Lecture Notes in Computer Science 3402

Commenced Publication in 1973
Founding and Former Series Editors:
Gerhard Goos, Juris Hartmanis, and Jan van Leeuwen

Michel Daydé   Jack J. Dongarra
Vicente Hernández   José M.L.M. Palma (Eds.)

High Performance Computing for Computational Science – VECPAR 2004

6th International Conference
Valencia, Spain, June 28-30, 2004
Revised Selected and Invited Papers

Volume Editors

Michel Daydé
ENSEEIHT
2, Rue Camichel, 31071 Toulouse Cedex 7, France
E-mail: dayde@enseeiht.fr

Jack J. Dongarra
University of Tennessee, TN 37996-1301, USA
E-mail: dongarra@cs.utk.edu

Vicente Hernández
Universidad Politecnica de Valencia
Camino de Vera, s/n, 46022 Valencia, Spain
E-mail: vhernand@dsic.upv.es

José M.L.M. Palma
Universidade do Porto
Faculdade de Engenharia
Rua Dr. Roberto Frias s/n, 4200-465 Porto, Portugal
E-mail: j.palma@fe.up.pt

Library of Congress Control Number: Applied for

CR Subject Classification (1998): D, F, C.2, G, J.2, J.3

ISSN 0302-9743
ISBN-10 3-540-25424-2 Springer Berlin Heidelberg New York
ISBN-13 978-3-540-25424-9 Springer Berlin Heidelberg New York

Springer is a part of Springer Science+Business Media

springeronline.com

Printed in Germany

Typesetting: Camera-ready by author, data conversion by Scientific Publishing Services, Chennai, India
Printed on acid-free paper SPIN: 11403937 06/3142 5 4 3 2 1 0

Preface

VECPAR is a series of international conferences dedicated to the promotion and advancement of all aspects of high-performance computing for computational science, as an industrial technique and academic discipline, extending the frontier of both the state of the art and the state of practice. The audience for and participants in VECPAR are seen as researchers in academic departments, government laboratories and industrial organizations. There is now a permanent website for the series, http://vecpar.fe.up.pt, where the history of the conferences is described.

The sixth edition of VECPAR was the first time the conference was celebrated outside Porto – at the Universitad Politecnica de Valencia (Spain), June 28–30, 2004. The whole conference programme consisted of 6 invited talks, 61 papers and 26 posters, out of 130 contributions that were initially submitted. The major themes were divided into large-scale numerical and non-numerical simulations, parallel and grid computing, biosciences, numerical algorithms, data mining and visualization.

This postconference book includes the best 48 papers and 5 invited talks presented during the three days of the conference. The book is organized into 6 chapters, with a prominent position reserved for the invited talks and the Best Student Paper. As a whole it appeals to a wide research community, from those involved in the engineering applications to those interested in the actual details of the hardware or software implementations, in line with what, in these days, tends to be considered as computational science and engineering (CSE).

Chapter 1 is concerned with large-scale computations; the first paper in the chapter, and in the book, was also the opening talk at the conference. Tetsuya Sato gives an overview of the greatest computations being performed on the Earth Simulator in Japan, followed by a series of 7 papers on equally large problems. Patrick Valduriez authors an invited talk on data management in large P2P systems; the companion to 4 more papers put together in Chapter 2.

The Grid technology in the roughly 5 years since it emerged has become one of the major driving forces in computer and also computational science and engineering. Fabrizio Gagliardi and Vincent Breton in the two first papers (invited talks) in chapter 3 present, respectively, the EGEE European Grid Infrastructure and applications of the Grid technology in medical applications. The 8 remaining papers in the chapter are a further example of the impact that the Grid is making in many areas of application.

Chapter 4 is the largest of all and the 12 papers in this chapter are an indication of the importance of cluster computing. Parallel and distributed computing is the title of chapter 5, which, despite its similarity with the previous chapter, includes papers where the emphasis has been put on the physical modelling and not so much on the strictly computing aspects of the simulations. The invited talk by Michael Heath opens chapter 5 and is a good example of how complex the computer simulation of real-life engineering systems can be. Since its early editions, linear algebra has occupied a relatively large proportion of the conference programme; linear algebra was the topic we chose to bring this book to a closure – Chapter 6.

Best Student Paper

There were 10 papers of high quality registered for the Best Student Paper competition. The laureate of the prize was German Molto for the paper:

- Three-Dimensional Cardiac Electrical Activity Simulation on Cluster and Grid Platforms, by German Molto, and also co-authored by Jose M. Alonso, Jose M. Ferrero, Vicente Hernandez, Marta Monserrat and Javier Saiz, all at Universidad Politecnica de Valencia.

To conclude, we would like to state in writing our gratitude to all the members of the Scientific Committee and the additional referees. Their opinions and comments were essential in the preparation of this book and the conference programme. We hope that the knowledge and the experience of many contributors to this book can be useful to a large number of readers.

December 2004

Michel Daydé,
Jack Dongarra,
Vicente Hernández,
José M.L.M. Palma

VECPAR is a series of conferences organized by the Faculty of Engineering of Porto (FEUP) since 1993

Acknowledgments

The sixth edition of the VECPAR conference brought new organizational challenges. The work was split between people in different countries. While the conference was held in Valencia (Spain), both the Web-based submission systems and the conference page were maintained at the Faculty of Engineering of the University of Porto (Portugal). Vitor Carvalho, once again, was responsible for the conference webpage and did an excellent job. Miguel Caballer, Gemma Cabrelles and Gabriel Garcia at the University of Valencia, under the guidance of Vicente Hernandez, did invaluable work on the Organizing Committee.

João Correia Lopes took care of the VECPAR databases and the Web-based submission systems; his collaboration was precious.

Committees

Organizing Committee

Vicente Hernandez (Chairman)
Antonio Vidal
Vicente Vidal
Victor García
Enrique Ramos
Ignacio Blanquer
Jose Roman
Jose Miguel Alonso
Fernando Alvarruiz
Jesus Peinado
Pedro Alonso
João Correia Lopes (Webchair)

Steering Committee

José Laginha Palma (Chairman), Universidade do Porto, Portugal
Jack Dongarra, University of Tennessee, USA
José Fortes, University of Purdue, USA
Álvaro Coutinho, Universidade Federal do Rio de Janeiro, Brazil
Lionel Ni, Hong Kong University of Science and Technology, Hong Kong, China

Scientific Committee

M. Daydé (Chairman)	ENSEEIHT-IRIT, France
O. Coulaud	INRIA, France
J.C. Cunha	Univ. Nova de Lisboa, Portugal
I.S. Duff	Rutherford Appleton Lab., UK and CERFACS, France
N. Ebecken	Univ. Federal do Rio de Janeiro, Brazil
W. Gentzsch	SUN, USA
A. George	Univ. of Florida, USA
L. Giraud	CERFACS, France
R. Guivarch	ENSEEIHT-IRIT, France
D. Knight	Rutgers Univ., USA
J. Koster	Bergen Center for Comp. Science, Norway
V. Kumar	Univ. of Minnesota, USA
R. Lohner	George Mason Univ., USA
O. Marques	Lawrence Berkeley National Laboratory, USA

A. Nachbin	Inst. Matemática Pura e Aplicada, Brazil
A. Padilha	Univ. do Porto, Portugal
B. Plateau	Lab. Informatique et Distribution, France
T. Priol	IRISA/INRIA, France
R. Ralha	Univ. do Minho, Portugal
H. Ruskin	Dublin City Univ., Ireland
E. Seidel	Louisiana State University, USA
A. Sousa	Univ. do Porto, Portugal
M. Stadtherr	Univ. of Notre Dame, USA
F. Tirado	Univ. Complutense de Madrid, Spain
B. Tourancheau	École Normale Supérieure de Lyon, France
M. Valero	Univ. Politécnica de Catalunya, Spain
E. Zapata	Univ. de Malaga, Spain

Invited Lecturers

– Tetsuya Sato
  Earth Simulator Center, Japan

– Patrick Valduriez
  INRIA and IRIN, Nantes, France

– Fabrizio Gagliardi
  EGEE, CERN, Switzerland

– Vincent Breton
  LPC Clermont-Ferrand, CNRS-IN2p3, France

– Michael T. Heath
  Computational Science and Engineering
  University of Illinois at Urbana-Champaign, USA

Sponsoring Organizations

The Organizing Committee is very grateful to the following organizations for their support:

UPV - Universidad Politécnica de Valencia
UP - Universidade do Porto
FEUP - Faculdade de Engenharia da Universidade do Porto
GVA - Generalitat Valenciana Conselleria de Empresa, Universidad y Ciencia

Additional Referees

Alberto Pascual
Albino dos Anjos Aveleda
Aleksandar Lazarevic
Alfredo Bautista
B. Uygar Oztekin
Boniface Nkonga
Bruno Carpentieri
Byung Il Koh
Carlos Balsa
Carlos Silva Santos
Christian Perez
Christine Morin
Congduc Pham
Daniel Ruiz
David Bueno
Eric Eilertson
Eric Grobelny
Eugenius Kaszkurewicz
Fernando Alvarruiz
Gabriel Antoniu
Gaël Utard
Germán Moltó
Gerson Zaverucha
Gregory Mounie
Guillaume Huard
Gérard Padiou
Helge Avlesen
Hui Xiong
Hung-Hsun Su
Ian A. Troxel
Ignacio Blanquer
Inês de Castro Dutra
J. Magalhães Cruz
J.M. Nlong
Jan-Frode Myklebust
Jean-Baptiste Caillau
Jean-Louis Pazat
Jorge Barbosa
Jose M. Cela
Jose Roman
Joseph Gergaud
José Carlos Alves
José L.D. Alves
José Miguel Alonso
João Manuel Tavares
João Tomé Saraiva
K. Park
Kil Seok Cho
Luis F. Romero
Luis Piñuel
Miguel Pimenta Monteiro
M.C. Counilh
Manuel Prieto Matias
Manuel Próspero dos Santos
Marc Pantel
Marchand Corine
Matt Radlinski
Michael Steinbach
Myrian C.A. Costa
Nicolas Chepurnyi
Nuno Correia
Olivier Richard
Pascal Henon
Paulo Lopes
Pedro Medeiros
Pierre Ramet
Ragnhild Blikberg
Rajagopal Subramaniyan
Ramesh Balasubramanian
Sarp Oral
Uygar Oztekin
Yves Denneulin
Yvon Jégou

Table of Contents

Chapter 1: Large Scale Computations

Chapter 2: Data Management and Data Mining

Chapter 3: Grid Computing Infrastructure

Chapter 4: Cluster Computing

Chapter 5: Parallel and Distributed Computing

Chapter 6: Linear and Non-Linear Algebra

Large Scale Simulations

Sato Tetsuya

The Earth Simulator Center,
Japan Agency for Marine-Earth Science and Technology, Japan

1 Introduction

Computer simulation has the history of 50 years. Its contribution to Science is indeed enormous and fruitful. Above all, significant is the fact that physical processes of nonlinear phenomena now become rather easily elucidated. Increase of simulation results that demonstrate nonlinear processes of observed phenomena in terms of fundamental laws has contributed decisively to finalizing modern western science based on reductionism. Nevertheless, simulation has not yet acquired real civil right as a self-supporting scientific methodology. This is mainly because the very tool of simulation, computer, remains not satisfactorily big enough. Because of this restriction of the capacity of available tools, simulations are used principally to explain only the direct physical causality of an individual observed phenomenon, not the global environmental evolution that bears the phenomenon. In other words, simulation can never be regarded as an independent, self-contained methodology of science that can cultivate a new field by itself, but remains an auxiliary and passive tool to observation and theory.

The Earth Simulator, which was born in March of 2002, manifested such a striking performance as 65% of the theoretical peak performance [1]. Not only its high record as 65% achieved by the AFES code (Atmospheric global simulation code for Earth Simulator), but also the average performance of all programs running on the Earth Simulator keeps 30% of the peak performance This indicates that the Earth Simulator is indeed an unexpectedly good simulator in its practical application.

Far more important and meaningful than this is the fact that one no longer needs to cut out only a minimal necessary region from the entire system that bears the phenomenon of concern, but can treat the entire system that includes simultaneously the concerned region and surroundings because of the large main memory.

This is a really revolutionary gift to humankind. Without any exaggeration one can say that humans have finally possessed a scientific tool to know not only the way nature evolves, such as climate changes and earth's interior variations, but also to design advanced technologies and products, such as fusion reactors and new automobiles.

M. Daydé et al. (Eds.): VECPAR 2004, LNCS 3402, pp. 1–9, 2005.

2 Examples of the Earth Simulator Project

Let us see some examples we have obtained by the Earth Simulator in these two years since the start of the Earth Simulator Project.

The Earth Simulator Project is carried out by the proposals from the public that are selected by the Selection Committee based on the Action Plan that the Director-General of the Earth Simulator Center adopts following the report of the Mission Definition Committee. The primary task is to contribute to the human life in making prediction of global environmental changes as reliable as possible. Nevertheless, in order to open the utilization to a wide field as much as possible and to make the maximal contribution to human society, each project is selected with a rule that a consortium or a society is assembled as broad as possible in the same field. About 40 projects are running this year.

The activity of each project is annually reported to public and the annual report is published. The reports for 2002 and 2003 have already been published [2]. In addition, a new journal called "Journal of the Earth Simulator" is published very recently [3]. Therefore, interested readers should refer to those publications.

In this paper, only a few results are chosen among all with an intention of introducing how large scale simulations the Earth Simulator has enabled us to do, rather than presenting the scientific interest.

2.1 Atmosphere-Ocean-Sea Ice-Land Coupled Simulation

No matter how good atmospheric circulation code one uses, one cannot make a sound, reliable prediction of future climate change, because the oceanic activity is equally important in the climate evolution. In the future prediction, the system to be dealt with in the simulation code must be as wide as possible so that all involved physical processes in time and space can be included. We have developed a climate code in which the atmospheric dynamics and the ocean dynamics are coupled with inclusion of the sea-ice and the land shape influence (CFES).

Figure 1 is a snapshot on a summer day of a preliminary run by the CFES code [2]. One can see cumulonimbus type of cloud formation and three-dimensional oceanic circulation in this figure. Although this is a preliminary run done one year ago, the code has improved largely since then and a practical production is now being progressed.

2.2 Global Seismic Wave Propagation

Earthquake calamity is usually localized, because the active seismic energy is strongly attenuated in relatively short distance and time compared to the global dimension. In the seismic wave propagation simulation, therefore, it might be reasonable to make a regional simulation in order to have a code as highly resolved as possible. However, we have never had a practical scientific tool to

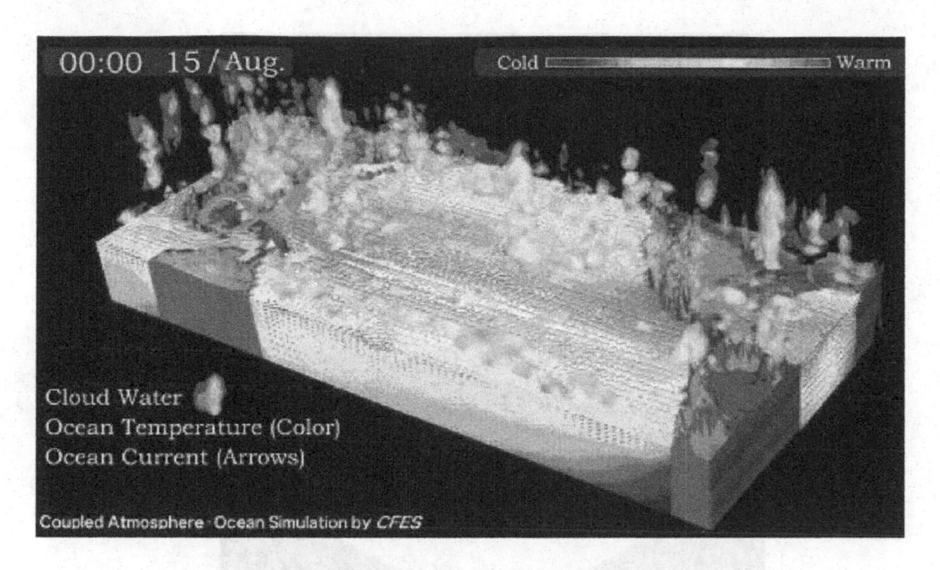

Fig. 1. Sapshot on a summer day of a preliminary run by the CFES code

study deep sea earthquakes or to reveal directly the global interior structure of the solid earth.

Figure 2 is a snapshot of the global seismic wave propagation caused by an earthquake in an Alaskan area [2]. In this particular simulation the interior structure of the globe is assumed, though used are as scientifically sound data as possible. This indicates that in collaboration with seismic wave observation the global simulation can greatly contribute to revealing details of the global interior structure, besides the prediction of global seismic disasters.

Reversals of the earth's dipole magnetic field in the dynamo region might be influenced by the mantle convection, though the time and space scales are by several orders of magnitude different. The Earth Simulator Center team is now developing a mantle/dynamo region coupled code that is optimized to the Earth Simulator [2]. What I want to emphasize here particularly is that humans have now invented a revolutionary tool to deal with the whole thing at all, thus enabling to make prediction of the future evolution, or, design future innovative technologies and products.

2.3 Carbon Nanotube (CNT)

The conventional simulator could of course deal with a nanometer material. But it was limited only to a fraction of the nanometer material, say, the size of a few nanometers, because of its smallness of capacity. Therefore, it was impossible to make design of new nanometer material by such a simulator, because it was too small to deal with the whole material at once. In contrast, the Earth Simulator is big enough to deal with several hundred nanometers of carbon nanotubes (CNT) at once.

Fig. 2. A snapshot of the global seismic wave propagation

Figure 3 is one example of the CNT stretching destruction [2]. In addition to this stretching, material properties such as the thermal conductivity are elucidated and new materials such as super-diamond are proposed. This example indicates that designing a new material by large scale simulation becomes practical and productive.

2.4 Lattice QCD

Let us introduce another special example that is not a dynamic simulation but a static numerical integration. In the QCD research the conventional approach is to introduce physical models and obtain approximated properties of the QCD equation, because the conventional computer is too small to solve it with enough accuracy. However, the Earth Simulator is big enough to attack the QCD equation in an exact way. The first solution obtained with a certain resolution has discovered that the quark mass is significantly smaller than the solution obtained by the approximated model [2]. At present a more accurate solution is being challenged with using higher resolution.

3 Encouragement of Simulation Culture

One may have seen from the above examples that simulation of the entire system could largely contribute to the security and welfare of human society. It is truly significant that the Earth Simulator has in practice presented the viability and

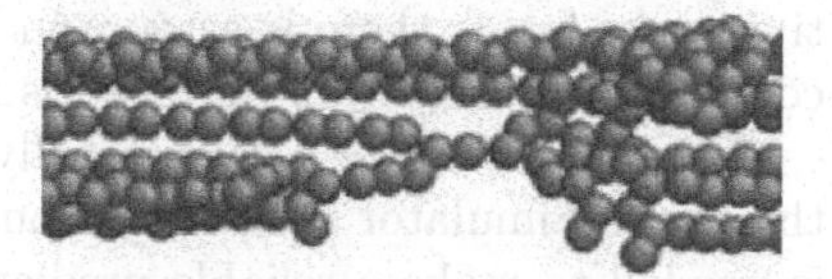

Fig. 3. The elasticity simulation of carbon nanotube

realization of knowing the future evolution of nature and designing the future technologies and products that can secure the sustainable human life.

Recently, the United States has announced the urgent necessity of developing large simulators exceeding the Earth Simulator. This is a rightful and natural direction to go beyond the stagnation of the modern science based on reductionism.

Simulations can easily present us several candidates so that we can choose at every necessary timing the best choice that could lead humankind and nature to the best relationship and also can revise the choice at any point, which I would call "Simulation Culture". There would be no better way at the moment to keep up a sustainable life than the quick spread of simulation culture.

4 Holistic Simulator

The bottleneck of spreading the simulation culture over the academia, industry, and general society lies in the developing speed of simulators. Petaflops simulators might be developed within several years, probably in the United States. Question, however, remains about the applicability and practicability of such simulators. If the actual performance for application remains one or few percent of the peak value that is quite likely for such a high peak performance parallel computer, then there would be no essential difference from the Earth Simulator., even though a peta-flops computer is realized.

Let us then change our strategy of developing large simulators. Quantitatively speaking, the Earth Simulator has made it possible to deal with a system with scale range of (1–5) $\times 10^3$ in the usual three-dimensional simulation, which corresponds to 10^{10} grid points in three dimension. This resolution may be good enough to deal with the entire system with reasonable accuracy. This would be true, though, only if the system is describable by one physical process. In the case of climate simulation, the resolution as small as 10 km becomes possible. However, the climate dynamics is governed by cloud formation describable in terms of the state change equation system as well as the atmospheric dynamics governed by the Navier-Stokes equation. Therefore, the climate scientist invokes so-called parameterization procedure for cloud formation, which is not based on the rigorous scientific law but to some extent based on the scientist's subjectivity. As far as there is a room in the simulation model such subjectivity can come in, the prediction obtained is not fully reliable.

In order to make a scientifically sound and truly reliable prediction, even the Earth Simulator is not big enough. Let us then consider how big simulator is needed to make a reliable prediction. In the case of climate simulation, the minimum scale of the system may be the scale of water drops, which is of the order of millimeter. Since the largest scale is the global size of the order of 10^7m, the total scale range can be as large as 10^{10}. For bio-system, the minimum scale can be of protein or amino acid, which is of the order of nanometer, and the largest one is an individual body of the order of meter. Therefore, the total scale range can again be as large as 10^9. Fusion plasma system and many other systems have also the scale range of 10^{10}.

The above simple consideration leads us to a conclusion that only simple technological progress of semi-conductor chips is a long way from a simulator that enables us to make a real prediction of the sustainable life.

In this paper, I would like to propose my own concept that can remove the difficulty of the high barrier of the scale range. The very point of removing the scale difficulty is to learn from the teachings of nature. "Nature is not homogeneous, but heterogeneous". Conversely speaking, heterogeneity means that nature is active and alive. Accordingly, the best way to grasp the way nature works is to assimilate the natural system. Viewing an organizing system in the heterogeneous nature, the interior of the system is also heterogeneous and governed by several different physical processes. My proposal is to design a simulator that can deal with one organizing system in a self-consistent way in which plural physical processes are weakly but firmly interacting to each other.

Suppose, for simplicity, a system that two processes, microscopic and macroscopic, are mutually interacting. Weakness of the mutual interaction allows us to make simulations of both processes rather independently. Firmness of the mutual interaction, in contrast, requires mutual information exchange as firmly as possible. We have already technologies at hand to simulate one physical process with enough resolution like the Earth Simulator. Therefore, the crucial point is whether or not we have a clever and smart way to fulfill the necessary information (data) exchange between the microscopic and macroscopic processes. From the standpoint of the macroscopic process, data are not necessary at every microscopic unit time step to be transferred from the microscopic part, because there would be no appreciable microscopic change of state in such a tiny time scale. On the other hand, from the microscopic standpoint, data are not necessary to be transferred at every microscopic unit time step from the macroscopic part, because in such a short time scale the macroscopic state would not suffer any meaningful change at all. A reasonably small macroscopic time would be a reasonable data exchange period, because the microscopic process could be appreciably influenced by an appreciable change in the macroscopic state in a small macroscopic time step and also the macroscopic process could be influ-

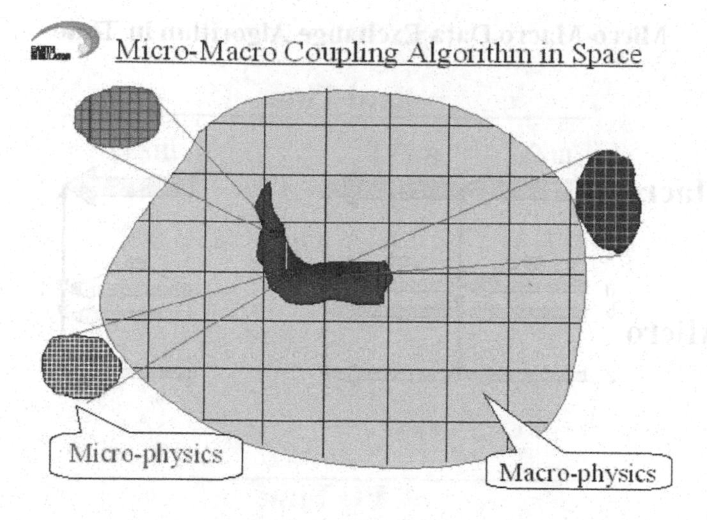

Fig. 4. Micro-Macro Coupling Algorithm inSpace

enced by an appreciable change in the microscopic state in that time step (see, Fig.4). By this way one can drastically reduce the data transfer rate.

Next, we must consider how drastically we can reduce the load of microscopic simulation compared with that of macroscopic simulation. Since we are concerned with the evolution of the entire system, the macroscopic simulation is the primary target. So, we assume that the entire system is divided into macroscopically reasonable grid system, say, 10^3 grid points in each direction. If we need to know the microscopic states at all grid points, it requires tremendously huge task for microscopic simulation. However, in most cases we do not need to make simulations at all macroscopic grid points. Suppose the macroscopic state is almost homogeneous. Then the microscopic states at all grid points must be the same. Thus, picking up only one grid point is enough for microscopic simulation. Usually, however, an interesting case where the microscopic process plays an important role in macroscopic evolution is when the macroscopic (environmental) state is changed drastically at local regions, consequently the microscopic process also being strongly activated (see, Fig.5). Thus in the macroscopic time scale both processes are strongly fed back mutually. This consideration tells us that we do not have to make microscopic simulations at all grid points, but that it would be enough to pick up several featuring grid points of the macroscopic grid system. Thus, we can reduce drastically the number of grid points that should be subjected to microscopic simulation.

In conclusion, it is quite possible to reduce the formidable task of microscopic simulation for both in time and space domains. One example of such a holistic simulator that can deal with the whole system as self-consistently as possible is presented in Fig. 6.

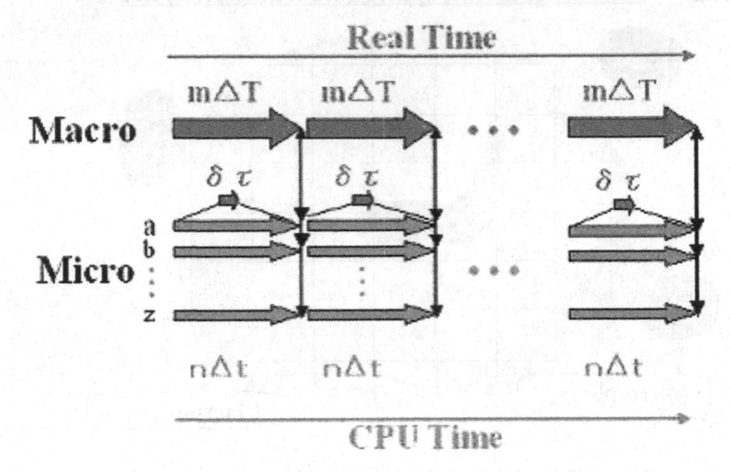

Fig. 5. Micro-Macro Data Exchange Algorithm in Time

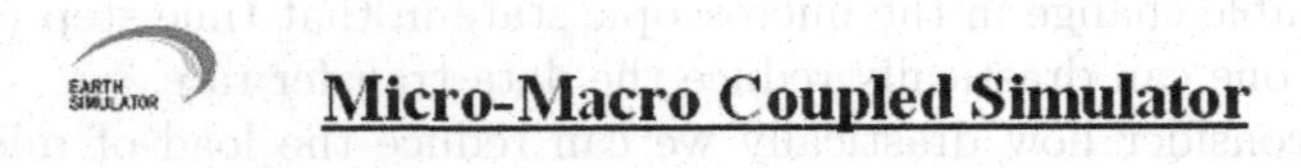

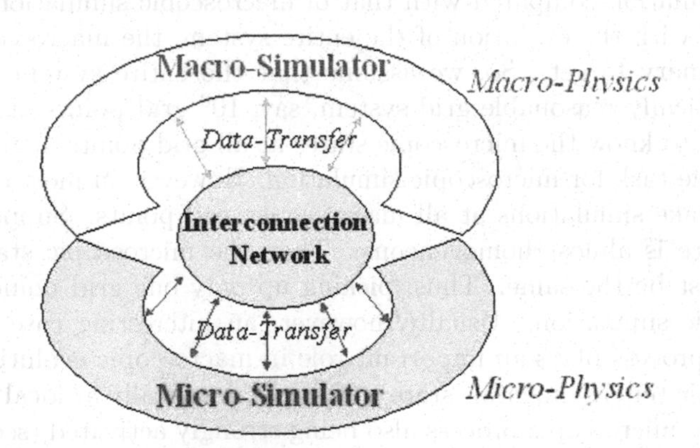

Fig. 6. Micro-Macro Coupled Simulator

5 Realization of Sustainable Life

Once such a holistic simulator would be in existence, then prediction of the earth's environmental changes such as climate changes and earthquakes would become far more reliable, since both microscopic and macroscopic processes could be simulated in a self-consistent way. Furthermore, symbiotic technologies and products such as eco-cars and securely synthesized biochemical medicines could be designed easily with low developing cost and time. Thus, sustainable human life could be realized securely and firmly.

Acknowledgments

I would like to thank for all who are participating in the Earth Simulator Project. Special thanks are given to all members of the Earth Simulator Center of JAMSTEC, in particular, to H. Hirano, K. Watanabe, K. Takahashi, W. Ohfuchi, A. Kageyama and E. Sasaki for their efforts in promoting the Earth Simulator Project.

References

[1] For instance, T. Sato, *We Think, Therefore We Simulate*, Look Japan, 32–34, **48**; T. Sato, *One Year Old Earth Simulator*, Proc. of ISC03, June, 2003; T. Sato, H. Murai, and S. Kitawaki, *How can the Earth Simulator Impact on Human Activities*, Proc. the Eighth Asia-Pacific Computer Systems and Architecture Conference (ACSAC'2003), Aizu, Japan, 2003; T. Sato, *The Earth Simulator: Roles and Impacts*, Proc. of the XXI International Symposium on Lattice Field Theory, Tsukuba, Japan, Sept. 2003.
[2] Annual Report of the Earth Simulator Center for 2002 and 2003.
[3] Journal of the Earth Simulator, the first issue, Apr. 2004. 9

Development and Integration of Parallel Multidisciplinary Computational Software for Modeling a Modern Manufacturing Process

Brian J. Henz[1], Dale R. Shires[1], and Ram V. Mohan[2]

[1] U.S. Army Research Laboratory, Attn: AMSRD-ARL-CI-HC,
High Performance Computing Division, Aberdeen Proving Ground, MD 21005, USA
Phone No.: (410) 278-6531 Fax No.: (410) 278-4983
{bhenz, dshires}@arl.army.mil
http://www.arl.army.mil/celst/

[2] Dept. of Mechanical Eng., Center for Adv. Materials and Smart Structures,
North Carolina A&T State University, Greensboro, NC 27411, USA
rvmohan@ncat.edu

Abstract. Modern multidisciplinary computational tools are required in order to model complex manufacturing processes. The liquid composite molding process, namely resin transfer molding (RTM) for the manufacture of composite material structures, includes models for fluid flow, heat transfer, and stress analysis during the process. In the past, these tools were developed independently, and an engineer would utilize each tool in succession, often in different environments with different data file formats, graphical user interfaces (if available), and sometimes different computer operating systems. The Simple Parallel Object-Oriented Computing Environment for the Finite Element Method (SPOOCEFEM) developed at the U.S. Army Research Laboratory provides a framework for development of multidisciplinary computational tools. A virtual manufacturing environment for the RTM process has been developed that integrates the coupled, multiphysics, multiscale physical models required for analyzing the manufacturing process. The associated code developments using SPOOCEFEM are discussed in this paper.

1 Introduction

Modeling of manufacturing processes such as liquid composite molding (LCM), namely resin transfer molding (RTM), requires integration of multidisciplinary numerical analysis tools. These tools can be combined in many ways including common data file formats or development in a common programming framework. For legacy codes that have been developed, optimized, and debugged over time, it may be easier to implement a common data format such as the eXtensible Data Model and Format (XDMF) from the U.S. Army Research Laboratory [1,2]. For new software tools being developed or parallelized, the authors have created the Simple Parallel Object-Oriented Computing Environment for the Finite Element Method (SPOOCEFEM) [3,4]. In this paper the SPOOCEFEM framework will be briefly discussed. Highlighted are three codes for flow, thermal, and stress analysis in a composite manufacturing process developed with SPOOCEFEM.

M. Daydé et al. (Eds.): VECPAR 2004, LNCS 3402, pp. 10–22, 2005.

These applications are coupled to show how SPOOCEFEM is used for the development and integration of multidisciplinary scientific software codes.

2 The SPOOCEFEM Framework

The SPOOCEFEM framework was developed expressly for the purpose of reducing the amount of time required for scientific software development and to promote code growth and maturation. The library avoids being overly generic but successfully compartmentalizes many of the functions common to the finite element method (FEM) approach. It has been developed in the C++ language and has been ported to numerous computing platforms including Intel-based Linux clusters, the SGI Origin, and the IBM Power series machines. While the initial ramp-up time to develop the fully functioning SPOOCEFEM library was significant, the library is being successfully used in several codes.

SPOOCEFEM uses numerous technologies to provide a complete development environment. At the lower levels are the high level language compilers (C++, C, and Fortran). On top of this are the OpenGL visualization libraries, the Message-Passing Interface (MPI) library for parallel interprocessor communication, and linear algebra packages such as BLAS (Basic Linear Algebra Subprograms) and LAPACK (Linear Algebra PACKage). Next are the visualization packages found in the Visualization Toolkit (VTK), binary and text file storage utilities, and linear system solver libraries. SPOOCEFEM sits atop all of this, along with specialized matrix and vector templates as shown in Figure 1.

SPOOCEFEM utilizes many of today's newest technologies to provide a complete integrated data file format for multidisciplinary applications. It uses XDMF for data storage [1]. This data format is extensible and designed for storing data used and generated by computational software. SPOOCEFEM utilizes only a subset of the larger XDMF framework for unstructured grid problems as encountered in the FEM. Data is segregated into descriptive text data (stored in the eXtensible Markup Language [XML]) and large binary data sets (stored in the Hierarchical Data Format [HDF]). Using XDMF guarantees the ability to quickly use visualization tools already available such as Paraview from Kitware (www.kitware.com). More details on SPOOCEFEM are outside the scope of this paper but are referenced elsewhere [3, 4].

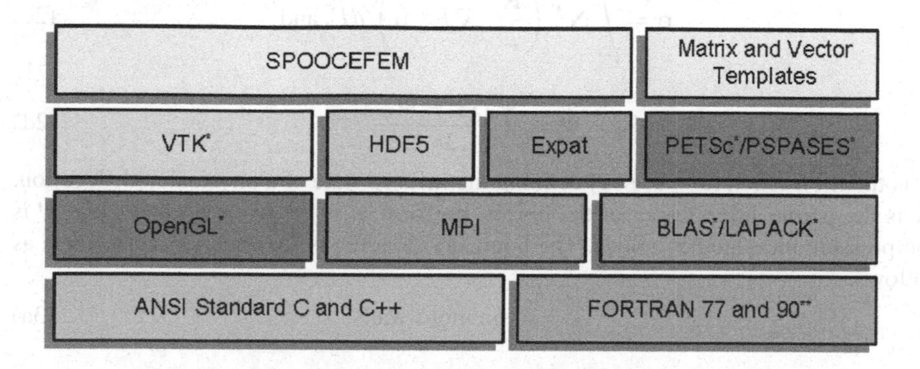

Fig. 1. SPOOCEFEM building block framework

3 Physical Modeling of RTM Composite Manufacturing Process

Physical modeling of the RTM process for composite manufacturing includes fluid flow, heat transfer, and stress analysis. Each of these models will now be briefly described, followed by a discussion of their integration with the SPOOCEFEM framework.

3.1 Resin Flow in Porous Media

Initially, the RTM process may be modeled as isothermal flow of resin through a porous media, namely a fibrous preform. In many instances, this model may be accurate enough. An implicit algorithm for this purpose based on the Pure Finite Element Method was developed by Mohan et al. [5] and Ngo et al. [6]. This algorithm was then implemented and parallelized in three incarnations. The first was a High Performance Fortran (HPF) version utilizing data parallelism. The second version was developed with Fortran 90 and MPI for message-passing parallelism [7, 8]. The current version utilizes SPOOCEFEM and is also based on message-passing parallelism, but with much of the book keeping routines managed by the SPOOCEFEM libraries [4]. This software is called the Composite Manufacturing and Process Simulation Environment (COMPOSE).

The transient mass conservation of the infusing resin is considered with a state variable defining the infused state of a node. Application of the Galerkin weighted residual formulation to the physical model equation leads to a semi-discrete equation of the form shown in Equation 1. Further details can be found in Mohan et al. [5].

$$\mathbf{C}\dot{\boldsymbol{\Psi}} + \mathbf{K}\mathbf{P} = \mathbf{q}, \tag{1}$$

where

$$\mathbf{C} = \int_{\Omega} \mathbf{N}^{\mathbf{T}}\mathbf{N}\, d\Omega, \tag{2a}$$

$$\mathbf{K} = \int_{\Omega} \mathbf{B}^{\mathbf{T}} \frac{\bar{\mathbf{K}}}{\mu} \mathbf{B}\, d\Omega, \tag{2b}$$

$$\mathbf{q} = \int_{\Gamma} \mathbf{N}^{\mathbf{T}} \left(\frac{\bar{\mathbf{K}}}{\mu} \cdot \nabla P \cdot \mathbf{n} \right) d\Gamma, \text{ and} \tag{2c}$$

$$\dot{\boldsymbol{\Psi}} = \frac{\boldsymbol{\Psi}_{n+1} - \boldsymbol{\Psi}_n}{\Delta t}. \tag{2d}$$

Ψ is the fill fraction ($0 \leq \Psi \leq 1$) defining the infused state of a physical node location, $\bar{\mathbf{K}}$ is the permeability tensor of the fibrous preform, μ is the resin viscosity, and P is the pressure measured at a node. The boundary conditions for equation 1 are given as follows:

$$\frac{\partial P}{\partial n} = 0 \text{ on mold walls,} \tag{3a}$$

$$P = 0 \text{ at flow front, and} \tag{3b}$$

$$P = P_0 \text{ prescribed pressure at inlet}$$
$$\text{or} \tag{3c}$$
$$q = q_0 \text{ prescribed flow rate at inlet,}$$

where P_0 and q_0 represent prescribed pressure and flow rate at the inlet(s), respectively. Initially,

$$\Psi = 1 \text{ at inlet and}$$
$$\Psi(t = 0) = 0 \text{ elsewhere.} \tag{4}$$

In equation 2, **N** is the elemental shape function and $\mathbf{B} = \nabla\mathbf{N}$. Equation 1 is solved in the structured FEM software COMPOSE utilizing element-based domain decomposition. In the present parallel developments, the same equation is solved utilizing SPOOCEFEM with node-based domain decomposition; otherwise, everything else including element types (three-noded triangular, four-noded quadrilateral, and four-noded tetrahedral), the stiffness matrix storage format (Compressed Sparse Row), and the iterative conjugate gradient (CG)-based solver and preconditioner remain the same.

Software Development. Utilizing the object-oriented design (OOD) of SPOOCEFEM the FEM problem is divided into classes. Most of the computational work, excluding the linear equation solvers in COMPOSE, takes place in the various element classes. A diagram of the `ComposeElement2dTri3` class hierarchy is shown in Figure 2. COMPOSE is capable of modeling the RTM process with 2.5-D or full 3-D analysis. For composite components typically seen in the aerospace industry, the 2.5-D model is most heavily utilized. The 2.5-D element types in COMPOSE are three-noded triangles and four-noded quadrilaterals. Multiple inheritance is used in the COMPOSE element classes in order to make code development more efficient. The `ComposeElement` class provides functionality common to all COMPOSE element types such as storage of viscosity data. The `ComposeElement2d` class provides mass matrix and local permeability computations. The functionality unique for each COMPOSE element type

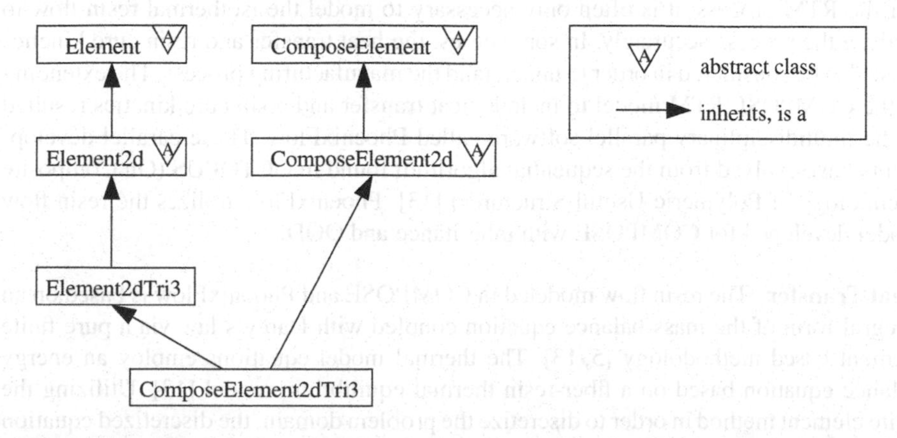

Fig. 2. Hierarchy of the `ComposeElement2dTri3` class showing multiple inheritance

Table 1. Function-oriented and OOD runtime comparison (in seconds)

Operation	FORTRAN 77	C++
Linear solver	2766	2694
Calculate pressure	114	128
Update	22	71
Core total time	2902	2893

such as stiffness matrix computation is available in the `ComposeElement2dTri3` class. More details on the development of the SPOOCEFEM-based COMPOSE software is available in Henz and Shires [9].

The code development time has decreased with the utilization of OOD from 12 months for the Fortran 90 version of COMPOSE to about 3 months for the C++ version [10]. In addition, code reuse of the OOD library is evident as discussed in later sections when thermal and cure components are added. Use of OOD in scientific computing always brings with it the performance concerns associated with the runtime system [11, 12]. Our own experience shows that OOD is not necessarily a sign of poor performance to come. The COMPOSE software that was originally developed in FORTRAN 77 was re-implemented with OOD in C++. Table 1 shows a break out of the three most time consuming routines in COMPOSE, the first of which is the linear solver with approximately 96% of the total runtime. The faster execution of the C++ version of the solver is most likely due to small enhancements performed during translation. The other routines, calculate pressure and update, are slower in C++ than FORTRAN (most noticeably in the update routine). The three fold increase in runtime for update is due to the numerous accesses to values contained in the vector templates. The advantage of these templates is that they allow quick use of numerous solvers and therefore this overhead is not critical in many instances and considered a good trade-off when compared to the increase in software development productivity.

3.2 Convection, Conduction, and the Exothermic Resin Curing Process

For the RTM process, it is often only necessary to model the isothermal resin flow to analyze the process accurately. In some cases, the heat transfer and resin cure kinetics must also be considered in order to understand the manufacturing process. The extending of the COMPOSE RTM model to include heat transfer and resin cure kinetics resulted in the multidisciplinary parallel software called PhoenixFlow. These parallel developments have evolved from the sequential algorithm found in OCTOPUS (On Composite Technology of Polymeric Useful Structures) [13]. PhoenixFlow utilizes the resin flow model developed for COMPOSE with inheritance and OOD.

Heat Transfer. The resin flow modeled in COMPOSE and PhoenixFlow is based on an integral form of the mass balance equation coupled with Darcy's law via a pure finite element-based methodology [5, 13]. The thermal model equations employ an energy balance equation based on a fiber-resin thermal equilibrium model [13]. Utilizing the finite element method in order to discretize the problem domain, the discretized equation for the heat transfer model in PhoenixFlow is given as

$$\mathbf{C}\dot{\mathbf{T}} + (\mathbf{K}_{ad} + \mathbf{K}_{cond})\,\mathbf{T} = \mathbf{Q}_q + \mathbf{Q}_{\dot{G}}, \tag{5}$$

where

$$\mathbf{C} = \int_{\Omega} \mathbf{W}^T \rho c_p \mathbf{N}\, d\Omega, \tag{6a}$$

$$\mathbf{K}_{ad} = \int_{\Omega} \mathbf{W}^T \left(\rho c_p\right)_r \left(\mathbf{u}\cdot\mathbf{B}_N\right) d\Omega, \tag{6b}$$

$$\mathbf{K}_{cond} = \int_{\Omega} \mathbf{B}_W^T k \mathbf{B}_N \, d\Omega, \tag{6c}$$

$$\mathbf{Q}_q = \int_{\Gamma} \mathbf{W}^T \left(-\mathbf{q}\cdot\mathbf{n}\right) d\Gamma, \text{ and} \tag{6d}$$

$$\mathbf{Q}_{\dot{G}} = \int_{\Gamma} \mathbf{W}^T \phi \dot{G}\, d\Gamma. \tag{6e}$$

$\mathbf{B}_W$ is the spatial derivative of the weighting function, $\mathbf{W}$, and $\mathbf{B}_N$ is the spatial derivative of the finite element shape function, $\mathbf{N}$. ϕ is porosity, ρ is density, c_p is specific heat, k is the thermal conductivity, $\mathbf{u}$ is the resin velocity, and $\dot{G}$ is the rate of heat generation by chemical reactions or other sources. The subscript r denotes resin properties. Physical quantities without the subscript denote the weighted average properties of the resin and the fiber.

The boundary conditions required to solve equation 5 are given by

$$T = T_w \text{ at the mold wall,} \tag{7a}$$

$$T = T_{r0} \text{ at the mold inlet,} \tag{7b}$$

$$k\frac{\partial T}{\partial n} = \left(1-\phi\right)\rho_r c_{pr} \mathbf{u}\cdot\mathbf{n}\left(T_{f0} - T\right) \text{ at the resin front, and} \tag{7c}$$

$$T\left(t=0\right) = T_w \text{ initially.} \tag{7d}$$

Resin Cure Model. In addition to the temperature calculations, the degree of resin cure is also computed in the current developments. The discretized system of equations is given as

$$\mathbf{C}\dot{\alpha} + \mathbf{K}\alpha = \mathbf{Q}_{\mathbf{R}_\alpha}, \tag{8}$$

where

$$\mathbf{C} = \int_{\Omega} \phi \mathbf{W}^T \mathbf{N}\, d\Omega, \tag{9a}$$

$$\mathbf{K} = \int_{\Omega} \phi \mathbf{W}^T \left(\mathbf{u}\cdot\mathbf{B}_N\right) d\Omega, \text{ and} \tag{9b}$$

$$\mathbf{Q}_{\mathbf{R}_\alpha} = \int_\Gamma \phi \mathbf{W}^T R_\alpha \, d\Gamma. \tag{9c}$$

$\mathbf{B}_N$ in equation 9b is the spatial derivative of the finite element shape function, $\mathbf{N}$. α is the degree of cure, and R_α is the rate of chemical reaction.

The boundary conditions for equation 8 are given as

$$\alpha = 0 \text{ at mold inlet, and} \tag{10a}$$

$$\alpha\,(t = 0) = 0 \text{ initially.} \tag{10b}$$

The resin curing process is exothermic. The heat generated, $\dot{G}$, is given by [14, 15, 16]

$$\dot{G} = H_R R_\alpha, \tag{11}$$

where H_R is the heat of reaction per unit volume for the pure resin.

Software Development. The first step in extending COMPOSE for nonisothermal analysis is to develop a new element type that includes routines for computing and storing stiffness matrices for the additional models. Graphically, this extension is shown in Figure 3. The `PhoenixFlowElement` class contains routines required by all PhoenixFlow element types. These routines include calculating effective material properties and the elemental viscosities. The `PhoenixFlowElement2d` class does not contain any functionality at this time but is used for convenience in the class hierarchy. The `PhoenixFlowElement2dTri3` class provides functionality for computing elemental stiffness matrices for heat transfer and resin cure.

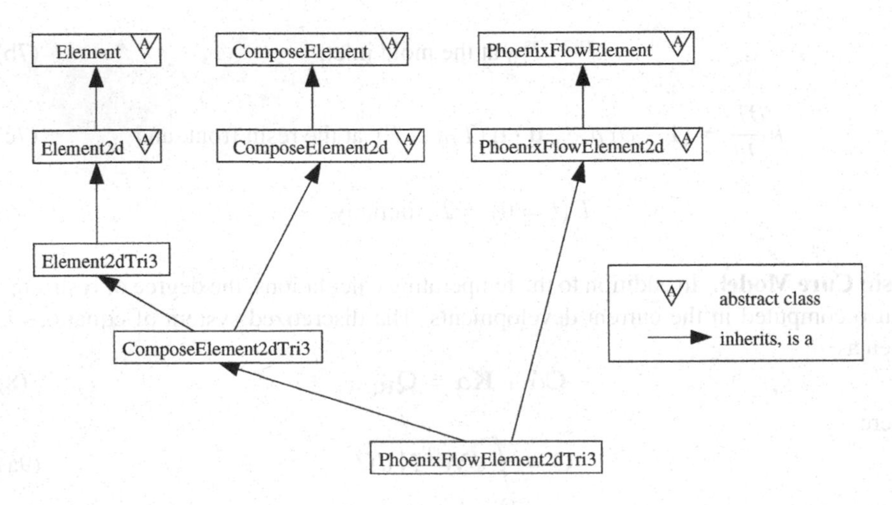

Fig. 3. Hierarchy of the `PhoenixFlowElement2dTri3` class showing multiple inheritance

3.3 Multiscale Residual Thermal Stress Analysis

Composite components manufactured with processes such as RTM often contain residual stresses induced during the process. These residual stresses are typically produced during the cool-down phase after the resin has cured. The model used to compute residual stresses in the current developments is based on the Multiscale Elastic Thermal Stress (METS) software developed by Chung [17]. The model also predicts homogeneous material properties for heterogeneous materials through the Asymptotic Expansion Homogenization (AEH) method [17].

The thermal residual stresses are computed from the following equation:

$$\left[\int_{\Omega} \mathbf{B}^T \mathbf{D} \mathbf{B} \, d\Omega\right] \Delta \mathbf{u}_{n+1} = \int_{\Omega} \mathbf{B}^T \mathbf{D} \Delta \epsilon_{n+1}^{th} \, d\Omega + \int_{\Gamma} \mathbf{N}^T \tau_{n+1} \, d\Gamma + \int_{\Omega} \mathbf{N}^T \mathbf{F}_{n+1} d\Omega - \int_{\Omega} \mathbf{B}^T \sigma_n \, d\Omega, \tag{12}$$

where σ is the stress experienced by the body, $\mathbf{F}$ is the body force per unit volume, $\mathbf{B}$ is the derivative of the shape function $\mathbf{N}$, $\tau = \sigma \cdot \mathbf{n}$, $\mathbf{u}$ is the nodal displacement, and ϵ is the elemental strain. A detailed derivation of the preceding equation can be found in Ngo [13].

The software to compute the residual thermal stresses developed at ARL utilizes a baseline code from Ngo [13]. The software from Ngo was developed with FORTRAN 77 and does not contain a common data file format or output besides simple elemental results in text files. This code was also serial and has been parallelized with the use of SPOOCEFEM. The new SPOOCEFEM-based software developed at ARL is called MSStress (MultiScale thermal residual Stress analysis).

Software Development. Since the MSStress software is currently coupled with COMPOSE and PhoenixFlow through common data files (XDMF), no inheritance of classes from previous applications is used. The key portion of the software development for MSStress is the implementation within the SPOOCEFEM framework so that it can be parallelized and have a GUI available for pre- and postprocessing capabilities. Some interesting details of the software development for MSStress are provided. The microscale model in MSStress is a repeating unit cell material configuration of the composite component which is used to compute the homogenized material properties for the macroscale model. As was soon discovered, the microscale model was much more difficult to parallelize because of the symmetry boundary conditions required in the unit cell model analysis. After a careful analysis, it was concluded that the microscale model was typically somewhat smaller than the macroscale model and could therefore be computed in serial on each processor used in the parallel analysis. Then the macroscale problem is subsequently solved in parallel with the computed material properties. The adverse affect on performance from not parallelizing the microscale model is not enough to offset the extra effort that would be required to parallelize the symmetric boundary condition routines at this time. A more detailed analysis of the MSStress software development is not in the scope of this work and will appear in future publications.

4 Computational Model Integration

In this section is a discussion of some of the issues encountered developing tightly coupled codes (such as COMPOSE and PhoenixFlow) and when coupling loosely through data I/O. The first section discusses the tightly coupled flow model in COMPOSE with the thermal and cure models in PhoenixFlow. The second section provides details about using XDMF in order to couple the thermal results with the multiscale model in MSStress.

4.1 Tightly Coupled Flow/Thermal/Cure

In modeling the RTM process, the fluid flow and thermal models must be tightly coupled. This coupling is accomplished through software engineering. The COMPOSE fluid flow model is implemented in the SPOOCEFEM framework [9] as a stand-alone code. The thermal and cure models in PhoenixFlow then take advantage of the COMPOSE software through inheritance and OOD.

An example of this OOD is the three-noded triangular element in PhoenixFlow. The COMPOSE three-noded triangular element has storage, computation, and assembly routines for the fluid flow stiffness matrix. The `PhoenixFlowElement` class inherits these routines and adds similar routines for thermal and cure analysis. In COMPOSE, the viscosity of an element remains constant, whereas the viscosity is computed at each flow time step in PhoenixFlow. For this, the storage of the `ComposeElement` class is utilized, and new routines for viscosity computation are added to the `PhoenixFlowElement` class.

In addition to the element classes, the `PhoenixFlowFEM` class inherits functionality from the `ComposeFEM` class. This functionality includes the one-shot filling routine [18] and sensitivity analysis for temperature independent parameters [19]. The main loop in the COMPOSE software computes filling at each time step until the mold is filled.

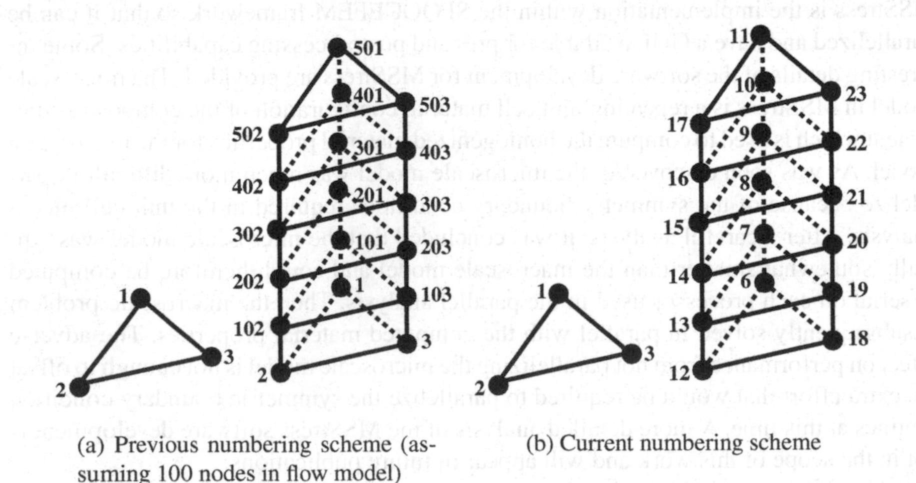

(a) Previous numbering scheme (assuming 100 nodes in flow model)

(b) Current numbering scheme

Fig. 4. Stack of six-noded wedge elements for PhoenixFlow thermal/cure analysis

Since temperature and degree of cure must be computed in between filling time steps, a new routine in PhoenixFlow has been developed that computes a single filling time step and then proceeds to thermal and cure analysis before continuing.

Another issue in the thermal analysis includes heat transfer in the through-thickness direction. This requires three-dimensional elements to be built up on top of the two-dimensional elements as seen in Figure 4. These extra elements and nodes increase the complexity of the linear equations many-fold in PhoenixFlow. As such, careful attention is paid to the nodal numbering and elemental connectivity in order to improve performance and reduce code complexity.

The additional thermal nodes and their numbering should preserve the connectivity graph of the original FEM nodal mesh configuration. The present numbering scheme as shown in Figure 4(b) preserves the shape of the nodal connectivity graph (Figures 5(c) and 5(d)). In contrast, the previous numbering algorithm (Figure 4(a)), did not preserve

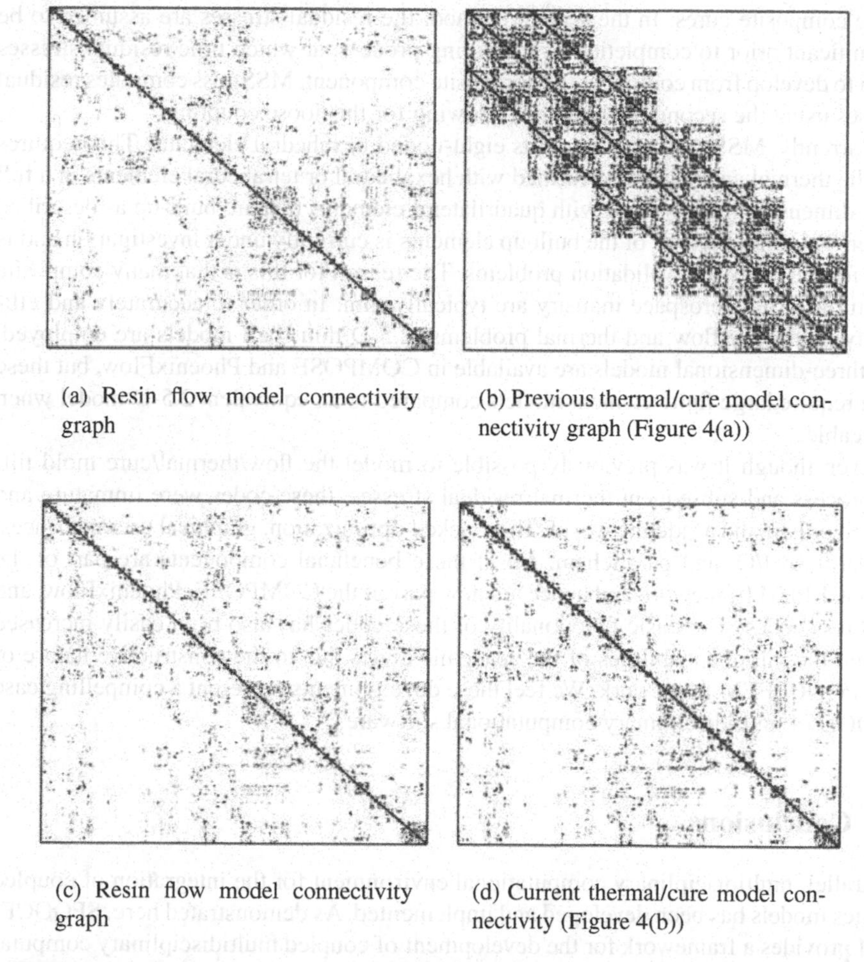

(a) Resin flow model connectivity graph

(b) Previous thermal/cure model connectivity graph (Figure 4(a))

(c) Resin flow model connectivity graph

(d) Current thermal/cure model connectivity (Figure 4(b))

Fig. 5. Comparison of previous connectivity graphs for resin flow vs. thermal/cure models the nodal connectivity graph (Figures 5(a) and 5(b)). Therefore, if a reverse Cuthill McKee (RCM) [20] pass is performed on the connectivity graph of the resin flow model, this optimization would be currently maintained for the thermal/cure model. Minimization of the connectivity graph bandwidth can lead to improvements in runtime and cache efficiency [21].

4.2 Coupled Flow/Thermal/Cure/Stress Analysis

The thermal residual stress analysis software (MSStress) requires temperature profiles from the nonisothermal RTM software (PhoenixFlow). The integration of the MSStress software with PhoenixFlow is accomplished through the XDMF file format. This format standardizes the storage and description of data so that MSStress can read the temperature results from PhoenixFlow.

There are two ways in which the residual stresses can be computed in the RTM process. In the first, the stresses are computed incrementally at each thermal time step as the composite cures. In the second method, the residual stresses are assumed to be insignificant prior to completion of the curing process, at which time residual stresses begin to develop from cooling of the composite component. MSStress computes residual stresses using the second method, thus allowing for this loose coupling.

Currently, MSStress only supports eight-noded hexahedral elements. This requires that the thermal analysis be performed with hexahedral or tetrahedral elements in a full three-dimensional analysis or with quadrilateral elements that are built-up as described earlier. This later method of the built up elements is currently under investigation and is used for the MSStress validation problems. The reason for this is that many composite structures in the aerospace industry are typically thin. In order to accurately and efficiently model the flow and thermal problems, 2.5-D thin shell models are employed. Full three-dimensional models are available in COMPOSE and PhoenixFlow, but these often require large finite element models compared to an equivalent 2.5-D model when applicable.

Even though it was previously possible to model the flow/thermal/cure mold filling process and subsequent thermal residual stresses, these codes were immature and resided only in the academic world. They lacked optimization, graphical user interfaces, standardized I/O, and parallelism. All of these beneficial components are part of the SPOOCEFEM framework and hence are now part of the COMPOSE, PhoenixFlow, and MSStress codes. The basic functionality of these codes has also been easily increased beyond the initial capabilities of the academic codes due to the constructive nature of the SPOOCEFEM framework. We feel these developments represent a compelling case for OOD of multidisciplinary computational software.

5 Conclusions

A parallel, multidisciplinary computational environment for the integration of coupled physics models has been developed and implemented. As demonstrated here, SPOOCEFEM provides a framework for the development of coupled multidisciplinary computational codes. In conjunction with the XDMF, SPOOCEFEM facilitates the coupling of developmental and legacy scientific codes. This coupling of codes is required in order to model several coupled physical phenomena in complex processes such as the RTM process. In addition, the software developed with SPOOCEFEM can be easily parallelized and ported in order to take advantage of high-performance computers including clusters of personal computers (PCs). The combination of COMPOSE/PhoenixFlow/MSStress models the RTM manufacturing process from resin injection through post-fill curing analysis, resulting in a computational model composite component with residual stress. These computed process induced initial residual stresses are then used by codes such as DYNA3D [22] or P-DINOSAURUS [23] as input for dynamic loading analysis. This coupling process is becoming more important as multiscale and multiphysics problems are analyzed.

Acknowledgments

This work was supported in part by a grant of computer time and resources by the Department of Defense High Performance Computing Modernization Program Office. Additional support was provided by the U.S. Army Research Laboratory Major Shared Resource Center (MSRC).

References

1. J. A. Clarke and R. R. Namburu. A Distributed Comuting Environment for Interdisciplinary Applciations. *Concurrency and Computation: Practice and Experience*, 14:1–14, 2002.
2. J. A. Clarke, C. E. Schmitt, and J. J. Hare. Developing a Full Featured Application from an Existing Code Using the Distributed Interactive Computing Environment. In *DoD High Performance Computing Modernization Program Users Group Conference*, Houston, TX, June 1998.
3. B. J. Henz, D. R. Shires, and R. V. Mohan. A Composite Manufacturing Process Simulation Environment (COMPOSE) Utilizing Parallel Processing and Object-Oriented Techniques. In *International Conference on Parallel Processing*, Vancouver, B.C. Canada, August 2002.
4. D. R. Shires and B. J. Henz. An Object-Oriented Approach for Parallel Finite Element Analysis. In *International Conference on Parallel and Distributed Processing Techniques and Applications*, Las Vegas, NV, June 2003.
5. R. V. Mohan, N. D. Ngo, and K. K. Tamma. On a Pure Finite Element Methodology for Resin Transfer Mold Filling Simulations. *Polymer Engineering and Science*, 39(1):26–43, January 1999.
6. N. D. Ngo, R. V. Mohan, P. W. Chung, and K. K. Tamma. Recent Developments Encompassing Non-Isothermal/Isothermal Liquid Composite Molding Process Modeling/Analysis: Physically Accurate, Computationally Effective and Affordable Simulations and Validations. *Journal of Thermoplastic Composite Materials*, 11(6):493–532, November 1998.
7. D. R. Shires, R. V. Mohan, and A. Mark. Optimization and Performance of a Fortran 90 MPI-Based Unstructured Code on Large Scale Parallel Systems. In *International Conference on Parallel and Distributed Processing Techniques and Applications*, Las Vegas, NV, June 2001.
8. D. R. Shires and R. V. Mohan. An Evaluation of HPF and MPI Approaches and Peformance in Unstructured Finite Element Simulations. *Journal of Mathematical Modelling and Algorithms*, 1:153–167, 2002.

9. B. J. Henz and D. R. Shires. Development and Performance Analysis of a Parallel Finite Element Application Implemented in an Object-Oriented Programming Framework. In *International Conference on Parallel and Distributed Processing Techniques and Applications*, Las Vegas, NV, June 2003.
10. D. R. Shires and B. J. Henz. Lessons Learned and Perspectives on Successful HPC Software Engineering and Development. In *International Conference on Software Engineering Research and Practice*, Las Vegas, NV, June 2004.
11. D. Post and R. Kendall. Software Project Management and Quality Engineering Practices for Complex, Coupled Multi-Physics Massively Parallel Computational Simulations: Lessons Learned from ASCI. Technical Report LA-UR-03-1274, Los Alamos Labratory, 2003.
12. T. L. Veldhuizen and M. Ed Jurnigan. Will C++ be Faster than FORTRAN? In *First International Scientific Computing in Object-Oriented Parallel Environments (ISCOPE)*, 1997.
13. N. D. Ngo. *Computational Developments for Simulation Based Design: Multi-Disciplinary Flow/Thermal/Cure/Stress Modeling, Analysis, and Validation for Processing of Composites*. PhD thesis, University of Minnesota, 2001.
14. S. Sourour and M. R. Kamal. *SPE Technical Paper*, 18(93), 1972.
15. M. R. Kamal and S. Sourour. Integrated Thermo-Rheological Analysis of the Cure of Thermosets. *SPE Technical Paper*, 18(187), 1973.
16. M. R. Kamal and S. Sourour. *SPE Technical Paper*, 13(59), 1973.
17. P. W. Chung, K. K. Tamma, and R. R. Namburu. Asymptotic Expansion Homogenization for Heterogeneous Media: Computational Issues and Applications. *Composites - Part A: Applied Science and Manufacturing*, 32(9):1291–1301, 2001.
18. V. R. Voller and Y. F. Chen. Prediction of Filling Times of Porous Cavities. *international Journal for Numerical Methods in Fluids*, 23:661–672, 1996.
19. B. J. Henz, K. K. Tamma, R. Kanapady, N. D. Ngo, and P. W. Chung. Process Modeling of Composites by Resin Transfer Molding: Sensitivity Analysis for Isothermal Considerations. In *AIAA-2002-0790, 40 th Aerospace Sciences Meeting*, Reno, NV, January 2002.
20. E. Cuthill and J. McKee. Reducing the Bandwidth of Sparse Symmetric Matrices. In *Proceedings of the 24th National Conference*, pages 157–172. Association for Computing Machinery, 1969.
21. G. Kumfert and A. Pothen. Two Improved Algorithms for Envelope and Wavefront Reduction. *BIT*, 37(3):559 – 590, 1997.
22. J. I. Lin. *DYNA3D: A Nonlinear, Explicit, Three-Dimensional Finite Element Code for Solid and Structural Mechanics*. Lawrence Livermore National Laboratory, 1998.
23. R. Kanapady and K. K. Tamma. *P-DINOSAURUS: Parallel Dynamic INtegration Operators for Structural Analysis Using Robust Unified Schemes*. Department of Mechanical Engineering, University of Minnesota, 2002.

Automatically Tuned FFTs for BlueGene/L's Double FPU

Franz Franchetti[†], Stefan Kral[†], Juergen Lorenz[†], Markus Püschel[‡], and Christoph W. Ueberhuber[†]

[†] Institute for Analysis and Scientific Computing,
Vienna University of Technology, Wiedner Hauptstrasse 8–10, A-1040 Wien, Austria
franz.franchetti@tuwien.ac.at
http://www.math.tuwien.ac.at/ascot

[‡] Dept. of Electrical and Computer Engineering,
Carnegie Mellon University, 5000 Forbes Avenue, Pittsburgh, PA 15213
pueschel@ece.cmu.edu
http://www.ece.cmu.edu/~pueschel

Abstract. IBM is currently developing the new line of BlueGene/L supercomputers. The top-of-the-line installation is planned to be a 65,536 processors system featuring a peak performance of 360 Tflop/s. This system is supposed to lead the TOP 500 list when being installed in 2005 at the Lawrence Livermore National Laboratory. This paper presents one of the first numerical kernels run on a prototype BlueGene/L machine. We tuned our formal vectorization approach as well as the Vienna MAP vectorizer to support BlueGene/L's custom two-way short vector SIMD "double" floating-point unit and connected the resulting methods to the automatic performance tuning systems SPIRAL and FFTW. Our approach produces automatically tuned high-performance FFT kernels for BlueGene/L that are up to 45 % faster than the best scalar SPIRAL generated code and up to 75 % faster than FFTW when run on a single BlueGene/L processor.

1 Introduction

IBM's BlueGene/L supercomputers [1] are a new class of massively parallel systems that focus not only on performance but also on lower power consumption, smaller footprint, and lower cost compared to current supercomputer systems. Although several computing centers plan to install smaller versions of BlueGene/L, the most impressive system will be the originally proposed system at Lawrence Livermore National Laboratory (LLNL), planned to be in operation in 2005. It will be an order of magnitude faster than the Earth Simulator, which currently is the number one on the TOP 500 list. This system will feature eight times more processors than current massively parallel systems, providing 64k processors for solving new classes of problems. To tame this vast parallelism, new approaches and tools have to be developed. However, tuning software for this machine starts by optimizing computational kernels for its processors. And

M. Daydé et al. (Eds.): VECPAR 2004, LNCS 3402, pp. 23–36, 2005.

BlueGene/L systems come with a twist on this level as well. Their processors feature a custom floating-point unit—called "double" FPU—that provides support for complex arithmetic.

Efficient computation of fast Fourier transforms (FFTs) is required in many applications planned to be run on BlueGene/L systems. In most of these applications, very fast one-dimensional FFT routines for relatively small problem sizes (in the region of 2048 data points) running on a single processor are required as major building blocks for large scientific codes. In contrast to tedious hand-optimization, the library generator SPIRAL [23], as well as the state-of-the-art FFT libraries FFTW [12] and UHFFT [22], use empirical approaches to automatically optimize code for a given platform.

Floating-point support for complex arithmetic can speed up large scientific codes significantly, but the utilization of non-standard FPUs in computational kernels like FFTs is not straightforward. Optimization of these kernels leads to complicated data dependencies that cannot be mapped directly to BlueGene/L's custom FPU. This complicacy becomes a governing factor when applying automatic performance tuning techniques and needs to be addressed to obtain high-performance FFT implementations.

Contributions of this Paper. In this paper we introduce and describe an FFT library and experimental FFT kernels (FFTW no-twiddle codelets) that take full advantage of BlueGene/L's double FPU by means of short vector SIMD vectorization. The library was automatically generated and tuned using SPIRAL in combination with formal vectorization [8] as well as by connecting SPIRAL to the Vienna MAP vectorizer [15, 16]. The resulting codes provide up to 45 % speed-up over the best SPIRAL generated code not utilizing the double FPU, and they are up to 75 % faster than off-the-shelf FFTW 2.1.5 ported to BlueGene/L without specific optimization for the double FPU.

We obtained FFTW no-twiddle codelets for BlueGene/L by connecting FFTW to the Vienna MAP vectorizer. The resulting double FPU no-twiddle codelets are considerably faster than their scalar counterparts and strongly encourage the adaptation of FFTW 2.1.5 to BlueGene/L [20]. Experiments show that our approach provides satisfactory speed-up while the vectorization facility of IBM's XL C compiler does not speed up FFT computations. Our FFT codes were the first numerical codes developed outside IBM that were run on the first BlueGene/L prototype systems DD1 and DD2.

2 The BlueGene/L Supercomputer

The currently largest prototype of IBM's supercomputer line BlueGene/L [1] is DD1, a machine equipped with 8192 custom-made IBM PowerPC 440 FP2 processors (4096 two-way SMP chips), which achieves a LINPACK performance of R_{max} = 11.68 Tflop/s, i. e., 71 % of its theoretical peak performance of R_{peak} = 16.38 Tflop/s. This performance ranks the prototype machine on position 4 of the TOP 500 list (in June 2004). The BlueGene/L prototype machine is roughly 1/20th the physical size of machines of comparable compute power that exist today.

The largest BlueGene/L machine, which is being built for Lawrence Livermore National Laboratory (LLNL) in California, will be 8 times larger and will occupy 64 full racks. When completed in 2005, the Blue Gene/L supercomputer is expected to lead the TOP 500 list. Compared with today's fastest supercomputers, it will be an order of magnitude faster, consume 1/15th of the power and, be 10 times more compact than today's fastest supercomputers.

The BlueGene/L machine at LLNL will be built from 65,536 PowerPC 440 FP2 processors connected by a 3D torus network leading to 360 Tflop/s peak performance. BlueGene/L's processors will run at 700 MHz, whereas the current prototype BlueGene/L DD1 runs at 500 MHz. As a comparison, the Earth Simulator which is currently leading the TOP 500 list achieves a LINPACK performance of $R_{max} = 36$ Tflop/s.

2.1 BlueGene/L's Floating-Point Unit

Dedicated hardware support for complex arithmetic has the potential to accelerate many applications in scientific computing.

BlueGene/L's floating-point "double" FPU was obtained by replicating the PowerPC 440's standard FPU and adding crossover data paths and sign change capabilities to support complex multiplication. The resulting PowerPC 440 CPU with its new custom FPU is called PowerPC 440 FP2. Up to four real floating-point operations (one two-way vector fused multiply-add operation) can be issued every cycle. This double FPU has many similarities to industry-standard two-way short vector SIMD extensions like AMD's 3DNow! or Intel's SSE 3. In particular, data to be processed by the double FPU has to be aligned on 16-byte boundaries in memory.

However, the PowerPC 440 FP2 features some characteristics that are different from standard short vector SIMD architectures; namely (*i*) non-standard fused multiply-add (FMA) operations required for complex multiplications; (*ii*) computationally expensive data reorganization within two-way registers; and (*iii*) efficient support for mixing of scalar and vector operations.

Mapping Code to BlueGene/L's Double FPU. BlueGene/L's double FPU can be regarded either as a complex FPU or as a real two-way vector FPU depending on the techniques used for utilizing the relevant hardware features.

Programs using complex arithmetic can be mapped to BlueGene/L's custom FPU in a straight forward manner. Problems arise when the usage of `real` code is unavoidable. However, even for purely complex code it may be beneficial to express complex arithmetics in terms of real arithmetic. In particular, switching to real arithmetic allows to apply common subexpression elimination, constant folding, and copy propagation on the real and imaginary parts of complex numbers separately. For FFT implementations this saves a significant number of arithmetic operations, leading to improved performance.

Since the double FPU supports all classical short vector SIMD style (inter-operand, parallel) instructions (e. g., as supported by Intel SSE 2), it can be used to accelerate real computations if the targeted algorithm exhibits fine-grain parallelism. Short vector SIMD vectorization techniques can be applied to speed up real codes.

The main challenge when vectorizing for the double FPU is that one data reorder operation within the two-way register file is as expensive as one arithmetic two-way FMA operation (i. e., four floating-point operations). In addition, every cycle either one floating-point two-way FMA operation or one data reorganization instruction can be issued. On other two-way short vector SIMD architectures like SSE 3 and 3DNow! data shuffle operations are much more efficient. Thus, the double FPU requires tailoring of vectorization techniques to these features in order to produce high-performance code.

Tools for Programming BlueGene/L's Double FPU. To utilize BlueGene/L's double FPU within a numerical library, three approaches can be pursued: *(i)* The implementation of the numerical kernels in C such that IBM's VisualAge XL C compiler for BlueGene/L is able to vectorize these kernels. *(ii)* Directly implement the numerical kernels in assembly language using double FPU instructions. *(iii)* Explicitly vectorize the numerical kernels utilizing XL C's proprietary language extension to C99 that provides access to the double FPU on source level by means of data types and intrinsic functions.

The GNU C compiler port for BlueGene/L supports the utilization of no more than 32 temporary variables when accessing the double FPU. This constraint prevents automatic performance tuning on BlueGene/L using the GNU C compiler.

In this paper, we provide vectorization techniques for FFT kernels tailored to the needs of BlueGene/L. All codes are generated automatically and utilize the XL C compiler's vector data types and intrinsic functions to access the double FPU to avoid utilizing assembly language. Thus, register allocation and instruction scheduling is left to the compiler, while vectorization (i. e., floating point SIMD instruction selection) is done at the source code level by our approach.

3 Automatic Performance Tuning of Numerical Software

Discrete Fourier transforms (DFTs) are, together with linear algebra algorithms, the most ubiquitously used kernels in scientific computing.

The applications of DFTs range from small scale problems with stringent time constraints (for instance, in real time signal processing) to large scale simulations and PDE programs running on the world's largest supercomputers. Therefore, the best possible performance of DFT software is of crucial importance. However, algorithms for DFTs have complicated structure, which makes their efficient implementation a difficult problem, even on standard platforms. It is an even harder problem to efficiently map these algorithms to special hardware like BlueGene/L's double FPU.

The traditional method for achieving highly optimized numerical code is hand coding, often in assembly language. However, this approach requires a lot of expertise and the resulting code is error-prone and non-portable.

Automatic Performance Tuning. Recently, a new paradigm for software creation has emerged: the automatic generation and optimization of numerical libraries. Examples include ATLAS [27] in the field of numerical linear algebra, and FFTW [12] which introduced automatic performance tuning in FFT libraries. In the field of digital signal processing (DSP), SPIRAL [23] automatically generates tuned codes for large classes of DSP transforms by utilizing state-of-the-art coding and optimization techniques. All these systems feature code generators that generate ANSI C code to maintain portability. The generated kernels consist of up to thousands of lines of code.

To achieve top performance in connection with such codes, the exploitation of special processor features such as short vector SIMD or FMA instruction set architecture extensions (like the double FPU) is a must. Unfortunately, approaches used by vectorizing compilers to vectorize loops [2, 3, 24, 29] or basic blocks [18, 19, 21] lead to inefficient code when applied to automatically generated codes for DSP transforms. In case of these algorithms, the vectorization techniques entail large overhead for data reordering operations on the resulting short vector code as they do not have domain specific knowledge about the codes' inherent parallelism. The cost of this overhead becomes prohibitive on an architecture like BlueGene/L's double FPU where data reorganization is very expensive compared to floating-point operations. This implies that vectorization techniques have to be adapted to BlueGene/L's FPU architecture.

FFTW. FFTW implements the Cooley-Tukey FFT algorithm [26] recursively. This allows for flexible problem decomposition that is the key for FFTW's adaptability to a wide range of different computer systems.

In an initialization step FFTW's *planner* applies dynamic programming to find a problem decomposition that leads to fast FFT computation for a given size on the target machine. Whenever FFTs of planned sizes are to be computed, the *executor* applies these problem decompositions to perform the FFT computation. At the leafs of the recursion the actual work is done in code blocks called *codelets*. These codelets come in two flavors (twiddle codelets and no-twiddle codelets) and are automatically generated by a program generator named `genfft` [10, 11].

SPIRAL. SPIRAL [23] is a generator for high performance code for discrete signal transforms including the discrete Fourier transform (DFT), the discrete cosine transforms (DCTs), and many others. SPIRAL uses a mathematical approach that translates the implementation and optimization problem into a search problem in the space of structurally different algorithms and their possible implementations for the best match to a given computing platform.

SPIRAL represents the different algorithms for a signal transform as formulas in a concise mathematical language, called SPL, that is an extension of the Kronecker product formalism [14, 28]. The SPL formulas expressed in SPL are automatically generated by SPIRAL's *formula generator* and automatically translated into code by SPIRAL's special purpose SPL compiler. Performance evaluation feeds back runtime information into SPIRAL's search engine which closes the feedback loop enabling automated search for good implementations.

4 Generating Vector Code for BlueGene/L

In this section we present two approaches to vectorizing FFT kernels generated by SPIRAL and FFTW targeted for BlueGene/L's double FPU.

- Formal vectorization exploits structural information about a given FFT algorithm and extends SPIRAL's formula generator and special-purpose SPL compiler.
- The Vienna MAP vectorizer extracts two-way parallelism out of straight-line code generated by SPIRAL or FFTW's codelet generator `genfft`.

Both methods are tailored to the specific requirements of BlueGene/L's double FPU. We explain the methods in the following.

4.1 Formal Vectorization

In previous work, we developed a formal vectorization approach [6] and applied it successfully across a wide range of short vector SIMD platforms for vector lengths of two and four both to FFTW [4, 5] and SPIRAL [7, 8, 9]. We showed that neither original vector computer FFT algorithms [17, 25] nor vectorizing compilers [13, 18] are capable of producing high-performance FFT implementations for short vector SIMD architectures, even in tandem with automatic performance tuning [9].

In this work we adapt the formal vectorization approach to the specific features of BlueGene/L's double FPU and the BlueGene/L system environment, and implement it within the SPIRAL system.

Short Vector Cooley Tukey FFT. Our DFT specific formal vectorization approach is the short vector Cooley-Tukey recursion [9]. For m and n being multiples of the machine's vector length ν, the short vector Cooley-Tukey FFT recursion translates a DFT_{mn} recursively into DFT_m and DFT_n such that any further decomposition of the smaller DFTs yields vector code for vector lengths ν. It adapts the Cooley-Tukey FFT recursion [26] to short vector SIMD architectures with memory hierarchies while resembling the Cooley-Tukey FFT recursion's data flow as close as possible to keep its favorable memory access structure.

In contrast to the standard Cooley-Tukey FFT recursion, all permutations in the short vector Cooley-Tukey FFT algorithm are either block permutations shuffling vectors of length ν or operate on $k\nu$ consecutive data items with k being small. For all current short vector SIMD architectures, the short vector Cooley-Tukey FFT algorithm can be implemented using solely vector memory access operations, vector arithmetic operations and data shuffling within vector registers.

Algorithm 1 (Short Vector Cooley-Tukey FFT)

ShortVectorFFT $(mn, \nu, y \leftarrow x)$
  BlockStridePermutation$(t_0 \leftarrow y, mn/\nu, m/\nu, \nu)$
  for $i = 0 \ldots m/\nu - 1$ **do**
    VectorFFT$(\nu \times \mathrm{DFT}_n, t_1 \leftarrow t_0[i\nu n \ldots (i+1)\nu n - 1])$
    StridePermutationWithinBlocks$(t_2 \leftarrow t_1, n/\nu, \nu^2, \nu)$
    BlockStridePermutation$(t_3[i\nu n \ldots (i+1)\nu n - 1] \leftarrow t_2, n, \nu, \nu)$
  endfor
  ApplyTwiddleFactors$(t_4 \leftarrow t_3)$
  for $i = 0 \ldots n/\nu - 1$ **do**
    VectorFFT$(\nu \times \mathrm{DFT}_m, y[i\nu m \ldots (i+1)\nu m - 1] \leftarrow t_4[i\nu m \ldots (i+1)\nu m - 1])$
  endfor

Algorithm 1 describes the short vector Cooley-Tukey FFT algorithm in pseudo code. It computes a DFT_{mn} by recursively calling VectorFFT() to compute vectors of $\nu \times \mathrm{DFT}_m$ and $\nu \times \mathrm{DFT}_n$, respectively. BlockStridePermutation() applies a coarse-grain block stride permutation [1] to reorder blocks of ν elements. StridePermutationWithinBlocks() applies n/ν fine-grain stride permutations to shuffle elements within blocks of ν^2 elements. ApplyTwiddleFactors() multiplies a data vector by complex roots of unity called twiddle factors. In our implementation the functions BlockStridePermutation(), StridePermutationWithinBlocks(), and ApplyTwiddleFactors() are combined with the functions VectorFFT() to avoid multiple passes through the data.

We extended Spiral's formula generator and SPL compiler to implement the short vector Cooley-Tukey recursion. For a given DFT_{mn}, Spiral's search engine searches for the best choice of m and n as well as for the best decomposition of DFT_m and DFT_n.

Adaptation to BlueGene/L. The short vector Cooley-Tukey FFT algorithm was designed to support all short vector SIMD architectures currently available. It is built on top of a set of C macros called the portable SIMD API that abstracts the details of a given short vector SIMD architecture. To adapt Spiral's implementation of the short vector Cooley-Tukey FFT algorithm to BlueGene/L we had to *(i)* implement the portable SIMD API for BlueGene/L's double FPU (with $\nu = 2$), and *(ii)* utilize the double FPU's fused multiply-add instructions whenever possible.

Spiral's implementation of the portable SIMD API mandates the definition of vector memory operations and vector arithmetic operations. These operations were implemented using the intrinsic and C99 interface provided by IBM's XL C compiler for BlueGene/L. Fig. 1 shows a part of the portable SIMD API for BlueGene/L.

A detailed description of our formal vectorization method and its application to a wide range of short vector SIMD architectures can be found in [6, 7, 9].

[1] A stride permutations can be seen as a transposition of a rectangular two-dimensional array [14].

```
/* Vector arithmetic operations and declarations*/
#define VECT_ADD(c,a,b) (c) = __fpadd(a,b)
#define VECT_MSUB(d,a,b,c) (d) = __fpmsub(c,a,b)
#define DECLARE_VECT(vec) _Complex double vec

/* Data shuffling */
#define C99_TRANSPOSE(d1, d2, s1, s2) {\
    d1 = __cmplx (__creal(s1), __creal(s2));\
    d2 = __cmplx (__cimag(s1), __cimag(s2)); }

/* Memory access operations */
#define LOAD_VECT(trg, src) (trg) = __lfpd((double *)(src))

#define LOAD_INTERL_USTRIDE(re, im, in) {\
      DECLARE_VECT(in1); DECLARE_VECT(in2);\
      LOAD_VECT(in1, in); LOAD_VECT(in2, (in) + 2);\
      C99_TRANSPOSE(re, im, in1, in2); }
```

Fig. 1. Macros included in SPIRAL's implementation of the portable SIMD API for IBM's XL C compiler for BlueGene/L

4.2 The Vienna MAP Vectorizer

This section introduces the Vienna MAP vectorizer [6, 15, 16] that automatically extracts two-way SIMD parallelism out of given numerical straight-line code. A peephole optimizer directly following the MAP vectorizer additionally supports the extraction of SIMD fused multiply-add instructions.

Vectorization of Straight-Line Code. Existing approaches to vectorizing basic blocks originate from very long instruction word (VLIW) or from SIMD digital signal processors (DSP) compiler research [3, 18, 19], and try to find either an efficient mix of SIMD and scalar instructions to do the required computation or insert data shuffling operations to allow for parallel computation. Due to the fact that SIMD data reordering operations are very expensive on IBM BlueGene/L's double FPU and FFT kernels have complicated data flow, these approaches are rendered suboptimal for the vectorization of FFT kernels for the double FPU.

MAP's vectorization requires that `all` computation is performed by SIMD instructions, while attempting to keep the SIMD reordering overhead reasonably small. The only explicit data shuffling operations allowed are swapping the two entries of a two-way vector register. In addition, MAP requires inter- and intra-operand arithmetic operations. On short vector extensions not supporting intra-operand arithmetic operations, additional shuffle operations may be required. When vectorizing complex FFT kernels for the double FPU, no data shuffle operations are required at all.

The Vectorization Algorithm. The MAP vectorizer uses depth-first search with chronological backtracking to discover SIMD style parallelism in a scalar

Table 1. Two-way Vectorization. Two scalar add instructions are transformed into one vector vadd instruction

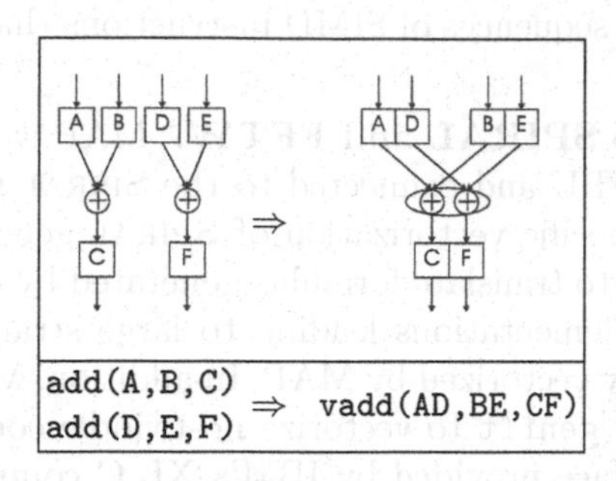

`add(A,B,C)` `add(D,E,F)`	⇒	`vadd(AD,BE,CF)`

code block. MAP's input is scalar straight-line codes that may contain array accesses, index computation, and arithmetic operations. Its output is vector straight-line code containing vector arithmetic and vector data shuffling operations, vector array accesses, and index computation. MAP describes its input as scalar directed acyclic graph (DAG) and its output as vector DAG.

In the vectorization process, pairs of scalar variables `s`, `t` are combined, i. e., *fused*, to form SIMD variables `st = (s,t)` or `ts = (t,s)`. The arithmetic instructions operating on `s` and `t` are combined, i. e., *paired*, to a sequence of SIMD instructions. This vectorization process translates nodes of the scalar DAG into nodes of the vector DAG. Table 1 shows an example of fusing two additions into a vector addition.

As initialization of the vectorization algorithm, store instructions are combined non-deterministically by fusing their respective source operands. The actual vectorization algorithm consists of two steps.

(*i*) Pick `I1 = (op1,s1,t1,d1)` and `I2 = (op2,s2,t2,d2)`, two scalar instructions that have not been vectorized, with `(d1,d2)` or `(d2,d1)` being an existing fusion.

(*ii*) Non-deterministically pair the two scalar operations `op1` and `op2` into one SIMD operation. This step may produce new fusions or reuse a existing fusion, possibly requiring a data shuffle operation.

The vectorizer alternatingly applies these two steps until either the vectorization succeeds, i. e., thereafter all scalar variables are part of at most one fusion and all scalar operations have been paired, or the vectorization fails. If the vectorizer succeeds, MAP commits to the first solution of the search process.

Non-determinism in vectorization arises due to different vectorization choices. For a fusion `(d1,d2)` there may be several ways of fusing the source operands `s1,t1,s2,t2`, depending on the pairing `(op1,op2)`.

Peephole Based Optimization. After the vectorization, a local rewriting system is used to implement peephole optimization on the obtained vector DAG. The rewriting rules are divided into three groups. The first group of rewriting rules aims at (*i*) minimizing the number of instructions, (*ii*) eliminating redundancies and dead code, (*iii*) reducing the number of source operands, (*iv*) copy propagation, and (*v*) constant folding. The second group of rules is used to extract SIMD-FMA instructions. The third group of rules rewrites unsupported SIMD instructions into sequences of SIMD instructions that are available on the target architecture.

Connecting MAP to SPIRAL and FFTW. MAP was adapted to support BlueGene/L's double FPU and connected to the SPIRAL system as backend to provide BlueGene/L specific vectorization of SPIRAL generated code. SPIRAL's SPL compiler was used to translate formulas generated by the formula generator into fully unrolled implementations leading to large straight-line codes. These codes were subsequently vectorized by MAP. In addition, MAP was connected to FFTW's code generator `genfft` to vectorize no-twiddle codelets. MAP's output uses the intrinsic interface provided by IBM's XL C compiler for BlueGene/L. This allows MAP to vectorize computational kernels but leaves register allocation and instruction scheduling to the compiler.

5 Experimental Results

We evaluated both the formal vectorization approach and the Vienna MAP vectorizer in combination with SPIRAL, as well as the Vienna MAP vectorizer as backend to FFTW's `genfft`. We evaluated 1D FFTs with problem sizes $N = 2^2$, $2^3, \ldots, 2^{10}$ using SPIRAL, and FFT kernels (FFTW's no-twiddle codelets) of sizes $N = 2, 3, 4, \ldots, 16, 32, 64$. All experiments were performed on one PowerPC 440 FP2 processor of the BlueGene/L DD1 prototype running at 500 MHz. Performance data is given in pseudo G op/s ($5N \log_2(N)$/runtime) or as speed-up with respect to the best code without double FPU support (scalar code).

For vector lengths $N = 2^2, \ldots, 2^{10}$, we compared the following FFT implementations: (*i*) The best vectorized code found by SPIRAL utilizing formal vectorization; (*ii*) the best vectorized code found by SPIRAL utilizing the Vienna MAP vectorizer; (*iii*) the best scalar FFT implementation found by SPIRAL (XL C's vectorizer and FMA extraction turned off); (*iv*) the best vectorizer FFT implementation found by SPIRAL using the XL C compiler's vectorizer and FMA extraction; and (*v*) FFTW 2.1.5 ported to BlueGene/L (without double FPU support).

In the following we discuss the results summarized in Fig. 2. The best scalar code generated by SPIRAL without compiler vectorization (thin line, crosses) serves as baseline, as it is the fastest scalar code we obtained. Turning on the vectorization and FMA extraction provided by IBM's XL C compiler for BlueGene/L (dashed line, x) actually slows down SPIRAL generated FFT code by up to 15 %. FFTW 2.1.5 (thin line, asterisk) runs at approximately 80 % of the performance provided by the best scalar SPIRAL generated code. Utilizing the Vienna MAP vectorizer as backend for SPIRAL (bold dashed line, hollow squares) produces the fastest code for problem sizes $N = 2^2$, 2^3, 2^4, for which it yields speed-ups of up to 55 %. For larger sizes MAP suffers from problems with the XL C compiler's register allocation; for problem sizes larger than 2^7, straight-line code becomes unfeasible. Formal vectorization (bold line, full squares) produces

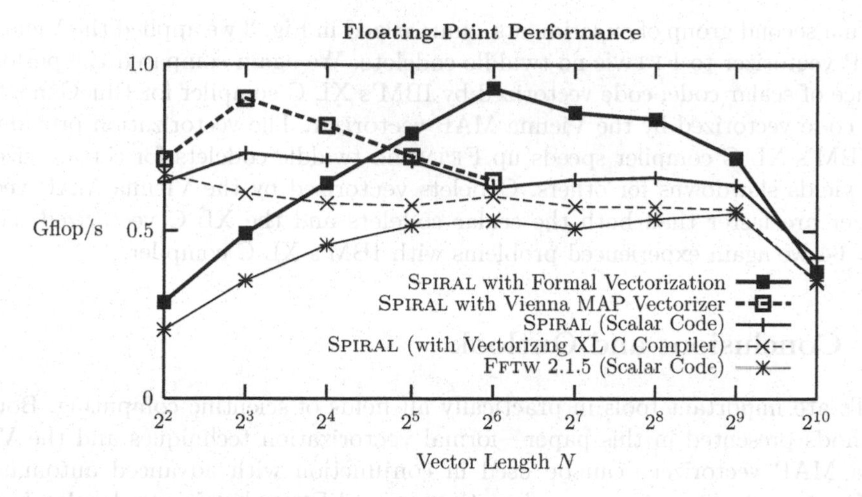

Fig. 2. Performance comparison of generated FFT scalar code and generated FFT vector code obtained with the MAP vectorizer and formal vectorization compared to the best scalar code and the best vectorized code (utilizing the VisualAge XLC for BG/L vectorizing compiler) found by SPIRAL. Performance is displayed in *pseudo Gflop/s* ($5N \log_2(N)$/runtime with N being the problem size)

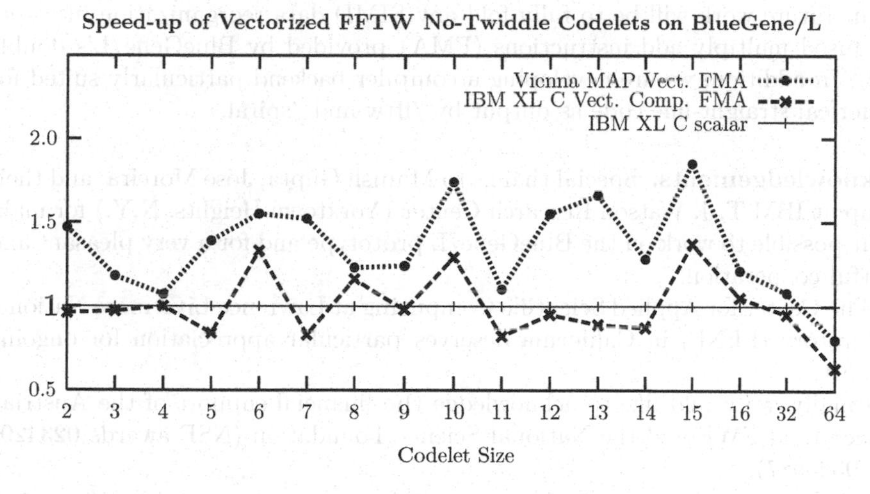

Fig. 3. Speed-up of the vectorization techniques applied by the MAP vectorizer compared to scalar code and code vectorized by IBM's VisualAge XL C compiler

the fastest code for $N = 2^5, \ldots, 2^9$, speeding up the computation by up to up to 45 %. Smaller problem sizes are slower due to the overhead produced by formal vectorization; for $N = 2^{10}$ we again experienced problems with the XL C compiler. In addition, all codes slow down at $N = 2^{10}$ considerably due to limited data-cache capacity.

In a second group of experiments summarized in Fig. 3 we applied the Vienna MAP vectorizer to FFTW's no-twiddle codelets. We again compared the performance of scalar code, code vectorized by IBM's XL C compiler for BlueGene/L, and code vectorized by the Vienna MAP vectorizer. The vectorization provided by IBM's XL C compiler speeds up FFTW no-twiddle codelets for certain sizes but yields slowdowns for others. Codelets vectorized by the Vienna MAP vectorizer are faster than both the scalar codelets and the XL C vectorized. For $N = 64$ we again experienced problems with IBM's XL C compiler.

6 Conclusions and Outlook

FFTs are important tools in practically all fields of scientific computing. Both methods presented in this paper—formal vectorization techniques and the Vienna MAP vectorizer—can be used in conjunction with advanced automatic performance tuning software such as SPIRAL and FFTW helping to develop high performance implementations of FFT kernels on the most advanced computer systems.

Performance experiments carried out on a BlueGene/L prototype show that our two vectorization techniques are able to speed up FFT code considerably.

Nevertheless, even better performance results will be yielded by improving the current BlueGene/L version of the Vienna MAP vectorizer. An integral part of our future work will be to fully fold any SIMD data reorganization into special fused multiply add instructions (FMA) provided by BlueGene/L's double FPU. In addition, we are developing a compiler backend particularly suited for numerical straight-line code as output by /fftw and /spiral.

Acknowledgements. Special thanks to Manish Gupta, José Moreira, and their group at IBM T. J. Watson Research Center (Yorktown Heights, N.Y.) for making it possible to work on the BlueGene/L prototype and for a very pleasant and fruitful cooperation.

The Center for Applied Scientific Computing at Lawrence Livermore National Laboratory (LLNL) in California deserves particular appreciation for ongoing support.

Finally, we would like to acknowledge the financial support of the Austrian science fund FWF and the National Science Foundation (NSF awards 0234293 and 0325687).

References

[1] G. Almasi et al., "An overview of the BlueGene/L system software organization," Proceedings of the Euro-Par '03 Conference on Parallel and Distributed Computing LNCS 2790, pp. 543–555, 2003.

[2] R. J. Fisher and H. G. Dietz, "The SCC Compiler: SWARing at MMX and 3DNow," in *Languages and Compilers for Parallel Computing (LCPC99)*, LNCS 1863, pp.399–414, 2000.

[3] ——, "Compiling for SIMD within a register," in *Languages and Compilers for Parallel Computing*, LNCS 1656, pp. 290–304, 1998.
[4] F. Franchetti, "A portable short vector version of FFTW," in *Proc. Fourth IMACS Symposium on Mathematical Modelling (MATHMOD 2003)*, vol. 2, pp. 1539–1548, 2003.
[5] F. Franchetti, H. Karner, S. Kral, and C. W. Ueberhuber, "Architecture independent short vector FFTs," in *Proc. ICASSP*, vol. 2, pp. 1109–1112, 2001.
[6] F. Franchetti, S. Kral, J. Lorenz, and C. W. Ueberhuber, "Efficient Utilization of SIMD Extensions," *IEEE Proceedings Special Issue on Program Generation, Optimization, and Platform Adaption, to appear.*
[7] F. Franchetti and M. Püschel, "A SIMD Vectorizing Compiler for Digital Signal Processing Algorithms," in *Proc. IPDPS*, pp. 20–26, 2002.
[8] ——, "Short vector code generation and adaptation for DSP algorithms." in *Proceedings of the International Conerence on Acoustics, Speech, and Signal Processing. Conference Proceedings (ICASSP'03)*, vol. 2, pp. 537–540, 2003.
[9] ——, "Short vector code generation for the discrete Fourier transform." in *Proceedings of the 17th International Parallel and Distributed Processing Symposium (IPDPS'03)*, pp. 58–67, 2003.
[10] M. Frigo, "A fast Fourier transform compiler," in *Proceedings of the ACM SIGPLAN '99 Conference on Programming Language Design and Implementation.* New York: ACM Press, pp. 169–180, 1999.
[11] M. Frigo and S. Kral, "The Advanced FFT Program Generator GENFFT," in AURORA *Technical Report TR2001-03*, vol. 3, 2001.
[12] M. Frigo and S. G. Johnson, "FFTW: An Adaptive Software Architecture for the FFT," in *ICASSP 98*, vol. 3, pp. 1381–1384, 1998.
[13] Intel Corporation, "Intel C/C++ compiler user's guide," 2002.
[14] J. Johnson, R. W. Johnson, D. Rodriguez, and R. Tolimieri, "A methodology for designing, modifying, and implementing Fourier transform algorithms on various architectures," *IEEE Trans. on Circuits and Systems*, vol. 9, pp. 449–500, 1990.
[15] S. Kral, F. Franchetti, J. Lorenz, and C. Ueberhuber, "SIMD vectorization of straight line FFT code," Proceedings of the Euro-Par '03 Conference on Parallel and Distributed Computing LNCS 2790, pp. 251–260, 2003.
[16] ——, "FFT compiler techniques," Proceedings of the 13th International Conference on Compiler Construction LNCS 2790, pp. 217–231, 2004.
[17] S. Lamson, "SCIPORT," 1995. [Online]. Available: **http://www.netlib.org/scilib/**
[18] S. Larsen and S. Amarasinghe, "Exploiting superword level parallelism with multimedia instruction sets," *ACM SIGPLAN Notices*, vol. 35, no. 5, pp. 145–156, 2000.
[19] R. Leupers and S. Bashford, "Graph-based code selection techniques for embedded processors," *ACM Transactions on Design Automation of Electronic Systems.*, vol. 5, no. 4, pp. 794–814, 2000.
[20] J. Lorenz, S. Kral, F. Franchetti, and C. W. Ueberhuber, "Vectorization Techniques for BlueGene/L's Double FPU," *IBM Journal of Research and Development, to appear.*
[21] M. Lorenz, L. Wehmeyer, and T. Draeger, "Energy aware compilation for DSPs with SIMD instructions," *Proceedings of the 2002 Joint Conference on Languages, Compilers, and Tools for Embedded Systems & Software and Compilers for Embedded Systems (LCTES'02-SCOPES'02).*, pp. 94–101, 2002.
[22] D. Mirkovic and S. L. Johnsson, "Automatic Performance Tuning in the UHFFT Library," in *Proc. ICCS 01*, pp. 71–80, 2001.

[23] M. Püschel, B. Singer, J. Xiong, J. M. F. Moura, J. Johnson, D. Padua, M. Veloso, and R. W. Johnson, "SPIRAL: A generator for platform-adapted libraries of signal processing algorithms," *Journal on High Performance Computing and Applications*, special issue on Automatic Performance Tuning, Vol. 18, pp. 21–45, 2004.
[24] N. Sreraman and R. Govindarajan, "A vectorizing compiler for multimedia extensions," *International Journal of Parallel Programming*, vol. 28, no. 4, pp. 363–400, 2000.
[25] P. N. Swarztrauber, "FFT algorithms for vector computers," *Parallel Comput.*, vol. 1, pp. 45–63, 1984.
[26] C. F. Van Loan, *Computational Frameworks for the Fast Fourier Transform*, ser. Frontiers in Applied Mathematics. Philadelphia: Society for Industrial and Applied Mathematics, 1992, vol. 10.
[27] R. C. Whaley, A. Petitet, and J. J. Dongarra, "Automated empirical optimizations of software and the ATLAS project," *Parallel Comput.*, vol. 27, pp. 3–35, 2001.
[28] J. Xiong, J. Johnson, R. Johnson, and D. Padua, "SPL: A Language and Compiler for DSP Algorithms," in *Proceedings of the Conference on Programming Languages Design and Implementation* (PLDI), pp. 298–308, 2001.
[29] H. Zima and B. Chapman, *Supercompilers for Parallel and Vector Computers.* New York: ACM Press, 1991.

A Survey of High-Quality Computational Libraries and Their Impact in Science and Engineering Applications

L.A. Drummond[1], V. Hernandez[2], O. Marques[1], J.E. Roman[2], and V. Vidal[2]

[1] Lawrence Berkeley National Laboratory,
One Cyclotron Road, MS 50F-1650, Berkeley, California 94720, USA
{ladrummond, oamarques}@lbl.gov

[2] Universidad Politécnica de Valencia,
DSIC, Camino de Vera, s/n – 46022 Valencia, Spain
{vhernand, jroman, vvidal}@dsic.upv.es

Abstract. Recently, a number of important scientific and engineering problems have been successfully studied and solved by means of computational modeling and simulation. Many of these computational models and simulations benefited from the use of available software tools and libraries to achieve high performance and portability. In this article, we present a reference matrix of the performance of robust, reliable and widely used tools mapped to scientific and engineering applications that use them. We aim at regularly maintaining and disseminating this matrix to the computational science community. This matrix will contain information on state-of-the-art computational tools, their applications and their use.

1 Introduction

In recent years, a number of important scientific and engineering problems, ranging in scale from the atomic to the cosmic, have been successfully studied and solved by means of computational modeling and simulation. Many of these computational models and simulations benefited from the use of available software tools and libraries to achieve high performance. These tools have facilitated code portability across many computational platforms while guaranteeing the robustness and correctness of the algorithmic implementations.

Software development time includes the time required for individual tasks, such as coding and debugging, and the time needed for group coordination, such as reporting bugs, synchronizing source code, and preparing releases. The inefficiency of ad hoc approaches to coordinating group activities has already become a significant hidden cost in many large scientific computing projects. Software tools greatly reduce the software development and maintenance costs because other specialized software development teams have taken on the hardy task of guaranteeing portability, scalability and robustness. In addition, software

M. Daydé et al. (Eds.): VECPAR 2004, LNCS 3402, pp. 37–50, 2005.

tools offer a variety of solutions to a problem through interfaces. Nowadays, there are several initiatives and approaches for making tool interfaces interoperable. Interoperability aggregates benefits to the use of tools because interoperable tools facilitate code changes, that are due to an increase in the complexity of the problem being addressed or algorithmic formulation.

As a result of all the advancements in tool development and deployment, there is nowadays a growing need and interest among computational scientists for identifying software tools that have been successfully used in the solution of computational problems that are similar to theirs. In this article, we highlight some of the most successful and robust numerical tool development projects and example of their applications. For the centerpiece of this work, we propose the creation of a matrix of software tools, their most relevant applications in science and engineering, with comments on tool performance and functionality with reference to other available similar tools. Further, we also propose to regularly update this table with feedback and results from the international community of computational scientists, as well as with software developments from other teams. We based our first efforts on current work with the U.S. Department of Energy (DOE) Advanced CompuTational Software (ACTS) Collection project [1] and the multidisciplinary research carried out by members of the High Performance Networking and Computing group of the Valencia University of Technology in Spain. Our goal is to promote the reuse of robust software tools and at the same time provide guidance on their use. Additionally, this reference matrix will be instrumental in the dissemination of state of the art software technology, while providing feedback not only to users but also to tool developers, software research groups and even to computer vendors. In Section 2, we briefly introduce the functionality of the selected computational tools. Section 3 provides an introduction to the different applications used in the performance matrix presented in Section 4. Lastly, conclusions are presented in Section 5.

2 Software Tools

In the computational science community, the complexity of today's scientific and engineering applications has led to new computational challenges that stress the need for more flexible, portable, scalable, reliable and high performing code developments. In turn, these challenges along with changes in the computer technology suggest that we need to move to a higher-level programing style, where library tools, and perhaps to some extent the compilers, take care of the details of the implementations [2,3]. Several examples of these state-of-the-art software developments will be studied in this article. First we take a look at some of the leading tool development projects. For the sake of space, we focus mostly on computational tools that provide implementation of numerical algorithms.

2.1 Tools from the ACTS Collection

The ACTS Collection comprises a set of computational tools developed primarily at DOE laboratories, sometimes in collaboration with universities and other funding agencies (NSF, DARPA), aimed at simplifying the solution of common and important computational problems. A number of important scientific problems have been successfully studied and solved by means of computer simulations built on top of ACTS Tools [4, 5, 6]. Here, we will focus on a subset of these tools. Namely, we will review some high performing applications that use ScaLAPACK, SuperLU, PETSc, SUNDIALS, and Global Arrays. We begin with brief descriptions of these tools. The reader is referred to the ACTS Information Center [7] for more details on these and other tools available in the Collection.

ScaLAPACK [8] is a library of high-performance linear algebra routines for distributed-memory message-passing multiple instruction multiple data computers and networks of workstations. It is complementary to the LAPACK library which provides analogous software for workstations, vector supercomputers, and shared-memory parallel computers. The ScaLAPACK library contains routines for solving systems of linear equations, least squares, eigenvalue problems and singular value problems. It also contains routines that handle many computations related to those, such as matrix factorizations or estimation of condition numbers.

SuperLU [9] is a general purpose library for the direct solution of large, sparse, nonsymmetric systems of linear equations. The library is written in C and is callable from either C or Fortran. SuperLU performs an LU decomposition with numerical pivoting and triangular system solves through forward and back substitution. The LU factorization routines can handle non-square matrices but the triangular solves are performed only for square matrices. Before the factorization step, the matrix columns may be preordered through library or user provided routines. Working precision iterative refinement subroutines are provided for improved backward stability. Routines are also provided to equilibrate the system, estimate the condition number, calculate the relative backward error, and estimate error bounds for the refined solutions.

PETSc (**P**ortable, **E**xtensible **T**oolkit for **S**cientific **c**omputation) [10] provides sets of tools for the parallel, as well as serial, numerical solution of PDEs that require solving large-scale, sparse linear and nonlinear systems of equations. PETSc includes nonlinear and linear equation solvers that employ a variety of Newton techniques and Krylov subspace methods. PETSc provides several parallel sparse matrix formats, including compressed row, block compressed row, and block diagonal storage. PETSc has a growing number of application users and development projects that re-use some of its functionality (see Section on SLEPc). The PETSc team has long worked on the development of interoperable interfaces with other ACTS tools, as well as other widely used tools.

SUNDIALS (**SU**ite of **N**onlinear and **DI**fferential/**AL**gebraic equation **S**olvers) [11] refers to a family of four closely related solvers: CVODE, for systems of ordinary differential equations; CVODES, which is variant of CVODE for sensitivity analysis; KINSOL, for systems of nonlinear algebraic equations; and IDA, for systems of differential-algebraic equations. These solvers have some code modules in common, primarily a module of vector kernels and generic linear system solvers, including one based on a scaled preconditioned GMRES method. All of the solvers are suitable for either serial or parallel environments.

Global Arrays (GA) [12] facilitates the writing of parallel programs that use large arrays distributed across different processing elements. GA offers a shared-memory view of distributed arrays without destroying their non-uniform memory access characteristics. Originally developed to support arrays as vectors and matrices (one or two dimensions), currently it supports up to seven dimensions in Fortran and even more in C. GA also comes with a visualizer that uses trace files to animate array access patterns. Its main purpose is to analyze the impact of distribution on performance. Additionally, GA offers some interfaces to numerical libraries like ScaLAPACK, TAO and the eigensolver library Peigs [13]. GA was originally developed for use in the NWChem computational chemistry package but has also been used in other chemistry packages such as GAMESS-UK and Columbus, as well as in other scientific fields.

2.2 Other Software Tools

The following tools have also made relevant contributions to computational sciences and are also included in the reference matrix.

SLEPc (**S**calable **L**ibrary for **E**igenvalue **P**roblem **c**omputations) [14], is a software library for computing a few eigenvalues and associated eigenvectors of a large sparse matrix or matrix pencil. It has been developed on top of PETSc and enforces the same programming paradigm. The emphasis of the software is on methods and techniques appropriate for problems in which the associated matrices are sparse, for example, those arising after the discretization of partial differential equations. SLEPc provides basic methods as well as more sophisticated algorithms and interfaces to some external software packages such as PARPACK [15]. It also provides built-in support for spectral transformations such as the shift-and-invert technique. SLEPc is a general library in the sense that it covers standard and generalized eigenvalue problems, both Hermitian and non-Hermitian, using either real or complex arithmetic.

SLICOT provides implementations of numerical algorithms for computations in systems and control theory [16]. The basic ideas behind SLICOT are usefulness of algorithms, robustness, numerical stability and accuracy, performance with respect to speed and memory requirements, portability and reusability, standardization and benchmarking. The current version of SLICOT consists of about 400 user-callable and computational routines in various domains of systems and control. Almost all of these routines have associated on-line documentation. About

200 routines have associated example programs, data and results. Reusability of the software is obtained by using standard Fortran 77, so as SLICOT can serve as the core for various existing and future computer aided control system design platforms and production quality software. SLICOT routines can be linked to MATLAB through a gateway compiler. P-SLICOT is based on ScaLAPACK and provides parallel versions of some SLICOT's routines.

3 Applications

Cosmic Microwave Background Data Analysis. A Microwave Anisotropy Dataset Computational Analysis Package (MADCAP) [17] has been developed to determine the two-point angular correlation functions (or power spectra) of the Cosmic Microwave Background (CMB) temperature and polarization modes, which are both a complete characterization of the CMB if the fluctuations are Gaussian, and are statistics which can be predicted by candidate cosmological models. MADCAP recasts the extraction of a CMB power spectrum from a map of the sky into a problem in dense linear algebra, and employs ScaLAPACK for its efficient parallel solution. The first implementation of MADCAP on a 512-processor Cray T3E used Cholesky decomposition of a symmetric positive definite matrix, which corresponds to the correlation matrix, followed by repeated triangular solves. It typically achieved 70-80% of theoretical peak performance. Porting to a 6086 processor IBM SP3 was trivial, but the initial performance was disappointing: 30% of peak. To reduce the communication overhead on this less tightly-coupled system, the Cholesky decomposition and triangular solves were replaced by explicit inversions and matrix-matrix multiplications. The cohesion of the ScaLAPACK library made this a trivial programming task, leading to a performance increase to around 50-60% of peak. MADCAP has recently been rewritten to exploit the parallelism inherent in the need to perform matrix-matrix multiplications, using ScaLAPACK's block cyclic distribution. The result has been near-perfect scaling, with $50 \pm 2\%$ of peak being achieved analyzing the same data on 1024, 2048, 3072 and 4096 processors.

Collisional Breakup in a Quantum System of Three Charged Particles. The complete solution of electron-impact ionization of hydrogen is the simplest nontrivial example of atomic collision theory. This problem remained open and challenged scientists for decades until it was successfully solved by the exterior complex scaling technique described in [6]. This technique employs a two-dimensional radial grid finite difference approximations for derivatives, and requires the solution of large, complex nonsymmetric linear systems. In this context, SuperLU was used to construct preconditioners for a conjugate gradient squared algorithm [18]. This approach allowed for the solution of order up to 736,000 on 64 processors of a Cray T3E, and up to 8.4 million on 64 processors of an IBM SP.

NWChem [19], developed at the Pacific Northwest National Laboratory, is a computational chemistry package that aims to be scalable both in its ability to treat large problems efficiently, and in its usage of available parallel computing resources. NWChem provides many methods to compute the properties of molecular and periodic systems using standard quantum mechanical descriptions of the electronic wavefunction or density. In addition, it has the capability to perform classical molecular dynamics and free energy simulations. These approaches may be combined to perform mixed quantum-mechanics and molecular-mechanics simulations. NWChem uses GA for the manipulation of data and computation in a distributed computing environment. The use of GA gives NWChem the necessary flexibility to scale the size of the problem and the number of processors used in the computation without any direct changes to the code. NWChem has been inside many computational chemistry applications and it has exhibited very good scaling (at least 70% efficiency [20]).

FUN3D is a code developed at NASA-Langley for investigating algorithmic issues related to the calculation of turbulent flows on unstructured grids. It is a tetrahedral vertex-centered unstructured grid code for compressible and incompressible Euler and Navier-Stokes equations. FUN3D is a valuable legacy code that has been promptly parallelized with the help of the PETSc library [21]. The developers of the FUN3D-PETSc code won the Gordon Bell Prize, 1999 in "Special Category" for achieving 0.23 Tflop/s on 3072 nodes of ASCI Red for unstructured mesh calculation.

Parallel Full-Potential Linearized Augmented Plane-Wave. Electronic structure calculations have become increasingly important for describing and predicting properties of materials. Developments in theory, modeling and simulation, resulted in new opportunities for the study of quantum dots, nanotubes, nanoparticles, nano-interfaces, etc. The study of such systems, however, usually require hundreds of atoms and can only be realized by using high-end computers and efficient algorithmic implementations. P-FLAPW [4] is a parallel implementation of the FLAPW (full-potential linearized augmented plane-wave) method for electronic structure calculations. This is one of the most accurate and heavily used method in materials science for determining structural, electronic and magnetic properties of new materials. The method is based on a density functional approach to approximating and solving Schrödinger's equation. The main part of the calculation involves the construction of the Hamiltonian matrix and then its diagonalization using ScaLAPACK to determine the lowest 5-10% of the eigenvalues and eigenfunctions which corresponds to the wavefunctions and energies of the electrons in the system being studied. The construction of the Hamiltonian and its diagonalization take approximately equal times. The use of ScaLAPACK has allowed the simulation of systems containing up to 700 atoms [4] on machines such as the IBM SP and the CRAY T3E. This corresponds to matrix sizes of about 35,000, which run efficiently on up to 512 processors on the SP and T3E. ScaLAPACK is also used for eigenvalue calculations in other LAPW codes such as the widely used WEIN2K code.

NIMROD. The NIMROD Project [22] is developing a modern computer code suitable for the study of long-wavelength, low-frequency, nonlinear phenomena in fusion reactor plasmas. These phenomena involve large-scale changes in the shape and motion of the plasma and severely constrain the operation of fusion experiments. NIMROD uses a high-order finite element representation of the poloidal plane and a finite Fourier series representation of the toroidal direction. The time advance is semi-implicit. SuperLU was incorporated in NIMROD as an alternative for a PCG solver. As a result, for each time-step of the linearized problem, SuperLU is more than 100 times and 64 times faster than PCG on 1 and 9 processors, respectively, of an IBM SP. The improvement in solve time for nonlinear problems is 10-fold. In this case, SuperLU is used to precondition matrix-free CG solves, in addition to solving other linear systems, and the overall improvement in code performance is 5-fold. In summary, the performance improvements are dramatic and equivalent to 3-5 years progress in hardware.

M3D [23] is multi-institutional code that implements a Multilevel, 3D, parallel, plasma simulation code. It is suitable for performing linear and non-linear calculations of plasmas in toroidal topologies including tokamaks and stellarators. Recent tokamak studies conducted with M3D have addressed such topics as high-beta disruptions, two-fluid effects on internal modes, Toroidicity-induced Alfven Eigenmodes (TAE modes) using a hybrid model, pellet injection, and the formation of current holes in discharges with off-axis current drive. M3D uses PETSc for the manipulation of unstructured meshed problems and solution of linear systems of equations. Non-linear solvers from PETSc are currently being studied for integration into the M3D code. The M3D developers reported that PETSc not only reduced the development time of M3D but also allowed them to easily test with different already parallelized preconditioners present in PETSc and also in another ACTS tool called Hypre. The analysis of steady state electromagnetic waves in cavities has many applications such as the design of industrial microwave ovens, particle accelerators or semi-anechoic chambers for electromagnetic compatibility studies.

Omega3P is a parallel distributed-memory code intended for the modeling and analysis of accelerator cavities. This code calculates cavity mode frequencies and field vectors by solving a generalized eigenvalue problem $Kx = \lambda Mx$ arising from a finite element discretization of Maxwell's equations. In this problem, K is the symmetric positive semi-definite stiffness matrix and M is the positive definite mass matrix. Both matrices are very sparse. The eigenvalues of interest are in the interior of the spectrum and usually tightly clustered. The original inexact shift-invert Lanczos algorithm [24] implemented in Omega3P showed poor convergence in applications where the matrices are ill-conditioned and the eigenvalues are not well separated, as in the case of long accelerator structures with many cells. A parallel exact shift-invert eigensolver was then devised, where SuperLU was incorporated with PARPACK [15], which implements the implicit restarted Arnoldi algorithm. This strategy allowed for the solution of a problem of order 7.5 million with 304 million nonzeros.

CAVIRES is another code for electromagnetic analysis of cavities. The underlying mathematical model is the eigenvalue problem solving the Maxwell equations in a bounded volume. The source-free wave equation is discretized with finite elements, which provide enough flexibility to handle strong variations in the solution as in the case of inhomogeneous cavities with several materials. Edge elements are used in order to avoid so-called spurious modes. All the element contributions yields a generalized algebraic eigenvalue problem in which the matrices can be complex and non-Hermitian. The dimension of the problem can be in the order of millions in research applications. In [25], the solution of the associated eigenvalue problem is carried out with SLEPc. The code computes the eigenvectors corresponding to a few of the smallest positive nonzero eigenvalues, avoiding the unwanted high-dimensional null-space. The efficiency obtained on 32 processors of an SGI Altix was 57 % for a problem of size 150,000.

Lambda Modes. Lambda modes can be used to study the steady state neutron flux distribution inside the reactor core (to detect sub-critical modes responsible for regional instabilities) as well as for transient analysis. The lambda modes equation is a differential eigenvalue problem derived from the neutron diffusion equation. The discretization of the problem through a collocation method leads to a generalized algebraic eigensystem whose dominant eigenmodes provide useful information about the neutron flux distribution and the safety of the reactor [26]. The matrices that are obtained have a block structure and the number of blocks depends on how many levels are considered when discretizing the energy. If upscattering effects are not considered the problem can be reduced to a standard eigenvalue problem in which the inverses of the diagonal blocks make it necessary to solve several linear systems in each step of the iterative eigensolver. The problem has been tackled with SLEPc for solving the eigenvalue problem combined with the linear system solvers provided by PETSc. Several benchmark reactors have been used for validation, with typical problem sizes ranging from 20,000 to 500,000. When computing only the dominant mode, A 55% efficiency was achieved with a problem size of 125,000, on 16 procs of an SGI cluster.

Time-Dependent Neutron Diffusion Equation. Complex space and time phenomena related with the neutron flux oscillations have been observed in several boiling water reactors. Furthermore, rapid transients of reactor power caused by a reactivity insertion due to a postulated drop or abnormal withdrawal of control rods from the core have strong space dependent feedback associated with them. Therefore, a fast transient analysis code is needed for simulating these phenomena. The solution is based on discretizing the spatial domain, obtaining a large system of linear stiff ordinary differential equations (ODEs). FCVODE, which is part of the SUNDIALS suite (see Section 2.1), has been used to solve this problem [27]. The test case used is a simplified two-dimensional reactor known as Seed Blanket or TWIGL. For a finite difference discretization with 133*133 cells the dimension of the ODE system is 35,378 if two energy groups are considered. When running the code on a cluster of 10 PC's linked with Gigabit Ethernet a superlinear speedup is obtained because in this particular application increasing the number of processors enhances the convergence of the process.

Information Retrieval. Many information retrieval techniques are based on vector space models, thus involving algebraic computations [28]. A matrix is built containing the frequencies (or weights) at which terms appear in the documents. A preprocessing step reduces the size of this terms-by-documents matrix by running a stemming algorithm and also eliminating stop words, i.e. terms with low semantic relevance such as articles, conjunctions, etc. Then, a query can be evaluated by computing the cosine of its angle with respect to each document. In techniques such as the Latent Semantic Indexing, the large terms-by-documents matrix are usually approximated by low-rank matrices, obtained with a singular value decomposition, so that a query evaluation is sped up while maintaining a similar precision-recall ratio. A code based on SLEPc has been developed for computing a small percentage of the largest singular values and vectors. The validation of the code has been carried out with the collection DOE-TREC [29]. It contains DOE's abstracts, over 225,000 documents. When computing the largest 400 singular values, the code achieves 54% efficiency on a 16-processor cluster.

Model Reduction in Control Engineering. High order systems appear frequently in control problems, providing better precision of results required in a control or simulation process. However, a larger system involves a higher computational cost. For this reason, reducing the order of the model is often a necessity. Reduction of a linear control system model consists in obtaining a lower order model with similar behavior. The reduced model decreases the computational requirements in the control or simulation processes but obtaining such a model is an expensive operation. Parallel subroutines for model reduction methods based on balancing techniques (appropriate for stable systems) have been developed [30]. These routines are based on ScaLAPACK and are included in the parallel version of SLICOT. They scale up to a moderate number of processors and allow, for instance, the reduction of a model of order 2000 in 400 seconds with 5 PC's connected with Gigabit Ethernet.

Cardiac Simulation. Cardiac arrhythmias are among the leading causes of mortality in developed countries. The triggering mechanisms, development and termination of these arrhythmias are not clearly understood yet. Mathematical models of the propagation of cardiac electrical activity are increasingly being considered as a helpful tool for better understanding of the mechanisms involved in the development of ventricular arrhythmias. By combining the mathematical formulation of membrane ion kinetics and the structural complexity of the heart, it is possible to simulate the electrical propagation in cardiac tissues. To study the complex dynamics underlying the formation and development of ventricular tachycardias and ventricular fibrillation, a 3D model of the tissue is required. In addition, the analysis of the evolution of ischemic conditions require the simulation of tissues over a period of minutes. These requirements make this kind of simulations a computational challenge. A simulator based on PETSc has been developed [31], allowing the simulation of a 100x100x100 cells cardiac tissue for a period of 250 ms in about 2.5 hours on a 32-processors cluster.

Simulation of Automotive Silencers. One of the design requirements of automotive exhaust silencers is the acoustic attenuation performance at the usual frequencies. Analytical approaches such as the plane-wave propagation model can be considered for studying the acoustic behavior of silencers with rectangular and circular geometries, as well as with elliptical cross-section. Then, the designer can study the effects of different design parameters such as length of the chamber, eccentricity of the ellipse, or location of the inlet and outlet ports. However, numerical simulation provides many advantages with respect to the analytic study. The presence of sudden area changes between pipes of different cross-section is one of the most remarkable features in exhaust silencers. In the neighborhood of sudden area changes, wave propagation is three-dimensional even in the low-frequency range, due to the generation of evanescent higher order modes at the discontinuity. A numerical simulator based on the finite element method has been developed [32], which uses PETSc to solve a linear system of equations per excitation frequency. The code obtains the results for a problem with 880,000 unknowns in the 20-3000 Hz frequency range in about 3 hours using a 32-processors cluster.

4 A Matrix of Applications, Tools and Their Performance

The software tools and computational applications introduced in the previous sections are only a small sample of the increasing number of scientific applica-

Table 1. Summary of computational applications and the software tools that they use

	ScaLAPACK	SuperLU	PETSc	SUNDIALS	GA	SLEPc	SLICOT
MADCAP	X						
3-Charged Particles		X					
NWChem					X		
FUN3D		X					
P-FLAPW	X						
NIMROD		X					
M3D			X				
Omega3P		X					
CAVIRES			X			X	
Lambda Modes			X			X	
Neutron Diffusion				X			
Info Retrieval						X	
Model Reduction	X						X
Cardiac Simulation			X				
Auto Silencers			X				

Table 2. List of applications and relevant computational problems and performance

Application	Computational Problem	Highlights
MADCAP	Matrix factorization and triangular solves.	Currently, 50% peak performance on an IBM SP, and nearly perfect scalability on 1024, 2048, 3072 and 4096 processors.
3-Charged Particles	Solution of large, complex nonsymmetric linear systems.	Solves systems of equations of order 8.4 million on 64 processors in 1 hour of wall clock time; 30 Gflops.
NWChem	Distribute large data arrays, collective operations.	Very good scaling for large problems.
FUN3D	Parallelization of legacy code, unstructured grids, compressible and incompressible Euler and Navier-Stokes equations.	Gordon Bell price, 0.23 Tflops on 3072 processors of ASCI Red.
P-FLAPW	Eigenvalue problems.	Study of systems up to 700 atoms corresponding to matrix sizes of about 35,000. Enables new scientific studies such impurities and disordered systems.
NIMROD	Quad and triangular high order finite elements, semi-implicit time integration, conjugate gradient based sparse matrix solvers.	Use of SuperLU in precondition matrix free CG results in code improvement of 5 fold, equivalent to 3-5 years progress in computing hardware.
M3D	Triangular linear finite elements, linear system solvers.	Currently, scales to 256 IBM SP processors. Many intrinsically 3D phenomena in laboratory plasmas have been modeled and understood in isolation.
Omega3P	Generalized eigenvalue problem.	Solution of a problem of order 7.5 million with 304 million nonzeros. Finding 10 eigenvalues required about 2.5 hours on 24 processors of an IBM SP.
CAVIRES	General eigenvalue problems, complex and nonsymmetric sparse matrices.	57% efficiency with 32 processors for a problem of size 150,000. Scalable to millions of unknowns.
Lambda Modes	Eigenvalue problems, implicit matrix.	55% efficiency on 16 processors for a problem of size 125,000.
Neutron Diffusion	System of linear stiff ODEs, finite differences, implicit time integration.	Very good scaling for large problems. Superlinear speedup. Tested with problem size 35,378.
Information Retrieval	Singular Value Decomposition of sparse rectangular matrices.	54% efficiency, solving problems of size 225,000.
Model Reduction	Dense matrices, solves Sylvester and Lyapunov equations.	48% efficiency solving problems of size 2,000.
Cardiac Simulation	Finite differences, sparse symmetric linear systems.	Very good scaling, tested with 100x100x100 cells tissues with 32 processors.
Automotive Silencers	3D finite elements, multiple sparse linear systems.	Good efficiency with 32 processors. Scalable to millions of unknowns.

tions that now benefit from the use of advanced computational software tools. We propose the creation and update of a reference matrix to be available through the World Wide Web. In this section, we present this information in two tables. Table 1 lists applications and the software tools that they use while Table 2 lists applications, computational problems and performance. Our goal is to present these data in a single table with hyperlinks to more information on the applications, their scientific or engineering relevance, general performance information, list of software tools used and relevant information of the implementation.

5 Conclusions

The realization of large complex, scalable interdisciplinary applications is ubiquitous in the agenda of many fields in computational sciences. Examples of these are all the emerging work on computational frameworks, problem solving environments, open source software development and cyber-collaborations. Many instances of such applications have led to remarkable discoveries in their domain fields. However, one of the great challenges we face today is how to best integrate and carry our achievements in science and engineering, as well as their corresponding software technology, to a solid base that will support and facilitate scientific discovery in the future. Furthermore, the ever evolving nature of the hardware and increase in the number of computational resources to support high end computing pose great penalties to software developments that are either monolithic, not scalable, contain hard-coded platform dependent tuning, or combinations of these. Quality general purpose tools can greatly benefit large and small software developments projects by providing core services that make applications less dependent on the hardware and programming trends. On high end computing, the use of parallel software tools offers more benefits for both legacy code and new code developments than compilers and language directives because the expertise and feedback from the use of tools can be embedded into the tools, thus providing further services to the computational science community. By contrast, the use of automatic parallelization through compilers and programming languages only gives, at best, parallel expression to algorithms known to the applications programmer.

Another essential aspect of this technology is user education. In building an infrastructure for today and future computational discoveries, we must first identify robust, scalable and portable software tools that have helped meet today's computational challenges. Then, we need to look for the missing functionality to tackle emerging problems. These efforts demand the creation of references to documentation on the tools and their applications.

The reference matrix presented in this article, and its more comprehensive version available on-line, is one of our mechanisms for disseminating information on state-of-the-art tool development projects and their functionality. The applications are used as a reference to new users on the areas where these tools have proved beneficial, and sometimes essential. The computational problems are listed to guide potential users in the tool selection process. We are aiming at collecting information on the best general purpose software tools available which will help us bring todays discoveries to support tomorrow research and development through high end computing.

References

[1] Drummond, L.A., Marques, O.: The Advanced Computational Testing and Simulation Toolkit (ACTS): *What can ACTS do for you?* Technical Report LBNL-50414, Lawrence Berkeley National Laboratory (2002) Available at http://acts.nersc.gov/documents.

[2] Press, W.H., Teukolsky, S.A.: Numerical Recipes: Does this paradigm have a future? http://www.nr.com/CiP97.pdf (1997)

[3] Henderson, M.E., Anderson, C.R., Lyons, S.L., eds.: Object Oriented Methods for Interoperable Scientific and Engineering Computing. SIAM, Philadelphia (1999)

[4] Canning, A., Mannstadt, W., Freeman, A.J.: Parallelization of the FLAPW method. Computer Physics Communications **130** (2000) 233–243

[5] Mitnik, D.M., Griffin, D.C., Badnel, N.R.: Electron-impact excitation of Ne5+. J. Phys. B: At. Mol. Opt. Phys. **34** (2001) 4455–4473

[6] Rescigno, T., Baertschy, M., Isaacs, W., McCurdy, W.: Collisional breakup in a quantum system of three charged particles. Science **286** (1999) 2474–2479

[7] Marques, O., Drummond, L.A.: The DOE ACTS Information Center. http://-acts.nersc.gov (2001)

[8] Blackford, L.S., Choi, J., Cleary, A., D'Azevedo, E., Demmel, J.W., Dhillon, I., Dongarra, J.J., Hammarling, S., Henry, G., Petitet, A., Stanley, K., Walker, D., Whaley, R.C.: ScaLAPACK User's Guide. SIAM, Philadelphia, PA (1997)

[9] Demmel, J.W., Gilbert, J.R., Li, X.: SuperLU Users' Guide. (2003)

[10] Balay, S., Buschelman, K., Gropp, W.D., Kaushik, D., Knepley, M., McInnes, L.C., Smith, B.F., Zhang, H.: PETSc users manual. Technical Report ANL-95/11 - Revision 2.2.0, Argonne National Laboratory (2004)

[11] Hindmarsh, A.C., Brown, P., Grant, K., Lee, S., Serban, R., Shumaker, D., Woodward, C.: SUNDIALS: Suite of Nonlinear and Differential/Algebraic Equation-Solvers. ACM TOMS (2004) (to appear)

[12] Nieplocha, J., Harrison, J., Littlefield, R.J.: Global Arrays: A portable 'shared-memory' programming model for distributed memory computers. In: Supercomputing'94. (1994) 340–349

[13] PNNL US: PeIGS, A Parallel Eigensolver Library. http://www.emsl.pnl.gov-/docs/global/peigs.html (2000)

[14] Hernández, V., Román, J.E., Vidal, V.: SLEPc: Scalable Library for Eigenvalue Problem Computations. Lecture Notes in Computer Science **2565** (2003) 377–391

[15] Lehoucq, R.B., Sorensen, D.C., Yang, C.: ARPACK Users' Guide, Solution of Large-Scale Eigenvalue Problems by Implicitly Restarted Arnoldi Methods. SIAM, Philadelphia, PA (1998)

[16] Benner, P., Mehrmann, V., Sima, V., Huffel, S.V., Varga, A.: SLICOT: A subroutine library in systems and control theory. In Datta, B.N., ed.: Applied and Computational Control, Signal and Circuits. Volume 1., Birkhauser (1999)

[17] Borrill, J.: MADCAP: the microwave anisotropy dataset computational analysis package. Available at ArXiv: http://xxx.lanl.gov/ps/astro-ph/9911389 (1999)

[18] Baertschy, M., Rescigno, T., Isaacs, W., Li, X., McCurdy, W.: Electron-impact ionization of atomic hydrogen. Physical Review A **63** (2001)

[19] High Performance Computational Chemistry Group: NWChem, a computational chemistry package for parallel computers, version 4.5. Pacific Northwest National Laboratory. Richland, Washington 99352, USA (2003)
[20] High Performance Computational Chemistry Group: NWChem Benchmarks. http://www.emsl.pnl.gov/docs/nwchem/index.html (2003)
[21] Anderson, W.L., Gropp, W.D., Kaushik, D.K., Keyes, D.E., Smith, B.F.: Achieving high sustained performance in an unstructure mesh CFD application. In: Proceedings from SC'99. (1999) also ANL/MCS-P776-0899.
[22] Sovinec, C.R., Barnes, D.C., Gianakon, T.A., Glasser, A.H., Nebel, R.A., Kruger, S.E., Schnack, D.D., Plimpton, S.J., Tarditi, A., Chu, M.S., the NIMROD Team: Nonlinear magnetohydrodynamics with high-order finite elements. Journal of Computational Physics (2003) submitted.
[23] Park, W., Belova, E.V., Fu, G.Y., Tang, X., Strauss, H.R., Sugiyama, L.E.: Plasma simulation studies using multilevel physics models. Physics Plasmas **6** (1999)
[24] Sun, Y.: The Filter Algorithm for Solving large-Scale Eigenproblems from Accelerator Simulations. PhD thesis, Stanford University (2003)
[25] Roman, J.E.: Software Portable, Escalable y Extensible para la Resolución de Problemas de Valores Propios Dispersos de Gran Dimensión. PhD thesis, Universidad Politécnica de Valencia (2002)
[26] Hernandez, V., Roman, J.E., Vidal, V., Verdu, G., Ginestar, D.: Resolution of the neutron diffusion equation with SLEPc, the Scalable Library for Eigenvalue Problem Computations. In: Nuclear Mathematical and Computational Sciences: A Century in Review, A Century Anew, Gatlinburg, TN, ANS (2003)
[27] García, V.M., Vidal, V., Verdu, G., Garayoa, J., Miro, R.: Parallel resolution of the two-group time-dependent neutron difussion equation with public domain ODE codes. 6th VECPAR Conference (2004)
[28] Berry, M.W., Browne, M.: Understanding Search Engines: Mathematical Modeling and Text Retrieval. SIAM, Philadelphia, PA (1999)
[29] NIST: TREC: Text retrieval conference. http://trec.nist.gov (2003)
[30] Guerrero, D., Hernandez, V., Roman, J.E.: Parallel SLICOT model reduction routines: The Cholesky factor of Grammians. In: 15th Triennal IFAC World Congress, Barcelona, Spain (2002)
[31] Alonso, J.M., Hernandez, V., Molto, G., Monserrat, M., Saiz, J.: Three-dimensional cardiac electrical activity simulation on cluster and grid platforms. 6th VECPAR Conference (2004)
[32] Alonso, J.M., Garcia, G., Hernandez, V., Denia, F.D., Fuenmayor, F.J.: A parallel computing approach for solving the Helmholtz equation in 3D domains: Application to the study of acoustic behaviour of silencers. For ECCOMAS (2004)

A Performance Evaluation of the Cray X1 for Scientific Applications

Leonid Oliker[1], Rupak Biswas[2], Julian Borrill[1], Andrew Canning[1], Jonathan Carter[1], M. Jahed Djomehri[2], Hongzhang Shan[1], and David Skinner[1]

[1] CRD/NERSC, Lawrence Berkeley National Laboratory, Berkeley, CA 94720
{loliker, jdborril, acanning, jtcarter, hshan, deskinner}@lbl.gov
[2] NAS Division, NASA Ames Research Center, Moffett Field, CA 94035
{rbiswas, djomehri}@nas.nasa.gov

Abstract. The last decade has witnessed a rapid proliferation of superscalar cache-based microprocessors to build high-end capability and capacity computers primarily because of their generality, scalability, and cost effectiveness. However, the recent development of massively parallel vector systems is having a significant effect on the supercomputing landscape. In this paper, we compare the performance of the recently-released Cray X1 vector system with that of the cacheless NEC SX-6 vector machine, and the superscalar cache-based IBM Power3 and Power4 architectures for scientific applications. Overall results demonstrate that the X1 is quite promising, but performance improvements are expected as the hardware, systems software, and numerical libraries mature. Code reengineering to effectively utilize the complex architecture may also lead to significant efficiency enhancements.

1 Introduction

The last decade has witnessed a rapid proliferation of superscalar cache-based microprocessors to build high-end capability and capacity computers [12]. This is primarily because their generality, scalability, and cost effectiveness convinced computer vendors, buyers, and users that vector architectures hold little promise for future large-scale supercomputing systems.

Two major elements have dramatically affected this common perception. The first is the recent availability of the Japanese Earth Simulator [3] that, based on NEC SX-6 vector technology, is the world's most powerful supercomputer both in terms of peak (40.96 TFlops/s) and sustained LINPACK (87.5% of peak) performance. The second is the constant degradation in sustained performance (now typically less than 10% of peak) for scientific applications running on SMP clusters built from processors using superscalar technology. The Earth Simulator, on the other hand, has demonstrated sustained performance of almost 65% of peak for a production global atmospheric simulation [11].

In order to quantify what a vector capability entails for scientific communities that rely on modeling and simulation, it is critical to evaluate it in the context

M. Daydé et al. (Eds.): VECPAR 2004, LNCS 3402, pp. 51–65, 2005.

of demanding computational algorithms. In earlier work [9], we conducted an evaluation of the SX-6 with the NAS Parallel Benchmarks and a suite of six applications. Here, we compare the performance of the Cray X1 vector testbed system at Oak Ridge National Laboratory [8] with that of the SX-6, and the superscalar cache-based IBM Power3 and Power4 architectures for scientific codes.

We first compare memory bandwidth and scatter/gather hardware behavior of a triad summation using regularly and irregularly strided data. Next, we use two synthetic benchmarks: a fast Fourier transform and an N-body solver to further compare the four target architectures. Finally, we present performance results for three scientific applications from cosmology (MADCAP), materials science (PARATEC), and fluid dynamics (OVERFLOW-D). For each of these benchmarks and applications, we examine the effort required to port them to the X1. Both intra- and inter-node parallel performance (in GFlops/s per processor) is reported for our application suite when running a small and a large test case. X1 results are generally promising; however, we expect performance improvements as the machine stabilizes to production mode. Note that none of the applications have been specifically optimized for the X1, so appropriate reengineering to utilize the architecture effectively would also be beneficial.

2 Target Architectures

We give a brief description of the salient architectural features of the X1, as well as those of the other machines that we examined. Table 1 presents a summary of their node characteristics and measured inter-node MPI latency [8, 13]. Observe that the X1 and the SX-6 are designed to sustain a higher bytes/flop ratio.

2.1 Cray X1

The recently-released X1 is designed to combine traditional vector strengths with the generality and scalability features of superscalar cache-based parallel systems (see schematic in Fig. 1). The computational core, called the single-streaming processor (SSP), contains two 8-stage vector pipes running at 800 MHz. Each SSP contains 32 vector registers holding 64 double-precision words, and operates at 3.2 GFlops/s peak. All vector operations are performed under a bit mask, allowing loop blocks with conditionals to compute without the need for scat-

Table 1. Architectural specifications of the X1, Power3, Power4, and SX-6 nodes

Node Type	CPU/ Node	Clock (MHz)	Peak (GFlops/s)	Mem. BW (GB/s)	Peak Bytes/Flop	MPI Latency (μsec)
X1	4	800	12.8	34	2.7	7.3
Power3	16	375	1.5	1.0	0.67	16.3
Power4	32	1300	5.2	6.4	1.2	17.0
SX-6	8	500	8.0	32	4.0	5.6

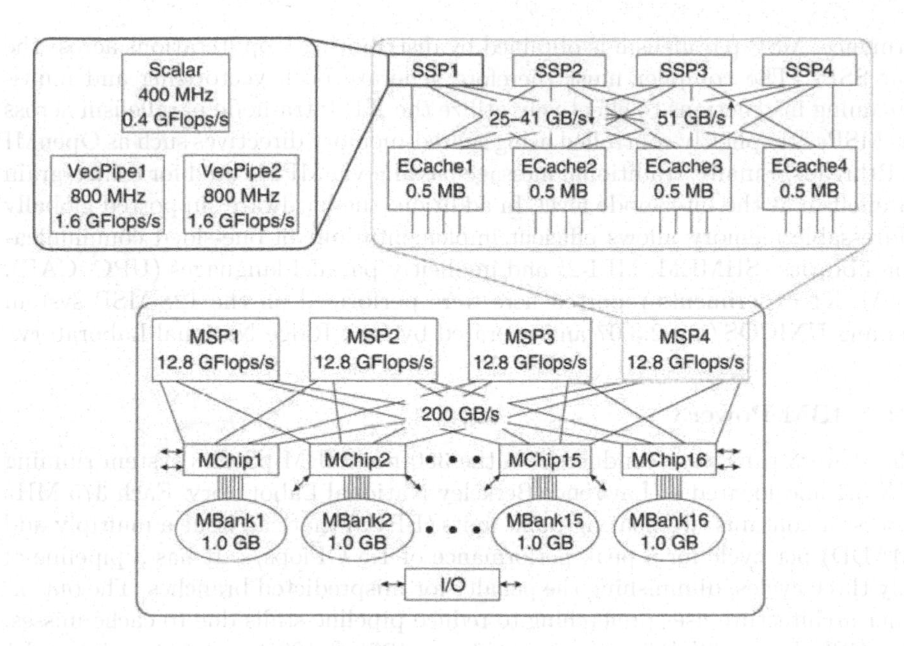

Fig. 1. Architecture of the X1 node (in red), MSP (in green), and SSP (in blue)

ter/gather operations. Each SSP can have up to 512 addresses simultaneously in flight, thus hiding the latency overhead for a potentially significant number of memory fetches. The SSP also contains a two-way out-of-order superscalar processor running at 400 MHz with two 16KB caches (instruction and data). The scalar unit operates at an eighth of the vector performance, making a high vector operation ratio critical for effectively utilizing the underlying hardware.

The multi-streaming processor (MSP) combines four SSPs into one logical computational unit. The four SSPs share a 2-way set associative 2MB data Ecache, a unique feature for vector architectures that allows extremely high bandwidth (25–51 GB/s) for computations with temporal data locality. An X1 node consists of four MSPs sharing a flat memory through 16 memory controllers (MChips). Each MChip is attached to a local memory bank (MBank), for an aggregate of 200 GB/s node bandwidth. Additionally, MChips can be used to directly connect up to four nodes (16 MSPs) and participate in remote address translation. To build large configurations, a modified 2D torus interconnect is implemented via specialized routing chips. The torus topology allows scalability to large processor counts with relatively few links compared with the crossbar interconnect of the IBM Power systems or the Earth Simulator; however, the torus suffers from limited bisection bandwidth. Finally, the X1 is a globally addressable architecture, with specialized hardware support that allows processors to directly read or write remote memory addresses in an efficient manner.

The X1 programming model leverages parallelism hierarchically. At the SSP level, vector instructions allow 64 SIMD operations to execute in a pipeline fashion, thereby masking memory latency and achieving higher sustained performance. MSP parallelism is obtained by distributing loop iterations across the four SSPs. The compiler must therefore generate both vectorizing and multistreaming instructions to effectively utilize the X1. Intra-node parallelism across the MSPs is explicitly controlled using shared-memory directives such as OpenMP or Pthreads. Finally, traditional message passing via MPI is used for coarse-grain parallelism at the inter-node level. In addition, the hardware supported globally addressable memory allows efficient implementations of one-sided communication libraries (SHMEM, MPI-2) and implicitly parallel languages (UPC, CAF).

All X1 experiments reported here were performed on the 128-MSP system running UNICOS/mp 2.3.07 and operated by Oak Ridge National Laboratory.

2.2 IBM Power3

The Power3 runs were conducted on the 380-node IBM pSeries system running AIX 5.1 and located at Lawrence Berkeley National Laboratory. Each 375 MHz processor contains two floating-point units (FPUs) that can issue a multiply-add (MADD) per cycle for a peak performance of 1.5 GFlops/s. It has a pipeline of only three cycles, diminishing the penalty for mispredicted branches. The out-of-order architecture uses prefetching to reduce pipeline stalls due to cache misses. The CPU has a 32KB instruction cache, a 128KB 128-way set associative L1 data cache, and an 8MB four-way set associative L2 cache with its own private bus. Each SMP node consists of 16 processors connected to main memory via a crossbar. Multi-node configurations are networked via the Colony switch.

2.3 IBM Power4

The Power4 experiments were performed on the 27-node IBM pSeries 690 system running AIX 5.1 and operated by Oak Ridge National Laboratory. Each 32-way SMP consists of 16 Power4 chips (organized as four MCMs), where a chip contains two 1.3 GHz processor cores. Each core has two FPUs capable of a fused MADD per cycle, for a peak of 5.2 GFlops/s. The superscalar out-of-order architecture can exploit instruction level parallelism through its eight execution units. Branch prediction hardware minimizes the effects of the relatively long pipeline (six cycles) necessitated by the high frequency design. Each processor has its own private L1 cache (64KB instruction, 32KB data) with prefetch hardware; however, both cores share a 1.5MB unified L2 cache. The L3 can operate as a stand-alone 32MB cache, or be combined with other L3s on the same MCM to create a 128MB interleaved cache. Multi-node configurations employ the Colony interconnect, but future systems will use the lower latency Federation switch.

2.4 NEC SX-6

The SX-6 results were obtained on the single-node system running SUPER-UX at the Arctic Region Supercomputing Center. The 500 MHz processor contains an 8-way replicated vector pipe capable of issuing a MADD per cycle, for a peak performance of 8.0 GFlops/s. The processors contain 72 vector registers, each holding 256 64-bit words. For non-vectorizable instructions, the SX-6 has a 500 MHz scalar processor with 64KB instruction and data caches, and 128 general-purpose registers. The 4-way superscalar unit has a peak of 1.0 GFlops/s and supports branch prediction, data prefetching, and out-of-order execution.

Unlike conventional architectures, the SX-6 vector unit lacks data caches. It therefore masks memory latencies by overlapping pipelined vector operations with memory fetches. The SX-6 uses high speed SDRAM with a peak bandwidth of 32 GB/s per CPU; enough to feed one operand per cycle to each of the replicated pipe sets. The nodes can be used to build large-scale multi-processor systems; for instance, the Earth Simulator contains 640 SX-6 nodes (but with a slightly faster memory), connected through a single-stage crossbar.

3 Microbenchmarks

We are developing a microbenchmark suite (called Xtreams) that measures low-level machine characteristics such as memory subsystem behavior and scatter/gather hardware support. Fig. 2(a) presents asymptotic memory bandwidth of the triad summation: $a(i) = b(i) + s \times c(i)$ using various strides for the four architectures in our study. The SX-6 achieves the best bandwidth: up to one, two, and three orders of magnitude better than the X1, Power4, and Power3, respectively. Observe that certain strides impact X1 and SX-6 bandwidth quite pronouncedly, by an order of magnitude or more. Analysis shows that strides containing multiples of two worsen performance due to increased DRAM bank conflicts. On the Power architectures, a precipitous drop in the transfer rate occurs for small strides, due to loss of cache reuse. This drop is more severe on the Power4, because of its more complicated cache structure. Note that the purpose of this benchmark is to measure the expected memory behavior via simple Fortran code fragments; however, machine theoretical performance can be

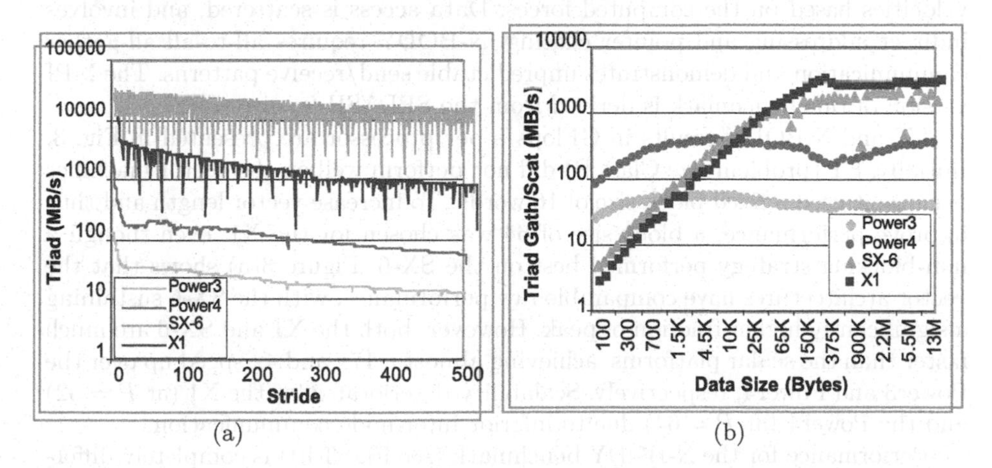

Fig. 2. Single-processor Xtreams triad performance (in MB/s) using (a) regularly strided data of various strides, and (b) irregularly strided data of various sizes

approached by using architecture-specific source transformations. For example, the X1 average bandwidth can be increased by up to a factor of 20x through F90 syntax and compiler directives for non-caching data and loop unrolling.

Fig. 2(b) presents single-processor memory bandwidth of indirect addressing through vector triad gather/scatter operations of various data sizes. For smaller sizes, the cache-based architectures show better data rates for indirect access to memory; but the X1 and the SX-6 effectively utilize their gather/scatter hardware to outperform the cache-based systems for larger data sizes. The X1 memory bandwidth behavior is comparable to that of the SX-6 (asymptotic performance is marginally higher); however, the X1 processor is more than 50% faster, indicating a poorer architectural balance between computational rate and memory subsystem performance (see Table 1).

4 Synthetic Benchmarks

We next use two synthetic benchmarks to compare and contrast the performance of the four target architectures. The first is FT, a fast Fourier transform (FFT) kernel, from the NAS Parallel Benchmarks (NPB) suite [7]. FFT is a simple evolutionary problem in which a time-dependent partial differential equation is transformed to a set of uncoupled ordinary differential equations in phase space. The bulk of the computational work consists of a set of discrete multi-dimensional FFTs, each accessing the same single large array multiple times, but with different strides every time.

The second synthetic benchmark is N-BODY, an irregularly structured code that simulates the interaction of a system of particles in 3D over a number of time steps, using the hierarchical Barnes-Hut algorithm. There are three main stages: building an octree to represent the distribution of the particles; browsing the octree to calculate the forces; and updating the particle positions and velocities based on the computed forces. Data access is scattered, and involves indirect addressing and pointer chasing. N-BODY requires all-to-all all-gather communication and demonstrates unpredictable send/receive patterns. The MPI version of this benchmark is derived from the SPLASH-2 suite [14].

FT and N-BODY results in GFlops/s per processor are presented in Fig. 3. Initially, FT (problem size Class B) did not perform well on the vector machines because it used a fixed block size of 16 words. To increase vector length and thus improve performance, a block size of 64 was chosen for the X1, even though a non-blocking strategy performed best on the SX-6. Figure 3(a) shows that the vector architectures have comparable raw performance, with the SX-6 sustaining a significantly higher fraction of peak. However, both the X1 and SX-6 are much faster than the scalar platforms, achieving almost a 17x and 4x speed up over the Power3 and Power4, respectively. Scalability deteriorates for the X1 (at $P = 32$) and the Power4 (at $P = 64$) due to inferior inter-node communication.

Performance for the N-BODY benchmark (see Fig. 3(b)) is completely different in that the scalar processors are far superior than the vector processors. For example, the Power3 achieves 7% of peak while the X1 runs at 0.3%. The dom-

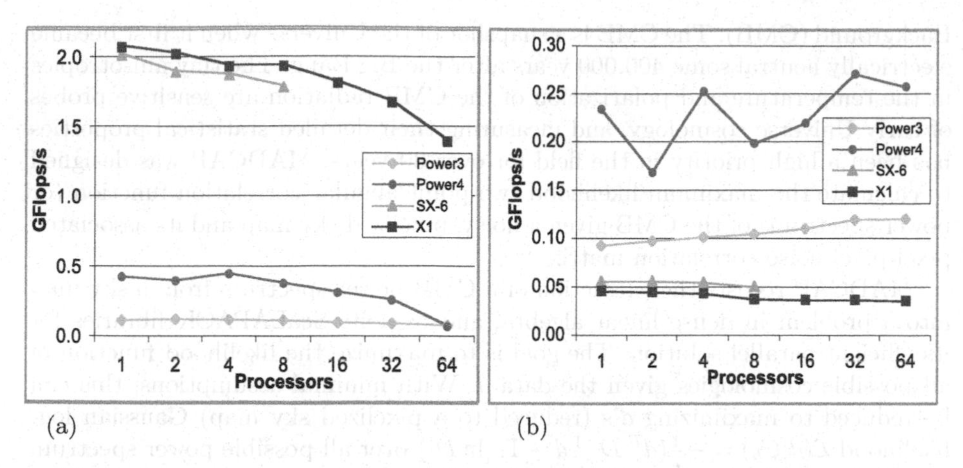

(a) (b)

Fig. 3. Per-processor performance (in GFlops/s) for (a) FT and (b) N-BODY

inant part of the code is browsing the octree and computing the inter-particle forces. Octree browsing involves significant pointer chasing, which is a very expensive operation on vector machines and cannot be vectorized well. We tried to improve vectorization by separating the octree browsing and force calculation phases; however, results were only negligibly better. These results demonstrate that although the vector architectures are equipped with scalar units, non-vectorized codes will perform extremely poorly on these platforms.

5 Scientific Applications

Three scientific applications were chosen to measure and compare the performance of the X1 with that of the SX-6, Power3, and Power4. The applications are: MADCAP, a cosmology code that extracts tiny signals from extremely noisy observations of the Cosmic Microwave Background; PARATEC, a first principles materials science code that solves the Kohn-Sham equations to obtain electronic wavefunctions; and OVERFLOW-D, a computational fluid dynamics production code that solves the Navier-Stokes equations around complex aerospace configurations. One small and one large case are tested for each application. SX-6 results are presented for only one node ($P \leq 8$) while those on Power3 ($P > 16$), Power4 ($P > 32$), and X1 ($P > 4$) are for multi-node runs. Performance is shown in GFlops/s per processor, and were obtained with `hpmcount` on the Power systems, `ftrace` on the SX-6, and `pat` on the X1.

5.1 Cosmology

From the application area of cosmology, we examined the Microwave Anisotropy Dataset Computational Analysis Package (MADCAP) [1]. MADCAP implements the current optimal general algorithm for extracting the most useful cosmological information from total-power observations of the Cosmic Microwave

Background (CMB). The CMB is a snapshot of the Universe when it first became electrically neutral some 400,000 years after the Big Bang. The tiny anisotropies in the temperature and polarization of the CMB radiation are sensitive probes of early Universe cosmology, and measuring their detailed statistical properties has been a high priority in the field for over 30 years. MADCAP was designed to calculate the maximum likelihood two-point angular correlation function (or power spectrum) of the CMB given a noisy, pixelized sky map and its associated pixel-pixel noise correlation matrix.

MADCAP recasts the extraction of a CMB power spectrum from a sky map into a problem in dense linear algebra, and exploits ScaLAPACK libraries for its efficient parallel solution. The goal is to maximize the likelihood function of all possible cosmologies given the data d. With minimal assumptions, this can be reduced to maximizing d's (reduced to a pixelized sky map) Gaussian log-likelihood $\mathcal{L}(d|C_b) = -\frac{1}{2}(d^T D^{-1} d - \mathrm{Tr}\,[\ln D])$ over all possible power spectrum coefficients C_b, where $D = \langle dd^T \rangle$ is the data correlation matrix.

Using Newton-Raphson iteration to locate the peak of $\mathcal{L}(d|C_b)$ requires the evaluation of its first two derivatives with respect to C_b. This involves first building D as the sum of the experiment-specific noise correlations and theory-specific signal correlations, and then solving a square linear system for each of the $\mathcal{N}_b$ spectral coefficients. To minimize communication overhead, this step is computed through explicit inversion of D (symmetric positive definite) and direct matrix-matrix multiplication. These operations scale as $\mathcal{N}_b \mathcal{N}_p^3$ for a map with $\mathcal{N}_p$ pixels, typically 10^4–10^5 for real experiments.

To take advantage of large parallel computer systems while maintaining high performance, the MADCAP implementation has recently been rewritten to exploit the parallelism inherent in performing $\mathcal{N}_b$ independent matrix-matrix multiplications. The analysis is split into two steps: first, all the processors build and invert D; then, the processors are divided into gangs, each of which performs a subset of the multiplications (gang-parallel). Since the matrices involved are block-cyclically distributed over the processors, this incurs the overhead of redistributing the matrices between the two steps.

X1 Porting. Porting MADCAP to vector architectures is straightforward, since the package utilizes ScaLAPACK to perform dense linear algebra calculations. However, it was necessary to rewrite the dSdC routine that computes the pixel-pixel signal correlation matrices, scales as $\mathcal{N}_b \mathcal{N}_p^2$, and does not have any BLAS library calls. The basic structure of dSdC loops over all pixels and calculates the value of Legendre polynomials up to some preset degree for the angular separation between these pixels. On superscalar architectures, this constituted a largely insignificant amount of work, but since the computation of the polynomials is recursive, prohibiting vectorization, the cost was appreciable on both the X1 and the SX-6. At each iteration of the Legendre polynomial recursion in the rewritten dSdC routine, a large batch of angular separations was computed in an inner loop. Compiler directives were required to ensure vectorization for both the vector machines.

Table 2. Per-processor performance of MADCAP on target machines

	Power3		Power4		SX-6		X1	
P	GFlops/s	% Peak	GFlops/s	% Peak	GFlops/s	% Peak	GFlops/s	% Peak
Small synthetic case: $\mathcal{N}_p = 8192$, $\mathcal{N}_b = 16$, without gang-parallelism								
1	0.844	56.3	1.98	38.1	4.55	56.9	5.36	41.9
4	0.656	43.7	1.23	23.7	2.66	33.2	4.26	33.3
9	0.701	46.7	1.11	21.3	—	—	3.22	25.2
16	0.520	34.7	0.731	14.1	—	—	2.48	19.4
25	0.552	36.8	0.628	12.1	—	—	1.84	14.4
Large real case: $\mathcal{N}_p = 14996$, $\mathcal{N}_b = 16$, with gang-parallelism								
4	0.831	55.4	1.90	36.5	4.24	53.0	5.66	44.2
8	0.688	45.9	1.62	31.2	3.24	40.5	5.22	40.8
16	0.615	41.0	1.26	24.2	—	—	4.09	32.0
32	0.582	38.8	0.804	15.5	—	—	3.35	26.2
64	0.495	33.0	0.524	10.1	—	—	2.15	16.8

Performance Results. Table 2 first shows the performance of MADCAP for a small synthetic dataset with $\mathcal{N}_p = 8192$ and $\mathcal{N}_b = 16$ on each of the four architectures under investigation, without the use of gang-parallelism. Note that for maximum efficiency, the processor sizes are set to squares of integers. As expected, the single processor runs achieve a significant fraction of the peak performance since the analysis is dominated by BLAS3 operations (at least 60%). However, as we use more processors with a fixed data size, the density of the data on each processor decreases, the communication overhead increases, and performance drops significantly. Observe that the X1 attains the highest raw performance, achieving factors of 6.5x, 3.5x, and 1.6x speedup compared with the Power3, Power4, and SX-6 respectively on 4 processors. With increasing processor counts, the fraction of time spent in BLAS3 computations decreases, causing performance degradation. The average vector length was about 62 per SSP (vector length 64) for all processor counts. It is therefore surprising that MADCAP did not attain a higher percentage of peak on the X1. We believe this is due, in part, to overhead associated with MADCAP's I/O requirements, which is currently under investigation.

Table 2 also shows the performance of the gang-parallel implementation of MADCAP for a larger real dataset from the Millimeter-wave Anisotropy Experiment Imaging Array (MAXIMA) [5], with $\mathcal{N}_p = 14996$ and $\mathcal{N}_b = 16$. MAXIMA is a balloon-borne experiment with an array of 16 bolometric photometers optimized to map the CMB anisotropy over hundreds of square degrees. The gang parallel technique is designed to preserve the density of data during matrix-matrix multiplication and produce close to linear speedup. In this regard, the results on the Power3 and Power4 are somewhat disappointing. This is because the other steps in MADCAP do not scale that well, and the efficiency of each gang suffers due to increased memory contention. The X1 shows significantly faster run times compared with the Power3/4 (about a factor of 4x); however, its sustained performance declines precipitously with increasing numbers of processors. This is due to the non-vectorizable code segments (such as I/O and data redistribution), which account for higher fractions of the overall run time.

5.2 Materials Science

The application that we selected from materials science is called PARATEC: PARAllel Total Energy Code [10]. The code performs ab-initio quantum-mechanical total energy calculations using pseudopotentials and a plane wave basis set. The pseudopotentials are of the standard norm-conserving variety. Forces can be easily calculated and used to relax the atoms into their equilibrium positions. PARATEC uses an all-band conjugate gradient (CG) approach to solve the Kohn-Sham equations of Density Functional Theory (DFT) and obtain the ground-state electron wavefunctions. DFT is the most commonly used technique in materials science, having a quantum mechanical treatment of the electrons, to calculate the structural and electronic properties of materials. Codes based on DFT are widely used to study properties such as strength, cohesion, growth, magnetic, optical, and transport for materials like nanostructures, complex surfaces, and doped semiconductors.

In solving the Kohn-Sham equations using a plane wave basis, part of the calculation is carried out in real space and the remainder in Fourier space using specialized parallel 3D FFTs to transform the wavefunctions. The code spends most of its time in vendor-supplied BLAS3 (~30%) and 1D FFTs (~30%) on which the 3D FFTs are built. For this reason, PARATEC generally obtains a high percentage of peak performance on different platforms. The code exploits fine-grained parallelism by dividing the plane wave components for each electron among the different processors [4].

X1 Porting. PARATEC is an MPI package designed primarily for massively parallel computing platforms, but can also run on serial machines. Since much of the computation involves FFTs and BLAS3, an efficient vector implementation of the code requires these libraries to vectorize well. While this is true for the BLAS3 routines on the X1, the standard 1D FFTs run at a low percentage of peak. Some code transformations were therefore required to convert the 3D FFT routines to use simultaneous ("multiple") 1D FFT calls. Compiler directives were also inserted in certain code segments to force vectorization on the X1 in loops where indirect indexing is used.

Performance Results. The results in Table 3 are for 3 CG steps of 250 and 432 Si-atom bulk systems and a standard LDA run of PARATEC with a 25 Ry cut-off using norm-conserving pseudopotentials. A typical calculation would require between 20 and 60 CG iterations to converge the charge density.

Results show that PARATEC vectorizes well on the SX-6 for the small test case. The X1 is significantly slower, even though it has a higher peak speed. One reason is that, on the X1, the code spends a much smaller percentage of the total time in highly optimized 3D FFTs and BLAS3 libraries than any of the

Table 3. Per-processor performance of PARATEC on target machines

	Power3		Power4		SX-6		X1	
P	GFlops/s	% Peak	GFlops/s	% Peak	GFlops/s	% Peak	GFlops/s	% Peak
Small case: 250 Si-atom bulk system								
1	0.915	61.0	2.29	44.0	5.09	63.6	2.97	23.2
2	0.915	61.0	2.25	43.3	4.98	62.3	2.82	22.0
4	0.920	61.3	2.21	42.5	4.70	58.8	2.65	20.7
8	0.911	60.7	2.09	40.2	4.22	52.8	2.48	19.4
16	0.840	56.0	1.57	30.2	—	—	1.99	15.5
32	0.725	48.3	1.33	25.6	—	—	1.67	13.0
Large case: 432 Si-atom bulk system								
32	0.959	63.9	1.49	28.7	—	—	3.04	23.8
64	0.848	56.5	0.75	14.4	—	—	2.82	22.0
128	0.739	49.3			—	—	1.91	14.9

other machines. This lowers sustained performance as other parts of the code run well below peak. Some of the loss in X1 scaling for $P > 4$ is also due to inter-node communication and shorter vector lengths. The parallel 3D FFTs require a transformation of the distributed grid which results in global communication. The average vector length drops from 49 for the uni-processor run to 44 on 16 processors. Performance for the large test case is poor for $P = 128$, primarily due to low average vector length (35) and increased communication overhead. PARATEC runs efficiently on the Power3, but performance on the Power4 is affected by the relatively poor ratio of memory bandwidth to peak performance. Nonetheless, the Power systems obtain a much higher percentage of peak than the X1 for all runs. Faster FFT libraries and some code rewriting to increase vectorization in conjunction with compiler directives could significantly improve the performance of PARATEC on the X1. Initial tests of systems larger than 432 Si-atoms show better scaling performance across all architectures.

5.3 Fluid Dynamics

In the area of computational fluid dynamics (CFD), we selected the NASA Navier-Stokes production application called OVERFLOW-D [6]. The code uses the overset grid methodology [2] to perform high-fidelity viscous simulations around realistic aerospace configurations. It is popular within the aerodynamics community due to its ability to handle complex designs with multiple geometric components. OVERFLOW-D, unlike OVERFLOW [2], is explicitly designed to simplify the modeling of problems when components are in relative motion. The main computational logic at the top level of the sequential code consists of a time-loop and a nested grid-loop. Within the grid-loop, solutions to the flow equations are obtained on the individual grids with imposed boundary conditions. Overlapping boundary points or inter-grid data are updated from the previous time step using a Chimera interpolation procedure. Upon completion of the grid-loop, the solution is automatically advanced to the next time step by the time-loop. The code uses finite differences in space, with a variety of implicit/explicit time stepping.

The MPI version of OVERFLOW-D takes advantage of the overset grid system, which offers a natural coarse-grain parallelism. A bin-packing algorithm clusters individual grids into groups, each of which is then assigned to an MPI process. The grouping strategy uses a connectivity test that inspects for an overlap between a pair of grids before assigning them to the same group, regardless of the size of the boundary data or their connectivity to other grids. The grid-loop in the parallel implementation is subdivided into two procedures: a group-loop over groups, and a grid-loop over the grids within each group. Since each MPI process is assigned to only one group, the group-loop is executed in parallel, with each group performing its own sequential grid-loop. The inter-grid boundary updates within each group are performed as in the serial case. Inter-group boundary exchanges are achieved via MPI asynchronous communication calls.

X1 Porting. The MPI implementation of OVERFLOW-D is based on the sequential version that was designed to exploit early Cray vector machines. The same basic program structure is used on all four target architectures except for a few minor changes in some subroutines to meet specific compiler requirements. In porting the code to the X1, three binary input files also had to be converted from the default IEEE format to 64-bit (`integer*8` and `real*8`) data types. Furthermore, for data consistency between FORTRAN and C, the `int` data in all C functions were converted to `long int`. The main loops in compute-intensive sections of the code were multi-streamed and vectorized automatically by the compiler; however, modifications were made to allow vectorization of certain other loops via pragma directives.

Performance Results. The results in Table 4 are for 100 time steps of an OVERFLOW-D simulation of vortex dynamics in the complex wake flow region around hovering rotors. A typical calculation would require several thousand time steps. Note that the current MPI implementation of OVERFLOW-D does not allow uni-processor runs. The grid system for the small test case consists of 41 blocks and about 8 million grid points, while that for the large case has 857 blocks and 69 million grid points. The Cartesian off-body wake grids surround the curvilinear near-body grids with uniform resolution, but become gradually coarser upon approaching the outer boundary of the computational domain.

Results for the small case demonstrate that the X1 is about a factor of 2x slower than the SX-6 and 1.5x (4x) faster than the Power4 (Power3), based on GFlops/s per processor. The SX-6 outperforms the other machines; e.g. its execution time for $P = 8$ (1.6 secs) is 90%, 50%, and 21% of the $P = 16$ timings for the X1, Power4, and Power3. The SX-6 also consistently achieves the highest percentage of peak performance while the X1 is the lowest. Scalability is similar for all machines except the Power4, with computational efficiency decreasing for a larger number of MPI tasks primarily due to load imbalance. Performance is particularly poor for $P = 32$ when 41 zones have to be distributed to 32

Table 4. Per-processor performance of OVERFLOW-D on target machines

	Power3		Power4		SX-6		X1	
P	GFlops/s	% Peak	GFlops/s	% Peak	GFlops/s	% Peak	GFlops/s	% Peak
Small case: 8 million grid points								
2	0.133	8.9	0.394	7.6	1.13	14.1	0.513	4.0
4	0.117	7.8	0.366	7.0	1.11	13.9	0.479	3.7
8	0.118	7.9	0.362	7.0	0.97	12.1	0.445	3.5
16	0.097	6.5	0.210	4.0	—	—	0.433	3.4
32	0.087	5.8	0.115	2.2	—	—	0.278	2.2
Large case: 69 million grid points								
16	0.115	7.7	0.282	5.4	—	—	0.456	3.6
32	0.109	7.3	0.246	4.7	—	—	0.413	3.2
64	0.105	7.0	0.203	3.9	—	—	0.390	3.0
96	0.100	6.7	0.176	3.4	—	—	0.360	2.8
128	0.094	6.3	0.146	2.8	—	—	0.319	2.5

processors. On the X1, a relatively small average vector length of 24 per SSP explains why the code achieves an aggregate of only 3.6 GFlops/s on 8 processors (3.5% of peak). Although the same general trends hold for the large case, performance is slightly better (for the same number of processors) because of higher computation-to-communication ratio and better load balance. Reorganizing OVERFLOW-D would achieve improve vector performance; however, extensive effort would be required to modify this production code.

6 Summary and Conclusions

This paper presented the performance of the Cray X1 vector testbed system, and compared it against the cacheless NEC SX-6 vector machine and the superscalar cache-based IBM Power3/4 architectures, across a range of scientific kernels and applications. Microbenchmarking showed that the X1 memory subsystem behaved rather poorly, given its large peak memory bandwidth and sophisticated hardware support for non-unit stride data access. Table ?? summarizes the performance of the X1 against the three other machines, for our synthetic benchmarks and application suite. All our codes, except N-BODY, were readily amenable to vectorization via minor code changes, compiler directives, and the use of vendor optimized numerical libraries. Results show that the X1 achieves high raw performance relative to the Power systems for the computationally intensive applications; however, the SX-6 demonstrated faster runtimes for several of our test cases. Surprisingly, the X1 did not achieve a high fraction of peak performance compared with the other platforms, even though reasonably large vector lengths were attained for the applications. The N-BODY kernel is an irregularly structured code poorly suited for vectorization, and highlights the extremely low performance one should expect when running scalar codes on this architecture.

Table 5. Summary overview of X1 performance

Application Name	Scientific Discipline	Lines of Code	X1 Speedup vs.						
			P	Power3	Power4	SX-6	P	Power3	Power4
FT	Kernel	2,000	8	17.3	5.1	1.1	64	18.4	19.4
N-BODY	Kernel	1,500	8	0.4	0.2	0.7	64	0.3	0.1
MADCAP	Cosmology	5,000	4	6.5	3.5	1.6	64	4.3	4.1
PARATEC	Mat. Sc.	50,000	8	2.7	1.2	0.6	64	3.3	3.8
OVERFLOW-D	CFD	100,000	8	3.8	1.2	0.5	128	3.4	2.2

Overall, the X1 performance is quite promising, and improvements are expected as the hardware, systems software, and numerical libraries mature. It is important to note that X1-specific code optimizations have not been performed at this time. This complex vector architecture contains both data caches and multi-streaming processing units, and the optimal programming methodology is yet to be established. For example, increasing cache locality through data blocking will lower the memory access overhead; however, this strategy may reduce the vector length and cause performance degradation. Examining these tradeoffs will be the focus of our future work.

Acknowledgements

The authors thank ORNL for providing access to the Cray X1. All authors from LBNL are supported by Office of Advanced Science Computing Research in the U.S. DOE Office of Science under contract DE-AC03-76SF00098. M.J. Djomehri is an employee of CSC and supported by NASA under contract DTTS59-99-D-00437/A61812D with AMTI/CSC.

References

1. J. Borrill, MADCAP: The Microwave Anisotropy Dataset Computational Analysis Package, in: *Proc. 5th European SGI/Cray MPP Workshop* (Bologna, Italy, 1999) astro-ph/9911389.
2. P.G. Buning *et al.*, Overflow user's manual version 1.8g, *Tech. Rep.*, NASA Langley Research Center, 1999.
3. Earth Simulator Center, See URL *http://www.es.jamstec.go.jp*.
4. G. Galli and A. Pasquarello, First-principles molecular dynamics, *Computer Simulation in Chemical Physics*, Kluwer, 1993, 261–313.
5. S. Hanany *et al.*, MAXIMA-1: A measurement of the Cosmic Microwave Background anisotropy on angular scales of 10′–5°, *The Astrophysical Journal*, 545 (2000) L5–L9.
6. R. Meakin, On adaptive refinement and overset structured grids, in: *Proc. 13th AIAA Computational Fluid Dynamics Conf.* (Snowmass, CO, 1997) Paper 97-1858.
7. NAS Parallel Benchmarks, See URL *http://www.nas.nasa.gov/Software/NPB*.
8. Oakridge National Laboratory Evaluation of Early Systems, See URL *http://www.csm.ornl.gov/evaluation/*.

9. L. Oliker *et al.*, Evaluation of cache-based superscalar and cacheless vector architectures for scientific computations, in: *Proc. SC2003* (Phoenix, AZ, 2003).
10. PARAllel Total Energy Code, See URL *http://www.nersc.gov/projects/paratec.*
11. S. Shingu *et al.*, A 26.58 Tflops global atmospheric simulation with the spectral transform method on the Earth Simulator, in: *Proc. SC2002* (Baltimore, MD, 2002).
12. Top500 Supercomputer Sites, See URL *http://www.top500.org.*
13. H. Uehara *et al.*, MPI performance measurement on the Earth Simulator, *Tech. Rep.*, NEC Research and Development, 2003.
14. S.C. Woo *et al.*, The SPLASH-2 programs: Characterization and methodological considerations, in: *Proc. 22nd Intl. Symp. on Computer Architecture* (Santa Margherita Ligure, Italy, 1995) 24–36.

Modelling Overhead of Tuple Spaces with Design of Experiments*

Frederic Hancke, Tom Dhaene, and Jan Broeckhove

University of Antwerp, Campus Middelheim - Middelheimcampus, Building G,
Middelheimlaan 1, B-2020 Antwerp, Belgium
{frederic.hancke, tom.dhaene, jan.broeckhove}@ua.ac.be

Abstract. In the world of high performance distributed computing new, better and faster distributed platforms evolve quite rapidly. Building application dependent performance test cases in order to make platform comparisons, is not satisfying anymore. This paper presents the use of a statistical technique, called Design of Experiments (DoE), to model the performance of two closely related distributed platforms, JavaSpaces and GigaSpaces, with as few tests as possible. All tests will be based on Tunable Abstract Problem Profiles (TAPPs), a technique for problem independent simulations.

1 Introduction

Currently, many distributed platforms provide the resources required to execute a scientific application efficiently. The choice of which specific platform to use is up to the application programmer and is mostly based on the availability (open source), user-friendliness, etc. Often, these applications require high performance. This means that decent performance evaluation techniques are needed.

On the one hand, much research has been done in the field of performance prediction in parallel and distributed computing [17, 3]. On the other hand, only few research results on performance evaluation can be found. Evaluation and comparison of distributed platforms are often realised through execution of some computationally expensive, complex but well-known applications, such as the calculation of Pi [2], generation of prime numbers [15], etc. Results of these experiments on different platforms are then compared using standard statistical techniques.

The aim of this contribution is to apply a predetermined set of statistical techniques to model the performance of a distributed platform. Design of Experiments (DoE) [6, 11, 16] is the covering name for this set of techniques. The idea of DoE is to obtain the best possible characterisation while executing as few experiments as possible. This technique has been widely and successfully used in the engineering world.

* Special thanks go to GigaSpaces Technologies for their technical support and for providing the appropriate license keys.

M. Daydé et al. (Eds.): VECPAR 2004, LNCS 3402, pp. 66–77, 2005.

The work presented in this paper builds upon previous results [4], where DoE was applied on JavaSpaces [7], an implementation of the Linda Tuple Spaces paradigm [8]. The remainder of this contribution focusses on the characterisation of GigaSpaces [1] exactly the same way JavaSpaces was modelled, which makes easy comparison possible.

First, an overview of the JavaSpaces and GigaSpaces implementations will be given. Then, the part of DoE that is used in the experiment will be discussed, followed by a brief description of the setup of the experiment. Finally the results of each experiment with the comparison of both models will be presented.

2 Tuple Spaces

The concept of `Tuple Spaces` was introduced in 1982 by the language `Linda` [8]. Linda is commonly described as a coordination language, developed to manage and control the coordination aspects of an application or algorithm (usually parallel). Linda does not constitute a full programming language in itself, but is designed to expand the semantics of an existing programming language in order to form a `parallel programming language`. Examples of this are the C-Linda [13], Fortran-Linda and Prolog-Linda [20] languages, respectively augmentations of the C, Fortran and Prolog programming languages.

The Linda language provides only a set of four simple operations needed to operate on the contents of a virtual shared memory, the Tuple Space. The data structures contained in this shared memory are called `tuples`, which can be manipulated by the aforementioned operations. The three most important operations for the rest of this paper are:

- **out(t)** which writes a tuple `t` into the Tuple Space,
- **in(t)** which returns a tuple from the Tuple Space that matches a given template `t` (and thereby deletes `t` from the Tuple Space), and
- **rd(t)** which performs essentially the same actions as **in**, but keeps `t` in the Tuple Space.

These operations call for the definition of the format of the datastructures of the Tuple Space, the tuples themselves. Basically, a tuple is an ordered sequence of `fields`, each having a determined type. Additionally, a Tuple Space does not require all tuples to have the same composition. This allows for the definition of `template tuples`: meta-tuples which describe the possible type-based composition of normal tuples.

Linda implementations must guarantee the atomicity of the previously mentioned operations. In the rest of this section an overview of two implementations, JavaSpaces and GigaSpaces, of the Tuple Space paradigm will be given. Both are based on the JavaSpaces specifications [19].

Sun Microsystems' `JavaSpaces` [7] specification builds upon Jini [5] and RMI [18]. The goal of JavaSpaces is to allow creation of robust distributed systems in general, and distributed algorithms in particular. These systems are modelled as a set of distributed components communicating by passing messages through a space. A message or entry is represented by a Java object that can be stored in the space (a distributed shared memory). JavaSpaces provides the basic operations read, write and take.

GigaSpaces [1] is a commercial implementation of the JavaSpaces specification [19]. While JavaSpaces is a reference implementation of the specification, GigaSpaces provides some performance improving features. As the experiments in this paper compare both platforms using only the two basic operations write and take, the extra features of GigaSpaces over JavaSpaces will not be discussed further.

3 Design of Experiments

Design of Experiments (DoE) is the covering term for a set of statistical techniques. These statistical techniques are Experimental Design, Analysis of Variance (ANOVA) and Regression Analysis (see Fig. 1). Each of these techniques will be described in the rest of this section. Experimental Design [11, 12] is a technique for describing the process to plan an experiment for a certain system. After the execution of this experiment it is possible to analyse the results with Analysis of Variance and Regression Analysis [14]. These techniques allow for a Response Surface Model (RSM) [21] to be built.

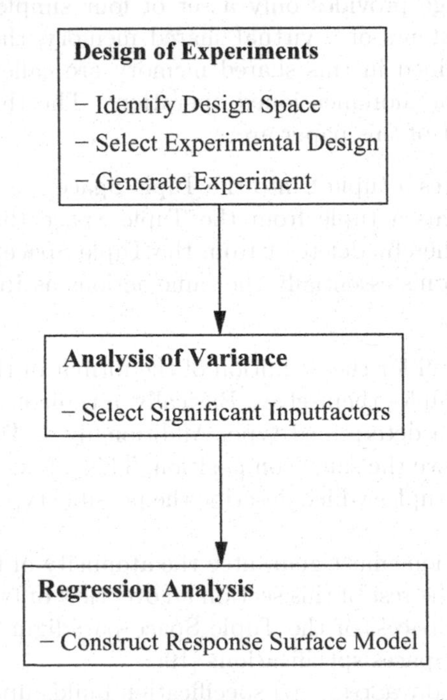

Fig. 1. Overview of Design of Experiments (DoE)

3.1 Experimental Design

Experimental Design provides the experiment to execute, in order to model a system. This system is referred to as a process [22]. A process is a combination of machines, methods, fysical reactions, human beings, etc. that, given an input and a set of inputfactors, generates an output where the values of a set of outputfactors (also called responses) can be derived. Among the inputfactors, one can distinguish controllable and uncontrollable inputfactors. Controllable inputfactors are changeable at any time, independent of other factors. Uncontrollable inputfactors are not independently changeable.

An *experiment* refers to a set of tests to be executed. The levels of the inputfactors are changed appropriately such that changes in values of outputfactors can be observed and identified. Therefore, these factors must be measurable. For each of these tests, each controllable inputfactor is assigned a level. The goal of experimental design is to set up as few tests as possible in an experiment. The choice of these tests is based on the fixed levels for inputfactors, replication and randomisation. Replication of every test in an experiment leads to an experimental error. This error is caused by the unknown uncontrollable input factors. The order of execution of all tests with its replicates are randomized. This way, influences of unknown uncontrollable factors on the outputfactors can be avoided.

Every experiment is generated using a *model*. This model must be consistent with the target of the experiment and therefore must not be chosen too simple (e.g. linear when the output is expected to be quadratic). On the other hand, a model chosen too complex leads to a larger amount of tests to execute. A linear model is always a good model to start with. Using this, linear effects and interaction effects of the inputfactors on the outputfactors may be detected and analysed. If the linear model is not satisfying, more tests can be defined for a more complex model, such as a quadratic model.

The *Central Composite Circumscribed* (CCC) design [11] is a quadratic model extending the linear 2^n *Full Factorial* design [11] (including the central point) with tests on the starpoints. In a full factorial design all inputfactors are assigned two levels: *high* (or $+1$) and *low* (or -1). Optionally the central point can be tested, which is called the *central* level (or 0). The CCC design extends this linear model with two levels: $-\delta$ and $+\delta$ ($\delta = (2^n)^{1/4}$). These levels lie beyond the levels -1 and $+1$. An experiment generated using this CCC design contains $m(2^n + 2n) + m_c$ number of tests (with m the number of replicates, n the number of controllable inputfactors, and m_c the number of replicates of the central point). Figure 2 illustrates the CCC design for $n = 2$.

Using the CCC design, a model can be built. For $n = 2$, x_A the value of inputfactor A, and x_B the value of inputfactor B, this model is given by

$$y = \gamma_0 + \gamma_1 x_A + \gamma_2 x_B + \gamma_{11} x_A^2 + \gamma_{22} x_B^2 + \gamma_{12} x_A x_B + \epsilon \tag{1}$$

This allows to predict outputfactor Y with respect to inputfactors A and B. The regression analysis will find the best possible fit to this model.

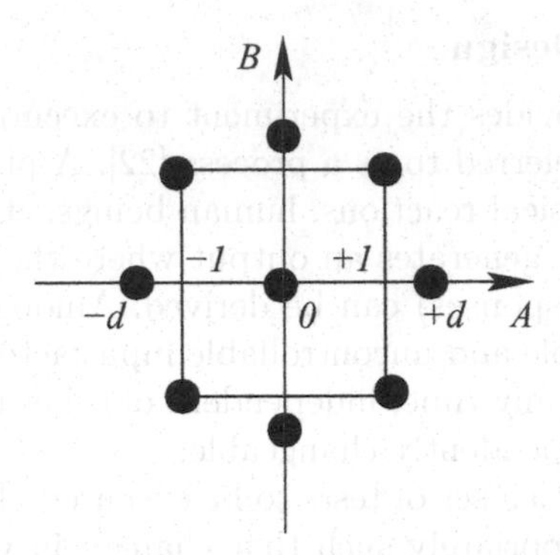

Fig. 2. Representation of the testpoints of a Central Composite Circumscribed design with inputfactors A and B

3.2 Analysis of Variance and Regression Analysis

After the designed experiment has been executed, the results can be fitted in the model and the model parameters can be obtained. To fit this model, the Analysis of Variance (ANOVA) with a Regression Analysis can be used [14].

ANOVA will test hypotheses for each (combination of) inputfactor(s). H_0 supposing the factor to be 0, H_1 supposing the factor not to be 0. If H_0 holds, the inputfactor A is considered to have no influence on outputfactor Y. If not (meaning the alternative H_1 holds), A is considered to influence Y at some testpoint. To test whether H_0 holds, a Level of Significance (LoS) must be chosen. The LoS indicates the risk of concluding H_0 is false, while being true in reality. Usually the LoS is chosen to be five percent. ANOVA is able to select the most significant inputfactors (for which H_0 is rejected). Using these factors a Response Surface Model (RSM) can be built by performing a regression analysis with the least squares method. This regression will result in approximations of the γ_i parameters in (1).

4 Processmodel Parameters

In order to model the performance of JavaSpaces and GigaSpaces, a number of parameters for the processmodel must be defined which can then be used as inputfactors for the Experimental Design phase. This processmodel is based on the Master-Slave pattern, which is widely used in distributed systems. It is based on the idea of one master program controlling a number of slave programs, each running on a different node of the distributed system. The number of slaves will be indicated by **tWo**. The master distributes **tTa** tasks to its slaves and collects their returned results. A task is defined as a problem taking some time to solve on a linear system. Suppose all tasks need equal solving time **tMe**. Then **tTi** = tTa $*$ tMe, the total theoretical time needed to solve all tasks on a linear system. In order to obtain full control of this process the idea of Tunable Abstract Problem Profiles (TAPPs) was developed [10]. These kind of sleep (dummy) tasks put the slaves' processor to sleep for a specified amount of time rather than doing any computation. This allows for a much better estimation of the overhead of a distributed platform. The wallclock time **pWCT** is the time needed for the master to distribute the first task until the time the last result is returned by the slaves. Specifically, for JavaSpaces this time will be pWCT.JS, for GigaSpaces pWCT.GS. Parameters starting with the letter t are either predefined or calculated from predefined parameters, those starting with p are either measured or calculated using measured and/or predefined parameters.

Prior to running the experiment, a set of XML-based tasklists is generated [10]. Each such tasklist contains tTa tasks, assigning each task an `id` and the `duration` (in ms) to sleep[1]. The TAPPs based application for the experiment [9] assigns each slave an `id`. At the master side, an XML-based tasklist is processed and distributed to a space, using tasks in the form of Java objects. Such a task is defined solely by the `id` and duration of the task from the tasklist. At the slave side, upon finishing a task, a result (also a Java object) is returned into the space containing again both `id` and duration of the task, plus the `id` of the slave processing the task.

The definition of the inputfactors for the processmodel is trivial: tWo indicates the number of slaves, tTa the number of tasks, and tMe the time to `solve` one task[2].

Besides the wallclock time pWCT, another outputfactor, the `speedup` **pSp**, will complete the processmodel:

$$\mathrm{pSp} = \frac{\mathrm{tTi}}{\mathrm{pWCT}}$$

Note that always $\mathrm{pSp} \leq \mathrm{tWo}$.

5 The Experiment

5.1 Setup

As discussed in the previous section the experiment contains three inputfactors. Table 1 summarizes the level values for each of these factors[3]. As the purpose of using a CCC design is to fit a quadratic model with as few tests as possible, it must be noted that using a larger design space automatically leads to less accuracy in the model. For the experiment, 23 Intel powered PC's participated, all running `SuSE Linux 8.2`. Implementation specific processes for the tuple space as well as the master process ran on an `IntelP4 2.0GHz (512MB RAM)` powered PC (named `queen`). The 22 computers (named `drones` in our case) running the different slaves can be divided into three categories. `drone01` to `drone04` are `Intel`

[1] Generating tasklists with equal durations is only one of the possibilities of the application. For more details, see [10].

[2] Remark that all tasks are supposed to be independent.

[3] Notice these must be rounded values.

Table 1. Level values for each inputfactor (tWo: number of slaves; tTa: number of tasks; tMe: execution time of one task in ms)

Inputfactor	$-\delta$	-1	0	$+1$	$+\delta$
tWo	2	6	12	18	22
tTa	46	230	500	770	954
tMe	46	433	1000	1567	1954

P4 1.7GHz (512MB RAM) computers, drone05 to drone14 are Intel P4 2.0GHz (512MB RAM) computers, and drone15 to drone22 are Intel P4 2.66GHz (1GB RAM) computers. Every participating PC is configured with Java SDK 1.4.1, Jini 1.2.1 and GigaSpaces 3.0. The PC's are connected with a 100Mbps fully switched ethernet network.

The complete experiment was designed using the CCC design with 10 replications of each test, resulting in $10(2^3 + 2 * 3) + 10 = 150$ tests. All tests were ordered randomly using the randomized block design with the replications as block variable [14]. This means that all tests were put in blocks (groups) per replication (10 blocks). All blocks contain different permutations of the same set of 15 tests, guaranteeing complete randomness of the experiment.

5.2 Results

The complete experiment resulted in 150 values of pWCT for each platform. These were then used for the Analysis of Variance (ANOVA), with the factors shown in the first column of Table 2. This results in the determination of the factors having a significant influence on the output factor pWCT (non-significant

Table 2. Results of the ANOVA and Regression Analysis for the outputfactors pWCT and pSp

Factor	pWCT.JS	pWCT.GS	pWCT.GS.JS	pSp.JS	pSp.GS	pSp.GS.JS
1	102030.5226	101522.2301	−6963.1944	−2.0716	−4.8326	−1.1523
tWo	−20355.4534	−20613.0193	–	3.2728E-1	7.5789E-1	4.0909E-1
tTa	107.9262	105.7999	–	−1.3248E-3	–	−4.3832E-4
tMe	49.3258	49.6197	10.7961	8.8639E-3	9.7387E-3	−7.3005E-4
tWo^2	932.7474	936.1124	–	4.4760E-3	7.4472E-3	–
tTa^2	–	–	−0.0007	0.0000E-4	0.0086E-4	–
tMe^2	–	–	−0.0042	−0.0437E-4	−0.0375E-4	0.0057E-4
tWo:tTa	−9.5961	−9.5133	–	2.6999E-4	–	−2.0565E-4
tWo:tMe	−4.7760	−4.6459	–	2.6362E-4	–	−2.0319E-4
tTa:tMe	0.1107	0.1108	–	–	–	0.0181E-4
R^2	0.8839	0.8920	0.5475	0.9550	0.9534	0.8511

factors are marked with – in the table). Then, Regression Analysis was applied to obtain a quadratic Response Surface Model (RSM), whose coefficients are shown in the second (JavaSpaces) and third (GigaSpaces) columns of Table 2. The same was done for the difference **pWCT.GS.JS** = pWCT.GS – pWCT.JS between both platforms in each data point. The results are shown in the fourth column. Similarly, the last three columns give the results for the speedup for both platforms as well as the difference **pSp.GS.JS** = pSp.GS – pSp.JS. The last row in the table shows the value of the R^2 statistic, which indicates the accuracy of the fitted model [14]. The closer to 1, the better the fit. Values of 0.9 or above are considered to be good.

Both models for the wallclock time pWCT for JavaSpaces and GigaSpaces result in detection of the same significant factors, and almost the same values for the coefficients and R^2. However, the model for the difference is worse, with only 54.75% accuracy. A possible reason is that the complexity of pWCT.GS.JS is too high for a quadratic fit, which is a drawback of using a CCC design.

The models for the speedup pSp are all better, as can be seen in the values of the R^2. It should be noted that the coefficients of some factors turn out to be very small (some even smaller than 1E-8), even for factors that are clearly selected as being significant by the ANOVA.

In order to compare both platforms directly, all 150 values used for the construction of the model, for pWCT.GS.JS (negative values mean GigaSpaces performs better) and pSp.GS.JS (positive values mean GigaSpaces performs better) have been plotted in Figure 3. They are shown in increasing order to give an idea of the distribution. From the plot for the wallclock time it can be seen that

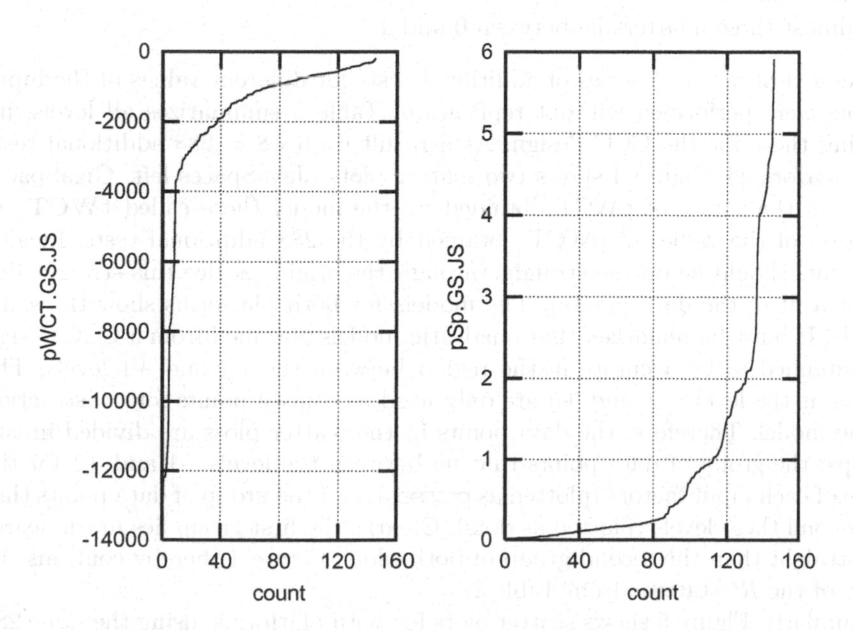

Fig. 3. Plot of pWCT.GS.JS in ms (left) and pSp.GS.JS (right)

Table 3. The verification level values for each inputfactor (tWo: number of slaves; tTa: number of tasks; tMe: execution time of one task in ms), the levels for the CCC design are marked in bold

Inputfactor	$-\delta$			**-1**			**0**			**+1**			$+\delta$
tWo	**2**		4	**6**	8	10	**12**	14	16	**18**	20		**22**
tTa	**46**		138	**230**	320	410	**500**	590	680	**770**	862		**954**
tMe	**46**	175	304	**433**	622	811	**1000**	1189	1378	**1567**	1696	1825	**1954**

(1) all values are negative, meaning GigaSpaces performs better than JavaSpaces for every test, and
(2) almost three quarters lie between 0 and −2000 ms.

The drop down in favour of GigaSpaces at the left of the plot comes from the tests for many tasks and short durations, such as for tTa = 954 and tMe = 46. The tuple space implementation of JavaSpaces seems to be less efficient than the implementation of GigaSpaces, especially for a relatively high number of tasks and small durations. Thus, the JavaSpaces implementation rapidly becomes a bottleneck. This is why JavaSpaces falls short in comparison to GigaSpaces. The plot on the right in Figure 3 for the speedup shows the same trend in favour of JavaSpaces:

(1) all values are positive, meaning GigaSpaces overclasses JavaSpaces for every test, and
(2) almost three quarters lie between 0 and 1.

As a verification, a series of additional tests for different values of the input factors were performed without replication. Table 3 summarizes all levels, including those for the CCC design. As a result $6 * 6 * 8 = 288$ additional tests were performed. Figure 4 shows two scatter plots (JavaSpaces left, GigaSpaces right) of the values of pWCT obtained by the model (here called **tWCT**) in function of the values of pWCT obtained by the 288 additional tests. Ideally, all points should lie on the straight through the origin (notice this straight line is not a fit of the data points). The models for both platforms show the same trends. It must be remarked that quadratic models obtained from a CCC design are intended to be accurate in the region between the −1 and +1 levels. The points at the levels $-\delta$ and $+\delta$ are only used as control points for construction of the model. Therefore, the data points in the scatter plots are divided in two groups: the group of data points that lie between the levels −1 and +1 for the value of each input factors (plotted as crosses), and the group of data points that lie beyond these levels (plotted as dots). Clearly, the first group lies much nearer the straight than the second group in both plots. Figure 4 thereby confirms the value of the R^2 statistic from Table 2.

Similarly, Figure 5 shows scatter plots for both platforms, using the same 288 additional data points. Again, **tSp** denotes the value obtained by the model for

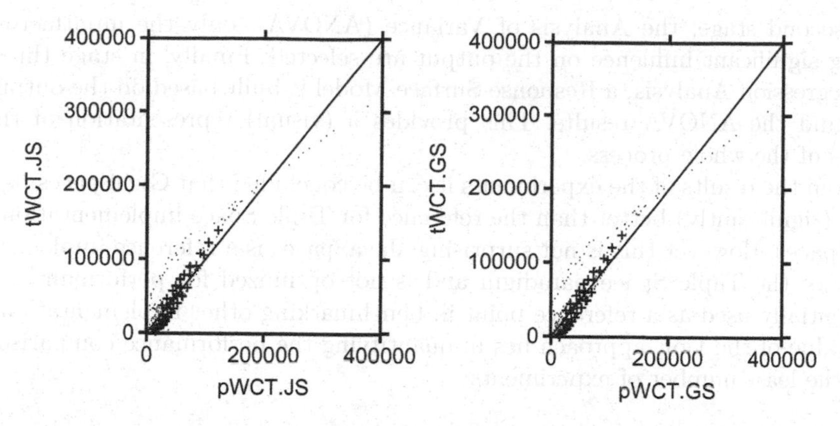

Fig. 4. Scatter plots of tWCT in function of pWCT for the additional data points (the crosses represent the data points between the levels −1 and +1; the dots represent the data points lying beyond these levels) for JavaSpaces (left) and GigaSpaces (right)

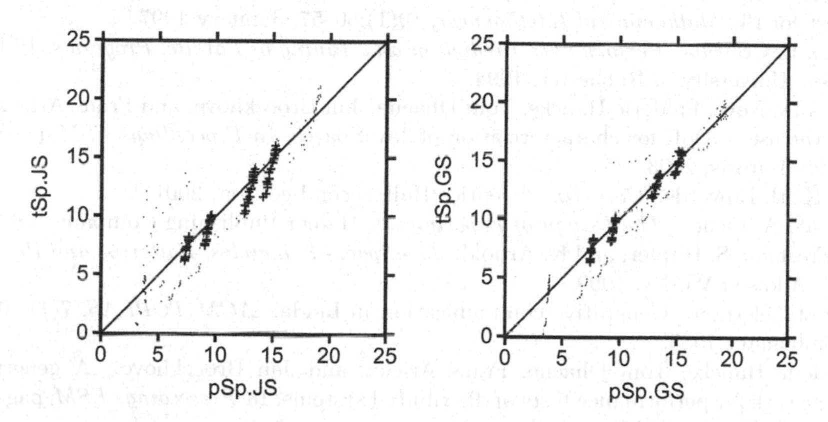

Fig. 5. Scatter plots of tSp in function of pSp for the additional data points for JavaSpaces (left) and GigaSpaces (right)

pSp. Here, the model for GigaSpaces shows higher accuracy than JavaSpaces, especially for the first group (crosses). Remark that the first group consists of four clusters of data points. This is due to the number of workers used, from left to right 8, 10, 14 and 16.

6 Conclusions

The aim of this paper has been to model and compare overhead of JavaSpaces and GigaSpaces, both Tuple Space variants, using a statistical technique called Design of Experiments (DoE). DoE consists of three stages. In a first stage, the Experimental Design, an experiment is constructed, with as few tests as possible.

In a second stage, the Analysis of Variance (ANOVA), only the inputfactors having significant influence on the output are selected. Finally, in stage three, the Regression Analysis, a Response Surface Model is built based on the output data and the ANOVA results. This provides a (visual) representation of the results of the whole process.

From the results of the experiments it can be concluded that GigaSpaces performs (significantly) better than the reference for Tuple Space implementations: JavaSpaces. However this is not surprising: JavaSpaces is a reference implementation of the Tuple Space paradigm and is not optimized for performance. It is essentially used as a reference point in benchmarking other implementations. The value of the DoE approach lies in quantifying the performance comparison with the least number of experiments.

References

1. GigaSpaces. http://www.gigaspaces.com.
2. David H. Bailey, Jonathan M. Borwein, Peter B. Borwein, and Simon Plouffe. The quest for Pi. *Mathematical Intelligencer*, 19(1):50–57, January 1997.
3. Mark E. Crovella. *Performance Prediction and Tuning of Parallel Programs.* PhD thesis, University of Rochester, 1994.
4. Hans De Neve, Frederic Hancke, Tom Dhaene, Jan Broeckhove, and Frans Arickx. On the use of DoE for characterization of JavaSpaces. In *Proceedings ESM*, pages 24–29. Eurosis, 2003.
5. W. Keith Edwards. *Core Jini.* Prentice Hall, second edition, 2001.
6. Ronald A. Fisher. *The Design of Experiments.* Hafner Publishing Company, 1971.
7. E. Freeman, S. Hupfer, and K. Arnold. *JavaSpaces Principles, Patterns, and Practice.* Addison Wesley, 1999.
8. David Gelernter. Generative communication in Linda. *ACM TOPLAS*, 7(1):80–112, January 1985.
9. Frederic Hancke, Tom Dhaene, Frans Arickx, and Jan Broeckhove. A generic framework for performance tests of distributed systems. In *Proceedings ESM*, pages 180–182. Eurosis, 2003.
10. Frederic Hancke, Gunther Stuer, David Dewolfs, Jan Broeckhove, Frans Arickx, and Tom Dhaene. Modelling overhead in JavaSpaces. In *Proceedings Euromedia*, pages 77–81. Eurosis, 2003.
11. Charles R. Hicks and Kenneth V. Jr. Turner. *Fundamental Concepts in the Design of Experiments.* Oxford University Press, Inc., 1999.
12. Douglas C. Montgomery. *Design and Analysis of Experiments.* John Wiley and Sons Inc., New York, 2001.
13. James Narem. An informal operational semantics of C-Linda v2.3.5. Technical report, Yale University, December 1990.
14. John Neter, Michael H. Kutner, Christopher J. Nachtsheim, and William Wasserman. *Applied Linear Statistical Models.* WCB/McGraw-Hill, fourth edition, 1996.
15. Michael S. Noble and Stoyanka Zlateva. Scientific computation with JavaSpaces. Technical report, Harvard-Smithsonian Center For Astrophysics and Boston University, 2001.
16. Ranjit K. Roy. *Design of Experiments using the Taguchi Approach.* John Wiley & Sons, Inc., 2001.

17. Jennifer M. Schopf. *Performance Prediction and Scheduling for Parallel Applications on Multi-User Clusters.* PhD thesis, University of California, 1998.
18. Inc. Sun Microsystems. Java RMI: A new approach to distributed computing. Technical report, Sun Microsystems, Inc., 1999.
19. Inc. Sun Microsystems. JavaSpaces service specification 1.2. Technical report, Sun Microsystems, Inc., 2001.
20. G. Sutcliffe, J. Pinakis, and N. Lewins. Prolog-Linda: An embedding of Linda in muProlog. Technical report, University of Western Australia, 1989.
21. Patrick N. Koch Timothy W. Simpson, Jesse D. Peplinski and Janet K. Allen. On the use of statistics in design and the implications for deterministic computer experiments. Technical report, Woodruff school of Engineering, Atlanta, Georgia, USA, September 1997.
22. Ledi Trutna, Pat Spagon, Dr. Enrique del Castillo, Terri Moore, Sam Hartley, and Arnon Hurwitz. *Engineering Statistics Handbook*, chapter 5 (Process Improvement). National Institute of Standards and Technology, 2000.

Analysis of the Interaction of Electromagnetic Signals with Thin-Wires Structures. Multiprocessing Issues for an Iterative Method*

E.M. Garzón[1], S. Tabik[1], A.R. Bretones[2], and I. García[1]

[1] Dept. Computer Architecture and Electronics,
University of Almería, Almería 04120, Spain
{ester, siham, inma}@ace.ual.es

[2] Dept. Electromagnetism and Physics of the Matter, Faculty of Sciences,
University of Granada, Severo Ochoa s/n, Granada 18071, Spain
arubio@ugr.es

Abstract. In this paper we analyse the computational aspects of a numerical method for solving the Electric Field Integral Equation (*EFIE*) for the analysis of the interaction of electromagnetic signals with thin-wires structures. Our interest concerns with the design of an efficient parallel implementation of this numerical method which helps physicist to solve the Electric Field Integral Equation for very complex and large thin-wires structures. The development of this parallel implementation has been carried out on distributed memory multiprocessors, with the use of the parallel programming library MPI and routines of PETSc (Portable, Extensible Toolkit for Scientific Computation). These routines can solve sparse linear systems in parallel. Appropriate data partitions have been designed in order to optimize the performance of the parallel implementation. A parameter named *relative efficiency* has been defined to compare two parallel executions with different number of processors. This parameter allows us to better describe the superlinear performance behavior of our parallel implementation. Evaluation of the parallel implementation is given in terms of the values of the speep-up and the relative efficiency. Moreover, a discussion about the requirements of memory versus the number of processors is included. It will be shown that memory hierarchy management plays a relevant role in the performance of this parallel implementation.

1 Introduction

The analysis of the radiation of broadband electromagnetic signals or pulses is of great interest in the fields of high-resolution radar, broadband radio communications or in the study of the vulnerability of electronic systems to electromagnetic pulses [1, 2, 3].

* This work was supported by the MCyT of Spain under contracts TIC2002-00228 and TIC2001-2364-C03-03.

M. Daydé et al. (Eds.): VECPAR 2004, LNCS 3402, pp. 78–89, 2005.

Maxwell's equations are the basis for studying all electromagnetic problems. Their solutions either involve time-harmonic analysis coupled with Fourier inversion in the frequency domain, or a direct formulation of the equations in the time domain and its subsequent solution using a time stepping algorithm. In some cases, the computational cost of the solutions in the time domain can be less than in the frequency domain. However, especially if the physical size of the structure compared to the wavelength (electrical size) to analyze is large, the computational cost in the time domain is still very high, since equations of both implicit spatial and temporal variables must be solved. Furthermore, the analysis in the time domain can be advantageous, or may even be the only alternative for solving some types of problems [4]. The development of high-performance computing systems allows us the application of these methods to large electrical dimension problems.

Time domain numerical techniques are based either on a description of the problem in terms of differential equations [5, 6] or on an integral-equation formulation [3]. The integral-equation forms are appropriate for the analysis of perfectly electric conducting geometries and have the advantage of including an explicit radiation condition, given that the computational space is generally limited to the surface of the structure. However, the integral-equation based method may become computationally very expensive or even impractical when the body length is large compared to the wavelength of the highest significant frequency of the excitation signal, i.e. for very short pulses. Although, it is necessary to mention that the combination with PWTD algorithms ('Plane Wave Time Domain') has allowed a decrease in time of computation and to extend the range of electrical sizes that is possible to analyze using integral-equations [7].

The focus of this paper is to describe and evaluate a parallel implementation of the DOTIG1 code [8]. This code solves the "Electric Field Integral Equation" (EFIE) for the analysis of the interaction of arbitrary transient electromagnetic signals with thin-wire structures. Due to the saving in run-time, the parallelization of DOTIG1 allows us to extend its application to the study of more complex geometries and larger electrical size structures.

At present, there are several models of architecture and software platforms for high-performance computing. On one hand, two major architectures exist: clusters of Cray-style vector supercomputers and clusters of superscalar uni/multiprocessors. The use of clusters of superscalar processors is more and more extended because of its excellent cost/performance ratio [9]. On the other hand, the existing software for automatic parallelization does not generate efficient parallel codes. Nevertheless, a wide range of high performance parallel software for the solution of linear algebra problems is available [10].

So, it is reasonable to develop the parallelization of DOTIG1 on distributed memory multiprocessors, applying the parallel programming library MPI, and using routines that are included in the PETSc library (free distribution) [11]. This library has been selected because it includes parallel routines for solving sparse systems of linear equations. In this library a wide diversity of iterative methods and preconditioners is available. These routines are especially appropriate for the DOTIG1 numerical problem. In this way, the parallel implementation of DOTIG1 is portable on clusters of uni/multiprocessors. The evaluation of the above mentioned implementation has been carried out on a Linux cluster of 12-nodes processors Xeon HT 2.4 Ghz.

The development of this paper is summarized as follows: in Section 2, a general description of the numerical method implemented by DOTIG1 is developed; in Section 3, details about parallel implementation are described; in Section 4, results of the performance evaluation of the parallel implementation of DOTIG1 are shown; and finally, in Section 5 the main conclusions are summarized.

2 Numerical Method

The DOTIG1 code computes the values of the currents induced on a structure modeled by thin wires, in the time domain, when it is excited by an arbitrary electromagnetic signal. The structure can be an antenna (when the source of excitation is inside the structure) or a scatterer (when the source of excitation is outside the structure). The Moments Method, which analyzes the thin-wire geometry by solving the Electric Field Integral Equation (EFIE), is applied [2]. The EFIE for the unknown current, $I(s,t)$, induced on the surface of the wires is formulated enforcing the boundary condition on the total tangential electric field along the wire axis. In this sense, the EFIE can be written as follows:

$$\hat{s}.\boldsymbol{E}^{in}(s,t) = \frac{\hat{s}}{4\pi\varepsilon_0}\int_{c(s\,)}[\frac{\hat{s}'}{C^2R}\partial_t\, I(s',t') + \frac{\boldsymbol{R}}{CR^2}\partial_s\, I(s',t') - \frac{\boldsymbol{R}}{R^3}q(s',t')]ds' \quad (1)$$

where $\hat{s}$ y $\hat{s}'$ are tangent vectors to the wire axis of contour $c(s')$ at positions $\boldsymbol{r}$ y $\boldsymbol{r}'$ respectively, see Figure 1; $q(s',t')$ is the charge distribution per unit length that can be expressed in terms of the unknown current by using the equation of continuity; $t' = t - R/C$ is the retarded time with $R = |\boldsymbol{r} - \boldsymbol{r}'|$ and C is the speed of light; $\boldsymbol{E}^{in}(s,t)$ is the known incident field on the wire in the case

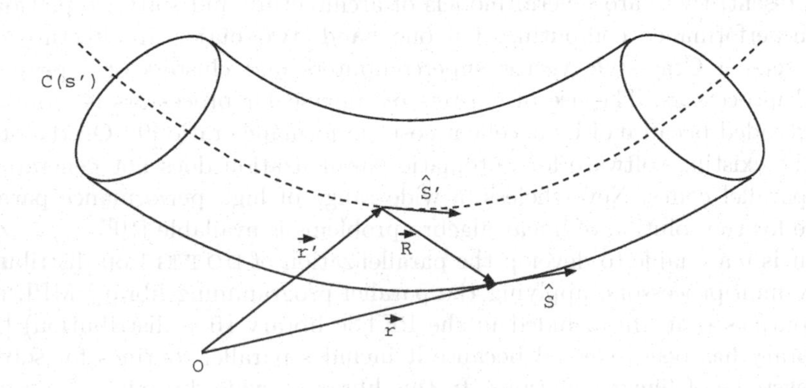

Fig. 1. Geometry of thin-wire

of a plane wave excitation or, in the antenna case, the voltage per unit length applied at the feeding points on the antenna.

To solve Equation (1), time is divided into n_t temporal intervals of equal duration Δt and the wire is approximated by n_s straight segments of equal length Δ. Then, a time stepping procedure is applied, which is based on the application of the point matching form of the Moments Method using delta type weighting functions. The Lagrange interpolation is used as temporal and spatial basis functions to expand the unknown function, $I(s,t)$. Next, several parameters are defined:

$$Z_{u,i} = \sum_{l=-1}^{+1} \sum_{p=0}^{+1} \delta(r_{i-l,u} - p)\mathcal{F}_{i-l,u,l,p} \tag{2}$$

$$E_u^{in}(\hat{s}_u, t_v) = \hat{s}.\boldsymbol{E}^{in}(\hat{s}_u, t_v) \tag{3}$$

$$E_u^s(\hat{s}_u, t_v) = -\sum_{i=1}^{n_s} \sum_{l=-1}^{+1} [\sum_{m=n; n \neq r_{i,u}}^{n+2} \mathcal{F}_{i,u,l,m} I_{i+l,v-r_{i,u}+m} + \sum_{m=n}^{n+2} \mathcal{G}_{i,u,l,m} \mathcal{Y}_{i+l,v-r_{i,u}-1+m}] \tag{4}$$

where δ is the delta of Dirac function; $\mathcal{Y}_{i+l,v-r_{i,u}-1+m} = \sum_{k=1}^{v-r_{i,u}-1} I_{i+l,k+m}$; $R_{i,u}$ is the discretized version of the magnitude of vector $\boldsymbol{R}$; the value of the integer n is always equal to -1, except when $R_{i,u}/C\Delta t < 1/2$, in this case $n = -2$; $r_{i,u}$ is the closest integer value of $R_{i,u}/C\Delta t$; $\mathcal{F}_{i,u,l,p}$ and $\mathcal{G}_{i,u,l,m}$ are the elements of the matrices of interaction. The dimension of these matrices is $n_s \times n_s \times 3 \times 2$, they only depend on the geometry, on the electromagnetic characteristics of the structure and on the coefficients used in the Lagrangian interpolation applied. Their specific dependence on the above mentioned parameters is given in [8].

According to the previous definitions, the Integral-Equation (1) can be expressed as follows:

$$\sum_{i=1}^{n_s} Z_{u,i} I_{i,v} = E_u^s(\hat{s}_u, t_v) + E_u^{in}(\hat{s}_u, t_v); \; u = 1 \dots n_s; \; v = 1 \dots n_t \tag{5}$$

where $\{I_{i,v}\}_{1 \leq i \leq n_s}$ are the unknown coefficients of the current expansion at time step t_v. Equation (5) can be expressed in a matricial form as follows:

$$\mathcal{Z} I_v = b_v; \qquad v = 1 \dots n_t \tag{6}$$

where $b_v = \{E_u^s(\hat{s}_u, t_v) + E_u^{in}(\hat{s}_u, t_v)\}_{1 \leq u \leq n_s}$ and $I_v = \{I_{i,v}\}_{1 \leq i \leq n_s}$. The matrix $\mathcal{Z} \in \mathcal{R}^{n_s \times n_s}$ is time independent; the value of $b_v \in \mathcal{R}^{n_s}$ depends on $E_u^s(\hat{s}_u, t_v)$ and $E_u^{in}(\hat{s}_u, t_v)$ (the field created at t_v by the wire currents during the n_s previous $\{t_k\}_{v-n_s \leq k \leq v-1}$, and the electric field incident at t_v). Therefore, the variation of the currents in time can be described by the Expression (6) in function of the antenna current and fields in n_s previous time steps. More details about the numerical method are described in [8].

Given a specific geometric structure, which determines matrix $\mathcal{Z}$, and the characteristics of the source of electromagnetic excitation (which allows to evaluate b_v), Equation (6) is fully defined. In order to solve Equation (6) two kinds of numerical methods can be implemented to compute the induced currents (I_v): (i) methods based on the iterative solutions of linear systems (called iterative methods); and (ii) methods based on the factorization of $\mathcal{Z}$ to accelerate the solution of linear systems, called direct method. Taking into account that $\mathcal{Z}$ is a sparse matrix, and that there exist sparse direct and iterative solvers for distributed memory message passing systems [10], many feasible parallel solutions can be considered for solving Equation (6). However, in this paper an iterative method will be analysed. Direct methods will be the scope of future works.

Iterative methods include the computation of n_t successive solutions of the linear systems $\mathcal{Z}I_v = b_v; v = 1 \ldots n_t$. The matrix $\mathcal{Z}$ is invariant and the right-hand-side vector b_v is calculated as a function of (i) the solutions (the currents) of the previous n_s linear equation systems (columns of the matrix I are the vectors I_v) and (ii) the interaction matrices $\mathcal{F}$ and $\mathcal{G}$. This procedure is described by Algorithm 1, which determines the induced currents in a structure of thin wires.

Algorithm 1

1. Computation of the time invariant parameters:
   $\mathcal{F}, \mathcal{G}, \mathcal{Z}$
2. FOR ($v = 1, v \leq n_t, v++$) 'time loop'
   Updating $b_v(\mathcal{F}, \mathcal{G}, I)$ $(O(12n_s^2))$
   Solving $\mathcal{Z}I_v = b_v$ $(O(n_{iter} \cdot nz))$

The theoretical computational complexities of the most costly stages for Algorithm 1 are shown on the right hand side. For real life problems, the values of n_s and n_t are large and the computational cost of the algorithm is also huge. An estimation of the computational cost to solve $\mathcal{Z}I_v = b_v$ by an iterative method is $O(n_{iter} \cdot nz)$; where n_{iter} is the number of iterations and nz the total number of non zero elements of the sparse matrix $\mathcal{Z}$. (As an example, for the applications treated in this paper, typical values of nz are $nz \approx n_s \cdot n_s \cdot 10^{-3}$).

3 Parallel Implementation

In Algorithm 1, linear equation systems are solved applying an iterative method. Notice that in this application, the coefficient matrix, $\mathcal{Z}$, is real, sparse and non-symmetric. The GMRES method is considered as one of the most effective non-stationary iterative methods for real non-symmetric linear systems solver [12]. This is the main reason why the correspondent parallel solver, included in PETSc library, was chosen for our application. Additionally, the convergence of the solution is accelerated by the Jacobi preconditioner (available in PETSc).

The memory requirements for Algorithm 1 are large, therefore, its parallel implementation on distributed memory multiprocessors is appropriate. Parallel implementation has been carried out using a Single Program Multiple Data (SPMD) programming model. In this context, details about the parallel implementation of Algorithm 1 are described in the following paragraphs.

The first stage of Algorithm 1 includes the computation of the time invariant parameters, where every processor computes the submatrices $\mathcal{F}_{local}$, $\mathcal{G}_{local}$ and $\mathcal{Z}_{local}$. In this sense, if P is the total number of processors and the label of every processor is named $q = 1, \ldots, P$ then processor q computes the following submatrices:

$$\begin{aligned} \mathcal{Z}_{local} &= \{Z_{u,i}\}_{s \le u \le e;\ 1 \le i \le n_s} \\ \mathcal{F}_{local} &= \{\mathcal{F}_{i,u,l,m}\}_{s-1 \le i \le e+1;\ 1 \le u \le n_s;\ -1 \le l \le 1;\ -2 \le m \le 1} \\ \mathcal{G}_{local} &= \{\mathcal{G}_{i,u,l,m}\}_{s-1 \le i \le e+1;\ 1 \le u \le n_s;\ -1 \le l \le 1;\ -2 \le m \le 1} \end{aligned} \quad (7)$$

where $s = \lceil \frac{n_s}{P} \rceil (q-1) + 1$ and if $q \ne P$ then $e = \lceil \frac{n_s}{P} \rceil q + 1$; if $q = P$ then $e = n_s$. Thus, the first stage of Algorithm 1 is carried out in parallel without communication processes.

It must be emphasized that the parallel computation of the time invariant matrices was designed in such a way that the resulting data distribution is optimal for the parallel computation of the main stage of Algorithm 1 (step 2). Moreover, the consequence of the slight data redundancy that has been included in the parallel data distribution of matrices $\mathcal{F}_{local}$ and $\mathcal{G}_{local}$ is to diminish the communication processes carried out during the last stage of the parallel algorithm.

The second stage of Algorithm 1 includes the time loop. The principle of the causality prevents the parallelization of this loop. Therefore, only computations included in every time step can be parallelized. Every time step consists of: (i) Updating b_v and (ii) Solving the linear system of equations $\mathcal{Z} I_v = b_v$.

Updating b_v consists of computing the following expression:

$$b_v = (E_u^{in}(\hat{s}_u, t_v) + E_u^{s}(\hat{s}_u, t_v))_{u=1 \ldots n_s} \quad (8)$$

where $E_u^{in}(\hat{s}_u, t_v)$ and $E_u^{s}(\hat{s}_u, t_v)$ are given by Equations (3) and (4), respectively. The input data to compute b_v are the matrices $\mathcal{F}$, $\mathcal{G}$ and I. Columns of I are the set of n_s vectors computed during the last previous n_s time steps, $\{I_k\}_{v-n_s \le k \le v-1}$.

The parallel computation of b_v consists of evaluating Equation (4) with the data in the local submatrices. In this sense, for every index u ($1 \le u \le n_s$) the processor q computes:

$$-\sum_{i=s-1}^{e+1} \sum_{l=-1; s \le i+l \le e}^{+1} [\sum_{m=n; n \ne r_{i,u}}^{n+2} \mathcal{F}_{i,u,l,m} I_{i+l, v-r_{i,u}+m} + \sum_{m=n}^{n+2} \mathcal{G}_{i,u,l,m} \mathcal{Y}_{i+l, v-r_{i,u}-1+m}] \quad (9)$$

Moreover, as it will be explained, the parallel implementation of the solution of the linear systems defines the partition of matrix I by rows. So that the local memory of each processor stores the following submatrix of I:

$$I_{local} = \{I_{i,j}\}_{s \le i \le e;\ 1 \le j \le n_s} \quad (10)$$

According to the parallel data distribution applied to matrices $\mathcal{F}$, $\mathcal{G}$ and I, every processor computes Expression (9) for the local submatrices without any communication process among processors.

To complete the computation of $E_u^s(\hat{s}_u, t_v)$ with $1 \leq u \leq n_s$ it is necessary to include a reduction sum by a collective communication. When the communication process is finished, the local memory of every processor stores $E_u^s(\hat{s}_u, t_v)$, with $1 \leq u \leq n_s$. To finish this phase, each processor updates the local vector b_v adding the values of $E_u^{in}(\hat{s}_u, t_v)$, with $s \leq u \leq e$. Therefore, it is necessary to explicitly include calls to the message passing routines in the parallel code. Notice that as a result of this stage, vector b_v is distributed among processors.

Then, the sparse linear system must be solved to finish a time step. The parallel routines of PETSc define a block partition by rows of the matrix $\mathcal{Z}$ and an homogeneous partition of the vectors b_v and I_v. Considering that I_v is a column of I and every processor stores a sub-vector of I_v, so matrix I is distributed by rows at the local memory of each processor. Consequently, the distribution of the matrix I has the same data distribution that those considered in the parallel processes of updating b_v.

In order to carry out Algorithm 1, n_t time steps must be completed. Therefore, n_t linear equation systems are successively solved, where the only difference is the value of the right hand side, because the matrix $\mathcal{Z}$ is invariant, thus, the preconditioner set-up operation is only carried out once (before of the time loop).

The above described details about parallel implementation of Algorithm 1 can be summarized as follows:

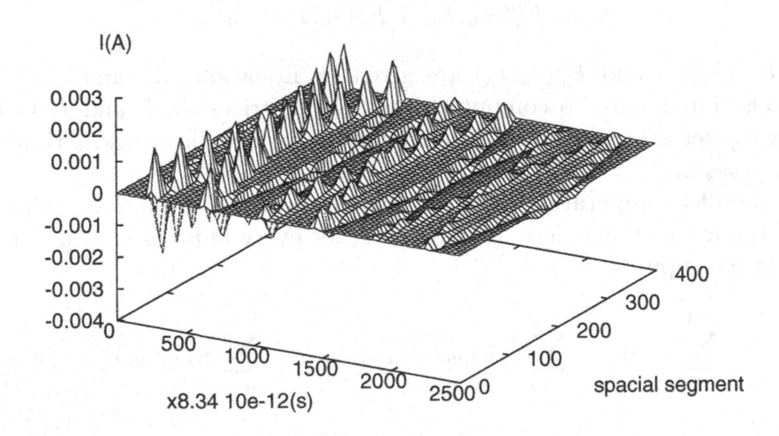

Fig. 2. Currents at $n_s = 400$ segments of the antenna versus time, for an antenna with two wires of 90° angle, 2 mm radius and 1 m length. The excitation at the vertex point is $E^{in} = Ae^{-g^2(t-t_{max})^2}$, where $A = 1.5$ V/m, $g = 10^9 s^{-1}$ and $t_{max} = 2.5 \times 10^{-10} s$

Parallel Algorithm 1

1. Computation of time invariant parameters:
   $\mathcal{F}_{local}$, $\mathcal{G}_{local}$, $\mathcal{Z}_{local}$
2. Jacobi preconditioner set-up operation
3. FOR ($v = 1$, $v \leq n_t$, $v + +$) 'time loop'
   Updating b_v (by MPI routines)
   Solving $\mathcal{Z}I_v = b_v$ (by PETSc routines)

Figure 2 shows an illustrative example of the numerical results obtained by the parallel code. The structure of the antenna considered in the example consists of two connected wires with a 90° angle, a 2 mm radius and 1 m of length. The excitation is a Gaussian pulse at the vertex point, $E^{in}(t) = Ae^{-g^2(t-t_{max})^2}$, where $A = 1.5$ V/m, $g = 10^9 s^{-1}$ and $t_{max} = 2.5 \times 10^{-10} s$. The temporal evolution of the current generated at the $n_s = 400$ segments of antenna has been drawn in Figure 2.

In the following section, results of the evaluation of our parallel implementation will be described.

4 Evaluation

Results to be analyzed in this section were obtained on a Linux cluster of 12-nodes processor Xeon Dual HT 2.4 Ghz with 1 GB RAM, 40 GB hard disk. Nodes are interconnected by a 1 Gigabit Ethernet network. The parallel code has been developed with SPMD programming model, applying MPI message passing interface [13] and PETSc routines (Portable, Extensible Toolkit for Scientific Computation) [11].

In order to evaluate our parallel implementation, the parameter speed-up has been obtained for a set of examples and several number of processors, $1 \leq P \leq 12$. Figure 3 shows the speed-up versus the number of processors, for different values of n_s and n_t. It can be seen that the results of the speed-up obtained by our parallel implementation include superlinear values. Moreover, the values of the speed-up increase as the dimension of matrix $\mathcal{Z}$ increases (large values of n_s).

There are many references about examples with superlinear speed-up results (see [14, 15, 16]). Theoretical foundations to model the superlinear speed-up have been developed by Gustafson [17] and Akl [18].

Two parameters, speed-up and efficiency, are the most used to evaluate the performance of a parallel implementation that solves a certain problem; these parameters compare the sequential and the parallel run times. But, when the speed-up and the efficiency are superlinear, it is interesting to compare two parallel executions with different number of processors. In this case, other parameters such as the *slowdown* defined by Akl [15] and the *incremental − speedup* defined by Garzón [14] can be used. Next, a new parameter related to the efficiency is defined to compare two parallel executions.

Let a computation be performed with P and Q processors in T_P and T_Q units of time, respectively; being $2 \leq Q < P$. The parameter named relative efficiency is defined as follows:

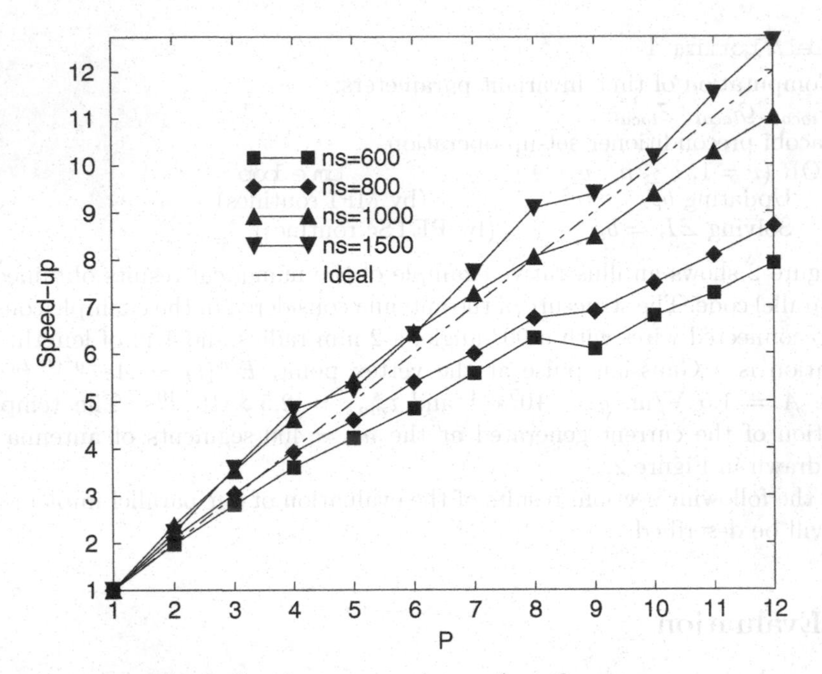

Fig. 3. Speed-up versus the number of processors

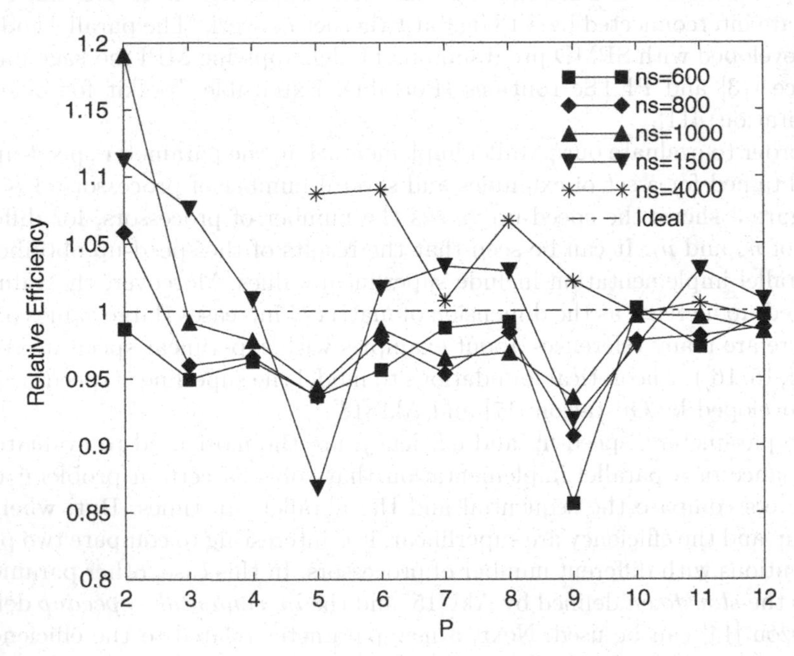

Fig. 4. Values of $E_r(P–1, P)$ obtained by the parallel implementation of *DOTIG1* on a cluster of 12 nodes Xeon HT 2.4 Ghz 1 GB RAM

$$E_r(Q,P) = \frac{Q \cdot T_Q}{P \cdot T_P} \tag{11}$$

The value of $Q \cdot T_Q$ can be considered as an evaluation of the parallel execution cost for Q processors. The parameter $E_r(Q,P)$ is appropriate to compare the parallel computational costs of two executions using Q and P processors. Notice that $E_r(P,P{+}M) = E_r(P,P{+}1) \cdot E_r(P{+}1,P{+}2) \cdots E_r(P{+}M{-}1,P{+}M)$. For most of the cases, $Q \cdot T_Q \leq P \cdot T_P$ because of the penalties of the parallel computation and so the relative e ciency usually verifies $E_r(Q,P) \leq 1$. Moreover, this parameter can be obtained, also, when it is not possible to measure the sequential run-time.

Figure 4 graphically shows the values of $E_r(P{-}1,P)$ for some examples evaluated in Figure 3 and, additionally, an example with very large requirement memory ($n_s = 4000$) is included in these measures. This example is not included in the analysis of speed-up because its large memory requirements prevent the sequential execution. However, parallel run-times for this example can be measured for larger values of number of processors $P > 3$. Each point represents the ratio of parallel execution cost with $P{-}1$ processors to the cost with P processors ($E_r(P{-}1,P) = (P{-}1) \cdot T_{P-1}/P \cdot T_P$). This parameter can explicitly show the increasing of the communication time and/or the decreasing of the access to data time when the number of P processors increases.

It can be seen from Figure 4 that for large matrices $E_r(P=2) > 1$, then $T_1 > 2 \cdot T_2$. Moreover, for large values of P ($P \geq 3$), the values of E_r are greater than 1 only for large values of n_s however, for $n_s < 1500$ the values of E_r are less

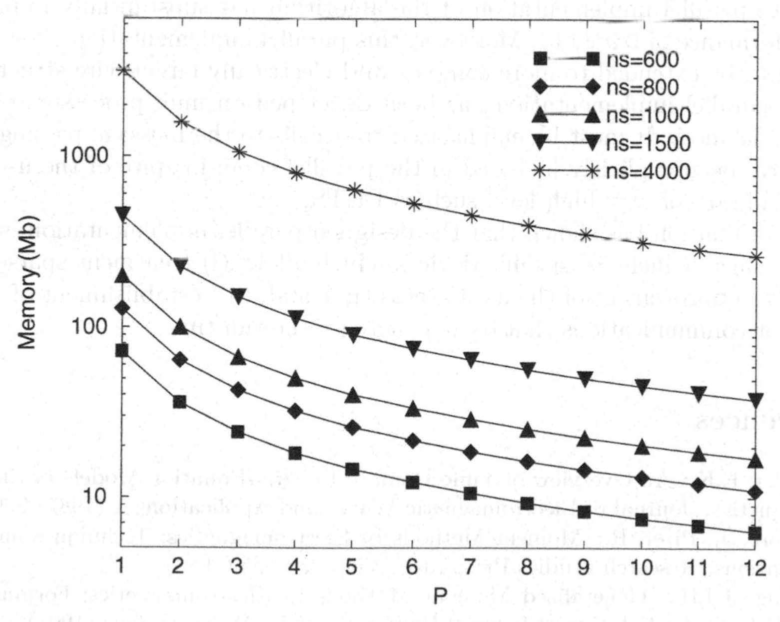

Fig. 5. Memory requirements of the parallel implementation of *DOTIG1* versus the number of processors

than 1. These values show the penalties of the parallel run-time (synchronization and communication times). These penalties do not become relevant, since the values of $E_r(P–1, P)$ are close to 1.

In order to analyze the reason of values of E_r to be greater than 1, we have done measurements of the memory resources used by each processor during a parallel execution. Figure 5 shows a log plot of the memory requirements of the algorithm versus number of processors. As it can be seen the memory requirements of the algorithm implemented are very large. Moreover, notice that the slope of the curves of the memory requirements show a sharp decrease when P is less and especially, for large values of n_s. Thus, it can be concluded that the inequality $T_1 > 2 \cdot T_2$ for large matrices is due to the improvement of the memory management.

Consequently, it can be said that the described parallel implementation substantially improves the management of the memory hierarchy and that the overhead of the communication processes is not relevant, specially when the dimension n_s is larger.

5 Conclusions

In this work, DOTIG1 [8], a numerical method that solves the Electric Field Integral Equation, has been implemented in a parallel computing system. This method is useful for analyzing the interaction of electromagnetic pulses with perfectly conducting electric thin-wire structures. The development of an appropriate parallel implementation of the algorithm has substantially improved the performance of DOTIG1. Moreover, this parallel implementation allows this method to be extended to more complex and electrically larger wire structures.

The parallel implementation has been developed on multiprocessors of distributed memory. It must be emphasized that calls to the message passing routines have been explicitly included in the parallel code, in spite of the use of a parallel library of very high level such as PETSc.

The evaluation has shown that the designed parallel implementation is very efficient since it includes specific devices which allow: (i) treatment sparse matrices, (ii) improvement of the data access time and, (iii) establishment of interprocessor communications that do not increase the run time.

References

1. Miller, E.K.: An Overview of Time-Domain Integral-Equation Models in Electromagnetics. Journal of Electromagnetic Waves and Applications **1** (1987) 269–293
2. Moore, J., Pizer, R.: Moment Methods in Electromagnetics: Techniques and applications. Research Studies Press Ltd, Wiley & Sons (1983)
3. Wang, J.J.H.: Generalized Moment Methods in Electromagnetics: Formulation and Computer Solution of Integral Equations. John Wiley & Sons (1991)
4. Landt, J., Miller, E., Deadrick, F.: Time domain modeling of nonlinear loads. IEEE Transactions on Antennas and Propagation **31** (1983) 121–126

5. Geuzaine, C., Meys, B., Beauvois, V., Legros, W.: A FETD Approach for the Modeling of Antennas. IEEE Transactions on Magnetics **36** (2000) 892–896
6. Nguyen, S., Zook, B., Zhang, X.: Parallelizing FDTD methods for solving electromagnetic scattering problems. In Astfalk, G., ed.: Applications on Advanced Architecture Computers. SIAM, Philadelphia (1996) 287–297
7. Ergin, A.A., Shanker, B., Michielssen, E.: Fast Evaluation of Three-dimensional Transient Wave Fields Using Diagonal Translation Operators. Journal of Computational Physics **146** (1998) 157–180
8. Bretones, A.R., Martin, R.G., Salinas, A.: DOTIG1, a Time-Domain Numerical Code for the Study of the Interaction of Electromagnetic Pulses with Thin-Wire Structures. COMPEL–The International Journal for Computation and Mathematics in Electrical and Electronic Engineering **8** (1989) 39–61
9. Bell, G., Gray, J.: What's Next in High-Performance Computing? Communications of ACM **45** (2002) 91–95
10. Dongarra, J.: Freely Available Software for Lineal Algebra on the Web. URL:http://www.netlib.org/utk/people/ JackDongarra/la-sw.html (april, 2003)
11. Balay, S., Buschelman, K., Gropp, W., Kaushik, D., Knepley, M., McInnes, L.C., Smith, B., Zhang, H.: PETSc Users Manual.Revision 2.1.5. Argonne National Laboratory, ANL-95/11 (2002)
12. Saad, Y.: Iterative Methods for Sparse Linear Systems. 2nd edn. SIAM (2003)
13. Snir, M., Otto, S., Huss-lederman, S., Walker, D., Dongarra, J.: MPI:The Complete Reference. MIT Press Boston (1998)
14. Garzón, E.M., García, I.: A parallel implementation of the eigenproblem for large, symmetric and sparse matrices. In: Recent advances in PVM and MPI. Volume 1697 of LNCS., Springer-Verlag (1999) 380–387
15. Akl, S.G.: Superlinear performance in real-time parallel computation. In: Proceedings of the Thirteenth Conference on Parallel and Distributed Computing and Systems, Anaheim, California,. (August 2001) 505 – 514
16. Nagashima, U., Hyugaji, S., Sekiguchi, S., Sato, M., Hosoya, H.: An experience with super-linear speedup achieved by parallel computing on a workstation cluster: Parallel calculation of density of states of large scale cyclic polyacenes. Parallel Computing **21** (1995) 1491–1504
17. Gustafson, J.L.: Reevaluating Amdahl's Law. Communications of the ACM **31** (1988) 532–533
18. Akl, S.G.: Parallel Computation: Models and Methods. Prentice Hall (1997)

A Performance Prediction Model for Tomographic Reconstruction in Structural Biology*

Paula Cecilia Fritzsche[1], José-Jesús Fernández[2], Ana Ripoll[1], Inmaculada García[2], Emilio Luque[1]

[1] Computer Science Department, University Autonoma of Barcelona. Spain
[2] Computer Architecture Department, University of Almería. Spain

Abstract. Three-dimensional (3D) studies of complex biological specimens at subcellular levels have been possible thanks to electron tomography, image processing and 3D reconstruction techniques. In order to meet computing requirements demanded by the reconstruction of large volumes, parallelization strategies with domain decomposition have been applied. Although this combination has already proved to be well suited for electron tomography of biological specimens, a performance prediction model still has not been derived. Such a model would allow further knowledge of the parallel application, and predict its behavior under different parameters or hardware platforms. This paper describes an analytical performance prediction model for BPTomo - a parallel distributed application for tomographic reconstruction-. The application's behavior is analyzed step by step to create an analytical formulation of the problem. The model is validated by comparison of the predicted times for representative datasets with computation times measured in a PCs cluster. The model is shown to be quite accurate with a deviation between experimental and predicted times lower than 10%.

1 Introduction

Analytical performance prediction models allow attaining fuller knowledge of parallel applications, prediction of their behavior under different parameters or hardware platforms, and better exploitation of the computation power of parallel systems. These models can provide answers to questions such as 'what happens to the performance if data and machine sizes are modified?', 'how many machines are needed to finish the work within a time?', 'how much performance can be gained by changing parameters?', 'is it possible to work with bigger data?'. The methodology based on prediction models allows evaluation of performance of the application, with no need of its implementation. This paper addresses the derivation and assessment of an analytical performance prediction model

* This work was supported by the MCyT under contracts 2001-2592 and 2002-00228 and partially sponsored by the Generalitat de Catalunya (Grup de Recerca Consolidat 2001SGR-00218).

M. Daydé et al. (Eds.): VECPAR 2004, LNCS 3402, pp. 90–103, 2005.

for BPTomo, a parallel application for tomographic reconstruction in structural biology that follows the Single Program Multiple Data (SPMD) paradigm.

Electron tomography, image processing and 3D reconstruction techniques allow structural analysis of complex biological specimens at sub-cellular levels [1, 2, 3]. Tomographic reconstruction allows 3D structure determination from a series of electron microscope images. These images are acquired at different orientations, by tilting the specimen around one or more axes. This structural information is critical to understanding biological function [4]. It is important to note that the methodology for tomographic analysis can be used in other disciplines ranging from engineering to earth system science [5].

The reconstruction files in electron tomography are usually large and in consequence the processing time needed is considerable. Parallelization strategies with data decomposition provide solutions to this kind of problem. The standard reconstruction method in electron tomography is called weighted back projection -WBP-. However, series expansion reconstruction methods improve the reconstructions using another basis function and do not have problems related with limited angle data or treatment of noise conditions, and thus they are being paid attention lately [6].

BPTomo is a parallelization of series expansion methods using blobs as basis functions for its application in electron tomography of biological specimens. Work has been carried out on parallelization in image reconstruction, but most of this either focuses on two-dimensional (2D) image reconstruction [7], or deals with 3D reconstruction using voxels as basis functions [5].

This paper presents an analytical performance prediction model for BPTomo. Due to nature and regularity of the application, an analytical prediction is more appropriated than simulation. In addition, the time needed to evaluate the performance is reduced. The behavior of application is analyzed step by step to create an analytical formulation of the problem. The model is validated by comparison of predicted times with computation times measured in a cluster.

The work reaches the conclusion that a performance prediction model proposed for BPTomo is quite accurate. The model allows beforehand the analysis of the optimum number of processors for the application, depending on its parameters. It also provides answers to the questions formulated at the beginning of the section. Moreover, the analytical model open gates not only to people who work in Biosciences but also general researchers that need to analyze 3D structures in any other discipline.

The rest of the article is organized as follows. Section 2 analyzes the BPTomo application. Section 3 derives the analytical model for the application. Section 4 presents the use of the model and the experimental validation. Section 5 summarizes and draws the conclusion of the work.

2 BPTomo Application

BPTomo is a parallel application of tomographic reconstruction that works with spherically symmetrical volume elements - blobs - [8]. Voxels and blobs are ba-

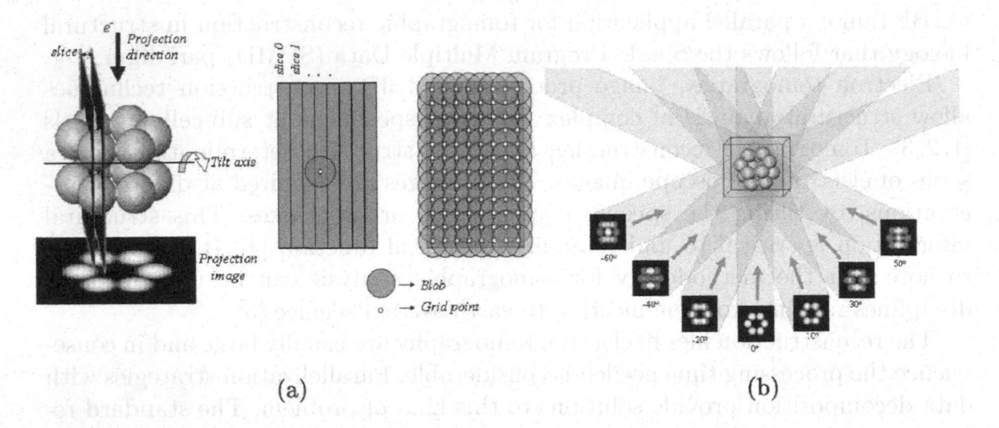

Fig. 1. (a) Acquisition of 2D data while the object is tilting around one axis. (b) 3D reconstruction from projections

sis functions used in the field of image reconstruction. Blobs have a smooth spherical shape while voxels have a cubic form. Blobs can also produce better reconstructions because they involve an implicit regularization mechanism [6].

2.1 Analysis

BPTomo uses an iterative series expansion reconstruction method. The method receives as an input a series of 2D projection images and through an iterative process, it produces a series of reconstructed slices representing a 3D image. Fig. 1(a) shows the acquisition of projection images from the unknown 3D object. The slices to be reconstructed are those planes orthogonal to the tilt axis. Interdependence between adjacent slices due to the blob extension makes the structure represented by a pseudo-continuous 3D density distribution. Fig. 1(b) sketches the 3D reconstruction process.

BPTomo implements an iterative reconstruction technique (IRT) that assumes that the 3D object or function f to be reconstructed can be approximated by a linear combination of a finite set J of known and fixed basis functions b_j.

$$f(r, \phi_1, \phi_2) \approx \sum_{j=1}^{J} x_j b_j(r, \phi_1, \phi_2) \tag{1}$$

(where (r, ϕ_1, ϕ_2) are spherical coordinates), and that the aim is to estimate the unknown x_j. IRT also assumes an image formation model where the measurements depend linearly on the object in such a way that:

$$y_i \approx \sum_{j=1}^{J} l_{i,j} x_j \tag{2}$$

where y_i denotes the i-th measurement of f and $l_{i,j}$ the value of the i-th projection of the j-th basis function.

Under those assumptions, the image reconstruction problem can be modeled as the inverse problem of estimating the x_j's from the y_i's by solving the system of linear equations given by Eq. (2). It may be subdivided into B blocks each of size S, and in consequence the iterative process can be described by its own iterative steps from the k to $k+1$ by

$$x_j^{k+1} = x_j^k + \lambda_k \sum_{s=1}^{S} \underbrace{\frac{y_i - \sum_{v=1}^{J} l_{i,v} x_v^k}{\sum_{v=1}^{J} s_v^b (l_{i,v})^2} l_{i,j}}_{Term(1)} \tag{3}$$

where λ_k denotes the relaxation parameter, $b = (k \bmod B)$ the index of the block, $i = bS + s$ represents the i-th equation of the whole system, and s_v^b the number of times that the component x_v of the volume contributes with nonzero value to the equations in the b-th block [8].

The Term (1) may be broken down into the following three stages: the computation of the forward projection of the object ($FProj$), the computation of the error between the experimental and calculated projections ($EComp$), and the refinement of the object by means of error-back projection ($BProj$).

$$FProj = \sum_{v=1}^{J} l_{i,v} x_v^k \tag{4}$$

$$EComp = y_i - FProj \tag{5}$$

$$BProj = \frac{EComp}{\sum_{v=1}^{J} s_v^b (l_{i,v})^2} l_{i,j} \tag{6}$$

Eq. (3) can then be rewritten as Eq. (7) shows, where the process passes through these stages for each block of equations and for every iterative step.

$$x_j^{k+1} = x_j^k + \lambda_k \sum_{s=1}^{S} BProj \tag{7}$$

A conceptual scheme of IRT can be seen in Fig. 2. It exhibits how IRT works giving the relationships between the three stages above mentioned.

2.2 SPMD Computational Model

The application has been designed following an SPMD computational model. The fact that blob basis functions are overlapping makes the adjacent slices interdependent. For data distribution, the slices are clustered into as many slabs as the number of nodes, as shown in Fig. 3(a). Each slab is composed of its corresponding data sub domain plus redundant slices from the neighboring nodes. Then, each node executes the same program over its own slab.

The slices in each slab are divided into these categories, see Fig. 3(b):

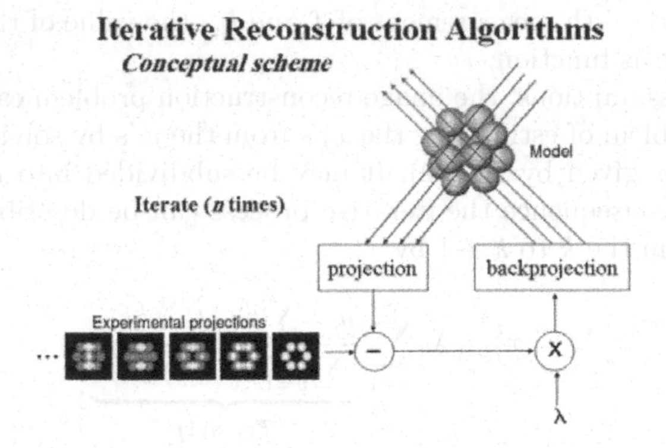

Fig. 2. IRT conceptual scheme

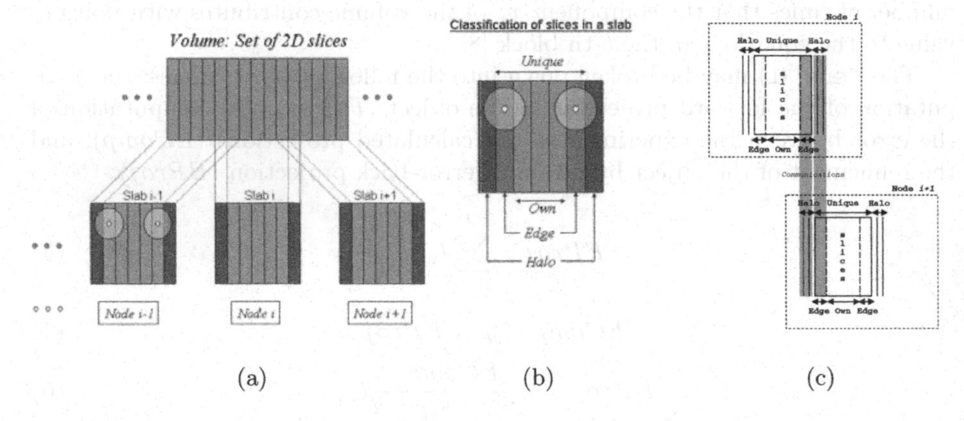

(a) (b) (c)

Fig. 3. (a) Distribution of slabs of slices. (b) Classification of slices in a slab. (c) The halo slices are modified with edge slices from the neighbor

Halo These slices are only needed by the node to reconstruct some of its own slices. They are redundant slices and coming from neighboring nodes where they are reconstructed.

Unique These slices are reconstructed by the node. These slices are further divided into the following subcategories:

Edge Slices that require information from the halo slices coming from the neighboring node.

Own Slices that do not require any information from the halo slices. These slices are independent from those in the neighbor nodes.

The interdependence between neighboring slices implies that, in order to compute either the forward projection Eq. (4) or the error-back projection Eq. (6) for a given slab, there must be a proper exchange of information between neighboring nodes. Fig. 3(c) shows this exchange of information.

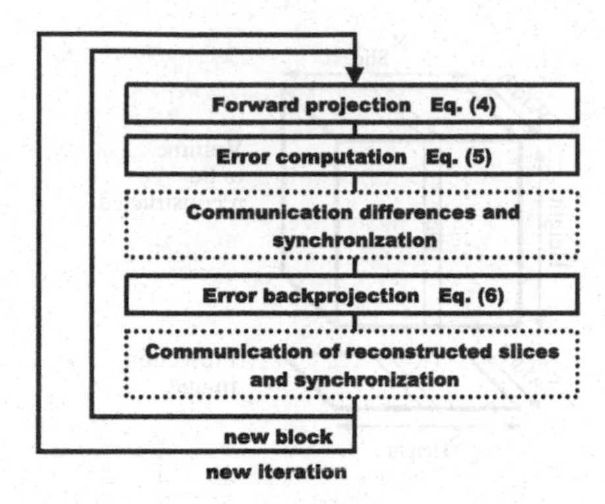

Fig. 4. IRT stages in a SPMD model

Therefore, two new stages appear in the IRT process, as Fig. 4 shows. One is the communication of differences between the experimental and calculated projections and the other is the communication of reconstructed slices which come from error-back projection stage. Both new stages involve synchronization where the nodes must wait for their neighbors.

3 BPTomo Analytical Model

The objective of the analytical model is to provide an estimation of the amount of computation and communication time per node required to reach the solution of an application that is executed in a parallel system. It is assumed that the parallel system is homogeneous for the hardware, the network and processing elements.

The total estimated execution time (T_{Est}) for a node can be approximated by the IRT time for a node (T_{IRT}) due to the negligible values of the data distribution, reading, and writing times; i.e. $T_{Est} \approx T_{IRT}$.

Table 1. Notation used in the model

Notation	Meaning
$AbyB$	Angles by block $\simeq$ Number of projection images per block. In Eq. (3) denoted by S.
$Blocks$	Number of blocks in which the complete set of projections is subdivided.
$Height$	Height of one slice.
$Width$	Width of one slice.
$Iter$	Number of iterations. It is related to the desired accuracy of the solutions.
N_{Slices}	Number of total slices to be processed.
S_{Edge}	Number of edge slices to be communicated.
S_{Height}	Height of one projection image. Note that $S_{Height} = N_{Slices}$.
S_{Width}	Width of one projection image.
T_{Comm}	Time required to transmit one pixel.
$Nodes$	Number of nodes or processors.

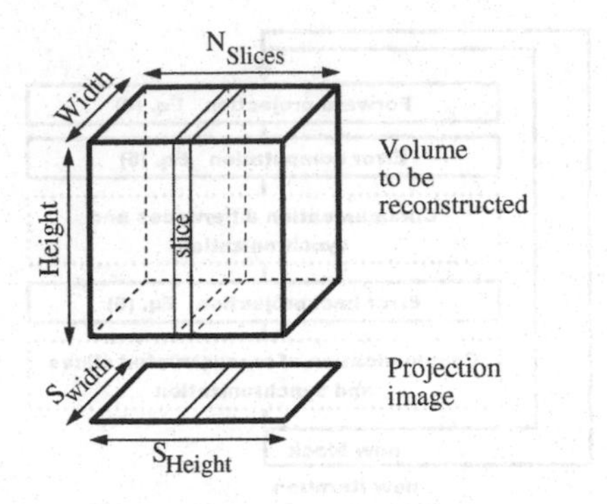

Fig. 5. Scheme of volume dimensions to reconstruct and project images

According to Fig. 4, T_{IRT} for a node can be estimated by:

$$T_{IRT} = T_{FProj} + T_{EComp} + T_{CommDiff} + T_{BProj} + T_{CommSlices} \quad (8)$$

where T_{FProj} is the forward projection time, T_{EComp} is the error computation time, $T_{CommDiff}$ is the differences communication time, T_{BProj} is the error-back projection time and $T_{CommSlices}$ is the reconstructed slices communication time. T_{FProj}, T_{EComp} and T_{BProj} are estimated values of the computation times, while $T_{CommDiff}$ and $T_{CommSlices}$ are estimated values of the communication times. For the sake of clarity, in Table 1 a description of the variables used for estimating the computation and communication times is given. Some of those variables are sketched in Fig. 5.

3.1 Estimated Computation Time

With the notation of Table 1 and based on Eq. (4)-Eq. (7), the estimation of the computational time required (per node) for evaluating the forward projection (T_{FProj}), the error computation (T_{EComp}) and the error-back projection (T_{BProj}) is analyzed in this subsection.

The size of every slice is $Height \cdot Width$, and each slice is projected $AbyB$ times per block. The projection image size is $S_{Height} \cdot S_{Width}$. In practice, $Height = Width = N_{Slices} = S_{Height} = S_{Width}$. There are $Nodes$ slabs and $N_{Slices}/Nodes$ slices in each slab. According to Eq. (3) the expressions are processed for each block of equations ($Blocks$) and for every iterative step ($Iter$). So, the forward projection time, for one node, can be estimated as:

$$T_{FProj} = Height \cdot Width \cdot AbyB \cdot (N_{Slices}/Nodes) \cdot Blocks \cdot Iter \cdot T_{PFProj} \quad (9)$$

where T_{PFProj} is the time required to "forward project" one pixel of one slice in the direction of one angle.

The time per node necessary to evaluate error computation (T_{EComp}), as given by Eq. (5), can be estimated taking into account that the number of pixels of one projection (one angle) per node is $S_{Width} \cdot N_{Slices}/Nodes$ and that there are *AbyB* projections per block. Again, this is done for each block of equations and for every iterative step. Therefore, the error computation time, for one node, can be estimated by:

$$T_{EComp} = S_{Width} \cdot AbyB \cdot (N_{Slices}/Nodes) \cdot Blocks \cdot Iter \cdot T_{PEComp} \quad (10)$$

where T_{PEComp} is the time required for the error computation of one projection of one pixel of one slice.

Error-back projection Eq. (6) exhibits a computational complexity equal to the forward projection; thus, an estimation of the time for evaluating the error-back projection is given by:

$$T_{BProj} = Height \cdot Width \cdot AbyB \cdot (N_{Slices}/Nodes) \cdot Blocks \cdot Iter \cdot T_{PBProj} \quad (11)$$

where T_{PBProj} is the time required to "backward project" the error in one projection of one pixel of one slice.

The T_{PFProj}, T_{PEComp}, and T_{PBProj} times have to be accurately determined. They can be weighted according to previous statistics obtained in [8] or can be measured running the application once. These values are platform-dependent, so for another environment with different machines and networking, such values have to be re-investigated.

3.2 Estimated Communication Time

Before giving expressions for estimating communication times, it is essential to investigate how IRT works when communicating. In order to optimize performance, communication stages are overlapped with computation stages. Therefore, communications are hidden in a total or partial way due to the order in which the slices are analyzed. As Fig. 6 shows, first, the left-edge slices are processed and sent immediately to the neighbor processor. The communication of the left-edge slices is overlapped with the processing of right-edge slices. Similarly, the communication of right-edge slices is overlapped with the processing of its own slices. This strategy is applied in both communication points; before and after error-back projection stage [8].

The transmission of differences between the experimental and calculated projection of the left-edge slices, for one node, involves the following amount of data:

$$Data_{DiffLeft} = S_{Width} \cdot AbyB \cdot S_{Edge} \cdot Blocks \cdot Iter \quad (12)$$

Similarly, the transmission of differences of right-edge slices ($Data_{DiffRight}$), for one node, involves the same amount of data of Eq. (12).

While the communication of differences of the left-edge slices is made, the processing of forward projection and error computation stages of the right-edge and of its own slices takes place. Similarly, the communication of differences of the right-edge slices is made while the processing of forward projection and error computation stages of its own slices takes place. Therefore, the estimated time is the following:

$$T_{CommDiff} = \begin{cases} T_{ExpCommDiff} & \text{if } T_{ExpCommDiff} \geq 0 \\ 0 & \text{if } T_{ExpCommDiff} < 0 \end{cases}$$

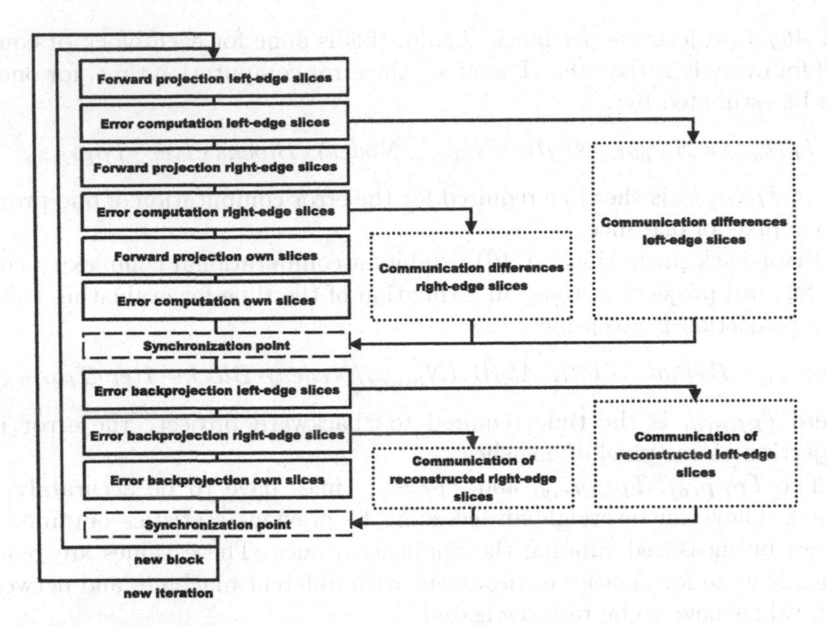

Fig. 6. Overlapping communication and computation. The continuous boxes represent steps of the iterative algorithm, the dotted boxes denote the transmission of the data already processed, and the dashed boxes mean synchronization points

where

$$T_{ExpCommDiff} = (Data_{DiffLeft} + Data_{DiffRight}) \cdot T_{Comm} - \\ -Height \cdot Width \cdot AbyB \cdot (N_{Slices}/Nodes - S_{Edge}) \cdot Blocks \cdot Iter \cdot T_{PFProj} \\ -S_{Width} \cdot AbyB \cdot (N_{Slices}/Nodes - S_{Edge}) \cdot Blocks \cdot Iter \cdot T_{PEComp}$$

and T_{Comm} is the time required to communicate a value through the network.

The transmission of reconstructed slices of left-edge slices, for one node, involves the following amount of data:

$$Data_{RSlicesLeft} = Width \cdot Height \cdot S_{Edge} \cdot Blocks \cdot Iter \quad (13)$$

Similarly, the transmission of the reconstructed slices of the right-edge slices involves the same amount of data of Eq. (13), ($Data_{RSlicesLeft} = Data_{RSlicesRight}$).

While the communication of the reconstructed slices of the left-edge slices is made, the processing of error-back projection of the right-edge and the own slices takes place. In the same way, the communication of the reconstructed slices of the left-edge slices is made while the processing of the error-back projection of the own slices takes place. So, the estimated time expression is the following:

$$T_{CommRSlices} = \begin{cases} T_{ExpCommRSlices} & \text{if } T_{ExpCommRSlices} \geq 0 \\ 0 & \text{if } T_{ExpCommRSlices} < 0 \end{cases}$$

where

$$T_{ExpCommRSlices} = (Data_{RSlicesLeft} + Data_{RSlicesRight}) \cdot T_{Comm} - \\ -Height \cdot Width \cdot AbyB \cdot (N_{Slices}/Nodes - S_{Edge}) \cdot Blocks \cdot Iter \cdot T_{PBProj}$$

4 Model Use

To validate the analytical model derived in the previous section, BPTomo[1] has been applied over two datasets with different sizes. The sizes are representatives of the current structural biology studies by electron tomography. The execution times obtained for both datasets were then compared to those times predicted by the model. The percentage of deviation has been used to show accuracy. Tables and graphics help in expressing the conclusions.

Dataset 1.
A first dataset consisting of 70 projection images of 64×64 pixels taken from the object tilted in the range [-70°, +68°] at intervals of 2° has been used to test the analytical model. In order to obtain times, it is essential to give values to some important parameters. We took 10 iterations for IRT, 1 block with 70 angles per block, and 1 or 2 edge slices to be communicated between neighboring nodes (denoted by $es = 1$ or $es = 2$). For T_{PFProj}, T_{PEComp}, and T_{PBProj} times, there are two ways to assign values. The first of these consists of using percentages 0.49%,0.01%, and 0.50%, respectively, according to the values used in a previous work [8]. The second consists of running the application once and measuring those times. T_{Comm} is related with overhead network value.

In this work, values of T_{PFProj}, T_{PEComp}, T_{PBProj}, and T_{Comm} were measured in a single node. The values obtained were 8.39E-07 sec., 3.95E-08 sec., 8.48E-7 sec., and 3.05E-5 sec., respectively.

Table 2 shows the scalability of the application using different number of nodes (first column): 1, 5, 10, 15, 20, and 25. The second, third, and fifth columns represent predicted computation times (T_{FProj}, T_{EComp} and T_{BProj}, respectively) while the fourth and sixth columns represent predicted communication times ($T_{CommDiff}$ and $T_{CommSlices}$, respectively). For the predicted communication times, there are two sub columns that correspond to the number of edge slices that was used ($es = 1$ or $es = 2$). The last column in the table is the predicted IRT time, resulting from the sum of the columns second to sixth. Accordingly, this last column also has two sub columns that refer to the number of edge slices ($es = 1$ or $es = 2$).

The main observation in Table 2 is that the predicted computation times decrease as the amount of nodes increases. Also, when the number of edge slices is 1, predicted communication times are 0, meaning that the communication is completely overlapped (i.e. hidden) with computation. If the number of edge slices is 2, there are some cases in which predicted communication times are greater

[1] BPTomo application is written in C+MPI

Table 2. Predicted times values for dataset 1 (in sec.)

$Nodes$	T_{FProj}	T_{EComp}	$T_{CommDiff}$		T_{BProj}	$T_{CommSlices}$		T_{IRT}	
			$es = 1$	$es = 2$		$es = 1$	$es = 2$	$es = 1$	$es = 2$
1	154.00	0.11	0.00	0.00	155.51	0.00	0.00	309.62	309.62
5	30.80	0.02	0.00	0.00	31.10	0.00	0.00	61.92	61.92
10	15.40	0.01	0.00	0.00	15.55	0.00	0.00	30.96	30.96
15	10.27	0.01	0.00	0.10	10.37	0.00	0.00	20.64	20.66
20	7.70	0.01	0.00	2.58	7.78	0.00	2.08	15.48	20.15
25	6.16	0.00	0.00	4.12	6.22	0.00	3.64	12.38	20.15

Table 3. Predicted times values for dataset 2 (in sec.), 10 iterations

$Nodes$	T_{FProj}	T_{EComp}	$T_{CommDiff}$	T_{BProj}	$T_{CommSlices}$	T_{IRT}
1	1231.95	0.45	0.00	1244.14	0.00	2476.54
5	246.39	0.09	0.00	248.83	0.00	495.31
10	123.20	0.05	0.00	124.41	0.00	247.65
15	82.13	0.03	0.00	82.94	0.00	165.10
20	61.60	0.02	0.00	62.21	0.00	123.83
25	49.28	0.02	0.00	49.77	0.00	99.06

than 0, instead. Thus, communication is not totally hidden by computation. As the estimated IRT time (T_{IRT}) is the sum of computation and communication times, when the number of edge slices is 1, the values monotonically decrease. For 2 edge slices, the estimated IRT times stop decreasing in one point. This is around 15 nodes, where communications begin playing an important role. The performance would therefore not improve when adding more nodes beyond 15.

Dataset 2.
A second dataset consisting of 70 projection images of 128×128 pixels taken from the object tilted in the range [-70°, +68°] at intervals of 2° has also been used to test the analytical model. The parameters and times for T_{PFProj}, T_{PEComp}, T_{PBProj}, and T_{Comm} are the same as used for Dataset 1.

The goal of using this new dataset is to quantify the influence of the size of the input images. In this dataset, the estimated communication times are 0 with either number of edge slices, 1 or 2, (see Table 3). Therefore, communication is totally hidden. This is due to the fact that the size of dataset 2 is greater than that of dataset 1. Consequently, longer computation time is need, and this time is sufficient to hide the communications completely. In order to analyze the influence of the number of iterations, for this dataset, predictions were carried out for 20 and 30 iterations as well.

Experimental Validation
The analytical model has been validated with a 25 node homogeneous PC Cluster (Pentium IV 1.8GHz., 512Mb DDR266, fast Ethernet) at the Computer Science Department, Universitat Autonoma de Barcelona. All the communications have

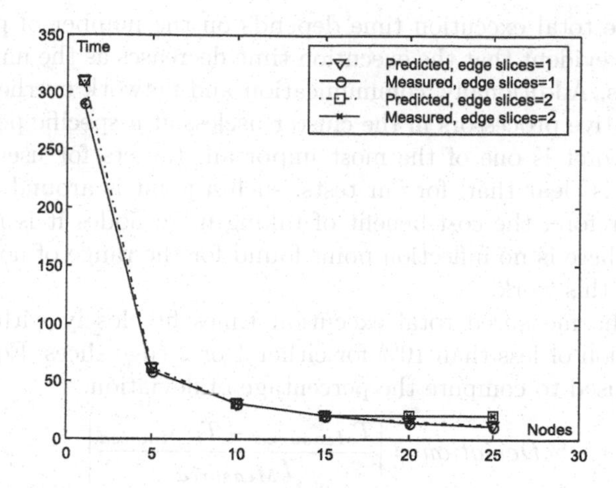

Fig. 7. Predicted versus measured total execution times (sec.) for dataset 1

been accomplished using a switched network with a mean distance between two communication end-points of two hops. The switches enable dynamic routes in order to overlap communication.

The predicted times from Table 2 and Table 3 are compared with execution times measured from BPTomo runs for the previously described datasets and parameters. The predicted and measured value times are shown in Fig. 7 for dataset 1, using edge slices equal to 1 and 2. The predicted and measured value times are shown in Fig. 8 for dataset 2, using iterations equal to 10, 20 and 30.

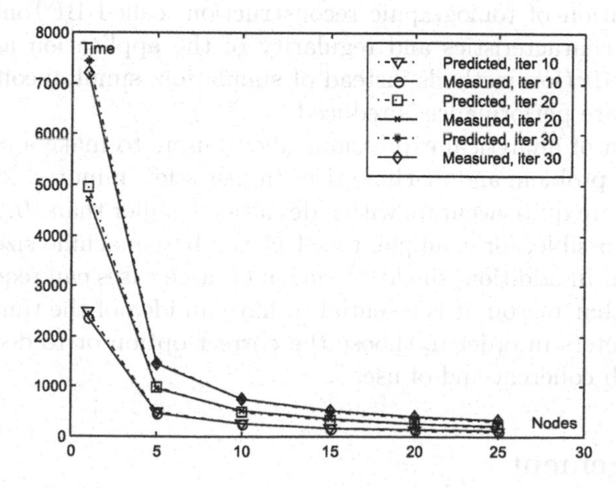

Fig. 8. Predicted versus measured total execution times (sec.) for dataset 2

Basically, the total execution time depends on the number of processors in the cluster. It is evident that the execution time decreases as the number of processors increases. Additionally, communication and network overheads increase making more active processors in the cluster useless at a specific point. Finding this inflection point is one of the most important targets for users of parallel applications. It is clear that, for our tests, such a point is around 15 nodes for Dataset 1. Therefore, the cost-benefit of taking more nodes it is not justified. For dataset 2, there is no inflection point found for the range of nodes that has been studied in this work.

In Fig. 7, the measured total execution times fit closely with predictions having a deviation of less than 10% for either 1 or 2 edge slices. Eq. (14) shows the expression used to compute the percentage of deviation.

$$\%Deviation = \left| \frac{T_{Measured} - T_{Estimated}}{T_{Measured}} \right| \tag{14}$$

Again, in Fig. 8, the predicted and measured execution are quite accurate, with a deviation smaller than 10% for a different number of iterations.

5 Conclusion

Obtaining good performance on distributed parallel applications is not at all a simple task. For example, many applications do not scale linearly on the number of nodes, and so a more powerful cluster configuration (e.g. more nodes) may not improve overall application performance. Furthermore, the execution of measurement sessions may be complex and time-consuming. In light of all the above, the availability of predicted performance figures proves to be helpful.

In this paper, an analytical performance prediction model for a parallel distributed application of tomographic reconstruction, called BPTomo, has been described. The characteristics and regularity of the application allow the use of analytical prediction methods instead of simulation; simultaneously the time needed to evaluate performance is reduced.

The adoption of analytical expressions allows users to make a complex analysis of different problem and machine sizes in just a few minutes. Moreover, the predicted times are quite accurate with a deviation smaller than 10% in all cases. This makes it possible, for example, to select the best machine size for a given input dimension. In addition, single execution of large cases can require hours or even days. For that reason, it is essential to have an idea of the time taken with different parameters in order to choose the correct option or to decide whether our work is both coherent and of use.

Acknowledgement

This work is the result of fruitful collaboration with the PhD lecturers at the Universitat Autonoma de Barcelona. Special acknowledgement is made of help received from the PhD-student staff at the UAB.

References

[1] J. Frank. Electron tomography. three-dimensional imaging with the transmission electron microscope. *Plenum Press*, 1992.

[2] A.J. Koster, R. Grimm, D. Typke, R. Hegerl, A. Stoschek, J. Walz, and W. Baumeister. Perspectives of molecular and cellular electron tomography. *J. Struct. Biol.*, 120:276–308, 1997.

[3] B.F. McEwen and M. Marko. The emergence of electron tomography as an important tool for investigating cellular ultrastructure. *J. Histochem. Cytochem.*, 49:553–564, 2001.

[4] A. Sali, R. Glaeser, T. Earnest, and W. Baumeister. From words to literature in structural proteomics. *Nature*, 422:216–225, 2003.

[5] S.T. Peltier, A.W. Lin, D. Lee, S. Mock, S. Lamont, T. Molina, M. Wong, L. Dai, M. E. Martone, and M. H. Ellisman. The telescience portal for tomography applications. *J. Paral. Distr. Comp.*, 63:539–550, 2003.

[6] J.J. Fernandez, A.F. Lawrence, J. Roca, I. Garcia, M.H. Ellisman, and J.M. Carazo. High performance electron tomography of complex biological specimens. *J. Struct. Biol.*, 138:6–20, 2002.

[7] Y. Censor, D. Gordon, and R. Gordon. Component averaging: An efficient iterative parallel algorithm for large and sparse unstructured problems. *Parallel Computing*, 27:777–808, 2001.

[8] J.J. Fernandez, J. M. Carazo, and I. Garcia. Three-dimensional reconstruction of cellular structures by electron microscope tomography and parallel computing. *J. Paral. Distr. Comp.*, 64:285–300, 2004.

Data Management in Large-Scale P2P Systems[1]

Patrick Valduriez and Esther Pacitti

Atlas group, INRIA and LINA, University of Nantes – France
Patrick.Valduriez@inria.fr
Esther.Pacitti@lina.univ-nantes.fr

Abstract. Peer-to-peer (P2P) computing offers new opportunities for building highly distributed data systems. Unlike client-server computing, P2P can operate without central coordination and offer important advantages such as a very dynamic environment where peers can join and leave the network at any time; direct and fast communication between peers, and scale up to large number of peers. However, most deployed P2P systems have severe limitations: file-level sharing, read-only access, simple search and poor scaling. In this paper, we discuss the issues of providing high-level data management services (schema, queries, replication, availability, etc.) in a P2P system. This implies revisiting distributed database technology in major ways. We illustrate how we address some of these issues in the APPA data management system under development in the Atlas group.

1 Introduction

Data management in distributed systems has been traditionally achieved by distributed database systems [19] which enable users to transparently access and update several databases in a network using a high-level query language (e.g. SQL). Transparency is achieved through a global schema which hides the local databases' heterogeneity. In its simplest form, a *distributed database system* is a centralized server that supports a global schema and implements distributed database techniques (query processing, transaction management, consistency management, etc.). This approach has proved effective for applications that can benefit from centralized control and full-fledge database capabilities, e.g. information systems. However, it cannot scale up to more than tens of databases. Data integration systems [30] extend the distributed database approach to access data sources on the Internet with a simpler query language in read-only mode. Parallel database systems [31] also extend the distributed database approach to improve performance (transaction throughput or query response time) by exploiting database partitioning using a multiprocessor or cluster system. Although data integration systems and parallel database systems can scale up to hundreds of data sources or database partitions, they still rely on a centralized global schema and strong assumptions about the network.

[1] Work partially funded by project MDP2P (Massive Data in P2P) [15] of the ACI "Masses de Données" of the French ministry of research.

M. Daydé et al. (Eds.): VECPAR 2004, LNCS 3402, pp. 104–118, 2005.

In contrast, peer-to-peer (P2P) systems adopt a completely decentralized approach to data sharing. By distributing data storage and processing across autonomous peers in the network, they can scale without the need for powerful servers. Popular examples of P2P systems such as Gnutella [8] and Kaaza [13] have millions of users sharing petabytes of data over the Internet. Although very useful, these systems are quite simple (e.g. file sharing), support limited functions (e.g. keyword search) and use simple techniques (e.g. resource location by flooding) which have performance problems. To deal with the dynamic behavior of peers that can join and leave the system at any time, they rely on the fact that popular data get massively duplicated.

Initial research on P2P systems has focused on improving the performance of query routing in the unstructured systems which rely on flooding. This work led to structured solutions based on distributed hash tables (DHT), e.g. CAN [24] and CHORD [27], or hybrid solutions with super-peers that index subsets of peers [32]. Although these designs can give better performance guarantees, more research is needed to understand their trade-offs between fault-tolerance, scalability, self-organization, etc.

Recently, other work has concentrated on supporting advanced applications which must deal with semantically rich data (e.g., XML documents, relational tables, etc.) using a high-level SQL-like query language, e.g. ActiveXML [2], Edutella [17], Piazza [29], PIER [9]. As a potential example of advanced application that can benefit from a P2P system, consider the cooperation of scientists who are willing to share their private data (and programs) for the duration of a given experiment. For instance, medical doctors in a hospital may want to share some patient data for an epidemiological study. Medical doctors may have their own, independent data descriptions for patients and should be able to ask queries like "age and last weight of the male patients diagnosed with disease X between day1 and day2" over their own descriptions.

Such data management in P2P systems is quite challenging because of the scale of the network and the autonomy and unreliable nature of peers. Most techniques designed for distributed database systems which statically exploit schema and network information no longer apply. New techniques are needed which should be decentralized, dynamic and self-adaptive.

In this paper, we discuss the main issues related to data management in large-scale P2P systems. we first recall the main principles behind data management in distributed systems and the basic techniques needed for supporting advanced functionality (schema management, access control, query processing, transaction management, consistency management, reliability and replication). Then we review P2P systems and compare the various architectures along several dimensions important for data management. we also discuss the state-of-the-art on data management in P2P systems. Finally, we illustrate how some of these issues (schema management, replication and query processing) are addressed in the context of APPA (*Atlas Peer-to-Peer Architecture*), a P2P data management system which we are building.

The rest of the paper is organized as follows. Section 2 recalls the main capabilities of distributed database systems. Section 3 discusses and compares P2P systems from the perspective of data sharing. Section 4 discusses data management in P2P systems. Section 5 introduces data management in the APPA system. Section 6 concludes.

2 Data Management in Distributed Systems

The fundamental principle behind data management is *data independence*, which enables applications and users to share data at a high conceptual level while ignoring implementation details. This principle has been achieved by *database systems* which provide advanced capabilities such as schema management, high-level query languages, access control, automatic query processing and optimization, transactions, data structures for supporting complex objects, etc.

A *distributed database* is a collection of multiple, logically interrelated databases distributed over a computer network. A *distributed database system* is defined as the software system that permits the management of the distributed database and makes the distribution *transparent* to the users [19]. Distribution transparency extends the principle of data independence so that distribution is not visible to users.

These definitions assume that each site logically consists of a single, independent computer. Therefore, each site has the capability to execute applications on its own. The sites are interconnected by a computer network with loose connection between sites which operate independently. Applications can then issue queries and transactions to the distributed database system which transforms them into local queries and local transactions (see Figure 1) and integrates the results. The distributed database system can run at any site s, not necessarily distinct from the data (i.e. it can be site 1 or 2 in Figure 1).

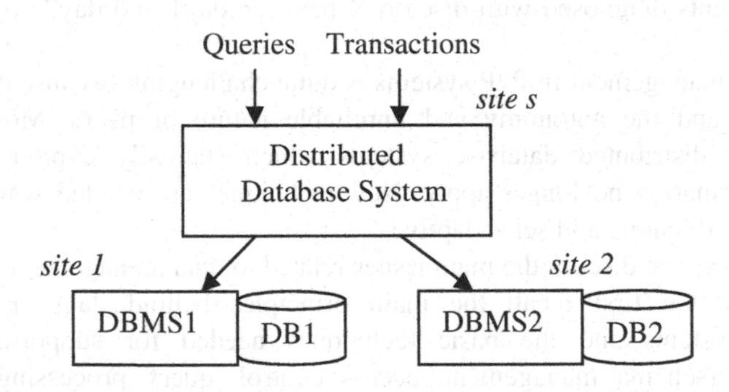

Fig. 1. A distributed database system with two data sites

The database is physically distributed across the data sites by fragmenting and replicating the data. Given a relational database schema, for instance, fragmentation subdivides each relation into partitions based on some function applied to some tuples' attributes. Based on the user access patterns, each of the fragments may also be replicated to improve locality of reference (and thus performance) and availability.

The functions provided by a distributed database system could be those of a database system (schema management, access control, query processing, transaction support, etc). But since they must deal with distribution, they are more complex to implement. Therefore, many systems support only a subset of these functions.

When the data and the databases already exist, one is faced with the problem of providing integrated access to heterogeneous data. This process is known as *data integration*: it consists in defining a *global schema* over the existing data and *mappings* between the global schema and the local database schemas. Data integration systems have received several names such as federated database systems, multidatabase systems and, more recently, mediators systems. In the context of the Web, mediator systems [30] allow general access to autonomous data sources (such as files, databases, documents, etc.) in read only mode. Thus, they typically do not support all database functions such as transactions and replication.

When the architectural assumption of each site being a (logically) single, independent computer is relaxed, one gets a *parallel database system* [31], i.e. a database system implemented on a tightly-coupled multiprocessor or a cluster. The main difference with a distributed database system is that there is a single operating system which eases implementation and the network is typically faster and more reliable. The objective of parallel database systems is high-performance and high-availability. High-performance (i.e. improving transaction throughput or query response time) is obtained by exploiting data partitioning and query parallelism while high-availability is obtained by exploiting replication.

The distributed database approach has proved effective for applications that can benefit from centralized control and full-fledge database capabilities, e.g. information systems. For administrative reasons, the distributed database system typically runs on a separate server and this reduces scale up to tens of databases. Data integration systems achieve better scale up to hundreds of data sources by restricting functionality (i.e. read-only querying). Parallel database systems can also scale up to large configurations with hundreds of processing nodes by relying on a single operating system. However, both data integration systems and parallel database rely on a centralized global schema.

3 P2P Systems

Peer-to-peer (P2P) systems adopt a completely decentralized approach to resource management. By distributing data storage, processing and bandwidth across all peers in the network, they can scale without the need for powerful servers. P2P systems have been successfully used for sharing computation, e.g. SETI@home [25], communication [11] or data, e.g. Gnutella [8] and Kaaza [13]. The success of P2P systems is due to many potential benefits: scale-up to very large numbers of peers, dynamic self-organization, load balancing, parallel processing, and fault-tolerance through massive replication. Furthermore, they can be very useful in the context of mobile or pervasive computing. However, existing systems are limited to simple applications (e.g. file sharing), support limited functions (e.g. keyword search) and use simple techniques which have performance problems. Much active research is currently on-going to address the challenges posed by P2P systems in terms of high-level data sharing services, efficiency and security. When considering data management, the main requirements of a P2P system are [7]:

- **Autonomy**: an autonomous peer should be able to join or leave the system at any time without restriction. It should also be able to control the data it stores and which other peers can store its data, e.g. some other trusted peers
- **Query expressiveness**: the query language should allow the user to describe the desired data at the appropriate level of detail. The simplest form of query is key look-up which is only appropriate for finding files. Keyword search with ranking of results is appropriate for searching documents. But for more structured data, an SQL-like query language is necessary.
- **Efficiency**: the efficient use of the P2P system resources (bandwidth, computing power, storage) should result in lower cost and thus higher throughput of queries, i.e. a higher number of queries can be processed by the P2P system in a given time.
- **Quality of service**: refers to the user-perceived efficiency of the system, e.g. completeness of query results, data consistency, data availability, query response time, etc.
- **Fault-tolerance**: efficiency and quality of services should be provided despite the occurrence of peers' failures. Given the dynamic nature of peers which may leave or fail at any time, the only solution is to rely on data replication.
- **Security**: the open nature of a P2P system makes security a major challenge since one cannot rely on trusted servers. Wrt. data management, the main security issue is access control which includes enforcing intellectual property rights on data contents.

There are many different architectures and network topologies that are possible for P2P systems. Depending on the architecture, the above requirements are more or less difficult to achieve. For simplicity, we consider three main classes: unstructured, structured and super-peer. Unstructured and structured systems are also called "pure" P2P while super-peer systems are qualified as "hybrid". Pure P2P systems consider all peers equal with no peer providing special functionality.

In *unstructured systems*, the simplest ones, each peer can directly communicate with its neighbors. Figure 2 illustrates a simple unstructured system, each peer supporting the same *p2p* software. Autonomy is high since a peer only needs to know its neighbors to log in. Searching for information is simple: it proceeds by flooding the network with queries, each peer processing and redirecting the incoming queries to its neighbors. There is no restriction on the expressiveness of the query language which could be high. Such query routing based on flooding is general but does not scale up to large numbers of peers. Also, the incompleteness of the results can be high since some peers containing relevant data or their neighbors may not be reached because they are either off-line. However, since all peers are equal and able to replicate data, fault-tolerance is very high.

Initial research on P2P systems has focused on improving the performance of unstructured systems and led to *structured systems* based on distributed hash tables

(DHT), e.g. CAN [24] and CHORD [27]. A DHT system provides a hash table interface with primitives put(key,value) and get(key), where key is typically a file name and each peer is responsible for storing the values (file contents) corresponding to a certain range of keys. There is an overlay routing scheme that delivers requests for a given key to the peer responsible for that key. This allows one to find a peer responsible for a key in *O(log n)*, where *n* is the number of peers in the network. Because a peer is responsible for storing the values corresponding to its range of keys, autonomy is limited. Furthermore, DHT queries are typically limited to exact match keyword search. Active research is on-going to extend the capabilities of DHT systems to deal with more general queries such as range queries and join queries [9].

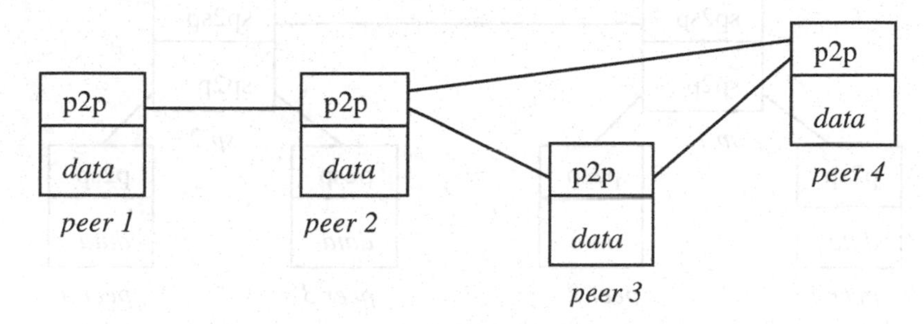

Fig. 2. P2P unstructured network

DHT overlay routing

$h(k_1)=p_1$ $h(k_2)=p_2$ $h(k_3)=p_3$ $h(k_4)=p_4$

p2p $d(k_1)$ *peer 1*

p2p $d(k_2)$ *peer 2*

p2p $d(k_3)$ *peer 3*

p2p $d(k_4)$ *peer 4*

Fig. 3. DHT network

Super-peer P2P systems are hybrid between pure systems and client-server systems. Unlike pure systems, peers are not all equal. Some peers, the super-peers, act as dedicated servers for some other peers and can perform complex functions such as indexing, query processing, access control and meta-data management. Using only one super-peer reduces to client-server with all the problems associated with a single server. For instance, Napster [16] which became famous for exchanging pirated music files used a central super-peer which made it easier to shut it down. Super-peers can also be organized in a P2P fashion and communicate with one another in sophisticated ways. Thus, unlike client-server systems, global information is not necessarily centralized and can be partitioned or replicated across all super-peers. Figure 4 illustrates a super-peer network that shows the different communication paths peer2super-peer (p2sp) and super-peer2super-peer (sp2sp). The main advantage of super-peer is efficiency and quality of service. A requesting peer simply sends the request, which can be expressed in a high-level language, to its responsible super-peer which can then find the relevant peers either directly through its index or indirectly using its neighbor super-peers. Access control can also be better enforced since directory and security information can be maintained at the super-peers. However, autonomy is restricted since peers cannot log in freely to any super-peer. Fault-tolerance is typically low since super-peers are single points of failure for their sub-peers.

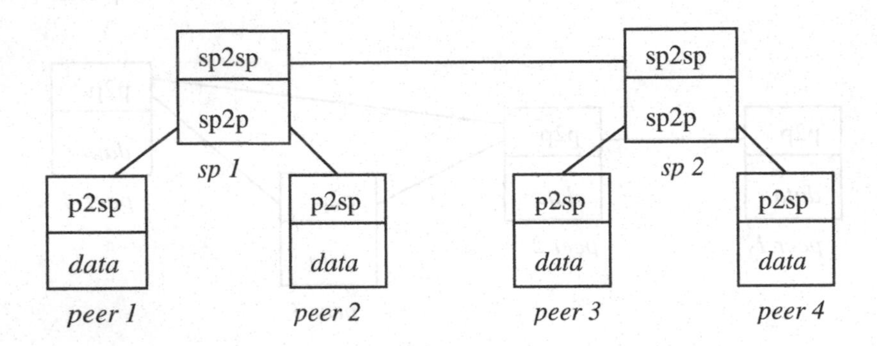

Fig. 4. Super-peer network

Table 1 summarizes how the requirements for data management are possibly attained by the three main classes of P2P systems. This is a rough comparison to understand the respective merits of each class. For instance, "high" means it can be high. Obviously, there is room for improvement in each class of systems. For instance, fault-tolerance can be made higher in super-peer by relying on replication and fail-over techniques.

Table 1. Comparison of P2P systems

Requirements	Unstructured	Structured	Super-peer
Autonomy	high	low	average
Query expressiveness	"high"	low	"high"
Efficiency	low	high	high
QoS	low	high	high
Fault-tolerance	high	high	low
Security	low	low	high

4 Data Management in P2P Systems

Advanced P2P applications must deal with semantically rich data (e.g., XML documents, relational tables, etc.). To address these applications, we need functions similar to those of distributed database systems. In particular, users should be able to use a high-level query language to describe the desired data. But the characteristics of P2P systems create new issues. First, the dynamic and autonomous nature of peers makes it hard to give guarantees about result completeness and makes static query optimization impossible. Second, data management techniques need to scale up to high numbers of peers. Third, the lack of centralized authority makes global schema management and access control difficult. Finally, even when using replication, it is hard to achieve fault-tolerance and availability in the presence of unstable peers. Most of the work on sharing semantically rich data in P2P systems has focused on schema management, and query processing and optimization. However, there has been very little work on replication, transactions and access control.

Schema management and query processing are generally addressed together for a given class of P2P system. Peers should be able to express high-level queries over their own schema without relying on a centralized global schema. Thus the main problem is to support decentralized schema mapping so that a query on one peer's schema can be reformulated in a query on another peer's schema. In PeerDB [18], assuming an unstructured network, schema mapping is done on the fly during query processing using information retrieval techniques. Although flexible, this approach limits query expressiveness to keyword search. Furthermore, query routing relies on flooding which can be inefficient. In PIER [9], a DHT network, the focus is on scaling up query processing to very large configurations assuming that de-facto standard schemas exist. However, only exact-match and equijoin queries are supported. In Edutella [17], a hybrid system, RDF-based schema descriptions are provided by super-peers. Thus, SQL-like query processing can be done by super-peers using distributed database techniques. Piazza [29] proposes a more general, network-independent, solution to schema management that supports a graph of pair-wise mappings between heterogeneous schema peers. Algorithms are proposed to reformulate a query in Xquery on a peer's schema into equivalent queries on the other peers' schemas. ActiveXML [2] is a general P2P system based on active XML documents, i.e. XML documents with embedded Web service calls in XQuery. Query processing in ActiveXML relies on a cost model which helps evaluating distributed queries and deciding which data and services to replicate.

Data replication in the presence of updates and transactions remains an open issue. The data sharing P2P systems like Gnutella and Kaaza deal with static, read-only files (e.g. music files) for which update is not an issue. Freenet [6] partially addresses updates which are propagated from the updating peer downward to close peers that are connected. However, peers that are disconnected do not get updated. ActiveXML [2] supports the definition of replicated XML fragments as Web service calls but does not address update propagation. Update is addressed in P-Grid [1], a structured network that supports self-organization. The update algorithm uses rumor spreading to scale and provides probabilistic guarantees for replica consistency. However, it only considers updates at the file level in a mono-master mode, i.e. only one (master) peer can update a file and changes are propagated to other (read-only) replicas.

Advanced applications are likely to need more general replication capabilities such as various levels of replication granularity and multi-master mode, i.e. whereby the same replica may be updated by several (master) peers. For instance, a patient record may be replicated at several medical doctors and updated by any of them during a visit of the patient, e.g. to reflect the patient's new weight. The advantage of multi-master replication is high-availability and high-performance since replicas can be updated in parallel at different peers. However, conflicting updates of the same data at different peers can introduce replica divergence. Then the main problem is to assure replica consistency. In distributed database systems [19], synchronous replication (e.g. Read-Once-Write-All) which updates all replicas within the same transaction enforces mutual consistency of replicas. However, it does not scale up because it makes use of distributed transactions, typically implemented by 2 phase commit. Preventive replication [22] can yield strong consistency, without the constraints of synchronous replication, and scale up to large configurations. However, it requires support for advanced distributed services and a high speed network with guaranteed maximum time for message reception as is the case in cluster systems. This assumption does not hold for P2P systems. A more practical solution is optimistic replication [20] which allows the independent updating of replicas and divergence until reconciliation. However, existing optimistic replication solutions do not address important properties of P2P systems such as self-organization.

5 Data Management in the APPA System

To illustrate data management in large-scale P2P systems, we introduce the design of the APPA system [3]. The main objectives of APPA are scalability, availability and performance for advanced applications. APPA has a layered service-based architecture. Besides the traditional advantages of using services (encapsulation, reuse, portability, etc.), APPA is network-independent so it can be implemented over different P2P networks (unstructured, DHT, super-peer, etc.). The main reason for this choice is to be able to exploit rapid and continuing progress in P2P networks. Another reason is that it is unlikely that a single P2P network design will be able to address the specific requirements of many different applications. Furthermore, different P2P networks could be combined in order to exploit their relative advantages, e.g. DHT for key-based search and super-peer for more complex searching.

There are three layers of services in APPA: P2P network, basic services and advanced services. The P2P network layer provides network independence with services that are common to all P2P networks : peer id assignment, peer linking and key-based storage and retrieval. The basic services layer provides services for peer management and communication over the network layer:

- **P2P data management:** stores and retrieves P2P data (e.g. meta-data, index data) by key in the P2P network.
- **Peer management:** provides support for peer joining (and rejoining) and for storage, retrieval and removal of peer ids.
- **Peer communication:** enables peers to exchange messages (i.e. service calls) even with disconnected peers using a persistent message queue.

- **Group membership management:** allows peers to join an abstract *group*, become *members* of the group and send and receive membership notifications. This is similar but much weaker than group communication [5].
- **Consensus module:** allows a given set of peers to reach agreement on a common value despite failures.

The advanced services layer provides services for semantically rich data sharing including schema management, replication, query processing, caching, security, etc. using the basic services. To capitalize on Web service standards, the shared data are in XML format (which may be interfaced with many data management systems) and the query language is XQuery. In addition, we assume each peer has data management capabilities (e.g. a DBMS) for managing its local XML data, possibly through a traditional wrapper interface.

The APPA services are organized differently depending on the underlying P2P network. For instance, in the case of a DHT network, the three service layers are completely distributed over all peers. Thus, each peer needs to manage P2P data in addition to its local data. In the case of a super-peer network, super-peers provide P2P network services and basic services while peers provide only the advanced services. APPA is being implemented using the JXTA framework [12] which provides a number of abstractions to P2P networks. In particular, JXTA provides global information sharing and group management on top of unstructured and DHT networks. Furthermore, it allows to organize some peers as super-peers.

To deal with semantically rich data, APPA supports decentralized schema management. Our solution takes advantage of the collaborative nature of the applications we target. We assume that peers that wish to cooperate, e.g. for the duration of an experiment, are likely to agree on a *Common Schema Description* (CSD). Our experience with scientific applications taught us this assumption is realistic [28]. Given a CSD, a peer schema can be specified using views. This is similar to the local-as-view approach in data integration [14] except that, in APPA, queries at a peer are expressed against the views, not the CSD. The peer schemas are stored as P2P data using the key-based storage and retrieval module, where the key is a combination of attribute and relation.

To focus on collaborative applications, we follow the small world assumption [10] which allows us to deal with groups of peers of manageable size. In this context, we can assume that the P2P system is self-organized [4] which yields fast communication between peers of the same group. Our replication model is based on the *lazy multi-master* scheme of distributed databases [21] which we transpose here to P2P. Multi-master replication allows a group of peers to update the same replicas, thus improving availability and performance. However, conflicting updates of the same data at different peers can introduce replica divergence. To solve this problem, we adapt log-based reconciliation [23] to address the properties of P2P systems. The original solution works as follows. Assuming each peer in the group holds a replica *r*, users locally execute *tentative actions* on *r* which respect some constraints and, record these actions in a local replication log. Periodically, all the logs are merged together by some peer in a *global log L*. Then the *reconcile* algorithm can be executed by that peer using *L* in order to produce a *best schedule* (an interleaved order of actions from different peers) which respects all the defined constraints. The best schedule is then applied at each peer to update *r*, possibly undoing some local tentative actions. This solution assures eventual consistency among replicas: if all users stop submitting actions then mutual consistency is achieved among all peers holding *r* [25]. However, this centralized solution is not well suited for P2P systems because decisions must be taken in a distributed way in order to preserve peers' autonomy and eventual consistency must be assured. Furthermore, it does not consider the case of peers joining or leaving a group which may impact the scheduling of actions.

In our solution, we use the P2P data management service (henceforth common storage) to log the tentative actions executed by each peer that updates *r*. The reconcile algorithm works as follows. Whenever a peer *p* updates *r*, the effects of the tentative action is immediately reflected locally and the corresponding tentative action (henceforth action) is logged in the common storage in the *action log* (see Figure 5). Thus, all actions may be eventually seen by the other peers of the group, even those that may be disconnected.

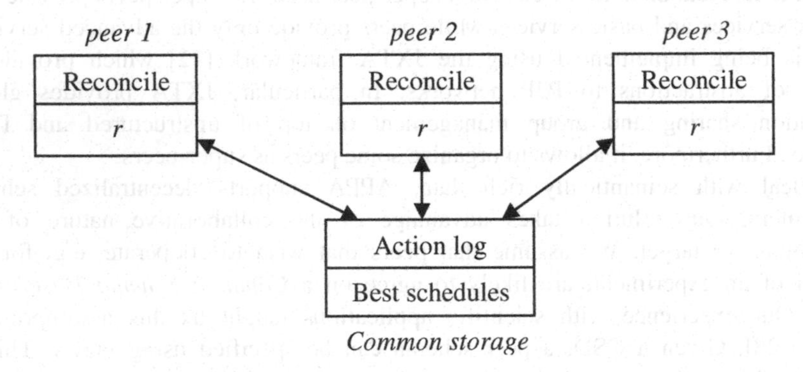

Fig. 5. Distributed Reconciliation in APPA

To manage the action log, the peers of the involved group agree (using the consensus module) on a time interval Δ in which the log actions must be grouped together to form a *log unit*. Hence, a log unit *l* holds an unordered set of actions performed by any peer of the group to update *r* during a time interval Δ Thus, the action log keeps a set of log units that are reconciled on demand (a peer reconciles the log unit he is involved) and, whenever log units do not conflict (for instance updates on different objects), they may be reconciled in parallel. Thus, whenever a peer *p* wishes to reconcile its actions *wrt*. the other peers' actions, it locally executes the reconcile algorithm using a complete log unit (stored in the action log) as input and produces the corresponding *best schedule unit*.

Our replication solution guarantees eventual consistency among replicas [26,25]. It is completely distributed and respects the autonomy of peers. Furthermore, information about replicas is systematically captured and can be exploited for other services, e.g. query processing.

Query processing in APPA deals with schema-based queries and considers data replication. Given a user query Q on a peer schema, the objective is to find the minimum set of relevant peers (query matching), route Q to these peers (query routing), collect the answers and return a (ranked) list of answers to the user. Since the relevant peers may be disconnected, the returned answers may be incomplete. Depending on the QoS required by the user, we can trade completeness for response time, e.g. by waiting for peers to get connected to get more results.

Query processing proceeds in four main phases: (1) query reformulation, (2) query matching, (3) query optimization and (4) query decomposition and execution. Phases 1, 2 can be done using techniques found in other P2P systems. However, phases 3 and 4 are new because of data replication. The optimization objective is to minimize the amount of redundant work (on replicas) while maximizing load balancing and query parallelism. This is done by statically selecting an optimal set of relevant replicas from which on-line peers are dynamically selected at run time based on load. Query decomposition and execution exploits parallelism using intermediate peers. Since some relevant peers may have only subsets of relations in Q, query decomposition produces a number of subqueries (not necessarily different), one for each peer, together with a composition query to integrate, e.g. through join and union operations, the intermediate results [30]. Finally, Q is sent to each peer which (if connected) reformulates it on its local schema (using the peer mappings), executes it and sends back the results to the sending peer which integrates the results. Result composition can also exploit parallelism using intermediate peers. For instance, let us consider relations r_1 and r_2 defined over CSD r and relations s_1 and s_2 defined over CSD r, each stored at a different peer, and the query *select * from r, s where r.a=s.a and r.b=2 and s.c=5* issued a peer q. A parallel execution strategy for Q is shown in Figure 6.

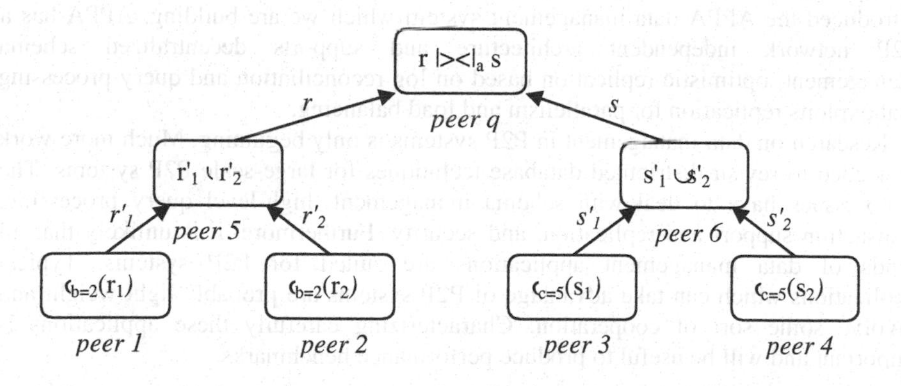

Fig. 6. Example of parallel execution using intermediate peers

This strategy exhibits independent parallelism between peers 1-4 (the select (σ) operations can all be done in parallel) and peers 5-6 (the union operations can be done in parallel). It can also yield pipelined parallelism. For instance, if the left-hand operand of an intermediate peer is smaller than the right-hand operand, then it would be entirely transferred first so the other operand could be pipelined thus yielding parallelism between peers 2-5-*q* and peers 4-6-*q*. Parallel execution strategies improve both the query response time and the global efficiency of the P2P system.

6 Conclusion

P2P systems adopt a completely decentralized approach to data sharing. By distributing data storage and processing across autonomous peers in the network, they can scale without the need for powerful servers. Although very useful, these systems are too simple and limited for advanced applications.

Advanced P2P applications such as scientific cooperation must deal with semantically rich data (e.g., XML documents, relational tables, etc.). Supporting such applications requires significant revisiting of distributed database techniques (schema management, access control, query processing, transaction management, consistency management, reliability and replication). When considering data management, the main requirements of a P2P system are autonomy, query expressiveness, efficiency, quality of service, and fault-tolerance. Depending on the P2P network architecture (unstructured, structured DHT, or hybrid super-peer), these requirements are more or less difficult to achieve. Unstructured networks have better fault-tolerance but can be quite inefficient because they rely on flooding for query routing. Hybrid systems have better potential to satisfy high-level data management requirements. However, DHT systems are best for key-based search and could be combined with super-peer networks for more complex searching.

Most of the work on sharing semantically rich data in P2P systems has focused on schema management, and query processing. However, there has been very little work on replication, transactions and access control. To illustrate some of these issues we introduced the APPA data management system which we are building. APPA has a P2P network independent architecture and supports decentralized schema management, optimistic replication based on log reconciliation and query processing that exploits replication for parallelism and load balancing.

Research on data management in P2P systems is only beginning. Much more work is needed to revisit distributed database techniques for large-scale P2P systems. The main issues have to deal with schema management, high-level query processing, transaction support and replication, and security. Furthermore, it is unlikely that all kinds of data management applications are suited for P2P systems. Typical applications which can take advantage of P2P systems are probably light-weight and involve some sort of cooperation. Characterizing carefully these applications is important and will be useful to produce performance benchmarks.

Acknowledgements

We wish to thank R. Akbarinia, V. Martins for their many inputs and fruitful discussions in the context of the APPA project, and S. Abiteboul and I. Manolescu for fruitful discussions in the context of the MDP2P project.

References

1. K. Aberer et al. P-Grid: a self-organizing structured P2P system. *SIGMOD Record, 32(3),* 2003.
2. S. Abiteboul et al. Dynamic XML documents with distribution and replication. *SIGMOD Conf.,* 2003.
3. R. Akbarinia, V. Martins, E. Pacitti, P. Valduriez. Replication and query processing in the APPA data management system. Submitted for publication, 2004.
4. E. Anceaume, M. Gradinariu, M. Roy. Self-organizing systems case study : peer-to-peer systems. *DISC Conf.,* 2003.
5. G. Chockler, I. Keidar, R. Vitenberg. Group communication specifications: a comprehensive study. *ACM Computing Surveys*, 33(427-469), 2001.
6. Clarke et al. Protecting Free Expression Online with Freenet. *IEEE Internet Computing, 6(1),* 2002.
7. N. Daswani, H. Garcia-Molina, B. Yang. Open problems in data-sharing peer-to-peer systems. *Int. Conf. on Database Theory*, 2003.
8. Gnutella. http://www.gnutelliums.com/.
9. R. Huebsch et al. Querying the Internet with PIER. *VLDB Conf.,* 2003.
10. Iamnitchi, M. Ripeanu, I. Foster. Locating data in (small world?) peer-to-peer scientific collaborations. *Int. workshop on P2P Systems (IPTPS),* 2002.
11. ICQ. http://www.icq.com/.
12. JXTA. http://www.jxta.org/.
13. Kazaa. http://www.kazaa.com/.
14. Levy, A. Rajaraman, J. Ordille. Querying heterogeneous information sources using source descriptions. *VLDB Conf.,* 1996.
15. MDP2P. http://www.sciences.univ-nantes.fr/info/recherche/ATLAS/MDP2P/.
16. Napster. http://www.napster.com/.
17. W. Nejdl, W. Siberski, M. Sintek. Design issues and challenges for RDF- and schema-based peer-to-peer systems. *SIGMOD Record, 32(3),* 2003.
18. B. Ooi, Y. Shu, K-L. Tan. Relational data sharing in peer-based data management systems. *SIGMOD Record, 32(3),* 2003.
19. T. Özsu, P. Valduriez. *Principles of Distributed Database Systems.* 2nd Edition, Prentice Hall, 1999.
20. E. Pacitti, O. Dedieu. Algorithms for optimistic replication on the Web. *Journal of the Brazilian Computing Society, 8(2),* 2002.
21. E. Pacitti, E. Simon: Update propagation strategies to improve freshness in lazy master replicated databases. *The VLDB Journal, 8(3-4),* 2000.
22. E. Pacitti, T. Özsu, C. Coulon. Preventive multi-master replication in a cluster of autonomous databases, *Euro-Par Conf.,* 2003), 2003.
23. N. Preguiça, M. Shapiro, C. Matheson. Semantics-based reconciliation for collaborative and mobile environments. *CoopIS Conf.*, 2003.
24. S. Ratnasamy et al. A scalable content-addressable network. *Proc. of SIGCOMM*, 2001.
25. SETI@home. http://www.setiathome.ssl.berkeley.edu/.
26. M. Shapiro. A simple framework for understanding consistency with partial replication. Technical Report, Microsoft Research, 2004.
27. Stoica et al. Chord: A scalable peer-to-peer lookup service for internet applications. *Proc. of SIGCOMM*, 2001.
28. Tanaka, P. Valduriez. The Ecobase environmental information system: applications, architecture and open issues. *ACM SIGMOD Record,* 3(5-6), 2000.

29. Tatarinov et al. The Piazza peer data management project. *SIGMOD Record 32(3)*, 2003.
30. Tomasic, L. Raschid, P. Valduriez. Scaling access to heterogeneous data sources with DISCO. *IEEE Trans. on Knowledge and Data Engineering*, 10(5), 1998.
31. P. Valduriez: Parallel Database Systems: open problems and new issues. *Int. Journal on Distributed and Parallel Databases*, 1(2), 1993.
32. Yang, H. Garcia-Molina. Designing a super-peer network. *Int. Conf. on Data Engineering*, 2003.

A High Performance System for Processing Queries on Distributed Geospatial Data Sets

Mahdi Abdelguerfi[1], Venkata Mahadevan[1], Nicolas Challier[1], Maik Flanagin[1], Kevin Shaw[2,*], and Jay Ratcliff[3]

[1] Department of Computer Science, University of New Orleans, New Orleans, Louisiana 70148, USA
Voice (504) 280-7076
{mahdi, venkata, nchallie, maik}@cs.uno.edu
http://www.cs.uno.edu

[2] Naval Research Laboratory, Stennis Space Center, Mississippi 39529, USA
shaw@nrlssc.navy.mil

[3] U.S. Army Corps of Engineers, New Orleans, Louisiana 70118, USA
jay.j.ratcliff@mvn02.usace.army.mil

Abstract. The size of many geospatial databases has grown exponentially in recent years. This increase in size brings with it an increased requirement for additional CPU and I/O resources to handle the querying and retrieval of this data. A number of proprietary systems could be ideally suited for such tasks, but are impractical in many situations because of their high cost. On the other hand, Beowulf clusters have gained popularity for providing such resources in a cost-effective manner. In this paper, we present a system that uses the compute nodes of a Beowulf cluster to store fragments of a large geospatial database and allows for the seamless viewing, querying, and retrieval of desired geospatial data in a parallel fashion i.e. utilizing the compute and I/O resources of multiple nodes in the cluster. Experimental results are provided to quantify the performance of the system and ascertain its feasibility versus traditional GIS architectures. Keywords: Parallel and distributed computing, Geospatial database, Beowulf cluster.

1 Introduction

Geospatial data is constantly being gathered by a variety of sensors, satellites, and other devices. The quantity of data collected by these units is staggering – NASA's Earth Observing System (EOS), for example, generates 1 terabyte of data on a daily basis [1]. The design and implementation of GIS databases have attracted considerable interest and research efforts in recent years [18, 19] and the performance problems associated with large databases have been widely documented by researchers for many years and several techniques have been devised to cope with this. One of the techniques that have gained popularity in recent years is parallel processing of geo-

* The work of Kevin Shaw was partially funded by the National Guard Bureau with program manager Major Mike Thomas.

M. Daydé et al. (Eds.): VECPAR 2004, LNCS 3402, pp. 119–128, 2005.

spatial database operations. Indeed, geospatial database operations are often time-consuming and can involve a large amount of data, so they can generally benefit from parallel processing [2, 3, 4].

There are 3 broad classes of parallel machines: shared memory systems, shared disk systems, and shared nothing systems. In a shared memory architecture, every processor has direct access to the memory of every other processor. In other words, there is only one physical memory address space for the entire system. A shared disk architecture, as its name suggests, provides shared access to the disks in the system from any of the processors via the communication network. Memory in this architecture is managed locally at each processor. Finally, in a shared nothing architecture, each processor has its own local memory and disk – nothing is shared and all communication between the processors is accomplished via the communication network. The shared nothing architecture is the most widely used design for building systems to support high performance databases, primarily due to its relatively low cost and flexible design.

Beowulf compute clusters are probably the best-known example of shared nothing machines in existence today. A number of factors can be attributed to this including the emergence of relatively inexpensive but powerful off-the-shelf desktop computers, fast interconnect networks such as Fast Ethernet and Gigabit Ethernet, and the rise of the GNU/Linux operating system. A compute cluster, for all intents and purposes, can be defined as a "Pile-of-PCs" [5] interconnected via some sort of network. Each node in the cluster usually has its own processor, memory, and optionally an I/O device such as a disk. However, it is important to note that a cluster is not simply a network of workstations (NOWs) – in a cluster, the compute nodes are delegated only for cluster usage and nodes typically have a dedicated, “cluster-only” interconnect linking them. In the Beowulf paradigm, all of the compute nodes (also called slave nodes) are isolated on a high-speed private network that is not directly visible to the outside world. A single computer called the “master” or “head-end” provides a single entry point to the cluster from the external network. This machine is sometimes referred to as the login or submit node. Essentially, the master node is a system with 2 network interfaces – one connected to the private Beowulf network and the other connected to the regular LAN. Users of the cluster will typically log in only to the master node. From here, they can spawn processes that will execute on the slave nodes. Beowulf clusters are typically applicable to any area of research where a speedup in program execution time is possible by splitting a large job into several sub-tasks that can run concurrently on the compute nodes. This capability provides an intriguing setting for evaluating the use of such a cluster for hosting large GIS databases, where there is an ever-increasing need for greater computing power to process and query the massive amounts of information being stored.

Partitioned parallelism is the main source of parallelism in a parallel shared-nothing system [16]. This is achieved through the declustering of the spatial data across the different computer nodes. The spatial database operation is then run locally on each node. Of course, the clustering scheme should distribute the data evenly across the computer nodes, and the design of the parallel spatial operation should be such that each node uses only the data fragment stored locally. Data clustering techniques implemented in our system includes hashing, round robin, spatial tiling [17], Hilbert space-filling Curve [14], and R*-tree based [13].

The work presented in this paper attempts to improve the performance aspects of geospatial querying as they pertain to a large geospatial database distributed across the compute nodes of a Beowulf cluster and is organized as follows. In Section 2 we describe the fundamental design of our Beowulf cluster and the environment in which geospatial data is stored and organized. Section 3 describes the functionality and design of the viewer used to query and retrieve geospatial data from the cluster. Section 4 discusses the test data selected for use in the process of evaluating overall system performance and explains the methodology used to experimentally gauge geospatial querying performance. An analysis of the performance results is provided in Section 5. In Section 6, we conclude with a discussion of the results and suggestions for future work.

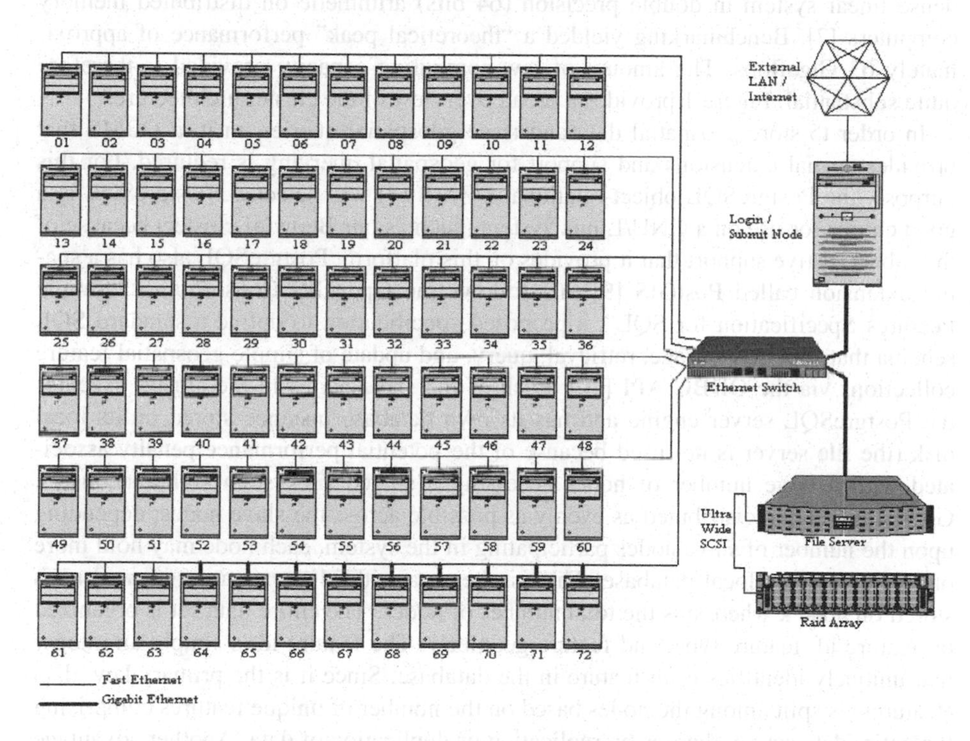

Fig. 1. Overview of 72-node Beowulf Cluster

2 Overview of System Organization

The Beowulf cluster housed at the Department of Computer Science at the University of New Orleans consists of 72 compute nodes, 1 login/submit node, and 1 file server. 63 of the slave nodes are 2.2 GHz Intel Pentium IV systems with 1 GB of memory, 20 GB of local disk storage, and Fast Ethernet networking; the remaining 9 nodes are 2.4 GHz Intel Pentium IV systems with 1GB of memory, 20 GB of local disk storage, and

Gigabit Ethernet networking. The file server is a dual 1.4 GHz SMP Intel Xeon system with 2 GB of memory, 500 GB of RAID-1 disk storage, and Gigabit Ethernet networking. Last, but not least, is the master node which is a dual 2.2 GHz SMP Intel Xeon system with 2 GB of memory, 300GB of disk storage, and 2 Ethernet interfaces. The interface that links the cluster with the private Beowulf network is Gigabit Ethernet; the external interface is 100BaseTX. All of the systems in the cluster are networked together by a Cisco Catalyst 4000 series switch with 10/100/1000 autosensing ports and a 12-Gbps backplane. Redhat Linux 7.3 is the operating system for all nodes in the cluster and the Warewulf Clustering System [6] is used to provide support utilities for cluster monitoring and maintenance. This cluster was recently benchmarked using HPL 1.0, a portable, freely available implementation of the standard High Performance Computing Linpack Benchmark. HPL solves a random, dense linear system in double precision (64 bits) arithmetic on distributed memory computers [7]. Benchmarking yielded a "theoretical peak" performance of approximately 63 Gigaflops. The amount of raw computing capacity provided is therefore quite substantial. Figure 1 provides a broad overview of the cluster architecture.

In order to store geospatial data and issue geospatial queries on it, a DBMS that provides spatial extensions and support for geospatial querying is required. For this purpose, the PostgreSQL object-relational DBMS [8] was selected. PostgreSQL is a good choice for use on a GNU/Linux system (such as our Beowulf cluster) because of the robust, native support that it provides on this platform. PostgreSQL also has a spatial extension called PostGIS [9] that follows the OpenGIS Consortium's "Simple Features Specification for SQL", a proposed specification to define a standard SQL schema that supports storage, retrieval, query, and update of simple geospatial feature collections via the ODBC API [10]. Each of the slave nodes in the cluster executes the PostgreSQL server engine and has its own database instance stored on its local disk (the file server is not used because of the potential performance penalty associated with a large number of nodes accessing a shared filesystem simultaneously). Geospatial data is distributed as evenly as possible across the slave nodes; depending upon the number of slave nodes participating in the system, each node may hold more or less data in its local database. That is, the data with id k = (feature-id mod n) is stored on node k where n is the total number of nodes. The entire data set is organized by feature id, feature type, and feature geometry. The feature id is simply an integer that uniquely identifies each feature in the database. Since it is the primary key, data (features) is split among the nodes based on the number of unique features comprising the entire data set i.e. there is no replication or duplication of data. Another advantage of this method is that if we know the id of a feature, we can easily locate the node in which the feature is stored. The feature type may be one of 7 different types specified by the "Simple Features" specification of the OpenGIS Consortium [10]. These include point, linestring, polygon, multipoint, multilinestring, multipolygon, and geometrycollection. Finally, the feature geometry is the set of geospatial coordinates comprising a particular feature. It is represented in the database as human readable text and the coordinates may be either 2-dimensional or 3-dimensional.

3 Functionality and Design of Data Viewing, Querying, and Retrieval Component

The system we have implemented is designed to exploit the parallel nature of the Beowulf cluster in order to provide fast response times when performing geospatial queries on a large data set. In particular, the "shared-nothing" parallel I/O and memory architecture of the Beowulf cluster is expected to provide a measurable performance gain in geospatial querying performance and data retrieval operations versus a single processor architecture. The functionality of the system can be broken down into 3 parts, namely: (i) geospatial query / search (ii) retrieval / download and (iii) visualization / display of data. Each of these steps is further described in detail below.

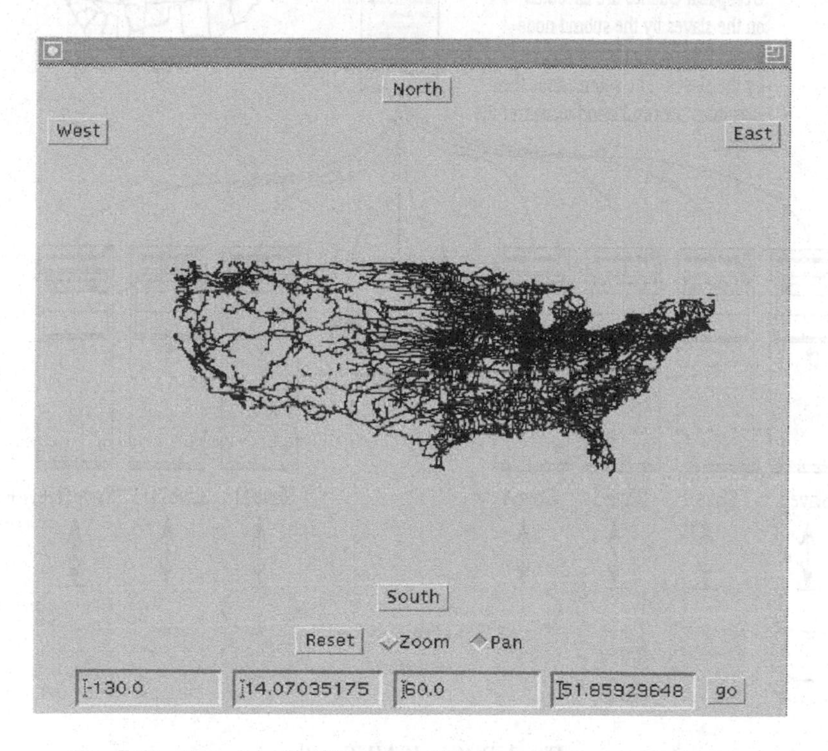

Fig. 2. User Interface for Viewing and Querying Geospatial Databases

Several types of geospatial queries are supported by the system's user interface that we implemented (see Figure 2 for a screen capture). These include area of interest (bounding box) queries, range / distance queries, and combination queries. The user interface itself runs on the master node and is responsible for spawning a separate thread to query each node in the system for relevant data encompassed within the geospatial query i.e. each query is executed in parallel across the selected nodes in the cluster that contain fragments of the geospatial database. This design makes good use of the independent disk and memory subsystems of each slave node. Once a node has finished searching its data fragment, the results (if any) are returned to the master node for visualization. Figure 3 succinctly depicts the entire process. In addition to the above functionality, the user interface also supports distributed panning of a large map – if the user pans the view and the new map data has not yet been retrieved, it is queried for on the fly by the interface.

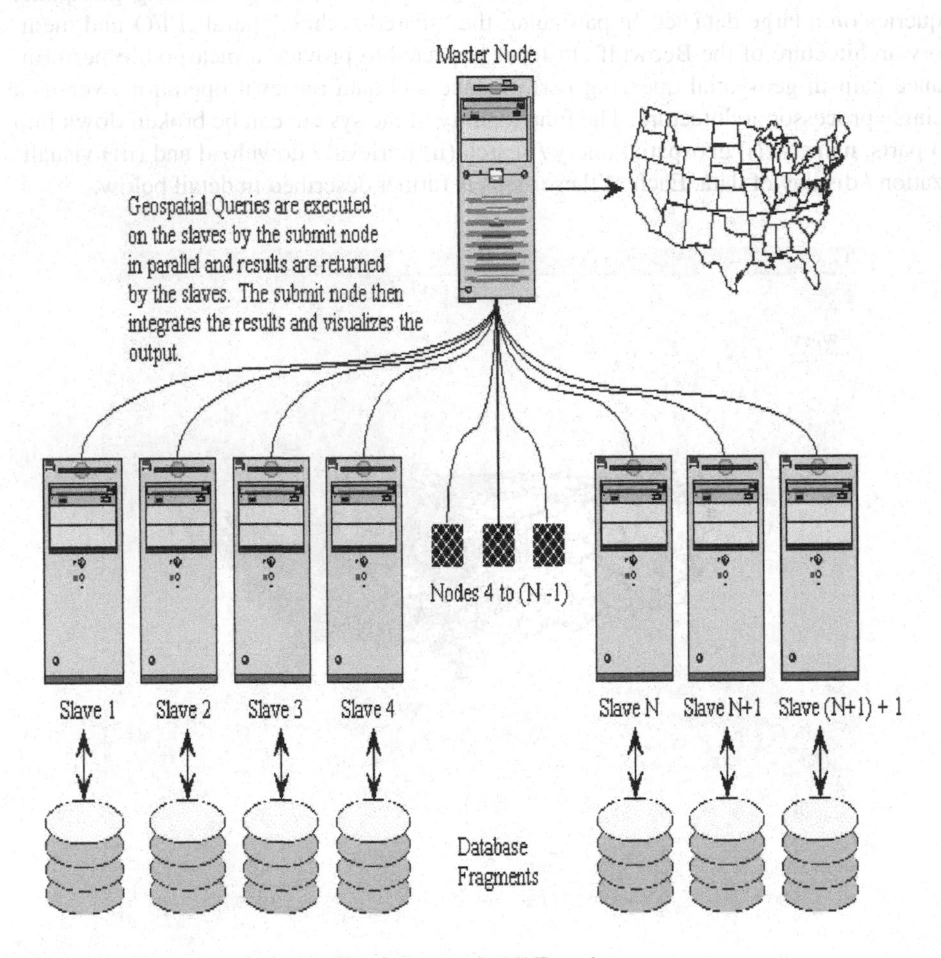

Fig. 3. Putting It All Together

4 Description of Test Data and Benchmarking Methodology

A large geospatial data set was obtained from the Bureau of Transportation Statistics (BTS): the 2002 National Transportation Data Hydrographic Features. The hydrographic features are a state-by-state database of both important and navigable water features creating a single, nationwide hydrography network containing named arcs and polygons [11]. The spatial domain of this data set is West: -179.252851, East: 179.830622, North: 71.441056 and South: 17.831509. The size of this data set is approximately 310 MB in compressed ESRI Shapefile format. When imported into a PostgreSQL database, the size is about 300 MB with 517537 unique features. The geometry of the features in this data set is represented as object type *multilinestring* from the OpenGIS Specification aforementioned. Due to the large number of unique features present in the data set and the sufficient degree of complexity present in each feature, distribution of the data across multiple nodes is expected to be beneficial.

The first step in the benchmarking process was to load the test data into the various database instances dispersed across the cluster nodes. This was accomplished as follows: a system script executes at the master node which batch loads fragments of the total data set into each database instance in the cluster – the number of fragments of course depends on the total number of slave nodes participating in the system. Once the test data has been loaded, queries can be executed on the slave nodes. The geospatial queries are similar to standard Structured Query Language (SQL) queries but provide a richer set of functions for dealing with geospatial data. For example, it is possible to express queries such as "return all features within 100 units of a particular point" using the PostGIS extensions to PostgreSQL. The following queries were selected for use in our evaluations:

Distance Query:
```
select gid, the_geom from [table_name]
where distance (the_geom, GeometryFromText(
`POINT(X Y)`, -1)) < [distance];
```

Window Query:
```
select gid, the_geom from [table_name]
where the_geom && GeometryFromText(
`BOX3D (X1 Y1, X2 Y2)`::box3d, -1);
```

The first query specifies that the feature id and feature geometry that lie within a particular distance from a point should be returned; the second query is a bounding box query – it simply specifies that all features that lie with the specified bounding box should be returned. It is conceivable that the distance function will be a computationally intensive operation within the queries because a separate distance calculation is required for finding the distance between the specified point and each point in the table (in other words, a separate distance calculation for each row of the table is needed).

The queries were tested under 3 different cluster configurations: 8 slave nodes, 16 slave nodes, and 32 slave nodes. The reference system was the master node which holds a complete copy of the data set – it was benchmarked separately. The benchmarking process itself works as follows: the master process on the master node executes each query concurrently on each of the slaves and forks a sub-process to keep track of the time it takes to retrieve the desired data. Each slave processes the query independently and returns the results to the master. When the last slave in the group has finished processing, the master process combines the results and its timing subprocess computes the time in seconds that the entire operation took to complete. In the next section, we will compare and contrast the benchmark results obtained under different cluster configurations as well as those of the reference system.

5 Performance Results

Table 1 summarizes the results for the bounding box query obtained from benchmarking using 8, 16, and 32 compute nodes (the references system is also included for comparison purposes). Each benchmark was run several times for all configurations; only averages are reported. In addition, the time reported is the total time it took from the start of the query to the visualization of the final output.

Table 1. Summary of results for bounding box (-95:30:10:10)

Number of nodes	Time (in seconds)
1 (reference system)	942.096
8	25.272
16	24.203
32	23.999

The results presented in the above table show that there is certainly a tangible performance gain that is associated with distributing and querying the data set across multiple nodes. Even with only 8 nodes, the query executed in less than a minute versus about 15.7 minutes for a single processor (the reference system). With more than 8 nodes, the difference is less marked, although there is a slight improvement with 16 and 32 nodes. This is probably attributable to the size of the data set versus the increase in communication overhead when more nodes are involved in the querying process. That is, network communication overhead caused by a large number of nodes returning query results to the master negates any processing gains achieved by having more processors and disks in the overall system. However, a larger data set may warrant the use of an even greater number of nodes.

Table 2 summarizes the results for the distance query obtained from benchmarking using 8, 16, and 32 compute nodes. As in the previous table, the time reported is the total time it took from the start of the query to the visualization of the final output.

Table 2. Summary of results for distance query (-100:35) distance < 5

Number of nodes	Time (in seconds)
1 (reference system)	846.126
8	17.741
16	16.453
32	15.729

The results for the distance query are in line with what was achieved with the bounding box query – a sharp difference in performance between the parallel executions of the queries versus those of the single processor execution. Once again, however, there was only a slight improvement in performance when the number of nodes was doubled or even quadrupled. It is our conjecture that this is due to the size of the data set used – we expect that a larger data set will benefit more from additional nodes.

6 Conclusions and Future Work

A scheme for viewing, querying, and retrieving geospatial data distributed across the compute nodes of a Beowulf cluster was discussed in this paper. The performance of this scheme under various cluster configurations was benchmarked, and the results certainly advocate using a compute cluster for mining geospatial data – large data sets such as the hydrographic data can benefit greatly from a reduction in query processing time when distributed across multiple nodes. However, it is expected that such a cluster will really make its mark when handling even larger data sets, on the order of tens of gigabytes or even terabytes of data, volumes that are simply unsustainable on even the most powerful non-parallel systems.

Current work in progress includes the parallel implementation of multi-scan queries such as the computationally expensive spatial join operation [12]. Towards this end, a semi-dynamic clustering scheme in which one of the two relations (usually the largest of the two) involved in the spatial join is declustered using an R*-tree [13] and the tree leaves placement policy is based on the well-known Hilbert curve [14]. This new approach has the advantage of dramatically reducing the replication of tuples among the computer nodes. Preliminary experiments indicate that this newly devised algorithm compares well with existing ones such as the parallel clone spatial join algorithm described in [15]. Underway is the development of an analytical model that will permit the study of the characteristics of the devised parallel join algorithm.

There is still a great deal of work that can be done in addition to that presented in this paper. In particular, we would like to make the system we have presented more modular in order to accommodate different types of data storage back-ends and visualization capabilities. Methods to handle both 2D and 3D visualization in parallel are of particular interest to us. Also, we are looking into adaptive clustering (adaptive to the size and type of data) and load balancing techniques to enhance the scalability of the system. In addition, we are interested in implementing a high-level web services application that would use this system as its back-end for providing rapid access to large data sets via a web browser interface.

Acknowledgements

We would like to thank the developers of the wonderful GeoTools Java GIS Toolkit (http://www.geotools.org) for their product, which helped considerably in developing the viewing components of the user interface.

References

1. Shashi Shekhar, Sanjay Chawla, "Spatial Databases: A Tour", Prentice Hall, 2002
2. Hector Garcia-Molina, Jeffrey D. Ullman, and Jennifer Widom, "Database System Implementation", Prentice Hall, 2000
3. Mahdi Abdelguerfi, Kam-Fai Wong (Eds.), "Parallel Database Techniques", IEEE Computer Society, 1998
4. Mahdi Abdelguerfi, Simon Lavington (Eds.), "Emerging Trends in Database and Knowledge-Base Machines", IEEE Computer Society, 1995
5. Daniel Ridge, Donald Becker, Phillip Merkey, Thomas Sterling, "Beowulf: Harnessing the Power of Parallelism in a Pile-of-PCs", Goddard Space Flight Center, CACR Caltech, Proceedings IEEE Aerospace, 1997
6. Greg Kurtzer, "The Warewulf Clustering System" http://www.runlevelzero.net/greg/warewulf/stuff/lbnl-pres/
7. Jack J. Dongara, Piotr Luszczek, Antoine Petitet, "The LINPACK Benchmark: Past, Present, and Future", 2001
8. John Worsley, Joshua Drake, Andrew Brookins (Ed.), Michael Holloway (Ed.), "Practical PostgreSQL", Command Prompt Inc., 2002
9. Refractions Research Inc., "PostGIS Manual" http://postgis.refractions.net/docs/, Refractions Research Inc., 2003
10. OpenGIS Consortium Inc., "OpenGIS Simple Features Specification for SQL", OpenGIS Consortium Inc., 1999
11. Bureau of Transportation Statistics (BTS), "2002 National Transportation Atlas Data Hydrographic Features" http://www.bts.gov/gis/download_sites/ntad02/metadata/hydrolin.htm, BTS, 2002
12. Geospatial Database Queries in a Parallel Environment, August 2003,
13. http://www.cs.uno.edu/Facilities/vcrl/RapportV3.doc
14. Beckmann, N., Kreigel, R., Seeger, B., The R*-tree: An Efficient and Robust Access method for Points and rectangles, Proceedings of the ACM SIGMOD International Conference on management of Data, pp.322-331, 1990
15. Moon, B., Jagadish, H., Faloutsos, C., and Saltz, J.H., Analysis of the Clustering Properties of Hilbert Space-filling Curve, Technical Report UMIACS-TR-96-20, CS Department.
16. Patel, J., DeWitt, D., Clone Join and Shadow Join: Two Parallel Spatial Join Algorithms, ACM GIS Conference, Washington D.C., 2000.
17. DeWitt, D., Gray, J., Parallel Database Systems: The future of Database Processing or a passing Fad? Communications of the ACM, June, 1992.
18. Patel, J., *et al.*, Building A Scalable GeoSpatial Database System: Technology, Implementation, and Evaluation , SIGMOD 1997.
19. R. Wilson, M. Cobb, F. McCreedy, R. Ladner, D. Olivier, T. Lovitt, K. Shaw, F. Petry, M. Abdelguerfi, "Geographical Data Interchange Using XML-Enabled Technology within the GIDB™ System", invited in edited manuscript: A B. Chaudhri (ed), *XML Data Management,* John Wiley & Sons, 2003.
20. Digital Mapping, Charting and Geodesy Analysis Program (DMAP) Team, http://dmap.nrlssc.navy.mil

Parallel Implementation of Information Retrieval Clustering Models

Daniel Jiménez and Vicente Vidal

Department of Computer Systems and Computation,
Polytechnic University of Valencia. Spain
{djimenez, vvidal}@dsic.upv.es

Abstract. Information Retrieval (IR) is fundamental nowadays, and more since the appearance of the Internet and huge amount of information in electronic format. All this information is not useful unless its search is efficient and effective. With large collections parallelization is important because the data volume is enormous. Hence, usually, only one computer is not sufficient to manage all data, and more in a reasonable time. The parallelization also is important because in many situations the document collection is already distributed and its centralization is not a good idea.

This is the reason why we present parallel algorithms in information retrieval systems. We propose two parallel clustering algorithms: α-Bisecting K-Means and α-Bisecting Spherical K-Means. Moreover, we have prepared a set of experiments to compare the computation performance of the algorithms. These studies have been accomplished in a cluster of PCs with 20 bi-processor nodes and two different collections.

1 Introduction

Briefly, the development of a total IR system is constituted by three phases:

1. Preprocessing: This part considers all processes once the collection is given and the weighted matrix of terms by documents is created. For example: lexical analysis, stemming, etc.
2. System Modeling: This part considers all necessary processes to prepare the collection for the queries. Modeling with LSI techniques, clustering methods, statistical techniques, etc.
3. System use: This part considers the set of queries given by the user.

This paper is centered in the second part. We study the parallelization of several models of IR systems, more specifically clustering, included in the algebraic paradigm. We assume a weighted matrix of terms by documents already distributed. Finally we stop when the system model is parallelized.

In this paper we present the parallelization of the α-Bisecting K-Means and the α-Bisecting Spherical K-Means algorithms. We also compare the performance between the sequential and parallel versions. This paper is organized as follows:

M. Daydé et al. (Eds.): VECPAR 2004, LNCS 3402, pp. 129–141, 2005.

In section 2 we present the parallelized IR models including all parallel algorithms developed. Next, in section 3, we explain the experiments executed and the collections used. Furthermore, we describe the cluster of PC's where the experiments have been executed. And finally we present the results of the experiments. In the last section, section 4, we present all the conclusions reached.

2 Parallelized IR Models

Our algorithm is based upon IR clustering models, concretely the K-Means algorithm family [1, 2, 3, 4, 5]. The main reasons are their good performance in IR, and their natural tendency to parallelization.

In all algorithms, we have used the following syntax: `M` represents the weighted matrix of terms by documents; M_i represents the i-th sub-matrix of `M` ; m_j represents the `j-th` column of a matrix (`j-th` document); `m` represents the number of terms; and `n` represents the number of documents.

2.1 Bisecting K-Means Algorithm

The α-Bisecting K-Means [3, 6] is a variant of the well known algorithm K-Means [7, 3], which has been parallelized many times [8, 9, 10]. The α-Bisecting K-Means uses the α-Bisection algorithm to perform all bisections required. In the α-Bisection, the α parameter depends of the document collection and number of clusters searched. This parameter is used to guarantee that all documents in a cluster have affinity, and all clusters have the same level of affinity.

Algorithm 1 Sequential α-Bisection

Input: $M \in \Re^{m \times n}$, α, `tol`, `maxiter`.

Output: $\pi_l \in \Re^{m \times nl}$ and $\pi_r \in \Re^{m \times nr}$ where π_l is the left cluster, π_r is the right cluster and `nl+nr=n`.

Step-1: Select a normalized concept vector, $c \in \Re^m$, of a set of documents.

Step-2: Divide `M` into two sub-clusters π_l and π_r according to:

$$m \in \pi_l \text{ if } m^{\mathrm{T}} \bullet c \geq \alpha$$
$$m \in \pi_r \text{ if } m^{\mathrm{T}} \bullet c < \alpha$$

Step-3: Calculate the new normalized concept vector of π_l, `nc`, defined as:

$$t = \frac{1}{nl} \sum_{m \in \pi_l} m \; ; \; nc = \frac{t}{\|t\|}$$

Step-4: If the stopping criterion ($\|nc - c\| \leq tol$) has been fulfilled or `maxiter=0` then finalize, else decrement `maxiter`, `c=nc` and go to step-2.

A collection could be seen as a sphere . The sphere's volume is determined by its radius raised to the document's dimension (number of terms). But we attempt to create k clusters, therefore we need to divide the collection in k sub-spheres with the same approximate volume. The α's value represents the radio of each k sub-sphere. Although we have considered the theoretical aspects indicated, also we have refined the formula that represents the α parameter experimentally. Finally we define α as follows:

$$\alpha = 1 - \frac{\overline{x} - \sigma}{\sqrt[m]{k}} \tag{1}$$

where $\overline{x}$ is the mean cosine distance, σ the standard deviation among each document of the collection and the centroid of the collection, and m is the number of terms.

In our implementation, step-1 of algorithm 1 is omitted because it is received as a parameter. With algorithm 1, we obtain two clusters, however we want to build k clusters $\{\pi_j\}_{j=1}^{k}$. Here is where appears algorithm 2, the α-Bisecting K-Means.

Algorithm 2 Sequential α-Bisecting K-Means

Input: $M \in \Re^{m \times n}$, $k \in \aleph$.
Output: k disjoint clusters$\{\pi_j\}_{j=1}^{k}$.
Step-1: Select M as the cluster to split, set t=1 and set α using expression (1).
Step-2: Find two clusters π_t and π_{t+1} with the α-Bisection, algorithm 1.
Step-3: If t=k-1 then finalize else select π_{t+1} as the cluster to split, increment t and go to step-2.

In the implementation, when algorithm 2 calls the α-Bisection algorithm (step-2) it also sends the required concept vector. Actually, the concept vector selected is a document of the collection chosen randomly.

In our parallel version of the α-Bisecting K-Means (algorithm 2), developed in this paper, we have assumed a distribution by documents as show in figure 1.

In others words, each processor has a little number of documents from the collection, achieving the necessary communications to all processors so they have the same normalized concept vector. Hence, we have developed a parallel algorithm to create a normalized concept vector (algorithm 3), a parallel version of

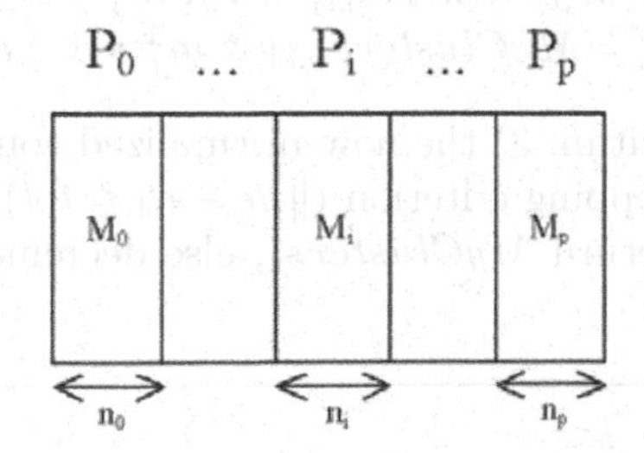

Fig. 1. Distribution of the collection

the α-Bisection (algorithm 4) and finally the parallel version of the α-Bisecting K-Means algorithm (algorithm 5).

Algorithm 3 Parallel Normalized Concept Vector

Note: This code is run by the i-th processor.
Input: $M_i \in \Re^{m \times ni}$, $idx_i \in \Re^{nidx_i}$.
Output: $ncv \in \Re^m$.
Step-1: $ncv_i = \sum_{j=1}^{nidx_i} m_{idx_i(j)} \; : \; m_{idx_i(j)} \in M_i$.
Step-2: Execute the global reduction adding all $nidx_i$ in N. Now N, in all processors, is equal to the sum of all $nidx_i$.
Step-3: Execute the global reduction adding all ncv_i in ncv. Now ncv, in all processors, is equal to the sum of all processed documents.
Step-4: $ncv = \frac{ncv}{N}$
Step-5: $ncv = \frac{ncv}{\|ncv\|}$

We should remark three things. First, that algorithm 3 calculates the normalized concept vector of several columns of the input matrix, actually the columns included in the index vector, idx. Second, step-2 and step-3 are optimized in the implementation. And third, in the implementation of the ncv 2-norm, it is also achieved in parallel, distributing the ncv vector and using a global reduction.

When we look for a cluster, we do not build explicitly the clusters. Instead, we build a vector idx. Which includes the cluster index of each document. The vector idx is used in the parallel Normalized Concept Vector algorithm and the parallel version of the α-Bisection.

Algorithm 4 Parallel α-Bisection

Note: This code is run by the i-th processor.
Input: $M_i \in \Re^{m \times ni}$, $idx_i \in \Re^{nidx_i}$, α, tol, maxiter.
Input/Output: $MyClusters_i \in \Re^{ni}$.
Step-1: Select a set of local documents and calculate in parallel the normalized concept vector $c \in \Re^m$ (algorithm 3).
Step-2: Divide M_i into two sub-clusters according to:

$$\left.\begin{array}{r} MyClusters_i[j] = MyClusters_i[j] + 1 \text{ if } m_j^{\mathrm{T}} \bullet c \geq \alpha \\ MyClusters_i[j] = MyClusters_i[j] \text{ if } m_j^{\mathrm{T}} \bullet c < \alpha \end{array}\right\} \begin{array}{l} j = idx_i(l) \\ l = 1 \ldots nidx_i \end{array}$$

Step-3: Calculate (algorithm 3) the new normalized concept vector, nc.
Step-4: Evaluate the stopping criterion ($\|nc - c\| \leq tol$). If it has been fulfilled or maxiter=0 then return $MyClusters_i$, else decrement maxiter, c=nc and go to step-2.

We should remark that, in our implementation, the 2-norm of $\|nc - c\|$ (step-4, algorithm 4) is achieved in parallel, distributing the vector and using a global

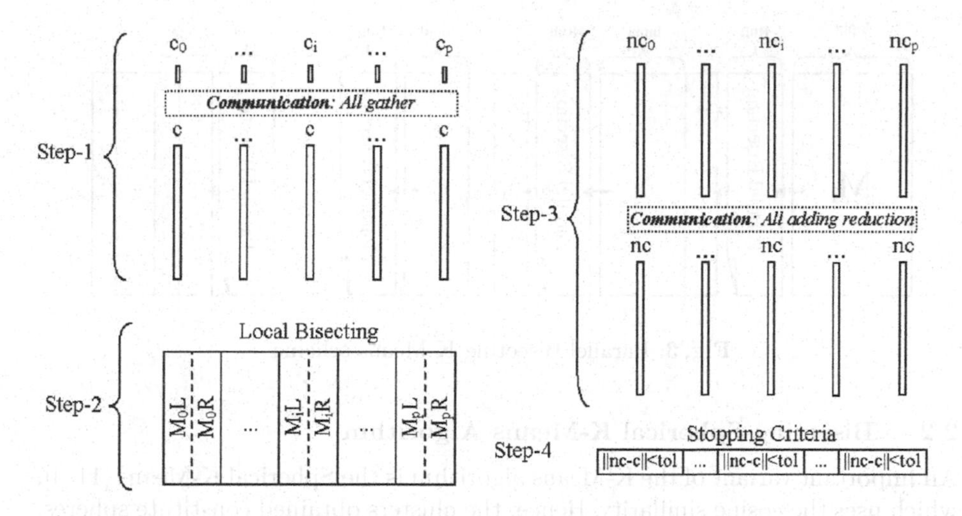

Fig. 2. Parallel α-Bisection scheme

reduction, analogous to the calculus of the ncv 2-norm in algorithm 3. Additionally, as done in the sequential version, the implementation of step-1 of algorithm 4 is omitted because it is received as a parameter.

We show in figure 2 a diagram depicting the parallel version of the α-Bisection.

And finally we present the parallel Bisecting K-Means algorithm and its diagram (algorithm 5 and figure 3):

Algorithm 5 Parallel α-Bisecting K-Means

Note: This code is run by the i-th processor.
Input: $M_i \in \Re^{m \times ni}$, $k \in \aleph$.
Output: k disjoint clusters$\{\pi_j\}_{j=1}^{k}$.
Step-1: Select M_i as the cluster to split, set t=1 and set α using expression (1).
Step-2: Find clusters π_t and π_{t+1} with the parallel α-Bisection, algorithm 4.
Step-3: If t is equal to k-1 then return all clusters else select π_{t+1} as the cluster to split, increment t and go to step-2.

The calculation of α is performed in parallel (step 1 algorithm 5). Concretely the parallelization is achieved to obtain the mean and the standard deviation. Each processor works with its own part of the collection, then all processors perform a global reduction adding the partial local sums. Just to remark, that each processor selects randomly a document of its sub-collection, then all processors calculate the normalized concept vector of all the documents chosen.

An schematic diagram depicting the parallel α-Bisecting K-Means algorithm is showed in figure 3.

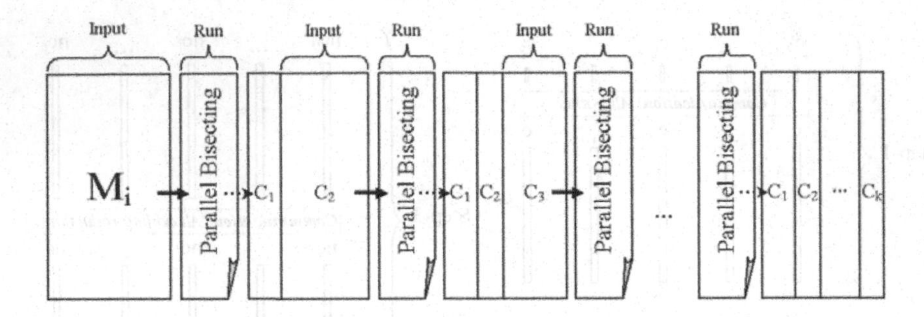

Fig. 3. Parallel Bisecting K-Means scheme

2.2 Bisecting Spherical K-Means Algorithm

An important variant of the K-Means algorithm is the Spherical K-Means [11, 4], which uses the cosine similarity. Hence, the clusters obtained constitute spheres, accordingly to its name. This algorithm seeks k disjoint clusters $\{\pi_j\}_{j=1}^{k}$ maximizing the quality of the obtained partitioning. This quality is given by the following objective function:

$$f\left(\{\pi_j\}_{j=1}^{k}\right) = \sum_{j=1}^{k} \sum_{x \in \pi_j} x^{\mathrm{T}} \bullet c_j \tag{2}$$

Consequently, the stopping criterion normally used is:

$$\left| f\left(\{\pi_j^t\}_{j=1}^{k}\right) - f\left(\{\pi_j^{t+1}\}_{j=1}^{k}\right) \right| \leq \epsilon \bullet f\left(\{\pi_j^{t+1}\}_{j=1}^{k}\right) \tag{3}$$

Algorithm 6 Sequential α-Bisecting Spherical K-Means

Input: $M \in \Re^{m \times n}$, tol, maxiter, $k \in \aleph$.

Output: k disjoint clusters$\{\pi_j\}_{j=1}^{k}$.

Step-1: Calculate k initial clusters $\left\{\pi_j^{(0)}\right\}_{j=1}^{k}$ applying the α-Bisecting K-Means, algorithm 2 and its normalized concept vectors $\left\{c_j^{(0)}\right\}_{j=1}^{k}$. Initialize t=0.

Step-2: Build a new partition $\left\{\pi_j^{(t+1)}\right\}_{j=1}^{k}$ induced by $\left\{c_j^{(t)}\right\}_{j=1}^{k}$ according to:

$$\pi_j^{(t+1)} = \left\{\forall x \in M : x^{\mathrm{T}} \bullet c_j^{(t)} > x^{\mathrm{T}} \bullet c_l^{(t)}\ ,\ 1 \leq l \leq k\ ,\ j \neq l\right\},\ 1 \leq j \leq k$$

Step-3: Calculate the new concept vector $\left\{c_j^{(t+1)}\right\}_{j=1}^{k}$ of the new partition.

Step-4: Finalize if the stopping criterion, expression 3, is fulfilled ($\epsilon = tol$) or t is equal to maxiter, else increment t and go to step-2.

The modification of the Spherical K-Means includes the α-Bisecting K-Means as the first step. The α-Bisecting Spherical K-Means, was developed in [5]. Furthermore, in this paper, we improve it because we have optimized the α-Bisecting K-Means changing the initial selection of the concept vectors and defining the α parameter in terms of the collection and number of clusters. The idea of the α-Bisecting Spherical K-Means is to use the α-Bisecting K-Means algorithm to obtain a first approximated solution and later refine it with the Spherical K-Means, taking advantage of both.

In this paper, we develop a parallel version of the α-Bisecting Spherical K-Means. We use a distribution by documents (see figure 1), the same distribution used in the parallel α-Bisecting K-Means (algorithm 5).

Algorithm 7 Parallel α-Bisecting Spherical K-Means

Note: This code is run by the i-th processor.
Input: $M_i \in \Re^{m \times ni}$, tol, m axiter, $k \in \aleph$.
Output: k disjoint clusters$\{\pi_j\}_{j=1}^{k}$.
Step-1: Calculate k clusters $\left\{\pi_j^{(0)}\right\}_{j=1}^{k}$ applying the parallel α-Bisecting K-Means, algorithm 5 and in parallel its concept vectors $\left\{c_j^{(0)}\right\}_{j=1}^{k}$, algorithm 3. Initialize t=0.
Step-2: Build a new partition $\left\{\pi_j^{(t+1)}\right\}_{j=1}^{k}$ induced by $\left\{c_j^{(t)}\right\}_{j=1}^{k}$ according to:

$$\pi_j^{(t+1)} = \left\{\forall x \in M_i : x^{\mathrm{T}} \bullet c_j^{(t)} > x^{\mathrm{T}} \bullet c_l^{(t)} \ , \ 1 \leq l \leq k \ , \ j \neq l\right\}, \ 1 \leq j \leq k$$

Step-3: Calculate in parallel the new normalized concept vector $\left\{c_j^{(t+1)}\right\}_{j=1}^{k}$ associated to the new partition, with algorithm 3.
Step-4: Finalize if the stopping criterion, expression 3, is fulfilled ($\epsilon = tol$) or t is equal to m axiter, else increment t and go to step-2.

The evaluation of the stopping criterion also occurs in parallel. In fact, the objective function evaluation is done in parallel. Basically, each processor calculates the partial sum with the documents that it contains; then all processors perform a global reduction adding the partial sums, obtaining the final result.

3 Experiments

3.1 Document Collections

In the study we use two test collections with different characteristics, what follows next is a concise description of both collections.

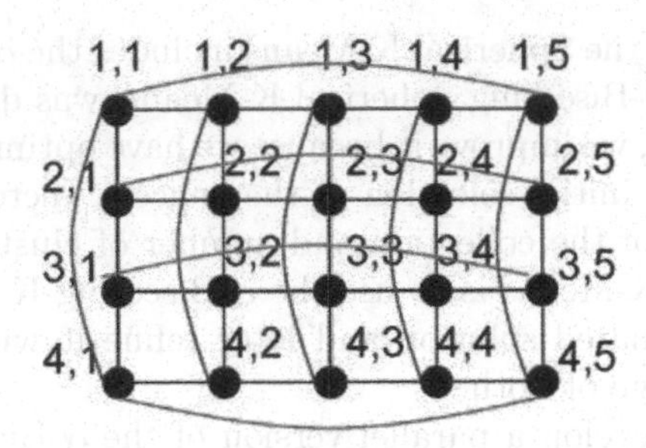

Fig. 4. Topology of the Cluster

Times Magazine of 1963: This collection contains articles from a 1963 Times Magazine and it has been obtained from ftp://ftp.cs.cornell.edu/pub/smart/time/ site. The collection language is English. A total number of 425 documents are included in the collection. Each document have an average of 546 words and 53 lines. The contents referred to world news, especially politics frequently mentioning words which in fact, reminded us of the typical news contents available in the cold war era.

DOE-TREC: This collection has been obtained from TREC (Text REtrieval Conference) site [12]. It contains a lot of abstracts of the U.S. Department of Energy (DOE). In this case, the collection language is English, too. The DOE-TREC collection is larger than the Times Magazine collection, presenting over 225000 documents.

3.2 Cluster of PCs

All performance studies of the parallel algorithms developed have been performed in a cluster of PCs with the following characteristics: the cluster has 20 bi-processor nodes (Pentium Xeon at 2 Ghz). These nodes have a toroidal 2D topology with SCI interconnection, as shown in figure 4. The main node is constituted by 4 Ultra SCSI hard disk of 36 Gbytes at 10000 rpm. The others nodes have an IDE hard disk of 40 Gbytes at 7200 rpm. All nodes have the same amount of RAM, concretely 1 GB.

Actually, we only include the experiments performed into 10 processors. We have always selected processors located into different nodes to optimize the performance. Two processors of the same node have to share an unique net card. Besides, the results obtained with 10 processors show clearly the performance evolution, and therefore we avoid presenting complex performance evolution graphs.

3.3 Description of Experiments

We compare the final parallel versions with their respective sequential versions. We study mainly the total time, efficiency and scalability. In all experiments, we use the two document collections described above, and tests were performed in the cluster of PCs commented above, too.

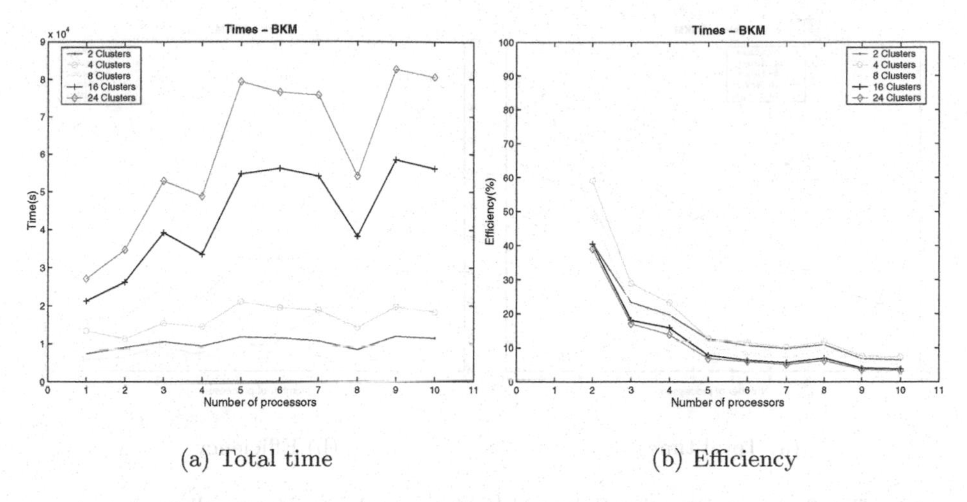

(a) Total time (b) Efficiency

Fig. 5. The α-Bisecting K-Means with the Times collection

In summary, the experiments to be carried out are indicated below:

- Parallel α-Bisecting K-Means vs. Sequential α-Bisecting K-Means: total time, efficiency and scalability.
- Parallel α-Bisecting Spherical K-Means vs. Sequential α-Bisecting Spherical K-Means: total time, efficiency and scalability.

All studies have been achieved with mean times, because the random selection of first centroids cause different number of required iterations and, hence, different times. So, we measure 11 times each sample (each method with a number of clusters and a number of processors).

3.4 Results of Experiments

Next, we present a series of graphs the summarizing results of our experiments. In figure 5, we show the study of the α-Bisecting K-Means method with the Times collection, we present the total time consumed by the algorithm and its efficiency. Analogously, in figure 6, we present the study of the α-Bisecting Spherical K-Means, also with the Times collection. In the same way, in figures 7 and 8, we depict the study of both methods, the α-Bisecting K-Means and the α-Bisecting Spherical K-Means algorithms, using as data the DOE collection.

In figure 5.a we observe that, in general, when we increase the number of processors, independently of the number of clusters, the total time increases, too. The reason of this behavior is the collection size. Due to the fact that the Times collection has so few documents, the communication times obtained are much bigger than the time reduction obtained when using more processors. But, we can highlight that when we work with four or eight processors performance improves. This occurs in all cases, independently of the number of clusters that we have used as a parameter. We suppose this behavior is caused by cache memory or a

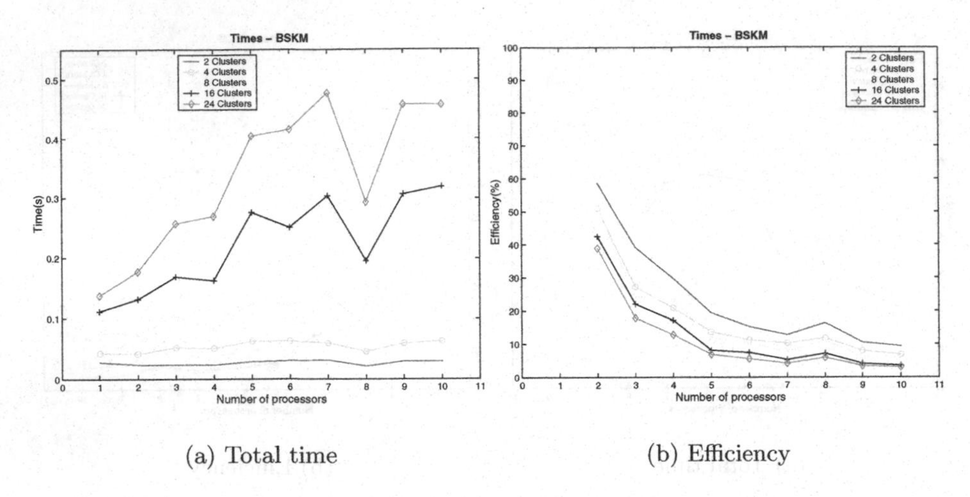

(a) Total time (b) Efficiency

Fig. 6. The α-Bisecting Spherical K-Means with the Times collection

favorable selection of the Cluster's nodes that improve the communications, as figure 4 shows depending on the nodes selected, communication costs can vary.

On the other hand, working with four and eight clusters and two processors we improve a little the total time. But the difference is insignificant. We can see this in figure 5.b, the efficiency study. All curves have a poor efficiency, and when we increase the number of processors efficiency decreases. Only with four and eight clusters and two processors we obtain a relatively acceptable efficiency (approximately 60% and 50% respectively).

The behavior of the α-Bisecting Spherical K-Means method (figure 6) is analogous to the α-Bisecting K-Means method (figure 5), both with the Times col-

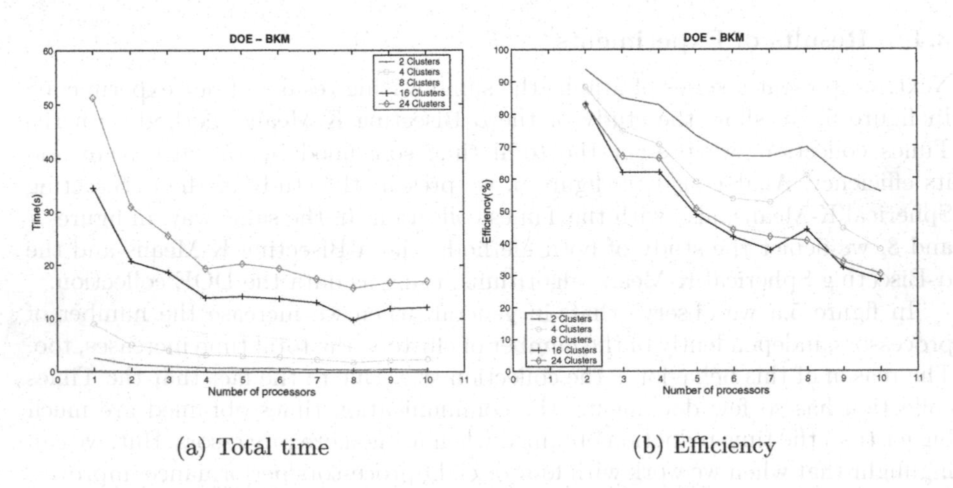

(a) Total time (b) Efficiency

Fig. 7. The α-Bisecting K-Means with the DOE-TREC collection

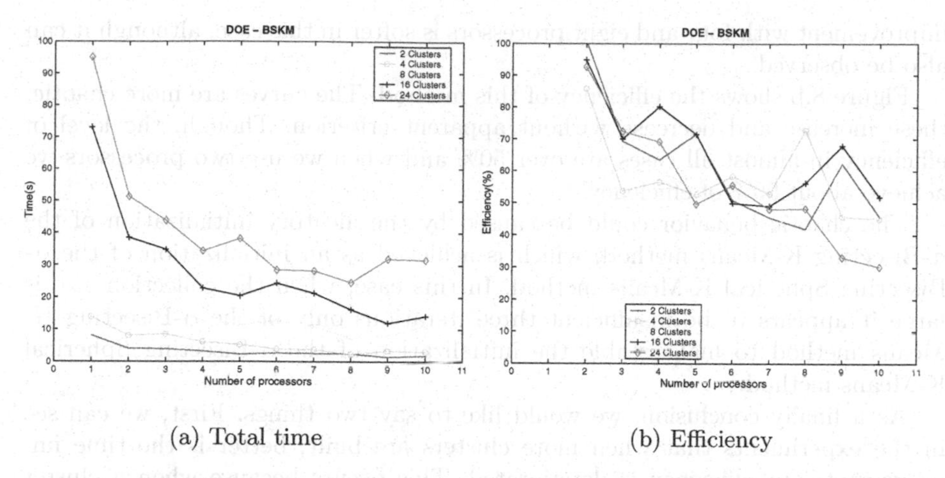

(a) Total time (b) Efficiency

Fig. 8. The α-Bisecting Spherical K-Means with the DOE-TREC collection

lection. When the number of processors increase the total time increases (figure 6.a), except with four and eight processors that decreases a little. But in this case, the curves obtained in small cluster groups (i.e. two, four and eight) are softer. And, the efficiency found in two clusters has improved much (figure 6.b), although the others are maintained approximately equal.

When the study is performed with a bigger collection, the parallelism benefits appear. In figure 7.a we show the total time needed to cluster the DOE collection with the α-Bisecting K-Means method. We observe, independently of the number of document clusters built, the total time improves when we use more processors. Logically, this improvement is more distinguishable when more document clusters are formed. We highlight the times obtained with four and eight processors. In both cases the performance relatively improves more than in the rest of cases. This situation is the same that in the study with the Times collection. Furthermore we assume that the reason is also the same, either the cache memory or the document distribution of the collection into the nodes. Seeing figure 7.a, we deduce that for the DOE collection size performance improves acceptably using up-to four processors. If we use more than four processors performance falls drastically.

This can be observed better in the efficiency study (figure 7.b). We observe upper efficiency, approximately, 60% when we work with up-to four processors. And with five or more processors efficiency falls down to 50% or lower. Though, it can be seen that the graph curve for four clusters delays its descent whereas the graph curve for two clusters maintains a level above 60% almost all the time, only dropping down to 55% with ten processor. Figure 7.b also presents small improvements in efficiency corresponding with four and eight processors.

Finally, we have studied the α-Bisecting Spherical K-Means method with the DOE collection. It can be seen in figure 8. In general, we observe a similar behavior to the α-Bisecting K-Means method with respect to the total time study (figure 8.a), but the curves, in this case, are more irregular. The relative improvement with four and eight processors is softer in this case, although it can also be observed.

Figure 8.b shows the efficiency of this method. The curves are more chaotic, these increase and decrease without apparent criterion. Though, the level of efficiency in almost all cases are over 50% and when we use two processors we achieve about 90% of efficiency.

The chaotic behavior could be caused by the aleatory initialization of the α-Bisecting K-Means method, which is achieved as an initialization of the α-Bisecting Spherical K-Means method. In this case, when the collection size is large it appears to be insufficient three iterations only of the α-Bisecting K-Means method to make stable the initialization of the α-Bisecting Spherical K-Means method.

As a finally conclusion, we would like to say two things. First, we can see in the experiments that when more clusters are built, better is the time improvement, but efficiency is deteriorated. This occurs because when a cluster is built the methods work with a subcollection of the original collection, and so on. Then as these subcollections are smaller than the original collection the performance is deteriorated. And second, these methods are scalable because as the collection size increases, it is possible to use more processors to improve performance.

4 Conclusions

The principal conclusion is with the parallelization of these algorithms we reduce the necessary time to build the required clusters. Of course, the collection size has to be enough large to obtain an improvement of performance, but nowadays the actual collections are usually very large and these will grow in the future. Besides, the parallel algorithms showed have a good efficiency, and good scalability, too. On the other hand we did not achieve great efficiency because the collection matrix is very sparse (0.0270 in the Times collection and 0.00053 in the DOE collection), and hence the number of operations to execute is small. We use functions optimized for sparse data (for example, the SPARSKIT library [13]).

Another conclusion is that the distribution of the collection used in the parallel algorithms presented in this paper cause good performance, a conclusion which we might indicate as simple and natural. As a matter of fact when a real collection is classified as distributed, what we would normally assume is that its documents are distributed as well.

We can also conclude that the most important operation in the algorithms is the calculation of the normalized centroid. Therefore, its optimization is fundamental to obtain good performance.

Finally, we want to indicate three future works in this research. First, we should improve the selection of the initial centroids in the α-Bisecting K-Means method, because the solution of the algorithms is more dependent and varies according to the chosen centroids. We should eliminate the randomness or modify the algorithm to grant independence respect to the initial selection.

Second, we should look for some mechanism to improve the calculation of the normalized centroid. Alternatively, we should develop a new version of both parallel algorithms to replace in some iterations the global parallel calculation of the concept vector by a local calculation.

And third, we should study what is the optimum number of iterations to use in the α-Bisecting K-Means method to make stable the initialization of the α-Bisecting Spherical K-Means method in the most of cases.

Acknowledgments

This paper has been partially supported by the Ministerio de Ciencia y Tecnología and project FEDER DPI2001-2766-C02-02

References

1. Savaresi, S., Boley, D.L., Bittanti, S., Gazzaniga, G.: (Choosing the cluster to split in bisecting divisive clustering algorithms)
2. Halkidi, M., Batistakis, Y., Vazirgiannis, M.: On clustering validation techniques. Journal of Intelligent Information Systems **17** (2001) 107–145
3. Steinbach, M., Karypis, G., Kumar, V.: A comparison of document clustering techniques. KDD-2000 Workshop on Text Mining (2000)
4. Dhillon, I.S., Fan, J., Guan, Y.: Efficient clustering of very large document collections. In R. Grossman, C. Kamath, V.K., Namburu, R., eds.: Data Mining for Scientifec and Engineering Applications. Kluwer Academic Publishers (2001) Invited Book Chapter.
5. Jiménez, D., Vidal, V., Enguix, C.F.: A comparison of experiments with the bisecting-spherical k-means clustering and svd algorithms. Actas congreso JOTRI (2002)
6. Savaresi, S.M., Boley, D.L.: On the performance of bisecting k-means and pddp. First Siam International Conference on Data Mining (2001)
7. Hartigan, J.: Clustering Algorithms. Wiley (1975)
8. Dhillon, I.S., Modha, D.S.: A data-clustering algorithm on distributed memory multiprocessors. In: Large-Scale Parallel Data Mining, Lecture Notes in Artificial Intelligence. (2000) 245–260
9. Kantabutra, S., Couch, A.L.: Parallel k-means clustering algorithm on nows. NECTEC Technical Journal **1** (2000) 243–248
10. Xu, S., Zhang, J.: A hibrid parallel web document clustering algorithm and its performance study. (2003)
11. Dhillon, I.S., Modha, D.S.: Concept decompositions for large sparse text data using clustering. Technical report, IBM (2000)
12. http://trec.nist.gov/: (Text retrieval conference (trec))
13. Saad, Y.: SPARSKIT: A basic tool kit for sparse matrix computations. Technical Report 90-20, NASA Ames Research Center, Moffett Field, CA (1990)

Distributed Processing of Large BioMedical 3D Images

Konstantinos Liakos[1,2], Albert Burger[1,2], and Richard Baldock[2]

[1] Heriot-Watt University, School of Mathematical and Computer Sciences, Ricarton Campus, EH14 4S, Edinburgh, Scotland, UK

[2] MRC Human Genetics Unit, Western Hospital, Crewe Road, EH4 2XU, Edinburgh, UK

Abstract. The Human Genetics Unit (HGU) of the Medical Research Council (MRC) in Edinburgh has developed the Edinburgh Mouse Atlas, a spatial temporal framework to store and analyze biological data including 3D images that relate to mouse embryo development. The purpose of the system is the analysis and querying of complex spatial patterns in particular the patterns of gene activity during embryo development. The framework holds large 3D grey level images and is implemented in part as an object-oriented database. In this paper we propose a layered architecture, based on the mediator approach, for the design of a transparent, and scalable distributed system which can process objects that can exceed 1GB in size. The system's data are distributed and/or declustered, across a number of image servers and are processed by specialized mediators.

1 Introduction

The Edinburgh Mouse Atlas [1, 2] is a digital atlas of mouse development, created at the MRC Human Genetics Unit (HGU), Edinburgh, to store, analyze and access mouse embryo development. The stored objects contain 3D images that correspond to conventional histological sections as viewed under the microscope and can be digitally re-sectioned to provide new views to match any arbitrary section of the experimental embryos [2]. The volume of the processed 3D objects can exceed 1GB and typical lab based machines fail to efficiently browse such large bio-medical image reconstructions, particularly since the users may wish to access more than one reconstruction at a time. This paper addresses that problem by proposing a layered distributed system design that provides scalable and transparent access to the image data; by transparency we mean that the system hides from the end user the connection among the different servers and the distribution of the data.

The proposed architecture has been influenced by the mediator approach. Mediators were first described by Wiederfold [3] to provide a coherent solution to the integration of heterogeneous and homogenous data by "abstracting, reducing, merging and simplifying them". In more detail mediators are "modules occupying

M. Daydé et al. (Eds.): VECPAR 2004, LNCS 3402, pp. 142–155, 2005.

an explicit active layer between the user applications and the data resources. They are a software module that exploits encoded knowledge about certain sets or subsets of data to create information for a higher layer of application" [3]. They encapsulate semantic knowledge about a number of distributed sources providing to the user-applications a unique representation of the distributed database context.

The middleware initially adopted for our purposes is the Common Object Request Broker Architecture (CORBA) [4, 5]. The CORBA framework provides the means to develop open, flexible and scalable distributed applications that can be easily maintained and updated. Data integration can be partially resolved via the use of the Interface Definition Language (IDL) that separates the data model from its implementation. In the past CORBA has been proposed as an efficient middleware solution for bioinformatics projects [6, 7, 8, 9]. In addition, the European Bioinformatics Institute (EBI) [10] has adopted CORBA as a middleware for a number of its bioinformatics systems. A competing technology to CORBA, also adopted by EBI, and gained a lot of attention in general, is Web Services[11]. However, Web Services are less efficient than CORBA/IIOP over slow connected networks [12]. As a consequence they currently cannot give optimum solutions to applications that require fast processing and transmission of large amounts of data. It should be noted though, that our proposed design is independent of the middleware platform. Web services are evolving rapidly and become the preferred option in many bioinformatics applications.

A property of this application is that a query on the data requests virtual objects. These objects are analogous to a view in database terms that represent 2D section images that are computed at run time from the original 3D voxel models stored in the database.

Our layered design is a modification of the mediator approach and is based on the requirement for easy re-configuration for performance reasons. Simple client-server designs such as that implemented in [2] for a genome-mapping prototype are inadequate to handle the requirements derived from large voxel image files. Although smaller size voxel images can be efficiently processed, larger images result in an unacceptably slow response, due to memory paging and CPU processing time. Furthermore integration of additional object resources becomes impractical. The volume of our current as well as our anticipated future objects, in addition to the requirement of providing very fast response times introduce the necessity of distributing the cost of processing by declustering biological images in order to provide a scalable solution, minimize the cost of the overall query response time, and make an optimum usage of the available hardware resources. Our design adopts an n-tier solution by distributing the cost of processing to a number of image servers. Such a task is accomplished by declustering and distributing the image data across different image servers, processing them in parallel. The image server processing is hidden by the use of one or more mediator layers, which are responsible to provide transparent access and to monitor image servers so that user requests are directed appropriately. In addition, mediators are designed to provide other services, such as query processing-decomposition and re-assembly, query optimization and user-behavior prediction or "look-ahead". This latter service enables pre-computation of predicted requests in order to accelerate the response time of the system. Finally an important aspect of our design is its ability to vary the number mediators and allow dynamically re-configure the system to optimize performance.

The emphasis of this paper is on the description and the efficiency of the prototype system to illustrate the advantages gained by data distribution and parallel processing. Issues such as the study of optimum declustering and placement approaches and the evaluation of particular prediction techniques will be reported elsewhere.

The remainder of this paper is organized as follows: Section 2 provides a brief discussion of the current Mouse Atlas system covering in more detail image processing issues and the notion of a virtual object. Section 3 provides a detailed description of the proposed distributed architecture, section 4 presents the query processing design and in section 5 some initial performance results to process large reconstructions are provided. Finally section 6 discusses future issues and concludes the paper.

2 Mouse Atlas

2.1 The Mouse Atlas System

The MRC Human Genetics Unit, Edinburgh, has developed the Edinburgh Mouse Atlas [1, 2] based on an object-oriented architecture in order to store and analyze biological data, including 3D images that relate to mouse embryo development [2]. The embryo framework for the database is represented as a set of voxel models (3D images), initially at each development stage defined by Theiler [13], but with the possibility of extension to finer time-steps especially at earlier stages. The voxel models correspond to conventional histological sections as viewed under the microscope and can be digitally re-sectioned to provide new views to match any arbitrary section of an experimental embryo.

The purpose of the system is the analysis and querying of complex spatial patterns, in particular the patterns of gene activity during embryo development. The underlying image processing and manipulation uses the Woolz image-processing library [1, 14, 15] that was developed for automated microscope slide scanning and is very efficient for binary set and morphological operations. The users navigate in the 3D space via the use of user interface components that correspond to particular viewing parameters, to define 2D sections (fig.1) at any orientation. At any given time only one such component can alter its value to generate a new query. For efficient browsing it is necessary to get the entire voxel image into the main memory so disk accesses are avoided. For very large reconstructions, i.e. images of $1-3GB$, this is impractical for typical laboratory based machines, particularly since users may wish to access many such reconstructions concurrently. Even with the entire image in memory most CPUs will be slow. While CPU and memory specifications will be steadily improved image processing

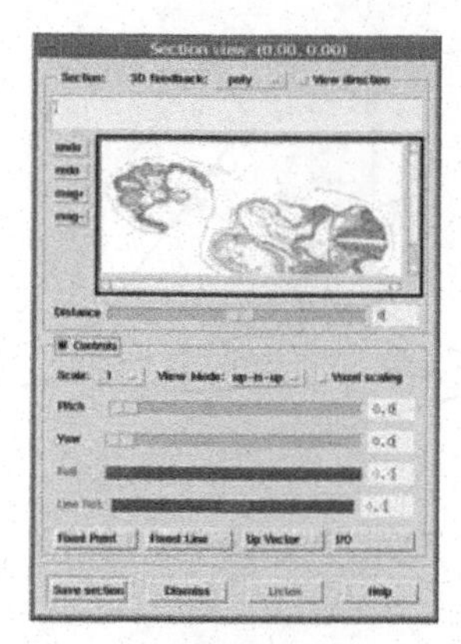

Fig. 1. GUI to process 3D Woolz images. Specialized components enable the rotation within the 3D space. The requested sections can be scaled, magnified and saved to disk

requirements are also expected to increase; e.g. future image volumes might well reach $10 - 100GB$.

There are many aspects of the Mouse Atlas, which are not directly relevant to the discussion that follows and therefore omitted from this paper. The interested reader is referred to [1, 2, 14, 15].

2.2 Virtual Objects and Woolz Image Processing

The Woolz library that has been developed by the MRC HGU performs all image-processing operations of this system. Woolz uses an interval coding data structure so that only grey-values corresponding to foreground regions (the embryo) are held in memory. For this type of image this can result in $30 - 50\%$ reduction in memory footprint in comparison with conventional image formats. While detailed information on various other aspects can be found in previous publications [1, 2, 14, 15], our emphasis here is on the efficient computation of requested 2D sections. These are virtual objects, objects that are not saved in the data sources, but are computed during run-time by Woolz library functions associated with the original 3D voxel data. This is analogous to a view in database terms. The rotation within the original 3D space that results in the generation of a section view is determined as follows.

Given an original coordinate $r = (x, y, z)^T$, the viewing plane (fig.2) is defined as a plane of constant z in the new coordinates $r' = (x', y', z')^T$, i.e. the new z-axis is the line-of-sight This axis is fully determined by defining a single fixed-point f that is the new coordinate origin, and a 3D rotation. The actual view plane is then defined to be perpendicular to this axis and is determined by a scalar distance parameter d along the new axis. In this way the transformation between the original r and viewing coordinates, r', is determined by a 3D rotation and translation with the viewing plane defined as a plane of constant $z' = d$. A 3D rotation can be defined in terms of Eulerian angles [16] with full details given

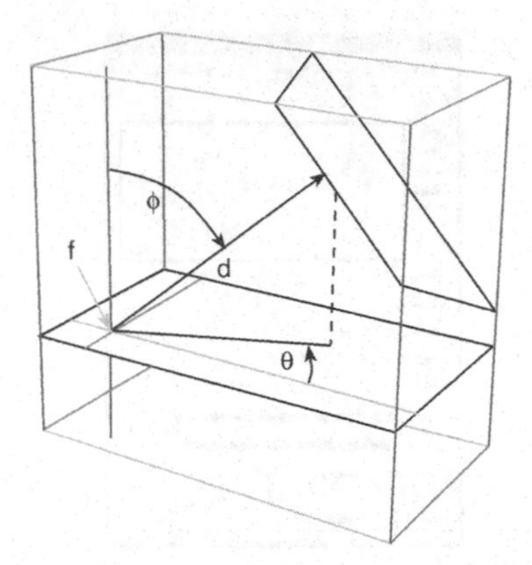

Fig. 2. The viewing plane is defined to be perpendicular to the viewing direction given by angles *phi* and *theta*. The actual plane is distance d from the fixed-point f. For the user interface *phi* is termed *pitch* and *theta* is *yaw*

in [2]. These mathematical details are not essential to the understanding of the rest of the paper, but highlight the point that the computation of a 2D section requested by a user query involves the retrieval of 3D data from the database as well as some image processing on that data.

3 The System's Architecture

The proposed design consists of four main components: client, directory, mediator and image servers (fig.3). The client consists of a GUI that provides the stage selection menus, controls for section parameters and is responsible for the generation of user queries while the mediators provide a transparent view onto the image server data sources. The mediators encapsulate the knowledge of the data sources (metadata), such as the objects which are stored in addition to their structural properties and the image servers where objects can be found, providing transparent access to the image data. Image servers provide direct access to the underlying image storage e.g. files and databases. It is only an image server that loads and directly queries a Woolz object. Due to the volume of the 3D image data, declustering is expected to be introduced. By declustering we mean the process of cutting an original 3D voxel into smaller dimensions placing them at different sites. A key difference in the concept of declustering between our approach and the one that is normally in use is that our declustering performance relates to the CPU speed up that can be gained; current declustering schemes are concerned with the disk IO speed up [17, 18]. Our system assumes that for each request the declustered 3D image data can be loaded into the main memory of each image server. Finally directory servers enable access to mediators and image servers by providing their connection details. Note that clients can only acquire information about a mediator while a mediator can acquire connection details related both to other mediators and image servers.

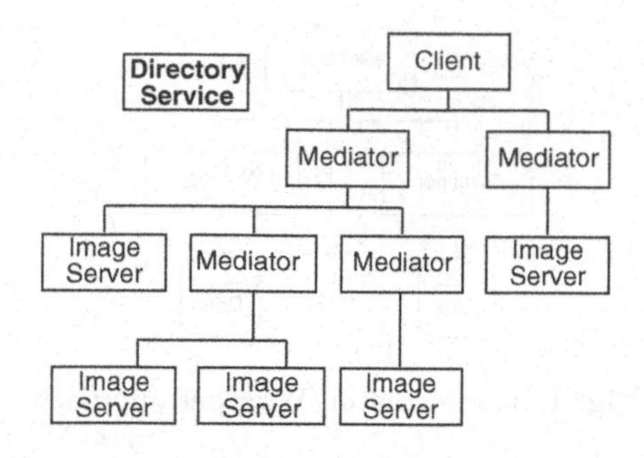

Fig. 3. Layered design based on the system's requirements

A mediator's task is to provide a unified view and transparent access to the declustered distributed Woolz objects, monitoring of the user access patterns and prediction of future requests so as to pre-compute related sections. The successful completion of these tasks depends on a metadata model that links image servers to Woolz image objects and Woolz objects to object properties, such as their structure and size. An image server on the other hand provides a logical view onto the Woolz data and a Woolz query-processing model to access them. The object oriented database in which Woolz are saved can be only accessed by appropriate image servers.

In a mediator/wrapper approach (fig.4) a mediator encapsulates a global data model to serve a number of heterogeneous data sources. Such an approach has been adopted by many distributed systems such as DISCO [19], GARLIC [20] and Informia [21]. In comparison to other bioinformatics implementations where the mediator approach is used to integrate heterogeneous data sources (Raj and Ishii [8], and Kemp et al [9]) our design seems similar because it also encapsulates a schema of its integrated image servers, however in contrast to most of them, where a centralized mediator serves a number of data sources, our mediators are fully scalable (fig.3). A hierarchy of mediators can be dynamically created according to the performance needs of the system at run time. As a consequence our design can result in both centralized and distributed topologies depending on performance requirements. More precisely a mediator might not only control a number of image servers but also a number of mediators as well. Hence a mediator's schema might be based on the schema of its lower-level mediators. The concept of such a composable approach is similar to AMOS II [22, 23], a lightweight OODBMS, where mediators regard interconnecting mediators as data sources. However the purpose of AMOS II and its implementation are different from ours. AMOS II is a federated multi-database system that combines heterogeneous data sources and views. On the other hand our system's main aim is to reduce the time needed to fetch, process and display large image objects. Because the processed Woolz objects are homogeneous, the emphasis is not given to the integration of diverse data sources. The mediator approach is used to distribute the cost of processing, exploit data parallelism and enable processing of very large reconstructions. Apart from that our proposal is based on standard

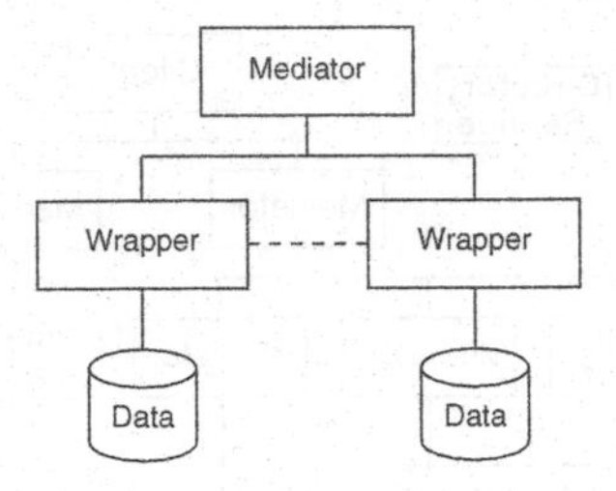

Fig. 4. The Mediator/Wrapper approach

and flexible solutions that facilitate scalability and maintenance, and is language and platform independent.

Furthermore our mediators perform additional tasks such as the monitoring of the user access patterns and the pre-computation of future requested sections so as to improve the overall performance of the system. These facilities can be seen as an extension of our query processing mechanism that is described in the following section. Details of this speculative computation model are beyond the scope of this paper and will be presented elsewhere. Under speculative pre-computation we mean that a mediator requests image servers to precompute possible future requested sections. Until the actual user request is received those predicted sections are not transmitted from the image servers to the mediator, where they are further processed, minimizing the cost of processing predicted sections that are not eventually needed.

The client component consists of a GUI that provides the tools to query and process Woolz image objects. Initially clients are unaware about the connection details of the mediators they are connected to. Such information is acquired from the directory server (fig.3). Another important issue is that clients are not aware of the underlying Woolz technology [2]. Their only intelligence relates to the display of the 2D sections, which are acquired as bitmap arrays, and their understanding of the 3D bounding box of the saved Woolz objects. An important property of the technology underlying the GUI is that it results in asking only the visible part of every reconstruction. As a consequence only a small part of the whole virtual object is requested, minimizing the data that needs to be transmitted over the net. The client GUI is intended to be as light as possible.

We are aware of the overhead that a layered approach might introduce. At this prototype level, our emphasis was on functionality. A performance study of the impact of multiple mediators is planned. The focal measures of such a study will be the tradeoff between the network and processing overhead and the potential improvement of the overall response time.

4 Query Processing

Kossman [24] has extensively described query processing optimization techniques. Assuming that Woolz internal image processing is optimal, our efforts

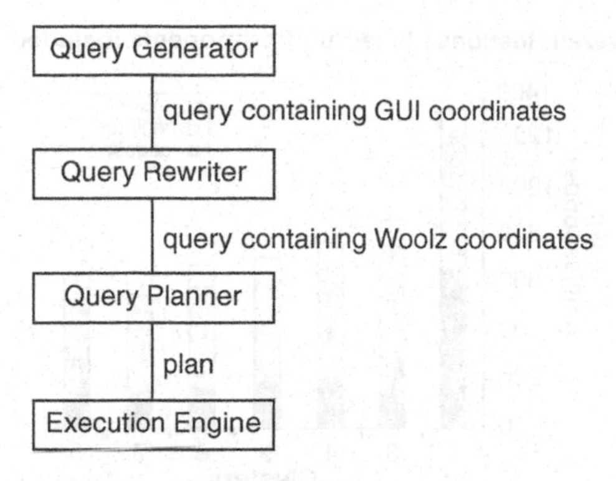

Fig. 5. Query Processing Architecture

are focused on accessing the Woolz data sources as fast as possible producing the adequate query for them. Our query processing architecture comprises of a query generator, a query rewriter, a query planner and a query execution module (fig.5). A query generator is found on the client side and involves the generation of a Woolz query based on the visible coordinate system of the GUI. The query rewriter module transforms the query from GUI coordinates into Woolz coordinates. The generation of a query can be also introduced on the mediator's side as a consequence of the pre-computation process algorithm that monitors user activity. The query planner, that includes a query rewriter module, selects only the data servers that can respond positively to a particular query. To optimize system performance only the image servers that can respond to queries are sent modified user requests. Only after the data servers are selected, a query-rewriting process starts so as to split the initial Woolz query and request only the subregion to which each data server can respond. The query execution model that is found both in the client and mediator is used to transform either a user or a mediator query into a compatible IDL query. On the other hand query execution on the image server side relates to the use of Woolz libraries so as to physically execute a query.

5 Experiments

To test our approach, a prototype system has been developed and used for a number of experiments. The experiments were run on a SUN Netra array, which includes 10 servers running SOLARIS 8. Each server has the same capabilities and the interconnection between them is provided via the use of dedicated switches whose speed is 100Mbps. Every server has 500MB of RAM, 500 MHz of UltraSPARC-II CPU and 256 Kbytes of e-cache. The developed programs were written according to the requirements of the system's design in Java [25]. A benchmark was developed that requests all 2D sections for each reconstruction for a given viewing orientation. Only after a client acquires a requested 2D section, a generation of a new query takes place. The window size is set up to

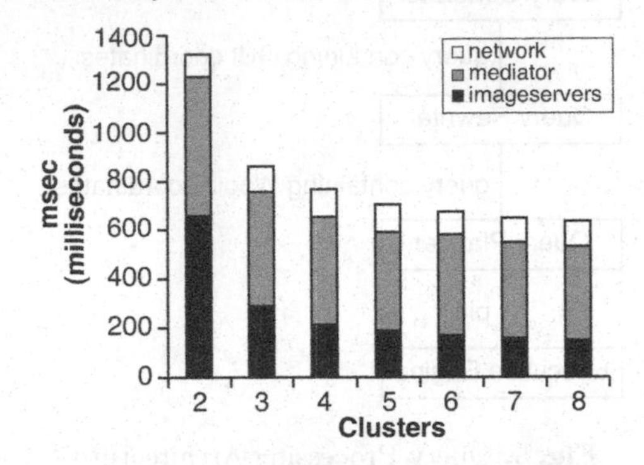

Fig. 6. Detailed analysis of the times spend throughout the experiment. Declustering and parallelism result in decreasing the system's average response time when 8 clusters are used

$800x400$ pixels while a centralized mediator is used throughout the experiments. Every image server responds to a mediator query by sending only the part of the section that it can generate. The requested parts are determined by the window size of the clients GUI and the section parameters.

The aim of the experiments was to establish the capability of our distributed design to handle large reconstructions and to understand and model the effects of data declustering and parallelism. For these tests the 3D reconstruction of a mouse embryo at its developmental stage Theiler stage 20 (TS20), a voxel image object of size 685.5 MB, was used. Parallelism is achieved by the use multiple processes communicating using CORBA and within a single process, e.g. the mediator or image server by the use of java programming threads [25]. Data declustering is the process of partitioning an object into a number of fragments.

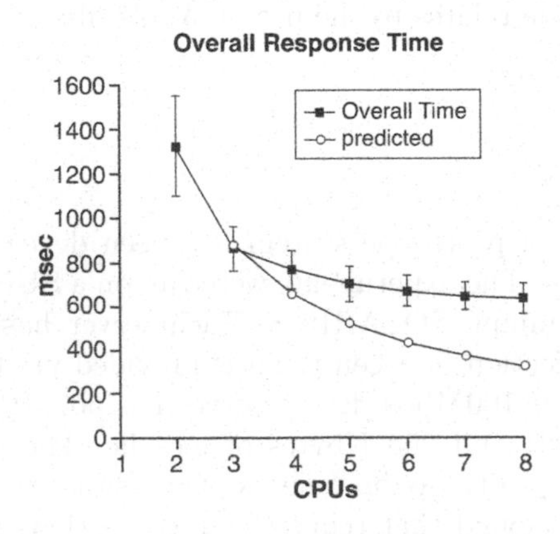

Fig. 7. Overall response time. The average response time of the system is reduced by the use of 8 clusters

The first experiment examines the overall advantage gained by the introduction of data parallelism and declustering. TS20 was cut into a number of clusters (2 – 8) of approximately equal size and placed in different locations over the network. The benchmark described previously was used to measure the system's performance. In this case the declustering was chosen so that each image server would respond so that the overall potential benefit of the parallelism could be established.

As expected the system's performance greatly improved through the use of the parallel processing design. In fig.6 an estimation of the time spent at each component of the system is given, while in fig.7 a more statistical analysis of the system's performance is provided. The displayed curve represents optimal performance assuming linear speedup with respect to the two-computer case i.e.: $\frac{2AVG(2)}{clusterNo}$ where $clusterNo$ is the number of clusters and $AVG(2)$ is the average

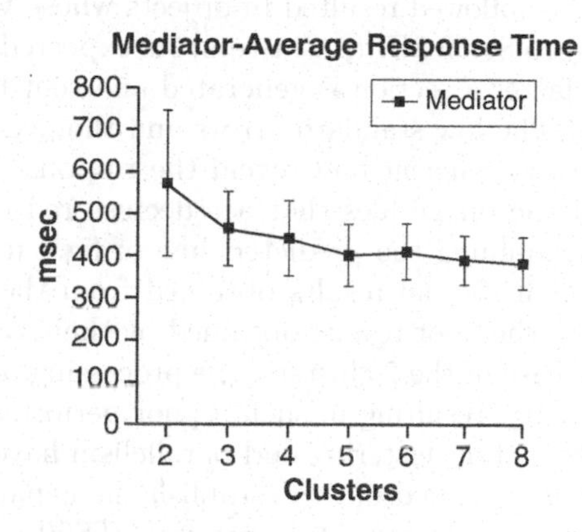

Fig. 8. Estimated Time spent on the mediator

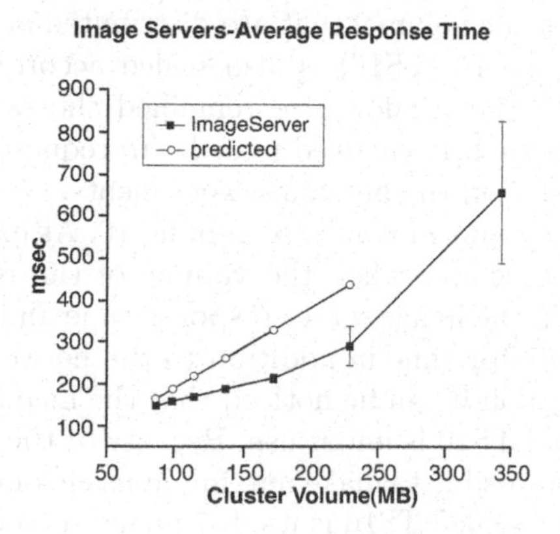

Fig. 9. Declustering effects. The process in which object were cut along their $Y-axis$ and placed in different image servers results in improving the image server's response times when the number of clusters increases

response time when 2 clusters are used.The difference between the curve and the measured points presumably reflects the performance of our declustering approach (fig.9), and the cost of processing many clusters simultaneously in the mediator (fig.8). The most important module of our experiment is the mediator since it monitors, directs and finally processes user requests. The use of one CPU in our mediator server results in similar performance results for all the experiments that were carried out (fig.8). The multi-threaded model adopted for the mediator can further improve its performance results provided that a multi CPU machine will be used. Under the current implementation java threads add an additional cost that is however negligible. The derived variations under the use of a small number of clusters are a consequence on variations on the response times of the individual image servers.

Finally the effect of declustering is studied by examining the performance of the image servers when the volume of their data sources decreases. The declustering approach that was followed resulted in objects whose volume varied from 342.75 MB (2 Clusters) to 85.69 MB (8 Clusters) As expected (fig.9) the smaller the image volume the faster a section is generated and sent back to the mediator for further process. The low standard errors and the average response time results are considered very efficient and reveal the response times that can be achieved for larger volume image files that are declustered to a larger number of image servers. By examining the predicted line of fig.9 it is noticeable that it performs worse than the actual results obtained from the experiment. Such behavior is explained by the poor results obtained by the usage of 2 clusters. To be more precise, in the case of the 2 clusters, the processing of them may exceed the available main memory, resulting in such a poor performance.

While both the effects of declustering and parallelism have been described in detail, a last experiment is carried out to establish the capability of our design to handle very large reconstructions. More precisely TS20 is scaled by a factor of 2 across its $X-Y-Z$ axis resulting in an object sizing 5.5 GB. As previously, the objects are declustered across their Y coordinate system, producing 8 sub objects of roughly the same volume, that are distributed across 8 image servers. In addition Theiler Stage 16 (TS16) is also scaled accordingly resulting in an object sizing 1.47 GB. The window size remained the same ($400x800$ pixels) while our benchmark is slightly changed in order to request the same amount of sections we had requested in the previous experiments.

The results of the experiment can be seen in fig.10. As expected, the response time of the system gets slower when the volume of the requested objects get larger. The gap between the image server response time and the system response time represent the mediator time in addition to the network and idle time. By better examining the graph it can be noticed that the mediator's time is slightly bigger when the original TS20 is under use. Because of the volume of the scaled TS16 and TS20 it is impossible to make maximum usage of the image servers. As a consequence when the scaled TS16 is used, 7 image servers respond in parallel while only 4 of them respond in parallel when the scaled TS20 is requested. As a result there is an overhead related to network requests and thread processing that

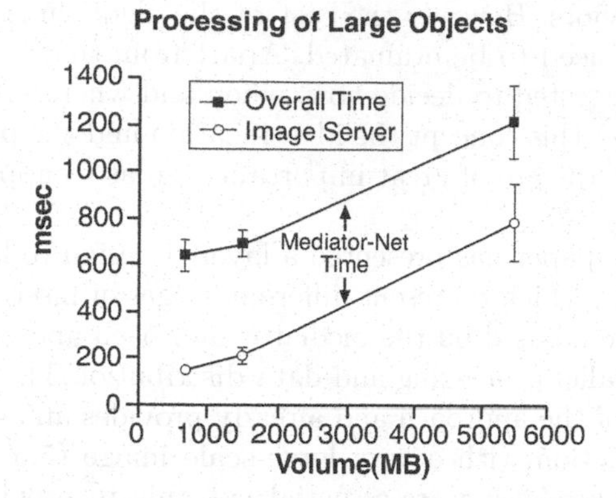

Fig. 10. The larger the volume and the dimensions of the objects are, the slowest their response time. The gap between the 2 diagrams represents the mediator time in addition to network and idle time

is not provided in these experiments. This overhead corresponds to the additional requests and threads that are generated when all image servers are involved in processing a user request, that result in increasing the cost of processing at the mediator side. All experiments are expected to reveal almost the same response time at the mediator side, given the fact that the same window size is used and the volume of the requested sections is roughly the same. Moreover such a feature also proposes further experimentations to define data declustering optimization techniques.

6 Future Work and Conclusions

This paper discussed the necessity of a distributed approach to efficiently process large biomedical 3D image data. In the Edinburgh Mouse Atlas, user queries do not directly correspond to the 3D objects saved in the database but rather to 2D sections, which need to be calculated at run-time from the 3D image data. This creates an interesting context in which new analysis related to different declustering schemes is required.

Some preliminary experiments reveal that user navigation patterns strongly affect the optimum cluster generation procedure and hence further details concerning optimum declustering and placement approaches is required. The required analysis will examine different placement approaches to identify the one that minimizes the cost of distribution and optimizes the performance of the system.

In addition our mediators will be enhanced with prediction capabilities to pre-compute future requested queries based on the monitoring of the user access patterns. Initial studies here suggest that time series prediction approaches such as the Autoregressive Integrated Moving Average (ARIMA) [26, 27] and the Exponential Weighed Moving Average (EWMA) [28] adapt fast and efficiently to changing user behaviors. However the cost of the algorithms and the improvement of the system need to be evaluated. Apart from that a unified prediction policy should be suggested to decide how, when and where precomputation will be introduced. Under this concept the identification and the properties of different user access patterns are of great importance, so as to apply prediction only when needed.

In summary, this paper has presented a layered design to handle large bioinformatics 3D images, which represent different stages of mouse embryo development. Such a design is based on the mediator approach and exploits the advantages gained by parallel processing and data distribution.The work thus far has shown the validity of the approach and already provides an acceptable response time for user interaction with a very large-scale image (5.5 GB). Data distribution and parallel processing are essential not only to provide faster response times but also to enable processing of large objects unable to fit into the main memory of a single machine. Such features have been presented not only by processing a very large scale biomedical image but also by examining the results of declustering and data distribution for a smaller scale object (685.5 MB) in which the overall response time has been decreased more than 100% when all 8 available distributed CPUs have been used.

Acknowledgments

The MRC HGU in Edinburgh has sponsored the presented work.

References

1. R. A. Baldock, F. J. Verbeek and J. L. Vonesch, 3-D Reconstructions for graphical databases of gene expression, Seminars in Cell and Developmental Biology, 1997, pp. 499-507.
2. R. A Baldock, C. Dubreuil, B. Hill and D. Davidson, The Edinburgh Mouse Atlas: Basic Structure and Informatics, Bioinformatics Databases and Systems, Ed. S Levotsky (Kluwer Academic Press, 1999), pp. 102-115.
3. G. Wiederhold, Mediators in the Architecture of Future Information Systems, IEEE Computer, vol 25, no 3, 1992 pp: 38-49.
4. Object Management Group, OMG, www.omg.org .
5. R. Orfali, D.Harkey, Client/Server Programming with Java and CORBA, Second Edition, Wiley Computer Publishing, ISBN: 0-471-24578-X,1998.
6. A. Spiridou, A View System for CORBA-Wrapped Data Source, European Bioinformatics Institut, Proceedings of the IEEE Advances in Digital Libraries 2000 (ADL2000).
7. J. Hu, C. Mungall, D. Nicholson and A. L. Archibald, Design and implementation of a CORBA-based genome mapping system prototype, Bioinformatics, ,2nd ed, vol.14, 1998,pp. 112-120.
8. P. Raj, N Ishii, Interoperability of Biological Databases by CORBA, Nagoya Institute of Technology, Proceedings of the 1999 International Conference on Information Intelligence and Systems, March 31 April 03, Rockville, Maryland, 1999.

9. J. Graham,L. Kemp, N. Angelopoulos, P. M.D.Gray, A schema-based approach to building a bioinformatics database federation, Proceedings of the IEEE International Symposium on Bio-Informatics and Biomedical Engineering (BIBE 2000),Arlilngton,Virginia, USA, November 08-10, 2000.
10. European Bioinformatics Instritute, http://www.ebi.ac.uk/ .
11. World wide World Consortium, www.w3.org .
12. D. C. Schmidt, S. Vinoski, Objet Interconnections: CORBA and XML-Part3: SOAP and Web Servises, C/C++ Users Journal, www.cuj.com
13. K. Theiler, The House Mouse: Atlas of Embryonic Development, Springer-Verlag, 1989.
14. R. A. Baldock, J. Bard, M. Kaufman, D. Davidson, A real mouse for your computer. BioEssays, vol. 14 no 7, 1992, pp. 501-503.
15. M. Ringwald, R. A. Baldock, j. Bard, M. Kaufman, J. T. Eppig, J. E. Richardson, J. H. Nadeau, D. Davidson, A database for mouse development, Science 265, 1994, pp. 2033-2034.
16. H. Goldstein, Classical Mechanic, Addison-Wesley, Reading, MA, 2nd ed., 1950.
17. Y. Zhou, S. Shekhar, M. Coyle,Disk Allocation Methods for Parallelizing Grid Files, ICDE 1994, 1994, pp 243-252
18. C. Chen, R. Sinha, R. Bhatia, Efficient Disk Allocation Schemes for Parallel Retrieval of Multidimensional Grid Data, Thirteenth International Conference on Scientific and Statistical Database Management, Fairfax, Virginia, July 18 - 20, 2001
19. A. Tomasic, L. Raschid, P. Valduriez, Scaling access to heterogeneous data sources with disco, IEEE Transactions on Knowledge and Data Engineering, vol 10, no 5 ,September/October 1998, pp. 808-823.
20. L. Haas, D. Kossmann, E. L. Wimmers, J. Yang, Optimizing Queries across Diverse Data Sources, Proceedings of the Twenty-third International Conference on Very Large Databases, Athens Greece, August 1999.
21. M. L. Barja, T. A. Bratvold, J. Myllymki, G. Sonnenberger, Informia, A Mediator for Integrated Access to Heterogeneous Information Sources, In Gardarin, French, Pissinou, Makki, Bouganim (editors), Proceedings of the Conference on Information and Knowledge Management CIKM'98, Washington DC, USA, November 3-7, 1998, pages 234-241, ACM Press, 1998.
22. V. Josifovski, T. Katchaounov, T. Risch, Optimizing Queries in Distributed and Composable Mediators, Proceedings of the Fourth IECIS International Conference on Cooperative Information Systems, Edinburgh, Scotland, September 02-04 1999.
23. T.Risch, V.Josifovski, Distributed Data Integration by Object-Oriented Mediator Servers, Concurrency and Computation: Practice and Experience J. 13(11), John Wiley & Sons, September, 2001.
24. D. Kossman, The State of the Art in Distributed Query Processing, ACM Computing Surveys, September 2000.
25. Java, www.java.sun.com
26. C. Chatfield, Time Series Forecasting, Charman & Hall/CRC, ISBN 1-58488-063-5, 2001.
27. G. E. P. Box, G. M. Jenkins, Time Series Analysis: forecasting and control, ISBN: 0-8162-1104-3, 1976.
28. C. D. Lewis, Industrial Business Forecasting Methods, ISBN: 0408005599

Developing Distributed Data Mining Applications in the KNOWLEDGE GRID Framework

Giuseppe Bueti, Antonio Congiusta, and Domenico Talia

DEIS University of Calabria, 87036 Rende (CS), Italy
{acongiusta, talia}@deis.unical.it

Abstract. The development of data intensive and knowledge-based applications on Grids is a research area that today is receiving significant attention. One of the main topics in that area is the implementation of distributed data mining applications using Grid computing services and distributed resource management facilities. This paper describes the development process of distributed data mining applications on Grids by using the KNOWLEDGE GRID framework. After a quick introduction to the system principles and a description of tools and services it offers to users, the paper describes the design and implementation of two distributed data mining applications by using the KNOWLEDGE GRID features and tools and gives experimental results obtained by running the designed applications on real Grids.

Keywords: Grid Computing, Data Processing, Data Mining.

1 Introduction

The knowledge extraction and building process in a distributed setting involves collection/generation and distribution of data and information, followed by collective interpretation of processed information into "knowledge." Knowledge construction depends on data analysis and information processing but also on interpretation of produced models and management of knowledge models. The knowledge discovery process includes mechanisms for evaluating the correctness, accuracy and usefulness of processed datasets, developing a shared understanding of the information, and filtering knowledge to be kept in accessible organizational memory that often is distributed.

The KNOWLEDGE GRID framework [2] provides a middleware for implementing knowledge discovery services in a wide range of high performance distributed applications. The KNOWLEDGE GRID offers to users a high level of abstraction and a set of services based on the use of Grid resources to support all phases of the knowledge discovery process. Therefore, it allows end users to concentrate on the knowledge discovery process they must develop without worrying about Grid infrastructure and fabric details.

The framework supports data mining on the Grid by providing mechanisms and higher level services for

M. Daydé et al. (Eds.): VECPAR 2004, LNCS 3402, pp. 156 – 169, 2005.

- searching resources,
- representing, creating, and managing knowledge discovery processes, and
- composing existing data services and data mining services in a structured manner,

allowing designers to plan, store, document, verify, share and re-execute their applications as well as their output results.

In the KNOWLEDGE GRID environment, discovery processes are represented as workflows that a user may compose using both concrete and abstract Grid resources. Knowledge discovery workflows are defined using a visual interface that shows resources (data, tools, and hosts) to a user and offers mechanisms for integrating them in a workflow. Single resources and workflows are stored using an XML-based notation that represents a workflow as a data flow graph of nodes, each representing either a data mining service or a data transfer service. VEGA (Visual Environment for Grid Applications) [3] is a software prototype that embodies the main components of the KNOWLEDGE GRID environment, including services and functionalities ranging from information and discovery services to visual design and execution facilities.

This paper describes the development process of two distributed data mining applications on Grids by using the KNOWLEDGE GRID framework. Section 2 gives a brief introduction to the system components and functionalities, describing also tools and services offered to data miners. Section 3 presents the implementation of a distributed data mining application for producing an intrusion detection predictive model along with its performance results. Section 4 discusses the integration of the data mining system KDDML in the KNOWLEDGE GRID. KDDML [1] is a system for performing complex data mining tasks in the form of high level queries, in particular we discuss here the problems we tackled in migrating this system into our environment and some experimental results. Section 5 draws some conclusions about the performed experiences.

2 KNOWLEDGE GRID Services and Tools

The KNOWLEDGE GRID architecture is designed on top of mechanisms provided by Grid environments such as the Globus Toolkit [4]. It uses the basic Grid services such as communication, authentication, information, and resource management to build more specific parallel and distributed knowledge discovery (PDKD) tools and services [5].

The KNOWLEDGE GRID services are organized into two layers: the *Core K-Grid layer*, which is built on top of generic Grid services, and the *High level K-Grid layer*, which is implemented over the Core layer. Figure 1 shows a general view of the KNOWLEDGE GRID architecture.

- The Core K-Grid layer includes two basic services: the Knowledge Directory Service (*KDS*) and the Resources Allocation and Execution Management Service (*RAEMS*). The KDS manages the metadata describing the characteristics of relevant objects for PDKD applications, such as data sources, data mining software, results of computations, data and results manipulation tools, execution plans, etc. The information managed by the KDS is stored into three repositories: the metadata describing features of data, software and tools, coded in XML documents, are stored in a Knowledge Metadata Repository (*KMR*), the information about the knowledge discovered after a PDKD computation is stored in a Knowledge Base Repository (*KBR*), whereas the Knowledge Execution Plan Repository (*KEPR*) stores the *execution plans* describing PDKD applications over the grid. An *execution plan* is an abstract description in XML of a distributed data mining application, that is a graph describing the interaction and data flow between data sources, data mining tools, visualization tools, and result storage facilities. The goal of RAEMS is to find a mapping between an execution plan and available resources on the grid, satisfying user, data and algorithms requirements and constraints.

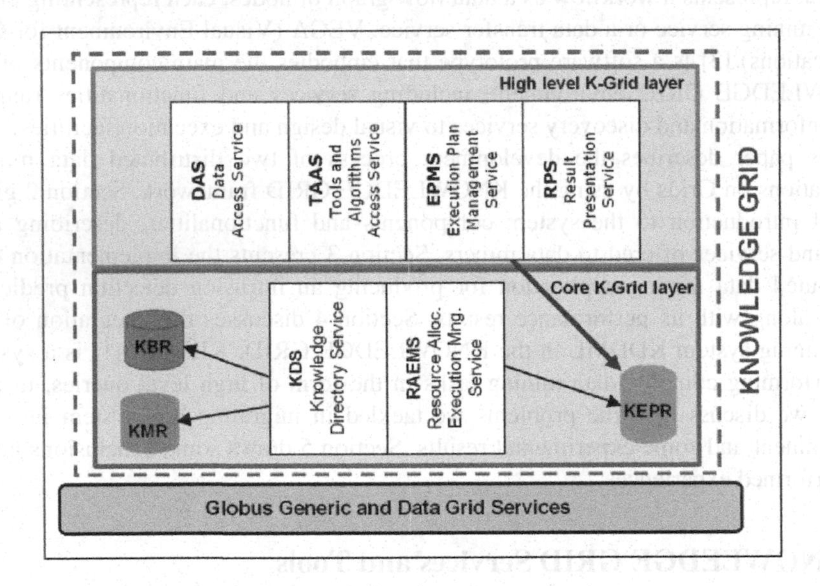

Fig. 1. Architecture of the KNOWLEDGE GRID

- The High level K-Grid layer includes the services used to build and execute PDKD computations over the Grid. The Data Access Service (*DAS*) is used for the search, selection, extraction, transformation and delivery of data to be mined. The Tools and Algorithms Access Service (*TAAS*) is responsible for search, selection, and download of data mining tools and algorithms. The Execution Plan Management Service (*EPMS*) is used to generate a set of different possible execution plans, starting from the data and the programs selected by the user. Execution plans are stored in the KEPR to allow the implementation of iterative knowledge discovery processes, e.g., periodical analysis of the same data sources varying in time. The Results Presentation Service (*RPS*) specifies how to generate, present and visualize the PDKD results (rules, associations, models, classification, etc.), and offers methods to store in different formats these results in the KBR.

The KNOWLEDGE GRID provides VEGA [3], a visual environment which allows a user to build an application and its execution plan in a semi-automatic way. The main goal of VEGA is to offer a set of visual functionalities that allows users to design distributed applications starting from a view of the present Grid configuration (i.e., available nodes and resources), and designing the different stages that compose an application within a structured environment. The high-level features offered by VEGA are intended to provide the user with easy access to Grid facilities offering a high level of abstraction, in order to leave her/him free to concentrate on the application design process. To fulfill this aim, VEGA builds a visual environment based on the component framework concept using and enhancing basic services offered by the KNOWLEDGE GRID and the Globus Toolkit. Fig. 2 shows a snapshot of the VEGA visual programming interface.

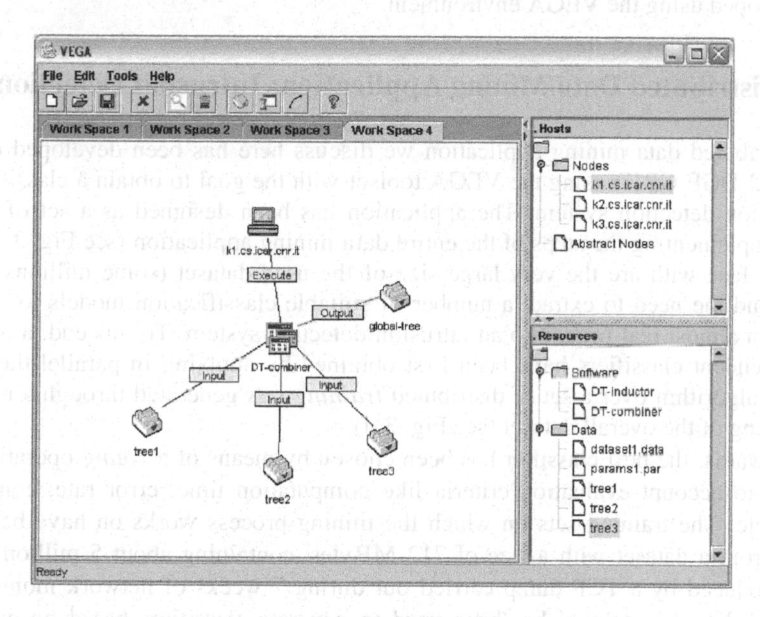

Fig. 2. The VEGA visual interface

VEGA overcomes the typical difficulties of Grid application programmers offering a high-level graphical interface. Moreover, by interacting with Knowledge Directory Service (KDS), VEGA shows available nodes in a Grid and retrieves additional information (metadata) about their published resources. Published resources are those made available for utilization by a Grid node owner by inserting specific entries in the KDS. The application design facility allows a user to build typical Grid applications in an easy, guided, and controlled way, having always a global view of the Grid status and the overall application that is being built.

Key concepts in the VEGA approach for designing a Grid application are the *visual language* used to describe in a component-like manner and through a graphical representation, the jobs constituting a knowledge discovery application, and the possibility to group these jobs in *workspaces* (on the left of Fig. 2) to form specific inter-dependent stages. Workspaces can be composed by using Grid nodes and resources shown on the corresponding panes (on the right of Fig. 2). A consistency checking module parses the model of the computation both while the design is in progress and prior to execute the application. Thus monitoring and driving the user actions as to obtain a correct and consistent graphical representation. Together with the workspace concept, VEGA makes available also the *virtual resource* abstraction. Thanks to this entity it is possible to compose applications working on data processed or generated in previous phases even if the execution has not been performed yet. VEGA includes an *execution service*, which gives the user the possibility to execute the designed application, monitor its status, and visualize results.

In the next two sections, we describe two distributed data mining applications that we developed using the VEGA environment.

3 A Distributed Data Mining Application: Intrusion Detection

The distributed data mining application we discuss here has been developed on the KNOWLEDGE GRID using the VEGA toolset with the goal to obtain a classifier for an intrusion detection system. The application has been designed as a set of workspaces implementing the steps of the entire data mining application (see Fig. 3). Main issues to face with are the very large size of the used dataset (some millions of records), and the need to extract a number of suitable classification models to be employed, in almost real-time, into an intrusion detection system. To this end, a number of independent classifiers have been first obtained by applying in parallel the same learning algorithm over a set of distributed *training sets* generated through a random partitioning of the overall dataset (see Fig. 3.d).

Afterwards, the best classifier has been chosen by means of a *voting* operation, by taking into account evaluation criteria like computation time, error rate, confusion matrix, etc. The training sets on which the mining process works on have been extracted from a dataset with a size of 712 MBytes, containing about 5 million of records produced by a TCP dump carried out during 7 weeks of network monitoring. The C4.5 data mining tool has been used to generate classifiers based on decision trees.

After the partitioning step, each training set has been moved to a node of the Grid providing the C4.5 data mining software (see Fig. 3.b). The induction of the decision trees has been performed in parallel on each node, followed by the validation of the models against a testing set. The results have been then moved back to the starting node to execute the voting operation.

3.1 Experimental Results

This section presents a comparison of the experimental performance results obtained on a single node (sequential case) with results obtained using 3 and 8 grid nodes.

In order to improve the transfers, datasets have been compressed before moving them to destination nodes (see Fig. 3.a); this permitted us to transfer files from 94% to 97% smaller in size.

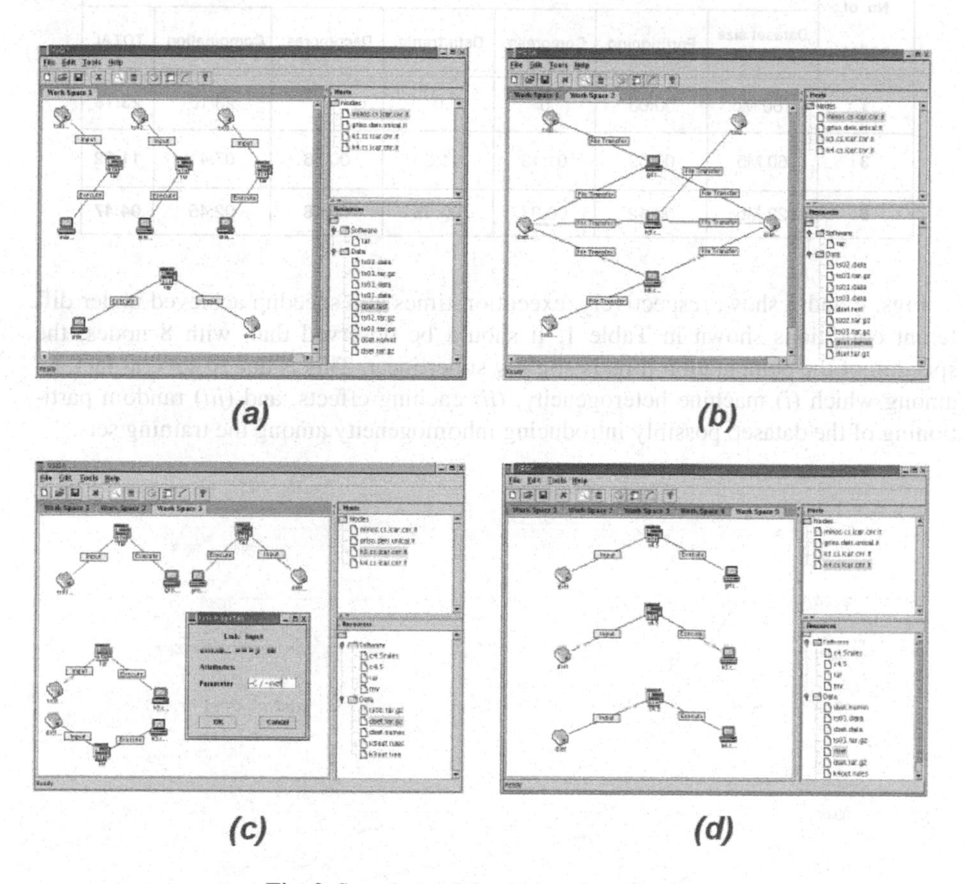

Fig. 3. Snapshots of the designed application

Table 1 shows the average execution times obtained from five runs for each configuration. Notice that compression, data transfer, and decompression steps were not needed in the execution on a single node. Moreover, it should be mentioned that, even when more than one node is used, partitioning and compression phases are still executed on a single machine, whereas decompression and computation phases are executed concurrently on different nodes.

The experiments were performed on Grid nodes of an early deployment of the SP3 Italian national Grid funded by MIUR. Machines hardware was ranging from dual Pentium III 800 MHz workstations to Pentium 4 1500 MHz PCs.

For each of the operations performed concurrently (data transfer, decompression, computation), the corresponding time is the maximum among the measured values on the different nodes.

Table 1. Execution times for different Grid configurations

No. of nodes	EXECUTION TIMES (mm:ss)						
	Dataset size per node	Partitioning	Compress.	Data transf.	Decompres.	Computation	TOTAL
1	180 Mb	00:08	0	0	0	23:10	**23:18**
3	60 Mb	00:09	01:13	01:33	00:56	07:41	**11:32**
8	23 Mb	00:12	00:37	00:55	00:18	02:45	**04:47**

Figs. 4 and 5 show, respectively, execution times and speedup achieved under different conditions shown in Table 1. It should be observed that, with 8 nodes, the speedup of the computation time is slightly superlinear. This is due to several factors, among which (*i*) machine heterogeneity, *(ii)* caching effects, and (*iii*) random partitioning of the dataset, possibly introducing inhomogeneity among the training sets.

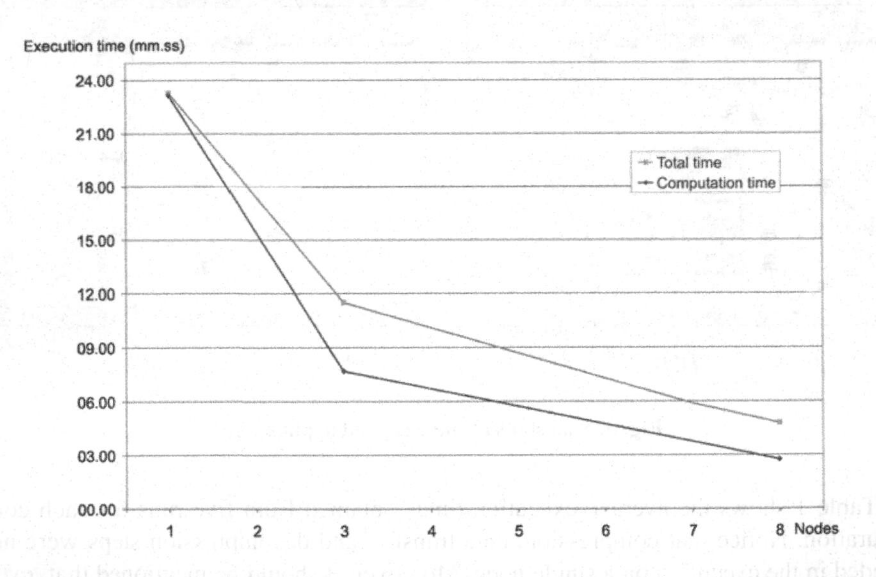

Fig. 4. Execution times on 1, 3 and 8 nodes

The total time suffers of the overhead added by the compression/decompression steps and data transfers. However, a total speedup factor of about 2 has been achieved employing three nodes and a speedup of about 5 has been obtained by using eight nodes.

These results confirm how the use of the KNOWLEDGE GRID may bring several benefits to the implementation of distributed knowledge discovery applications both in terms of data analysis distribution and scalability results.

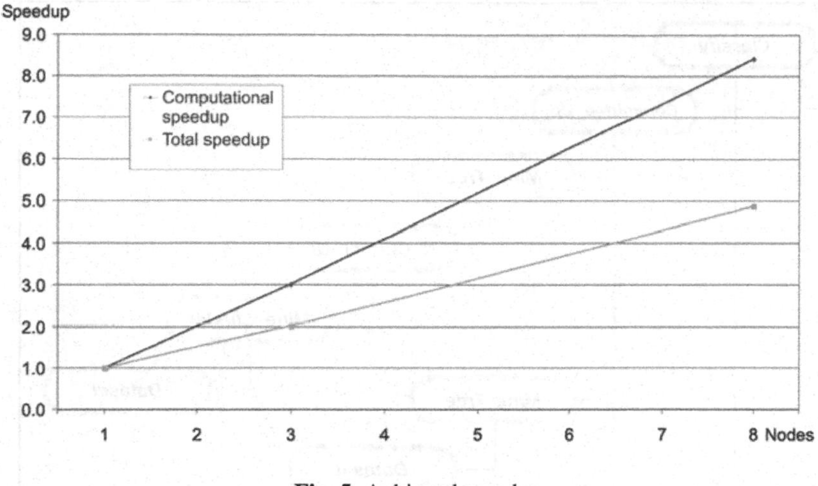

Fig. 5. Achieved speedup

4 KDD Markup Language and the Knowledge Grid

KDDML-MQL is an environment for the execution of complex Knowledge Discovery in Databases (KDD) processes expressed as high-level queries [1]. The environment is composed of two layers, the bottom layer called KDDML and the top one called MQL.

KDDML (KDD Markup Language) is an XML-based query language for performing knowledge extraction from databases. It combines, through a set of operators, data manipulation tasks such as data acquisition, preprocessing, mining and post-processing, employing specific algorithms from several suites of tools. Compositionality of the algorithms is obtained by appropriate wrappers between algorithm internal representations and the KDDML representation of data and models. The KDDML system is implemented in Java and consists of a graphical user interface for editing queries and an interpreter for their execution.

MQL (Mining Query Language) is an algebraic language, in the style of SQL. MQL queries are simpler to write and to understand than KDDML ones. The MQL system compiles an MQL query into one or more KDDML queries.

The overall system has been developed by the Computer Science department at University of Pisa (Italy); it has been tested on a variety of scenarios and it is gradually being updated with new features.

We are currently cooperating with KDDML developers so as to achieve the integration of the system into the KNOWLEDGE GRID environment. Aims of this effort are:

- combining the possibility to express a data mining application as a query with the advantages offered by the KNOWLEDGE GRID environment for executing the data mining query on a Grid;
- to extend the KNOWLEDGE GRID environment with a new class of knowledge discovery applications.

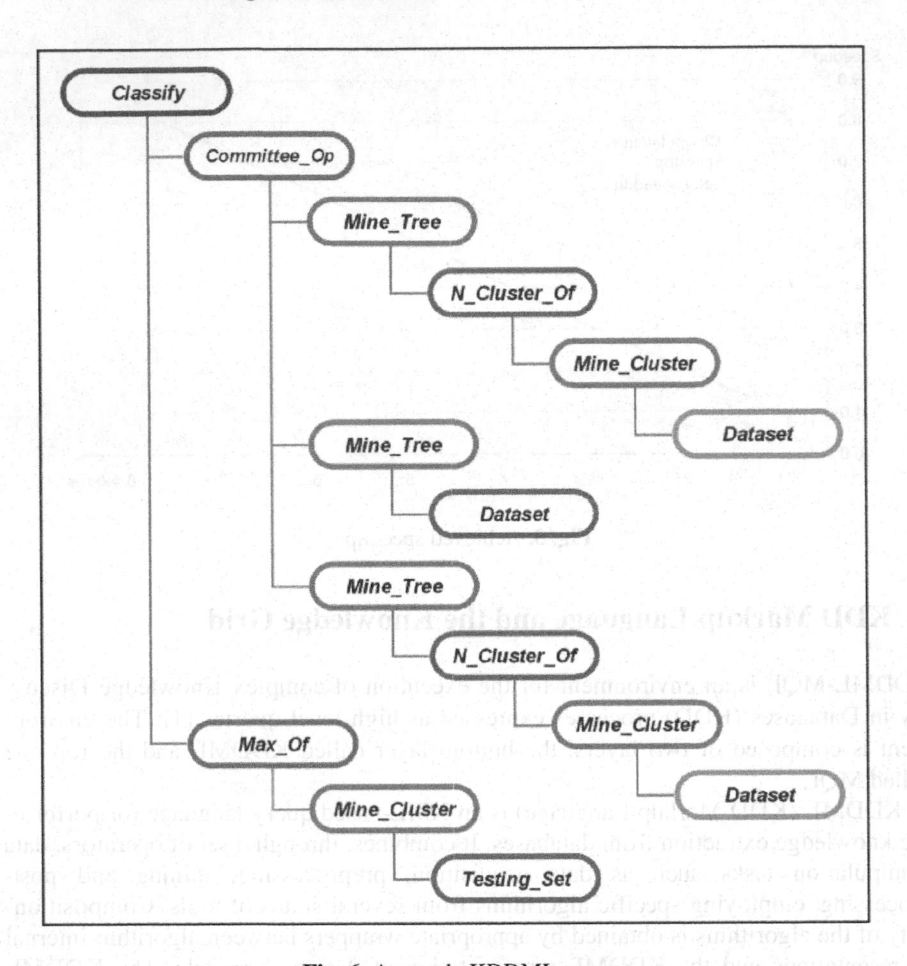

Fig. 6. A sample KDDML query

A KDDML query has the structure of a tree in which each node is a KDDML operator specifying the execution of a KDD task (pre-processing, classification, etc), or the logical combination (and/or operators) of results coming from lower levels of the tree. As, an example, Fig. 6 shows the structure of a query executed on the KNOWLEDGE GRID. Additional operators are provided for typical query specification (selection, join, etc.), along with more specific queries, such as: *Rule_Exception* that selects the records not verifying a certain set of rules, *Rule_Support* that selects the records verifying a certain set of rules, *Preserved_Rules* that finds the rules holding along a hierarchy. KDDML operators can be applied both to a dataset, a leaf in the tree, and to the results produced by another operator.

KDDML can support the parallel execution of the KDD application other than the sequential one, thanks to the possibility to split the query representing the overall KDD process into a set of sub-queries to be assigned to different processing nodes.

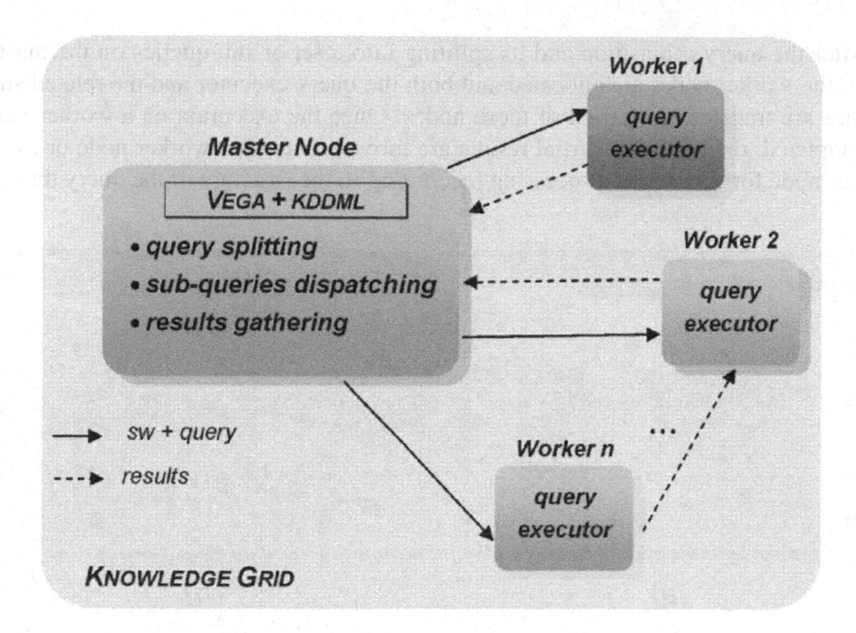

Fig. 7. Logical architecture of the distributed query execution

The parallel execution has been tested by KDDML developers exploiting a shared file system (on a cluster with a shared-disk architecture), used to retrieve datasets and exchange partial results. In order to enable such a system to execute on a Grid there are some problems to solve for ensuring the correct and reliable execution of the applications. First of all, the location of the data source is no more known a priori. Then the execution of the sub-queries need to be adequately coordinated, along with the exchange of the partial results for constructing the final answer.

The process of enabling the parallel and distributed execution of KDDML applications over a Grid setting required the modification of the structure of KDDML to transform it into a distributed application composed of three standalone components:

- *query entering and splitting*, this component includes the user interface for entering the query and the module that care for splitting it in sub-queries;
- *query executor*, executes a single sub-query;
- *results visualization.*

The distributed execution of KDDML has been modeled according to the master-worker paradigm (see Fig. 7). In particular, the *query entering* component acts as the master, while each instance of the *query executor* is a worker.

The main objectives of the execution of such queries on the Grid are:

- distributing them to a number of Grid nodes to exploit parallelism,
- avoiding an excessive fragmentation of the overall task, i.e. allocating a number of nodes so that the exchanges of intermediate (partial) results is minimized.

After the query acquisition and its splitting into a set of sub-queries on the master node, the worker nodes are allocated and both the query executor and the related sub-queries are transferred on each of these nodes. Once the execution on a worker node is completed, the produced partial results are moved to another worker node or to the master node for subsequent processing (according to the structure of the query tree).

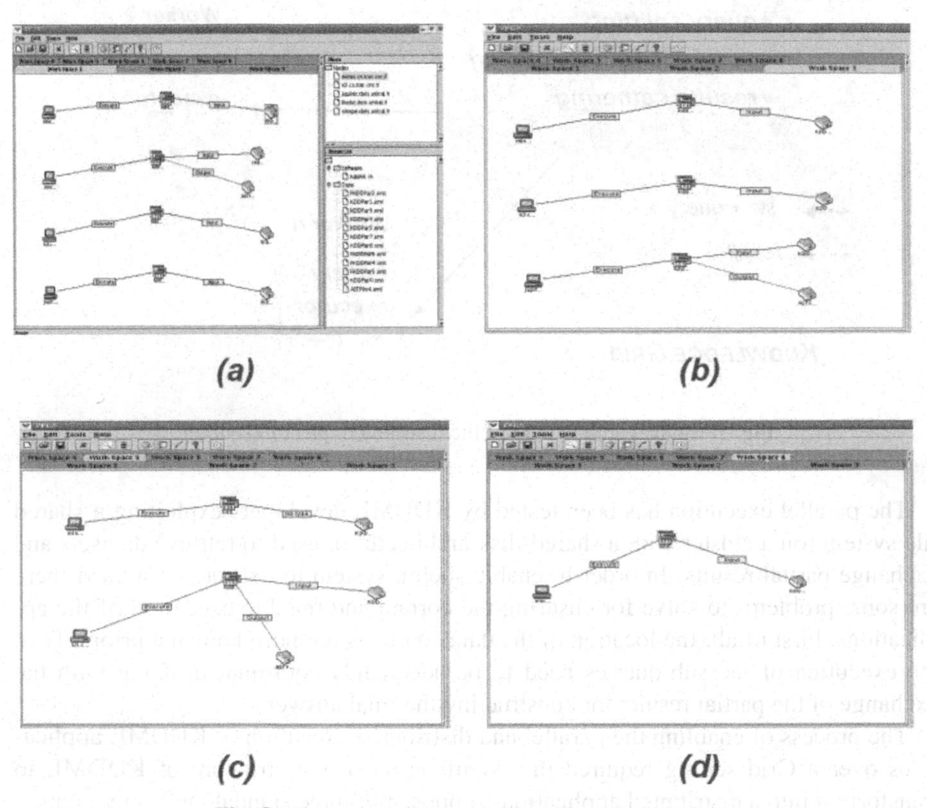

Fig. 8. Workspaces for query dispatch and execution

These requirements have been took into consideration in the formulation of an appropriate allocation policy. First of all, it is necessary to distinguish between two kind of sub-queries, *frontier queries* and *intermediate queries*: the former are those sub-queries that do not need to use results produced by other queries. They are thus at the lowest level in the tree. The latter are queries located between the lowest level and the root of the tree, they can be executed only when the underlying queries have produced their results.

The allocation policy takes into account both the requirements mentioned above and the structure of the query tree. It starts form the leafs of the tree and allocates each sub-query belonging to a branch to the same Grid node. The branches we consider here not always end on the tree root since they must not contain common paths. It should be noted that a branch individuates a set of sub-queries that operate on results produced by another sub-query of the same branch. Thus it is not fruitful to execute them in parallel, because there is an ordering to ensure in their execution, i.e. each parent query can start only after its descendant has produced its results. In addition, allocating them on the same Grid node brings the further advantage that partial results on which they operate do not need to be transferred prior to be used for subsequent processing.

For supporting the execution of KDDML applications, VEGA provides a template reflecting the structure described above. In particular, there is a first workspace for the query entering and decomposition. The other workspaces are dynamically configured on the basis of number and structure of the generated sub-queries, both in terms of number of worker nodes to be allocated and about the transfer of partial results. Fig. 8 shows some workspaces obtained after the instantiation of the general template based on the submitted query.

4.1 Preliminary Results

This section presents some preliminary results obtained executing a KDDML application on the KNOWLEDGE GRID with the goal of validating the approach and testing its feasibility.

The application aims at extracting a classification model starting from three different datasets (see Fig. 6). Two of them are first submitted to a clustering algorithm, then two clusters are selected among the obtained clusters (one per each dataset). Such clusters are then used to obtain two classification models through the C4.5 algorithm. The C4.5 algorithm is also (directly) applied to the third dataset. Afterwards, these three decision trees are fed into the *Committee_Op* operator, which merges them into the final classification model by applying a voting operation. The final step is constituted by the validation of the model against a set of records extracted from a testing set by applying a clustering algorithm and selecting the larger cluster.

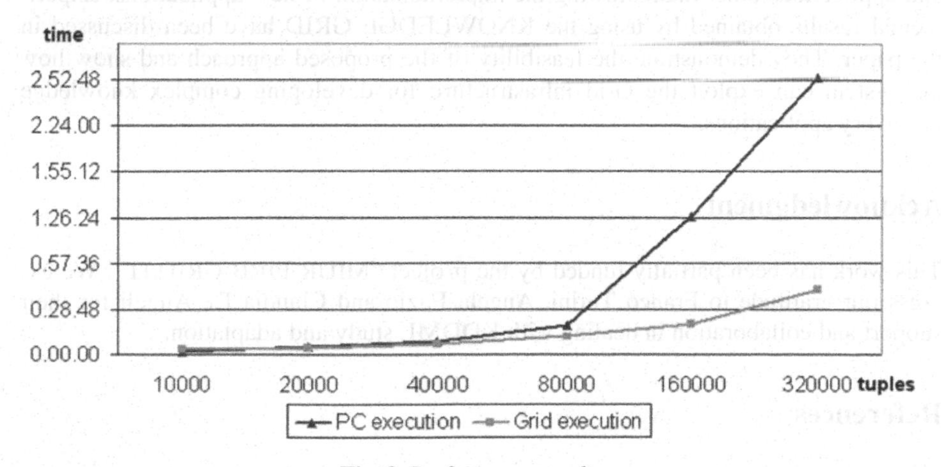

Fig. 9. Performance results

This application is represented by a query with a quite complex structure. The number of sub-queries is equal to the number of nodes of the corresponding tree (the query generates 11 sub-queries, 4 of which are frontier queries, as can be seen in Fig. 6). Since a branch in a tree always starts with a leaf, it follows that the number of nodes to be allocated is equal to the number of leafs, i.e. to the number of frontier queries.

The sub-queries executed on each of the allocated Grid nodes are those included in each of the four branches. When the computations of each branch are completed, the partial results they produced must be moved to another processing node. The number of transfers of partial results is equal to the number of branches in the tree less one. In our application only three transfers were needed.

Fig. 9 shows the execution time with different dataset sizes, both for the execution on the KNOWLEDGE GRID and for the sequential execution of all the queries. It can be noted that when the size of the dataset increases the advantages of the Grid execution became soon very profitable.

5 Conclusion

Complex applications can be developed on Grids according to a distributed approach. Among those, data mining applications may benefit from Grid resources and facilities. The KNOWLEDGE GRID system is a framework that offers the users high-level features for implementing distributed data mining applications. It integrates and completes data Grid services by supporting distributed data analysis, knowledge discovery, and knowledge management services.

We discussed two real data mining applications implemented on a Grid by using the KNOWLEDGE GRID framework and its visual programming environment offered by VEGA. In particular, the implementation of the KDDML system shows how the KNOWLEDGE GRID can be used to port on Grids, pre-existing data mining tools and applications other than allowing the implementation of new applications. Experimental results obtained by using the KNOWLEDGE GRID have been discussed in the paper. They demonstrate the feasibility of the proposed approach and show how the system can exploit the Grid infrastructure for developing complex knowledge discovery applications.

Acknowledgment

This work has been partially funded by the project "MIUR FIRB GRID.IT". We express our gratitude to Franco Turini, Angela Fazio and Claudia De Angeli for their support and collaboration in dealing with KDDML study and adaptation.

References

1. Alcamo, P., Domenichini F., Turini F.: An XML based environment in support of the overall KDD process. *Proc. 4th Intl. Conference on Flexible Query Answering Systems*. Physica-Verlag Heidelberg New York, (2000) 413-424

2. Cannataro, M., Talia, D.: The Knowledge Grid. *Communications of the ACM*, 46(1) (2003) 89-93
3. Cannataro, M., Congiusta A., Talia, D., Trunfio, P.: A Data Mining Toolset for Distributed High-Performance Platforms. *Proc. 3rd Intl. Conference Data Mining 2002*, Bologna, Italy, WIT Press, (2002) 41-50
4. Foster, I., Kesselman, C., Tuecke, S.: The Anatomy of the Grid: Enabling Scalable Virtual Organizations. *Intl. Journal of Supercomputer Applications*, 15(3) (2001)
5. Kargupta, H., Joshi, A., Sivakumar, K, Yesha, Y. (eds.) : *Data Mining: Next Generation Challenges and Future Directions*, MIT/AAAI Press, (2004)

Scaling Up the Preventive Replication of Autonomous Databases in Cluster Systems*

Cédric Coulon, Esther Pacitti, and Patrick Valduriez

INRIA and LINA, University of Nantes – France
{Cedric.Coulon, Esther.Pacitti}@lina.univ-nantes.fr
Patrick.Valduriez@inria.fr

Abstract. We consider the use of a cluster system for Application Service Providers. To obtain high-performance and high-availability, we replicate databases (and DBMS) at several nodes, so they can be accessed in parallel through applications. Then the main problem is to assure the consistency of autonomous replicated databases. Preventive replication [8] provides a good solution that exploits the cluster's high speed network, without the constraints of synchronous replication. However, the solution in [8] assumes full replication and a restricted class of transactions. In this paper, we address these two limitations in order to scale up to large cluster configurations. Thus, the main contribution is a refreshment algorithm that prevents conflicts for partially replicated databases. We describe the implementation of our algorithm over a cluster of 32 nodes running PostGRESQL. Our experimental results show that our algorithm has excellent scale up and speed up.

1 Introduction

High-performance and high-availability of database management has been traditionally achieved with parallel database systems [9], implemented on tightly-coupled multiprocessors. Parallel data processing is then obtained by partitioning and replicating the data across the multiprocessor nodes in order to divide processing. Although quite effective, this solution requires the database system to have full control over the data and is expensive in terms of software and hardware.

Clusters of PC servers provide a cost-effective alternative to tightly-coupled multiprocessors. They have been used successfully by, for example, Web search engines using high-volume server farms (e.g., Google). However, search engines are typically read-intensive which makes it easier to exploit parallelism. Cluster systems can make new businesses such as Application Service Providers (ASP) economically viable. In the ASP model, customers' applications and databases (including data and DBMS) are hosted at the provider site and need be available, typically through the Internet, as efficiently as if they were local to the customer site. To improve performance, applications and data can be replicated at different nodes so that users

* Work partially funded by the MDP2P project of the ACI "Masses de Données" of the French ministry of research.

M. Daydé et al. (Eds.): VECPAR 2004, LNCS 3402, pp. 170–183, 2005.

can be served by any of the nodes depending on the current load [1]. This arrangement also provides high-availability since, in the event of a node failure, other nodes can still do the work. However, managing data replication in the ASP context is far more difficult than in Web search engines since applications can be update-intensive and both applications and databases must remain autonomous. The solution of using a parallel DBMS is not appropriate as it is expensive, requires heavy migration to the parallel DBMS and hurts database autonomy.

In this paper, we consider a cluster system with similar nodes, each having one or more processors, main memory (RAM) and disk. Similar to multiprocessors, various cluster system architectures are possible: shared-disk, shared-cache and shared-nothing [9]. Shared-disk and shared-cache require a special interconnect that provide a shared space to all nodes with provision for cache coherence using either hardware or software. Shared-nothing (or distributed memory) is the only architecture that supports our autonomy requirements without the additional cost of a special interconnect. Furthermore, shared-nothing can scale up to very large configurations. Thus, we strive to exploit a shared-nothing architecture.

To obtain high-performance and high-availability, we replicate databases (and DBMS) at several nodes, so they can be accessed in parallel through applications. Then the main problem is to assure the consistency of autonomous replicated databases. An obvious solution is *synchronous replication* (e.g. Read-One-Write-All) which updates all replicas within the same (distributed) transaction. Synchronous (or eager) replication enforces mutual consistency of replicas. However, it cannot scale up to large cluster configurations because it makes use of distributed transactions, typically implemented by 2 phase commit [5].

A better solution that scales up is *lazy replication* [6]. With lazy replication, a transaction can commit after updating a replica, called *primary copy*, at some node, called *master* node. After the transaction commits, the other replicas, called *secondary copies*, are updated in separate refresh transactions at *slave* nodes. Lazy replication allows for different replication configurations [7]. A useful configuration is *lazy master* where there is only one primary copy. Although it relaxes the property of mutual consistency, strong consistency[1] is eventually assured. However, it hurts availability since the failure of the master node prevents the replica to be updated. A more general configuration is (lazy) *multi-master* where the same primary copy, called a *multi-owner copy*, may be stored at and updated by different master nodes, called *multi-owner* nodes. The advantage of multi-master is high-availability and high-performance since replicas can be updated in parallel at different nodes. However, conflicting updates of the same primary copy at different nodes can introduce replica divergence.

Preventive replication can prevent the occurrence of conflicts [8], by exploiting the cluster's high speed network, thus providing strong consistency, without the constraints of synchronous replication. The refresher algorithm in [8] supports multi-master configurations and its implementation over a cluster of 8 nodes shows good performance. However, it assumes that databases are fully replicated across all cluster

---

[1] For any two nodes, the same sequence of transactions is executed in the same order.

nodes and thus propagates each transaction to each cluster node. This makes it unsuitable for supporting heavy workloads on large cluster configurations. Furthermore, it considers only write multi-owner transactions. However it may arise that a transaction must read from some other copy to update a multi-owner copy.

In this paper, we address these two limitations in order to scale up to large cluster configurations. The main contribution is a multi-master refreshment algorithm that prevents conflicts for partially replicated databases. We also show the architectural components necessary to implement the algorithm. Finally, we describe the implementation of our algorithm over a cluster of 32 nodes running PostGRESQL and present experimental results that show that it yields excellent scale-up and speed up.

The rest of the paper is structured as follows. Section 2 presents the cluster system architecture in which we identify the role of the replication manager. Section 3 defines partially replicated configurations. Section 4 presents refreshment management including the principles of the refresher algorithm and the architectural components necessary for its implementation. Section 5 describes our prototype and show our performance model and experiments results. Finally, Section 6 concludes the paper.

2 Cluster System Architecture

In this section, we introduce the architecture for processing user requests against applications into the cluster system and discuss our general solutions for placing applications, submitting transactions and managing replicas. Therefore, the replication layer is identified together with all other general components. Our system architecture is fully distributed since any node may handle user requests.

Shared-nothing is the only architecture that supports sufficient node autonomy without the additional cost of special interconnects. Thus, we exploit a shared-nothing architecture. However, in order to perform distributed request routing and load balancing, each node must share (i) the user directory which stores authentication and authorization information on users and (ii) the list of the available nodes, and for each node, the current load and the copies it manages with their access rights (read/write or read only). To avoid the bottleneck of central routing, we use Distributed Shared Memory [4] to manage this shared information. Each cluster node has four layers (see Figure 1): Request Router, Application Manager, Transaction Load Balancer and Replication Manager.

A request may be a query or an update transaction for a specific application. The general processing of a user request is as follows. When a user request arrives at the cluster, traditionally through an access node, it is sent randomly to a cluster node. There is no significant data processing at the access node, avoiding bottlenecks. Within that cluster node, the user is authenticated and authorized through the Request Router, available at each node, using a multi-threaded global user directory service. Thus user requests can be managed completely asynchronously. Next, if a request is accepted, then the Request Router chooses a node j, to submit the request. The choice of j involves selecting all nodes in which the required application is available and copies used by the request are available with the correct access right, and, among these nodes, the node with the lightest load. Therefore, i may be equal to j. The Request Router then routes the user request to an application node using a traditional load balancing algorithm.

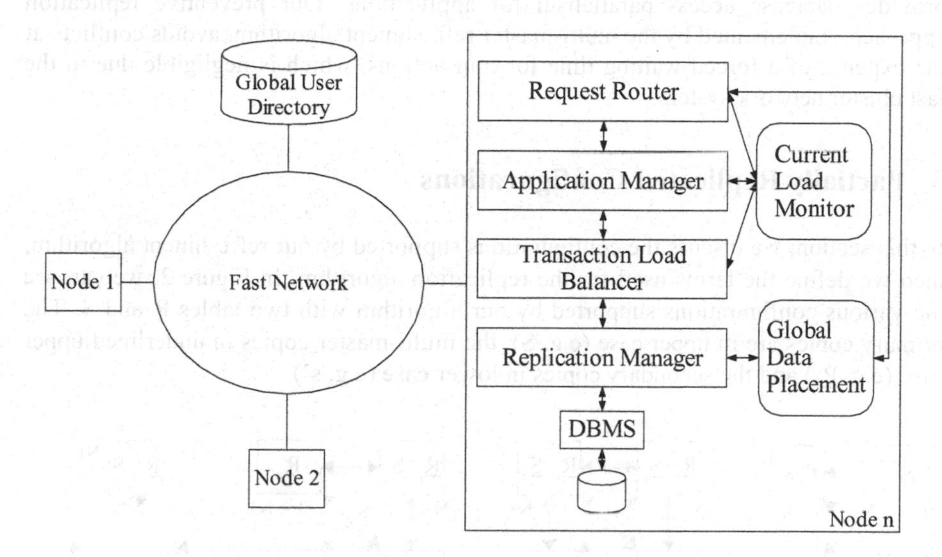

Fig. 1. Cluster system architecture

However, the database accessed by the user request may be placed at another node k since applications and databases are both replicated and not every node hosts a database system. In this case, the choice regarding node k will depend on the cluster configuration and the database load at each node.

Each time a new node is available, it must register to other nodes. To accomplish this, it uses the global shared memory that maintains the lists of available nodes. If a node is no more available, it is removed from the list either by itself, if it disconnects normally, or by the request router of another node, if it crashes.

A node load is specified by a current load monitor available at each node. For each node, the load monitor periodically computes application and transaction loads. For each type of load, it establishes a load grade and stores the grades in a shared memory. A high grade corresponds to a high load. Therefore, the Request Router chooses the best node for a specific request using the node grades.

The Application Manager is the layer that manages application instantiation and execution using an application server provider. Within an application, each time a transaction is to be executed, the Transaction Load Balancer layer is invoked which triggers transaction execution at the best node, using the load grades available at each node. The "best" node is defined as the one with lighter transaction load. The Transaction Load Balancer ensures that each transaction execution obeys the ACID (atomicity, consistency, isolation, durability) properties, and then signals to the Application Manager to commit or abort the transaction.

The Replication Manager layer manages access to replicated data and assures strong consistency in such a way that transactions that update replicated data are executed in the same serial order at each node. We employ data replication because it provides database access parallelism for applications. Our preventive replication approach, implemented by the multi-master refreshment algorithm, avoids conflicts at the expense of a forced waiting time for transactions, which is negligible due to the fast cluster network system.

3 Partially Replicated Configurations

In this section, we discuss the configurations supported by our refreshment algorithm, then we define the terms used for the replication algorithm. In Figure 2, we can see the various configurations supported by our algorithm with two tables R and S. The primary copies are in upper case (e.g. S), the multi-master copies in underlined upper case (e.g. $\underline{R}_1$) and the secondary copies in lower case (e.g. s').

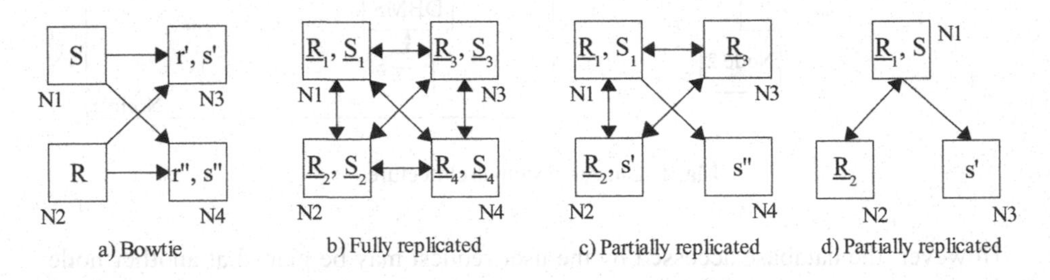

Fig. 2. Useful configurations

First, we have a bowtie lazy-master configuration (Figure 2a) where there are only primary copies and secondary copies. Second, we have a multi-master configuration (Figure 2b) where there are only multi-master copies and all nodes hold all copies,. Thus, this configuration is also called fully-replicated. Then we have two examples of partially-replicated configurations (Figure 2c and 2d) where all types of copies (primary, multi-master and secondary) may be stored at any node. For instance, in Figure 2c, node N1 carries the multi-owner copy R_1 and a primary copy S, node N2 carries the multi-owner copy R_2 and the secondary copy s', node N3 carries the multi-owner copy R_3 , and node N4 carries the secondary copy s''. Such a partial replicated configuration allows an incoming transaction at node N1, such as the one in Example 1, to read S_1 in order to update R_1:

Example 1: Incoming transaction at node N1

```
UPDATE R1 SET att1=value
WHERE att2 IN
    (SELECT att3 FROM S)
COMMIT;
```

This transaction can be entirely executed at N1 (to update R_1) and N2 (to update R_2). However it cannot be entirely executed at node N3 (to update R_3) which does not hold a copy of S.

A *Multi-Owner Transaction* (MOT) may be composed of a sequence of read and write operations followed by a *commit* (as produced by the above statement). This is more general than in [8] which only considers write operations. A *Multi-Owner Refresh transaction* (MOR) corresponds to the resulting sequence of write operations of an MOT, as written in the Log History [8]. The *origin node* is the node that receives a MOT (from the client) which updates a set of multi-owner copies. *Target nodes* are the nodes that carry multi-owner copies which are updated by the incoming MOT. Note that the origin node is also a target node since it holds multi-owner copies which are updated by MOT. For instance, in Figure 2b whenever node N1 receives a MOT that updates R_1, then N1 is the origin node and N1,N2, N3 and N4 are the target nodes. In Figure 2c, whenever N3 receives a MOT that updates R_3, then the origin node is N3 and the target node are N1, N2 and N3.

4 Preventive Refreshment for Partially Replicated Configurations

In this section, we present the preventive refreshment algorithm for the partially replicated configurations. Next, we introduce the architecture used for the Replication Manager layer to show how the algorithm works, using an example.

4.1 Refresher Algorithm

We assume that the network interface provides global First-In First-Out (FIFO) reliable multicast: messages multicast by one node are received at the multicast group nodes in the order they have been sent [3]. We denote by Max, the upper bound of the time needed to multicast a message from a node i to any other node j. We also assume that each node has a local clock. For fairness reasons, clocks are assumed to have a drift and to be ε-synchronized [2]. This means that the difference between any two correct clocks is not higher that ε (known as the precision).

To define the refresher algorithm, we need a formal correctness criterion to define strong consistency. In multi-master configurations, inconsistencies may arise whenever the serial orders of two multi-owner transactions at two nodes are not equal. Therefore, multi-owner transactions must be executed in the same serial order at any two nodes. Thus, global FIFO ordering is not sufficient to guarantee the correctness of the refreshment algorithm. Hence the following correctness criterion is necessary:

Definition 3.1 *(Total Order) Two multi-owner transactions MOT1 and MOT2 are said to be executed in Total Order if all multi-owner nodes that commit both MOT1 and MOT2 commit them in the same order.*

Proposition 3.1 *For any cluster configuration C that meets a multi-master configuration requirement, the refresh algorithm that C uses is correct if and only if the algorithm enforces total order.*

We now present the refresher algorithm. Each multi-owner transaction is associated with a chronological time stamp value. The principle of the preventive refresher algorithm is to submit a sequence of multi-owner transactions in chronological order at each node. As a consequence, total order is enforced since multi-owner transactions are executed in same serial order at each node. To assure chronological ordering, before submitting a multi-owner transaction at node i, the algorithm has to check whether there is any older committed multi-owner transaction enroute to node i. To accomplish this, the submission time of a new multi-owner transaction MOT_i is delayed by Max + ε (recall that Max is the upper bound of the time needed to multicast a message from a node to any other node). After this delay period, all older transactions are guaranteed to be received at node i. Thus chronological and total orderings are assured.

Whenever a multi-owner transaction MOT_i is to be triggered at some node i, the same node multicasts MOT_i to all nodes 1, 2, …n, of the multi-owner configuration, including itself. Once MOT_i is received at some other node j (notice that i may be equal to j), it is placed in the pending queue for the multi-owner triggering node i, called multi-owner pending queue (noted moq_i). Therefore, at each multi-owner node i, there is a set of multi-owner queues, moq_1, moq_2, …, moq_n. Each pending queue corresponds to a node of the multi-owner configuration and is used by the refresher to perform chronological ordering.

With partially replicated configurations, some of the target nodes may not be able to perform MOT_i because they do not hold all the copies necessary to perform the read set of MOT_i (recall the discussion on Example 1). However the write sequence of MOT_i (which corresponds to its MOR_i) needs to be ordered using MOT_i's timestamp value in order to ensure consistency. So MOT_i is delayed by Max + ε to prevent conflicts and waits for the reception of the corresponding MOR_i (the resulting sequence of write operations of MOT_i). At origin node i, when the commitment of MOT_i is detected, the corresponding MORi is produced and node i multicasts MOR_i towards the involved target nodes. Finally, upon reception of the MOR_i , at a target node j, the content of the waiting MOT_i is replaced with the content of the incoming MOR_i . Then, MOT_i is executed which enforces strong consistency.

A key aspect of our algorithm is to rely on the upper bound Max on the transmission time of a message by the global FIFO reliable multicast. Therefore, it is essential to have a value of Max that is not over estimated. The computation of Max resorts to scheduling theory (e.g. see [9]) and takes into account several parameters such as the global reliable network itself, the characteristics of the messages to multicast and the failures to be tolerated. In summary, our approach trades the use of a worst case multicast time at the benefit of reducing the number of messages exchanged on the network. This is a well known trade-off. This solution brings simplicity and ease of implementation.

4.2 Replication Manager Architecture

To implement the multi-master refresher algorithm, we add several components to a regular DBMS. Our goal is to maintain node autonomy since the implementation solution does not require the knowledge of system internals. Figure 3 shows the architecture for refreshing partially replicated databases in our cluster system. The Replica Interface receives MOTs, coming from the clients. The Propagator and the Receiver manage the sending and reception (respectively) of transactions (MOT and MOR) inside messages within the network. Whenever the Receiver receives a MOT, it places it in the appropriate pending queue, used by the Refresher. Next, the Refresher executes the refresher algorithm to ensure strong consistency. Each time a MOT is chosen to be executed by the Refresher, it is written in a *running queue* which is then read by the Deliverer in FIFO order. The Deliverer submits transactions, read from the running queue, to the DBMS.

For partially replicated configurations, when a MOT is composed of a sequence of reads and writes, the refresher at the target nodes must assure the correct ordering. However the MOT execution must be delayed until its corresponding MOR is received. This is because the MOR is produced only after the commitment of the corresponding MOT at the origin node. At the target node, the content of the MOT (sequence of read and write operations) is replaced by the content of the MOR (sequence of write operations). Thus, at the target node, when the receiver receives a MOR, it interacts with the Deliverer.

At the origin node, the Log Monitor checks constantly the content of the DBMS log to detect whether replicated copies have been changed. For partially replicated configurations, whenever a MOT is composed of reads and writes, the log monitor detects the MOT commitment in order to produce the corresponding MOR. Then, the propagator multicasts the MOR to the involved target nodes.

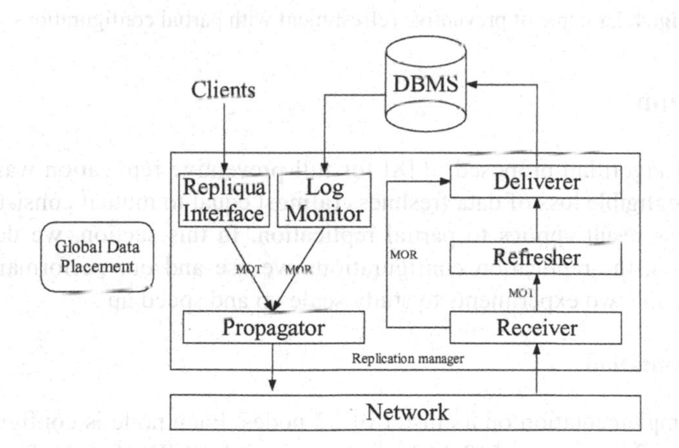

Fig. 3. Preventive Replication Manager Architecture

Let us now illustrate the refresher algorithm with an example of execution. In Figure 4, we assume a simple configuration with 3 nodes (N1, N2 and N3) and 2 copies (R and S). N1 carries a multi-owner copy of R and a primary copy of S, N2 a multi-owner copy of R, and N3 a secondary copy of S. The refreshment proceeds in 5 steps. In step 1, N1 (the origin node) receives MOT from a client, using the Replica Interface, which reads S and updates R_1. For instance, MOT can be the resulting read and write sequence produced by the transaction of Example 1. Then, in step 2, N1 multicasts MOT, using the Propagator, to the involved target nodes, i.e. N1 and N2. N3 is not concerned by MOT because it only holds a secondary copy s. In step 3, MOT can be performed using the Refresher algorithm at N1. At N2, MOT is also managed by the Refresher and then put in the running queue. MOT cannot yet be executed at the target node (N2 does not hold S) because the Deliverer needs to wait for its corresponding MOR (see step 4). In step 4, after the commitment of MOT at the origin node, the Log Monitor produces MOR and multicasts it to all involved target nodes. In step 5, N2 receives MOR and the Receiver replaces the content of MOT by that of MOR. The Deliverer can then submit MOT.

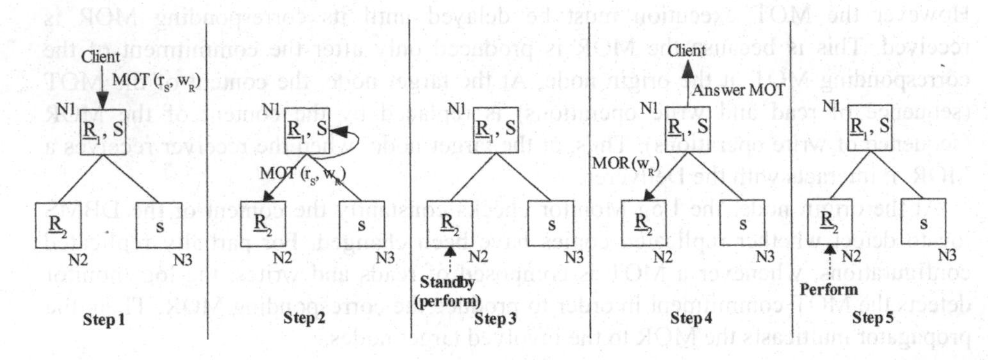

Fig. 4. Example of preventive refreshment with partial configurations

5 Validation

The refresher algorithm proposed in [8] for full preventive replication was shown to introduce a negligible loss of data freshness (almost equal to mutual consistency). The same freshness result applies to partial replication. In this section, we describe our implementation, the replication configurations we use and our performance model. Then we describe two experiments to study scale up and speed up.

5.1 Implementation

We did our implementation on a cluster of 32 nodes. Each node is configured with a 1GHz Pentium 3 processor, 512 MB of memory and 40GB of disk. The nodes are linked by a 1 Gb/s Myrinet network. We use Linux Mandrake 8.0/Java and Spread toolkit that provides a reliable FIFO message bus and high-performance message service among the cluster nodes that is resilient to faults. We use the PostGRESQL[2] Open Source DBMS at each node. We chose PostGRESQL because it is quite complete in terms of transaction support and easy to work with.

[2] http://www.postgresql.org

Our implementation has four modules: *Client*, *Replicator*, *Network* and *Database Server*. The Client module simulates the clients of the ASP. It submits multi-owner transactions randomly to any cluster node, via RMI-JDBC, which implement the Replica Interface. Each cluster node hosts a Database Server and one instance of the Replicator module. For this validation which stresses transaction performance, we implemented directly the Replicator module within PostGRESQL. The Replicator module implements all system components necessary for a multi-owner node: Replica Interface, Propagator, Receiver, Refresher and Deliverer. Each time a transaction is to be executed it is first sent to the Replica Interface which checks whether the incoming transaction writes on replicated data. Whenever a transaction does not write on replicated data, it is sent directly to the local transaction manager. Even though we do not consider node failures in our performance evaluation, we implemented all the necessary logs for recovery to understand the complete behavior of the algorithm. The Network module interconnects all cluster nodes through the Spread toolkit.

5.2 Configurations

We use 3 configurations involving 3 copies R, S and T with 5 different numbers of nodes (4, 8, 16, 24 and 32). This choice makes it easy to vary the placement of copies while keeping a balanced distribution. In the Fully Replicated configuration (FR), all nodes hold the 3 copies, i.e. *{R,S,T}*. In the Partially Replicated 1 configuration (PR1), one fourth of the nodes holds *{R,S,T}*, another holds *{R,S}*, another holds *{R,T}* and another holds *{S,T}*. In the Partially Replicated 2 configuration (PR2), one fourth of the nodes holds *{R,S,T}*, another holds *{R}*, another holds *{S}* and another holds *{T}*.

Table 1. Placement of copies per configuration

	Fully Replicated (FR)	**Partially Replicated 1 (PR1)**				**Partially Replicate 2 (PR2)**			
nodes	RST	RST	RS	RT	ST	RST	R	S	T
4	4	1	1	1	1	1	1	1	1
8	8	2	2	2	2	2	2	2	2
16	16	4	4	4	4	4	4	4	4
24	24	6	6	6	6	6	6	6	6
32	32	8	8	8	8	8	8	8	8

5.3 Performance Model

Our performance model takes into account multi-owner transaction size, arrival rates and the number of multi-owner nodes. We vary multi-owner transactions' arrival rates (noted). We consider workloads that are either low (=3s) or bursty (=200ms). We define two types of multi-owner transactions (denoted by */MOT/*), defined by the number of write operations. Small transactions have size 5, while long transactions have size 50. Multi-owner transactions are either Single Copy (SC), only one replica is updated, or Multiple Copy (MC), several replicas are involved. Thus, with our 3 copies, we have 3 SC transactions (on either *{R}*, *{S}* or *{T}*) and 7 MC transactions (on either *{R}*, *{S}*, *{T}*, *{R,S}*, *{R,T}*, *{S,T}* or *{R,S,T}*).

Table 2. Performance parameters

Param.	**Definition**	**Values**
/MOT/	Multi-owner transaction size	5; 50
Transaction Type	Involved a Single or Muli- Replicas Copies	SC (single), MC (multi)
Ltr	Long transaction ratio	66%
	Mean time between transactions (workload)	bursty: 200ms; low: 3s
Nb-Multi-Owner	Number of replicated PostGRESQL instances	4, 8, 16, 24, 32
M	Number of update transactions submitted for the tests	50
Max +	Delay introduced for submitting a MOT	100ms
Configuration	Replication of R, S and T	FR, PR1, PR2

To understand the behavior of our algorithm in the presence of short and long transactions, we set a parameter called *long transaction ratio* as *ltr* = 66 (66% of update transactions are long). Updates are done on the same attribute of a different tuple. Finally our copies R, S and T have 2 attributes (the first attribute is used as primary key and the second as data) and carry each 100000 tuples and no indexes. The parameters of the performance model are described in Table 2.

5.4 Scale Up Experiment

This experiment studies the algorithm's scalability. That is, for a same set of incoming multi-owner transactions, scalability is achieved whenever in augmenting the number of nodes the response times remain the same. We consider SC and MC transactions in *bursty* and *low* workloads (= 200ms and = 20s), varying the number of nodes for each configuration (FR, PR1 and PR2). SC and MC transactions randomly update R, S or T. For each test, we measure the average response time per transaction. The duration of this experiment is the time to submit 50 transactions.

The experiment results (see Figure 5) clearly show that for all tests scalability is achieved. In contrast with synchronous algorithms, our algorithm has linear response time behavior since for each incoming multi-owner transaction it requires only the multicast of *n* messages for nodes that carry all required copies plus *2n* messages for nodes that do not carry all required copies, where *n* corresponds to the number of target nodes.

The results also show the impact of the configuration on performance. First, on low workloads (see Figure 5.b and 5.d), the impact is very little (transaction response times are similar for all configurations) because at each node transaction executes alone. In bursty workloads (see Figure 5.a and 5.c), the number of transaction augments however several of them execute in parallel due to partial replication. Thus partial replication increases inter-transaction parallelism by allowing different nodes to process different transactions. The performance improvement of PR over FR is best for SC transactions (see Figure 5.a) and quite significant (a factor of 2.33 for PR1 and of 7 for PR2). For MC transactions (see Figure 5.c), the performance improvement is still a factor of 3.5 for PR1 but only of 1.6 for PR2. This is because FR performs relatively better for MC transactions than for SC transactions. For SC transactions, PR2 which stores individual copies of R, S and T at different nodes outperforms PR1. But for MC transactions, PR1 outperforms PR2. Thus the configuration and the placement of the copies should be tuned to selected types of transactions.

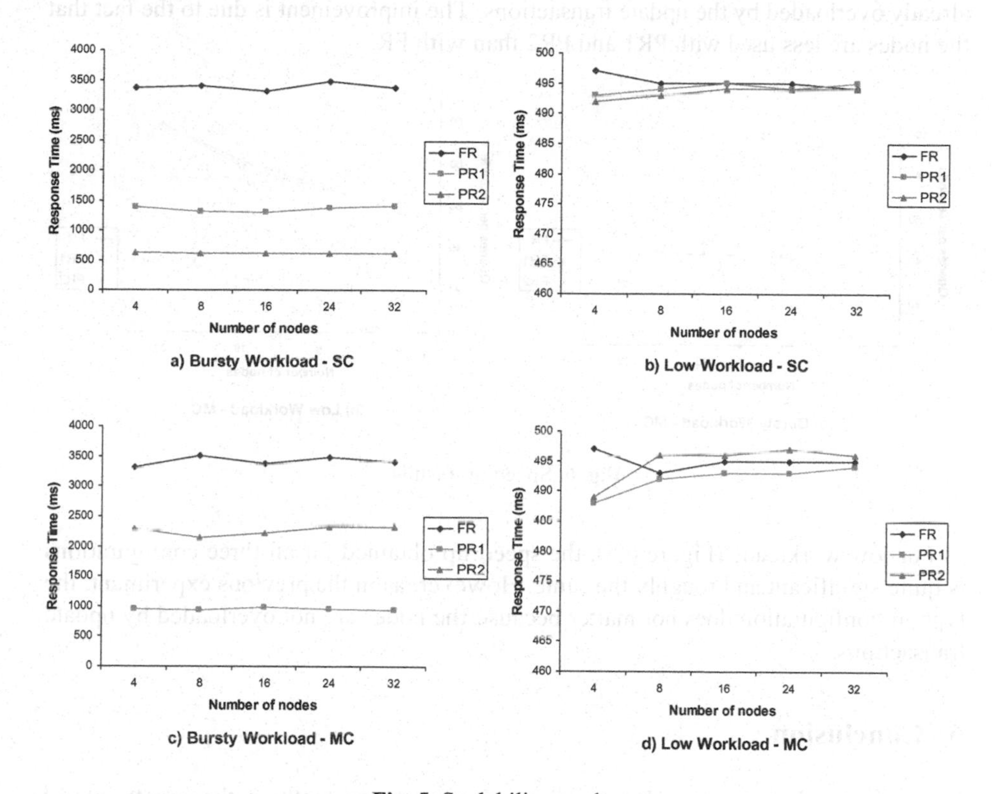

Fig. 5. Scalability results

5.5 Speed Up Experiment

The second experiment studies the performance improvement (speed up) for read queries when we increase the number of nodes. To accomplish this, we modified the previous experiment by introducing clients that submit queries. We consider MC read queries in bursty and low workloads (= 200ms and = 20s), varying the number of nodes for each configuration (FR, PR1 and PR2). Since each read query involves only one node, the performance difference between SC and MC queries is negligible. Thus, we only consider MC queries which are more general.

The number of read clients is 128. Each client is associated to one node and we produce an even distribution of clients at each node. Thus, the number of read clients per node is 128/number of nodes. The clients submit queries sequentially while the experiment is running. For each test, we measured the throughput of the cluster, i.e. the number of read queries per second. The duration of this experiment is the time of submitting 1000 read queries.

The experiment results (see Figure 6) clearly show that the increase of the number of nodes improves the cluster's throughput. For bursty workloads (Figure 6.a), the improvement obtained with partial replication is moderate because the nodes are already overloaded by the update transactions. The improvement is due to the fact that the nodes are less used with PR1 and PR2 than with FR.

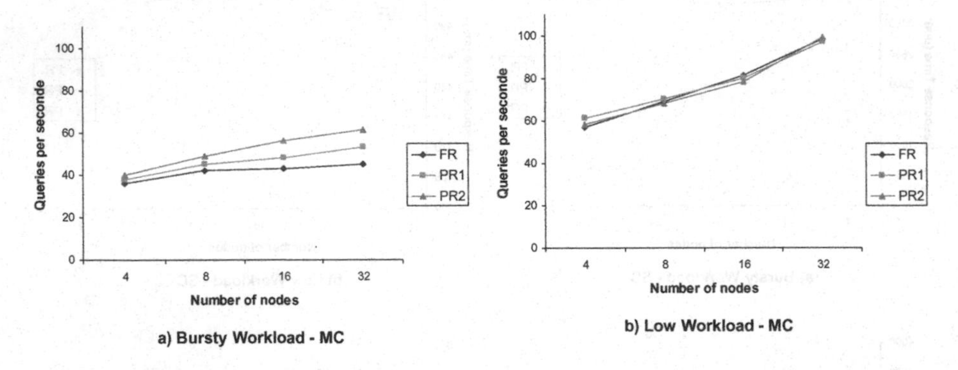

Fig. 6. Speed up results

For low workloads (Figure 6.b), the speed up obtained for all three configurations is quite significant and roughly the same. However, as in the previous experiment, the type of configuration does not matter because the nodes are not overloaded by update transactions.

6 Conclusion

Preventive replication provides strong copy consistency, without the constraints of synchronous replication. In this paper, we addressed the problem of scaling up preventive replication to large cluster configurations. The main contribution is a multi-master refreshment algorithm that prevents conflicts for partially replicated databases. The algorithm supports general multi-owner transactions composed of read and write operations. Cluster nodes can support autonomous databases that are considered as black boxes. Thus our solution can achieve full node autonomy.

We described the system architecture components necessary to implement the refresher algorithm. We validated our refresher algorithm over a cluster of 32 nodes running PostGRESQL. To stress transaction performance, we implemented directly our replicator module within PostGRESQL.

Our experimental results showed that our multi-master refreshment algorithm scales up very well. In contrast with synchronous algorithms, our algorithm has linear response time behaviour. We also showed the impact of the configuration on performance. For low workloads, full replication performs almost as well as partial replication. But for bursty workloads in partial replicated configurations an important amount of transactions are executed in parallel, thus the performance improvement of partial replication can be quite significant (by a factor up to 7).

The speed up experimental results also showed that the increase of the number of nodes can well improve the query throughput. However the performance gains strongly depend on the types of transactions (single versus multiple copy) and of the configuration. Thus an important conclusion is that the configuration and the placement of the copies should be tuned to selected types of transactions.

References

1. S. Gançarski, H. Naacke, E. Pacitti, P. Valduriez: Parallel Processing with Autonomous Databases in a Cluster System, *Int. Conf. of Cooperative Information Systems (CoopIS)*, 2002.
2. L. George, P. Minet: A FIFO Worst Analysis for a Hard Real Time Distributed Problem with Consistency Constraints, *Int. Conf. On Distributed Computing Systems (ICDCS)*, 1997.
3. V. Hadzilacos, S. Toueg: Fault-Tolerant Broadcasts and Related Problems. *Distributed Systems*, 2nd Edition, S. Mullender (ed.), Addison-Wesley 1993.
4. M.D. Hill et al. Cooperative Shared Memory: Software and Hardware for Scalable Multiprocessors. *ACM Trans. On Computer Systems*, 11(4), 1993.
5. T. Özsu, P. Valduriez: *Principles of Distributed Database Systems*. Prentice Hall, Englewood Cliffs, New Jersey, 2nd edition, 1999.
6. E. Pacitti, P. Valduriez: Replicated Databases: concepts, architectures and techniques, *Network and Information Systems Journal*, Hermès, 1(3), 1998.
7. E. Pacitti, P. Minet, E. Simon: "Replica Consistency in Lazy Master Replicated Databases". *Distributed and Parallel Databases, Kluwer Academic, 9(3), May 2001*.
8. E. Pacitti, T. Özsu, C. Coulon : "Preventive Multi-Master Replication in a Cluster of Autonomous Databases" , *Int. Conf. on Parallel and Distributed Computing (Euro-Par* 2003), Klagenfurt, Austria, 2003.
9. K. Tindell, J. Clark: Holistic Schedulability analysis for Distributed Hard Real-time Systems. *Micro-processors and Microprogramming*, 40, 1994.
10. P. Valduriez: Parallel Database Systems: open problems and new issues. *Int. Journal on Distributed and Parallel Databases*, 1(2), 1993.

Parallel Implementation of a Fuzzy Rule Based Classifier

Alexandre G. Evsukoff, Myrian C.A. Costa, and Nelson F.F. Ebecken

COPPE/Federal University of Rio de Janeiro,
P.O.Box 68506, 21945-970 Rio de Janeiro RJ, Brazil
Tel: (+55) 21 25627388, Fax: (+55) 21 25627392
evsukoff@coc.ufrj.br, myrian@nacad.ufrj.br
nelson@ntt.ufrj.br

Abstract. This works presents an implementation of a fuzzy rule based classifier where each single variable fuzzy rule based classifier (or a set of them) is assigned to a different processor in a parallel architecture. Partial conclusions are synchronized and processed by a master processor. This approach has been applied to a very large database and results are compared with a parallel neural network approach.

Keywords: fuzzy systems, data mining, parallel computing.

1 Introduction

The huge amount of data generated by Data Warehousing and transaction databases has pushed Data Mining algorithms through parallel implementations [1], [2]. One requirement of data mining is efficiency and scalability of mining algorithms. Therefore, parallelism can be used to process long running tasks in a timely manner.

Fuzzy reasoning techniques are a key for human-friendly computerized devices, allowing symbolic generalization of high amount of data by fuzzy sets and providing its linguistic interpretability [4].

Fuzzy rule based classifier [3] has shown reasonable results in serial implementation. The current approach to rule based classification is divided in two steps:

1. Derive a fuzzy rule based classifier to each variable.
2. Aggregate the partial conclusions of each classifier into a global conclusion.

This approach allows a simple parallel implementation, since each single variable fuzzy rule based classifier (or a set of them) is assigned to a different processor in a parallel architecture, and partial conclusions are synchronized and processed by a master processor. This works presents such an implementation describing the parallel processing scheme and how rule bases are computed in each processor. Results for the classification of a very large database are presented and compared with a parallel neural network approach.

The next section introduces the current approach. The third section describes the parallel implementation. The fourth section presents and discusses the results achieved by three classifiers built with this methodology. Finally, some concluding remarks are presented.

M. Daydé et al. (Eds.): VECPAR 2004, LNCS 3402, pp. 184–193, 2005.

2 Fuzzy Rule Based Classifier

Consider a pattern recognition problem where observations are described as a n-dimensional vector $\mathbf{x}$ in a variable space X^n and classes are represented by the set $\mathbf{C} = \{C_1 \ldots C_m\}$. The solution consists in assigning a class label $C_j \quad \mathbf{C}$ to an observation $\mathbf{x}(t) \quad X^n$. The problem of designing a fuzzy system for pattern recognition is to build a *classifier*, which should executes correctly the mapping $X^n \quad \mathbf{C}$.

2.1 Preliminaries

For many applications, the mathematical representation of natural language concepts through fuzzy sets is done by the definition of a *base variable* x, whose domain X is a subset of real numbers where concepts can be expressed as a fuzzy set, denoted as:

$$A = \{(x, \propto_A(x)), x \quad X\}. \tag{1}$$

where $\propto_A : X \quad [0,1]$ is the function that characterizes the *membership* of an element $x \quad X$ to the fuzzy set A.

Each input variable $x(t) \quad X$ is described using ordered linguistic terms in a *descriptor set* $\mathbf{A} = \{A_1, \ldots, A_k\}$. The *meaning* of each term $A_i \quad \mathbf{A}$ is given by a fuzzy set. The collection of fuzzy sets used to describe the input variable forms a *fuzzy partition* of the input variable domain.

Strong, normalized and triangular fuzzy partitions are completely determined by the location of the triangle vertices, which can be viewed as *prototypes* (best representatives) of the respective fuzzy set. Trapezoidal membership functions are used for the two fuzzy sets at each end of the domain, as shown in Fig. 1, to deal with off-limit points.

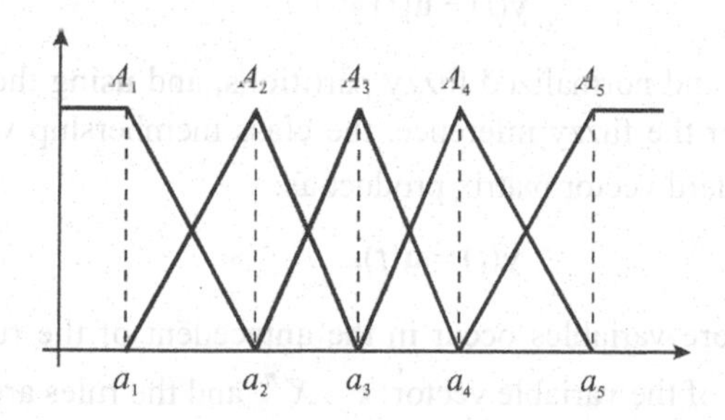

Fig. 1. A fuzzy partition

For a given input $x(t) \quad X$, the membership vector (or fuzzification vector) $\mathbf{u}(t)$ is computed by the fuzzy sets in the fuzzy partition of the input variable domain as:

$$\mathbf{u}(t) = \left(\propto_{A_1}(x(t)) \ldots \propto_{A_k}(x(t))\right). \tag{2}$$

In fuzzy systems models, fuzzy sets describing input variables are combined in fuzzy rules, as described next.

2.2 Fuzzy Rules

In pattern recognition applications, the rules' conclusions are the classes in the set $\mathbf{C} = \{C_1 \ldots C_m\}$. The fuzzy rule base has thus to relate input linguistic terms $A_i \quad \mathbf{A}$ to the classes $C_j \quad \mathbf{C}$, in rules such as:

$$if\ x(t)\ is\ A_i\ then\ class\ is\ C_j\ with\ cf = {}_{ij} \tag{3}$$

where ${}_{ij} \quad [0,1]$ is a confidence factor (*cf*) that represents the rule certainty.

The confidence factor weights for all rules define the symbolic fuzzy relation , defined on the Cartesian product $\mathbf{A} \cdot \mathbf{C}$. A value $\propto (A_i, C_j) = {}_{ij}$ represents how much the term A_i is related to the class C_j in the model described by the rule base. A value ${}_{ij} > 0$ means that the rule (i, j) occurs in the rule base with the confidence factor ${}_{ij}$. The rule base can be represented by the matrix $= \left[{}_{ij} \right]$, of which the lines $i = 1...n$ are related to the terms in the input variable descriptor set $\mathbf{A}$ and the columns $j = 1...m$ are related to classes in the set $\mathbf{C}$.

The output of the fuzzy system is the class membership vector (or the fuzzy model output vector) $\mathbf{y}(t) = \left(\propto_{C_1}(x(t)), \ldots, \propto_{C_m}(x(t)) \right)$, where $\propto_{C_j}(x(t))$ is the output membership value of the input $x(t)$ to the class C_j. The class membership vector $\mathbf{y}(t)$ can also be seen as a fuzzy set defined on the class set $\mathbf{C}$. It is computed from the input membership vector $\mathbf{u}(t)$ and the rule base weights in using the fuzzy relational composition:

$$\mathbf{y}(t) = \mathbf{u}(t) \circ \tag{4}$$

Adopting strong and normalized fuzzy partitions, and using the sum-product composition operator for the fuzzy inference, the class membership vector $\mathbf{y}(t)$ is easily computed as a standard vector matrix product as:

$$\mathbf{y}(t) = \mathbf{u}(t). \quad . \tag{5}$$

When two or more variables occur in the antecedent of the rules, input variables are the components of the variable vector $\mathbf{x} \quad X^n$ and the rules are written as:

$$if\ \mathbf{x}(t)\ is\ B_i\ then\ class\ is\ C_j\ with\ cf = {}_{ij} \tag{6}$$

Each term B_i in the multi-variable descriptor set $\mathbf{B} = \{B_1, \ldots, B_M\}$ represents a combination of terms in the input variables' descriptor sets. It is a symbolic term used for computation purposes only and does not necessarily need to have a linguistic interpretation. All combination must be considered in such a way that the model is complete, *i.e.* it produces an output for whatever input values. For instance consider a $n = 2$ variables problem, where the first variable is described by 3 fuzzy sets and the second by 4 fuzzy sets. The resulting multi-variable descriptor set have $M = 12$ elements as shown in Fig. 2.

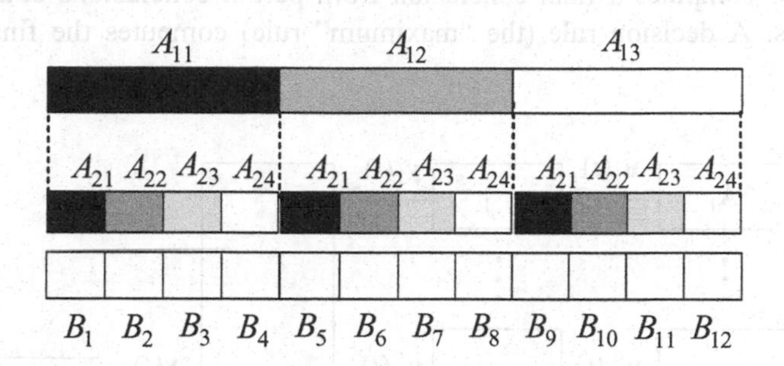

Fig. 2. Combination of description terms

The multi-variable fuzzy model described by a set of rules (6) is analogous to the single variable fuzzy model described by rules of type (3) and the model output $\mathbf{y}(t) = \left(\propto_{C_1}(\mathbf{x}(t)), \ldots, \propto_{C_m}(\mathbf{x}(t))\right)$ is computed as:

$$\mathbf{y}(t) = \mathbf{w}(t). \tag{7}$$

The combination of terms in multiple antecedent rules of type (6) implies in an exponential growth of the number of rules in the rule base. For problems with many variables, reasonable results [3] can be obtained considering the aggregation of partial conclusions computed by single antecedent rules of type (3).

The final conclusion is the membership vector $\mathbf{y}(t) = \left(\propto_{C_1}(\mathbf{x}(t)), \ldots, \propto_{C_m}(\mathbf{x}(t))\right)$, which is computed by the aggregation of partial conclusions $\mathbf{y}_i(t)$ as in multi-criteria decision-making [5]. Each component $\propto_{C_j}(\mathbf{x}(t))$ is computed as:

$$\propto_{C_j}(\mathbf{x}(t)) = \mathrm{H}\left(\propto_{C_j}(x_1(t)), \ldots, \propto_{C_j}(x_n(t))\right) \tag{8}$$

where $\mathrm{H} : [0,1]^N \quad [0,1]$ is an aggregation operator.

The best aggregation operator must be chosen according to the semantics of the application. Generally, a conjunctive operator, such as the "minimum" or the "product", gives good results to express that all partial conclusions must agree. A weighted operator like OWA may be used to express some compromise between partial conclusions.

The final decision is computed by a decision rule. The most usual decision rule is the "maximum rule", where the class is chosen as the one with greatest membership value. The "maximum rule" is often used to determine the most suitable solution. Nevertheless, other decision rules can be used including risk analysis and reject or ambiguity distances.

When all variables are used in single variable rules as (3) and their outputs are aggregated as (8), the fuzzy classifier behaves as a standard Bayesian classifier. The current approach is flexible enough so that some partial conclusions can be computed from the combination of two or three variables in multi-variable rules (6). An aggregation operator computes a final conclusion from partial conclusions obtained from all sub-models. A decision rule (the "maximum" rule) computes the final class as shown in Fig. 3.

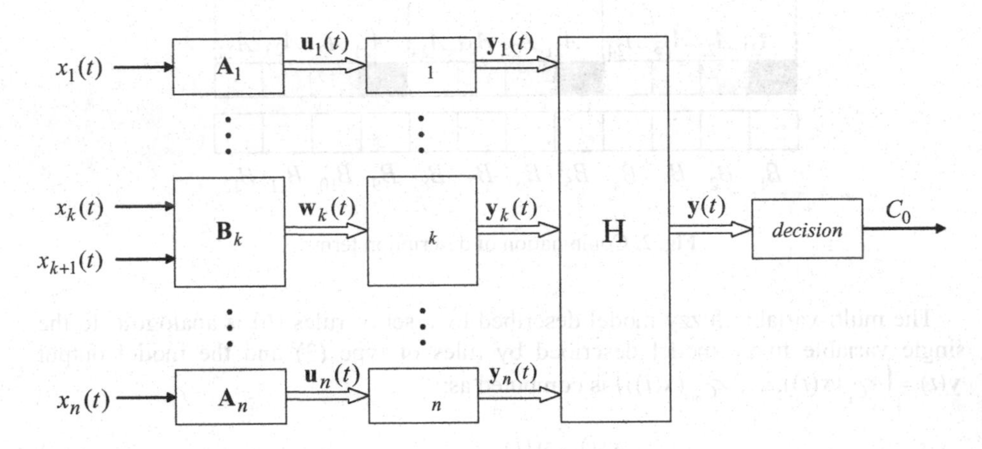

Fig. 3. The fuzzy rule based classifier

The rule base is the core of the model described by the fuzzy system and its determination from data is described next.

2.3 Rule Base Identification

The determination of the rule base is often called the "learning" phase. Consider a labeled data set T, where each sample is a pair $\left(\mathbf{x}(t), \mathbf{v}(t)\right), t = 1..N$, of which $\mathbf{x}(t)$ is the vector containing the variable values and $\mathbf{v}(t) = \left(v_1(t) \ldots v_m(t)\right)$ is the vector containing the assigned membership values of $\mathbf{x}(t)$ to each class.

For each fuzzy sub-model k, the rule base weights matrix each component $_{ij}^{k}$ of the rule base $_k$ is computed as:

$$_{ij}^{k} = \frac{\sum_{t=1..N} z_{ik}(t) v_j(t)}{\sum_{t=1..N} z_{ik}(t)} \tag{9}$$

where $\mathbf{z}_k(t) = \mathbf{u}_k$ if the sub-model k is described by single antecedent rules as (3) or $\mathbf{z}_k(t) = \mathbf{w}_k(t)$ if the sub-model is described by multiple antecedent rules as (6).

When the rule base weights are computed as (9), the single variable fuzzy classifier (of which the output is computed as (5) and (8)) behaves like a Bayesian naïve classifier, since class membership distribution is computed over the universe of each variable. The parallel implementation of the fuzzy rule base classifier is presented next.

3 Parallel Implementation

The parallel implementation of the algorithm is done according to the data fragmentation. There are at least two approaches for data fragmentation: vertical and horizontal. In the vertical fragmentation, the data is divided by columns, in such a way that all samples of one (or a sub-set of) variable(s) is (are) allocated to each processors. In the horizontal fragmentation, the data is divided by lines and a set of samples of all variables is allocated to each processor.

The horizontal fragmentation approach is suitable when multiple antecedent rules are considered or in the case of the neural network model. In this case, each processor computes a rule base for the subset of the training data set assigned to it. The final rule base is computed as the mean of the rule base computed by each processor.

Since is very difficult to stay whether the variables should be combined or not, in this work only the single variable fuzzy classifier was used with the vertical strategy. In the vertical fragmentation (Fig. 4), a master processor selects inputs and distributes the inputs thorough the available nodes. In the learning phase, each node computes a fuzzy rule base weight matrix for each sub-model from the input variables that was assigned to that node.

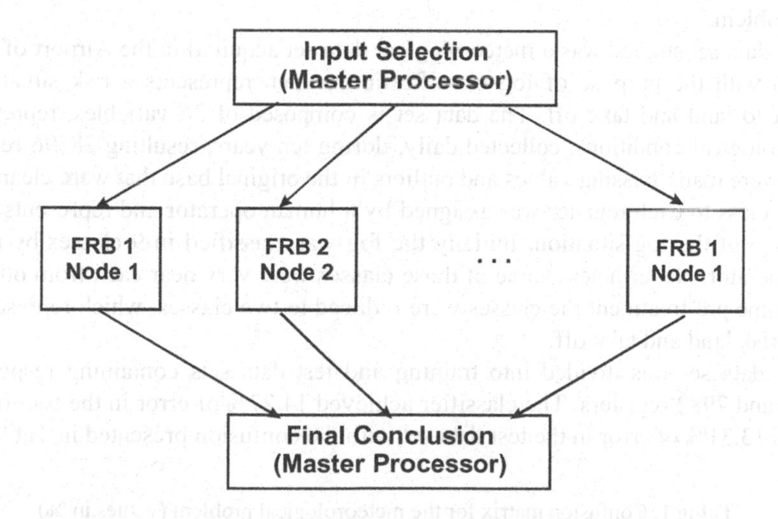

Fig. 4. The parallel implementation workflow

In the processing or testing phase, the rule base weight matrix for each sub model is used to compute a partial conclusion on class descriptions. All partial conclusions are synchronized in the master processor and final conclusions are computed by aggregation of partial conclusions of each sub-model (see Fig. 4).

4 Results and Discussion

The machine used for execution and analysis performance was the Cluster Mercury of the High Performance Computing Center (NACAD) of COPPE/UFRJ. The cluster has 16 dual Pentium III, 1 GHz, processor nodes interconnected via Fast Ethernet and Gigabit networks and 8GB of total memory. The execution is controlled by PBS (Portable Batch System) job scheduler avoiding nodes sharing during execution. The cluster runs Linux Red Hat on processing and administrative nodes powered by Intel compilers. The application was developed using MPICH as message passing environment.

The results were analyzed from different perspectives, as presented in next subsections. First of all, the classification performance was evaluated against one real world problem on serial version of the algorithm. Then, the size scalability was analyzed with synthetic data sets and the experiments were performed for the serial and the parallel versions of the algorithm, with similar results. Finally, the speed-up analysis that was also done using the synthetic data sets.

4.1 Classifier Performance Evaluation

The single variable fuzzy classifier has had reasonable results in serial implementation with benchmark problems, when compared to decision tree and neural network classifiers [3]. In this work the performance of the classifier was tested against a large real problem.

The data set studied was a meteorological data set acquired in the Airport of Rio de Janeiro with the purpose of fog classification, which represents a risk situation for aircraft to land and take off. The data set is composed of 36 variables, representing meteorological conditions, collected daily, during ten years, resulting 28596 registers. There were many missing values and outliers in the original base that were cleaned up.

The class to each register was assigned by a human operator and represents a classification of the fog situation. Initially the fog was classified in 9 classes by the human operator. Nevertheless some of those classes were very near each from other and after some pre-treatment the classes were reduced to two classes, which represent risk of not risk land and take off.

The data set was divided into training and test data sets containing respectively 18500 and 7982 registers. The classifier achieved 14.27% of error in the training data set and 13.31% of error in the test data set with the confusion presented in Table I.

Table 1. Confusion matrix for the meteorological problem (values in %)

	C_1	C_2
C_1	51.40	6,31
C_2	7.00	35.27

As it can be seen from Table I, the classifier performance is quite good for such a large problem. The neural network classifier had been obtained 13.32 % of error in the test data set, which have a much more complex learning algorithm [6].

4.2 Size Scalability Analysis

In order to evaluate the scalability of the algorithm for larger problems, a set of synthetic problems were generated. Each problem's data set was created using independent normally distributed random variables with different means for each class.

The number of classes $m = 2$ for all synthetic problems, distributed with the same *a priori* probability. The number of variables was varied as $\{16, 32, 64, 128, 256\}$ and the number of registers of the data set was varied as $\{10, 20, 50, 100\}$ thousands of registers. All combinations of number of variables and number of registers were considered, resulting into 20 synthetic data sets.

The processing time for each one of the synthetic data set, running in a single processor machine, is shown in Table II. In all experiments were used 5 membership functions for the fuzzy partition of each input variable.

Table 2. Size scalability analysis

	10	**20**	**50**	**100**
16	1.515	2.751	5.93	11.477
32	2.663	4.699	11.15	22.225
64	4.677	8.918	21.67	44.257
128	9.041	17.409	43.276	91.852
256	17.455	34.644	85.544	196.971

The processing time increases almost linearly with the number of variables and the number of registers. The time of each experiment is shown in Table I. For instance, the time for the 256 variables data set with 50 thousands registers was 85.544 seconds, while the same number of variables and 100 thousand registers was 196.971 seconds. The execution time for the 128 variables data set and 50 thousand registers was 43.276 seconds.

A closer look at the processing time reveals that most of the computing time is spend with I/O to open and read the training and test data. The learning algorithm itself is very efficient. This issue is stressed in the next section.

4.3 Speed Up Analysis

Considering the size scalability described above, the speed-up study was carried out with the 100 thousands registers datasets.

Several experiments were performed in order to evaluate the speed-up analysis. Only the computing part of the program were be used such this analysis, since most of the overall time is spent with I/O. Despite of the communicating overhead due to data distribution the performance of the algorithm is primarily influenced by the processing of base rule. Speed-up is calculated against time of serial execution of the algorithm. The results of the speed up analysis are shown in Fig. 5.

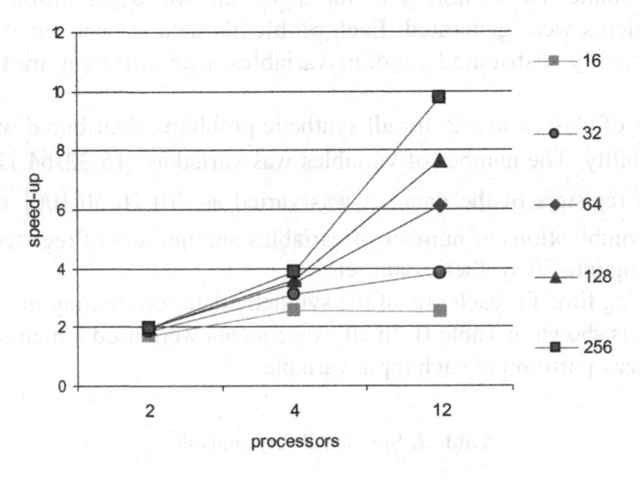

Fig. 5. Fuzzy classifier speed-up analysis

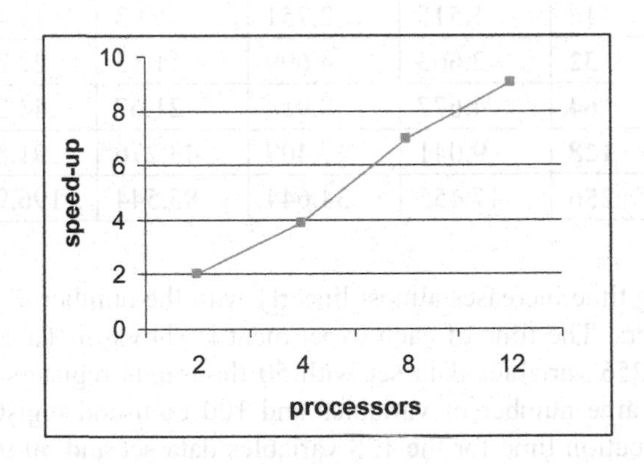

Fig. 6. Neural network classifier speed-up analysis

As it can be seen in Fig. 5, as the number of variables increases, more efficient is the effect of the parallel implementation. For 16 variables the efficiency speed-up does not justify the parallel implementation. However for 256 variables the speed-up efficiency increases almost linearly with the number of processors.

The same analysis was performed for the neural network classifier with similar speed-up results for the same number of processors, as can be seen in Fig. 6.

5 Conclusions

This work has presented a parallel implementation of fuzzy rule based classifier. The main advantage of this approach is its flexibility to combine single or multiple-variable fuzzy rules and to derive fuzzy rules from numerical or nominal variables.

Previous results have shown reasonable classification rates when applied in benchmark problems. The parallel implementation allows a faster execution of the classification, allowing its application to very large databases.

In the experiments performed, the communication and I/O time have been much greater than the processing time in the learning phase. This is due to option to consider only single variables rubes. A better classification performance could be expected if multi-variables rules would be considered. Nevertheless, the choice of which variables should be combined is a very difficult optimization problem, as should be very time consuming task.

The choice of variables to combine in multi-variable rule bases as so as the parameters of the fuzzy partitions for each variable are important design issues that can improve classification rates and have not been considered in this work. It is considered for further development of the symbolic approach developed in this work.

Acknowledgements

This work has been supported by the Brazilian Research Council (CNPq) under the grant 552052/2002-7. The authors are grateful to High Performance Computing Center (NACAD-COPPE/UFRJ) where the experiments were performed.

References

1. Sousa, M. S. R.; M. Mattoso and N. F.F. Ebecken (1999). Mining a large database with a parallel database server. *Intelligent Data Analysis* 3, pp. 437-451.
2. Coppola, M. and M. Vanneschi, (2002). High-performance data mining with skeleton-based structured parallel programming. Parallel Computing 28, pp. 783-813.
3. Evsukoff, A.; A. C. S. Branco and S. Gentil (1997). A knowledge aquisition method for fuzzy expert systems in diagnosis problems. *Proc. 6th IEEE International Conference on Fuzzy Systems – FUZZIEEE'97*, Barcelona.
4. Zadeh, L. (1996). Fuzzy logic = computing with words. *IEEE Trans. on Fuzzy Systems*, 4 (2), pp. 103-111.
5. Zimmermann, H.-J. (1996). *Fuzzy Set Theory and its Applications*, Kluwer.
6. Costa, M. C. A. and N. F. F. Ebecken (2000).Data Mining High Performance Computing Using Neural Networks. *Proc. of High Peformance Computing 2000, HPC 2000. Maui.* Wessex Institute of Technology.

The EGEE European Grid Infrastructure Project

Fabrizio Gagliardi

CERN, EGEE Project Direction, IT-DI, 1211 Geneva 23
Fabrizio.Gagliardi@cern.ch

Abstract. The state of computer and networking technology today makes the seamless sharing of computing resources on an international or even global scale conceivable. Extensive computing Grids that integrate large, geographically distributed computer clusters and data storage facilities have changed from representing a dream to becoming a vision and, with the Enabling Grids for E-science in Europe project (EGEE), a reality today. EGEE aims to provide a European Grid infrastructure for the support of many application domains. This infrastructure is built on the EU Research Network GEANT and exploits Grid expertise that has been generated by previous projects. EGEE is a EU funded project that involves 71 partners from Europe, Russia and the United States. The project started in April 2004 for a first phase of 2 years.

1 Introduction

The vision of Grid computing is that of an infrastructure which will integrate large, geographically distributed computer clusters and data storage facilities, and provide simple, reliable and round-the-clock access to these resources.

In recent years, a number of projects have demonstrated first results for various aspects of Grid computing. Europe has achieved a prominent position in this field, in particular thanks to the success of the European DataGrid (EDG) project, which managed to establish a functional Grid testbed comprising more than 20 centres in Europe. Other individual countries, such as the UK, France, and Italy, have developed comprehensive "e-Science" programs that rely on emerging national computing Grids. However, as yet, there were no real production-quality Grids that can offer continuous and reliable Grid services to a range of scientific communities.

The new European project EGEE (Enabling Grids for E-Science in Europe), launched in April 2004, aims to transform the vision into reality. Its goal is to integrate current national, regional and thematic Grid efforts, and create over the next two years a seamless Grid infrastructure for the support of scientific research.

1.1 The EGEE Vision

The vision of EGEE is to provide researchers in academia and industry with a common market of computing resources, enabling round-the-clock access to major computing resources independent of geographic location,. The resulting infrastructure will surpass the capabilities of local clusters and individual supercomputing centres in

M. Daydé et al. (Eds.): VECPAR 2004, LNCS 3402, pp. 194–203, 2005.

many respects, providing a unique tool for collaborative compute-intensive science ("e-Science") in the many research area. It supports distributed research communities, including relevant Networks of Excellence, which share common computing needs and are prepared to integrate their own distributed computing infrastructures and agree to common access policies.

The EGEE infrastructure is built on the EU research network GEANT and exploits Grid expertise that has been generated by projects such as the EU DataGrid project, other EU supported Grid projects and the national Grid initiatives such as UK e-Science, INFN Grid, Nordugrid and the US Trillium (cluster of projects). EGEE will preserve the current strong momentum of the international Grid community and the enthusiasm of hundreds of young researchers.

Mostly funded by European Union funding agencies, this project has a world-wide mission and receives important contributions from the US, Russia and other non EU partners. It will provide interoperability with other Grids around the globe, including the US NSF Cyberinfrastructure, and will substantially contribute to the efforts in establishing a "worldwide" Grid infrastructure.

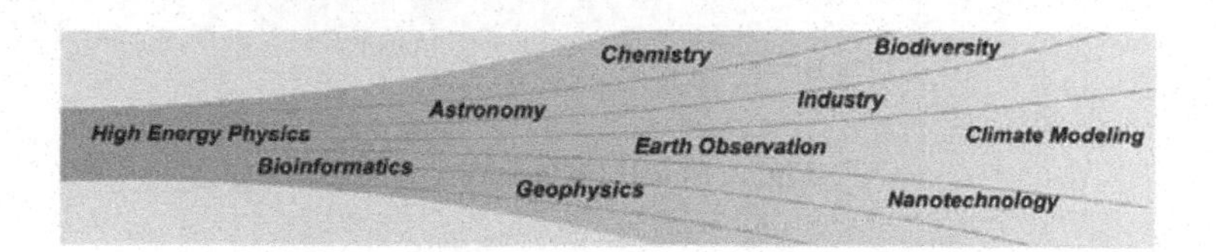

Fig. 1. Applications. The above figure shows a schema of the evolution of the European Grid infrastructure from two pilot applications in high energy physics and biomedical Grids, developing into an infrastructure serving multiple scientific and technological communities. The EGEE project covers Year 1 and 2 of a planned four year programme

EGEE was proposed by experts in Grid technologies representing the leading Grid activities in Europe. The international nature of the infrastructure has naturally lead to a structuring of the Grid community into twelve partner regions or "federations", which integrate regional Grid efforts. The federations represent 70 partner institutions, and cover both a wide range of scientific applications as well as industrial applications.

The project is conceived as a two-year project part of a four-year programme. Major implementation milestones after two years will provide the basis for assessing subsequent objectives and funding needs. In addition to demonstrating the added value of Grid technology quantitatively, the project aims to achieve qualitative improvement in terms of new functionality not previously available to the participating scientific communities.

Given the service oriented nature of this project, two pilot application areas have been selected to guide the implementation and certify the performance and functionality of the evolving European Grid infrastructure.

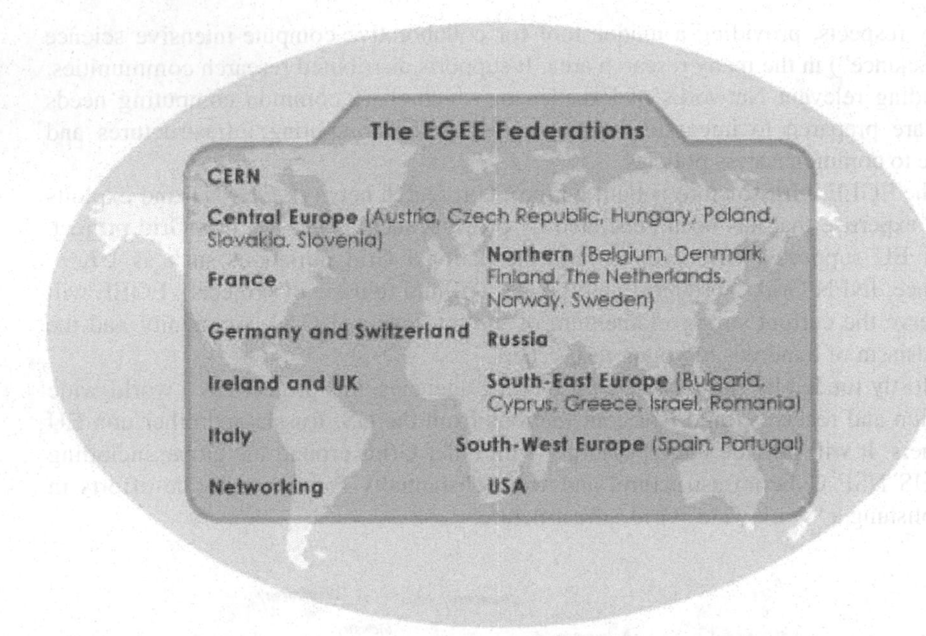

Fig. 2. EGEE Federations

One of these pilot initiatives is the Large Hadron Collider Computing Grid (LCG: www.cern.ch/lcg). It relies on a Grid infrastructure in order to store and analyse petabytes of real and simulated data from high-energy physics experiments at CERN in Switzerland. EGEE launched on April 1st already deploying production Grid service based on the infrastructure of the LCG project running second generation middleware. Based on the experiences made with the first generation of the middleware that was used before start of EGEE, the projects experts work on developing a "next generation" Grid facility. The new middleware has as main goal the delivery of solutions to the problems encountered in its first generation, some of which were configuration problems of several sites, and the vulnerability of the system to problem sites. A new core set of sites that would guarantee stable service and commit significant computing and storage resources is foreseen for EGEE. Only when the data challenges are started and stabilised can new sites be brought in one-by-one as each is verified to be correctly configured and stable.

The second pilot application is Biomedical Grids, where several communities are facing daunting challenges such as the data mining of genomic databases, and the indexing of medical databases in hospitals, which amount to several terabytes of data per hospital per year.

In addition to this endeavour, this application has added importance to a special concern about the project's technical aspects. Biomedical applications have significant security requirements such as confidentiality that need to be addressed early on and that represent a major task in the applications and security groups of EGEE.

1.2 First Steps

Given the rapidly growing scientific need for a Grid infrastructure, it was deemed essential for the EGEE project to "hit the ground running", by deploying basic services, and initiating joint research and networking activities before the project even started officially. To ensure that the project would ramp up quickly, project partners had agreed to begin providing their unfunded contribution prior to the official start of the project on April 1st this year. The LCG project provided basic resources and infrastructure already during 2003, and Biomedical Grid applications are in planning at this stage. The available resources and user groups are expected to rapidly expand throughout the course of the project.

EGEE builds on the integration of existing infrastructures in the participating countries, in the form of national GRID initiatives, computer centres supporting one specific application area, or general computer centres supporting all fields of science in a region.

1.3 The EGEE Mission

In order to achieve the vision outlined above, EGEE has an ambitious three-fold mission:

Efficient delivery of production level Grid services - Their essential elements are manageability, robustness, resilience to failure. Also, a consistent security model will be provided, as well as the scalability needed to rapidly absorb new resources as these become available, while ensuring the long-term viability of the infrastructure.

Professional Grid middleware re-engineering in support of the production services - This will support and continuously upgrade a suite of software tools capable of providing production level Grid services to a base of users which is anticipated to grow and diversify.

Clear outreach and training efforts - These can proactively market Grid services to new research communities in academia and industry, capture new e-Science requirements for the middleware and service activities, and provide the necessary education to enable new users to benefit from the Grid infrastructure.

Reflecting this mission, EGEE is structured in three main areas of activity: services, middleware re-engineering and networking. It is essential to the success of EGEE that the three areas of activity should form a tightly integrated "Virtuous Cycle", illustrated in the figure. In this way, the project as a whole can ensure rapid yet well-managed growth of the computing resources available to the Grid infrastructure as well as the number of scientific communities that use it. As a rule, new communities contribute new resources to the Grid infrastructure. This feedback loop is supplemented by an underlying cyclical review process. It covers overall strategy, middleware architecture, quality assurance and security status, and ensures a careful filtering of requirements, representing a coordinated prioritization of efforts and maintenance of production-quality standards.

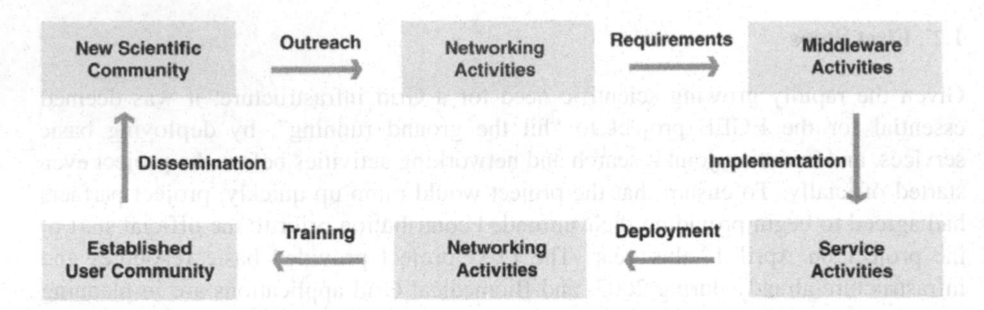

Fig. 3. EGEE Virtuous Cycle. The above figure shows the "Virtuous Cycle" for EGEE development. A new scientific community establishes first contact with the project through outreach events organized by Networking Activities. Follow-up meetings of applications specialists may lead to definition of new requirements for the infrastructure. If approved, the requirements are implemented by the Middleware Activities. After integration and testing, the new middleware is deployed by the Service Activities. The Networking Activities then provide appropriate training to the community in question, so that it becomes an established user

2 The Stakeholder's Perspective

In order to convey the scope and ambition of the project, this section presents the expected benefits for EGEE stakeholders and an outline of the procedure for the stakeholder to participate in EGEE. The key types of EGEE stakeholders are users, resource providers, and industrial partners.

2.1 EGEE Users

Once the EGEE infrastructure is fully operational, users will perceive it as one unified large scale computational resource. The complexity of the service organisation and the underlying computational fabric will remain invisible to the user. The benefits of EGEE for the user include:

Simplified access – Today most users have accounts on numerous computer systems at several computer centres. The resource allocation procedures vary between the centres and are in most cases based on complicated applications submitted to each centre or application area management. The overhead involved for a user in managing the different accounts and application procedures is significant. EGEE will reduce this overhead by providing means for users to join virtual organisations with access to a Grid containing all the operational resources needed.

On demand computing – By allocating resources efficiently, the Grid promises greatly reduced waiting times for access to resources.

Pervasive access – The infrastructure will be accessible from any geographic location with good network connectivity, thus providing even regions with limited computer resources access on an as-need basis to large resources.

Large scale resources – Through coordination of resources and user groups, EGEE will provide application areas with access to resources of a scale that no single computer centre can provide. This will enable researchers to address previously intractable problems in strategic application areas.

Sharing of software and data – With its unified computational fabric, EGEE will allow wide spread user communities to share software and databases in a transparent way. EGEE will act as the enabling tool for scientific collaborations, building and supporting new virtual application organisations.

Improved support – Counting with the expertise of all the partners, EGEE will be able to provide a support infrastructure that includes in depth support for all key applications and around the clock technical systems support for GRID services.

A potential user community will typically come into contact with EGEE through one of the many outreach events supported by the Dissemination and Outreach activity, and will be able to express their specific user requirements via the Applications Identification and Support activity. An EGEE Generic Applications Advisory Panel (EGAAP) will use criteria such as scientific interest, Grid added value and Grid awareness, and make its recommendations to the Project Executive Board. After negotiating access terms, which will depend, amongst other things, on the resources the community can contribute to the Grid infrastructure, users in the community will receive training from the User Training and Induction activity. From the user perspective, the success of the EGEE infrastructure will be measured in the scientific output that is generated by the user communities it is supporting.

2.2 Resource Providers

EGEE resources will include national GRID initiatives, computer centres supporting one specific application area, or general computer centres supporting all fields of science in a region. The motivation for providing resources to the EGEE infrastructure will reflect the funding situation for each resource provider. EGEE will develop policies that are tailored to the needs of different kinds of partners. The most important benefits for resource providers are:

Large scale operations – Through EGEE, a coordinated large scale operational system is created. This leads to significant cost savings and at the same time an improved level of service provided at each participating resource partner. With this new infrastructure, the critical mass needed for many support actions can be reached by all participating partners.

Specialist competence – By distributing service tasks among the partners, EGEE will make use of the leading specialists in Europe to build and support the infrastructure. The aggregate level of competence obtained is a guarantee for the success of the EGEE project. In this sense, the Grid connects distributed competence just as much it as connects distributed computational resources. Each participating centre and its users will thus have access to experts in a wide variety of application and support fields.

User contacts – The EGEE distributed support model will allow for regional adaptation of new users and bring them in close contact with other regional user communities. The existence of regional support is of fundamental importance when introducing new users and user communities with limited previous experience of computational techniques.

Collaborations among resource partners – It is foreseen that several partners within the EGEE framework will form close collaborations and launch new development and support actions not yet included in the predictions for the project. This will lead to cost sharing of R&D efforts among partners and in the longer run allow for specialization and profiling of participating partners to form globally leading centres of excellence within EGEE. These benefits motivate the many partners that support the project, representing aggregate resources of over 3.000 CPUs provided by 10 sites already at month 1. One of the many milestones throughout the project is the expansion of the number of CPUs that form the infrastructure to 10.000 CPUs provided by 50 sites (see Table 1) by the end of Year 2.

Table 1. EGEE CPU Deployment

Region	CPU nodes	Disk (TB)	CPU Nodes Month 15	Disk (TB) Month 15
CERN	900	140	1800	310
UK + Ireland	100	25	2200	300
France	400	15	895	50
Italy	553	60.6	679	67.2
North	200	20	2000	50
South West	250	10	250	10
Germany + Switzerland	100	2	400	67
South East	146	7	322	14
Central Europe	385	15	730	32
Russia	50	7	152	36
Totals	3084	302	8768	936

2.3 Industrial Partners

The driving forces for EGEE are scientific applications, and the bulk of the current partners represent different publicly funded research institutions and computer resource providers from across Europe. Nevertheless, it is foreseen that industry will benefit from EGEE in a great variety of ways:

Industry as partner – Through collaboration with individual EGEE partners, industry participates in specific activities where relevant skills and manpower are required and available, thereby increasing know-how on Grid technologies.

Industry as user – As part of the networking activities, specific industrial sectors will be targeted as potential users of the installed Grid infrastructure, for R&D applications. The pervasive nature of the Grid infrastructure should be particularly attractive to high-tech SMEs, because it brings major computing resources – once only accessible to large corporations – within grasp.

Industry as provider – Building a production quality Grid requires industry involvement for long-term maintenance of established Grid services, such as call centres, support centres and computing resource provider centres. The EGEE vision also has inspiring long-term implications for the IT industry. By pioneering the sort of comprehensive production Grid services which are envisioned by experts – but which at present are beyond the scope of national Grid initiatives – EGEE will have to develop solutions to issues such as scalability and security that go substantially beyond current Grid R&D projects. This process will lead to the spin off of new and innovative IT technology, which will offer benefits to industry, commerce and society that go well beyond scientific computing, in much the same way that the World Wide Web, initially conceived for science, has had a much broader impact on society. Major initiatives launched by several IT industry leaders in the area of Grids and utility computing already emphasize the economic potential of this emerging field.

Industry will typically come in contact with EGEE via the Industry Forum, as well as more general dissemination events run by the Dissemination and Outreach activity. The Industry Forum of EGEE liaises between the project partners and industry in order to ensure businesses get what they need in terms of standards and functionality and that the project benefits from the practical experience of businesses. Interested companies will be able to consult about potential participation in the project with the Project Director and with regional representatives on the EGEE Project Management Board. As the scope of Grid services expands during the second two years of the programme, it is envisaged that established core services will be taken over by industrial providers with proven capacity. The service would be provided on commercial terms, and selected by a competitive tender.

2.4 Service Activities

The Service Activities creates, operates, supports and manages a production quality Grid infrastructure which makes resources at many resource centres across Europe accessible to user communities and virtual organisations in a consistent way according to agreed access management policies and service level agreements, while maintaining an overall secure environment. These activities build on current national and regional initiatives such as the UK e- Science Grid, the Italian Grid, and NorduGrid, as well as infrastructures being established by specific user communities, such as LCG.

The structure of the Grid services comprises: EGEE Operations Management at CERN; EGEE Core Infrastructure Centres in the UK, France, Italy and at CERN, responsible for managing the overall Grid infrastructure; Regional Operations Centres, responsible for coordinating regional resources, regional deployment and support of services. The basic services that are offered are: middleware deployment and installation; a software and documentation repository; Grid monitoring and problem tracking; bug reporting and knowledge database; Virtual Organization (VO) Services; and Grid Management Services. Continuous, stable Grid operation represents the most ambitious objective of EGEE, and requires the largest effort.

2.5 Middleware Re-engineering Activities

The current state-of-the-art in Grid Computing is dominated by research Grid projects that aim to deliver test Grid infrastructures providing proofs of concept and opening opportunities for new ideas, developments and further research. Only recently has there been an effort to agree on a unified Open Grid Services Architecture (OGSA) and an initial set of specifications that set some of the standards in defining and accessing Grid services. Building a Grid infrastructure based on robust components is thus becoming feasible. However, this will still take a considerable integration effort in terms of making the existing components adhere to the new standards, adapting them to further evolution in these standards, and deploying them in a production Grid environment.

The middleware activities in EGEE focus primarily on re-engineering already existing middleware functionality, leveraging the considerable experience of the partners with the current generation of middleware. The project's experts work on developing a "next generation" Grid facility. The new middleware has as main goal the delivery of solutions to the problems encountered in its first generation, some of which were configuration problems of several sites, and the vulnerability of the system to problem sites. A new core set of sites that would guarantee stable service and commit significant computing and storage resources is foreseen for EGEE. Only when the data challenges are started and stabilised can new sites be brought in one-by-one as each is verified to be correctly configured and stable.

In addition to this endeavour, biomedical application has added importance to a special concern about the project's technical aspects. Biomedical applications have significant security requirements such as confidentiality that need to be addressed early on and that represent a major task in the applications and security groups of EGEE.

As experience has shown, geographic co-location of development staff is essential, and therefore these activities are based on tightly-knit teams concentrated in a few major centres with proven track records and expertise.

2.6 Networking Activities

The networking activities in EGEE strive to facilitate the induction of new users, new scientific communities and new virtual organisations into the EGEE community. The project develops and disseminates appropriate information to these groups proactively. It also identifies and takes into account their emerging Grid infrastructure needs. The goal is to ensure that all users of the EGEE infrastructure are well supported and to provide input to the requirements and planning activities of the project.

Specific activities of EGEE are: Dissemination and Outreach; User Training and Induction; Application Identification and Support; Policy and International Cooperation. The Application Identification and Support Activity has three components, two Pilot Application Interfaces – for high energy physics and biomedical Grids – and one more generic component dealing with the longer term recruitment of other communities.

3 Conclusions

EGEE is a two-year project conceived as part of a four-year programme and officially started on April 1st 2004. It deployed basic services, initiated middleware and dissemination activities already before the formal start of the project. The available resources and user groups will rapidly expand during the course of the project.

Several measures of quality of service will be used to assess the impact of this Grid infrastructure. In addition to demonstrating the added value of Grid technology quantitatively, the project aims to achieve qualitative improvement in terms of new functionality not previously available to the participating scientific communities.

A second two-year project is hopefully to follow on from EGEE, in which industry might progressively take up the operations and maintenance of a stable Grid infrastructure from the academic community.

Grid Technology for Biomedical Applications

Vincent Breton[1], Christophe Blanchet[2], Yannick Legré[1],
Lydia Maigne[1], and Johan Montagnat[3]

[1] LPC, CNRS-IN2P3 / Université Blaise Pascal,
Campus des Cézeaux, 63177 Aubière Cedex, France
{Breton, Legré, Maigne}@clermont.in2p3.fr
[2] IBCP, CNRS, 7, passage du Vercors, 69367 Lyon CEDEX 07, France
Christophe.Blanchet@ibcp.fr
[3] CREATIS, CNRS UMR5515- INSERM U630, INSA,
20 Ave. A. Einstein, Villeurbanne, France
johan@i3s.unice.fr

Abstract. The deployment of biomedical applications in a grid environment has started about three years ago in several European projects and national initiatives. These applications have demonstrated that the grid paradigm was relevant to the needs of the biomedical community. They have also highlighted that this community had very specific requirements on middleware and needed further structuring in large collaborations in order to participate to the deployment of grid infrastructures in the coming years. In this paper, we propose several areas where grid technology can today improve research and healthcare. A crucial issue is to maximize the cross fertilization among projects in the perspective of an environment where data of medical interest can be stored and made easily available to the different actors of healthcare, the physicians, the healthcare centres and administrations, and of course the citizens.

1 Introduction

Last summer, about 10000 elderly people died in one European country because of unusually long and severe hot weather. For two weeks, the overall increase of mortality rate in hospitals and healthcare centres remained unnoticed. To better handle this kind of situation, a strategy is to set up a monitoring service recording daily on a central repository the number of casualties in each healthcare centre. With the present telemedicine tools, such a monitoring service requires an operator in each healthcare centre to submit the information to the central repository and an operator to validate the information provided. In case of emergency, for instance if the monitoring service identifies an abnormal increase of the mortality rate, experts have to be called to analyze the information available at the central repository. If they want additional information, they need to require it from the operators in each healthcare centre. This extra request may introduce major delays and extra work on health professionals who are already overworked.

With the onset of grid technology, such a monitoring service would require much less manpower. Indeed, grid technology delivers today access in a secure way to data

M. Daydé et al. (Eds.): VECPAR 2004, LNCS 3402, pp. 204–218, 2005.

stored on distant grid nodes. Instead of having one operator in each centre in charge of transmitting information daily to the central repository, the information on the number of casualties is stored locally on a database which is accessible by the central repository. In case of emergency, the experts can access further to the healthcare centre database to inquire about the patient medical files. In this scenario, patient medical files stay in healthcare centres and the central monitoring service picks up only what is needed for its task.

2 Vision for a Grid for Health

The example used to introduce this paper illustrates the potential impact of grid technology for health. The grid technology is identified as one of the key technologies to enable the European research Area. Its impact is expected to reach much beyond eScience to eBusiness, eGouvernment, … and eHealth. However, a major challenge is to take the technology out of the laboratory to the citizen. A HealthGrid (figure 1) is an environment where data of medical interest can be stored and made easily available to the different actors of healthcare, the physicians, the healthcare centres and administrations, and of course the citizens. Such an environment has to offer all guarantees in terms of security, respect of ethics and regulations. Moreover, the association of post-genomics and medical data on such an environment opens the perspective of individualized healthcare.

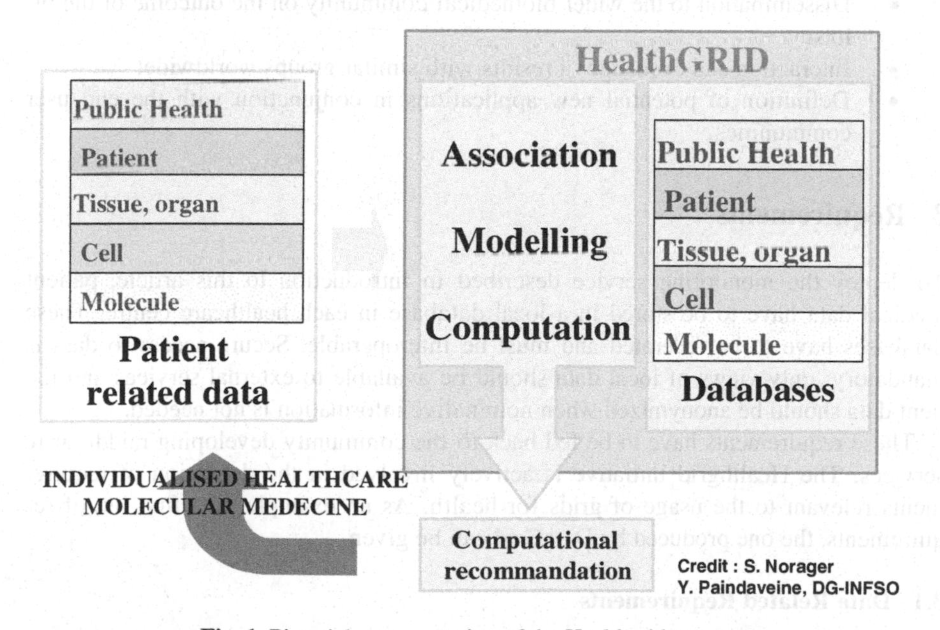

Fig. 1. Pictorial representation of the Healthgrid concept

While considering the deployment of life sciences applications, most present grid projects do not address the specificities of an e-infrastructure for health, for instance the deployment of grid nodes in clinical centres and in healthcare administrations, the connection of individual physicians to the grid and the strict regulations ruling the access to personal data,…[18].

Technology to address these requirements in a grid environment is under development and a pioneering work is under way in the application of Grid technologies to the health area.

In the last couple of years, several grid projects have been funded on health related issues at national and European levels. These projects have a limited lifetime, from 3 to 5 years, and a crucial issue is to maximize their cross fertilization. Indeed, the Healthgrid is a long term vision that needs to build on the contribution of all projects. The Healthgrid initiative, represented by the Healthgrid association (http://www.healthgrid.org), was initiated to bring the necessary long term continuity. Its goal is to collaborate with projects on the following activities:

- Identification of potential business models for medical Grid applications;
- Feedback to the Grid-development community on the requirements of the pilot applications deployed by the European projects;
- Dialogue with clinicians and people involved in medical research and Grid development to determine potential pilots;
- Interaction with clinicians and researchers to gain feedback from the pilots;
- Interaction with all relevant parties concerning legal and ethical issues identified by the pilots;
- Dissemination to the wider biomedical community on the outcome of the pilots;
- Interaction and exchange of results with similar groups worldwide;
- Definition of potential new applications in conjunction with the end user communities.

3 Requirements

To deploy the monitoring service described in introduction to this article, patient medical data have to be stored in a local database in each healthcare centre. These databases have to be federated and must be interoperable. Secure access to data is mandatory: only views of local data should be available to external services and patient data should be anonymized when nominative information is not needed.

These requirements have to be fed back to the community developing middleware services. The Healthgrid initiative is actively involved in the definition of requirements relevant to the usage of grids for health. As an example of a first list of requirements, the one produced by DataGrid can be given.

3.1 Data Related Requirements

- Provide access to biology and medical image data from various existing databases
- Support and improve existing databases import/export facilities

- Provide transparent access to data from the user point of view, without knowledge of their actual location
- Update databases while applications are still running on their own data versions

3.2 Security Related Requirements

- Grant anonymous and private login for access to public and private databases
- Guarantee the privacy of medical information
- Provide an encryption of sensitive message
- Fulfill all legal requirements in terms of data encryption and protection of patient privacy
- Enforce security policies without the need to modify applications

3.3 Administration Requirements

- Provide "Virtual Grids", ie the ability to define subgrids with a restricted access to data and computing power
- Provide "Virtual Grids" with customizable login policies

3.4 Network Requirements

- Provide batch computing on huge dataset
- Provide fast processing applications which transfer small set of images between different sites : storage site, processing site and physician site
- Provide interactive access to small data sets, like image slices and model geometry

3.5 Job Related Requirements

- Allow the user to run jobs "transparently", without the knowledge of the underlying scheduling mechanism and resource availability
- Manage jobs priorities
- Deal with critical jobs : if no resource are available, jobs with lowest priority will be interrupted to allow execution of critical jobs
- Permit to chain batch jobs into pipelines
- Provide a fault-tolerant infrastructure with fault detection, logging and recovery
- Notify an user when a job fails for a Grid independent reason
- Provide an interactive mode for some applications
- Support for parallel jobs
- Provide a message passing interface inside a local farm and also at the Grid level
- Offer Job monitoring and control, i.e. query job status, cancel queuing or running job
- Provide logs to understand system failures and intruders detection.

4 DataGrid

The deployment of biomedical applications in a grid environment has started about three years ago in several European projects and national initiatives. These applications have demonstrated that the grid paradigm was relevant to the needs of the biomedical community.

The European DataGrid (EDG) project [1] successfully concluded on 31 March 2004. It aimed at taking a major step towards making the concept of a world-wide computing Grid a reality. The goal of EDG was to build a test computing infrastructure capable of providing shared data and computing resources across the European scientific community. The budget for the project was around 10 million euros and 21 partner institutes and organizations across Europe were involved.

After a massive development effort involving seven major software releases over three years, the final version of EDG software is already in use in three major scientific fields: High Energy Physics, Biomedical applications and Earth Observations. At peak performance, the EDG test bed shared more than 1000 processors and more than 15 Terabytes of disk space spread in 25 sites across Europe, Russia and Taïwan. The software is exploited by several bio-medical applications in the area of bioinformatics and biomedical simulation.

4.1 Monte-Carlo Simulation for Nuclear Medicine and Radio/Brachytherapy

The principle of Monte-Carlo simulations is to reproduce radiation transport knowing the probability distributions governing each interaction of particles in the patient body and in the different equipments needed in nuclear medicine and brachy/radiotherapy: gamma-camera and PET for nuclear medicine, accelerator head for radiotherapy and internal radiation applicators for brachytherapy. Accuracy is therefore only limited by the number of particles generated. As a consequence, Monte-Carlo simulations are increasingly used in nuclear medicine to generate simulated PET and SPECT images to assess the performances of reconstruction algorithms in terms of resolution, sensitivity and quantification and to design innovative detectors. In external beam radiotherapy, Monte-Carlo simulations are needed for accelerator head modelling and computation of beam phase space [2]. Accelerator beam modelling is especially critical to reach high accuracy in gradient and shielded regions for Intensity Modulated Radiation Therapy. Measurement of dose deposit by applicators in brachytherapy is often difficult experimentally because of high gradient. In the case of electron sources, Monte-Carlo simulation is especially relevant provided electron transport is properly described.

All in all, the major limiting factor for the clinical implementation of Monte-Carlo dose calculations methods is the large computing time requested to reach the desired accuracy. Most of the commercial systems, named TPS (Treatment Planning Systems), for clinical routine use an analytic calculation to determine dose distributions and so, errors near heterogeneities in the patient can reach 10 to 20%. Such codes are very fast comparing to Monte Carlo simulations: the TPS computation time for an ocular brachytherapy treatment is lower than on minute, thus allowing its usage in clinical practice, while a Monte Carlo framework could take 2 hours.

To evaluate the impact of parallel and distributed Monte Carlo simulations, a radiotherapy treatment planning was performed on the EDG Testbed, from pre-processing and registration of medical images on the Storage Elements (SEs) of the grid to the parallel computation of Monte Carlo simulations GATE (Geant4 Application for Tomographic Emission [3].

The application framework is depicted in figure 3. Sets of 40 DICOM slices or so, 5122 pixels each, acquired by CT scanners are concatenated and stored in a 3D image format. Such image files can reach until 20 MB for our application. To solve privacy issues, DICOM headers are wiped out in this process.

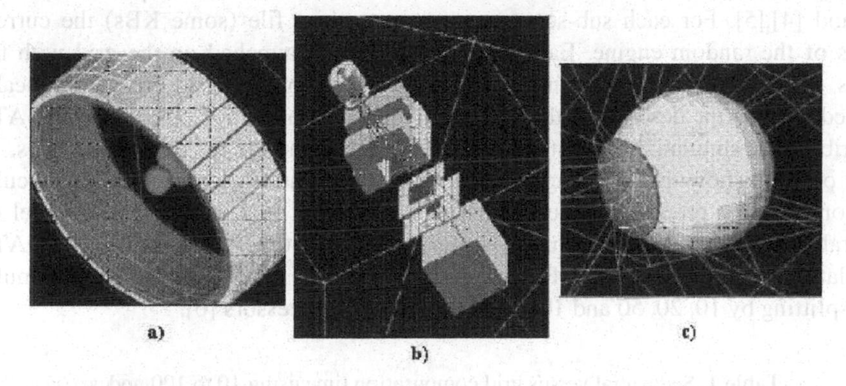

Fig. 2. GATE Monte Carlo simulations: a) PET simulation; b) Radiotherapy simulation; c) Ocular brachytherapy simulation

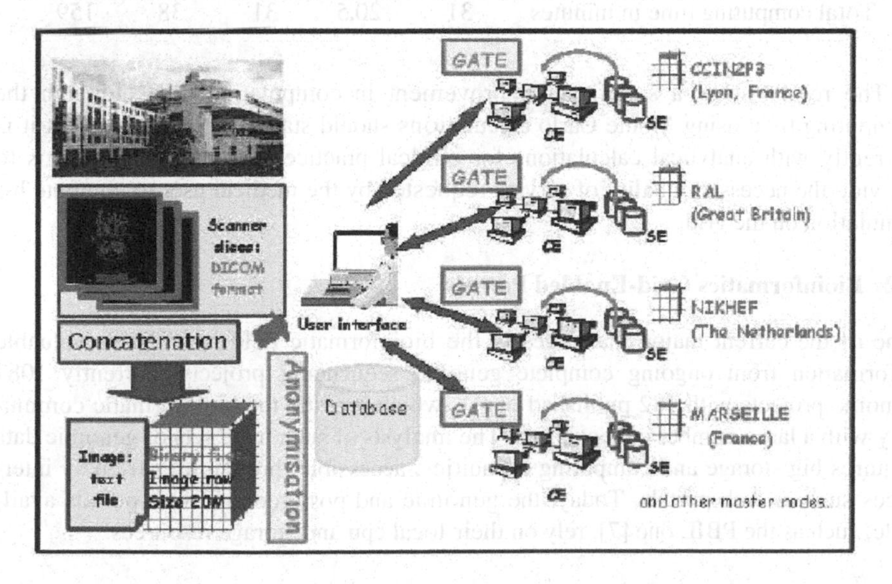

Fig. 3. Submission of GATE jobs on the DataGrid testbed

The 3D image files are then registered and replicated on the sites of the EDG testbed where GATE is installed in order to compute simulations (5 sites to date). During the computation of the GATE simulation, the images are read by GATE and interpreted in order to produce a 3D array of voxels whose value is describing a body tissue. A relational database is used to link the GUID of image files with metadata extracted from the DICOM slices on the patient and additional medical information. The EDG Spitfire software is used to provide access to the relational databases.

Every Monte Carlo simulation is based on the generation of pseudorandom numbers using a Random Numbers Generator (RNG). An obvious way to parallelize the calculations on multiple processors is to partition a sequence of random numbers generated by the RNG into suitable independent sub-sequences.

To perform this step, the choice has been done to use the Sequence Splitting Method [4],[5]. For each sub-sequences, we save in a file (some KBs) the current status of the random engine. Each simulation is then launched on the grid with the status file. All the other files necessary to run Gate on the grid are automatically created: the script describing the environment of computation, the macros GATE describing the simulations, the status files of the RNG and the job description files.

In order to show the advantage for the GATE simulations to partition the calculation on multiple processors, the simulations were split and executed in parallel on several grid nodes. Table 1 illustrates the computing time in minutes of a GATE simulation running on a single P4 processor at 1.5GHz locally and the same simulation splitting by 10, 20, 50 and 100 jobs on multiple processors [6].

Table 1. Sequential versus grid computation time using 10 to 100 nodes

Number of jobs submitted	10	20	50	100	Local
Total computing time in minutes	31	20,5	31	38	159

The results show a significant improvement in computation time although the computing time using Monte Carlo calculations should stay comparable to what it is currently with analytical calculations for clinical practice. The next challenge is to provide the necessary quality of service requested by the medical user to compute his simulation on the grid.

4.2 Bioinformatics Grid-Enabled Portals

One of the current major challenges in the bioinformatic field is to derive valuable information from ongoing complete genome sequencing projects (currently 1087 genome projects with 182 published ones), which provide the bioinformatic community with a large number of sequences. The analysis of such huge sets of genomic data requires big storage and computing capacities, accessible through user-friendly interfaces such as web portals. Today, the genomic and post-genomic web portals available, such as the PBIL one [7], rely on their local cpu and storage resources.

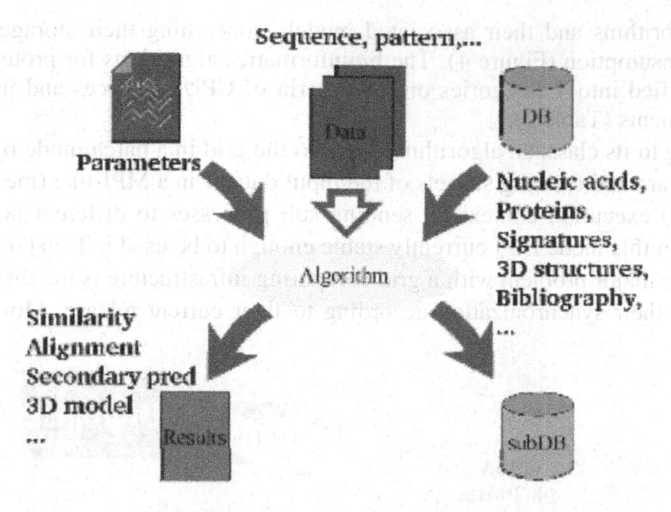

Fig. 4. Bioinformatics algorithm schema

Grid computing may be a viable solution to go beyond these limitations and to bring computing resources suitable to the genomic research field. A solution explored in the European DataGrid project was to interface the NPS@ web site [8] dedicated to protein structure analysis to the DataGrid infrastructure.

Table 2. Classification of the bioinformatics algorithms used in GPS@ according to their data and CPU requirements

		Input/Output DATA	
		Small	**Large**
CPU consummer	**Moderate**	Protein secondary structure prediction (GOR4, DPM, Simpa96…). Physicochemical profiles…	BLAST ProScan (protein pattern) …
	Intensive	Multiple alignement with CLUSTAL W or Multalin …	FASTA, SSEARCH PattInProt (protein pattern) Protein secondary structure predictions (SOPMA, PHD,…) CLUSTAL W (complete genomes)…

The bioinformatics sequence algorithms used on NPS@ web portal are of different types depending on the data analyses they aim to compute: sequence homology and similarity searching (*e.g.* BLAST [9]), patterns and signatures scanning (*e.g.* PattIn-Prot), multiple alignment of proteins (*e.g.* ClustalW [10]), secondary structure prediction and so on. The "gridification" of web portals for genomics has to deal with these different algorithms and their associated models concerning their storage and CPU resources consumption (Figure 4). The bioinformatics algorithms for protein analysis can be classified into 4 categories on the criteria of CPU resources and input/output data requirements (Table 2).

According to its class, an algorithm is sent to the grid in a batch mode by distributing the software and creating subsets of the input data or in a MPI-like (message passing interface) execution context by sending sub processes to different nodes of the grid (although this mode isn't currently stable enough to be used in DataGrid).

In fact, the major problem with a grid computing infrastructure is the distribution of the data and their synchronization according to their current release. Moving on the

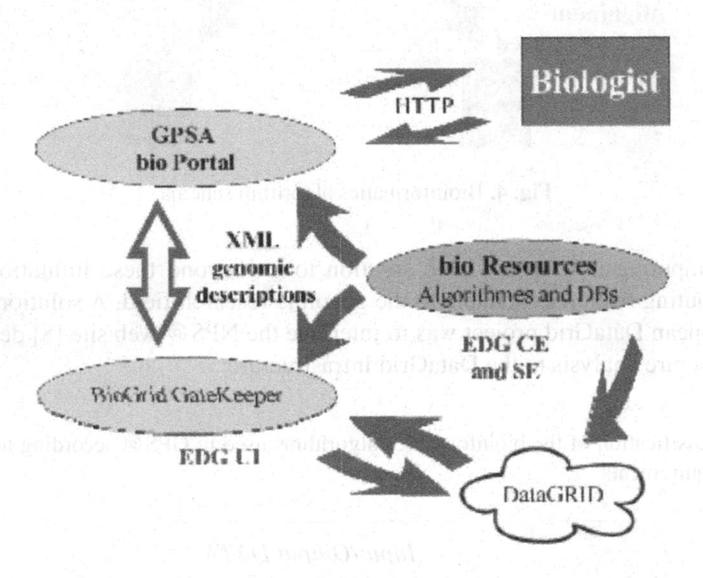

Fig. 5. Bioinformatic job processing on GPS@ web portal

grid a databank which size varies from tens of megabytes (*e.g.* SwissProt [11]) to gigabytes (*e.g.* EMBL [12]), requires a significant fraction of the network bandwith and therefore increases the execution time. One simple solution can be to split databanks into subsets sent in parallel to several grid nodes, in order to run the same query on each subset. Such approach requires a synchronization of all the output files at the end. A more efficient solution would be to update the databanks on several referenced nodes with the help of the Replica Manager provided by the DataGrid middleware and to launch the selected algorithms on these nodes. To summarize, the algorithms working on short dataset are sent at runtime with the data through the grid sandbox while the ones analyzing large datasets are executed on the grid nodes where the related databanks have been transported earlier by the DataGrid replica manager service.

GPS@ - Grid Protein Sequence Analysis. Depending on the algorithms, DataGrid submission process is different. Some of the algorithms on NPS@ portal have been adapted in agreement to the DataGrid context, and will be made available to the biologist end-user in a real production web portal in the future (they can be tested on the URL http://gpsa.ibcp.fr). GPS@ offers a selected part of the bioinformatic queries available on NPS@, with the addition of job submission on DataGrid resources accessible through a drag-and-drop mechanism and simply pressing the "submit" button. All the DataGrid job management is encapsulated into the GPS@ backoffice: scheduling and status of the submitted jobs (Fig. 5). And finally the result of the biological grid jobs are displayed into a new web page, ready for further analysis or for download.

4.3 Humanitarian Medical Development

The training of local clinicians is the best way to raise the standard of medical knowledge in developing countries. This requires transferring skills, techniques and resources. Grid technologies open new perspectives for preparation and follow-up of medical missions in developing countries as well as support to local medical centres in terms of teleconsulting, telediagnosis, patient follow-up and e-learning. To meet requirements of a development project of the French NPO Chain of Hope in China, a first protocol was established for describing the patient pathologies and their pre- and post-surgery states through a web interface in a language-independent way. This protocol was evaluated by French and Chinese clinicians during medical missions in the fall 2003 [13]. The first sets of medical patients recorded in the databases will be used to evaluate grid implementation of services and to deploy a grid-based federation of databases (see figure 6). Such a federation of databases keeps medical data distributed in the hospitals behind firewalls. Views of the data will be granted according to individual access rights through secured networks.

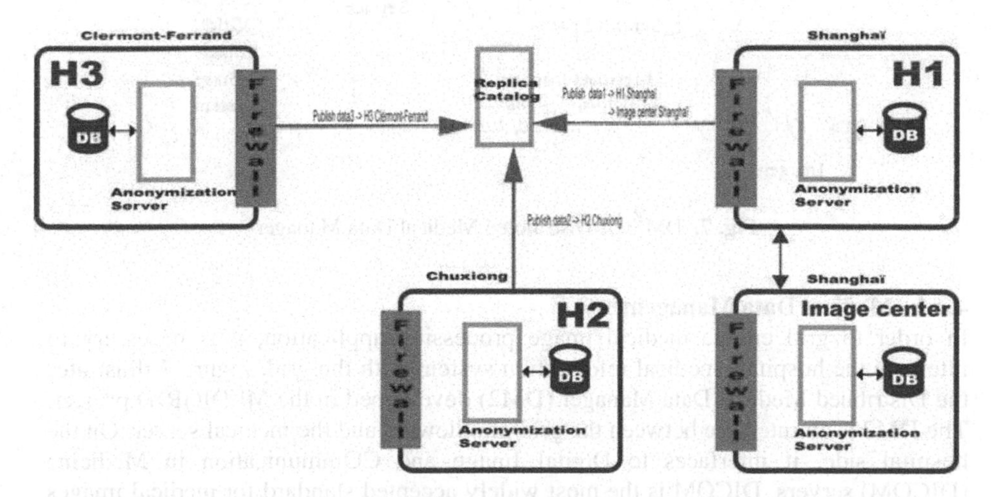

Fig. 6. Schematic representation of data storage architecture and information flow between three hospitals

4.4 Medical Image Storage, Retrieval and Processing

Medical images represent tremendous amounts of data produced in healthcare centers each day. Picture Archiving and Communication Systems (PACS) are proposed by imager manufacturers to help hospitals archiving and managing their images. In addition to PACS, Radiological Information Systems (RIS) are deploy to store medical records of patients. PACS and RIS are developped with an clinical objective of usability. However, they usually do not address the problems of large scale (cross-site) medical data exchange as most of them are proprietary manufacturers solution that do not rely on any existing standard for interoperability. Moreoever, they are usually weakly addressing the security problems arising in wide scale grid computing since they are usually bounded to each health center. Finally, they do not consider the increasing need for automated data analysis and they do not offer any interface to external computing resources such as a grid.

Grids are a logical extension of regional PACS and RIS. However, many technical issues as well as an adaptation of the medical world to Information Technology tools delay the wide applicability of grid technologies in this field. In the European DataGrid project and the French National MEDIGRID project [14], we have been working on medical data management and grid processing on medical data [15].

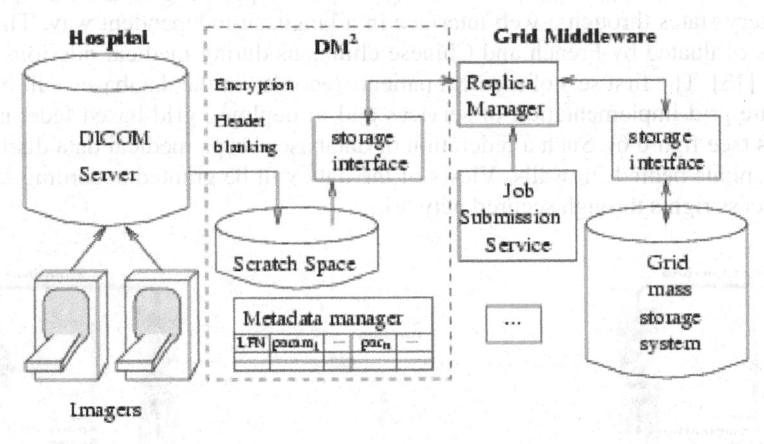

Fig. 7. DM^2: A Distributed Medical Data Manager

4.4.1 Medical Data Management

In order to grid enable medical image processing application, it is necessary to interface the hospitals medical information system with the grid. Figure 7 illustrates the Distribued Medical Data Manager (DM2) developped in the MEDIGRID project. The DM2 is an interface between the grid middleware and the medical server. On the hospital side, it interfaces to Digital Image and COmmunication in Medicine (DICOM) servers. DICOM is the most widely accepted standard for medical images storage and communication. Most recent medical imagers are DICOM compliant: they produce images that they can locally archive or transfer to a DICOM server. The DM2 can be conneted to one central or several distributed DICOM servers available in the hospital. On the grid side, the DM2 provides a grid storage interface. Each time a medical image is recored on the DICOM server, the DM2 registers a new file entry on the grid data management system. It also contains a database storing metadata (the medical record) associated to images. From the grid side, images recorded are accessible to authorized users as any file regularly registered. The image can be accessed by a grid node through the DM2 interface which translates the grid incoming request into a DICOM request and returns the desired file. However, the DM2 provide additional services such as automatic data anonymization and encryption to prevent critical data from being accessible by non accredited users. See [17] for details.

4.4.2 Medical Data Processing

Through an interface such as the DM2 a grid can offer an access to large medical image and metadata databases spread over different sites. This is usefull for many data intensive medical image processing applications for which large datasets are needed. Moreover, grids are well suited to handle computation resulting from exploring full databases. For instance, epidemiology is a field requiring to collect statistics over large population of patients. For rare pathologies, this can only be achieved if data providing from a large number of sites can be assembled. Building digital atlases also requires to assemble large image sets.

An example of a grid image content-based retrieval application is described in detail in [16] and depicted in figure 8. The typical scenario is a medical doctor diagnosing a medical image of one of his/her patient that has been previously registered (1,2). To confirm his/her diagnosis, he/she would like to confront it to similar known medical cases. The application allows the physician to search for recorded images similar to the source image he is interested in. Target candidates are first selected out of the database by querrying image metadata (3). Several similarity criterion may then be used to compare each image from the database to the source image. The similarity compuation algorithms return a score. Images of the database can be ranked according to this score and the physician can download highest score images corresponding to most similar cases. This application requires one similarity job to be started for each candidate image to be compared in the database (4). These computations are distributed over available grid nodes to speed up the search (5,6). Given the absence of dependencies between the computations, the computing time is therefore divided by the number of processor available, neglecting the grid overhead (network transmission of data and scheduling overhead). This application has been successfully deployed on the EDG testbed, processing up to hundreds of images (tens of images are processed in parallel).

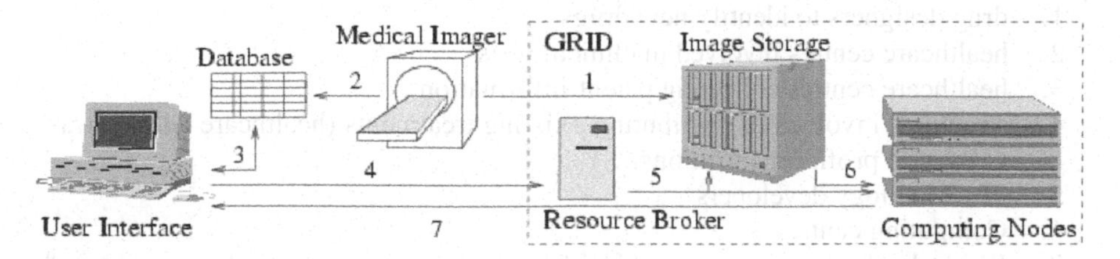

Fig. 8. Application synopsis

5 An Emerging Project: A Grid to Address a Rare Disease

Another area where grid technology offers promising perspectives for health is drug discovery. This chapter presents the potential interest of a grid dedicated to Research and development on a rare disease.

5.1 The Crisis of Neglected Diseases

There is presently a crisis in research and development for drugs for neglected diseases. Infectious diseases kill 14 million people each year, more than ninety percent of whom are in the developing world. Access to treatment for these diseases is problematic because the medicines are unaffordable, some have become ineffective due to resistance, and others are not appropriately adapted to specific local conditions and constraints. Despite the enormous burden of disease, drug discovery and development targeted at infectious and parasitic diseases in poor countries has virtually ground to a standstill, so that these diseases are de facto neglected. Of the 1393 new drugs approved between 1975 and 1999, less than 1% (6) was specifically for tropical diseases. Only a small percentage of global expenditure on health research and development, estimated at US$50-60 billion annually, is devoted to the development of such medicines. At the same time, the efficiency of existing treatments has fallen, due mainly to emerging drug resistance.

The unavailability of appropriate drugs to treat neglected diseases is among other factors a result of the lack of ongoing R&D into these diseases. While basic research often takes place in university or government labs, development is almost exclusively done by the pharmaceutical industry, and the most significant gap is in the translation of basic research through to drug development from the public to the private sector. Another critical point is the launching of clinical trials for promising candidate drugs. Producing more drugs for neglected diseases requires building a focussed, disease-specific R&D agenda including short-, mid- and long-term projects. It requires also a public-private partnership through collaborations that aims at improving access to drugs and stimulating discovery of easy-to-use, affordable, effective drugs.

5.2 The Grid Impact

The grid should gather:

1. drug designers to identify new drugs
2. healthcare centres involved in clinical tests
3. healthcare centres collecting patent information
4. structures involved in distributing existing treatments (healthcare administrations, non profit organizations,...)
5. IT technology developers
6. Computing centres
7. Biomedical laboratories searching for vaccines, working on the genomes of the virus and/or the parasite and/or the parasite vector

The grid will be used as a tool for:

1. Search of new drug targets through post-genomics requiring data management and computing
2. massive docking to search for new drugs requiring high performance computing and data storage
3. handling of clinical tests and patent data requiring data storage and management
4. overseeing the distribution of the existing drugs requiring data storage and management

A grid dedicated to research and development on a given disease should provide the following services:

1. large computing resources for search for new targets and virtual docking
2. large resources for storage of post genomics and virtual docking data output
3. grid portal to access post genomics and virtual docking data
4. grid portal to access medical information (clinical tests, drug distribution,…)
5. a collaboration environment for the participating partners. No one entity can have an impact on all R&D aspects involved in addressing one disease.

Such a project would build the core of a community pioneering the use of grid-enabled medical applications. The choice of a neglected disease should help reducing the participants reluctance to share information. However, the issue of Intellectual Property must be addressed in thorough details.

6 Conclusion

The grid technology is identified as one of the key technologies to enable the European research Area. Its impact is expected to reach much beyond eScience to eBusiness, eGouvernment, … and eHealth. However, a major challenge is to take the technology out of the laboratory to the citizen. A HealthGrid is an environment where data of medical interest can be stored and made easily available to the different actors of healthcare, the physicians, the healthcare centres and administrations, and of course the citizens. Such an environment has to offer all guarantees in terms of security, respect of ethics and regulations. Moreover, the association of post-genomics and medical data on such an environment opens the perspective of individualized healthcare.

The deployment of biomedical applications in a grid environment has started about three years ago in several European projects and national initiatives. These applications have demonstrated that the grid paradigm was relevant to the needs of the biomedical community. In this paper, we have reported on our experience within the framework of the DataGrid project.

References

1. Special issue of Journal of Grid Computing dedicated to DataGrid, to be published in 2004
2. Fraas B.A., Smathers J. and Deye J., Issues limiting the clinical use of Monte Carlo dose calculation algorithms, Med. Phys. 30, vol. 12, (2003) 3206-3216

3. Santin G., Strul D., Lazaro D., Simon L., Krieguer M., Vieira Martins M., Breton V. and Morel C. GATE, a Geant4-based simulation platform for PET and SPECT integrating movement and time management, IEEE Trans. Nucl. Sci. 50 (2003) 1516-1521
4. Traore, M. and Hill, D. (2001). The use of random number generation for stochastic distributed simulation: application to ecological modeling. In 13th European Simulation Symposium, Marseille, pages 555-559, Marseille, France.
5. Coddington, P., editor (1996). Random Number Generators For Parallel Computers, Second Issue. NHSE Review.
6. Maigne, L., Hill, D., Breton, V., and et al. (2004). Parallelization of Monte Carlo simulations and submission to a Grid environment. Accepted for publication in Parallel Processing Letters.
7. Perriere, G, Combet, C, Penel, S, Blanchet, C, Thioulouse, J, Geourjon, C, Grassot, J, Charavay, C, Gouy, M, Duret, L and Deléage, G.(2003). Integrated databanks access and sequence/structure analysis services at the PBIL. Nucleic Acids Res. 31, 3393-3399.
8. Combet, C., Blanchet, C., Geourjon, C. and Deléage, G. (2000). NPS@: Network Protein Sequence Analysis. Tibs, 25, 147-150.
9. 9 Altschul, SF, Gish, W, Miller, W, Myers, EW, Lipman, DJ (1990) Basic local alignment search tool. J. Mol. Biol. 215, 403-410
10. Thompson, JD, Higgins, DG, Gibson, TJ (1994) CLUSTAL W: improving the sensitivity of progressive multiple sequence alignment through sequence weighting, position-specific gap penalties and weight matrix choice. Nucleic Acids Res. 22, 4673-4680.
11. Bairoch, A, Apweiler, R (1999) The SWISS–PROT protein sequence data bank and its supplement TrEMBL in 1999. Nucleic Acids Res. 27, 49-54
12. Stoesser, G, Tuli, MA, Lopez, R, Sterk, P (1999) the EMBL nucleotide sequence database. Nucleic Acids Res. 27, 18-24.
13. V. J. Gonzales, S. Pomel, V. Breton, B. Clot, JL Gutknecht, B. Irthum, Y. Legré. Empowering humanitarian medical development using grid technology, proceedings of Healthgrid 2004, to be published in Methods of Information in Medecine
14. MEDIGRID project, French ministry for research ACI-GRID project, http://www.creatis.insa-lyon.fr/MEDIGRID/
15. J. Montagnat, V. Breton, I. E. Magnin, Using grid technologies to face medical image analysis challenges, Biogrid'03, proceedings of the IEEE CCGrid03, pp 588-593, May 2003, Tokyo, Japan.
16. J. Montagnant, H. Duque, J.M. Pierson, V. Breton, L. Brunie, I. E. Magnin, Medical Image Content-Based Queries using the Grid, HealthGrid'03, pp 138-147, January, 2003, Lyon, France
17. H. Duque, J. Montagnat, J.M. Pierson, L. Brunie, I. E. Magnin, DM2: A Distributed Medical Data Manager for Grids, *Biogrid'03, proceedings of the IEEE* CCGrid03, pp 606-611, May 2003, Tokyo, Japan.
18. Hernandez V, Blanquer I, The GRID as a healthcare provision tool, proceedings of Healthgrid 2004, to be published in Methods of Information in Medecine

Three-Dimensional Cardiac Electrical Activity Simulation on Cluster and Grid Platforms*

J.M. Alonso[1], J.M. Ferrero (Jr.)[2], V. Hernández[1], G. Moltó[1], M. Monserrat[2], and J. Saiz[2]

[1] Departamento de Sistemas Informáticos y Computación, Universidad Politécnica de Valencia. Camino de Vera s/n 46022 Valencia, Spain
{jmalonso, vhernand, gmolto}@dsic.upv.es
Tel. +34963877356, Fax +34963877359

[2] Departamento de Ingeniería Electrónica
Universidad Politécnica de Valencia. Camino de Vera s/n 46022 Valencia, Spain
{cferrero, monserr, jsaiz}@eln.upv.es
Tel. +34963877600, Fax +34963877609

Abstract. Simulation of action potential propagation on cardiac tissues represents both a computational and memory intensive task. The use of detailed ionic cellular models, combined with its application into three-dimensional geometries turn simulation into a problem only affordable, in reasonable time, with High Performance Computing techniques. This paper presents a complete and efficient parallel system for the simulation of the action potential propagation on a three-dimensional parallelepiped modelization of a ventricular cardiac tissue. This simulator has been integrated into a Grid Computing system, what allows an increase of productivity in cardiac case studies by performing multiple concurrent parallel executions on distributed computational resources of a Grid.

1 Introduction

Cardiac arrhythmias are one of the first causes of mortality in developed countries. Among them, ventricular tachycardias and ventricular fibrillation stand out because of triggering sudden cardiac death. In spite of intense research, the mechanisms of generation, maintenance and termination of these arrhythmias are not clearly understood.

Recently, mathematical models of the propagation of cardiac electrical activity are being considered as a powerful and helpful tool to better understand the mechanisms involved in the development of ventricular arrhythmias. The electrical activity of cardiac cells is described by detailed models of ion movements through the cell membrane. By combining the mathematical formulation

* The authors wish to thank the financial support received from (1) the Spanish Ministry of Science and Technology to develop the projects CAMAEC (TIC2001-2686) and GRID-IT (TIC2003-01318), and (2) the Universidad Politécnica de Valencia for the CAMAV project (20020418).

M. Daydé et al. (Eds.): VECPAR 2004, LNCS 3402, pp. 219–232, 2005.

of membrane ion kinetics and the structural complexity of the heart it is possible to simulate in computers the complex electrical propagation in cardiac tissues.

Earlier studies characterised this tissue as a simple one dimensional fiber. A more realistic model consisted of a thin sheet of myocardium, where one cell was connected to its four neighbours in a two-dimensional regular structure. However, to study the complex dynamics underlying the formation and development of ventricular tachycardias and ventricular fibrillation, a 3D model of a cardiac tissue with appropriate dimensions is required. This 3D tissue consists of a very large number of cardiac cells (typically hundreds of thousands), governed by time-consuming ionic models which require tenths of state variables for each cell. Besides, provided that the simulation periods are typically milliseconds and even seconds, integrated with time steps of a few μs, an action potential propagation simulation can last for several days or even weeks on a traditional serial platform.

In addition to all the computational requirements for a single simulation, there are many cardiac case studies that demand the execution of several simulations. For example, testing the effects of new drugs requires the execution of multiple parametric simulations, where for instance the drug concentration is changed. As another example, studying the effects of late ischemia, it is necessary to vary the junctional resistances in a determined interval and observe the evolution of the electrical activity of the tissue under different anisotropy conditions.

Therefore, to harness all this computational burden, we have integrated two different technologies. First of all, High Performance Computing techniques offer the possibility to reduce the execution time of a single simulation, as well as to enable the simulation of larger three-dimensional tissues during longer time by performing execution on a cluster of PCs. On the other hand, Grid Computing technology emerges as a solution for the collaborative usage of multi-organisational computational resources [1]. In this work, both techniques have been integrated into a system that allows concurrent parallel simulations of action potential propagation on remote, multi-organisational resources of a Grid infrastructure, based upon the public domain Globus Toolkit [2] and the commercial InnerGrid [3] middlewares.

The article is structured as follows. Section 2 presents the underlying mathematical model. Then, section 3 explains the parallelisation approach implemented in order to reduce the execution time of a single simulation. Next, section 4 discusses the performance results achieved. Later, section 5 describes the Grid Computing approaches with both middlewares. Section 6 presents a case study to analyse the Grid advantages and finally section 7 summarizes the main achievements.

2 Mathematical Model and Geometry

The action potential propagation on a monodomain modelization of a cardiac tissue can be described by the following partial derivative equation:

$$\nabla \cdot \sigma \nabla V_m = C_m \cdot \frac{\partial V_m}{\partial t} + I_{ion} + I_{st} \tag{1}$$

where σ represents the conductivity tensor, V_m is the membrane potential of the cells, C_m stands for the membrane capacitance, I_{st} represents an stimulus current to provoke an action potential and I_{ion} is the sum of ionic currents traversing the membrane of each cell, computed by the comprehensive and detailed Luo-Rudy Phase II ionic model [4].

The term action potential denotes a transient change of the membrane potential caused by the electrically excitable heart cells. When a stimulus applied to the cell leads to depolarization of resting membrane potential up to its threshold, then an action potential is induced. This response is characterised by an initially fast rise of the membrane potential followed by a slow recovery to the resting potential.

In our three-dimensional modelization, the ventricular tissue cardiac cells are linked with resistances within a parallelepiped geometry. Cardiac muscle fibers are assumed to have faster longitudinal than transversal or transmural conductivity, accounting for the anisotropy condition of a ventricular cardiac tissue.

Equation (1) is spatially discretized using a seven-point finite difference stencil and employing the Crank-Nicholson's semi-implicit method, what leads to the following algebraic equation:

$$G_L \cdot V_m^{t+1} = G_R \cdot V_m^t + I_{ion}^t + I_{st}, \forall t = 1, 2, ..., n \; . \tag{2}$$

The matrices G_L and G_R account for the conductivity along the cells of the tissue. The I_{ion} term encapsulates the cellular ionic model, requiring the resolution of several time-dependent ordinary differential equations. Thus, the simulation turns into an iterative process where the membrane potential of the cells is reconstructed through the resolution of a large sparse linear equation system for each simulation time step.

Even though there have been several parallel approaches to this computational problem [5], the good efficiency results achieved on a beowulf cluster, logically based on a distributed memory paradigm, together with appearing to be the first simulation system to approach both a parallel and a Grid Computing philosophy represent a step forward in the study of the electrical activity of the heart.

3 Parallel Solution

3.1 Parallelization Approach

The cells in our three-dimensional parallelepiped are numbered following a natural ordering and assigned to the processors by groups with contiguous numeration indexes of approximately the same size. This way, each processor is in charge of performing all the calculations corresponding to its part of the tissue.

3.2 Conductivity Matrix Generation

The conductivity matrices G_R and G_L are generated in parallel with no communication among the processors. These matrices, together with the ionic (I_{ion}), the stimulus (I_{st}) and membrane potential (V_m) vectors, have been partitioned among the processors following a rowwise block-striped distribution, what overcomes the memory constraints that may arise when simulating a large three-dimensional tissue on a single computer, thus enabling the simulation of larger tissues.

3.3 Cell Membrane Potential Calculation

In order to obtain the membrane potential of the cells, a large sparse linear equation system must be solved for each simulation time step. The G_L coefficient matrix is symmetric and positive definite, with a size equal to the number of cells in the tissue. For a 3D tissue of 1 million cardiac cells (100x100x100 cells), the coefficient matrix has dimension 1 million with 7 million nonzero elements.

Based upon the framework that the PETSc library [6] offers, two different strategies have been employed in order to solve this large sparse linear equation system. A parallel direct method, based on a Multifrontal Cholesky Factorization provided by the MUMPS library [7], integrated within PETSc, and a parallel iterative method, based on the Preconditioned Conjugate Gradient, have been tested. For the direct method, a previous step of ordering to reduce the fill-in is performed with the Multilevel Nested Dissection algorithm implemented in the METIS library [8].

3.4 Right-Hand Side Vector Generation

The right-hand side vector generation of the linear equation system has been fully parallelised. First of all, a distributed updating of the state of the cells and the computation of the I^t_{ion} term, via the Luo-Rudy Phase II model, take place. This represents the most time-consuming step of the simulation, where each processor only updates its local part of the tissue without any inter-process communication.

Then, a sparse matrix-vector product $G_R \cdot V^t_m$, with the data distribution described in section 3.2, must be carried out in parallel. Communications are needed in this stage taking into account the sparsity pattern of the G_R matrix, where each processor can demand the membrane potential of cells residing in its neighbour processors.

Next, the I_{st} stimulus vector is computed with no communications. Finally, the right-hand side vector is generated as a sum of three vectors, with no inherent communication cost.

4 Experimental Results of the Parallel Implementation

The simulations have been run on a cluster of 20 dual-processor 2 GHz Pentium Xeon with 1 GByte of RAM, interconnected with a 4x5 torus SCI network. Figure 1 shows the execution time of a single time step, comparing the direct and the iterative method, when simulating an action potential propagation on a 50x50x50 cells tissue.

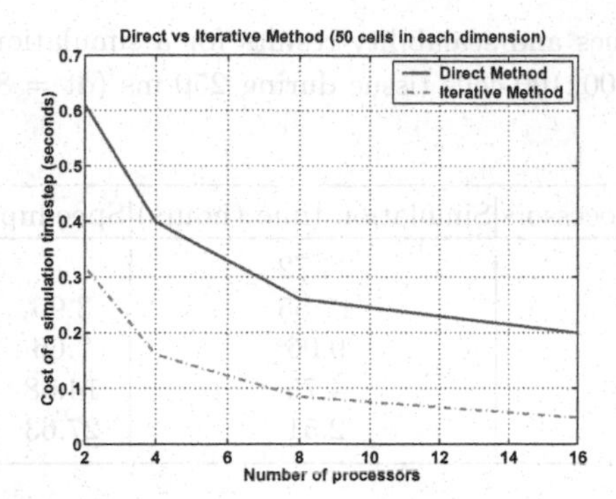

Fig. 1. Conjugate Gradient Method versus solution of the triangular systems after a Multifrontal Cholesky Factorization

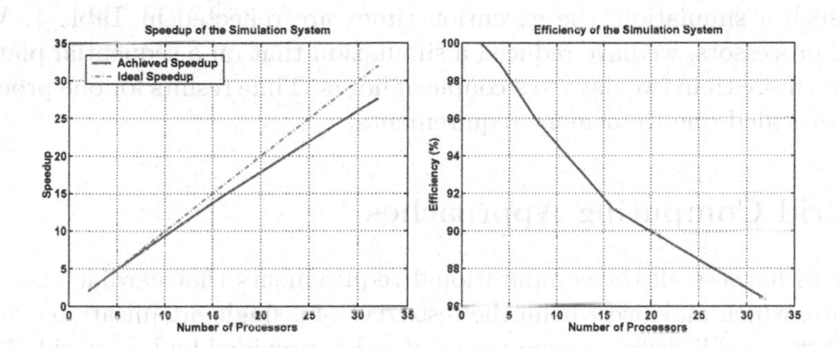

Fig. 2. Scalability of the simulation system

Regarding the direct method, the ordering, symbolic and numerical factorizations cost have been neglected because the coefficient matrix remains constant through the simulation and thus, they can be reused. Therefore, only the solution of the triangular systems is included in the cost of the direct method, as the time of the three steps would vanish in a long simulation. It should be taken into account that, for very large tissues, the factorization could exceed the available memory, what represents a handicap which an iterative method does not suffer from.

The conjugate gradient method with no preconditioning has probed to be the best iterative solver tested. Besides, as Fig. 1 reflects, it performs twice as fast as the resolution of the triangular systems. In fact, the coefficient matrix is well conditioned and convergence is obtained within few iterations with a good residual tolerance.

Figure 2 shows the speedup and efficiency of the whole simulation system when simulating an action potential propagation on a 100x100x100 cells cardiac tissue during 250 ms, using the conjugate gradient method with no precondi-

Table 1. Execution times and scalability results for a simulation of action potential propagation on a 100x100x100 cells tissue during 250 ms (dt = 8 μs), using up to 32 processors

Number of processors	Simulation time (hours)	Speedup	Efficiency
2	34.72	-	-
4	17.55	3.95	98.88
8	9.06	7.66	95.79
16	4.75	14.58	91.18
32	2.51	27.63	86.36

tioning, and employing up to 32 processors. Simulations have been performed running two processes per node. It should be pointed out that the simulation system scales quite linear with the number of processors.

For such a simulation, the execution times are reflected in Table 1. When using 32 processors, we have reduced a simulation that on a sequential platform would last more than two days to a couple of hours. Time results for one processor are not provided due to memory requirements.

5 Grid Computing Approaches

In order to harness all the computational requirements that cardiac case studies require, which may overwhelm the resources of a single organisation, we have deployed two middlewares, a commercial solution provided by InnerGrid [3] software and the wide adopted public domain industrial standard Globus Toolkit [2].

Innergrid is a multi-platform software that enables the collective sharing of computational resources, within an organisation, in order to enlarge the productivity of parametric sequential jobs. This software offers a fault-tolerance scheme that guarantees the execution of the tasks as long as there are living nodes in the Grid. InnerGrid exposes a web interface from which the configuration and the management of the tasks is performed.

On the other hand, the Globus Toolkit is an open source tool that allows the generation of inter-organisational Grids within a secure and transparent environment. It offers basic building tools for data transfer, parallel execution and integration with remote execution policies, among other features.

5.1 Enabling Portability

A distributed Grid infrastructure is, at first glance, an unknown pool of computational resources. It can not be assumed that the execution hosts will have available the required dynamic libraries that the simulation system depends on, neither the computational nor the system libraries dependences. Therefore, the simulation system should have no requirements of any external library. We have approached this problem by static linking the application, that is, introducing the code from all the dynamic libraries into a single executable with no external library dependences. The MPI message passing layer is also introduced into the executable by static linking with a standard MPICH [9][10] implementation, specially configured to disable shared memory communication between processes in the same node of a cluster, which is known to introduce memory leak problems because of relying on the System V IPC facilities [9].

In addition, all sort of platform-dependent optimised software should not be employed, such as the BLAS or LAPACK libraries implementations for a concrete architecture, as well as platform-dependent compiler optimization flags, such as `-march` or `-mcpu` which allow the compiler to emit specific code for a specific platform. This way, the simulation system will not execute any illegal instruction on the remote machine. Fortunately, traditional compiler optimization flags, i.e. `-O3`, can be used with no risk.

Therefore, it is possible to achieve a self-contained parallel simulation system that can be executed on different Linux versions. Besides, having integrated the MPI communication software, it can be executed in parallel on a remote cluster without depending on the MPI implementation of the execution host. This has been ensured by parallel simulations in a variety of machines of different architectures such as Pentium III, Pentium Xeon and even Intel Itanium 2 with different Linux flavours such as Red Hat Linux Advanced Server, Red Hat 8.0 and Debian GNU/Linux.

This process of adapting the simulation system to the Grid infrastructure results in a weighty executable file that can be lightened by discarding the symbols from its object files. Through compression, with a Lempel-Ziv coding scheme, the executable archive has been reduced, in our case, to a self-contained simulation system of less than 2 MBytes. This self-contained simulator performs on average 2% slower than the optimised counterpart.

It should be pointed out that such a simulator runs on compatibility mode on an Intel Itanium 2 (64 bit) platform and thus, it is up to 8 times slower than on an Intel Pentium Xeon (32 bit). Therefore, we have natively compiled on the Intel Itanium 2 platform in order to achieve comparable execution times on both architectures, and to be able to exploit Itanium Grid execution nodes. This results on two self-contained simulation systems, one for IA-32 and other for IA-64 platforms, an strategy that could be refined to target more architectures.

5.2 Grid Infrastructure

The available Grid infrastructure, shown in Fig. 3, is composed of local resources, belonging to our research group, the High Performance Networking and Computing Group (GRyCAP-UPV), and remote resources from the Distributed Systems Architecture & Security group (ASDS-UCM), at Universidad Complutense de Madrid. Table 2 summarises the main characteristics of the machines.

The Globus Toolkit version 2.4 [2] has been installed on all the machines of this testbed. Besides, provided that there is no Itanium version of InnerGrid yet, and this software is focused for single-organisational resources, this middleware has only been installed on Ramses and Kefren clusters.

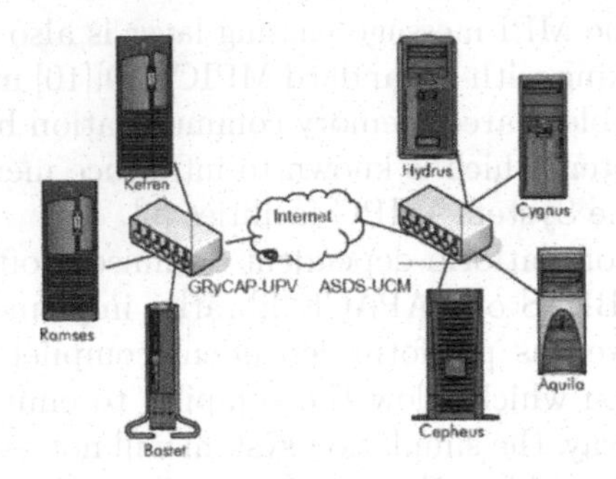

Fig. 3. Computational resources within the Grid infrastructure

Table 2. Detailed machine characteristics

Machine	Processors	Memory	Job Manager
kefren	10 (2 x Intel Xeon 2.0 Ghz)	1 GByte	pbs, fork
ramses	12 (2 x Intel Pentium III 866 Mhz)	512 MBytes	pbs, fork
bastet	2 x Itanium 2 (900 Mhz)	4 GBytes	fork
hydrus,cygnus	1 x Pentium 4 (2.53 Ghz)	512 MBytes	fork
aquila	1 x Pentium III (666 Mhz)	128 MBytes	fork
cepheus	1 x Pentium III (666 Mhz)	256 MBytes	fork

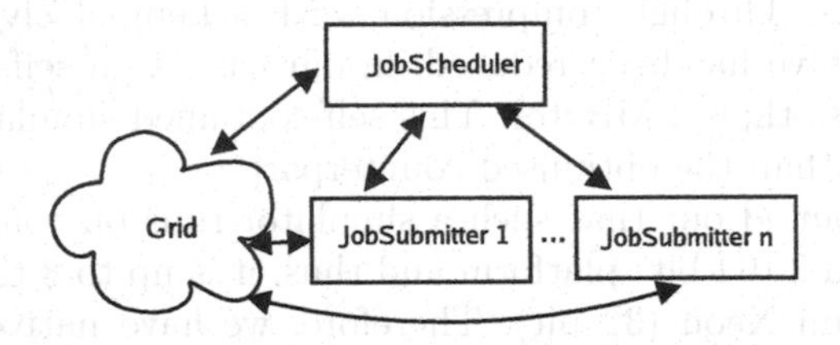

Fig. 4. Scheme of the Grid Computing system developed

5.3 Globus Developments

Figure 4 shows a conceptual view of the Grid Computing system developed. The `JobScheduler` is the module responsible for the allocation of simulations to the computational resources. It delegates into a `JobSubmitter` instance, for each simulation, which will be in charge of the proper execution of the task in the resource.

The JobScheduler Module. This module reads an input file with a parametric description of the multiple simulations that form a whole case study. For each simulation, it finds out the best available resource, from a predefined list of machines, by consulting their number of available nodes on the machine, via the Monitoring and Discovery Service (MDS) provided by Globus. Clusters with the Globus Resource Allocation Manager (GRAM) Reporter installed report the number of free computing nodes, delegating in the local queue manager

(LoadLeveler, PBS, etc). For workstations or sequential PCs, an estimate of the CPU usage during the last minute serves as an indicator of the availability of the resource. This strategy allows to customise a parallel execution to the number of available nodes in the host.

We have included basic quality of service capabilities by specifying the minimum and maximum number of processors on which a parallel simulation can be run. These numbers are thought to increase productivity and to ensure executions with some minimum requirements. Besides, a problem-dependent memory estimator of each simulation prevents the execution of tasks on machines that will otherwise become memory exhausted.

Then, this module selects an appropriate executable based on the architecture of the remote machine. The JobScheduler is also responsible for submitting the unassigned simulations and restarting the failed executions, delegating on an instance of the JobSubmitter module. If no available resources exist, it periodically checks their availability to continue submitting pending tasks.

Given that the amount of data generated by the simulator is quite large (hundreds of MBytes), the GridFTP protocol is employed for all the data transfers as opposed to the GASS (Global Access to Secondary Storage) service. This ensures high performance reliable transfers for high-bandwidth networks.

The JobSubmitter Module. Each instance of this module is in charge of the proper execution of a single simulation. First of all, the input files that the simulation system needs are staged in, via the GridFTP service, to the execution host. Through the Globus native interface, the remote machine is queried about its availability to run MPI jobs, so a serial or parallel execution can be selected. The execution of the simulation is integrated, if configured, with the queue manager of the remote node (PBS, LoadLeveler, etc), thus respecting the execution policies of the organisation.

While the simulation is running on the remote resource, a checkpoint job is periodically submitted by this module, which transfers, if not already done, a compressed image of the checkpoint data generated by the application to the local machine. Thus, the latest checkpoint files always will reside in the submission machine and a failed simulation can be automatically restarted on a new computational resource. A message digest mechanism ensures that the latest checkpoint data is only transferred once, thus saving bandwidth.

Once the execution has finished, all the result files are compressed, transferred back to the submission node and saved in the appropriate local folder created for this simulation. All the output files in the execution node are deleted, and finally, the JobSubmitter module annotates whether the simulation has finished correctly or not. This information will be used by the JobScheduler module to be able to restart or resume the failed simulations.

5.4 InnerGrid Developments

InnerGrid software has been tested as an alternative, easy-to-use middleware for Grid execution. InnerGrid automatic file staging capabilities, combined with its built-in task scheduler, dramatically simplifies the extra development in order to execute cardiac case studies. We have developed a new module that allows to specify the memory and disk requirements of each simulation. Besides, this module specifies the varying parameters of the study.

Then, a task can be seen as the instantiation of a module, and thus, new tasks, that define the range of variation of the parameters, can be created.

6 Case Study

In order to test the capabilities of the Grid Computing system developed, a real case study has been executed on the available Grid infrastructure.

6.1 Description

Myocardial ischemia is a condition in which oxygen deprivation to the heart muscle is accompanied by inadequate removal of metabolites because of reduced flood. The mechanisms of generation of ventricular tachycardia and ventricular fibrillation (the most mortal of arrhythmias) can be studied using a model of a regionally ischemic cardiac tissue (which would result from the occlusion of a coronary artery) in which certain part of the substrate is ischemic while the surrounding tissue remains in normal conditions [11].

In the ischemic zone, the values of several electrophysiological parameters suffer variations through time as ischemia progresses. Extracellular potassium concentration ($[K^+]_o$), in first place, changes from its normal value (5.4 mmol/L) to a value of 12.5 mmol/L in the first 5 minutes, reaching a plateau for the next 5 minutes of ischemia [12]. In second place, intracellular concentration of ATP ($[ATP]_i$) decreases almost linearly with time from a normal value of around 6.8 mmol/L to 4.6 mmol/L in the first 10 minutes of ischemia, while the intracellular concentration of ADP ($[ADP]_i$) increases from 15 μmol/L to 100 μmol/L in the same time interval [12, 13]. Finally, acidosis reduces the maximum conductance of sodium (I_{Na}) and calcium ($I_{Ca(L)}$) currents to around 75% of its normal values between the fifth and the tenth minute of the ischemic episode [14].

Thus, the four parameters mentioned (which are present in the model of the cardiac action potential) change differently with time during the first 10 minutes of myocardial ischemia. In the simulations presented here, the short-term electrical behaviour of the tissue was simulated in different instants of time after the onset of ischemia. Time was, therefore, the changing parameter of the set of simulations.

The simulated virtual 3D tissue comprised a central ischemic zone, consisting on a 20x20x20-element cube (which represents 2x2x2 mm) in which $[K+]_o$, $[ATP]_i$, $[ADP]_i$, I_{Na} and $I_{Ca(L)}$ changed with time, embedded in a 60x60x60-element cube. The electrophysiological parameters of the tissue that surrounds the central ischemic zone is maintained in their normal values.

This case study will analyse the influence in action potential propagation of different degrees of ischemic conditions that take place from 0 to 10 minutes from

Table 3. Distribution of the simulations in the testbed, for each machine. The number in parentheses indicates the number of processors involved in the execution

Machine	Simulations	Machine	Simulations
Kefren	4 (4 p.), 3 (3 p.), 2 (2 p.) 2 (1 p.)	Hydrus	2 (1 p.)
Ramses	1 (6 p.), 3 (5 p.), 1 (4 p.), 1 (1 p.)	Cygnus	2 (1 p.)

the onset of ischemia. Using a time increment of 0.5 minutes, this results in 21 independent parametric simulations that can be executed in a Grid infrastructure. Each execution will perform a simulation during 80 ms with a time step of 10 μs. A supra-threshold stimulus will be applied to all the cells at the bottom plane during the simulation time interval [50, 52] ms. A snapshot of the membrane potential of the tissue cells will be stored every 1 ms during the interval [40, 80] ms, resulting in a total 68 MBytes of raw data. Besides, a total 180 MBytes of RAM is required for the execution of each simulation on a sequential platform.

6.2 Execution Results

For the Globus-based Grid Computing system designed, Table 3 summarises the task distribution in the Grid. As maximum, parallel executions have been limited to a quarter the total available processors of the remote resource, implementing a polite policy that allows multiple concurrent simulations. The minimum number of processors where set to one, thus allowing serial executions. In the table, an entry like 3 (5 p.) indicates that three simulations were performed with five processors each one.

The machines Bastet and Cepheus do not appear in the table because they were heavily overloaded during the test and thus, the scheduler never chose them. Besides, the machine Aquila did not have enough RAM to host a simulation and thus, the scheduler did not considered it for execution.

The execution of the whole case study lasted for 10.27 hours in the Grid deployment. It can be seen that the scheduler dynamically assigned to each machine a number of simulations proportional to its computational power, thus resulting in a scheduling policy that takes into account Grid computers with little computational resources.

The execution of the case study via the InnerGrid middleware lasted for 22.57 hours, distributing the tasks among the nodes of the two clusters. Executing the same case study in a traditional sequential manner in one node of the cluster Kefren required 81.1 hours. Finally, performing a High Performance Computing approach, executing each simulation with 4 processors in the same cluster (thus allowing two concurrent executions), required a total 11.9 hours.

While InnerGrid seems more appropriate to take advantage of idle computers in single-organisational Grids, the Globus Toolkit is more focused on running on dedicated resources from different organisations. Therefore, a Globus-based solution is much more appropriate for the cardiac electrical activity simulation

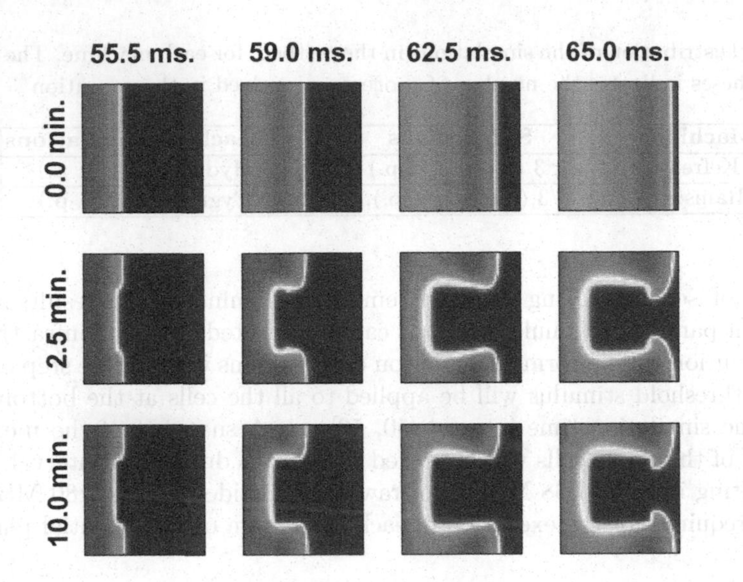

Fig. 5. Membrane potential at four simulation time instants under three degrees of ischemia. Propagation takes place from left to right. Colourbar is the same as shown in Fig. 6

problem, as it offers the possibility to access distant computational resources in a transparent manner.

6.3 Case Study Results

Figure 5 summarises the results obtained in the case study. It represents the membrane potential of the cells of a vertical tissue slab at the center of the cube, for three different ischemic conditions and at four simulation time instants. Ischemic conditions at 0 minutes introduce no perturbation in action potential

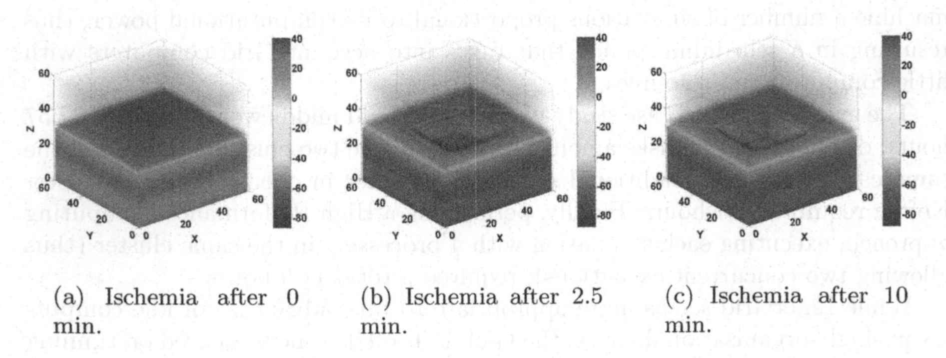

(a) Ischemia after 0 min.

(b) Ischemia after 2.5 min.

(c) Ischemia after 10 min.

Fig. 6. Three-dimensional representation of the membrane potential of the tissue at simulation time instant 59 (ms.)

propagation, while more severe degrees of ischemia provoke a slowdown in propagation within the tissue area affected.

Figure 6 offers a three-dimensional representation of the membrane potential of all the cells of the tissue at simulation time 59 ms, for three different degrees of ischemia. Again, it can be seen that the ischemic zone reduces the velocity of action potential propagation, which is more severe as the ischemia progresses in time.

7 Conclusions

This paper presents an efficient parallel implementation for the simulation of the action potential propagation on a three-dimensional monodomain ventricular cardiac tissue. The use of a High Performance Computing-based model has reduced the simulation times from days to few hours. Besides, larger 3D tissues can be simulated during longer time by only employing more processors, thus enlarging the global available memory.

In addition, to harness resource-starved cardiac case studies, a Globus-based Grid Computing system has been developed, what allows the integration of MPI parallel executions on multiprocessor machines within a Grid infrastructure. The system features important capabilities such as self-contained executable, dependencies migration, data compression, cross-linux portability, as well as the integration of parallel executions in the Grid.

Besides, InnerGrid commercial product has been tested as an alternative middleware for creating single-organisational Grids. As part of our work, a new InnerGrid module has been developed which allows varying several parameters and managing the execution of the parametric tasks from a web environment.

As High Performance Computing techniques are responsible for speedup, Grid Computing is responsible for productivity. The integration of both computational strategies has represented to be a key combination in order to enhance productivity when executing cardiac case studies.

References

1. Foster, I., Kesselman, C., Tuecke, S.: The Anatomy of the Grid: Enabling Scalable Virtual Organizations. International Journal of High Performance Computing Applications **15** (2001) 200–222
2. Foster, I., Kesselman, C.: Globus: A Metacomputing Infrastructure Toolkit. Intl. J. Supercomputer Applications **11** (1997) 115–128
3. GridSystems S.A.: InnerGrid Nitya Technical Specifications. (2003)
4. Luo, C.H., Rudy, Y.: A Dynamic Model of the Cardiac Ventricular Action Potential. I Simulations of Ionic Currents and Concentration Changes. Circulation Research **74** (1994) 1071–1096
5. Vigmond, E.J., Aguel, F., Trayanova, N.A.: Computational Techniques for Solving the Bidomain Equations in Three Dimensions. IEEE Transactions on Biomedical Engineering **49** (2002) 1260–1269

6. Balay, S., Buschelman, K., Gropp, W.D., Kaushik, D., Knepley, M., McInnes, L.C., Smith, B.F., Zhang, H.: PETSc User Manual. Technical Report ANL-95/11 - Revision 2.1.5, Argonne National Laboratory (2002)
7. Amestoy, P.R., Duff, I.S., L'Excellent, J.Y., Koster, J.: MUltifrontal Massively Parallel Solver (MUMPS Version 4.3) User's Guide. (2003)
8. Karypis, G., Kumar, V.: METIS. A Software Package for Partitioning Unstructured Graphs, Partitioning Meshes and Computing Fill-reducing Orderings of Sparse Matrices. University of Minnesota. Version 4.0. (1998)
9. Gropp, W.D., Lusk, E.: User's Guide for MPICH, a Portable Implementation of MPI. Mathematics and Computer Science Division, Argonne National Laboratory. (1996)
10. Gropp, W., Lusk, E., Doss, N., Skjellum, A.: A High-Performance, Portable, Implementation of the MPI Message Passing Interface Standard. Parallel Computing **22** (1996) 789–828
11. Coronel, R.: Heterogeneity in Extracellular Potassium Concentration During Early Myocardial Ischaemia and Reperfusion: Implications for Arrhythmogenesis. Cardiovasc. Res. **28** (1994) 770–777
12. Weiss, J.N., Venkatesh, N., Lamp, S.T.: ATP-sensitive k^+ Channels and Cellular k^+ Loss in Hypoxic and Ischaemic Mammalian Ventricle. J. Physiol. **447** (1994) 649–673
13. Ferrero (Jr.), J.M., Saiz, J., Ferrero, J.M., Thakor, N.: Simulation of Action Potentials from Metabolically Impaired Cardiac Myocytes. Role of ATP-sensitive k^+ Current. Circ. Res. **79** (1996) 208–221
14. Yatani, A., Brown, A.M., Akaike, N.: Effects of Extracellular pH on Sodium Current in Isolated, Single Rat Ventricular Cells. J. Membr. Biol. **78** (1984) 163–168

2DRMP-G: Migrating a Large-Scale Numerical Mathematical Application to a Grid Environment

T. Harmer[1], N.S. Scott[2], V. Faro-Maza[2], M.P. Scott[3], P.G. Burke[3], A. Carson[1], and P. Preston[1]

[1] Belfast eScience Centre, Queen's University Belfast, Belfast BT7 1NN, UK
Phone: +44 (0)28 9027 4626 Fax: +44 (0)28 9068 3890
t.harmer@qub.ac.uk

[2] School of Computer Science, Queen's University Belfast, Belfast BT7 1NN, UK
Phone: +44 (0)28 9027 4626 Fax: +44 (0)28 9068 3890
{ns.scott, v.faro-maza}@qub.ac.uk

[3] Department of Applied Mathematics and Theoretical Physics, Queen's University Belfast, Belfast BT7 1NN, UK
Phone: +44 (0)28 9027 3197
{m.p.scott, p.burke}@qub.ac.uk

Abstract. We report on the migration of a traditional, single architecture application to a grid application using heterogeneous resources. We focus on the use of the UK e-Science Level 2 grid (UKL2G) which provides a heterogeneous collection of resources distributed within the UK. We discuss the solution architecture, the performance of our application, its future development as a grid-based application and comment on the lessons we have learned in using a grid infrastructure for large-scale numerical problems.

Keywords: Large Scale Simulations in all areas of Engineering and Science; Numerical Methods; Grid Computing.

1 Introduction

Traditionally, large-scale numerical mathematical applications have focused on using supercomputers. Each new supercomputer deployment has resulted in significant development activities as legacy software is ported and tuned to extract the best possible performance on the new computational resource. This process is both difficult and time-consuming. However, in common with many traditional scientific applications, the application we are developing does not have any real-time constraints in its execution or urgency in obtaining results. There is, therefore, no pressing need to use the fastest available computational resource. Any collection of computational resources that can deliver results in a reasonable time is acceptable. In addition, we are interested in considering cases where data sizes would make use of a single computational resource difficult if not impossible. Our primary focus, therefore, is to investigate how a traditional atomic

M. Daydé et al. (Eds.): VECPAR 2004, LNCS 3402, pp. 233–246, 2005.

physics application can be migrated to a grid environment thereby extending its usefulness to experimentalists and theorists who need to compute atomic data but who have no access to supercomputer resources.

A grid infrastructure provides an environment for the co-operative use of distributed computational and storage resources. One application of a grid infrastructure is to use dynamic collections of computational resources in solving large-scale numerical mathematical problems. Thus, for example, an application can enquire what computational resources are currently available in the application's grid environment (usually called resource discovery) and direct work to those resources that are available. Grid middleware enables the secure use of remote resources and the transport of data within the grid infrastructure.

In our application we have concentrated on using available computational resources together with large collections of generic resources. In addition, we are interested in identifying those parts of our application that need very fast computational resources and those that can be assigned to more modest ones. Thus, our implementation will direct the most computationally intensive parts of the application to the most appropriate computational resources that are available for use. Our application has a long execution time and the grid implementation will automatically restart grid tasks that fail switching to other computational resources that are available.

In the following section we describe the atomic physics application. We outline the traditional computational method, illustrate how it is unsuitable for a grid environment and show how it can be recast to exploit a grid environment. The corresponding software suite, 2DRMP, is described in §3. A number of test-bed grids have been deployed across the globe. In this paper we report on the use of the UK e-Science Level 2 grid (UKL2G) which provides a heterogeneous collection of resources distributed within the UK [1]. Our application has been executing almost everyday on the UKL2G for one year. It has yet to fail to complete although its execution time varies as it is dependant on the number of available resources in the UKL2G. A brief overview of the UKL2G is presented in §4. The grid enabled package 2DRMP-G, including its resource task allocation and monitoring, is described in §5. In §6 we present and comment on 2DRMP application timings on the UKL2G and the UK's supercomputing resource HPCx. In the final section we present some concluding remarks.

2 The Atomic Physics Application

For over thirty years the R-matrix method has proved to be a remarkably stable, robust and efficient technique for solving the close-coupling equations that arise in electron collision theory [2]. We begin by sketching traditional 'one sector' R-matrix theory. Then we introduce a propagation approach that subdivides the internal region into a collection of smaller and partially independent sub-regions [4]. This novel approach has the advantage of extending the boundary radius of the internal region far beyond that which is possible using the traditional 'one-sector' technique. Currently this approach is restricted to electron scattering from hydrogen and hydrogen-like systems.

2.1 Overview of Traditional R-Matrix Theory

The basic idea behind the traditional method is to divide the configuration space describing the collision process into two regions by a sphere of radius $r = a$, where a is chosen so that the charge distribution of the target atom or ion is contained within the sphere. In the internal region ($r \leq a$) exchange and correlation effects between the scattering electron and the target electrons must be included, whereas in the external region such effects can be neglected thereby considerably simplifying the problem.

In the internal region the $(N+1)$-electron wavefunction at energy E is expanded in terms of an energy independent basis set, ψ_k, as

$$\Psi_E = \sum_k A_{Ek} \psi_k. \tag{1}$$

The basis states, ψ_k, are themselves expanded in terms of a complete set of numerical orbitals, u_{ij}, constructed to describe the radial motion of the scattered electron. The expansion coefficients of the u_{ij} set are obtained by diagonalizing the $(N+1)$-electron Hamiltonian matrix in the internal region.

In the outer region the equations reduce to coupled second-order ordinary differential equations that can be solved by a variety of means such as the 1-D propagation technique implemented in the FARM package [3].

The main goal of the computation is to calculate accurate scattering amplitudes for elastic and inelastic excitation processes. An example would be the inelastic $1s \rightarrow 7p$ process for electron scattering by atomic hydrogen. The larger the principal quantum number of the eigenstate involved in the transition of interest, the larger the region of configuration space that needs to be spanned and hence the larger the number of basis states that are required in the expansion of the scattering wavefunction.

A useful heuristic is that energy of the highest numerical orbital, u_{ij}, needs to be at least twice that of the scattering energy to ensure convergence of the basis set. It can be shown that the energy of the nth orbital is $O(\frac{n^2}{a^2})$.

From this it is clear that if we wish to increase a to accommodate eigenstates with higher principal quantum numbers and perform a calculation over the same range of scattering energies then n must increase in proportion to a. Also the number of basis functions in the expansion of the R-matrix basis states is approximately proportional to n^2. Thus, the dimension of the Hamiltonian matrix within the internal region is proportional to a^2.

To put this in perspective consider the calculation of electron scattering by atomic hydrogen involving the $n = 4$ and $n = 7$ eigenstates. A radius of $a = 60a.u.$ is needed to envelop the $n = 4$ eigenstates and a radius of $a = 200a.u.$ is needed to envelop the $n = 7$ eigenstates. Thus, the dimension of Hamiltonian matrix increases by a factor of 11 in moving from an $n = 4$ calculation to an $n = 7$ calculation. However, since matrix diagonalization is an n^3 problem the time required to diagonalize increases by a factor 1,371. Moving to $n = 10$ from $n = 7$, where the required radius is $350a.u.$, would increase the diagonalization time by a further factor of 30,000 to 40,000 times longer than an $n = 4$ calculation.

Data from electron collision processes are of crucial importance in the analysis of important physical phenomena in fields that include laser physics, plasma physics, atmospheric physics and astronomy. This is particularly true for excitation to highly excited states and for collisions with atoms in highly excited states e.g. laser-excited or metastable targets. However, because of the high principal quantum numbers involved the R-matrix radius is large. The corresponding dense Hamiltonian matrices, which are typically of the order of 50,000 to 100,000, clearly require supercomputer performance for solution. The latency involved in the distributed diagonalization of a large dense matrix over collections of heterogeneous resources renders the method unsuitable for a grid environment.

2.2 Overview of 2-D R-Matrix Propagation Theory

An alternative approach is to redesign the basic algorithms to exhibit a higher degree of embarrassing parallelism so that the considerable computational resources of grid environments can be exploited. This is one of the motivations behind the development of the 2-D R-matrix propagation technique the theory of which is sketched below.

As in traditional R-matrix theory the two-electron configuration space (r_1, r_2) is divided into two regions by a sphere of radius a centered on the target nucleus. However, in the 2-D variant the inner-region is further divided into sub-regions as illustrated in Figure 1.

Within each sub-region energy independent R-matrix basis states, $\theta_k^{LS\pi}(\mathbf{r}_1,\mathbf{r}_2)$, are expanded in terms of one-electron basis functions, ν_{ij}, whose radial forms are solutions of the Schrödinger equation. The expansion coefficients of the ν_{ij} set are obtained by diagonalizing the corresponding 2-electron Hamiltonian matrix. The expansion coefficients and the radial basis functions are then used

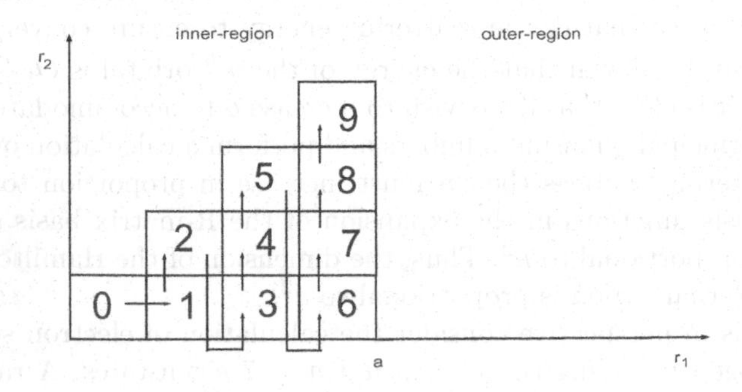

Fig. 1. Subdivision of the inner-region configuration space (r_1, r_2) into a set of connected sub-regions labelled 0..9

to construct surface amplitudes, $\omega_{inl_1l_2k}$, associated with each sub-region edge $i \in \{1,2,3,4\}$.

For each incident electron energy a set of local R-matrices (R_{ji}) can be constructed from the surface amplitudes as follows:

$$(R_{ji})_{n\ l_1l_2nl_1l_2} = \frac{1}{2a_i}\sum_k \frac{\omega_{jn\ l_1l_2k}\,\omega_{inl_1l_2k}}{E_k - E}, j,i \in \{1,2,3,4\} \quad (2)$$

Here a_i is the radius of the ith edge, E is the total energy of the two-electron system and E_k are the eigenenergies obtained by diagonalizing the two-electron Hamiltonian in the sub-region. By using the local R-matrices, the R-matrix on the boundary of the innermost sub-region can propagated across all sub-regions, working systematically from the r_1-axis at the bottom of each strip to its diagonal as illustrated in Figure 1, to yield the global R-matrix, $\Re$, on the boundary of the inner- and outer-region ($r_1 = a$).

Finally, the global R-matrix $\Re$ is transformed onto an appropriate basis for use in the outer-region. Here, a traditional 1-D propagation is performed on the transformed R-matrix to compute the scattering data of interest.

3 Overview of 2DRMP Package

The 2-D R-matrix propagation software package (2DRMP) consists of the seven stages, illustrated in Figure 2.

Each stage belongs to one of four functional blocks: A, B, C or D. The blocks must be performed sequentially. Blocks A and B are independent of the collision energy and need only be performed once while blocks C and D are dependent on the collision energy and must be repeated hundreds of times. Communication between stages is through files. The focus of this paper is on the energy independent stages A and B.

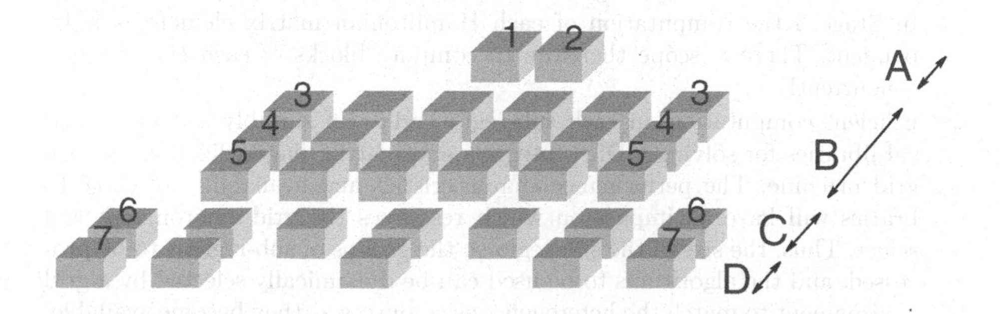

Fig. 2. The 2DRMP package. Blocks A and B are independent of the collision energy and need only be performed once while blocks C and D are dependent on the collision energy and must be repeated hundreds of times

Block A contains two independent stages, 1 and 2: Stage 1 constructs the atomic basis functions to be used in the transformation of the Global R-matrix, $\Re$, while Stage 2 computes radial integrals that can be used in the construction of the Hamiltonian matrix in off-diagonal sub-regions.

Block B consists of stages 3, 4 and 5. Each column of Block B corresponds to a sub-region. While the three stages of each column must be performed sequentially each column is independent and can be computed concurrently. Stage 3 constructs the sub-region Hamiltonian matrix, Stage 4 diagonalizes the matrix and Stage 5 constructs the corresponding surface amplitudes, $\omega_{inl_1l_2k}$.

Block C involves the propagation of a global R-matrix, $\Re$, across the sub-regions of the inner-region. This computation will be performed for each scattering energy.

Block D corresponds to a 1-D propagation code such as FARM [3]. Again this will be performed for each scattering energy, although FARM can process more than one energy at at time.

The energy independent phase (Blocks A and B) is embarrassingly parallel and has several features that make it attractive for investigating grid environments.

- There is considerable flexibility in the choice of sub-region size. Depending on grid resources available one can opt for a large number of small sub-regions or a small number of large sub-regions e.g. 1000 sub-regions with matrices of the order of 3000x3000 or 250 sub-regions with matrices of the order 10,000x10,000.
- The choice of basis set size within each sub-region can only be determined experimentally. Hence there is scope for computational steering at the start of the computation.
- Each sub-region computation is independent, therefore, sub-regions can be computed individually or in dynamically sized groups as resources become available. This distribution is further complicated by the different computational characteristics of diagonal, off-diagonal and axis sub-regions.
- In Stage 3 the computation of each Hamiltonian matrix element is independent. There is scope therefore to compute blocks of each Hamiltonian concurrently.
- Efficient computation on each sub-region relies on suitably tuned numerical libraries for solving ODEs, integrals and matrix diagonalization on each grid machine. The performance characteristics and availability of these libraries will have an impact on which resources the grid environment will select. Thus, the size of the sub-regions, the groups of sub-regions to be processed, and the algorithms to be used can be dynamically selected by a grid environment to match the heterogeneous resources as they become available.

The energy dependent Blocks C exhibits increasing parallelism as the computation proceeds and also has dynamic features which make it suitable to utilize a grid environment. However, this aspect of the computation will not be considered in this paper.

4 The UK Level 2 Grid: A Grid of Heterogeneous Resources

The UKL2G is a heterogenous collection of computational and storage resources that is intended to permit pilot grid applications to be developed. Some eight (plus) sites distributed throughout the UK provide resources to this grid infrastructure [1].

Each site in the grid provides a collection of computational and storage resources for use in the national grid resource. These resources remain under local control but are intended to be used as available resources to other sites on the grid. The network connectivity between the sites has been upgraded and is intended to enable the rapid distribution of data between computational and storage resources in this testbed grid environment.

The use of a resource is controlled using digital certificates that enable sites (automatically) to authenticate users submitting jobs to grid resources and enable accounting to be performed. These certificates permit resources to be used without logging on to a resource or providing passwords when the resource is to be used. The authorization to use a resource is still under local control by approval of a user and the registration of the user's digital certificate.

The UKL2G provides a dynamic database of available resources in the grid environment and thus provides a means of querying which computational resources are available and their types and hardware configuration. Each grid site runs application software to update the national resource database periodically. The existence of a resource in the national database does not indicate that any user can use that resource.

Some UK2LG resources can be used directly by starting a job on the machine. Other resources provide job queues that enable job scheduling and the specification of resources that are required in that job. This enables local resource managers to control resource use and, for example, provide high-throughput job queues or large-resource use job queues.

The range of resources on the UKL2G is wide. Most resources are commodity desktop machines, there are a few large-scale commodity resource pools, there are a number of closely coupled clusters and some supercomputers including the UK's national supercomputer HPCx. In addition, the UK e-Science programme has recently purchased two 20-processor data clusters and two 64-processor compute clusters[6]. These resources will be available for production use in late 2004 and are intended to enable large-scale grid-based pilot systems to be developed.

5 The Grid Enabled 2D R-Matrix Propagator: 2DRMP-G

The 2DRMP-G grid architecture uses the concept of a task pool to define the processing that is possible at any stage in its execution. A task is added to the task pool, and is thus a candidate for execution, when its prerequisite steps have completed. A work flow defines the tasks in the 2DRMP-G and the prerequisite of

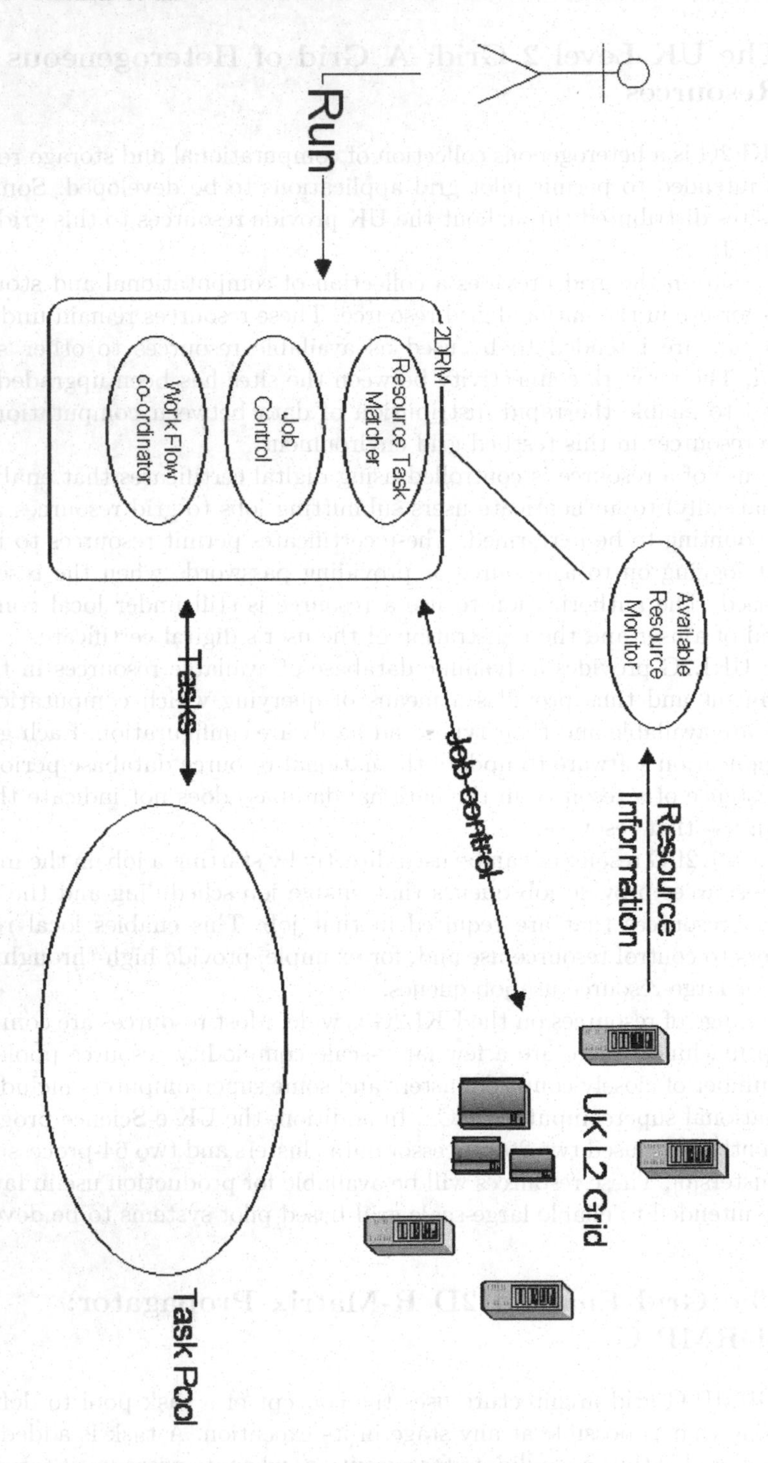

Fig. 3. 2DRMP-G Solution Architecture

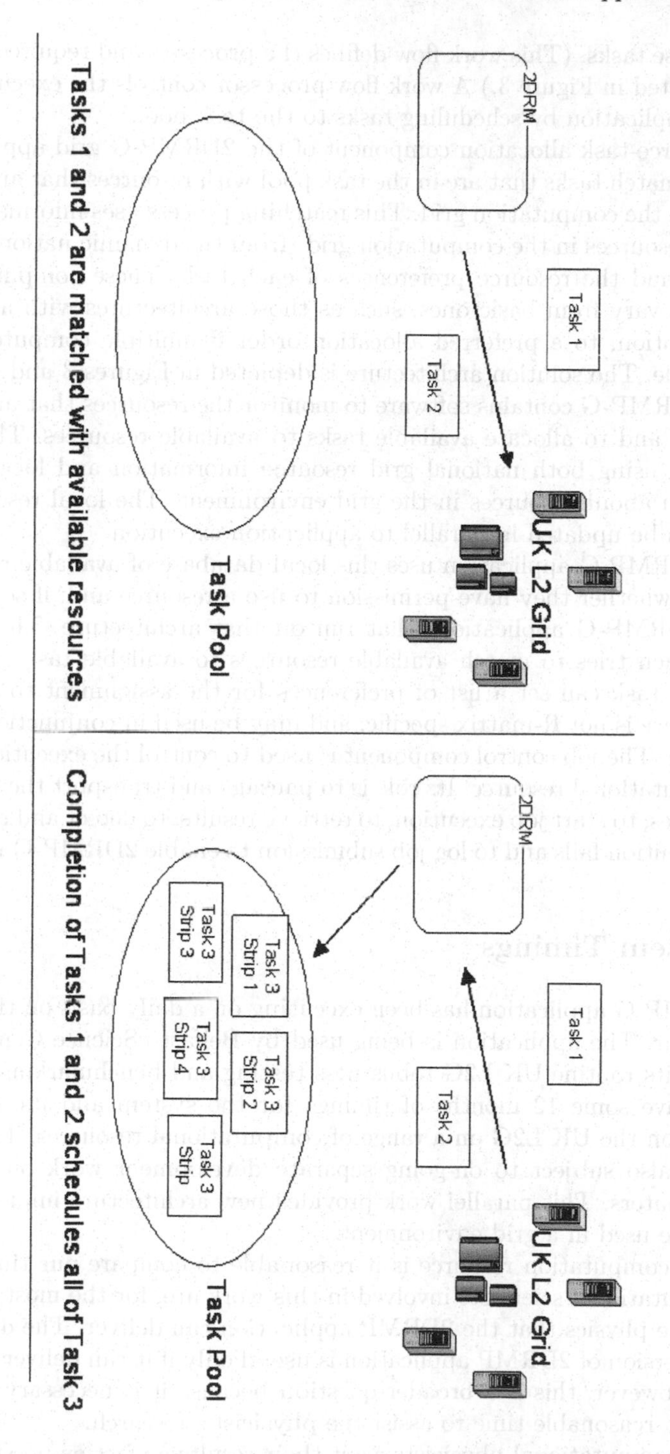

Fig. 4. 2DRMP-G Job Scheduling by Work Flow

each of these tasks. (This work flow defines the processes and required execution order depicted in Figure 3.) A work flow processor controls the execution of the 2DRMP application by scheduling tasks to the task pool.

A resource-task allocation component of the 2DRMP-G grid application attempts to match tasks that are in the task pool with resources that are currently available in the computation grid. This matching process uses information about available resources in the computation grid (from the dynamic national resource database) and the resource preferences of each task. These compute resource preferences vary from basic ones, such as those architectures with an available implementation, to a preferred allocation order if multiple compute resources are available. The solution architecture is depicted in Figures 3 and 4.

The 2DRMP-G contains software to monitor the resources that are available in the grid and to allocate available tasks to available resources. The resource monitoring using both national grid resource information and local historical information about resources in the grid environment. The local resource information can be updated in parallel to application execution.

The 2DRMP-G application uses this local database of available resources to determine whether they have permission to use a resource and, if so, search for specific 2DRMP-G applications that run on that architecture. The allocation software then tries to match available resources to available tasks in the task pool. Each task can set a list of preferences for the assignment to a resource. This software is not R-matrix specific, and may be used in conjunction with any application. The job control component is used to control the execution of a task on a computational resource. Its role is to package and transport the data that a task requires, to start job execution, to retrieve results, to detect and reschedule a task if execution fails and to log job submission to enable 2DRMP-G monitoring.

6 System Timings

The 2DRMP-G application has been executing on a daily basis on the UK L2G for one year. The application is being used by Belfast eScience Centre (BeSC) as part of its routine UK L2G robustness testing and benchmarking work. The authors have some 12 months of timings for the system and its components executing on the UK L2G on a range of computational resources. The 2DRMP system is also subject to on-going separate development work on a range of supercomputers. This parallel work provides new architecture implementations that can be used in a grid environment.

Which computation resource is it reasonable to compare our timings with? The computational scientists involved in this work are, for the most part, interested in the physics that the 2DRMP application can deliver. The development of a grid version of 2DRMP application is useful only if it can deliver meaningful physics. However, this is a broader question because it is necessary to produce results in a reasonable time to assist the physicist's research.

Many computational physicists want their results as fast-as-possible and assume very fast computation resources. Indeed, it has been a tradition in com-

Component (sub-region)	Grid Average Time(secs)	HPCx Average Time(secs)	Component (sub-region)	Grid Average Time(secs)	HPCx Average Time(secs)
NEWRD(0)	428	181	DIAG(0)	54	16
NEWRD(1)	13	0.54	DIAG(1)	379	79
NEWRD(2)	434	177	DIAG(2)	53	15.5
NEWRD(3)	12	0.44	DIAG(3)	311	80
NEWRD(4)	12	0.52	DIAG(4)	300	76
NEWRD(5)	441	196	DIAG(5)	62	16
NEWRD(6)	12	0.5	DIAG(6)	378	82
NEWRD(7)	12	0.43	DIAG(7)	290	74
NEWRD(8)	12	1.28	DIAG(8)	350	50.5
NEWRD(9)	444	203	DIAG(9)	54	9.5
AMPS(0)	4	0.6	AMPS(6)	5	1.5
AMPS(1)	4	1	AMPS(7)	5	1.8
AMPS(2)	4	1.5	AMPS(8)	4	1.4
AMPS(3)	4	1.4	AMPS(9)	5	0.6
AMPS(4)	5	1.3			
AMPS(5)	4	0.8			

Fig. 5. 2DRMP-G Timings and HPCx Timings. NEWRD, DIAG, and AMPS correspond to blocks B3, B4 and B5 respectively of Fig. 2

putational physics to tailor systems to each new generation of supercomputers as they become available. Some computational scientists do not see the point of work as described in this paper because it uses the commodity resources available on a heterogeneous grid. On the grid infrastructures that are currently available we will not deliver the results of the 2DRMP computation as fast as traditional supercomputer or closely coupled clusters. However, we are interested in demonstrating that a grid infrastructure can create a sustainable physics infrastructure where the physics is the focus of the work and not the large-scale tailoring of programs to computational environments.

We thus choose to compare the times we achieve in a grid environment with timings timings taken for the same components tailored for execution on HPCx, the UK's large-scale supercomputing resource. In figure 5 the average execution times for each of the Block B components of the 2DRMP application, illustrated in Fig. 2, are plotted. This is a significant and, perhaps, somewhat unfair comparison but the results are what would be expected. The HPCx components execute much faster than the components on a heterogenous grid infrastructure. However, the timings do illustrate the same pattern of timings.[1]

In a grid solution the execution time must be augmented with times that define the transfer of data to the remote resource, the transfer of results from the

[1] For example, because of symmetry the matrix size on the diagonal sub-regions(0,2,5 and 9) is half of that on the off-diagonal sub-regions. The matrix diagonalizations on the off-diagonal sub-regions (DIAG[1,3,4,6,7,8]) are therefore considerably longer than their diagonal sub-region counterparts (DIAG[0,2,5,9]). Due to the reuse of radial integrals on off-diagonal sub-regions the reverse is the case in NEWRD.

Component(sub-region)	Operation	Time(secs)
NEWRD(0)	Housekeeping	2.5
	Data Upload	27
	Execution	428
	Data Download	5
NEWRD(3)	Housekeeping	1.5
	Data Upload	26
	Execution	12
	Data Download	5.5

Fig. 6. Detailed Timings Sub-region Execution

remote resource and timings for necessary housekeeping on the remote machine. Detailed timings for two sub-regions are presented in figure 6. For a short-lived execution data transfer times dominate the execution time. For a long-lived execution the transfer times still represent a significant proportion of the execution time. These times reflect the relatively slow network connection to the job allocation server (100 megabit/sec) and will fall when data server is upgraded (to 1 Gigabit/sec) shortly. Nevertheless, these timings do show that care must be taken in design to ensure that remote job execution is desirable given the overhead of data transfers. It does also give rise to obvious optimizations where data is moved from its creation point to its next point of use rather than being moved from our central server to the remote computing resource. These optimizations are much in the spirit of the conventional processor-processor communication optimization that are common in tightly coupled and loosely coupled multiprocessors.

The execution of the individual components is significantly slower than the computation can be achieved on HPCx. In addition, remote job execution job involved significant housekeeping operations of the remote machines. Is a grid solution a reasonable solution for large scale physics computations? Firstly, it is likely that the resources on grid infrastructures will get better. Thus, for example, the UK L2G has recently been augmented with a number of utility 64-node processor clusters and a large-scale, high-performance data storage solution that might be use to host a temporary data repository. These additions will provide a more diverse utility processor infrastructure and enable faster execution times for component applications of 2DRMP-G. In addition, it is likely that for some large-scale computational problems a single supercomputer will not be capable of executing a complete application and some form of computation partitioning will be necessary. This is the case for 2DRMP-G application where the computation physicists are interested in tackling very large data sets.

So, is our 2DRMP-G application just a prototype which must wait for a more complete grid infrastucture to be in place? In Fig 7 the table lists the average time to run 2DRPM application tailored for the HPCx and on the 2DRPM-G on the UK L2G. These times are the start to finish times and include the queue time that is required on HPCx and any wait time for free resources on the UKL2G. Thus, with utility resources we can achieve better turn-around time in our executions that we can achieve using the UK's supercomputing resource.

Machine	Average Execution Time
HPCx	typically several hours
UK L2G	typically 30mins

Fig. 7. Executing Times for HPCx and UKL2G, including queuing time

(In relation to queuing timings for HPCx the interested reader should consult the statistics provided at the HPCx website [5]). This reflects the management strategy of HPCx that gives priority to applications that will make the most use of its large-scale resources and also how busy the machine is.

7 Concluding Remarks

We have reported on the migration of a traditional, single architecture application to a grid application using heterogeneous resources. We have described a solution architecture which has proved effective in deploying the 2DRMP G application on the UKL2G. The performance results for a modest but realistic electron scattering computation have been compared with those achieved on HPCx.

The application attempts to provide both useful physics data, be a robustness and reliability tool for the UK L2G, and be an active testbed for 2DRMP-G software development. As such, the application is still in active development and our interests include

- strategies for the allocation of computational tasks to available computing resources;
- defining the tasks as a collection of web/grid services that will enable a more mix-and-match style of R-matrix application development;
- computational steering to achieve better granularity of results in interesting areas by enabling visualization of interim results from grid tasks and the capability to re-configure sub-region sizes during execution.

The 2DRMP-G application has demonstrated that useful physics can be obtained from a grid environment that consists of utility resources and modest PC-based clusters. The UKL2G and other grids are likely to increase in number and quality of resources and should provide a good execution environment for large-scale computational physics applications. The availability of very high performance resources within a grid framework, such as HPCx in the UK, would make a significant difference to 2DRMP-G application as it would permit the computational significant tasks to be directed to these resources. However, it seems unlikely given current usage patterns that such resources would be made available in a fashion that would make their integration in to large-scale grid solutions practical.

We believe that the 2DRMP-G demonstrates that a grid infrastructure can provide a good environment for computational physics infrastructure and that is a 2DRMP-G sustainable infrastructure for R-matrix development.

Acknowledgements

The authors are grateful to the UK EPSRC for their support through the UK e-Science Programme and HPC grant GR/R3118/01 on Multiphoton, Electron Collisions and BEC. VFM acknowledges the receipt of an ESF/Queen's University postgraduate studentship.

References

1. R Alan et al, Building the e-Science Grid in the UK: Middleware, applications and tools deployed at Level 2, UK e-Science All Hands Meeting, September 3-5 2003, p556-561 (http://www.nesc.ac.uk/events/ahm2003/AHMCD/)
2. Burke, P.G., Berrington, K.A.(eds): Atomic and Molecular Processes and R-matrix Approach. IOP Publishing, Bristol (1993)
3. Burke,V.M. and Noble, C.J.: Comput. Phys. Commun. **85**(1995)471-500
4. Heggarty, J.W., Burke, P.G., Scott, M.P., Scott, N.S.: Compt. Phys. Commun. **114** (1998) 195-209
5. Monthly reports on HPCx's activities and performance (http://www.hpcx.ac.uk/projects/reports/monthly/)
6. UK National Grid Service - http://www.ngs.ac.uk

Design of an OGSA-Compliant Grid Information Service Using .NET Technologies

Ranieri Baraglia[1], Domenico Laforenza[1],
Angelo Gaeta[2], Pierluigi Ritrovato[2], and Matteo Gaeta[2]

[1] Information Science and Technologies Institute,
Italian National Research Council, Pisa, Italy
(ranieri.baraglia, domenico.laforenza)@isti.cnr.it

[2] CRMPA - University of Salerno, Salerno, Italy
(agaeta, ritrovato, gaeta)@crmpa.unisa.it

Abstract. The new trend of Grid is represented by the migration towards a model built on concepts and technologies inheriting from Grid and Web Services communities. By merging these two technologies, the new concept of Grid Service has emerged and has been formalized through the introduction of the Open Grid Services Architecture (OGSA). OGSA defines the semantics of Grid Service instance, such as, how it is created, how it is named, how its lifetime is determined. OGSA places no requirements on implementation aspects. No mandatory constraint is requested for implementing Grid Services; the developers can exploit the features that characterize their target implementation environment.

In the framework of the Grid Based Application Service Provision (GRASP) European project, we are developing a OGSA-compliant middleware by using .NET technologies. According to this goal we are investigating the design of a Grid Information Service (GIS) by exploiting these technologies, which we consider an interesting operating environment for Grid Services, due to the high level of integration of the provided services. Advantages and drawbacks of both technologies, and our solution are here pointed out[1].

Keywords: Grid, OGSA, Middleware, Grid Information Service, Web Services.

1 Introduction

A Grid Information Service (GIS) holds information about people, software, services and hardware that participate in a Grid, and more generally in a virtual

[1] This work is partially supported by the European Community under the Infomation Society Technologies (IST) - 5th Framework Programme for RTD - project GRASP, contract no.: IST-2001-35464. This document does not represent the opinion of the European Community, and the European Community is not responsible for any use that might be made of data appearing therein.

M. Daydé et al. (Eds.): VECPAR 2004, LNCS 3402, pp. 247–259, 2005.

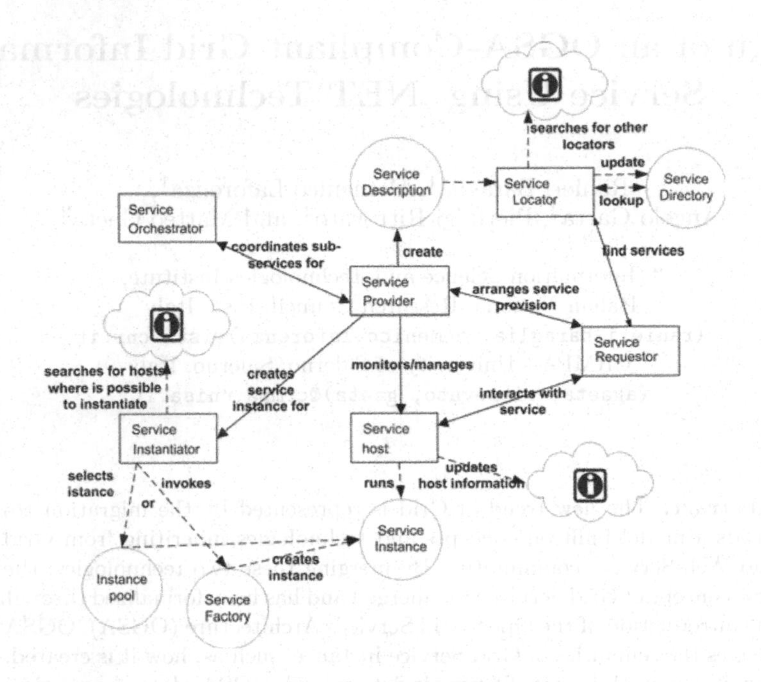

Fig. 1. Interactions between internal roles and entities of the GRASP environment

organization (VO). This wide correlated set of static and dynamic information is available upon request. A GIS should also provide protocols, methods, APIs, etc. for description, discovery, lookup, binding and data protection [1].

In a Grid environment there are many diversities in terms of nature of the information sources (e.g. physical characteristics, location, sensitivity), of demands placed on those information sources (e.g. access rates, query vs. subscribe, accuracy demands) and ways in which information is employed (e.g. discovery, brokering, monitoring).

Currently, the GIS architecture is based on two fundamental entities: highly distributed information providers and specialized aggregate directory services. Information providers form a common VO neutral infrastructure providing the access to detailed and dynamic information about Grid entities. Aggregate directories provide, often specialized, VO-specific views of federated resources, services and so on [2].

The goal of this paper is to investigate the design of a GIS by using .NET technologies, which we consider an interesting operating environment for Grid Services, due to the high level of integration of the provided services. This work was conducted within the activities related to the Grid Based Application Service Provision (GRASP) European project [3]. One of the main project objectives is to design and implement a layered architecture for Application Service Provision (ASP) using GRID technologies. The GRASP system environment internal roles and their interactions are shown in Figure 1. The internal roles are shaped like rectangles while the relations between internal roles and entities (e.g., service instance, service directory) are drawn using dashed lines. The Information Service, shown in Figure 1 as clouds containing the "i", is not a centralized one, but distributed inside the GRASP environment. Figure 1 also shows the interactions between internal roles and the Information Service.

The paper is structured as follows. Section 2 and Section 3 which describe "a service view" of an information service implemented in GT3, and the Microsoft (MS) technologies that can be exploited to implement an OGSA-compliant GIS, respectively. Section 4 describing a possible use of a GIS in a service-oriented environment. Finally, Section 5 which present some conclusions along with future directions.

2 The Current Service View of an Information Service

In our work we have made reference to the Information Service developed in the Globus Toolkit (GT3) [4]. GT3 is based on Open Grid Service Architecture (OGSA) specifications [5]. In GT3, Monitoring and Discovery Services (MDS) are tremendously changed with respect to the GT2 [6]. Here there is not a standalone service (like the MDS in GT2), but the Information Service is built upon Service Data and Service Data Definitions, which are part of the OGSA core, and specialized collective-layer Index Services that are delivered as part of GT3.

The OGSI [7] core in GT3 provides a generic interface for mapping Service Data queries and subscriptions for Service Data notification to service implementation mechanisms. In essence, this subsumes the role of the Grid Resource Information Service (GRIS) backend server module in MDS, while relying on more basic OGSA binding mechanisms for secure access to the query interface in place of the GSI-enabled LDAP protocol [8].

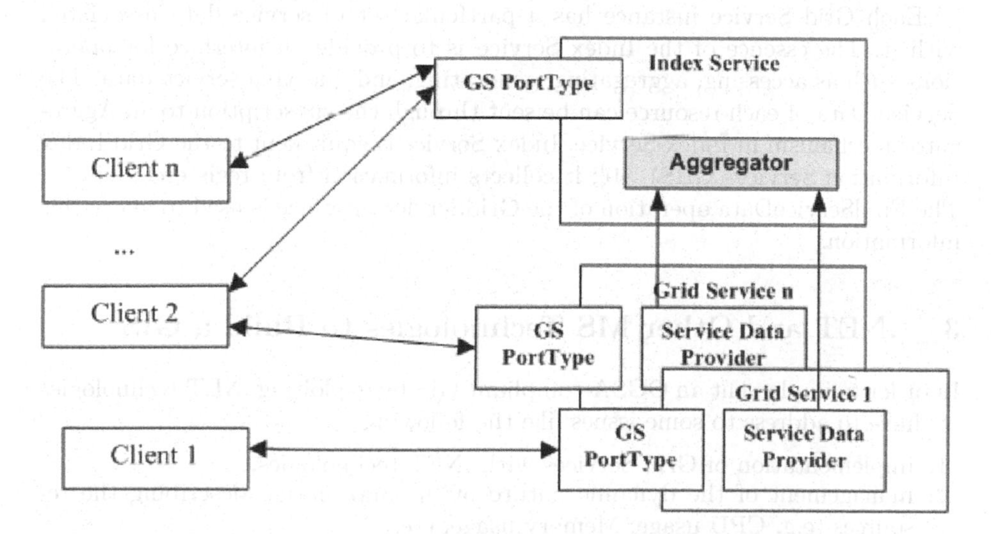

Fig. 2. An information service view implemented in GT3

GT3 will provide Service Data Providers as part of its base service implementation. These Service Data sources within the base service implementations replace the role of the host-information provider scripts in MDS. The base service implementations expose status information as well as probed, measured, or discovered platform information according to well-defined Service Data Definitions in their Service Type WSDL [9]. These Service Data Definitions in WSDL subsume the role of MDS schema written in the RFC2252 LDAP schema format, and are used to standardize the information exchange between providers and consumers within specific service domains.

The Index Service components in GT3 provide the following features: a) An interface for connecting external information provider programs to service instance, b) A generic framework to aggregate Service Data, c) A registry of Grid Services. Typically, there is one Index Service per VO. However, when multiple large sites compose a VO, each site can run its own Index Service that will index the resources available at the site level. In this case each Index Service would be included in the VO's Index Service.

With the GT3 Index Service, there will be multiple ways to obtain information: MDS-style query propagation, Cache prefetch (each source of information can be independently selected for cache prefetch), and Subscription/Notification. In Figure 2 a "service view" of the Information Service is shown.

According to the service oriented view, each resource is virtualized in a service. In Figure 2 the Grid Services representing resources are depicted. Each resource has a Service Data Provider that takes the role of the GRIS service of MDS. As mentioned above, GT3 offers some Service Data Providers but it is even possible to use customized and/or external providers. Like in MDS, a client can directly query a resource by invoking the FindServiceData operation of the Grid Service PortType.

Each Grid Service instance has a particular set of service data associated with it. The essence of the Index Service is to provide an interface for operations such as accessing, aggregating, generating, and querying service data. The Service Data of each resource can be sent through the subscription to an Aggregator mechanism in Index Service. Index Service is equivalent to the Grid Index Information Service (GIIS) [10]; it collects information from registered "child". The FindServiceData operation of the GridService interface is used to query this information.

3 .NET and Other MS Technologies to Build a GIS

In order to implement an OGSA-compliant GIS by exploiting .NET technologies we have to address to some issues like the following:

1. implementation of Grid Services with .NET technologies,
2. management of the dynamic nature of the information describing the resources (e.g. CPU usage, Memory usage, etc.),
3. extension of the MS directory service functionalities (e.g. Active Directory) in order to implement the OGSA Index Service functionalities.

The first issue is related to the implementation of Grid Service Specification prescribed in [7], in a MS .NET language [11], [12]. In the framework of the GRASP project, we have selected the implementation of Grid Service Specification provided by the Grid Computing Group of the Virginia University, named OGSI.NET [13].

To manage the dynamic nature of information describing the resources, GT3 leverages on Service Data Providers. In the MS environment, we rely on Performance Counters and Windows Management Instrumentation (WMI) architecture to implement the Service Data Providers. For each component of a MS system we have a performance object (e.g. Processor Object) gathering all the performance data of the related entity. Each performance object provides a set of Performance Counters that retrieves specific performance data regarding the resource associated to the performance object. For example, the "% Processor Time" is a Performance Counter of the Processor Object representing the percentage of time during which the processor is executing a thread. The performance counters are based on services at the operating system level, and they are integrated in the .NET platform. In fact, .NET Framework [11] provides a set of APIs that allows the management of the performance counters.

To perform the collection and provisioning of the performance data to an index service, we leverage on Windows Management Instrumentation (WMI) [14] architecture. WMI is a unifying architecture that allows the access to data from a variety of underlying technologies. WMI is based on the Common Information Model (CIM) schema, which is an industry standard specification, driven by the

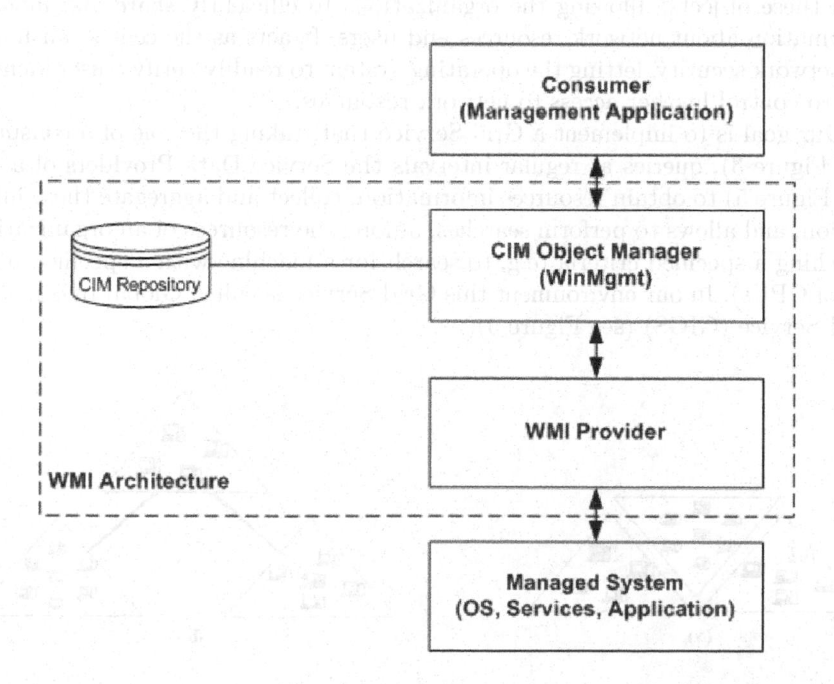

Fig. 3. WMI Architecture

Distributed Management Task Force (DMTF) [15]. WMI provides a three-tiered approach for collecting and providing management data. This approach consists of a standard mechanism for storing data, a standard protocol for obtaining and distributing management data, and a WMI provider. A WMI provider is a Win32 Dynamic-Link Library (DLL) that supplies instrumentation data for parts of the CIM schema. Figure 3 shows the architecture of WMI.

When a request for management information comes from a consumer (see Figure 3) to the CIM Object Manager (CIMON), the latter evaluates the request, identifies which provider has the information, and returns the data to the consumer. The consumer only requests the desired information, and never knows the information source or any details about the way the information data are extracted from the underlying API. The CIMOM and the CIM repository are implemented as a system service, called WinMgmt, and are accessed through a set of Component Object Model (COM) interfaces.

WMI provides an abstraction layer that offers the access to many system information about hardware and software and many functions to make calculations on the collected values. The combination of performance counters and WMI realizes a Service Data Provider that each resource in a VO should provide.

To implement an OGSA-compliant Index Service, we exploit some Active Directory (AD) features [16]. AD is a directory service designed for distributed networking environments providing secure, structured, hierarchical storage of information about interesting objects, such as users, computers, services, inside an enterprise network. AD provides a rich support for locating and working with these objects, allowing the organizations to efficiently share and manage information about network resources and users. It acts as the central authority for network security, letting the operating system to readily verify a user identity and to control his/her access to network resources.

Our goal is to implement a Grid Service that, taking the role of a consumer (see Figure 3), queries at regular intervals the Service Data Providers of a VO (see Figure 5) to obtain resources information, collect and aggregate these information, and allows to perform searches, among the resources of an organization, matching a specified criteria (e.g. to search for a machine with a specified number of CPUs). In our environment this Grid Service is called Global Information Grid Service (GIGS) (see Figure 5).

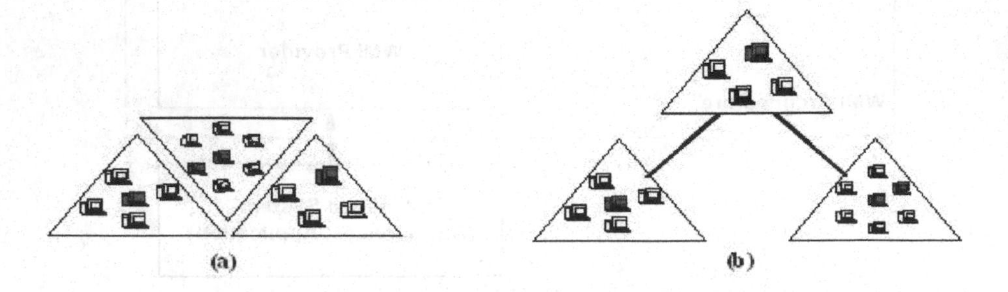

Fig. 4. Active Directory partitioning scheme

The hosts that run the GIGS have to be Domain Controllers (DC). A DC is a server computer, running on Microsoft WindowsNT, Windows2000, or Windows Server2003 family operating systems, that manages security for a domain. The use of a DC permits us to create a global catalog of all the objects that reside in an organization, that is the primary goal of AD services. This scenario is depicted in Figure 4 (a) where black-coloured machines run GIGS and the other ones are Service Data Providers. Obviously, black-coloured hosts could also be Service Data Providers.

In order to avoid that the catalog grows too big and becomes slow and clumsy, AD is partitioned into units, the triangles in Figure 4 (a). For each unit there is at least a domain controller. The AD partitioning scheme emulates the Windows 2000 domain hierarchy (see Figure 4 (b)). Consequently, the unit of partition for AD services is the domain. GIGS has to implement an interface in order to obtain, using a publish/subscribe method, a set of data from Service Data Providers describing an active directory object. Such data are then recorded in the AD by using Active Directory Service Interface (ADSI), a COM based interface to perform common tasks, such as adding new objects.

After having stored those data in AD, the GIGS should be able to query AD for retrieving such data. This is obtained exploiting the Directory Services

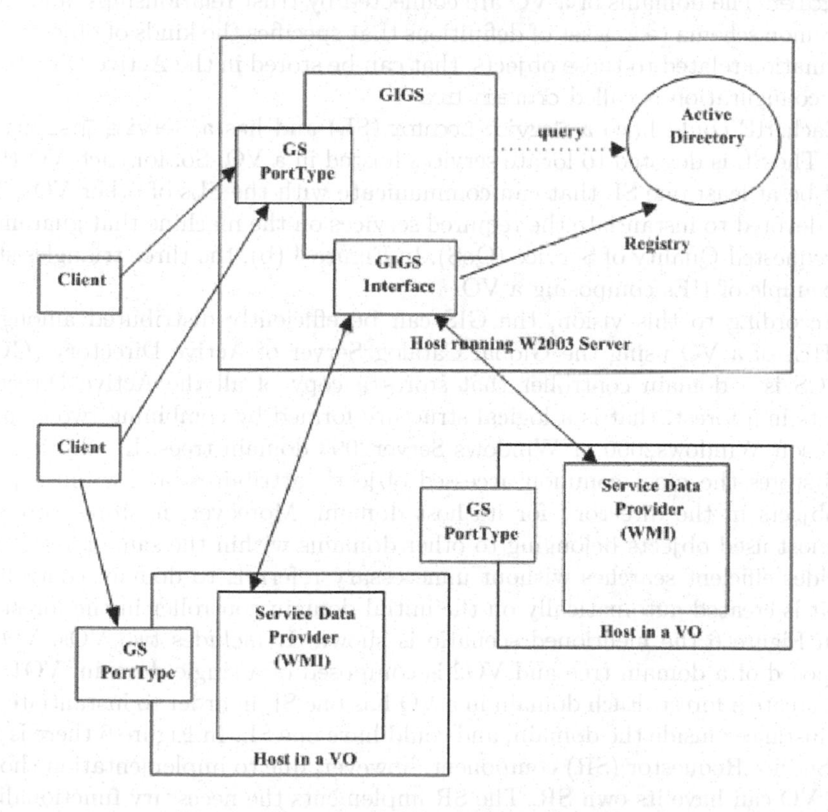

Fig. 5. Example of GIS in MS Environment

Markup Language (DSML) Services for Windows (DSfW). DSML provides a means of representing directory structural information and directory operations as an XML document. The purpose of DSML is to allow XML-based enterprise applications to leverage profile and resource information from a directory in their native environment. DSfW uses open standards such as HTTP, XML, and SOAP, so a high level of interoperability is possible. Furthermore, it supports DSML version 2 (DSMLv2), a standard supported by OASIS [19]. The solution is schematized in Figure 5, where each host is virtualized in a grid service.

4 Scenario

The scenario that we will show in this section will be developed in the framework of the GRASP project. It is presented in order to demonstrate a possible use of a GIS in a service oriented environment.

The scenario here considered is composed of two different VOs, each of them enclosing a set of Service Providers (SPs) that offer various services for their clients inside or outside VOs (see Figure 6). Different domains, called `Hosting Environments` (HEs), compose a VO and each domain has, at least, one `domain controller`. The domains of a VO are connected by trust relationships, and share a common schema (e.g. a set of definitions that specifies the kinds of objects, and information related to those objects, that can be stored in the Active Directory). This configuration is called `domain tree`.

Each HE could have a `Service Locator` (SL) and has a `Service Instantiator` (SI). The SL is devoted to locate services hosted in a VO. So, for each VO there must be at least one SL that can communicate with the SLs of other VOs. The SI is devoted to instantiate the required services on the machine that guarantees the requested Quality of Service (QoS). In Figure 4 (b), the three triangles show an example of HEs composing a VO.

According to this vision, the GIS can be efficiently distributed among all the HEs of a VO using the Global Catalog Server of Active Directory (GCS). A GCS is a domain controller that stores a copy of all the Active Directory objects in a forest, that is a logical structure formed by combining two or more Microsoft Windows2000 or Windows Server2003 domain trees. In addition, the GCS stores the most common accessed object's attributes, and a full copy of all objects in the directory for its host domain. Moreover, it stores copies of the most used objects belonging to other domains within the same forest. GCS provides efficient searches without unnecessary referrals to domain controllers, and it is created automatically on the initial domain controller in the forest.

In Figure 6 the mentioned scenario is shown. It includes two VOs: VO1 is composed of a domain tree and VO2 is composed of a single domain. VO1 and VO2 create a forest. Each domain in a VO has one SI, in order to instantiate service instances inside the domain, and could have one SL. In Figure 6 there is just one Service Requestor (SR) component, however, due to implementation choices each VO can have its own SR. The SR implements the necessary functionalities to invoke and operate a service.

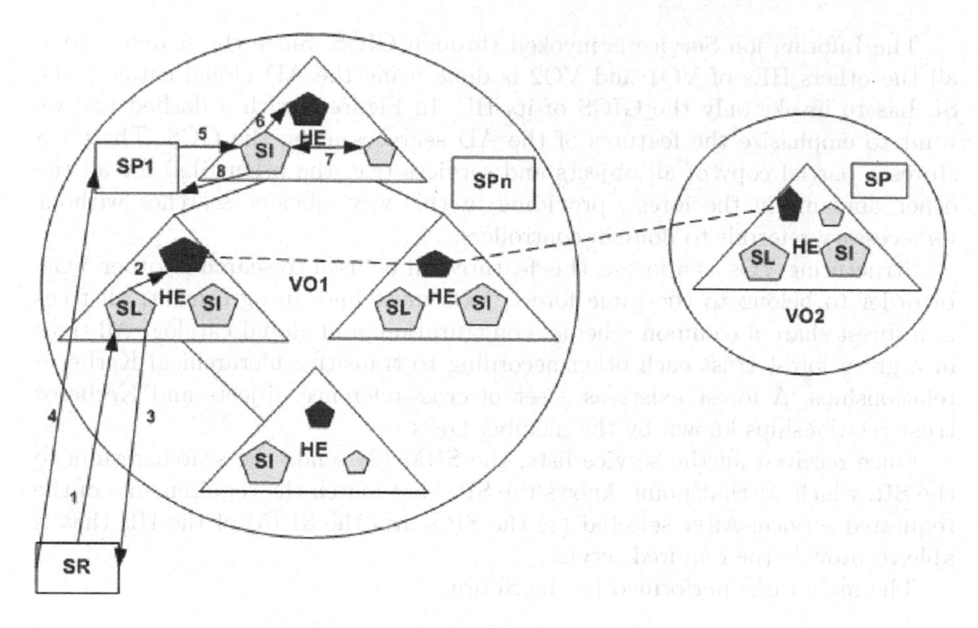

Fig. 6. Scenario

In Figure 6 the black-coloured machines are domain controllers and run the GIGS, and the numbers associated to each link represent the sequence of performed operations. When the SR issues a request for a service, it contacts (1) a SL of VO1 forwarding to it a structure that contains some attributes that identify the service being requested (e.g. QoS required, type of service). In order to execute more requests, the SL is made up of three processes that run in pipeline fashion:

1. `RequestRecipient` (RR): receives and processes all service location requests,
2. `RequestManager` (RM): accesses its related Service Directory to find the requested service and sends to all the other SLs a request for the same service,
3. `SearchResultManager` (SRM): receives the research results (a Service Lists), both local and remote. SRM, after receiving the original request from the SR, acts as master that collects the Service Lists from all the other SLs and sends them to the SR.

When the processed request arrives to the RM, it searches for the requested service inside its own service directory (UDDI) and contacts (2) the Information Service (invoking the GIGS belonging to its domain) in order to obtain the addresses of the other SLs. Once obtained the addresses of other SLs, the RM sends them the request to perform the same search simultaneously.

RM sends the number of SLs involved in the search to the SRM. The search of a service is considered as completed when SRM has received a number of Service Lists equal to the number of SLs involved in the search.

The Information Service is invoked through GIGS. Since the search among all the others HEs of VO1 and VO2 is done using the AD global catalog, the SL has to invoke only the GIGS of its HE. In Figure 6 with a dashed line we want to emphasize the features of the AD searches using the GCS. The GCS stores a partial copy of all objects and services (e.g. the other SLs) for all the other domains in the forest, providing in this way efficient searches without unnecessary referrals to domain controllers.

Structuring VOs as a forest, this feature can be used to search in other VOs. In order to belong to the same forest, VOs must find an agreement; all trees in a forest share a common schema, configuration, and global catalog. All trees in a given forest trust each other according to transitive hierarchical Kerberos relationships. A forest exists as a set of cross-reference objects and Kerberos trust relationships known by the member trees.

Once received all the service lists, the SRM (3) sends the search results to the SR, which, at that point, knows the SPs that match the requirements of the requested service. After selected (4) the SP starts the SI (5) of the HE that is able to provide the required service.

The main tasks performed by the SI are:

1. checks the SP rights in order to verify if it can create the requested service with the requested QoS,
2. chooses, between all the hosts of a HE that could run a service instance of the required type, the host able to guarantee the requested requirements (e.g., QoS, Price),
3. manages pools of idle instances.

To perform these tasks, the SI has to interact with other subsystems of the GRASP environment, such as the Information Service and Accounting. For example, to choose the right factory the SI has to be able to refine the information about QoS required in a set of host-figures (e.g. CPU rate, Memory required). Factory Manager, an object of the SI subsystem, does this operation. It accepts in input a level of QoS related to the service, and refines it into hosts data. The Factory Manager (6) performs a search in the Information Service, filtered by hosts data, price data and service type. A SI only queries the domain controller of its HE since it is able to instantiate inside its HE only. Information Service returns a list of hosts able to guarantee the required QoS. Factory Manager selects a host, eventually performing a more fine-grained selection on the basis of information received by other subsystems (e.g. the Monitoring subsystems of the selected hosts). By exploiting these two selection levels, we perform a load-balancing task inside the HE. Once obtained the right host, the SI (7) invokes the factory service on that host and obtains the service locator (a structure that contains zero or more GSHs and zero or more GSRs) [7] of the created instance. The service locator is then returned to the SP (8) and the SP returns it to the SR.

5 Conclusions

Our work was focused on providing a possible implementation of a Grid Information Service using .NET technologies, trying to take full advantage from the MS technologies features. This is a work in progress in the framework of the GRASP project. Then, due to the project requirements, it could change under way.

On the basis of our experience with the GRASP project, OGSA seems to be an ideal environment for development of dynamic, collaborative, multi-institutional and VO based scenarios, as the one described in this paper (See Section 4), where a global information model is fundamental. In our opinion, the generation of these kinds of scenarios, identifying key technologies and addressing user's needs in different fields such as eLearning, eHealth, eCommerce, is one of the greater challenges of ICT. OGSA tries to address this challenge by defining requirements and service to support systems and applications in both eScience and eBusiness [21]. This is not, obviously, a simple task and feedbacks have to come from international projects and initiatives, as the aforementioned GRASP. One potential risk, in our opinion, is related to *usability* of Grid. Grid architectures and systems must not be used only by IT skilled people but mainly by doctors, business managers and students, otherwise the big challenge of any new technology, that is to improve every user's life, will be dropped. While OGSA defines a solid foundation upon which IT professionals can build systems and applications, research must be done in order to define and develop intelligent user interfaces that should help to bridge the "semantic gap", that is the gap existing between user's demands (in a natural language) and how these demands are interpreted by the infrastructure, making Grid easy to handle and trasparent. These and other important properties are research topics of the so called Next Generation Grids (NGG)[20].

In our work, we want to leverage on MS technologies in order to:

1. validate the use of MS technologies to develop grid middleware and grid services,
2. verify advantages of MS technologies,
3. investigate interoperability between grid services developed with different technologies.

The OGSA semantic definition of Grid Service instance does not dictate a particular hosting environment for Grid Services and promotes the development of many more environments such as J2EE, Websphere, .NET, and Sun One. Currently, there are few projects going on this way. In our opinion, in order to obtain an integrated environment, composed by distributed, heterogeneous, dynamic "virtual organizations", formed by many and different resources, the development of different hosting environments should be investigated and encouraged.

The adopted approach has some advantages. The MS technologies described in the paper are fully integrated in all the new MS operating systems and the solution proposed to implement a GIS is based directly upon them. In this way, we rely on technologies optimized for .NET hosting environments, that means an improvement from a performance viewpoint.

For example, among the Service Data Providers provided by GT3 there is a Java-based host information data provider that produces the following types of data: CPU count, memory statistics, OS type, and logical disk volumes. Obviously, this solution can be used in Windows operating systems. On the other side, our solution for the development of Service Data Providers based on the Performance Counter and WMI architecture, represents a valid and feasible alternative that doesn't need integration with third party software.

Furthermore, the features of the Active Directory (AD) present several advantages when used to implement an Information Service. In particular, we have seen that the use of global catalog permits us to distribute the Information service among the domains of a Virtual Organization (VO), allowing more efficiently searches of information, avoiding the necessity to query all the domain controllers.

The conducted experience has shown that the AD is flexible, robust, efficient, extensible and scalable. However, if we want to take advantage from AD, the set of VO members must share a common schema, configuration, and global catalog.

The studies performed in this work have shown that the MS technologies and the MS operating systems features can be used to develop grid middleware service oriented. Following the new steps towards a refactoring of the OGSI in terms of Web Services [18], as future work we plan to investigate the features of the new Microsoft operating system named Longhorn [17]. It presents key innovations for Presentation Layer, Data Layer and Communications Layer. All these layers provide their functionalities in terms of Web Services and our effort will be focused on investigating how these functionalities can be extended to realize an infrastructure for the development of Grid Services. In this way, a developer can rely on an OGSA compliant infrastructure "embedded" in the operating system.

References

1. Foster I. and Kesselman C., *The Grid: Blueprint for a Future Computing Infrastructure*, Morgan Kaufmann Publishers, USA, 1999.
2. I. Foster, C. Kesselman, S. Tuecke. "The Anatomy of the Grid: Enabling Scalable Virtual Organizations" *International J. Supercomputer Applications*, 15(3), 2001.
3. http://www.eu-grasp.net
4. I. Foster, C. Kesselman, J. M. Nick, S. Tuecke, "The Physiology of the Grid: An Open Grid Services Architecture for Distributed Systems" Integration, *Open Grid Service Infrastructure WG, Global Grid Forum*, June 22, 2002.
5. http://www.globus.org/ogsa/releases/alpha/docs/infosvcs/indexsvcoverview.html
6. http://www.globus.org/mds/
7. S. Tuecke, K. Czajkowski, I. Foster, J. Frey, S. Graham, C. Kesselman, T. Maquire, T. Sandholm, D. Snelling, P. Vanderbilt, "Grid Service Specification", *Open Grid Service Infrastructure WG, Global Grid Forum*, Version 1, April 5, 2003. http://www.ggf.org/ogsi-wg
8. T. A. Howes, M. Smith. "A scalable, deployable directory service framework for the Internet" *Technical report*, Center for Information Technology Integration, University of Michigan, 1995.

9. L. Wall, A. Lader, *Web Services Building Blocks for Distributed Systems*, Prentice-Hall, Inc., 2002.
10. K. Czajkowski, I. Foster, N. Karonis, C. Kesselman, S. Martin, W. Smith, S. Tuecke, "A resource management architecture for metacomputing systems", in *Proceedings of the 4th Workshop on Job Scheduling Strategies for Parallel Processing*, 1998, pp. 4-18.
11. L. Wall, A. Lader, *Building Web Services and .Net Applications*, McGraw-Hill/Osborne, 2002.
12. http://www.microsoft.com/net/
13. http://www.cs.virginia.edu/gsw2c/ogsi.net.html
14. http://msdn.microsoft.com/library/enus/wmisdk/wmi/about_wmi.asp
15. http://www.dmtf.org/standards/standardcim.php
16. http://www.msdn.microsoft.com/windows2000/activedirectory/
17. http://msdn.microsoft.com/Longhorn/
18. http://www.globus.org/wsrf/default.asp
19. http://www.microsoft.com/windows2000/server/evaluation/news/bulletins/dsml.asp
20. Next Generation Grid(s) European Grid Research 2005-2010, Expert Group Report, 16 June 2003
21. I. Foster, D. Berry, H. Kishimoto et al. "The Open Grid Services Architecture Version 1.0", July 12, 2004. http://forge.gridforum.org/projects/ogsa-wg

A Web-Based Application Service Provision Architecture for Enabling High-Performance Image Processing

Carlos de Alfonso[1], Ignacio Blanquer[1], Vicente Hernández[1], and Damià Segrelles[1]

Universidad Politécnica de Valencia - DSIC,
Camino de Vera s/n, 46022 Valencia, Spain
{calfonso, iblanque, vhernand, dquilis}@dsic.upv.es
http://www.grycap.upv.es

Abstract. The work described in this paper presents a distributed software architecture for accessing high performance and parallel computing resources from web applications. The architecture is designed for the case of medical image processing, although the concept can be extended to many other application areas. Parallel computing algorithms for medical image segmentation and projection are executed remotely through Web services as being local. Ubiquity and security on the access and high performance tools have been the main design objectives. The implementation uses the Application Service Provision model, of increasing importance in the exploitation of software applications. The system is under clinical validation and considers all the industrial and de-facto standards in medical imaging (DICOM, ACR-NEMA, JPEG2000) and distributed processing (CORBA, Web Services, SOAP).

1 Introduction

The introduction of multislice Computer Tomography (CT) and higher-field Magnetic Resonance Imaging (MRI) has drastically increased the size of medical imaging studies. Resources needed for processing such large studies exceed those available in standard computers, and High Performance Computing (HPC) techniques have been traditionally applied to solve it.

Until the appearance of commodity-based clusters of computers, access to parallel computing resources was limited to research centers who can afford large investment in parallel computers. Cluster computing has leveraged the use of HPC, although it has increased the complexity of parallel programming. However, current parallel applications require dedicate resources and are designed as monolithic applications. The work presented in this article suggests the use of client/server models for accessing shared computing resources seamlessly and transparently. Data servers, computing resources and client workstations cooperate on a distributed environment.

Moreover, medical information is stored sparsely and medical records are unlikely up-to-date. The remote access to medical information is a very important

M. Daydé et al. (Eds.): VECPAR 2004, LNCS 3402, pp. 260–273, 2005.

issue that is undertaken by many different points of view, such as the standardization of health records (CEN ENV 13606), telemedicine, collaborative working, etc. Medical images need to be accessed and processed from outside of the hospitals. Dedicated networks are expensive, but public networks (such as the Internet) are unreliable and generally slower. Notwithstanding the difficulties, access through public networks is preferred, since they are widely available. The purpose of the work presented in this paper is the demonstration of the suitability of Internet as the communication media for combining high performance resources, access to medical images and workflow control for the implementation of a distributed and secure accessible image diagnosis system.

The objective is the design and implementation of a software architecture able to access efficiently high performance resources able to execute distributed memory parallel computing algorithms on large medical image datasets and transmit efficiently and securely the results to a standard computer, only requiring a web browser and a standard Internet connection on the client side.

The following sections describe the status of the technology, the design and implementation of the main components (parallel computing server and algorithms and web services based architecture) as well as the results obtained with such approaches. Conclusions and future work are finally presented.

2 Status of the Technology

The work presented in this paper combines medical imaging, high-performance computing and web-based access to computing and data resources. Next subsections present the status of the technology on each area.

2.1 Medical Imaging and High Performance Computing

Medical imaging is the projection of medical information on a plane. Medical imaging is generally related to anatomy, and image pixels present different intensities depending on the different tissues and structures (fat, air or specific organs). Tissues are identified using different physical properties. Radiation sources are used in X-Rays (2D) and Computer Tomography (3D) and non-invasive sources such as sound waves or magnetic fields are used in Ultrasound (2D or 3D) and Magnetic Resonance Imaging (3D). On the other side, functional imaging reflects dynamic properties and electric or magnetic activity or even chemical reactions (Single Proton Emission Computer Tomography or SPECT and Positron Emission Tomography or PET).

The use of Information and Communication Technologies (ICT) on Medical Imaging enable to process the large number of studies produced daily and with a high degree of repeatability. ICT provides with tools for complex operations such as tissue classification, 3D projection or fusion of different modalities.

Segmentation is the association of the voxels (3D pixels) of an image to tissues [1, 2]. Difficulties in segmentation arise from several factors: Artefacts are introduced during the acquisition and preprocessing, the homogeneity of organs and tissues is usually poor, the ratio of signal vs. noise is small. Moreover, in some cases is compulsory to have a previous knowledge of the size or shape of the structures. Manual segmentation methods are tedious and lack of repeatability, but automatic methods are not feasible for general purpose images.

Thus, semiautomatic segmentation is the most suitable solution for general images. Typical semiautomatic methods are Threshold classification and Region Growing. These methods rely on the homogeneity of the regions of interest to be segmented, defining the interval of intensity levels that relate to each tissue in the former and identifying by neighbourhood and similarity the voxels in the latter. Threshold segmentation is typically inaccurate since there is not a direct relation between intensity level and tissue. Region growing on the other side, lacks of repeatability due to the need of user interaction. However, a combination of both methods leads to good results, since Threshold segmentation can select homogeneous isolated regions that can be separated by Region Growing.

Semiautomatic methods can be improved computing the threshold levels according to standard Hounsfield units or by statistic clustering of the image. Gradients, image blurring or multiresolution processes can reduce segmentation path leakages or undesired holes. Parallel computing methods for segmentation are not deeply exploited: although parallel computing threshold segmentation is straightforward, parallel computing version for other segmentation methods are far more complex. 3D Medical images are usually projected in 3D for better diagnosis and surgery planning. Basic 3D projection consists either on the extraction of geometrical models of the organs and tissues (surface rendering [3]), or direct projection of the volumetric information (volume rendering [4]). Surface rendering require segmenting the image, computing and filtering a mesh of the surface of the organ, and computing the projection using standard graphics hardware. Surface rendering discards the volumetric information. Volume rendering instead, use directly the information available on the 3D image and provide images that are more relevant for medical diagnosis.

Two main approaches are considered for Volume Rendering: Back-to-front [5] and front-to-back projection [6]. A widely used back-to-front method is Splatting, which consists on projecting the voxels along a projection direction. Voxels are interpolated on the image plane by a function kernel known as 'splat'. Front-to-back methods consist on the projection of rays from the image to the volume, and interpolating the contribution of each voxel traversed. Both methods produce a bitmap with the projection of the 3D volume. Different types of images are obtained depending on the function that computes the contribution of each voxel. Maximum Intensity Projection (MIP) [7] images consider the value of the projected pixel as the maximum of the intensities of the voxels either traversed by the ray or projected on the pixel. Surface Shading Projection [8] considers the amount of light reflected from the surface boundaries of the voxels projected.

Parallel versions of volume rendering algorithm consist either on distributing the data volume among a set of processors (back-to-front methods) [9] or dividing the image plane among them (front-to-back methods) [10]. Parallel Splatting-like methods are more scalable in memory, although the postprocessing affects the scalability on computing cost. On the other side, Parallel Ray Casting-like methods require replication of the data volume (low memory scalability), but since postprocessing is not needed, it obtains good computing scalability.

2.2 Web-Based Access to High Performance Resources

Remote access to processes is a problem that has been extensively analyzed. Many solutions have been proposed to access remote objects and data. An object-oriented precursor standard for remote procedure call is CORBA (Common Object Request Broker Access). CORBA eases the remote execution of processes with independence of architecture and location. CORBA shares with GRID concepts such as resource discovering, integrity, ubiquity or platform independence. CORBA is supported by all main manufacturers for distributed interoperability of objects and data access. Nevertheless one of the core problems addressed by GRID has been security, authentication and accounting problems. These issues are not clearly solved in CORBA, although there are initiatives working on establishing a confident security platform.

The consolidation of the Web as a main infrastructure for the integration of information opened the door to a new way for interoperating through web-based protocols. Pioneer web services started using Web Interface Definition Language (WIDL), a first specification for describing remote Web services. The evolution of WIDL leaded to Simple Object Access Protocol (SOAP), which has support for message-oriented as well as procedural approaches. SOAP provides a simple and lightweight mechanism for exchanging structured and typed information between peers in a decentralized, distributed environment using XML. Last evolution in Web Services is the use of Web Services Description Language (WSDL). It is a simple way for service providers to describe the basic format of requests to their systems regardless of the underlying protocol (such as SOAP or XML) or encoding (such as MIME). WSDL is a key part of the effort of the Universal Description, Discovery and Integration (UDDI) initiative to provide directories and descriptions of services for electronic business.

However, all these technologies have been developed considering mainly interoperability, rather than High Performance Computing (HPC). Problems relevant to HPC such as load balancing, communication costs, scalability, process mapping or data reutilization have not addressed. Several efforts have been performed on other projects. Several approaches [11, 12, 13] treat the HPC components as black boxes. More complex approaches [14, 15, 16] extend the syntax and semantics of CORBA and similar ORB (Object Request Broker) technologies to deal with data distribution, process mapping and memory reusing. Unfortunately all these approaches have reached a moderate impact, incomparable to the success of conventional ORBs.

But the take up of HPC on the Internet has arrived with the GRID. As defined in [17], a GRID provides an abstraction for resource sharing and collaboration across multiple administrative domains. Resources are physical (hardware), informational (data), capabilities (software) and frameworks (middleware). The key concept of GRID is the Virtual Organization (VO). A VO [18] refers to a temporary or permanent coalition of geographically dispersed individuals, groups or organizational units (belonging or not to the same physical corporation) that pool resources, capabilities and information to achieve a common objective.

The VO concept is the main difference among GRID and other similar technologies. High throughput, High Performance Computing (HPC) and Distribute Databases are usually linked to GRID. High throughput applications can be executed locally in computing farms up to a scale only limited by the resources, not involving cooperation. HPC is more efficient on supercomputers or clusters than in GRID due to the overhead of encryption and the latencies of geographically distributed computers. Access to distributed data is efficiently provided by commercial databases. GRID is justified when access rights policies must be preserved at each site, without compromising the autonomy of the centers.

GRID technologies have evolved from distributed architectures. Recent trends on GRID are being directing towards Web service-like protocols. Open GRID Services Architecture (OGSA) represents an evolution towards a GRID system architecture based on Web services concepts and technologies. OGSA defines uniform exposed service semantics (the GRID service); defines standard mechanisms for creating, naming, and discovering transient GRID service instances; provides location transparency and multiple protocol bindings for service instances; and supports integration with underlying native platform facilities. The OGSA defines in terms of WSDL the mechanisms required for creating and composing sophisticated distributed systems.

Although OGSA is an appropriate environment, many problems need to be faced for achieving good performance, higher security and true interoperability. The aim of the work presented in this article has been to develop an architecture specific for the problem of medical image diagnosis, but considering conventional ORB, Web services' gate to HPC, GRID security issues and ASP models.

3 The ASP - HPC Architecture

The architecture presented in this work comprises two main structures. On one hand, a high performance server implements components that provide parallel image processing methods. This structure enables a platform independent access to the high-performance computing. On the other side, an ASP and Web service architecture provides the means for accessing to these services through web browsers and Java applications. Figure 1 shows a diagram of the architecture.

HPC is ensured by the use of a homogeneous and dedicated cluster of PCs, and the use of MPI and POSIX threads for the implementation of the parallel algorithms. Interoperability is achieved by the use of a CORBA front-end, which enables the access from Unix, MS Windows or any other flavor of Operating Systems. The web services provide an interface to applications compatible with current trends of GRID architectures. The ASP server implements all the necessary components for data encoding and decoding, information flow, security and data access. Finally, the web browsers execute the applets that show the information to the users. Next subsections describe both architectures.

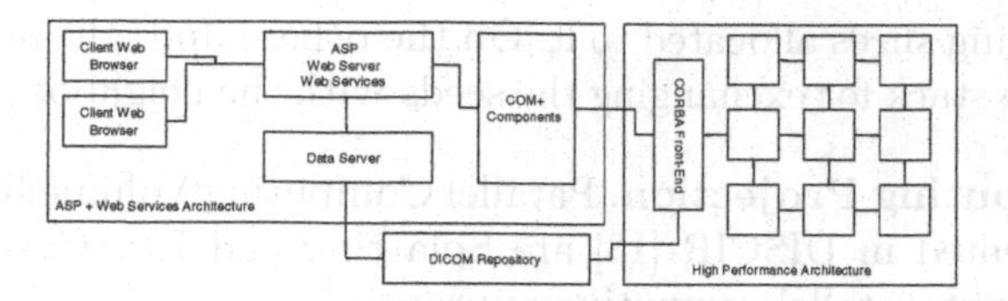

Fig. 1. Different components of the ASP architecture to HPC Image Processing

3.1 The High-Performance Architecture

The High Performance architecture used in this work is DISCIR [13]. DISCIR is component-based and includes several objects that implement parallel algorithms for segmentation and projection of 3D images. More details on the implementation of the algorithms can be found at [19]. The algorithms are designed considering a mixture of distributed and shared memory approaches (adequate to modern platforms such as clusters of biprocessor computers). Parallel approach is carried out using MPI and POSIX threads. Each processor on a node executes a separate thread, and each processor works on a different block of data.

Medical image slices are distributed according to an overlapped block-cyclic approach for improving load balancing (medical data is not uniform and computing cost depends mainly on the number of voxels with relevant information). Image data is compressed using a 3D variant of Compress Storage Row (CSR) that stores consecutively the different slices and introduces a fourth vector to point to the start of each slice [20]. Overlapping reduces communication costs when computing the normals to surfaces and Region Growing segmentation.

Parallel Computing Segmentation. Segmentation methods implemented are Threshold classification and Region Growing (RG). Parallel version of Threshold segmentation is straightforward, since no dependencies arise among the voxel classification function. RG however, is far more complex. It starts from an initial set of voxels and extracts the set of neighbor voxels that verify a homogeneity criterion (based on an interval around the mean intensity of the voxels).

No parallel versions of RG are available in the literature. Standard RG [1] lack of structured access, penalizing memory coherency and producing many unnecessary access. Span-based solutions were proposed for 2D approaches [21]. A 3D extension of Span methods has been implemented [19]. This extension considers 2 simultaneous Spans on the Z and X axis, with higher priority to the Z-span for spreading the seeds among the slices. A homogeneity criterion considers an adaptive heuristic function that penalizes the isolated voxels and promotes the inclusion of voxels surrounded of segmented voxels (with a maximum factor of two on the interval radius). This heuristic function reduces the undesired leaks and reduces holes within organs [19].

Different blocks of slices are processed on different computers. Seeds belonging to overlapped slices are exchanged between the neighboring processes, which start new segmentation actions. Two levels of seed stacks are used for reducing the contention accessing new seeds. On one hand, each thread owns a stack for the seeds involving slices allocated to it. On the other hand, all threads in a node share a common stack for exchanging the seeds with the neighbor processors [19].

Parallel Computing Projection. Parallel Computing Volume Rendering techniques implemented in DISCIR [13] are Splatting and Ray Casting. Each one requires a different parallel computing approach.

Splatting [5] is a back-to-front technique that works on the image space for projecting each voxel of the volume on the image plane, along the projection ray. Each voxel is projected and the Splatting kernel is applied to the pixels of the image plane where the projection aims at. There are no dependencies on the voxel projection, provided that a z-buffer is somehow computed to preserve the order of the data. Thus, distributed volume voxels can be independently projected on the image plane by different processes and threads. Postprocessing of different image planes generated by each process and thread is necessary to sum up the different contributions. This postprocessing is performed on each node combining all the results from the threads and globally by the root processor. Image plane is compressed using Run Length Encoding (RLE) or CSR [22] formats, to reduce both communication and processing time.

Ray Casting [6] is the front-to-back technique used in the HPC architecture. Ray Casting traces a ray along the projection direction from each pixel in the image plane. Thus, no dependencies appear at image plane level. However, volume data accessed by each ray depends on the point of view and cannot be initially forecasted. Volume replication is used to avoid the overhead of data redistribution. Image plane is distributed cyclically by blocks of rows among processes and threads for the shake of load balancing. No postprocessing is required, since final image is merely a composition of the different fragments. Trilinear interpolation is performed at voxel level for increasing the quality of the image. Acceleration methods, such as octrees or early ray termination are applied.

3.2 The Application Service Provision Architecture

The objective of this architecture is the development of a high performance image diagnostic system on an ASP, making profit of the advantages that the communications across the internet provide at low cost.

Main objectives for the design of the system are:

- Ubiquitous access from a web browser to applications for sending, receiving and processing medical image studies, both of small and of large volume.
- Efficient transmission of the images on non-dedicated networks (Internet). Images have been compresses using the JPEG2000 format that enable sending and receiving, overlapping communications and processing times.
- Access to high performance images processing tools (3D reconstruction, image segmentation, multiplanar reconstruction, etc.). Web services have been used to provide the access to high performance computers.
- Security in data transferences. The use of SSL (Secure Sockets Layers) communication protocols, digital signatures and client's certificates assures identity and confidentiality of the data.

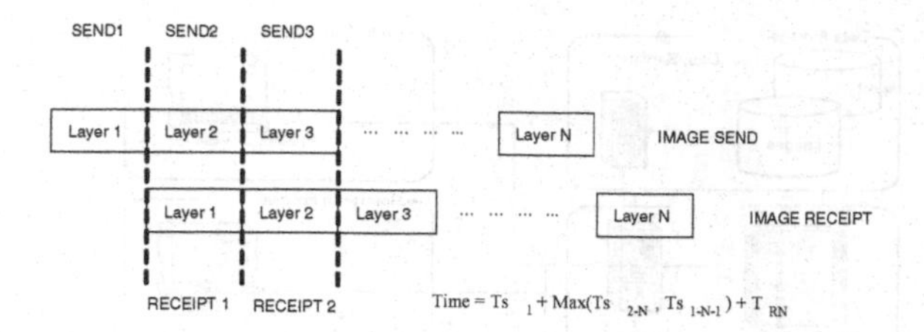

Fig. 2. Sending and receiving images using JPEG2000 compression and the overlapped pipeline. The scale is on the horizontal axis. In the ideal case, the total communication time is totally overlapped, except from the reception and sending of last and first layers

Image Compression and Overlapped Transmission. The image sending and report receiving application reads the DICOM studies produced in a centre by means of standard digital scanning devices (CTs, MRI, etc.). The application has access to the images and sends them to the data servers. Images are previously compressed in JPEG2000 format [23]. This format is based on Wavelets' transformation and enables splitting the image into layers. Each layer needs from the previous layers for its visualization [24], and the more layers are used the better the image quality is. JPEG2000 deals efficiently with 12-bit depth medical images, achieving compressions ratios above 2, higher than those obtained with standard image and data lossless compression formats. Besides, it enables the division of the images into layers for its progressive sending, and then the receiving peer can process and visualize progressively the information from less to more quality, as the layers are being received.

Layers are divided into blocks to optimize communications. Progressive sending and receiving of images allows overlapping both tasks. Moreover, once a layer has been sent, it can be immediately retrieved and decompressed by the recipient without waiting for that the rest of layers for visualizing the image.

It is important to emphasize that as images are being received progressively, the user can start the analysis and reporting of a study before it is completely downloaded. Therefore, analysis and diagnosis times can also be overlapped.

Architecture and High performance integration. The architecture that is used in the development of the applications corresponds to four clearly differentiated layers, which can be hosted in different physical locations. A TCP/IP connection is required for the connection among layers. Each layer is connected to the neighboring layer that precedes or follows it. The four used layer are:

- Data layer: It corresponds to the layer entrusted of saving the patient's demographic data and DICOM studies, stored in the data servers.
- Logic layer: It takes charge of the information process. For the case of high performance processing, it also provides the connection to the HPC resources. All these communications are performed using Web Services [25].
- Presentation layer: The interface offers a web environment for I/O.
- Client layer: It is constituted by a simple Web browser with a Sun's Java Virtual Machine, which interacts with the client and the presentation layer.

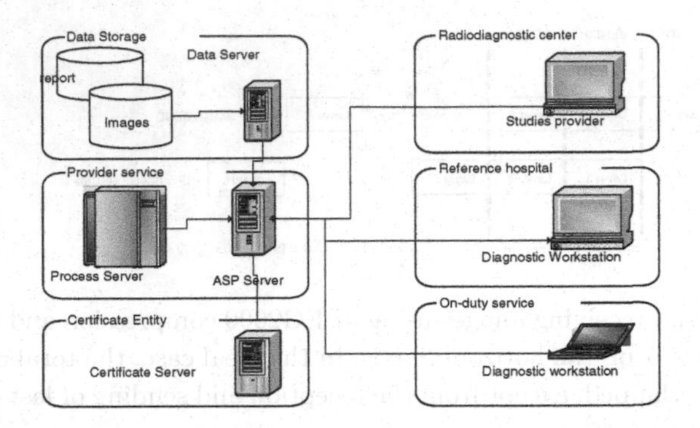

Fig. 3. Functional scheme of the different servers. Each box shows a different physical location. HPC, data and certificates services can be provided by external centers

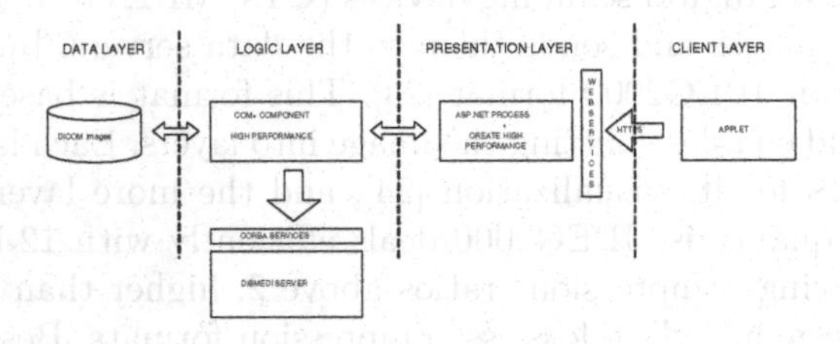

Fig. 4. Integrated scheme of HPC in the architecture ASP. Access to HPC services is provided to through COM+ object and using a CORBA interface. This scheme preserves the platform independence, and the web-based access. It also solves the problem of back of persistence for web services objects, needed for data distribution

The tools used for the development of the applications are Visual Studio .NET for the Web Services, logical and presentation layers and Sun's JDK for the applets executed in the Web browsers (such as the DICOM images diagnostic tool and the compression / decompression tool in JPEG2000 format).

As a response to the need for computing resources, the DISCIR high performance server [13] has been integrated in the platform. DISCIR is a parallel computing system for medical image segmentation and projection, based in cluster of PCs and whose high performance services are offered through CORBA. In the architecture defined in this paper, COM+ components are used in the logical layer to enable the communication to the DISCIR server through the CORBA interface. Clients request an image processing operation from an applet that is executed in a Web browser. This request is translated into a Web service request in the presentation layer, which creates a COM+ object for accessing to the HPC server (logical layer). The server executes in parallel the algorithms for attending the request, returning the results to the Web browser.

Security. The nature of the information requires a mechanism that ensures confidentiality when it is transferred across the Internet. Privacy and integrity of the information is ensured by using HTTPs communication protocols. HTTPs is based on the SSL (Secure Sockets Layers) protocol, which uses an encoding method based on a public and a private key. The information is encoded using the private key and can only be decoded on the other side using the corresponding public key. The authentication and authorization of the users is performed by means of client certificates. Every client needs an electronic certificate that allows it to connect to the web applications. Every client certificate is signed by a certificate authority (CA) trusted by the application and the server.

Since subprograms access critical parts of the computer (hard disc, communications ports, etc.), all subprograms are certified using digital signatures validated by the CA, thus avoiding malicious access by other applets.

In practice, a user must contact the web administrator to ask for a personal certificate which will be used for accessing the web server. The administrator will process the request and will provide the certificate to the user, and install the complementary in the server. The trusted CA will use the certificate to authenticate the access to the web server. Once the user connects to the web server, it will download an applet that has been previously signed by the server (for ensuring confidence). This applet use HTTPs communication protocol and the same personal certificate for ensuring his/her identity.

3.3 Test Cases

Six test cases extracted from daily practice have been used for testing the algorithms. Size, image dimensions and a brief description for each image are provided in table 1. The total size of the studies is the usual one for minimising patient harm and maximizing image information.

3.4 Compression Results

The compression ratio of the JPEG2000 format is better than other standard lossless compression formats. Classical medical image compression with JPEG or TIFF cannot deal with raw 12-bit DICOM images (required for diagnosis). Rar or ZIP formats present worse compression ratios. Moreover, JPEG2000 compression is progressive and considering an acceptable error ratio of 2%, the compression

Table 1. Test battery used in the work. Each case contains several images of 512x512 voxels. All the studies are CT scans of neuro and bone-joint

Study #	Images	Size	Description
CT1	31	512x512	Study of the nee of a male patient.
CT2	27	512x512	Study of the lumbar vertebras of a male patient.
CT3	45	512x512	Study of the right low limb of a male patient.
CT4	56	512x512	Neuronavigation study of the head of a male patient.
CT5	54	512x512	Study of the wrist of a male patient.
CT6	34	512x512	Study of the wrist of a male patient.

Table 2. Compression ratio for all the cases in the test battery. JPEG2000 compression ratio is better for all the cases

Image	Uncompressed	Compressed	JPEG2000 Ratio	Rar Ratio
CT1	524.288	264.815	1,98	1,74
CT2	524.288	223.246	2,35	2,01
CT3	524.288	220.010	2,38	1,64
CT4	524.288	183.052	2,86	1,70
CT5	524.288	164.433	3,19	2,14
CT6	524.288	172.030	3,04	2,13
CT7	524.288	215.662	2,47	1,88

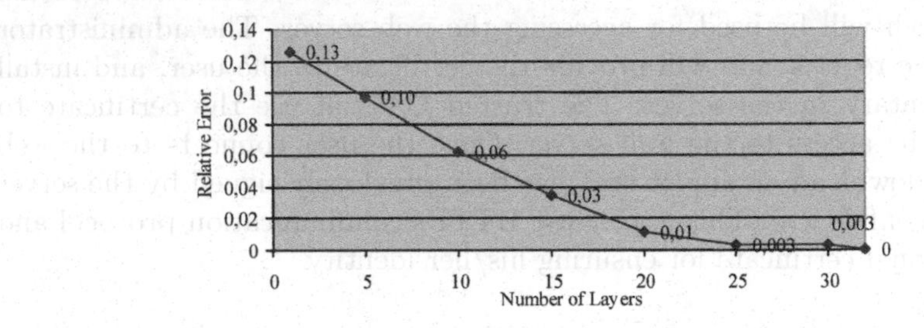

Fig. 5. Evolution of the relative error with respect to the original pixels of a JPEG2000 image. Error is reduced as more layers are included, reaching a ratio below 2% for 20 layers (equivalent in this case to the 25% of the total compressed image). This is the worst case, as the rest of images present lower error with less number of layers

ratio of JPEG raises above 8. As an example, figure 5 shows the evolution of the relative error depending of number of layers for the test case image CT1.

3.5 Performance Results

Results in terms of response time and speed-up are presented in this paper for the segmentation and projection parallel algorithms developed. The cases in the test battery have been used on a cluster of 9 biprocessor PCs, able to reach a sustained speed of 9,5 GFlops and with a total amount of 4,5 GBytes of RAM.

Segmentation. Threshold parallel segmentation is totally scalable and presents high efficiency (above 90% in all cases). Computing time required is of the order of milliseconds. Results are presented in the web interface by explicit request.

3D Projection. Response time and speed-up for parallel Splatting projection is presented in this section. It is important to outline that response time for configurations above 8 processors is less than one second, short enough for an interactive use of the application. Higher efficiency is obtained in larger cases.

Computing cost scalability of the Splatting algorithm [19] is reduced. Speed-up continues increasing until the postprocessing cost (which depends on the

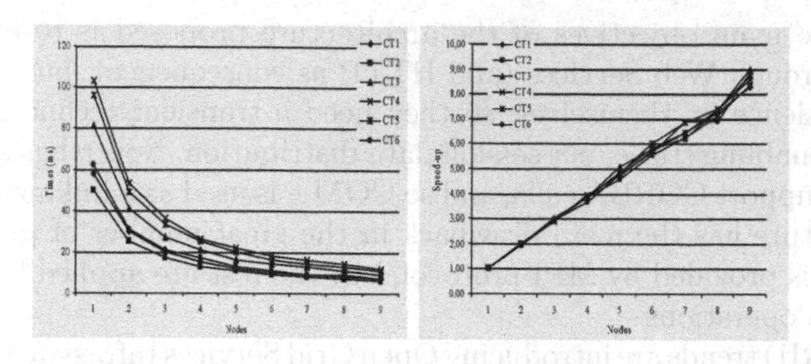

Fig. 6. Segmentation time for the six cases using from 1 to 9 bipro nodes. Response time (left) and speed-up (right). Process is highly parallel and Speed-up is linear

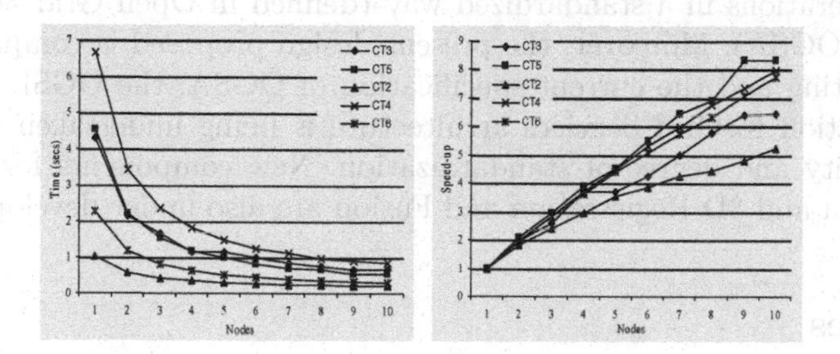

Fig. 7. Projection time using from 1 to 9 nodes. Left figure shows the response time and right figure the speed-up. Speed-up is almost linear for larger cases. Communication cost is the same in all cases, whereas computing costs varies on the number of voxels

number of processes) is comparable to the cost of projecting the whole volume. However, spatial scalability of Splatting is high, since data distribution do not require nor replication neither creating costly temporal structures.

4 Conclusions and Future Work

This work describes an integral architecture that combines high performance, medical image processing, distributed computing, web access and security components for the deployment of realistic applications on the web.

This approach benefits from the advantages of the individual components:

- HPC by the use of parallel computing components on a cluster of PCs.
- Ubiquitous access provided by the use of Internet.
- Security provided by Web-based SSL and certification processes.
- Multiplatform support, by the use of CORBA and web-based protocols.

The system is compliant with current standards such as CORBA, MPI, DICOM, SSL, HTTPs, and Web services. It does not need large requirements on the client side, constituting an ideal framework for challenging applications.

One of the main objectives of the architecture proposed is to enable HPC processing through Web Services (and HTTP as consequence), but they do not support transience by themselves, so they need a transient technology such as CORBA for enabling static, per session data distribution. Nevertheless, Web services do not support CORBA calls, and so COM+ is used as a linking technology. This architecture has the main drawback in the great number of protocols employed. HPC is provided by MPI protocol, but the rest are applied for obtaining transient web operations.

Recent GRID trends are introducing Open Grid Services Infrastructure (OGSI) as an extension of Web Services for enabling the so called Grid Services. These Grid Services are essentially Web Services that support Grid characteristics (authentication, security, encryption, authorization, resource sharing, etc.) and transient operations in a standardized way (defined in Open Grid Services Architecture - OGSA). Moreover, the present design proposed is compatible with Grid computing and the current specification of OGSA, the OGSI, and therefore a migration to Grid Services architecture is being undertaken to increase the portability and degree of standardization. New components for Advanced Segmentation and 3D Registration and Fusion are also under development.

References

1. Haralick M., Shapiro L.G., *"Image Segmentation Techniques"*, Comput. Vision, Graphics, Image Processing, Vol. 29, Num. 1, pp. 100-132, 1985.
2. Lundstrm C. *"Segmentation of Medical Image Volumes"*, Ph D Thesis Linkping University, 1997.
3. W.E. Lorensen, H. Cline, *"Marching Cubes: a High Resolution 3D Surface Construction Algorithm"*, Computer Graphics, Vol 21 (4) pp. 163-169, July 1994.
4. Drebin R. A., Carpenter L, Hanrahan P., *"Volume Rendering"*, Computer Graphics, Vol. 22(4), pp. 1988. 65-73, 1998.
5. Westover L., *"Footprint evaluation for volume rendering"*, Proc. SIGGRAPH '90, pp. 367-376, 1990.
6. Levoy M. *"Efficient Ray Tracing of Volume Data"*, ACM Transactions on Graphics Vol 9. Num.3, pp. 245-261, 1990.
7. Sato Y., Shiraga N., Nakajima S., et al. *"LMIP: Local Maximum Intensity Projection- A New Rendering Method for Vascular Visualization"*, Journal of Comp. Assist. Tomography, Vol. 22, Num. 6, pp. 912-917, Nov. 1998.
8. Yagel R., Cohen D., Kaufman A., *"Normal Estimation in 3D Discrete Space"*, The Visual Computer, Vol 8, Num. 5-6, pp 278-291, 1992.
9. A. Law, R. Yagel *"Exploitting Spatial, Ray and Frame Coherency for Efficient Parallel Volume Rendering"*, Proceedings of GRAPHICON'96, Vol. 2, St. Petesburg, Russia, July1996, pp. 93-101.
10. H. Ray, H., Pfiser, D. Silver, *"Ray Casting Architectures for Volume Visualisation"*, Internal Report, Rutgers State University.
11. Wierse A., Lang U., *"COVISE: COllaborative VIsualization and Simulation Environment"*, http://www.hlrs.de/organization/vis/covise/
12. T. Breitfeld, S. Kolibal, et. al, *"Java for Controlling and Configuring a Distributed Turbine Simulation System"*, Workshop Java for HPNC, Southampton 1998.

13. Alfonso C., Blanquer I., Gonzlez A., Hernndez V., *"DISMEDI: Una Arquitectura Multiusuario, Distribuida y de Altas Prestaciones para el Diagnstico por Imagen."*, Informed 2002, Libro de Ponencias, Comunicaciones y Posters, 2002.
14. T. Priol, *"The Cobra project"*, INRIA, http://www.irisa.fr/, 1997.
15. Keahey K., Gannon D., *"PARDIS: A Parallel Approach to CORBA"*, Proceedings of the 6th IEEE Int Symposium on High Perf. Distributed Computing, 1997
16. Keahey K., *"Requirements of Component Architectures for High-Performance Computing"*, http://www.acl.lanl.gov/cca-forum.
17. Expert Group Report, *"Next Generation GRID(s)"*, European Commission. http://www.cordis.lu/ist/grids/index.htm
18. Foster I., Kesselman C., *"The GRID: Blueprint for a New Computing Infrastructure"*, Morgan Kaufmann Pulishers, Inc., 1998.
19. Blanquer I., *"Sistemas Distribuidos de Componentes de Altas Prestaciones Orientados a Objetos para el Diagnóstico por Imagen"*, PhD Tesis, Universidad Politécnica de Valencia, DSIC, 2003.
20. Alfonso C., Blanquer I., Hernández V., *"Reconstruccin 3D en Paralelo de Imágenes Médicas de Gran Dimensión Con Almacenamiento Comprimido"*, Proceedings of the XIX CASEIB, 2001.
21. Foley J, et. al. *"Computer Graphics Principles and Practice"*, Addison-Wesley, Reading, Massachusets, 1990.
22. Saad Y. *"SPARSKIT: A Basic Toolkit for Sparse Matrix Computations"*, University of Minnesota Department of Computer Science and Engineering, http://www.cs.umn.edu/Research/arpa/SPARSKIT/paper.ps.
23. Joint Photograph Expert Group, *"JPEG 2000 Specification"*, http://www.jpeg.org
24. I. Blanquer, V. Hernndez, D. Segrelles, *"Diagnóstico por Imagen de Altas Prestaciones sobre Servicios ASP"*, Actas de las "XI Jornadas de Paralelismo", pp. 139. Madrid, Sept 2003.
25. World Wide Web Consortium, *"Web Services Description Requirements"*, http://www.w3.org/TR/ws-desc-reqs/.

Influence of Grid Economic Factors on Scheduling and Migration

Rafael Moreno-Vozmediano[1] and Ana B. Alonso-Conde[2]

[1] Dept. de Arquitectura de Computadores y Automática,
Universidad Complutense. 28040 - Madrid, Spain
Tel.: (+34) 913947615 - Fax: (+34) 913947527
rmoreno@dacya.ucm.es

[2] Dept. Economía Financiera, Contabilidad y Comercialización,
Universidad Rey Juan Carlos. 28032 - Madrid, Spain
Tel./Fax: (+34) 914887788
abac@fcjs.urjc.es

Abstract. Grid resource brokers need to provide adaptive scheduling and migration mechanisms to handle different user requirements and changing grid conditions, in terms of resource availability, performance degradation, and resource cost. However, most of the resource brokers dealing with job migration do not allow for economic information about the cost of the grid resources. In this work, we have adapted the scheduling and migration policies of our resource broker to deal with different user optimization criteria (time or cost), and different user constraints (deadline and budget). The application benchmark used in this work has been taken from the finance field, in particular a Monte Carlo simulation for computing the value-at-risk of a financial portfolio.

1 Introduction

A grid is inherently a dynamic system where environmental conditions are subjected to unpredictable changes: system or network failures, addition of new hosts, system performance degradation [1], variations in the cost of resources [2], etc. In such a context, resource broker becomes one of the most important and complex pieces of the grid middleware. Efficient policies for job scheduling and migration are essential to guarantee that the submitted jobs are completed and the user restrictions are met.

Most of the resource brokers dealing with job migration face up to the problem from the point of view of performance [3] [4] [5]. The main migration policies considered in these systems include, among others, performance slowdown, target system failure, job cancellation, detection of a better resource, etc.

However, there are hardly a few works that manage job migration under economic conditions [6] [7]. In this context, new job migration policies must be contemplated, like the discovery of a new cheaper resource, or variations in the resource prices during the job execution. There are a broad variety of

M. Daydé et al. (Eds.): VECPAR 2004, LNCS 3402, pp. 274–287, 2005.

reasons that could lead resource providers to dynamically modify the price of their resources, for example:

- Prices can change according to the time or the day. For example, the use of a given resource can be cheaper during the night or during the weekend.
- Prices can change according to the demand. Resources with high demand can increase their rates, and vice versa.
- A provider and a consumer can negotiate a given price for a maximum usage time or a maximum amount of resources consumed (CPU, memory, disk, I/O, etc.). If the user jobs violate the contract by exceeding the time or the resource quota, the provider can increase the price.

In this work we have extended the GridWay resource broker capabilities [1] [3] to deal with economic information and operate under different optimization criteria: time or cost, and different user constraints: deadline and/or budget.

This paper is organized as follows. Section 2 describes the GridWay framework. In Section 3 we analyze the extensions to the GridWay resource broker to support scheduling and migration under different user optimization criteria (time optimization and cost optimization), and different user constraints (cost limit and time limit). Section 4 describes the experimental environment, including the grid testbed and the application benchmark, taken from the finance field [12]. In Section 5 we show the experimental results. Finally, conclusions and future work are summarized in Section 6.

2 The GridWay Resource Broker

The GridWay (GW) framework [1] [3] is a Globus compatible environment, which simplifies the user interfacing with the grid, and provides resource brokering mechanisms for the efficient execution of jobs on the grid, with dynamic adaptation to changing conditions.

GW incorporates a command-line user interface, which simplifies significantly the user operation on the Grid by providing several user-friendly commands for submitting jobs to the grid ("gwsubmit") along with their respective configuration files (job templates), stopping/resuming, killing or re-scheduling jobs ("gwkill"), and monitoring the state and the history of the jobs ("gwps" and "gwhistory"). For a given job, which can entail several subtasks (i.e., an array job), the template file includes all the necessary information for submitting the job to the grid:

- The name of the executable file along with the call arguments.
- The name of the input and output files of the program.
- The name of the checkpoint files (in case of job migration).
- The optimization criterion for the job. In this implementation, we have incorporated two different optimization criteria: *time* or *cost*.
- The user constraints. The user can specify a time limit for its job (*deadline*), as well as a cost limit (*budget*).

The main components of the GW resource broker are the following:

2.1 Dispatch Manager and Resource Selector

The Dispatch Manager (DM) is responsible for job scheduling. It invokes the execution of the Resource Selector (RS), which returns a prioritized list of candidates to execute the job or job subtasks. This list of resources is ordered according to the optimization criterion specified by the user.

The DM is also responsible for allocating a new resource for the job in case of migration (re-scheduling). The migration of a job or job subtask can be initiated for the following reasons:

- A forced migration requested by the user.
- A failure of the target host.
- The discovery of a new better resource, which maximizes the optimization criterion selected for that job.

2.2 Submission Manager

Once the job has been submitted to the selected resource on the grid, it is controlled by the Submission Manager (SM). The SM is responsible for the job execution during its lifetime. It performs the following tasks:

- **Prolog**. The SM transfers the executable file and the input files from the client to the target resource
- **Submission**. The SM monitors the correct execution of the job. It waits for possible migration, stop/resume or kill events.
- **Epilog**. When the job execution finishes, the SM transfers back the output files from the target resource to the client.

3 Scheduling and Migration Under Different User Specifications

We have adapted the GW resource broker to support scheduling and migration under different user optimization criteria (time optimization and cost optimization), and different user constraints (budget limit and deadline limit). Next, we analyze in detail the implementation of these scheduling alternatives.

3.1 Time Optimization Scheduling and Migration

The goal of the time optimization criterion is to minimize the completion time for the job. In the case of an array job, this criterion tries to minimize the overall completion time for all the subtasks involved in the job.

To meet this optimization criterion, the DM must select those computing resources being able to complete the job – or job subtasks – as faster as possible, considering both the execution time, and the file transfer times (prolog and epilog). Thus, the RS returns a prioritized list of resources, ordered by a rank function that must comprise both the performance of every computing resource, and the file transfer delay. The time optimization rank function used in our RS implementation, is the following:

$$TR(r) = PF(r)(1 - DF(r)) \qquad (1)$$

Where

$TR(r)$	is the Time-optimization Rank function for resource r
$PF(r)$	is a Performance Factor for resource r
$DF(r)$	is a file transfer Delay Factor for resource r

The Performance Factor, $PF(r)$, is computed as the product of the peak performance (MHz) of the target machine and the average load of the CPU in the last 15 minutes. The Delay Factor, $DF(r)$, is a weighted value in the range [0-1], which is computed as a function of the time elapsed in submitting a simple job to the candidate resource – for example, a simple Unix command, like "date" – and retrieving its output.

The resource selector is invoked by the dispatch manager whenever there are pending jobs or job subtasks to be scheduled, and also at each resource discovery interval (configurable parameter). In this case, if a new better resource is discovered, which maximizes the rank function, the job can be migrated to this new resource. To avoid worthless migrations, the rank function of the new discovered resource must be at least 20% higher than the rank function of the current resource. Otherwise the migration is rejected. This condition prevents from reallocating the job to a new resource that is not significantly better than the current one, because in this case the migration penalty time (transferring the executable file, the input files, and the checkpoint files) could be higher than the execution time gain.

3.2 Cost Optimization Scheduling and Migration

The goal of the cost optimization criterion is to minimize the CPU cost consumed by the job. In case of an array job, this criterion tries to minimize the overall CPU consumption for all the job subtasks. In this model we have only considered the computation expense (i.e. the CPU cost). However, other resources like memory, disk or bandwidth consumption could be also incorporated to the model.

To minimize the CPU cost, the resource selector must know the rate of every available resource, which is usually given in the form of price (Grid $) per second of CPU consumed. These rates can be negotiated with some kind of trade server [8] [9], and different economic models can be used in the negotiation [10]. Once the resource selector has got the price of all the resources, it returns an ordered list using the following rank function:

$$CR(r) = \frac{PF(r)}{Price(r)} \qquad (2)$$

Where

$CR(r)$	is the Cost-optimization Rank function for resource r
$PF(r)$	is a Performance Factor for resource r (similar to the time-optimization scheduling)
$Price(r)$	is the CPU Price of resource r, expressed in *Grid \$ per (CPU) second*

It is important to point out that the CPU Price of the resource, *Price*(r), can not be considered by itself as an appropriate rank function, since the total CPU cost, which is the factor to be minimized, is given by the product of the CPU price and the execution time. In this way, a low-priced but very slow resource could lead to a higher CPU consumption than another more expensive but much faster resource. So the most suitable resource is that one that exhibits the best ratio between performance and price.

As in the previous case, the resource selector is invoked by the dispatch manager whenever there are pending jobs to be scheduled, an also at each resource discovery interval. If a new better resource is discovered, whose rank function exceeds more than 20% the current resource rank value, then the job is migrated to that new resource.

3.3 Scheduling and Migration Under User Constraints (Budget/Deadline Limits)

Independently of the optimization criterion selected (time optimization or cost optimization), the user can also impose a budget and/or a deadline limit. In this case, the resource broker must be able to minimize the specific optimization user criterion (time or cost), but without exceeding the budget or deadline limits. To implement this behavior, the RS uses the same rank functions - (1) or (2), depending on the optimization criterion - to get an ordered list of candidates, but in addition, it must be able to estimate if each candidate resource meets the user budget and/or deadline limits. Otherwise, the specific resource can not be eligible for executing the given task.

These estimations have been implemented exclusively in array jobs, since we make use of the known history of the first executed subtasks to estimate the cost or deadline of the pending subtasks for every available resource.

First we analyze the model developed for job scheduling and migration under budget limit. Let:

B_L	Budget Limit imposed by the user
$B_C(r,s,t)$	Budget Consumed by resource r to execute task s, at time t
$N_C(r,t)$	Number of tasks completed by resource r at time t
$N_P(t)$	Total Number of Pending tasks (not started) at time t
$N_S(t)$	Total Number of Started tasks, but not completed, at time t

The Budget Available at time t, $B_A(t)$, is

$$B_A(t) = B_L - \sum_{\forall s, \forall r} B_C(r,s,t) \tag{3}$$

Assuming that, in average, the started tasks have been half executed, the average Budget Available per pending task at time t, $\overline{B}_A(t)$, is estimated as

$$\overline{B}_A(t) = \frac{B_A(t)}{N_P(t) + \frac{N_S(t)}{2}} \tag{4}$$

On the other hand, the average Budget per task Consumed by resource r at time t, $\overline{B}_C(r,t)$, is

$$\overline{B}_C(r,t) = \frac{\sum_{\forall s} B_C(r,s,t)}{N_C(r,t)} \tag{5}$$

If the average budget per task consumed by resource r is higher than the average budget available per pending task, i.e.,

$$\overline{B}_C(r,t) > \overline{B}_A(t) \tag{6}$$

then it is assumed that resource r is not eligible for executing a new task at the current time t.

It is obvious that this budget estimation model can not be applied when $N_C(r,t) = 0$, so it is only useful for array jobs. In fact, when the first subtasks of the array are scheduled, there is no information about the average budget consumed by each resource, and hence every available resource is considered an eligible candidate. For the subsequent subtasks, the historical information about the average budget consumed by a given resource is used to estimate whether it is likely to violate the user budget limit. If so, the resource is excluded from the candidate list.

Similarly, we have developed a model for scheduling and migration under deadline limit. Let:

D_L Deadline Limit imposed by the user (expressed as maximum elapsed time, in seconds)
$T_C(r,s)$ Time consumed by resource r for completing task s
$N_C(r,t)$ Number of tasks completed by resource r at time t

The Remaining Time at time t, $T_R(t)$, is

$$T_R(t) = D_L - t \tag{7}$$

The average time per task consumed by resource r, $\overline{T}_C(r)$, is

$$\overline{T}_C(r) = \frac{\sum_{\forall s} T_C(r,s)}{N_C(r,t)} \tag{8}$$

If the average time per task consumed by resource r is higher than the remaining time, i.e.,

$$\overline{T}_C(r) > T_R(t) \tag{9}$$

then resource r is not eligible for executing a new task at the current time t.

As in the previous case, this time estimation model is only applicable for array jobs, when there is available historical information about the time consumed by a given resource, and hence $N_C(r,t) > 0$.

4 Experimental Environment

In this section we analyze the grid testbed and the application benchmark used in our experiments.

4.1 Grid Testbed

The main features of the computational resources employed in our grid testbed are summarized in Table 1.

Table 1. Characteristics of the machines in the experimental testbed

Hostname	Architecture / OS	Perf. Factor (peak MHz)	Delay Factor (x 100)	CPU Price (Gris $)	Slots
hydrus.dacya.ucm.es	i686 / Linux	2539	2	20	1
cygnus.dacya.ucm.es	i686 / Linux	2539	2	20	1
cepheus.dacya.ucm.es	i686 / Linux	650	3	5	1
aquila.dacya.ucm.es	i686 / Linux	662	3	5	1
belle.cs.mu.oz.au	i686 / Linux	2794	15	5	2

Resources hydrus, cygnus, cepheus, and aquila, which are located in Spain (Computer Architecture Dept., Univ. Complutense of Madrid), are uniprocessor systems, and it is assumed that they have only one available slot, i.e., only one task can be issued simultaneously to each computer. Resource belle, which is located in Australia (Computer Science Dept., Univ. of Melbourne), is a 4-processor system, and it is assumed that it has two available slots, i.e., up to two tasks can be issued simultaneously to this computer.

The client machine is located in Spain, so the delay factor for the belle system, located in Australia, is much higher than the delay factors for the systems located in Spain (hydrus, cygnus, cepheus, and aquila).

With regard to the CPU prices, it is assumed that the systems are used at European peak time, and Australian off-peak time, so that belle system exhibits a CPU price significantly lower than hydrus and cygnus systems. On the other hand, cepheus and aquila are low-performance systems, and hence their rates are also lower.

4.2 Application Benchmark

The experimental benchmark used in this work is based on a financial application, specifically, a Monte Carlo (MC) simulation algorithm for computing the Value-at-Risk (VaR) of a portfolio [12] [13]. We briefly describe this application.

The VaR of a portfolio can be defined as the maximum expected loss over a holding period, Δt, and at a given level of confidence c, i.e.,

$$Prob\{|\Delta P(\Delta t)| < VaR\} = 1 - c \tag{10}$$

where $\Delta P(\Delta t) = P(t + \Delta t) - P(t)$ is the change in the value of the portfolio over the time period Δt.

The Monte Carlo (MC) approach for estimating VaR consists in simulating the changes in the values of the portfolio assets, and re-evaluating the entire portfolio for each simulation experiment. The main advantage of this method is its theoretical flexibility, because it is not restricted to a given risk term distribution and the grade of exactness can be improved by increasing the number of simulations.

For simulation purposes, the evolution of a single financial asset, S(t), can be modelled as a random walk following a Geometric Brownian Motion [11]:

$$dS(t) = \mu S(t)dt + \sigma S(t)dW(t) \tag{11}$$

where $dW(t)$ is a Wiener process, μ the instantaneous drift, and σ the volatility of the asset.

Assuming a log-normal distribution, using the Itô's Lemma, and integrating the previous expression over a finite time interval, δt, we can reach an approximated solution for estimating the price evolution of $S(t)$:

$$S(t + \delta t) = S(t)e^{(\mu - \sigma^2/2))\delta t + \sigma\eta\sqrt{\delta t}} \tag{12}$$

where η is a standard normal random variable.

For a portfolio composed by k assets, $S_1(t)$, $S_2(t)$,..., $S_k(t)$, the portfolio value evolution can be modelled as k coupled price paths:

$$\begin{aligned}
S_1(t + \delta t) &= S_1(t)e^{(\mu_1 - \sigma_1^2/2))\delta t + \sigma_1 Z_1\sqrt{\delta t}} \\
S_2(t + \delta t) &= S_2(t)e^{(\mu_2 - \sigma_2^2/2))\delta t + \sigma_2 Z_2\sqrt{\delta t}} \\
&\dots \\
S_k(t + \delta t) &= S_k(t)e^{(\mu_k - \sigma_k^2/2))\delta t + \sigma_k Z_k\sqrt{\delta t}}
\end{aligned} \tag{13}$$

where $Z_1, Z_2, ..., Z_k$ are k correlated random variables with covariance $cov(Z_i, Z_j) = cov(S_i, S_j) = \rho_{ij}$

To simulate an individual portfolio price path for a given holding period Δt, using a m-step simulation path, it is necessary to evaluate the price path of all the k assets in the portfolio at each time interval:

$$S_i(t + \delta t), S_i(t + 2\delta t), ..., S_i(t + \Delta t) = S_i(t + m\delta t), \quad \forall i = 1, 2, ..., k \tag{14}$$

where δt is the basic simulation time-step ($\delta t = \Delta t/m$).

For each simulation experiment, j, the portfolio value at target horizon is

$$P_j(t + \Delta t) = \sum_{i=1}^{k} w_i S_{i,j}(t + \Delta t) \quad \forall j = 1, ..., N \tag{15}$$

where w_i is the relative weight of the asset S_i in the portfolio, and N is the overall number of simulations.

Then, the changes in the value of the portfolio are

$$\Delta P_j(\Delta t) = P_j(t + \Delta t) - P(t) \quad \forall j = 1, ..., N \tag{16}$$

Finally, the VaR of the portfolio can be estimated from the distribution of the N changes in the portfolio value at the target horizon, taking the $(1-c)$ percentile of this distribution, where c is the level of confidence.

The Monte Carlo solution for VaR estimation is inherently parallel, since different simulation experiments can be distributed among different computers on a grid, using a master-worker paradigm. In order to generate parallel random numbers we use the Scalable Parallel Random Number Generators Library [14].

To adapt the application to a grid environment and allow migration, the application must save periodically some kind of checkpoint information. This adaptation is very straightforward, since saving regularly the output file, which contains the portfolio values generated by the Monte Carlo simulation up to this moment, it is sufficient to migrate and restart the application on a different host from the last point saved. The checkpoint file is saved every 1,000 simulations. To save properly the checkpoint file, the application should never be interrupted while the file is being written to disk. To avoid this problem we have protected the checkpoint file update code section against interruption signals, by masking all the system signals before opening the file, and unmasking them after closing the file. The C code is the following:

```
/*************************************************/
/*  C code fragment for saving checkpoint file    */
/*************************************************/
/* Initialize signal group including all the signals */
sigfillset(&blk_group);
/* Mask all the system signals */
sigprocmask(SIG_BLOCK, &blk_group, NULL);
/* Open checkpoint file in append mode */
chkp = fopen("output_file","a");
/* Save checkpoint information */
.........................
/* Write the last 1,000 computed values to the file */
.........................
/* Close checkpoint file */
fclose(chkp);
/* Unmask system signals */
sigprocmask(SIG_UNBLOCK, &blk_group, NULL);
/*************************************************/
```

5 Experimental Results

In this section we analyze how this implementation of the GridWay resource broker behaves under different optimization criteria and user constraints, using the grid testbed and the application benchmark described in the previous section.

5.1 Cost Optimization and Time Optimization Scheduling

First we show the behavior of the resource broker under different optimization criteria (cost and time) without considering user constraints. We assume that the user submits a 4-subtask array job, each subtask performing 500,000 simulations.

Table 2 shows the ordered resource list returned by the resource selector when the user selects the CPU cost as optimization criterion.

Fig. 1.a shows the resulting cost-optimization scheduling for the four subtasks submitted, assuming that no job migration is allowed (remember that belle system has two execution slots). As we can observe, this scheduling completes all the four tasks in a period of 22:35 mm:ss, with a CPU expense of 20,090 Grid $.

If job migration is allowed, as shown in Fig. 1.b, the resource broker gets a much better scheduling, since the CPU cost is reduced to 16,770 Grid $, and elapsed time is also reduced to 20:57 mm:ss. So, these results highlight the relevance of job migration in grid scheduling.

Notice that, for a given task, the CPU cost is computed as the product of the execution time and the CPU Price of the resource. Prolog, epilog, and migration

Table 2. Resource ranking for cost-optimization scheduling

Ranking	Host	Cost-optimization rank function value
1	belle.cs.mu.oz.au	559
2	aquila.dacya.ucm.es	132
3	cepheus.dacya.ucm.es	130
4	cygnus.dacya.ucm.es	127
5	hydrus.dacya.ucm.es	127

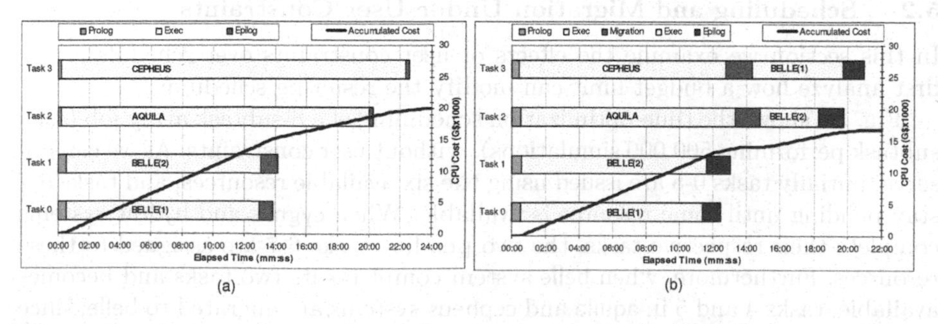

Fig. 1. Cost-optimization scheduling (no user constraints). a) Without migration. b) With migration

Table 3. Resource ranking for time-optimization scheduling

Ranking	Host	Time-optimization rank function value
1	cygnus.dacya.ucm.es	2488
2	hydrus.dacya.ucm.es	2488
3	belle.cs.mu.oz.au	2375
4	aquila.dacya.ucm.es	642
5	cepheus.dacya.ucm.es	630

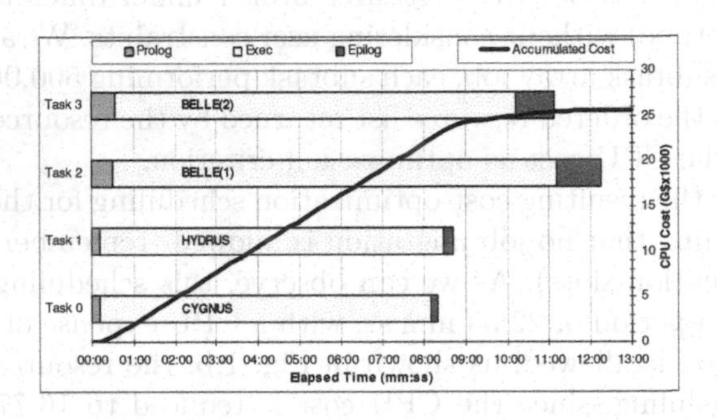

Fig. 2. Time-optimization scheduling (no user constraints)

times are not contemplated for computing the CPU cost, since these periods are used just for file transmission, and no CPU expense is considered.

Table 3 shows the ordered resource list returned by the resource selector when the user selects the time as optimization criterion, and Fig. 2 shows the resulting scheduling. As we can observe, the time-optimization criterion gets an important time reduction, since the four tasks are completed in 12:14 mm:ss, but at the expense of increasing the CPU cost up to 25,610 Grid $. In this case, scheduling with or without migration leads to similar results, since the rank functions of both cygnus and hydrus systems do not exceed more than 20% the rank function of belle system, so job migration is discarded.

5.2 Scheduling and Migration Under User Constraints

In this section we examine the effects of user constraints over scheduling. We first analyze how a budget limit can modify the resulting scheduling.

Fig. 3.a shows the time-optimization scheduling of a 8-subtask array job (each subtask performing 500,000 simulations), without user constraints. As we can observe, initially tasks 0-5 are issued using the six available resources, and tasks 6-7 stay pending until some resource is available. When cygnus and hydrus systems complete their respective tasks, the two pending tasks 6-7 are assigned to these resources. Furthermore, when belle system completes its two tasks and becomes available, tasks 4 and 5 in aquila and cepheus systems are migrated to belle, since it exhibits a higher rank function. This scheduling can be considered optimal in time, and it takes 17:55 mm:ss, consuming a CPU cost of 53,500 Grid $.

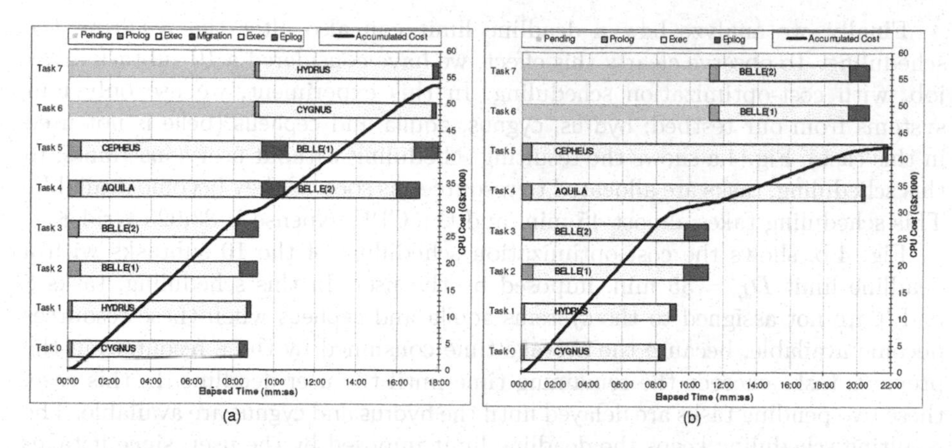

Fig. 3. Time-optimization scheduling. a) No user constraints. b) Scheduling under budget limit

Now, we are going to consider a budget limit $B_L = 45{,}000$ Grid \$. With this user constraint, the resulting scheduling is shown in Fig. 3.b. When cygnus and hydrus systems complete their respective tasks, they are not eligible to execute the two pending tasks 6 and 7, since the average budget consumed by these two systems exceeds the average available budget per task. Instead of selecting these two expensive systems, the resource broker waits until belle system is available, and then the two pending tasks are assigned to it. This scheduling keeps the budget limit imposed by the user – the overall CPU consumption is 42,190 Grid \$ –, but it is not optimal in time, since it takes 21:45 mm:ss to complete all the tasks.

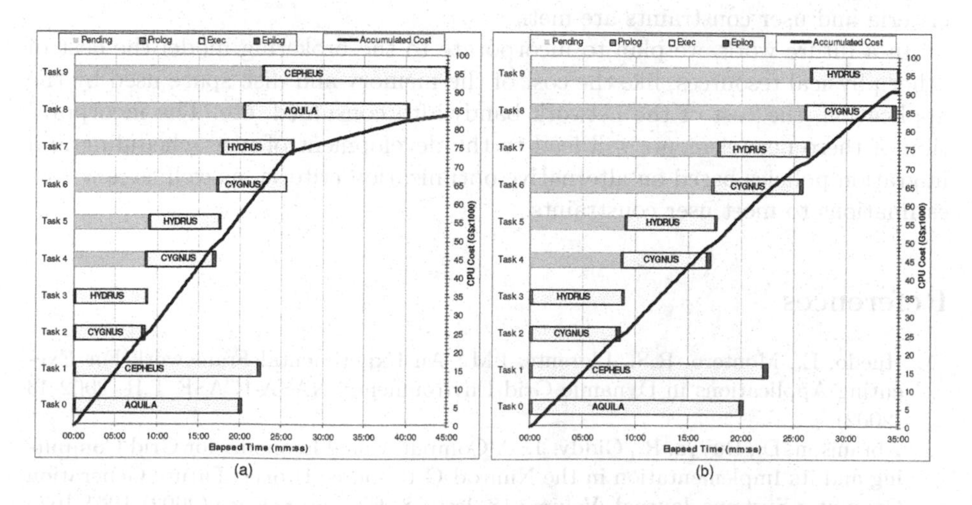

Fig. 4. Cost-optimization scheduling. a) No user constraints. b) Scheduling under deadline limit

Finally, we analyze how a deadline limit can also alter the resulting the scheduling. To observe clearly this effect, we have considered a 10-subtask array job, with cost-optimization scheduling. In this experiment, we use only four systems from our testbed: hydrus, cygnus, aquila and cepheus (belle is not used in this case). Fig. 4.a shows the resulting scheduling without user constraints. In this scheduling, tasks are allocated to resources as soon as they become available. This scheduling takes almost 45 min. and the CPU expense is 83,925 Grid $.

Fig. 4.b. shows the cost-optimization scheduling of the 10 subtasks with a deadline limit $D_L = 35$ min. imposed by the user. In this scheduling, tasks 8 and 9 are not assigned to the systems aquila and cepheus when these resources become available, because the average time consumed by these resources in the previous tasks exceeds the remaining time until the user deadline. In this case, these two pending tasks are delayed until the hydrus and cygnus are available. The resulting scheduling keeps the deadline limit imposed by the user, since it takes 35 min to complete all the subtasks, but the CPU cost goes up to 91,015 Grid $.

6 Conclusions and Future Work

In this paper we have adapted the scheduling and migration strategies of our resource broker to deal with economic information and support different user optimization criteria (time or cost), and different user constraints (deadline and budget). This implementation of the resource broker uses different rank functions to get an ordered list of resources according to the optimization criteria specified by the user, and performs different time and cost estimations based on historical information to discard those resources that are likely to violate the user constraints. The results also show that migration is essential to adapt resource mapping to changing conditions on the grid, and guarantee that optimization criteria and user constraints are met.

In a future work, we plan to incorporate to the brokering model the cost of other physical resources, like the cost of the memory and disk space used by the application, the cost of the network bandwidth consumed, etc. The incorporation of these new elements will lead to the development of new scheduling and migration policies based on alternative optimization criteria, as well as new cost estimations to meet user constraints.

References

1. Huedo, E., Montero, R.S., Llorente, I.M.: An Experimental Framework For Executing Applications in Dynamic Grid Environments. NASA-ICASE T.R. 2002-43 (2002)
2. Abramson, D., Buyya, R., Giddy, J.: A Computational Economy for Grid Computing and its Implementation in the Nimrod-G Resource Broker. Future Generation Computer Systems Journal, Volume 18, Issue 8, Elsevier Science (2002) 1061-1074
3. Huedo, E., Montero, R.S., Llorente, I.M.: An Framework For Adaptive Execution on Grids. Intl. Journal of Software - Practice and Experience. In press (2004)

4. Allen, G., Angulo, D., Foster, I., and others: The Cactus Worm: Experiments with Dynamic Resource Discovery and Allocation in a Grid Environment. Journal of High-Performance Computing Applications, Volume 15, no. 4 (2001)
5. Vadhiyar, S.S., Dongarra, J.J.: A Performance Oriented Migration Framework For The Grid. http://www.netlib.org/utk/people/JackDongarra/papers.htm (2002)
6. Moreno-Vozmediano, R., Alonso-Conde, A.B.: Job Scheduling and Resource Management Techniques in Economic Grid Environments. Lecture Notes in Computer Science (LNCS 2970) - Grid Computing (2004) 25-32
7. Sample, N., Keyani, P., Wiederhold, G.: Scheduling Under Uncertainty: Planning for the Ubiquitous Grid. Int. Conf. on Coordination Models and Languages (2002)
8. Buyya, R., Abramson, D., Giddy, J.: An Economy Driven Resource Management Architecture for Global Computational Power Grids. Int. Conf. on Parallel and Distributed Processing Techniques and Applications (2000)
9. Barmouta, A. and Buyya, R., GridBank: A Grid Accounting Services Architecture (GASA) for Distributed Systems Sharing and Integration. 17th Annual Int. Parallel and Distributed Processing Symposium (2003)
10. Buyya, R., Abramson, D., Giddy, J., and Stockinger, H.: Economic Models for Resource Management and Scheduling in Grid Computing The Journal of Concurrency and Computation, Volume 14, Issue 13-15 (2002) 1507-1542
11. Jorion, P.: Value at Risk: The Benchmark for Controlling Market Risk McGraw-Hill Education (2000)
12. Alonso-Conde, A.B., Moreno-Vozmediano, R.: A High Throughput Solution for Portfolio VaR Simulation. WSEAS Trans. on Business and Economics, Vol. 1, Issue 1, (2004) 1-6
13. Branson, K., Buyya, R., Moreno-Vozmediano, R., and others: Global Data-Intensive Grid Collaboration. Supercomputing Conference, HPC Challenge Awards (2003)
14. Mascagni, M., Srinivasan, A.: Algorithm 806: SPRNG: a scalable library for pseudorandom number generation. ACM Trans. on Mathematical Software (TOMS), Volume 26, Issue 3, September (2000) 436-461

Extended Membership Problem for Open Groups: Specification and Solution*

Mari-Carmen Bañuls and Pablo Galdámez

Instituto Tecnológico de Informática,
Universidad Politécnica de Valencia, 46022 Valencia (Spain)
{banuls, pgaldamez}@iti.upv.es

Abstract. A particular case of open group is that of a large–scale system where an unbounded and dynamically changing set of client nodes continuously connect and disconnect from a reduced and stable group of servers. We propose an extended specification of the membership problem, that takes into account the different consistency properties that both types of nodes require from the membership information in such a scenario. The specification is completely independent of any group communication layer, assuming only the presence of a local failure detector at each node. We also describe a membership service that satisfies the specification and sketch a proof for its properties.

Keywords: distributed systems, high availability, group membership, group communication.

1 Introduction

An open group is characterised by allowing connections from external nodes. This is a very general definition that suits diverse configurations. Our interest focuses on a particular scenario, typical of client-server architectures, consisting of a reduced, preconfigured group of nodes serving requests from a set of external clients whose identities and number are not predetermined, and which will typically connect over a WAN. From a practical point of view, this is the typical scenario that appears on the interaction between clusters devoted to offer highly available services and clients that connect to access to those services.

The problem involves thus the interaction between two node sets with very different properties. The lifetime of client–server connections is generally much shorter than that of server–server connections, which are expected to last long periods of time in order to implement the required services. Moreover, the failure rate that affect clients — nodes and communications — is expected to be at least one order of magnitude larger the failure rate observed in the group of servers.

* Partially supported by the EU grant *IST-1999-20997*, by *Spanish Research Council (CICYT)*, under grant *TIC2003-09420-C02-01* and by the Polytechnic University of Valencia, under grant *20030549*.

M. Daydé et al. (Eds.): VECPAR 2004, LNCS 3402, pp. 288–301, 2005.

In this paper we address the problem of specifying and solving the membership service for such an environment.

The group membership problem has been extensively discussed both from a theoretical [1, 2, 3] and a practical point of view [4, 5, 6]. Diverse specifications of the problem, together with their solutions, correspond to the requirements of particular applications. The most accepted specifications of the membership problem [7, 8] impose a series of well–defined consistency requirements on the information delivered to all group members. Nevertheless, when we deal with large–scale groups, specific questions arise depending on the required system features, and the requirements do not need to be the same for all the involved nodes.

Whereas it would be desirable to have an homogeneous membership service providing accurate membership information to every single node, the associated cost is clearly excessive when dealing with big enough groups. On the other hand, client–server interactions that occur in a setup as the one we envision do not generally require precise information about membership. In this paper we explore which semantics should an extended membership service have, so that it provides accurate membership information to the group of servers, while relaxing at some extent membership information regarding clients. The information about servers supplied to clients should enable client applications built on top of the service to react upon server failures. Analogously, servers should receive enough information to allow the reaction of server applications upon client failures, connections and disconnections. In both cases, membership information is required so that upper software components, such as group communication, replicated object support infrastructures or whichever considered middleware, can be designed using membership information as the starting point upon which reactions could be implemented. On one hand, the convenience of employing a unique membership service with a homogeneous interface for both servers and clients motivates our work, so that we could facilitate the deployment of upper software components that benefit from using a common interface. On the other hand, observations about client–server interactions and client–server software architectures, provide the means to relax the strict and classical membership semantics for client membership information. Specifically, clients do not need to be aware about the presence of other clients, but rather some notion about the server group configuration that enables them to contact server nodes as needed. In contrast, server nodes need precise information about the rest of the servers, and accurate but possibly delayed information about clients. In this way, server applications with fault tolerant capabilities which are common for cluster setups are enabled.

In the following section we give the specification of the extended membership problem. Then, in Sections 3 and 4 we present an algorithm designed to satisfy such requirements. Each section deals with one component of the proposed protocol. In Section 5 we sketch the proof of the different properties the protocol satisfies. Finally, in Section 6, conclusions and relation with existing work close the paper.

2 Specification of the Extended Membership Problem

2.1 System Model

The system is modelled as asynchronous and partitionable, and processes communicate by means of message passing. Failures may happen to nodes (stopping failures) or network channels, and partitions may also occur. After crashing, a process is allowed to recover and rejoin the group. The system is equipped with a local failure detector at each node.

Although failures may happen to every node and connection in the system, the scenario under study allows us to assume that client–server connections are subject to more frequent failures than those internal to the group of servers.

The strict liveness properties required by the classical membership problem specification are not achievable in a pure asynchronous system [2]. An eventually perfect failure detector, which would allow the system to escape that result [9] is neither implementable in this model. Nevertheless, in practical systems, such a component can be implemented [8] provided the system exhibits a reasonable behaviour. Therefore our system model includes an assumption on the dynamics, namely that within the server group there exist failure-free intervals of time long enough for the failure detector to stabilise and a full reconfiguration to take place. In the considered scenario this is a natural assumption, enough to make the service implementable. As we will see such an assumption is not required for the group of clients as this is not required to satisfy such strict guarantees.

2.2 Terminology

We specify the membership problem for the group of servers in the classical manner. The terminology used to describe this subproblem and the corresponding solution is that traditionally found in the context of membership services. Thus, the term *view* refers to an image of the group membership held by a particular node. The output of the service is membership information, delivered to the applications or components in the form of views. We also say that the view has been *installed* or *committed* by that node. From a (fictitious) global point of view, we say that a view is *confirmed* (or *committed*) when some node has installed it. Since we are modelling our system, and in particular the group of servers, as partitionable, there may be at times two or more disjoint subsets of servers that remain correct, i.e. operational, but mutually disconnected. The membership service should deliver different disjoint views of the group membership in each subgroup. In this case, there may exist a single *majority view* comprising more than one half of the preconfigured group of server nodes, and maybe multiple *minority views*.

To reflect the particular scenario we are concerned about, we will use the term *halo* to denote the unbounded and dynamically changing set of client nodes, and the term *core* when speaking about the central group of servers. Both groups require different membership information regarding each other. In that sense, we introduce the concepts of *halo view*, to denote the group of clients a given server node knows about, and that of *membership horizon*, to denote the knowledge a client has about the group of servers.

2.3 Problem Specification

The membership service must provide core nodes with views that contain the correct and connected servers, and such views must be delivered with some consistency guarantees. The specification of the core part of the service, following [7, 8], is contained in the four safety properties and the single liveness property every membership service must satisfy.

C.1 Self Inclusion. A core node is member of each view it installs.
C.2 Initial View Event. Each event occurs in some view context.
C.3 Local Monotonicity. Two committed views are never installed in different order by two different core nodes.
C.4 View Agreement. A majority view is never installed in a core node if there exists a previous committed majority view not known by this node.
C.5 Membership Precision. If there are stable components[1] in the core system, the same (correct) view is installed as the last one in every node of the same component.

The extension of the problem regards halo membership information, which concerns both core and halo nodes. Whenever a new client connects, the core group must reach agreement on the joining, and the client must receive information on which core nodes are capable to be contacted for its requests. The problem may be specified by the following list of properties.

H.1 Unicity. A client receives a different unique identifier each time it joins the group.
H.2 Validity. A client node does not include a core member in its horizon unless the core node included the client in its halo view.
H.3 Halo Consistency. Two different core members do not disagree on the identifier of a particular client.
H.4 Halo Liveness. If a client fails or disconnects, its identifier will eventually be invalidated.
H.5 Horizon Liveness. If a core member included in a client's horizon fails, the client will eventually update its horizon to exclude the node. Conversely, if a core member joins the group and both the client and the new member stay connected, the client will eventually include the core node in its horizon.

---

[1] Subsets of processes that remain correct and mutually connected, whereas they are disconnected from any other process not in the component.

3 Core Protocol Description

3.1 Basic Ingredients

The membership protocol we propose to solve the specification above consists of two separate parts, namely the core and halo components. The core component solves the general membership problem for any process group with local failure detectors which fulfils the dynamical assumption in section 2.1. It relies on some basic ingredients.

Identifiers. Each core node has a unique, preconfigured identifier within the cluster. This enables a deterministic Master choice among living members. Views do also hold unique identifiers with a well-defined order relation.

Protocol Phases. The description of the service is divided in a set of subprotocols, taking charge of the different phases of the algorithm, which are depicted in fig. 1. Each phase realises definite tasks, achieved by exchanging specific messages. During basic operation, **Setmem** messages include view proposals, while **Step** and **Ends** lead virtually synchronous steps. In the Master change phase, **ChangeM**, and **View** messages are used to collect information on the last proposals. Finally, synchronising of histories and partition merging use messages **Join**, **Joined**, **Ready** and **Update**.

State. Each core node records views in a local list as they are being proposed and confirmed, together with a label (`prepared`, `committed`, `released`) indicating its stage of confirmation. Two such lists exist, called strong and weak list and dedicated to majority and minority views, respectively. Furthermore, a consec-

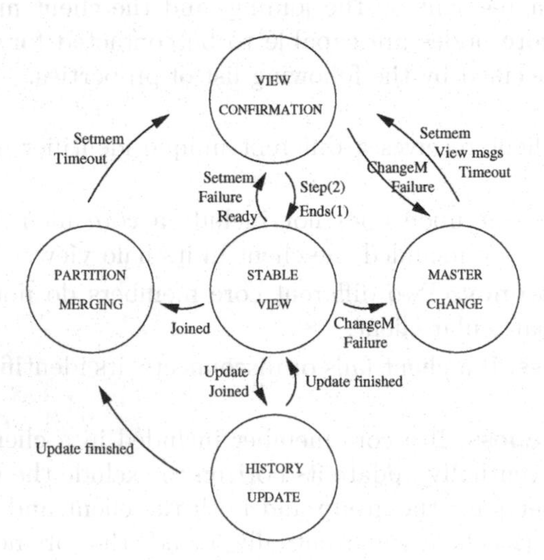

Fig. 1. Phases of the Core Membership Service and events or messages driving transitions

utive list of all majority views called `majority history` is kept in stable storage to be used during partition merging.[2]

3.2 Basic Operation

Each node starts in a predefined stable `zero` configuration, where they are the Master and only member, and broadcast **Join** messages trying to find a hypothetic majority partition to merge with. If such a group exists, the `Partition Merging` phase will be launched. Merging of two minority partitions is also allowed, playing the majority role the one that contains the would-be Master of the joint group.

`View Proposal and Confirmation.` New views are installed at Master's request, whenever a failure is detected or a new member is accepted. Confirmation of views is carried out as a three-phase commit. During the first phase, the Master proposes a new view V_p by sending **Setmem**(V_p,(V_c,confirm)) to all nodes in V_p, being V_c the last `committed` view in its strong list. At the same time, the pair (V_p,`prepared`) is appended to the end of the list. All other nodes will discard any `prepared` view still in the `strong` list, change V_c's label to `committed`, append (V_p,`prepared`) to the list and reply with an **Ends**(0) message, where the argument indicates the completed phase. During the second phase, after receiving all **Ends**(0) messages, the Master locally labels V_p as `committed`, setting any previous `committed` view to `released`, and sends **Step**(1) to the remaining nodes. They will also set it `committed` upon reception of the message, and reply with **Ends**(1). finally, after all **Ends**(1) messages have been received, V_p is labelled `released` by the Master, which sends **Step**(2) to the others. They will also release the view when receiving this message.

The procedure is restarted each time a new view composition is decided, due to either a failure or merging, as depicted by fig. 2 (`left`) for the case of a node crash.

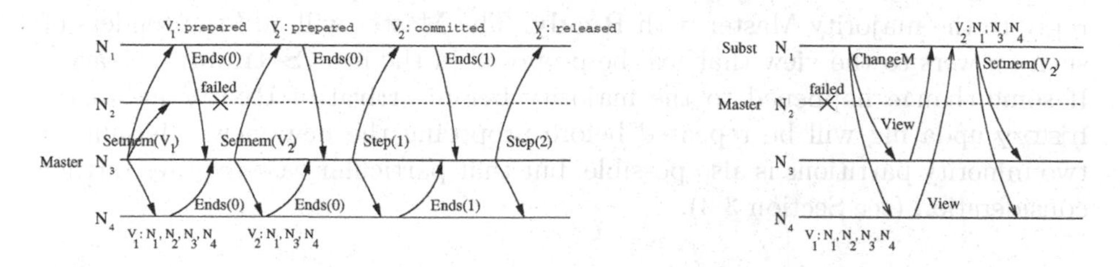

Fig. 2. Proposal and confirmation of a view (left) and Master Changing procedure (right)

[2] A simple subprotocol takes charge of occasionally cutting away the oldest part of the list (after all preconfigured nodes get simultaneously connected) to keep its size finite.

Master Change. If the Master itself fails, the pretended substitute, N_w, will start the Master Changing Protocol by interrogating possibly alive nodes with a **ChangeM**(V_ℓ) message, being V_ℓ the last view in its strong list. The remaining nodes will reply with a **View** message, as shown in fig. 2 (right), including information on the last `committed` and `prepared` views (respectively V_c and V_p) of their strong list.

1. If ($V_c > V_\ell$), **View**(V_c,`committed`) is sent. [3]
2. If ($V_c = V_\ell$) and ($V_p > V_\ell$), **View**((V_c,`committed`),(V_p,`prepared`)) is sent.
3. If ($V_c < V_\ell$), **View**(V_p,`prepared`) is sent.

When examining the answers, the would-be Master will take a different action depending on the collected information.

– Exclude itself from the group, if a $V_c > V_\ell$ has been reported or a $V_p > V_\ell$ exists with $N_w \notin V_p$ and all nodes in $V_p \cap V_\ell$ reporting it as `prepared`.[4]
– Prepare a new view V_n and propose it with **Setmem** in any other case. The message will include the order to confirm V_ℓ if some node had it locally `committed` or all nodes in $V_p \cap V_\ell$ have reported it as `prepared` with no later view `prepared` in any of them. Such confirmation is definitive if V_n is in majority, so that partitioned or failed nodes that were present in V_ℓ will have to be notified about its confirmation when they rejoin the majority.

Partition Joining and History Update. Any minority Master in a stable configuration probes for larger partitions with periodic **Join** broadcasts. The message includes the view composition and the last majority view it knows about. Merging is only allowed between partitions that share the majority history, therefore the first task during this phase is to update the history of the minority. When a majority Master gets that message, it sends a **Joined** message to each member of the minority group, containing the majority view identifier and the latest part of the majority history. Each node in minority will process the message and reply to the majority Master with **Ready**. The Master will add the senders of such answers to the view that will be proposed in the next **Setmem** message.[5] If some change happened to the majority before arrival of **Ready** messages, history updating will be repeated before proposing the new view. Merging of two minority partitions is also possible, but that particular case requires further consideration (see Section 3.3).

[3] In this case, N_w is not in V_c and will exclude itself.

[4] Reception of a **ChangeM** message after a vote request causes a node to dismiss any further messages from the old Master. A timeout is established so that a new node assumes mastership if N_w excludes itself.

[5] Such proposal will take place after a limited wait, or when a failure within the majority forces a reconfiguration.

3.3 Minority Operation

When nodes fail or network partitions, majority may be lost. In that case, special considerations apply to view proposal, Master change and partition merging. Here we enumerate the main features corresponding to this operation mode. Detailed descriptions can be found in [10].

- View proposal and confirmation uses the weak list, whereas the strong one is set aside without modification as soon as the majority is lost. Minority views that are `committed` do not have the definitive character of majority ones, and are not stored in the stable history. Higher level applications must be informed about the minority character of the group.
- During Master change, when the substitute Master turns out to be in minority, `prepared` views received in **View** messages cannot be discarded if some nodes were missing them.
- When two minority partitions merge, the one that has the most up to date majority history must share such information with the other, irrespective of which is the leading one. To this end, **Update** messages are sent to the obsolete partition whenever a **Join** is detected, before the real merging takes place. When majority is recovered after merging, only the latest among the pending `prepared` majority views will be confirmed.

4 Halo Protocol Description

4.1 Generalities

The halo component is specifically designed for maintaining information on a halo set such as the one described in the introduction. It is also devised so that it can be laid on top of an existing core membership service. The basic ingredients for this component are the following.

Node Identifiers. A unique identifier is dynamically assigned to each halo node as it joins the group.

Specific Messages. These are mainly **Horizon** (to notify a client about core members it can contact) and **Token**. The latter contains a list of entries (**Propose**, **Confirm**, **Update**, **Remove** or **Definitive**) indicating the action to take on a particular client.

State. The halo protocol requires that each core node maintains the halo view, containing information on the known clients. Each client is represented by an entry of the form [*id*, *address*, N_r, *status*, *view*, *updating*], where *id* is the client identifier inside the group, *address* represents its physical address, N_r is the identifier of its representative core member, and *view* is a view identifier whose meaning depends on the value of *status*, which may be *proposed*, *committed*, *definitive* or *removed*. The *updating* field holds view information only when the client information is being updated, being otherwise null. The halo view

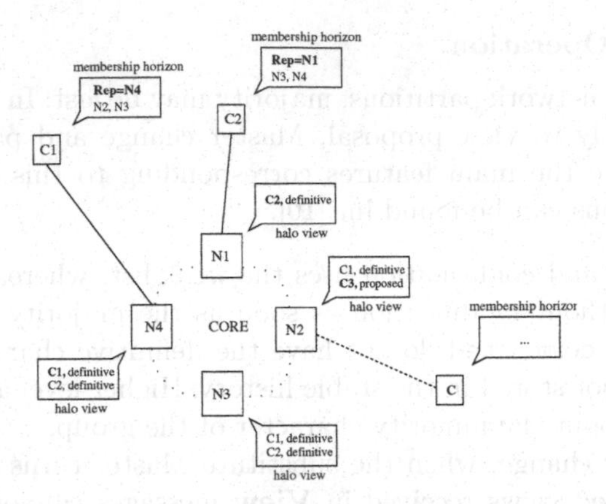

Fig. 3. Contents of halo membership information maintained by core and client nodes at a given time

is initially empty, and it is reset each time the node abandons the majority. Each core member keeps also a copy of the last received token. Conversely, the state of a halo node contains only its assigned identifier and its `membership horizon`, which holds a set of core nodes that are aware of this client's presence. The `horizon` is initialised when the core replies to a connection attempt, with a snapshot of the core membership when agreement was achieved on the identifier granted to this node. Further updates to the `horizon` do not necessarily reflect all core group changes, and thus it will in general not coincide with a core view. Figure 3 shows a possible instant configuration of halo membership, to illustrate these concepts.

4.2 Basic Operation

The protocol for maintaining `halo views` is based on the circulating token algorithm for mutual exclusion. The token is generated by the Master of the core group when the majority is attained, and it circulates along a logical ring formed by the view members. The token carries its own identifier (increased after each round), the first available client identifier and a vector of client entries.

`ClientAddition`. When a client tries to join the group, it contacts one or several core members, which will ask the client to confirm the request before proceeding.[6] The first core member to get the token after the client's confirmation, N_r, will act as its `representative` within the group, and propose its joining.

1. N_r retrieves the available client identifier, ID_f, from the token.[7] Then it adds an entry [*id*, *address*, N_r, *proposed*, V], to its `halo view`, being V the

[6] This prevents obsolete messages from launching a duplicate joining round once the client is definitively added to the group.

[7] This implies also increasing it in the next token.

current view identifier, and $id = ID_f$ the proposed client identifier. An entry **Propose**(N_r,*id*,*address*,N_r) is added to the token and this is passed on. The first identifier in such entry indicates N_r is the initiator of the token round.
2. As the token circulates, the other core nodes add the same local entry to their halo views. When N_r gets the token back, it locally promotes *status*(*id*) to *committed*, and replaces the token entry by **Confirm**(N_r,*id*). A message **Horizon**(*id*,*members*) is sent to the client, being *members* the list of core nodes in the current view, and the token is passed on.
3. Subsequent core members will conveniently update *status*(*id*) when receiving the token. At the end of the round, N_r will set *status*(*id*) to *definitive*, and replace the token entry by **Definitive**(N_r,*id*).
4. Finally, after the third round, *status*(*id*) is *definitive* at all core members, and N_r deletes the token entry.

Client Disconnection. Clients are removed from the group by a similar procedure. The representative N_r adds the entry **Remove**(N_r,*id*) to the token. Core members knowing about *id* will mark their halo view entry as *removed* when processing the token. After one round, N_r removes its local entry for *id* and changes the token entry to **Definitive**(N_r,*id*).

4.3 Core View Changes and Token Recovery

Token circulation is only allowed within stable core views, i.e. when the view has been `committed` by all members. Thus token processing must be suspended when a **Setmem**, **ChangeM** or **Change** message is received, keeping a record of the last ring composition to be used in the recovery phase. When the stable stage is reached again, the token must be recovered.

To this end, when a node reaches the `released` phase in a view V_{new} (or whenever it receives a recovery token), it sends back its last received token to its

Table 1. Token recovery from N_b's *halo view* and *reversed token* information

status(*id*)	*view*	*updating*	$T_r[id]$	**recovery entry**
proposed	V_{old}	*null*	not present	**Propose**(N_b,*id*,*address*,N_r)
			Propose(*id*)	**Update**(N_b,*id*,*address*,N_r,V_{old})
		V_u	**Update**(*id*,V_u)	**Update**(N_b,*id*,*address*,N_r,V_{old})
			anything else	**Update**(N_b,*id*,*address*,N_r,V_u)
committed	V_{old}	*null*	**Confirm**(*id*)	**Definitive**(N_b,*id*)
			anything else	**Update**(N_b,*id*,*address*,N_r,V_{old})
		$V_u > V_{old}$	**Update**(*id*,V_{old})	**Update**(N_b,*id*,*address*,N_r,V_u)
			anything else	**Update**(N_b,*id*,*address*,N_r,V_{old})
		$V_u = V_{old}$	**Confirm**(*id*)	**Definitive**(N_b,*id*)
			anything else	**Update**(N_b,*id*,*address*,N_r,V_{old})
removed			**Remove**(*id*)	**Definitive**(N_b,*id*)
			anything else	**Remove**(*id*)

preceding node in the former ring.[8] Each node will compare that reversed token identifier, T_r, with that of the last locally received token, T_{last}. Then, either for exactly one node, N_b, $T_r < T_{last}$, or $T_r = T_{last}$ with N_b following (in the ring order) the node that sent T_r. This indicates the limit of previous token advance. Thus N_b regenerates the token and sends it along the new ring with identifier $T_{last} + 1$. For each local entry [*id*,*address*, N_r,*status*,*view*,*updating*], N_b adds an entry to the new token, depending on which information the reversed token carries on *id* ($T_r[id]$), as indicated in table 1.

5 Properties

The correctness of the proposed algorithm can be shown by proving it satisfies the specification in 2.3. In this section, proofs for the various properties are only sketched. More detailed explanations can be found in [10].

5.1 Core Properties

Self Inclusion (C.1.) and Initial View Event(C.2.). Proof of both properties is immediate from the algorithm definition. Any node starts in a stable zero configuration, what guarantees Initial View Event. Such view contains the node itself as only member, and the node can only receive proposals that contain it, thus satisfying Self Inclusion.

Local Monotonicity (C.3.). If two different nodes (N_p, N_q) locally commit two common views, (V_i, V_j), both commit events follow in the same order in both nodes.

Proof: The three-phase commit assures that commission of a given view is only possible after the first preparation phase has concluded and acknowledgement of all **Setmem** messages has arrived to the Master. Therefore, the preparation of the view in a given node follows the corresponding event in the Master, and precedes itself the commit event in any other node. Besides, local view lists are handled in such a way that confirmation or discarding of pending majority views is realised before preparing new ones. Therefore, if N_p commits V_i before V_j, necessarily N_q prepares V_i before V_j. If the commission order in N_q was opposite, we would conclude that N_p prepares V_j before committing V_i. But in N_p, preparation of V_i must also precede that of V_j, so the commission order must be the same in both nodes.

View Agreement (C.4.). If a majority view V_j is committed in some node, then any other involved node must also commit it, before confirming any later majority view $V_k > V_j$.

Proof: First we observe that discarding of a committed V_j is not possible in any node that remains in the majority, since any messages from the Master will

[8] If the immediate predecessor is not alive, the token is sent to the previous one, and so on.

include confirmation of V_j until it is acknowledged by all remaining nodes. If a node included in V_j leaves the group before confirming it, the Partition Joining procedure ensures that it will update its majority history before being able to install any further majority views, and it will thus confirm V_j. Therefore, no node discards V_j.

If we consider two nodes, N_p, N_q, each one committing one of V_j, V_k and included in both views, then any of them executes both prepare actions, as the corresponding commit exists. But since $V_j < V_k$, if they were consecutive,[9] then there would exist a node in which commit of V_j preceded that of V_k. It is easy to show that all prepare actions for V_j must then precede commit of V_k in any other node. Finally, since prepare events are not consecutive in the algorithm and $V_j < V_k$, we conclude that commit of V_j must necessarily happen prior to preparation (and thus commission) of V_k in any node that commits this latter view.

Membership Precision (C.5.). If there exists a subset of nodes that stay connected and correct, all its components will eventually install the same final view.

PROOF: Fulfilling of this property is conditioned to the behaviour of the system. We thus assume that from a given moment on, a particular subset of nodes stays interconnected and correct, with no connection with other nodes. After such situation arises, local failure detectors assure that every node which is not in the connected component will be eventually excluded from any view that the correct nodes install. If two such nodes have installed different views, at least one of them will not be in the majority, and will try the Joining procedure. Since the network is correct, its messages will eventually reach a majority group and a common view will be installed. After this moment, since no more failures occur and the subsystem remains connected, no more changes will happen.

5.2 Halo Properties

Unicity (H.1.) and Validity (H.2.). The algorithm trivially satisfies Unicity property, since the token algorithm solves mutual exclusion problem, and guarantees that clients are added by only one core node at a time. Moreover, client identifiers are not reused. Validity condition is also fulfilled, since the **Horizon** message is only sent to the client after completing the first token round, in which all core members have added it to their halo views.

Halo Consistency (H.3.). Two core nodes do not disagree on the identifier of a given client.

PROOF: The token algorithm guarantees that two core nodes do not try to add the same client with different identifiers. The request for confirmation from the client whenever a joining attempt is detected by a core node, assures that an

[9] If V_j and V_k were not consecutive, the same result follows by induction on the number of intermediate majority views.

obsolete joining trial will not cause launching of a new proposal by a core node that has joined the group after the client's definitive confirmation.

Halo Liveness (H.4.). If a client fails or leaves, its identifier is eventually released.

PROOF: If the client voluntarily disconnects, its representative will be informed and it will correspondingly launch the **Remove** round. If the client fails, the representative failure detector is assumed to eventually notice it and cause an analogous effect.

Horizon Liveness (H.5.). If a core node fails, it will eventually disappear from all clients' horizons. Conversely, if new core members join the group, clients will eventually include them in their horizon, provided both client and core member remain correct.

PROOF: The first part of the property is guaranteed by each representative, which after a core view change will inform its represented clients about the failed nodes. The second part is provided by **Update** rounds referring each client, eventually started by the various representatives.

6 Conclusions

In this paper we have extended the membership problem specification to cover a large-scale system where an unbounded halo of client nodes, continuously connect and disconnect from a reduced group of core nodes, which remain correct and interconnected most of the time. Our protocol evolved from HMM [11], the membership protocol for the Hidra distributed architecture, as it was extended to provide client membership support. Moreover, the new specification applies for partitionable environments, and it has been made independent of the particular failure detection procedure.

There exists precedent work on large-scale groups. The membership protocol in [12] is also designed for a WAN, and based on a client-server architecture. The approach in such work is however completely different, as in [12] server nodes are distributed geographically and devoted to maintain clients membership, not including their own. In [13] an architecture is proposed for a similar large-scale group communication, which relies on the idea of multiple roles for group membership. In that proposal, as in later works on highly available distributed services [14], group communication guarantees are exploited to provide the different desired semantics for client and server nodes.

Different from many other solutions, our specification is completely independent of group communication services or other parts of the system. It thus attempts to be implementable in any distributed environment. The algorithm is carefully defined, to cover in detail some subtle aspects whose in depth treatment is not found in the literature, namely update of information on view history during partition merging and change from minority to majority status and vice-versa. The formal aspect of the work is completed by showing how the specifying properties can be proved.

Acknowledgements. We are grateful to F.D. Muñoz for his helpful comments and suggestions.

References

[1] F. Cristian. Reaching agreement on processor-group membership in synchronous distributed systems. *Distributed Computing*, 4(4):175–188, 1991.
[2] T.D. Chandra et al. On the impossibility of group membership. In *Proceedings of the 15th Annual ACM Symposium on Principles of Distributed Computing (PODC'96)*, pages 322–330, New York, USA, 1996. ACM.
[3] E. Anceaume et al. On the formal specification of group membership services. Technical Report TR95-1534, 25, 1995.
[4] C. P. Malloth and A. Schiper. View synchronous communication in large scale distributed systems. In *Proceedings of the 2nd Open Workshop of the ESPRIT project BROADCAST (6360)*, Grenoble, France, 1995.
[5] A. Ricciardi and K. P. Birman. Process membership in asynchronous environments. Technical Report NTR 92-1328, Department of Computer Science, Cornell University, Ithaca, New York, February 1993.
[6] Y. Amir et al. The Totem single-ring ordering and membership protocol. *ACM Transactions on Computer Systems*, 13(4):311–342, 1995.
[7] G. Chockler, I. Keidar, and R. Vitenberg. Group communication specifications: a comprehensive study. *ACM Computing Surveys*, 33(4):427–469, 2001.
[8] O. Babaoglu, R. Davoli, and A. Montresor. Group communication in partitionable systems: Specification and algorithms. *Software Engineering*, 27(4):308–336, 2001.
[9] T.D. Chandra and S. Toueg. Unreliable failure detectors for reliable distributed systems. *Journal of the ACM*, 43(2):225–267, 1996.
[10] M.C. Bañuls and P. Galdámez. Technical Report ITI-ITE-03/01, Instituto Tecnológico de Informática, Univ. Politécnica Valencia, 2003. http://www.iti.upv.es/sifi/en/publications/2003/index.html.
[11] F.D. Muñoz, P. Galdámez, and J.M. Bernabéu. HMM: A membership protocol for a multi-computer cluster. Technical Report ITI-DE-2000/02, Instituto Tecnológico de Informática, Univ. Politécnica Valencia, February 2000.
[12] I. Keidar et al. A client-server oriented algorithm for virtually synchronous group membership in WANs. In *International Conference on Distributed Computing Systems*, pages 356–365, 2000.
[13] O. Babaoglu and A. Schiper. On group communication in large-scale distributed systems. In *ACM SIGOPS European Workshop*, pages 17–22, 1994.
[14] C.T. Karamanolis and J.N. Magee. Client access protocols for replicated services. *Software Engineering*, 25(1):3–21, 1999.

Asynchronous Iterative Algorithms for Computational Science on the Grid: Three Case Studies

Jacques Bahi, Raphaël Couturier, and Philippe Vuillemin

LIFC, University of Franche-Comté, France**
{bahi, couturie, vuillemi}@iut-bm.univ-fcomte.fr
http://info.iut-bm.univ-fcomte.fr/

Abstract. In this paper we are interested in showing that developing asynchronous iterative algorithms for computational science is easy and efficient with the use of Jace: a Java-based library for asynchronous iterative computation. We experiment three typical applications in a heterogeneous cluster composed of distant machines. To illustrate the adequacy of our algorithms in this context, we compare the execution times of both synchronous and asynchronous versions.

Keywords: Asynchronous iterative algorithms, computational science problems, grid computing.

1 Introduction

Recent developments in networks and distributed systems have led scientists to design algorithms for the grid. Several architectures are available for the grid, if we consider homogeneous or heterogeneous machines with local or distant networks and homogeneous or heterogeneous networks. One purpose of this paper is to explain how to build efficient iterative algorithms to solve scientific applications using a cluster composed of distant heterogeneous machines with heterogeneous networks. The context of execution we consider is easy to obtain since it simply consists in connecting standard workstations with classical networks (including ADSL networks). Nevertheless, the heterogeneity of computers and the possible low bandwidth added to high latencies may be responsible for low performances. Traditional algorithms require several synchronizations during their execution in order to update a part or the whole of the data, or simply to start the next step of the computation. In this case, in a distant cluster context, synchronizations often penalize performances.

Roughly speaking, scientists categorize their problems in two kinds: direct algorithms and iterative ones. The former class consists in finding the problem solution in one step. The latter proceeds by successive approximations of the

** This work was supported by the “Conseil Régional de Franche-Comté”.

M. Daydé et al. (Eds.): VECPAR 2004, LNCS 3402, pp. 302–314, 2005.

solution. The class of problems that can be solved using iterative algorithms is large. It features, for example, sparse linear systems, ordinary differential equations, partial differential equations, Newton based algorithms, polynomial roots finders...

A solution to reduce or even to suppress synchronization steps in an iterative algorithm consists in desynchronizing communications and iterations thus obtaining an asynchronous iterative algorithm. Obviously, this is not always possible because if the convergence conditions are taken into consideration, the algorithm may not converge. Fortunately, the class of converging asynchronous iterative algorithms (we call them *AIAC* algorithms for Asynchronous Iterations with Asynchronous Communications) is large enough to provide an efficient alternative to synchronous iterative algorithms, especially in the context of distant grid computing [14].

Another purpose of this paper is to show that efficient asynchronous iterative algorithms for computational science could be designed for the grid. To do this, we have chosen three algorithms that we have implemented and run in a distant heterogeneous cluster environment. Programming environments to manage asynchronous iterations need only two simple features:

- threads to manage communications and computation asynchronously,
- and a thread-safe message passing library to implement communications.

In this paper, we have chosen to use Jace [13, 15] (a Java based library designed to build asynchronous iterative algorithms) for our experiments.

The following section recalls the principle of asynchronous iterative algorithms. Section 3 briefly presents the Jace programming environment. In Section 4, the three scientific problems are presented. The first one is a linear problem. Then a differential-equation problem is presented and applied to two different chemical-reaction applications. The first models a one-dimensional equation, whereas the second models a two-dimensional equation. Section 5 is devoted to experiments in a distant heterogeneous cluster context. Finally, we conclude this paper and give some perspectives.

2 Asynchronous Iterative Algorithms

As our goal is not to present asynchronous iterative algorithms theory formally, only the important points and criteria that characterize this class of algorithms will be reviewed. Interested readers can see, for example [2, 4, 12], to have more details about asynchronism and AIAC algorithm.

First we present the main differences between synchronous and asynchronous iterative algorithms. In the synchronous class, at a given time, either a processor is computing the same iteration as the others or it is idle (waiting for other processors to finish their current iteration). Communications can be synchronous or even asynchronous. In the latter case, idle times between two iterations can be reduced since asynchronous communications overlap a part of the computation. This is very important to ensure the convergence of these algorithms. Nevertheless, the advantage of asynchronous communications lies in the fact that the convergence behavior is strictly the same as the sequential iterative equivalent.

In the AIAC algorithm class, both communications and iterations are asynchronous. So, as soon as a processor finishes an iteration, it starts the next one with the last data received during a previous iteration, without waiting for any other processor. Consequently all idle times are suppressed and the number of iterations before the convergence is in general greater than in the synchronous case. Nevertheless, the execution time could be considerably reduced, especially in a distant cluster context. Another very interesting feature of this class of algorithms is its tolerance to long message delays. As a consequence, we actually think that this class is suitable for grid computing.

3 Jace for Asynchronous Iterative Algorithms

Jace [13] is a Java based environment designed to implement asynchronous algorithms as simply as possible. Communications are implemented with RMI[1] and asynchronism between computation and communication is performed by threads. Nevertheless it is possible to develop synchronous algorithms using Jace.

According to the kind of algorithms, synchronous or asynchronous, blocking or non-blocking receptions should respectively be used. In both cases, sendings are non blocking. More precisely, Jace uses a thread to send messages which are tagged to differentiate them. Messages are primarily stored in a queue and the thread tries to send them as soon as possible. Nonetheless, if the computation thread produces more messages than the network is able to send, then some old messages may be replaced by new ones. In the same way, a queue, managed by another thread, is used to receive messages. If the computation thread does not often get its messages, some older versions of a message can be replaced by a newer one.

3.1 Skeleton and Implementation Scheme for Parallel Iterative Algorithms

With Jace, parallel iterative algorithms always have the same skeleton, whatever the version (synchronous or asynchronous). Their canvas is as follows:

Algorithm 1 Parallel iterative algorithm skeleton

Initialize the dependencies
repeat
  Computation of local data
  Exchange of non-local data
  Convergence detection
until Global convergence is achieved

---

[1] Remote Method Invocation.

We start by initializing the dependencies (i.e. we detect which are the data to be exchanged with other processors during the overall execution, and which processors are concerned). After this initialization, we start the iteration loop: we first execute the computation phase which approximates local data. For this phase, different solvers can be used according to the problem. After the results have been computed, the dependency values are exchanged with the nodes concerned (synchronous algorithms would use blocking receptions and asynchronous ones would use non-blocking receptions).

The last phase of the iteration loop relates to the convergence detection. We proceed by centralizing the state of all nodes over one (the master node). We consider that the global convergence is achieved when the residual error on each node (defined as $\max_i |y_i^{t+1} - y_i^t|$ where y_i^t is the local vector computed at iteration t) is smaller than ϵ (a stopping criterion fixed for each problem). However, the process is different according to the version:

- For synchronous algorithms, each processor sends its local error to the master node after each iteration, and then stops its activity until it receives the convergence response from the master (this period corresponds to the idle times).
  When the master has received all the local errors, it computes the global error of the whole system and decides whether the global convergence is achieved or not.
  Then, the master node sends the convergence response to the others which keep on computing.
- For AIAC algorithms, the nodes only send a message to the master if their local convergence state changes and stays constant during several iterations (hence, there are no idle times between two iterations).
  The master decides that global convergence is reached when all the nodes are in a local convergence state at a given time. Then, it orders the other processors to stop computing. Interested readers can consult [11] for more information.

Finally, in both versions, as long as the convergence is not reached, the algorithm computes another iteration in order to get more accurate results.

3.2 Parallelization Methodology

In this section, we explain how we implement AIAC algorithms starting from sequential ones. This approach is divided in two parts:

1. This part consists in parallelizing a sequential algorithm in order to obtain a synchronous iterative algorithm. This step requires a strong attention to split the data, to compute dependencies and manage communications. In fact, the data distribution is frequently a delicate part because local indices must carefully be taken into consideration and one should ensure that they correspond to global indices. Moreover, exchanges of data are primordial in order to obtain the same results as in the sequential version.

2. The desynchronization of a synchronous algorithm requires only two adaptations (as described in section 3.1). The convergence detection is different in the context of AIAC and all blocking receptions are replaced by non-blocking ones so it is a quite easy task to realize.

4 Presentation of the Problems

In this section, the three scientific applications we experimented are described.

4.1 A Linear Problem

The block Jacobi method is the simplest iterative method to solve a linear system. Let $Ax = b$ be a linear system to be solved, and M a block diagonal matrix of A. Then, the solution of the system is computed by solving successively $x^{k+1} = Tx^k + M^{-1}b$ with $T = M^{-1}N$ and $A = M - N$ (the decomposition of A is described in figure 1 with the example of a 3x3 block structure).

The asynchronous convergence criterion is $\rho(|T|) < 1$ where $\rho(T)$ is the spectral radius of T:

$$\rho(T) = \max_i\{|\lambda_i|\}$$

where λ_i is an eigenvalue of T. Interested readers can see for example [1, 2, 8].

The matrices A were randomly generated so that the matrices T are nonnegative and diagonally dominant.

4.2 Differential Equation Problem

To solve an Ordinary Differential Equation (ODE) defined as

$$y'(t) = f(y(t), t),$$

computational scientists often use parallel synchronous methods. CVODE [5], is a well-known and efficient tool to solve such systems. It proceeds by using an implicit method for stiff problems (like BDF in case of CVODE, or Runge-Kutta or more simply Euler) to discretize the time. The resulting nonlinear system is generally solved using a Newton method. Consequently, at each iteration of the Newton method, a linear system should be solved. An iterative method is especially suited since the linear system is sparse.

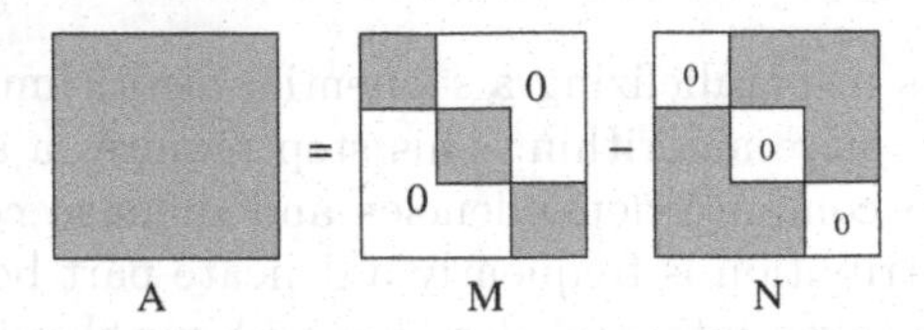

Fig. 1. Decomposition of A with a block diagonal matrix

An alternative to synchronizations in a parallel execution is possible since the convergence of nonlinear systems has been studied in the asynchronous context, see for example [7, 8]. So it is possible to build an asynchronous iterative solver for ODE. The main features of the solver we build are summarized in the following. To discretize the time, an implicit Euler method is used. Then, the obtained nonlinear system is linearized using the Newton method. To apply this method, the Jacobian matrix must be computed. Firstly, as this step is time consuming, it is efficient to compute this matrix, only once, for each time step. Secondly, as this matrix is very sparse, a relevant compressed row storage technique provides a huge speed up in the construction of the matrix and a considerable memory saving. Moreover, it is possible to use an asynchronous multisplitting method to solve this system; hence, all synchronizations during a time step computation can be suppressed. Only one synchronization is required for the first iteration of each time step. Finally, the linear system is solved using the GMRES method [6, 9, 10, 16].

Solving a One-Dimensional ODE System: A Chemical Reaction Problem. In this section, we present a problem which is a large stiff one-dimensional system of Partial Differential Equations (PDE).

This system models a chemical reaction mechanism which leads to an oscillating reaction. It deals with the conversion of two elements A and B into two others C and D by the following series of steps:

$$\begin{aligned} A &\rightarrow X \\ 2X + Y &\rightarrow 3Y \\ B + X &\rightarrow Y + C \\ X &\rightarrow D \end{aligned} \tag{1}$$

There is an autocatalysis and when the concentrations of A and B are kept constant, the concentrations of X and Y oscillate with time.

We want to determine the evolutions of the concentrations u and v of both elements X and Y according to time in each point of a given one-dimensional space. We discretize this space in function of time in order to obtain a one-dimensional ODE system. If the discretization is made with N points, the evolution of the u_i and v_i for $i = 1, ..., N$ is given by the following differential system:

$$\begin{aligned} u_i' &= 1 + u_i^2 v_i - 4u_i + \alpha(N+1)^2(u_{i-1} - 2u_i + u_{i+1}) \\ v_i' &= 3u_i - u_i^2 v_i + \alpha(N+1)^2(v_{i-1} - 2v_i + v_{i+1}) \end{aligned} \tag{2}$$

The boundary conditions are:

$$\begin{aligned} u_0(t) &= u_{N+1}(t) = \alpha(N+1)^2 \\ v_0(t) &= v_{N+1}(t) = 3 \end{aligned}$$

and initial conditions are:

$$u_i(0) = 1 + sin(2\pi x_i) \text{ with } x_i = \frac{i}{N+1},\ i = 1, ..., N$$
$$v_i(0) = 3$$

Here, we fix $\alpha = \frac{1}{50}$. N is a parameter of the problem. For further information about this problem and its formulation, we refer to [3].

For this problem, the u_i and v_i of the system are represented in a single vector as follows:

$$y = (u_1, v_1, ..., u_N, v_N)$$

with $u_i = y_{2i-1}$ and $v_i = y_{2i}$, $i \in \{1, ..., N\}$. So we obtain a one-dimensional ODE system as

$$y'(t) = f(y(t), t),$$

representing (2). The y_j functions, $j \in \{1, ..., 2N\}$ thereby defined will also be referred to as spatial components in the remaining of this section.

In order to exploit the parallelism, the y_j functions are initially homogeneously distributed over the processors. Since these functions are represented in a one-dimensional space (the state vector y), we have chosen to logically organize our processors in a linear way and map the spatial components (y_j functions) over them. Hence, each processor applies the Newton method (as described in part 4.2) over its local components using the needed data from other processors involved in its computations.

From the problem formulation (2) it arises that the processing of components y_p to y_q also depends on the two spatial components before y_p and the two spatial components after y_q. Hence, if we consider that each processor owns at least two functions y_j, the non-local data needed by each processor to perform its iterations come only from the previous processor and the following one in the logical organization.

Solving a Two-Dimensional ODE System: An Advection-Diffusion Problem. The problem is a large stiff two-dimensional system of PDE that models a chemical mechanism involving diurnal kinetics, advection and diffusion. We want to determine the evolution of the concentration of two chemical species in each point of a given two-dimensional space, according to time. We discretize this space in order to obtain a grid. The concentration evolution of the species in each $x-z$ point (mesh) of the resulting grid (x for the columns and z for the rows) is given by the following differential system:

$$\frac{\delta c^i}{\delta t} = K_h \frac{\delta^2 c^i}{\delta x^2} + V \frac{\delta c^i}{\delta x} + \frac{\delta}{\delta z} K_v(z) \frac{\delta c^i}{\delta z} + R^i(c^1, c^2, t), \tag{3}$$

with i= 1,2 (i is used to distinguish each chemical species) and where the reaction terms are given by:

$$R^1(c^1, c^2, t) = -q_1 c^1 c^3 - q_2 c^1 c^2 + 2q_3(t) c^3 + q_4(t) c^2$$
$$R^2(c^1, c^2, t) = q_1 c^1 c^3 - q_2 c^1 c^2 - q_4(t) c^2 \tag{4}$$

The constants and parameters for this problem are as follows: $K_h = 4.0 \times 10^{-6}, V = 10^{-3}, K_v(z) = 10^{-8}exp(z/5), q_1 = 1.63 \times 10^{-16}$, $q_2 = 4.66 \times 10^{-16}, c^3 = 3.7 \times 10^{16}$, and the diurnal rate constants $q_j(t)$ are given by:

$$\begin{aligned} q_j(t) &= exp[-a_j/sin(\omega t)] \; for \; sin(\omega t) > 0 \\ q_j(t) &= 0 \qquad\qquad\qquad\quad for \; sin(\omega t) \leqslant 0 \end{aligned} \tag{5}$$

where $j = 3,\ 4,\ \omega = \pi/43200,\ a_3 = 22.62,\ a_4 = 7.601$.

The time interval of integration is [0s, 7200s]. Homogeneous Neumann boundary conditions are imposed on each boundary and the initial conditions are:

$$\begin{aligned} c^1(x, z, 0) &= 10^6\alpha(x)\beta(z), \\ c^2(x, z, 0) &= 10^{12}\alpha(x)\beta(z) \\ \alpha(x) &= 1 - (0.1x - 1)^2 + (0.1x - 1)^4/2, \\ \beta(z) &= 1 - (0.1z - 1)^2 + (0.1z - 4)^4/2 \end{aligned} \tag{6}$$

With a discretization of the considered space, we can transform the two-dimensional PDE system to obtain an two-dimensional ODE system as

$$y'(t) = f(y(t), t),$$

representing (3).

In order to exploit the parallelism, there are three methods to distribute the data over the nodes: 1) to decompose the grid according to the z components (i.e. horizontal decomposition), 2) according to the x components (i.e. vertical decomposition), or 3) according to the z and the x components (a mix of the first two methods). We have chosen the first strategy because it is the most efficient[2].

To solve this two-dimensional system, we represent the $x - z$ grid into a single y vector composed of N elements ($N = 2 \times x \times z$ because the problem deals with two chemical species in each mesh of the grid). Then, we logically organize our processors in a linear way and homogeneously map the spatial components (the y_j functions with $j \in \{1, ..., N\}$) over them (this corresponds to the first strategy described before). Hence, each processor applies the Newton method (as described in part 4.2) over its local components using the needed data from other processors involved in its computation.

From the advection-diffusion problem formulation (3), it arises that the processing of c^i for a given $x - z$ mesh of the grid depends on c^1 and c^2 for the same mesh, but also depends of c^1 and c^2 for the North, East, West and South neighbors. Hence, after a computation, each processor must always send the same parts of the subgrid it computes: 1) for the bottom row of the subgrid the solution components are sent to the processor below it and 2) for the top row they are sent to the processor above it. Then, we can deduce what the corresponding components of y are and actually send them to the processors.

---

[2] In fact, dependencies between neighbor meshes of the $x - z$ grid are much larger horizontally than vertically.

5 Experiments

In this section, our asynchronous algorithms are compared to the synchronous versions (for each of the three problems described in the previous section), in order to evaluate the efficiency of asynchronism.

In the context of our experiments, we have studied the execution time of the synchronous and the asynchronous algorithms according to the size of the problem. Each time given in the following tables is the average of a series of executions. Each table also features the ratio of the execution times (*ratio* = *synchronous time/asynchronous time*).

Remark 1. The number of iterations is not mentioned in the following results, because, due to the conditions of our experimentations, no pertinent conclusion can be deduced from this kind of information. Indeed, our experimentations were made in non dedicated environnements and contrary to dedicated parallel environments such as dedicated clusters and parallel computers, the number of iterations dramatically changes according to unpredictable parameters such as the load of the network and the load of the workstations. In such environments, the number of asynchronous iterations is directly related to the unpredictable character of the conditions of experimentations. However, the number of iterations in the asynchronous mode is often (and almost always) higher than in the synchronous one. But, the whole execution times in asynchronous modes are lower, because the desynchronizations (of the iterations and the communications) suppress all the unpredictable idle-times and so reduce the whole execution times. Only the information provided by the mean value of a serie of the execution times have a relevant signification.

The experiments have been performed with the 1.2.2 version of the JDK[3] running the Linux Debian operating system. The virtual parallel machine used was composed of 6 workstations with at least 256MB RAM and using the following processors:

- 1 AMD Athlon(tm) XP 1600+,
- 1 Intel(R) Pentium(R) 4 CPU 1.70GHz,
- 2 Intel(R) Pentium(R) 4 CPU 1.60GHz,
- 1 AMD Athlon(tm) XP 1800+,
- 1 Intel(R) Pentium(R) 4 CPU 2.40GHz.

Five of those workstations were connected together by an Ethernet 100Mb/s connection and the sixth one was linked by an ADSL connection with a 512Kb/s download bandwidth and a 128Kb/s upload bandwidth.

This heterogeneous configuration (networks and processors) is typical in a distant grid context.

[3] Java Development Kit.

5.1 Jacobi Solver Results

For the Jacobi method experiments, we have studied the behavior of both synchronous and asynchronous versions according to S, the size of the linear system. This experiment has been performed with the following parameters: $\varepsilon = 10^{-5}$ (the stopping criterion) and 100 nonzero elements per row.

The results are given in the table 1 where the times are given in seconds. We can see that the execution times of the asynchronous algorithm are always lower than those of the synchronous version.

Table 1. Execution times and ratios of synchronous and asynchronous versions

S	Sync. time (in s)	Async. time (in s)	ratio
1,000	461	81	5.69
5,000	642	228	2.81
10,000	1123	270	4.16
15,000	1464	356	4.11
20,000	1947	454	4.28
30,000	2696	748	3.60
40,000	3546	969	3.66

Remark 2. The difference between the ratio values can be explained by the fact that each matrix is randomly generated for each experiment. If S is different, the matrix would be completely different too. Consequently, the number of iterations may change to solve the problem so the ratio changes too.

5.2 One-Dimensional Chemical Problem Results

For this experiment, we have studied the behavior of both synchronous and asynchronous versions according to the size of y vector. This experiment has been performed with the following parameters: $\varepsilon = 10^{-7}$ (the stopping criterion), h (the time step) $= 1.10^{-7}s$ and T (the time interval) $=$ [0s, 0.001s]. The results are given in table 2 where the times are given in seconds.

Table 2. Execution times and ratios of synchronous and asynchronous versions

Size of the y vector	Sync. time (in s)	Async. time (in s)	ratio
5,000	270	62	4.35
10,000	481	120	4.00
20,000	941	282	3.33
30,000	1607	398	4.04

With those results, we can also see that the execution times of the asynchronous version are always lower than the synchronous ones.

5.3 Advection-Diffusion Problem Results

For the Advection-Diffusion problem experiment, we have studied the behavior of both synchronous and asynchronous versions according to the size of $x - z$ grid. This experiment has been performed with the following parameters: $\varepsilon = 1$ (the stopping criterion), h (the time step) $= 180s$ and T (the time interval) $=$ [0s, 7200s]. The results are given in table 3 where the times are given in seconds

Table 3. Execution times and ratios of synchronous and asynchronous versions

Size of the grid	Sync. time (in s)	Async. time (in s)	ratio
30*30	285	29	9.83
60*60	463	59	7.85
90*90	767	135	5.68
120*120	1074	327	3.28
150*150	1445	503	2.87

For this experiment, we can also notice that the asynchronous execution times are lower than the synchronous ones, whatever the size of the problem.

Remark 3. In table 3, we can see that when the size of the problem increases, the ratio decreases. This is due to the following phenomenon: the larger the system is, the more computation times are significant compared to communication times. And when the ratio *computation time/communication time* increases , the ratio *synchronous time/asynchronous time* decreases because computation time is significant compared to communication times. Hence, for very large problems, it is necessary to involve even more processors in order to reduce computation times and preserve an efficient ratio *synchronous time/asynchronous time*.

We can conclude that in a grid context, it is very important to take the heterogeneity of both processors and networks into account.

6 Conclusion

Performances of asynchronous iterative algorithms (AIAC) are far better than synchronous ones in the context of distant heterogeneous cluster computing. AIAC algorithms are not penalized by synchronizations because they do not use blocking receptions and, hence they are much more efficient than synchronous ones in a heterogeneous context (i.e. with heterogeneous networks and heterogeneous processors). Moreover they are long-delay message tolerant. That is why we do think they are especially suitable for computational science in the context of grid computing.

Jace is fully suited to the implementation of AIAC algorithms because firstly, it makes it possible to implement parallel synchronous algorithms and secondly, to receive the last version of a message via non-blocking receptions. Moreover, the adaptation of synchronous algorithms, designed to be later asynchronous, to AIAC algorithms is simplified and does not require much time since only the receptions and the convergence detection procedures need modifications (as an example, it took only one hour to obtain the AIAC version of the Advection-Diffusion problem, the most complex of the three applications, whereas the synchronous version from sequential one required two weeks).

Currently, we think that a lot of iterative algorithms would be easily adaptable to the context of AIAC (after having first checked their convergence properties) and would provide considerable gains of execution time, especially in the context of grid computing.

In future works, we plan to study and implement three-dimensional PDE systems. Overlapping techniques which generally speed up the performances in multisplitting methods could be integrated. Finally, finite element methods could be considered too, as they are commonly used in computational science.

References

1. El Tarazi, M. N.: Some convergence results for asynchronous algorithms. Numerisch Mathematik **39** (1982) 325–340
2. Bertsekas, D., Tsitsiklis, J.: Parallel and Distributed Computation: Numerical Methods. Prentice Hall, Englewood Cliffs NJ (1989)
3. Hairer, E., Wanner, G.: Solving ordinary differential equations II: Stiff and differential-algebraic problems. Ser. Springer series in computational mathematics. Berlin: Springer-Verlag **14** (1991) 5–8
4. El Baz, D., Spiteri, P., Miellou, J.-C. and Gazen, D.: Asynchronous iterative algorithms with flexible communication for nonlinear network flow problems. Journal of Parallel and Distributed Computing, **38(1)** (1996) 1–15
5. Cohen, S. D., Hindmarsh, A. C.: CVODE, A Stiff/Nonstiff ODE Solver in C. Computers in Physics, **10(2)** (1996) 138–143
6. Bahi, J. M., Miellou, J.-C. and Rhofir, K.: Asynchronous multisplitting methods for nonlinear fixed point problems. Numerical Algorithms, **15(3,4)** (1997) 315–345.
7. Szyld, D. and Xu, J.J.: Convergence of some asynchronous nonlinear multisplitting methods. Research Report 99-6-30, Department of Mathematics, Temple University (1999)
8. Frommer, A., Szyld, D.: On asynchronous iterations. J. Comp. Appl. Math., **123** (2000) 201–216
9. Guivarch, R., Padiou, G. and Papaix, P.: Asynchronous Schwarz alternating method in observation based distributed environment. Réseaux et Systèmes Répartis - Calculateurs Parallèles. Editions Hermes **13(1)** (2001) 35–45
10. Arnal, J., Migallon, V. and Penadés, J.: Nonstationary parallel newton iterative methods for nonlinear problems. Lectures Notes in Comput. Science, VECPAR'2000 **1981** (2001) 380–394
11. Bahi, J., Contassot-Vivier, S., Couturier, R.: Asynchronism in a global computing environment. 16th Annual International Symposium on High Performance Computing Systems and Applications, HPCS'2002, (2002) 90–97.
12. Bahi, J., Contassot-Vivier, S., Couturier, R.: Coupling Dynamic Load Balancing with Asynchronism in Iterative Algorithms on the Computational Grid. 17th IEEE and ACM Int. Conf. on International Parallel and Distributed Processing Symposium, IPDPS 2003 (2003) 40a, 9 pages

13. Bahi, J., Domas, S.,Mazouzi, K.: Java Asynchronous Computation Environment. 12-th Euromicro Conference on Parallel, Distributed and Network based Processing, (2004)
14. Bahi, J., Contassot-Vivier, S., Couturier, R.: Performance comparison of parallel programming environments for implementing AIAC algorithms. 18th IEEE and ACM Int. Conf. on International Parallel and Distributed Processing Symposium, IPDPS 2004 (2004) 247b, 8 pages
15. Mazouzi, K., Domas, S. and Bahi, J.: Combination of java and asynchronism for the grid: a comparative study based on a parallel power method. 6th International Workshop on Java for Parallel and Distributed Computing, JAVAPDC workshop of IPDPS 2004, IEEE computer society press (2004) 158a, 8 pages
16. Bahi, J., Couturier, R. and Vuillemin, P.: Solving nonlinear wave equations in the grid computing environment: an experimental study. Journal of Computational Acoustics (to appear)

Security Mechanism for Medical Image Information on PACS Using Invisible Watermark

Guan-tack Oh[1], Yun-Bae Lee[1], and Soon-ja Yeom[2]

[1] Dept. of Computer Science, Graduate School, Chosun University,
375 Seosuk-Dong, Dong-ku, Gwangju 501-759, Korea
Tel : +82-062-230-7736, Fax : +82-062-233-6896
yblee@mina.chosun.ac.kr
[2] School of Computing, University of Tasmania, Australia

Abstract. The emerging digital revolution is impacting all areas of society, including medical areas. The era of digital hospital is coming and its success will depend on the rapid adaptation of emerging technology. Picture Archive and Communication System(PACS) could give the necessary competitive edge for medical institutions. Pacs digitizes captured images using the technique in Computed Radiography, stores them in massive storage devices, and transmits the patient's medical information(or images) to networked clinics and hospital wards, anywhere anytime. However, to be useful, the issues of possible fabrication and damages to images, and authentication have to be addressed. The watermark technique could provide some solution to these problems. This paper describes an improved secure medical image system addressing issues on PACS, accurate patient data, and privacy. Further, the proof of concept is demonstrated by experiment.

1 Introduction

Traditionally, the film has been used to capture, store, and display radiographic images in hospital. Today, many radiological modalities generate digital images that can be viewed directly on monitor displays. Picture Archiving and Communication System (PACS) consists of image acquisition archiving and retrieval, communication, image processing, distribution, and display of patient information from various imaging modalities [1]. Digital archival will eliminate the need for bulky film room and storage site and the possibility of losting films. Digital image acquisition requires interfacing the PACS to the digital imaging modalities such as CT, MRI, CR, DSA (Digital Subtraction Angiography), supersonic, endoscope, and film digitizer. The modality interfaces require that the devices to be used comply with the digital imaging and communication in medicine (DICOM) standard. DICOM represents an international standard of definitions and methods for transferring information and images between devices thus to advertise reliable interoperability among different modules [2].

The PACS system offers an efficient means of viewing, analyzing, and documenting study results, and furnishes a method for effectively communicating study results

M. Daydé et al. (Eds.): VECPAR 2004, LNCS 3402, pp. 315–324, 2005.

to the referring physicians [1]. Therefore, because of possibility of making wrong diagnosis, all the images from medical equipments must ensure at least same quality when they are transmitted through the public media as internet. In this paper, we introduce and prove some possibilities of medical image fabrication or modulation, and we provide a solution to protect medical images from malicious fabrication. The remainder of this paper is organized as follows. In section II, the brief introduction about DICOM and DICOM standards are presented. In section III, we illustrate the experimental evidence for possibilities of medical image fabrication. In section IV, we provide a method to prevent medical image from malicious modulation, and finally, the conclusion in section V.

2 DICOM

DICOM (Digital Image Communication in Medicine) is the industry standard for the transmission of digital images and other associated information between digital imaging equipments. The American College of Radiology (ACR) and the National Electrical Manufacturers Association (NEMA) formed a joint committee to develop a standard for DICOM. This DICOM Standard was developed according to the NEMA procedures. This standard is developed in liaison with other standardization organizations including CEN TC251 in Europe, and JIRA and MEDIS-DC in Japan, with review also by other organizations including IEEE, HL7 and ANSI in the USA [2].

DICOM enables digital communications among image acquisition devices, image archive, hardcopy devices and other diagnostic and therapeutic equipment that is connected into a common information infrastructure. In order to provide the support for connecting different devices from various manufacturers on standard networks, it is closely connected with network protocols like OSI and TCP/IP. It also defines data structures/formats for medical images and related data, network-oriented service such as image transmission and query of an image archive, formats for storage media exchange, etc.

The DICOM standard does not address issues of security policies, though clear adherence to appropriate security policies is necessary for any level of security [3]. The standard only provides mechanisms that can be used to implement security policies with regard to the interchange of DICOM objects between Application Entities (AE) using TCP/IP protocol that could not ensure security issues between AEs [3]. The brief view of DICOM supplements related with security issues are:

- Supplement 31: Security Enhancement One, Secure Transport Connection

This Supplement is to allow DICOM applications to exchange messages over a Secure Transport Connection. While this protects the data during transit, it did not provide any lifetime integrity checks for DICOM SOP Instances. It only addresses the authentication of network communication security. The TLS (Transport Layer Security) and ISCL (Integrated Secure Communication Layer) protocols are used for exchange medical image data securely on the communication network [3].

• Supplement 41: Security Enhancement Two, Digital Signature

This Supplement adds some mechanisms for adding Digital Signatures to DICOM SOP Instances as a first step toward lifetime integrity[1] checks [3].

• Supplement 51: Media Security

This supplement defines a framework for the protection of DICOM files for media interchange based on the ISO 7498-2[2] [3].

Application Entities may not trust the communications channel by which they communicate with other Application Entities. Thus, this Standard provides mechanisms for AEs to authenticate each other securely, AEs to detect any tampering with or alteration of messages exchanged and AEs to protect the confidentiality of those messages while they are traversing the communication channel. However, as mentioned in ISO7498-2, the DICOM standard does not provide the protection mechanism for the medical images [4].

3 Experimental Study

3.1 Background

The DICOM image file contains both a header, which includes text information such as patient's name, image size, medical record number, imaging modality, and other demographic information in the same file.

For purpose of image processing, extraction of pixel data value is needed. Thus, the only part of pixel data from DICOM image is needed, in the other word, the pixel data element tagged as (7FE0, 0010) can be extracted.

The Pixel Data Element (7FE0, 0010) and Overlay Data Element (60xx, 3000) shall be used for the exchange of encoded graphical image data. These elements along with additional Data Elements, specified as attributes of the image information entities defined in PS3.3-2001, shall be used to describe the way in which the Pixel Data and Overlay Data are encoded and shall be interpreted [5].

The Pixel Data Element and Overlay Data Element have a VR (Value Representation)[3] of OW (Other Word String: uncompressed) or OB (Other Byte String: compressed), depending on the negotiated Transfer Syntax[4]. The only difference between OW and OB being that OB, a string of bytes, shall be unaffected by Byte Ordering [6].

---

[1] Data Integrity - verifying that data within an object has not been altered or removed.

[2] ISO 7498-2 : Information Processing System – Open Systems Interconnection – Basic reference model – Part 2 : Security Architecture.

[3] Specifies the data type and format of the Values contained in the value field of a data element.

[4] Transfer Syntax : A set of encoding rules able to unambiguously represent one or more abstract syntaxes [6].

Notice that because of the most of medical image has more than 8 bit length in each pixel it cannot be extracted as BMP or JPEG. Therefore, it must be modulated as 8 bit pixel for easy processing.

The processing steps for modulating and demodulating DICOM image into JPEG image are:

1. Get medical image from MRI
2. Modulate DICOM standard image to raw data using Visual Gate in PI-View
3. Demodulate raw data into JPEG image
4. Fabricating JPEG image as needed
5. Demodulating fabricated JPEG image into DICOM standard image
6. Insert this fabricated image into original data.

It is not difficult to modulating DICOM image into BMP or JPEG using PACS. However, because the DICOM image has its header that contains Patient Name, ID and Study Date etc., modulating BMP or JPEG image into DICOM image is not easy. It is also difficult to decide the Transfer Syntax of DICOM image.

3.2 Medical Image Modulation and Diagnostic Results

The digitalized image has been modulated intentionally and testing the modulated images by some experts for the reason how these reading effects the diagnostic results. A M.R.I (Magnetic Resonance Imaging) of normal vertebra is modulated as HNP (Herniated Nucleus Pulpous).

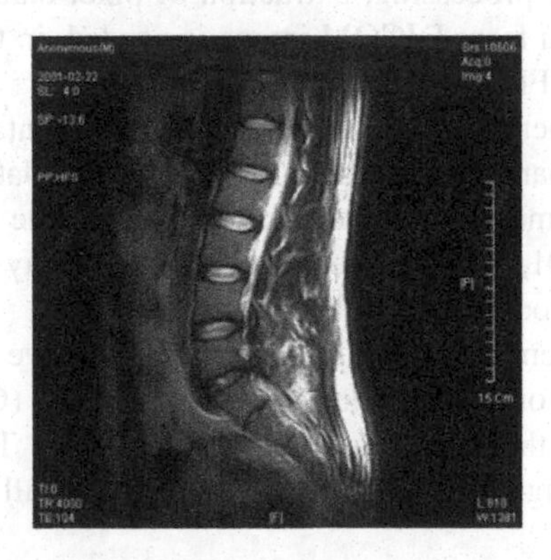

Fig. 1. MRI of normal patient's L-SPINE LATERAL

Then the modulated image is examined by two radiotherapists and three residents who work for the C university MRI laboratory in Gwangju, Korea. As the result of medical image modulation, a MRI of normal patient's L-spine lateral (shown in figure 1) is fabricated as shown in figure 2 that possibly results the completely different diagnosis.

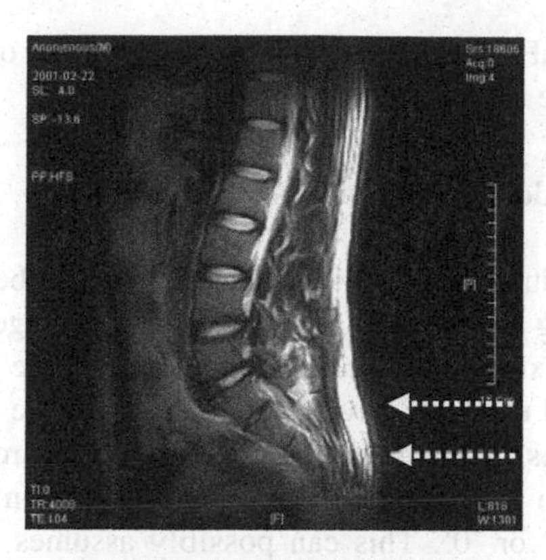

Fig. 2. Fabricated MRI

The study result from two radiotherapists and three residents of C university MRI laboratory in Gwangju was either Herniation of Intervertebral Disk or Bulging Disc. This wrong diagnostic result could cause serious problems, and the fabricated medical image must be verified in any reason. Therefore, we provide the invisible watermarking mechanism to verify fabricated medical images in the following section.

4 Security Mechanism

4.1 Digital Watermark

Compared to ordinary paper or film form medical image, digitized medical information including both image and text provides many advantages, such as easy and inexpensive duplication and re-use, less expensive and more flexible transmission either electronically (e.g. through the Internet) or physically (e.g. as CD-ROM). Furthermore, transferring such information electronically through network is faster and needs less effort than physical paper copying, distribution and update. However, these advantages also significantly increase the problems associated with enforcing copyright on the electronic information [7]. The use of digital imaging technique and digital contents based on theInternet has grown rapidly for last several years, and the needs of digital image protection become more important. For the purpose of medical image protection from modulation/fabrication, the invisible watermarking technique is adapted to the medical image. The fabrication or alteration of medical image is shown and verified by an experimental works.

The basic principle of watermarking methods is to add copyright information into the original data by modifying it in a way that the modifications are perpetually invisible and robust [8]. Digital Watermarking makes it possible to mark uniquely each image for every medical image. Watermarking techniques aim at producing watermarked images that suffer no little quality degradation and perceptually identical to the original versions [9]. However, there will be some traces on the watermark image, if the watermarked image is modulated.

4.2 Watermark Embedding Method Based on JPEG

In this section, we introduce the JPEG-based watermark embedding method, then the algorithm for embedding watermarks into the medical images and experimental results are shown in the next section. The basic principle of the JPEG-based embedding method is that quantized elements have a moderate variance level in the middle frequency coefficient ranges, where scattered changes in the image data should not be noticeably visible [7]. In general, the value of each pixel in a binary image can be represented as either '1' or '0'. This can possibly assumes that there is no 'noise' space which can be used for embedding additional information. In this paper, we use the Minerva's [9] algorithm to embedding watermark. The Minerva's algorithm for binary images is based on the ratio of '1' and '0' in a selected block. In Minerva's algorithm, they suppose '1' represent black bit and '0' represent white bit in the source binary image. According to Minerva's algorithm, the rate of blacks and number of 1s in the block b can be expressed as:

$$P_1(b) = \frac{N_1(b)}{64} \quad \text{-------------------------------------} \ (1)$$

Since $P_1(b)$ and $N_1(b)$ represent the probability of rate of blacks in the block b and the number of 1s in the block b respectively, the probability of rate of whites can be expressed as equation (2).

$$P_0(b) = 100 - P_1(b) \quad \text{-------------------------------------} \ (2)$$

A bit indicating black is embedded into the block of b if $P_1(b) \rangle t$ (where t is a certain threshold), otherwise the white bit is embed into the block b until given threshold is reached.

For a confirmation, we provide some practical examples and the visualized interfaces of the system that can insert or extract watermark into/from the original images. Table 1 shows a part of algorithm for applying watermark images into medical images. Also in figure 3, the visualized interface of the embedding watermark is illustrated. As shown in figure 3, Pic1 and Pic2 are merged as "Merge Image". From right upper side of figure 3, the value can be taken from 0 to 1, which indicates the brightness of the watermark image. The system used for embedding watermark image is:

Microsoft Windows™ NT/2000SEVER
Pentium IV 800 MHz CPU
128 MB RAM and 20GB hard disk space
Ethernet network card
Screen resolution of 1280*1024 32-bit color display

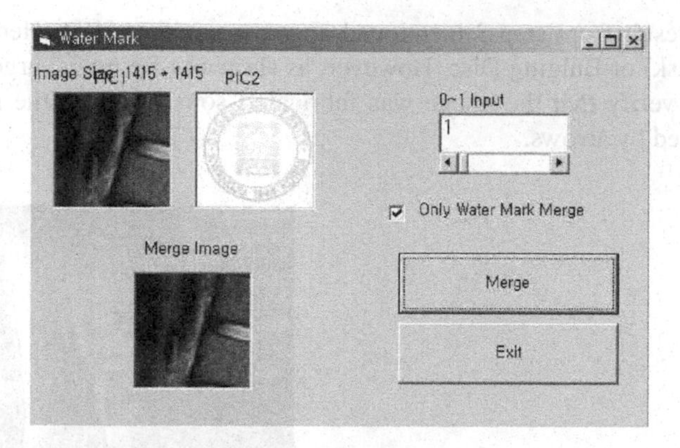

Fig. 3. Visual interface of watermark embedding system

Table 1. Algorithm used for embedding watermark

```
      Select Case Check1.Value
          Case 0
              Picture3.PSet (X, y), RGB(Int(R1 * intMerge + R2 *
              (1 - intMerge)), Int(G1 * intMerge + G2 * (1 - intMerge)),
              Int(B1 * intMerge + B2 * (1 - intMerge)))
          Case 1
              If R1 = 255 And G1 = 255 And B1 = 255
              Then Picture3.PSet (X, y), RGB(Int(R2), Int(G2), Int(B2))
              Else Picture3.PSet (X, y), RGB(Int(R1 * intMerge + R2 *
              (1 - intMerge)), Int(G1 * intMerge + G2 * (1 - intMerge)),
              Int(B1 * intMerge + B2 * (1 - intMerge)))
   End Select
```

4.3 Result

For the proof of identifying fabricated medical images, following four steps are tested:

Step 1: A watermark (shown in figure 4) is inserted into the original image (shown in figure 1).
Step 2: The watermarked image (shown in figure 5, which looks same as original image) is fabricated (shown in figure 6) on purpose.
Step 3: The fabricated image is studied from two radiotherapists and three residents of MRI laboratory in the university hospital.
Step 4: The watermark image is extracted (shown in figure 7) from the fabricated image to identifying whether the image was fabricated or not.

The study result from step 3 mentioned above was either HID (Herniation of Intervertebral Disk) or Bulging Disc. However, as shown in figure 7, surgeons or radiotherapists can verify that the image was fabricated somewhere in the middle of the image as marked by arrows.

Fig. 4. Watermark

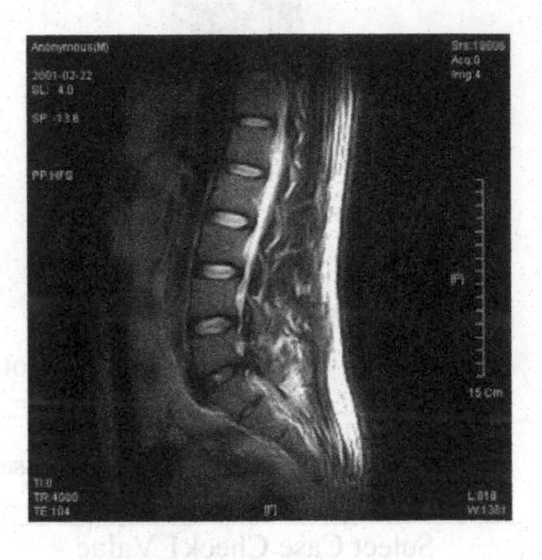

Fig. 5. Image with watermark

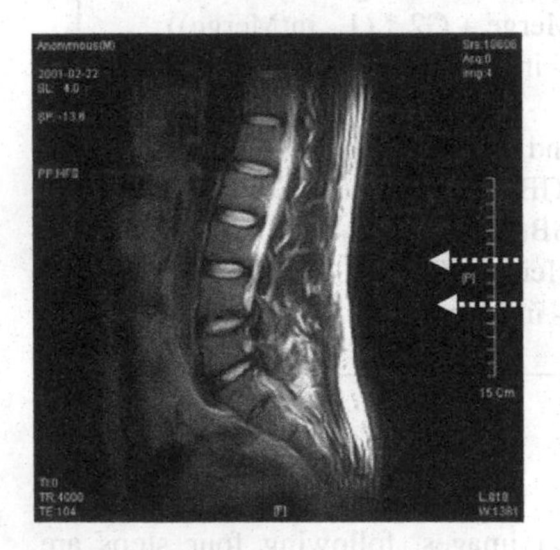

Fig. 6. Fabricated image

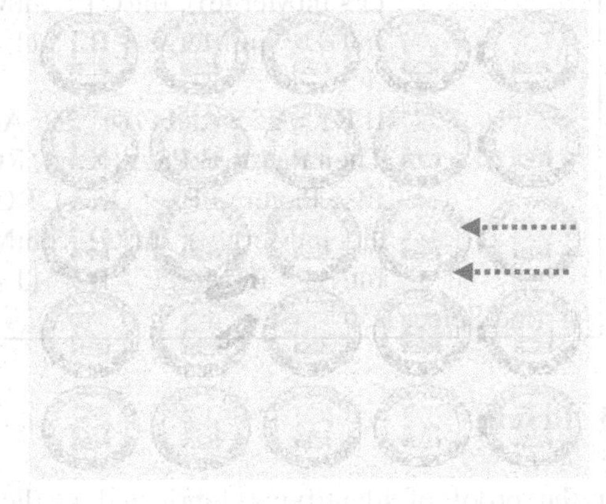

Fig. 7. Extracted watermark from fabicated image

5 Conclusion

Recently, the trend of PACS considers a scheme to apply copyright protection mechanism like watermarking for transferring medical information on the public network to prevent medical information from fabrication or modification of medical data. There are mainly two points of views for applying the digital watermarking into medical image. One is to strengthening the security issues of PACS, and the other one is to waiting until the global standardization of medical information protection scheme on the communication channel is accomplished. In particular, the group who has opposite opinion insists the security issues of PACS is insignificance since the medical image could not be shared on the Internet or external communication channels at all.

However, for the assumption of convenience adapting IT and mobile communication technology into medical information system, it is obvious that real time online retrieving of medical information will become reality soon. Thus, it is very important to presuppose and preventing the counter measurement of the problems of security issues, which can possibly occurred from transferring the medical information on the public communication channels. Hereafter, the secured transmission scheme of the medical image on the mobile communication network should studied seriously.

Acknowledgement. We are grateful to surgeon Dr. C. H. Jeung from the neurological department of Wooil Hospital in Gwangju, who examine and compare the medical images that were fabricated intentionally. We also want to thank two radiotherapists and three residents of C university MRI laboratory in Gwangju. This study was supported by research funds from Chosun University 2000 in Korea.

References

[1] Wayne T. Dejarnette, "Web Technology and its Relevance to PACS and Teleradiology," Applied Radiology, August 2000.
[2] DICOM (Digital Imaging and Communications in Medicine), Part 1(PS3.1-2001): Introduction and Overview, Published by National Electrical Manufacturers Association, 1300 N. 17th Street Rosslyn, Virginia 22209 USA, 2001 at http://medical.mena.org/ dicom/2003.html
[3] DICOM (Digital Imaging and Communications in Medicine), Part 15(PS3.15-2001): Security Profiles, Published by National Electrical Manufacturers Association, 1300 N. 17th Street Rosslyn, Virginia 22209 USA, 2001 at http://medical. mena.org/ dicom/2003.html
[4] DICOM (Digital Imaging and Communications in Medicine), Part 10(PS3.10-2001): Media Storage and File Format for Media Interchange, Published by National Electrical Manufacturers Association, 1300 N. 17th Street Rosslyn, Virginia 22209 USA, 2001 at http://medical. mena.org/ dicom/2003.html
[5] DICOM (Digital Imaging and Communications in Medicine), Part 3(PS3.3-2001): Information Object Definitions, Published by National Electrical Manufacturers Association, 1300 N. 17th Street Rosslyn, Virginia 22209 USA, 2001 at http://medical. mena.org/ dicom/2003.html
[6] DICOM (Digital Imaging and Communications in Medicine), Part 5(PS3.5-2001): Data Structures and Encoding, Published by National Electrical Manufacturers Association, 1300 N. 17th Street Rosslyn, Virginia 22209 USA, 2001 at http://medical. mena.org/ dicom/2003.html
[7] Zhao, J. & E. Koch, "Embedding Robust Labels into Images for Copyright Protection," Proceedings of the 1995 KnowRight Conference, pp. 242-251.

[8] Jian Zhao, "A WWW Service to Embed and Prove Digital Copyright Watermarks", Proc. of the European Conference on Multimedia Applications, Services and Techniques, Louvain-La-Neuve, Belgium, May 1996.
[9] Minerva M. Yeung & Sharath Pankanti, "Verification Watermarks on Fingerprint Recognition and Retrieval", Proc. of SPIE Conference on Security and Watermarking of Multimedia Contents, P.W Wong and E.J. Delp (eds), Vol. 3657, San Jose, January 1999.

Parallel Generalized Finite Element Method for Magnetic Multiparticle Problems

Achim Basermann[1] and Igor Tsukerman[2]

[1] NEC Europe Ltd., C&C Research Laboratories,
Rathausallee 10, D-53757 Sankt Augustin, Germany
basermann@ccrl-nece.de,
http://www.ccrl-nece.de/~basermann/
[2] Department of Electrical & Computer Engineering,
The University of Akron, OH 44325-3904, USA
itsukerman@uakron.edu,
http://coel.ecgf.uakron.edu/igor/public_html

Abstract. A parallel version of the Generalized Finite Element Method is applied to multiparticle problems. The main advantage of the method is that only a regular hexahedral grid is needed; the particles do not have to be meshed and are represented by special basis functions approximating the field behavior near the particles. A general-purpose parallel Schur complement solver with incomplete LU preconditioning (A. Basermann) showed excellent performance for the varying problem size, number of processors and number of particles. In fact, the scaling of the computational time with respect to the number of processors was slightly superlinear due to cache effects. Future research plans include parallel implementation of the new Flexible Local Approximation MEthod (FLAME) that incorporates desirable local approximating functions (e.g. dipole harmonics near particles) into the difference scheme.

1 Introduction

The Finite Element Method (FEM) – a very powerful numerical tool in a variety of application areas – is not very effective for problems with (multiple) moving boundaries, as complex meshing procedures are required. In addition, remeshing can cause numerical artifacts, e.g. spurious force values [15, 26]. For multiparticle problems, the disadvantages of the standard FEM are especially obvious, since mesh generation may become impractical if the number of particles is, say, a few dozen or more.

Since multiparticle problems are of great practical interest (see Section 4), alternative modelling techniques are well worth exploring. One obvious candidate would be the Fast Multipole Method (FMM) [13]; however, FMM is not as powerful for problems with electric or magnetic inhomogeneities and possibly nonlinearities, and with a moderate number of particles whose magnetic dipole moments are induced by the external field and therefore are not known *a priori*.

This paper explores a different approach and deals with computational methods where material interfaces are represented *algebraically*, by relevant additional ap-

M. Daydé et al. (Eds.): VECPAR 2004, LNCS 3402, pp. 325–339, 2005.

proximating functions, rather than *geometrically*, on complex conforming meshes. One class of such methods is the Generalized FEM that allows one to employ any reasonable approximating functions, not necessarily piecewise-polynomial as in the standard FEM. GFEM was previously applied by one of the authors (I.T., with co-workers) to electromagnetic problems in [17, 22, 23]; this paper deals with parallelization of the algorithm. We use the general-purpose parallel Schur complement solver developed by the other author (A.B.).

An alternative approach proposed by one of the authors [27] and named 'FLAME' (Flexible Local Approximation MEthod) is related to GFEM but does not employ partition of unity and avoids numerical volume quadratures, the main bottleneck in GFEM. These ideas are briefly reviewed in Section 8 and are left for future research.

One practical motivation for the simulation of electrostatic and magnetostatic multiparticle problems is self-assembly of nanoparticles (Section 4). It is of great interest in a number of nanotechnology applications, such as microarrays for high-resolution chemical or biological sensing and DNA sequencing, photonic bandgap materials, novel magnetic storage media.

Our research is thus driven by a combination of several important factors:

- New computational methods on simple geometrically nonconforming meshes.
- Parallelization techniques and parallel solvers for such methods.
- Practical applications in nanotechnology.

The paper is structured as follows: a brief overview of GFEM (Section 2) and its specific application to multiparticle problems (Section 3); a brief description of self-assembly of nanoparticles as our practical motivation (Section 4); parallelization of the algorithm (Section 5); validation of numerical results (Section 6); scaling with respect to the number of processors and particles (Section 7). The prospect of future work is outlined in Section 8.

2 The Generalized Finite Element Method

The idea of partition of unity (PU) in the context of Finite Element Analysis was introduced in [1, 7, 14, 20, 21]. A system of functions $\{\varphi_i\}$, $1 \le i \le n_{\text{patches}}$ forms a partition of unity over a set of overlapping 'patches' (subdomains) covering the computational domain Ω if

$$\sum\nolimits_{i=1}^{n_{\text{patches}}} \varphi_i \equiv 1 \quad \text{in } \Omega, \qquad \text{supp}(\varphi_i) = \Omega_i \; . \tag{1}$$

To construct a partition of unity in practice, one starts with a set of functions $\tilde{\varphi}_i$ satisfying the second part of condition (1), $\text{supp}(\tilde{\varphi}_i) = \Omega_i$, and then normalizes these functions to satisfy the first part of (1) [1, 14]:

$$\varphi_i = \tilde{\varphi}_i \,/\, \sum\nolimits_{k=1}^{n_{\text{patches}}} \tilde{\varphi}_k \; . \tag{2}$$

If solution u is approximated *locally*, i.e. over each patch Ω_i, by a certain function u_{hi}, these local approximations can be merged using the φ's as functional weights:

$$u = \sum_{i=1}^{n_{\text{patches}}} u\varphi_i \; ; \qquad u_h = \sum_{i=1}^{n_{\text{patches}}} u_{hi}\varphi_i \; . \tag{3}$$

This way of merging local approximations is the central point of the whole computational procedure, as it ensures that the global approximation accuracy over the whole domain is commensurate with the local accuracy over each subdomain.

It is this guarantee of the approximation accuracy that distinguishes the PU-based GFEM from various other 'meshless' methods in different engineering applications [6]. Thus, under fairly general assumptions, the global approximation essentially reduces, due to the partition of unity, to local approximations over each patch. Rigorous mathematical analysis is available in [1, 14, 20, 21].

Locally, within each patch Ω_i, the numerical solution u_{hi} is sought as a linear combination of some approximating functions $g_m^{(i)}$:

$$u_{hi} = \sum_m a_m^{(i)} g_m^{(i)} \; . \tag{4}$$

The final system of approximating functions $\{ \hat{\psi}_m^{(i)} \}$ is built using elements of the partition of unity φ_i as weight functions:

$$\psi_m^{(i)} = g_m^{(i)} \varphi_i \; . \tag{5}$$

The global approximation u_h is a linear combination of the ψ's,

$$u_h = \sum_{m,i} a_m^{(i)} \psi_m^{(i)} \; , \tag{6}$$

and the standard Galerkin procedure can be applied.

3 Application of GFEM to Multiparticle Problems

A class of problems under consideration consists in finding the distribution of the magnetic field around a system of magnetic particles being deposited on a magnetized substrate (see Section 4). The governing equation is

$$\nabla\cdot\mu\nabla u = \nabla\cdot(\mu \boldsymbol{M}) \tag{7}$$

where $\boldsymbol{M}$ is the given magnetization of permanent magnets and u is the magnetic scalar potential (so that the magnetic field $\boldsymbol{H} = \boldsymbol{M} - \nabla u$). The permeability μ can depend on coordinates but not, in the linear case under consideration, on the potential u. Parameter μ can also be discontinuous across material boundaries and, in particular, is in general different inside and outside magnetic particles. The computational domain Ω is either two- or three-dimensional, with the usual mathematical assumption of a Lipschitz-continuous boundary, and standard types of boundary conditions (e.g. homogeneous Dirichlet).

The weak form of (8), suitable for the application of FEM, is

$$(\mu\nabla u, \; \nabla u') \;\; = \; (\mu \boldsymbol{M}, \nabla u'), \qquad \forall u' \in H^1(\Omega) \tag{8}$$

where the parentheses denote the standard inner product in $L_2(\Omega)^3$.

Conventional FE analysis faces serious difficulties for multiple particles: meshes would be geometrically very complex and difficult to construct, and it would be almost impossible to avoid 'flat' elements with large approximation errors [24, 25]. Moreover, changing positions of the particles would lead to mesh distortion and related numerical artifacts, especially in force computation [15, 26].

In our implementation of GFEM [16], the particles are not meshed at all, and only a regular hexahedral background grid is introduced. A patch in GFEM consists of eight brick elements sharing the same (non-boundary) node of the background mesh. Each particle contributes not a patch but rather a system of approximating functions $g_m^{(i)}$ that get multiplied by the brick-element shape functions φ as in (6).

The local approximating functions $g_m^{(i)}$ consist of two separate subsets, the first one containing the standard trilinear basis functions for the background brick mesh, and the second one contributed by the particles. It is the second subset that accounts for the field jump on the particle boundary. In brick patches that intersect the boundary, these additional approximating functions are, in the simplest case,

$$g_j^p = R - R_j \quad \text{if } R > R_j \text{, and zero otherwise} \tag{9}$$

where R_j is the radius of the j-th particle, R is the distance from the particle center. Alternatively, the second subset of $g_m^{(i)}$ can be constructed by a coordinate mapping [2] or, better yet, by dipole approximations as explained in [27].

The stiffness matrix is assembled in the usual way, by integrating $\mu \nabla \psi_i \cdot \nabla \psi_j$ over each element and summing up the individual element contributions. The integrals involving trilinear functions are standard and computed analytically. Integrals containing the jump functions (9) (or, alternatively, dipole harmonics) require numerical quadratures. For numerical integration, we use a simple scheme described in [16]. Obviously, more sophisticated adaptive integration techniques should give better results; see [2, 20, 21] and especially [10, 11, 12].

4 Practical Applications

One important application where the motion and positioning of magnetic particles need to be controlled is assembly of particles into biochemical micro-arrays. G. Friedman proposed an assembly process driven by magnetic templates prepatterned on the substrate [30]. Nanoparticles can be used as carriers of different biochemical agents. This approach has several advantages: Magnetic forces are relatively long-range, easily tuneable, do not require any permanent supply of power and do not usually cause any adverse biological or chemical effects.

This procedure yields a substantially higher resolution than the existing methods: printing and lithography with sequential masking. In addition, the use of micro- or nanoparticles can improve the uniformity of biochemical distribution over the substrate. For illustration, Fig. 1 shows the optical image of microparticles in one of the experiments of Friedman's group.

5 Parallelization

5.1 Matrix Structures and Assembly

Parallelization of matrix assembly is conceptually straightforward, as usually is the case in the standard FEM as well: the global system matrix is a sum of separate element-wise contributions that can be computed independently. The overall procedure has the following stages:

- Setting up the graph of the system matrix. This involves examining the particle positions with respect to the grid cells and nodes. (Recall that the unknowns in the GFEM system are not only the nodal values of the potential but also the coefficients of the additional basis functions near the particles.)
- Once the matrix structure has been set up, partitioning of the nodal graph is performed by METIS (function Metis_PartGraphKway, weights on vertices only). As a result, each grid node is assigned to a certain processor.

Matrix assembly proceeds on an element-by-element basis. Each processor handles only those elements (grid cells) that have at least one node allocated to that processor.

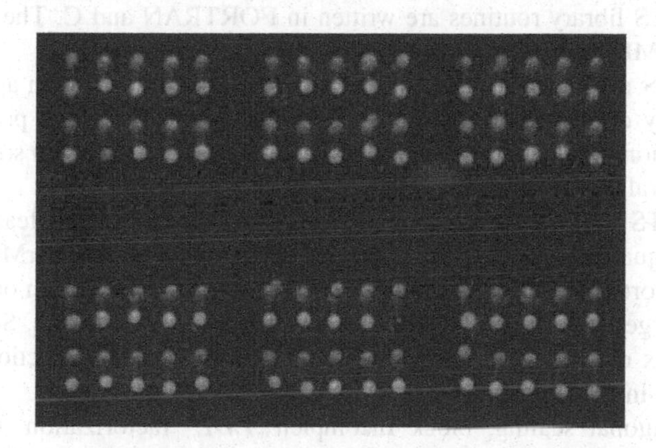

Fig. 1. (Courtesy G. Friedman and B. Yellen, Drexel University, Philadelphia.) Two differently colored (by fluorescent molecules) sets of particles (2.8 μm average diameter) assembled on an array of micromagnets. The first set of particles was assembled while applying an external bias field perpendicular to the substrate. Then the substrate was rinsed and the second set was assembled at the opposite poles of micromagnets after applying a bias field in the opposite direction

5.2 Parallel Solver

Our main objective has been to assess the performance of NEC's general-purpose suite of parallel iterative solvers with different preconditioners (PILUTS: Parallel Incomplete LU with Threshold preconditioned Solvers) for the GFEM problem. The scaling results in Section 7 show that these solvers perform quite well indeed. The use of regular meshes in GFEM, as opposed to complex unstructured meshes that would be needed in standard FEM, undoubtedly has an additional advantage of improving the performance of the solver. Iterative solvers with incomplete factorizations for preconditioning have been chosen since the fill-in in the complete factors of a direct method is huge and results in high memory requirements as well as long execution times. Moreover, parallelization of a direct solver for distributed memory machines (here: PC clusters) is considerably more difficult than for iterative algorithms.

The PILUTS library includes parallel sparse solvers for real symmetric positive definite (spd), general real symmetric and real non-symmetric matrices. As basic iterative methods, the **C**onjugate **G**radient (CG) algorithm, a **sym**metric variant of the **Q**uasi-**M**inimal **R**esidual method (symQMR), the **Bi**-**C**onjugate **G**radient **stab**ilised (BiCGstab) algorithm and the **F**lexible **G**eneralised **M**inimal **RES**idual (FGMRES) method are provided [3, 4, 5, 8, 18, 28]. CG is applied to equation systems with symmetric positive definite matrices, symQMR usually to systems with general symmetric matrices, and BiCGstab as well as FGMRES usually to systems with non-symmetric matrices. For convergence acceleration of the basic iterations, a selection of parallel preconditioning methods is available. The preconditioners include scaling methods, symmetric or non-symmetric incomplete block factorizations with threshold and **D**istributed **S**chur **C**omplement (DSC) algorithms [3, 4, 5, 18, 19].

The PILUTS library routines are written in FORTRAN and C. The parallelization is done using MPI.

All PILUTS methods can be called as parallel subprograms in an application code parallelized by domain partitioning. The parallel application code provides the distributed equation system data to a local PILUTS subroutine that then solves the matrix problem in parallel.

The PILUTS environment additionally provides the following features: first, the distributed equation system data can be re-partitioned by ParMETIS and re-distributed in order to reduce couplings between sub-domains and in order to accelerate the convergence of block factorizations or DSC preconditioners. Second, the distributed matrix data can be re-ordered by the METIS nested dissection algorithm to reduce the fill-in for local incomplete decompositions.

Simple diagonal scaling, block **I**ncomplete LDL^T factorization with **T**hreshold (ILDLT) with preceding diagonal scaling, DSC preconditioning with preceding diagonal scaling using ILDLT for local diagonal matrix blocks, or DSC using complete LDL^T decompositions of these blocks can precondition the current PILUTS symQMR solver [3, 4, 5, 18, 19]. The symQMR method with block ILDLT preconditioning is used in the numerical and performance tests in this paper [4].

6 Numerical Results: Validation

Before evaluating the performance of the parallel algorithm, one needs to be convinced in the accuracy of the simulation. Two test problems were used for validation purposes. The first problem, with only one particle, has a well-known analytical solution (the particle in a uniform external field behaves as a dipole). The second problem, with four particles, does not have a simple analytical solution but, due to the small number of particles, was easily solved by MATLAB / FEMLAB with second order tetrahedral elements. A more detailed description of the problems and numerical results follows.

A test problem with one particle. A magnetic particle with relative permeability $\mu = 10$ and radius $r_0 = 0.1$ is placed at the center of the unit cube. The boundary conditions correspond to the exact potential of the same particle in a uniform external field applied in the x-direction. **Fig. 2** – **Fig. 5** below compare numerical and analytical potential distributions, for different grid sizes, along the line $y = z = 0.5$ passing through the center of the particle.

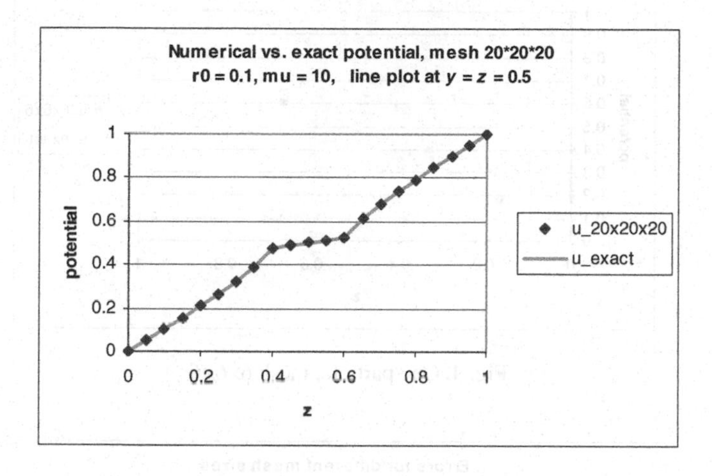

Fig. 2. One particle, mesh 20×20×20

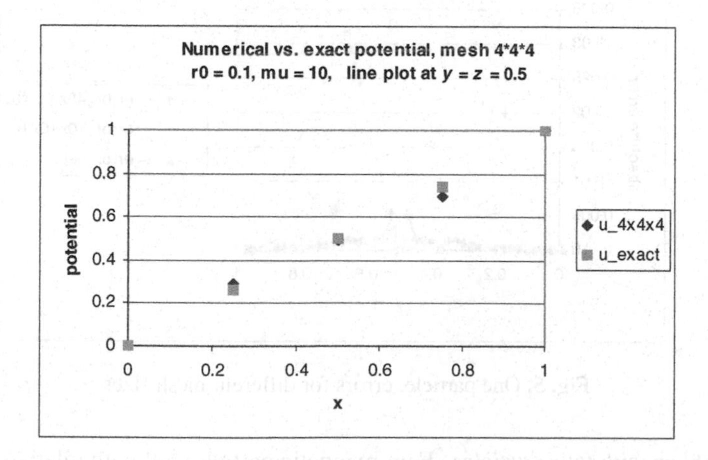

Fig. 3. One particle, mesh 4×4×4

For mesh size $h = 0.05$, i.e. only two grid points per particle radius, there is no appreciable discrepancy between the theoretical and numerical results (**Fig. 2**). More notable is the fact that the nodal values of the potential remain quite accurate when the grid size is comparable or even bigger than the particle radius – that is, when the grid is very far from resolving the geometry of the particle (**Fig. 3** and **Fig. 4**).

Error distributions for different mesh sizes, along the same line $y = z = 0.5$, are plotted in Fig. 5.

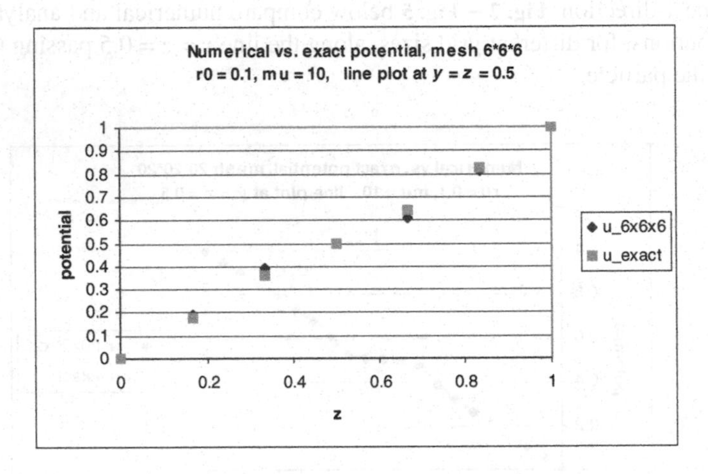

Fig. 4. One particle, mesh 6×6×6

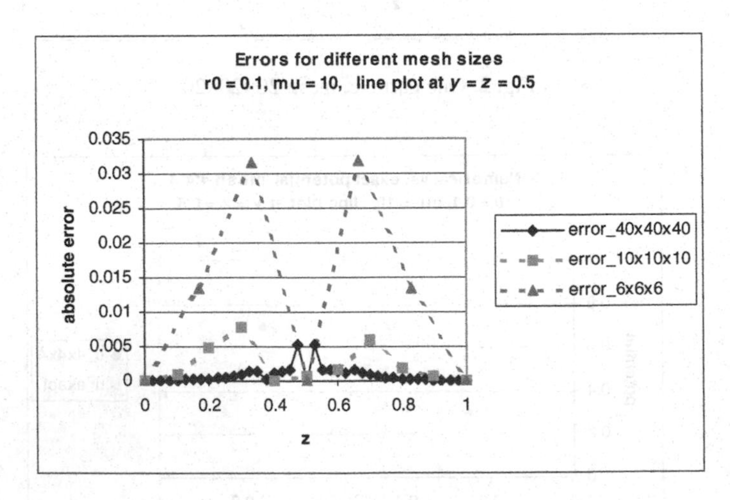

Fig. 5. One particle, errors for different mesh sizes

A test problem with four particles. Four magnetic particles, all with relative permeability $\mu = 10$ and radius $r_0 = 0.1$, are centered at (½ , ½, ½), (1/6, 1/6, 1/6), (1/6, 1/6, ½)

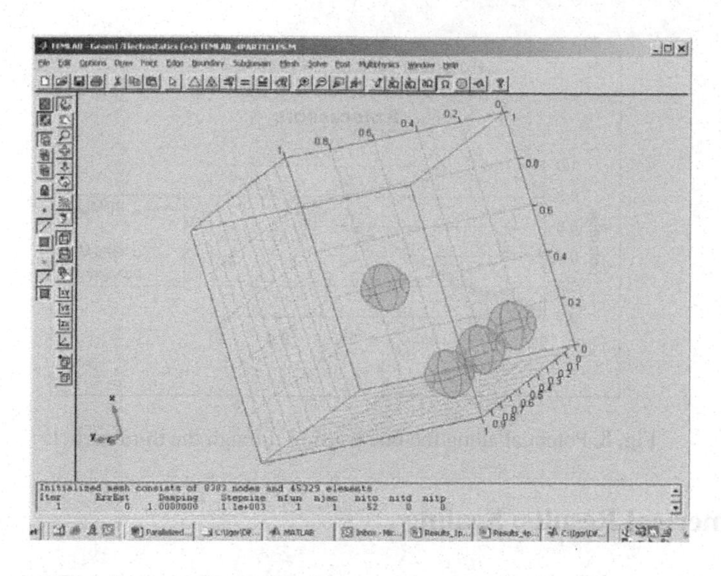

Fig. 6. A test problem with four particles (FEMLAB rendition)

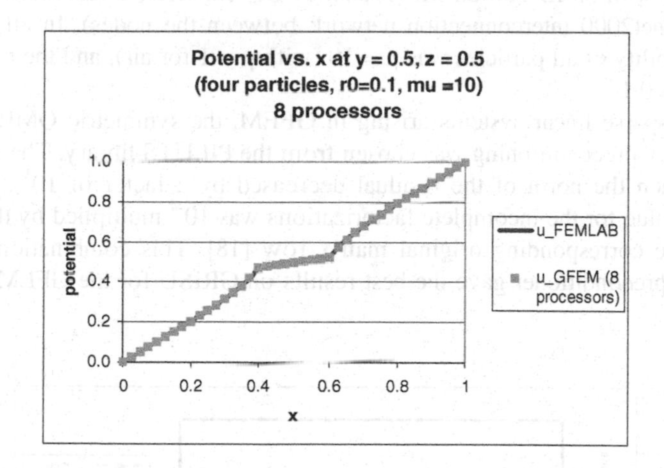

Fig. 7. Potential along the line passing through the first particle

and (1/6, 1/6, 5/6), respectively, within a unit cube (Fig. 6). The Dirichlet bound ary condition $u = x$ is applied (it corresponds *approximately* to a uniform external field in the x-direction; this correspondence would be more precise if the boundary were farther away). A regular $40 \times 40 \times 40$ mesh was used in GFEM. The FEMLAB mesh contained 45,329 *second order* elements, with 63,508 degrees of freedom (unknowns). In addition to the line $y = z = 0.5$ passing through the first particle, line $y = 0.15$, $z = 0.5$ crossing the third particle is also used for plotting. The agreement between the FEMLAB and GFEM results is excellent (see Fig. 7 and Fig. 8).

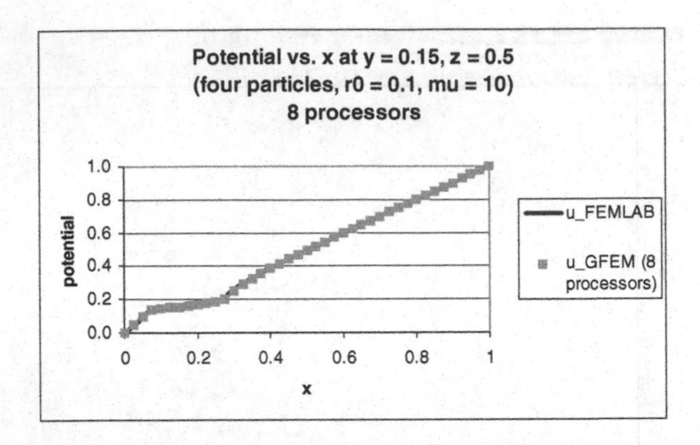

Fig. 8. Potential along the line passing through the third particle

7 Numerical Results: Scaling

The following experiments were performed on NEC's PC cluster GRISU (32 2-way SMP nodes with AMD Athlon MP 1900+ CPUs, 1.6 GHz, 1 GB main memory per node, Myrinet2000 interconnection network between the nodes). In all simulations, the permeability of all particles was $\mu = 10$ (with $\mu = 1$ for air), and the radius of each particle was 0.1.

For the sparse linear systems arising in GFEM, the symmetric QMR solver with block ILDLT preconditioning was chosen from the PILUTS library. The iteration was stopped when the norm of the residual decreased by a factor of 10^{10}. The relative threshold value for the incomplete factorizations was 10^{-4} multiplied by the Euclidean norm of the corresponding original matrix row [18]. This combination of iterative solver and preconditioner gave the best results on GRISU for the GFEM simulations performed.

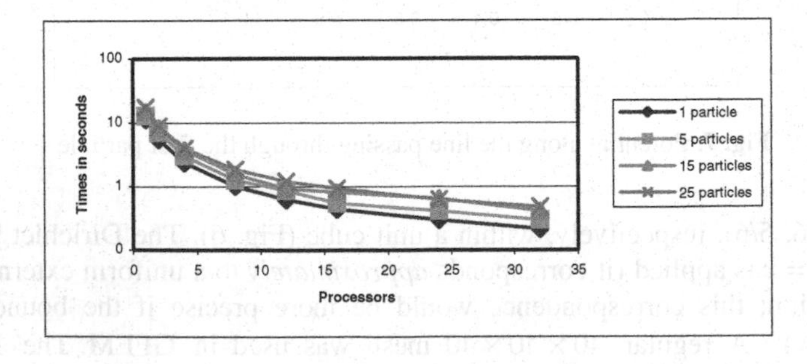

Fig. 9. Solver execution times for different particle numbers

Fig. 9 shows solver execution times for simulation runs with 1, 5, 15, and 25 particles on 1, 2, 4, 8, 12, 16, 24, and 32 processors of GRISU. A regular 30x30x30 mesh was chosen. This results in equation systems of order 30131 with 778617 nonzeros for one particle, of order 31446 with 875234 nonzeros for five particles, of order 34767 with 1119207 nonzeros for 15 particles, and of order 38064 with 1361494 nonzeros for 25 particles. The execution times in **Fig. 9** slightly increase with the increasing number of particles. The scaling behavior with respect to the number of processors is very favorable. **Fig. 10** even shows superlinear speedups in all cases due to cache effects. Even on 32 processors, the parallel efficiency of the solver is high, although the problem is only of medium size. The speedup here is defined as the sequential execution time divided by the parallel execution time. The sequential method is symmetric QMR with *global* ILDLT preconditioning. In the parallel case, block Jacobi preconditioning is applied with ILDLT factorization of the *local* diagonal blocks. Thus the preconditioner is different for the sequential and the parallel algorithm; it depends on the number of processors used (couplings between subdomains are neglected in the parallel preconditioner). Note that in these tests the construction of the ILDLT factors does not dominate the iterative solver time either in the sequential or in the parallel case. Significant cache effects on the Athlon cluster are known (in particular for structured finite element codes).

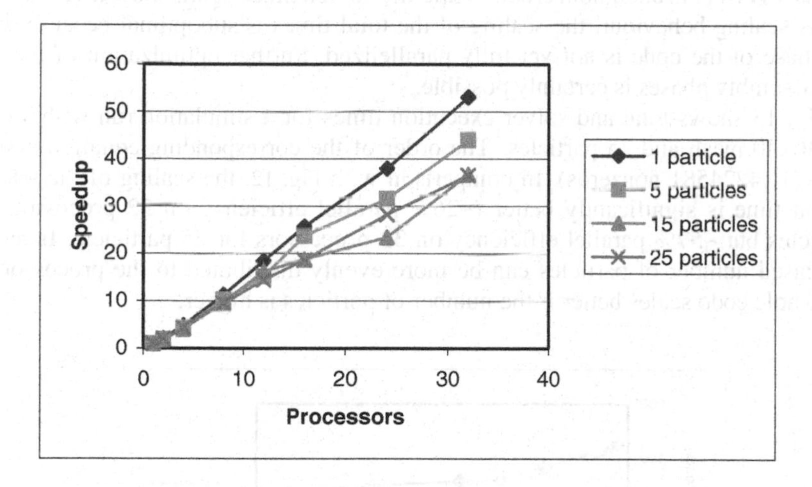

Fig. 10. Solver speedup for the varying number of particles

Fig. 11 displays the number of solver iterations to reach the stopping criterion. The number of iterations increases with the increasing number of processors, since the block preconditioner neglects couplings between subdomains and so the quality of the preconditioner reduces with the increasing number of processors. On the other hand, computational costs for construction and application of the preconditioner decrease as well with the increasing number of processors. Since the number of couplings between subdomains stays relatively small due to METIS partitioning (and due to an advantageous matrix structure), the increase in the number of iterations from 2 to 32 processors is not drastic and thus leads to the good scaling.

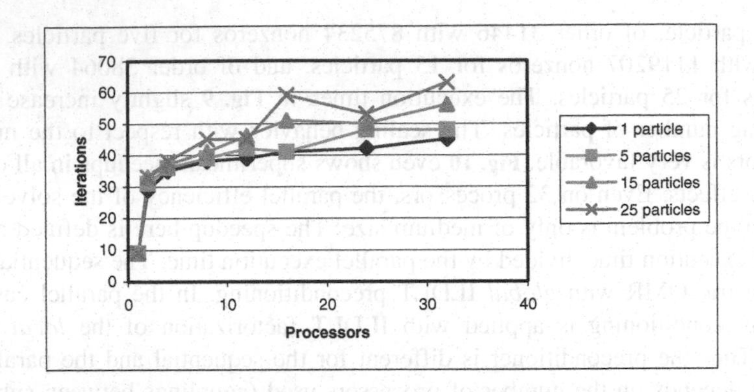

Fig. 11. Solver iterations for the varying number of particles

Fig. 12 displays the total and solver execution times for a simulation run with a regular 40x40x40 mesh and 4 particles. The order of the corresponding equation system is 74118 (2170600 nonzeros). While the solver times again show a very advantageous scaling behaviour, the scaling of the total times is suboptimal because the set-up phase of the code is not yet fully parallelized. Further optimization of the set-up and assembly phases is certainly possible.

Fig. 13 shows total and solver execution times for a simulation run with a regular 40x40x40 mesh and 25 particles. The order of the corresponding equation system is 100871 (4224581 nonzeros). In comparison with Fig. 12, the scaling of the total execution time is significantly better (~26% parallel efficiency on 32 processors for 4 particles but ~57% parallel efficiency on 32 processors for 25 particles). Indeed, the increased number of particles can be more evenly distributed to the processors, and the whole code scales better if the number of particles is higher.

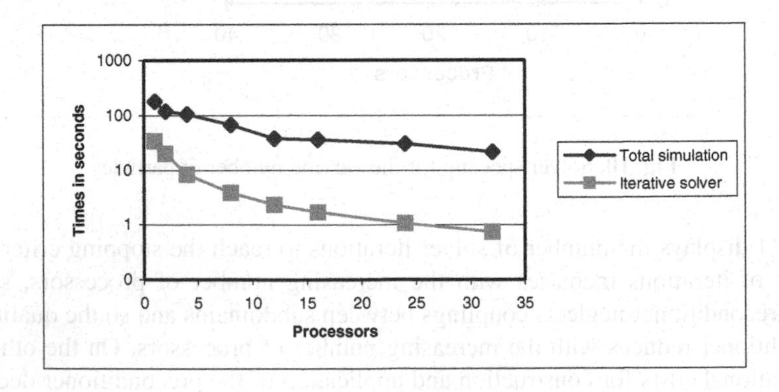

Fig. 12. Total execution times vs. solver execution times (mesh 40x40x40, 4 particles)

In the approach chosen, the treatment of particles in the matrix assembly phase is computationally expensive. Further optimization of this phase is possible, but it will

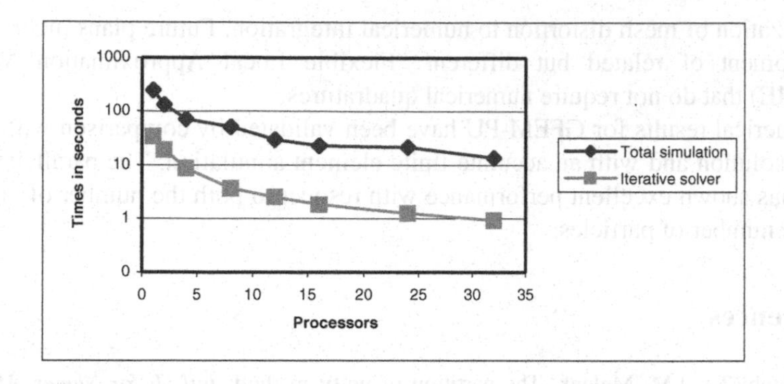

Fig. 13. Total execution times and solver execution times (mesh 40x40x40, 25 particles)

with high probability remain the most time-consuming part of the GFEM simulation. An alternative approach proposed very recently in [27] can substantially reduce this cost, while providing the same or higher accuracy.

8 Future Work: Difference Schemes with Flexible Approximation

Although GFEM-PU does provide almost complete freedom of choice of the basis, there is a substantial price to pay for it. Multiplication by the partition of unity functions makes the set of approximating functions more complicated, and possibly ill-conditioned or even linearly dependent [1, 14]. Another major difficulty is the high cost of the Galerkin quadratures that need to be computed numerically in geometrically complex 3D regions.

A new framework proposed in [27] provides a systematic way of constructing difference schemes with arbitrary approximating functions. Like GFEM, such schemes can in particular be used to represent slanted or curved material interfaces *algebraically*, by introducing special approximating functions rather than *geometrically*, on conforming meshes. At the same time, construction of these schemes is algorithmically and computationally simpler than GFEM. The variational versions of the new method require surface, rather than volume, integration; the Trefftz versions need no integration at all.

One interesting niche for the new schemes may be problems that admit a convenient *algebraic* description – this includes not only spherical or spheroidal particles but also algebraically varying material parameters as in some protein models [9, 29].

9 Conclusion

In the Generalized Finite Element – Partition of Unity (GFEM-PU) method for multiparticle problems in nanotechnology applications, only very simple regular meshes are needed and no mesh distortion occurs when particles change their positions. Implementation difficulties are shifted from generation of complex FE meshes and minimization of mesh distortion to numerical integration. Future plans include further development of related but different "Flexible Local Approximation MEthods" (FLAME) that do not require numerical quadratures.

Numerical results for GFEM-PU have been validated by comparison with the analytical solution and with an accurate finite element simulation. The parallelized algorithm has shown excellent performance with respect to both the number of processors and the number of particles.

References

1. I. Babuška, J.M. Melenk, The partition of unity method, *Intl. J. for Numer. Methods in Engineering*, vol. 40, No. 4, pp.727-758, 1997.
2. I. Babuška, G. Caloz, J.E. Osborn, Special finite-element methods for a class of 2nd-order elliptic problems with rough coefficients, *SIAM Journal on Numerical Analysis*, vol. 31, No. 4, pp. 945-981, 1994.
3. A. Basermann, QMR and TFQMR methods for sparse non-symmetric problems on massively parallel systems, *The Mathematics of Numerical Analysis*, series: *Lectures in Applied Mathematics*, vol. 32, pp. 59-76, 1996.
4. A. Basermann, Parallel Block ILUT/ILDLT Preconditioning for Sparse Eigenproblems and Sparse Linear Systems, *Num Linear Algebra with Applications*, vol. 7, pp. 635-648, 2000.
5. A. Basermann, U. Jaekel, and K. Hachiya, Preconditioning parallel sparse iterative solvers for circuit simulation, *Proceedings of The 8th SIAM Conference on Appl Lin Alg,* July 2003, Williamsburg, VA, USA (http://www.siam.org/meetings/la03/proceedings/basermaa.pdf).
6. T. Belytschko *et al.* Meshless methods: an overview and recent developments, *Computer Methods in Applied Mechanics and Engineering*, vol. 139, No. 1-4, pp. 3-47, 1996.
7. C.A. Duarte, I. Babuška, J.T. Oden, Generalized finite element methods for three-dimensional structural mechanics problems, *Computers & Structures*, vol. 77, No. 2, pp. 215-232, 2000.
8. R. W. Freund, N. M. Nachtigal, QMR: A quasi-minimal residual method for non-Hermitian linear systems, *Numerische Mathematik*, 60:315-339, 1991.
9. J. A. Grant, B. T. Pickup, A. Nicholls, A smooth permittivity function for Poisson–Boltzmann solvation methods, *J. Comp. Chem.*, vol. 22, No. 6, 608–640, 2001, and references therein.
10. M. Griebel and M. A. Schweitzer. A particle-partition of unity method for the solution of elliptic, parabolic and hyperbolic PDE. *SIAM J. Sci. Comp.*, 22(3):853-890, 2000.
11. M. Griebel and M. A. Schweitzer. A particle-partition of unity method-Part II: efficient cover construction and reliable integration. *SIAM J. Sci. Comp.*, 23(5):1655-1682, 2002.
12. M. Griebel and M. A. Schweitzer. A particle-partition of unity method-Part III: a multilevel solver. *SIAM J. Sci. Comp.*, 24(2):377-409, 2002.
13. L. Greengard, V. Rokhlin, H. Cheng, A fast adaptive multipole algorithm in three dimensions, *J. of Comp .Phys.*, vol. 155, No. 2, pp. 468-498, 1999; and references therein.
14. J.M. Melenk, I. Babuška, The partition of unity finite element method: Basic theory and applications, *Comput. Methods Appl. Mech. Engrg.*, vol. 139, pp. 289-314, 1996.
15. W. Müller, Comparison of different methods of force calculation, *IEEE Trans. Magn.*, vol. 25, No. 2, pp.1058-1061, 1990.

16. A. Plaks, I. Tsukerman, G. Friedman, B. Yellen, Generalized Finite Element Method for magnetized nanoparticles, *IEEE Trans. Magn.*, vol. 39, No. 3, pp. 1436–1439, 2003.
17. Leonid Proekt, Igor Tsukerman, Method of overlapping patches for electromagnetic computation, *IEEE Trans. Magn.*, vol. 38, No. 2, pp.741-744, 2002.
18. Y. Saad, *Iterative Methods for Sparse Linear Systems*. 2nd ed., SIAM, Philadelphia, 2003.
19. Y. Saad, M. Sosonkina, Distributed Schur complement techniques for general sparse linear systems, *SISC*, 21:1337-1356, 1999.
20. T. Strouboulis, I. Babuška, K.L. Copps, The design and analysis of the Generalized Finite Element Method, *Comp Meth in Appl Mech & Engrng*, vol. 181, No. 1-3, pp. 43-69, 2000.
21. T. Strouboulis, K. Copps, I. Babuska, The generalized finite element method, *Computer Methods in Applied Mech. and Engineering*, vol. 190, No. 32-33, pp. 4081-4193, 2001.
22. I. Tsukerman and L. Proekt, Generalized scalar and vector elements for electromagnetic computation, *XI Intern Symposium on Theoretical Electr Eng*, August 2001, Linz, Austria.
23. L. Proekt, S. Yuferev, I. Tsukerman, N. Ida, Method of overlapping patches for electromagnetic computation near imperfectly conducting cusps and edges, *IEEE Trans. Magn.*, vol. 38, No. 2, pp.649-652, 2002.
24. I.A. Tsukerman, Approximation of conservative fields and the element 'edge shape matrix', *IEEE Trans. Magn.*, vol. 34, pp.3248-3251, 1998.
25. I.A. Tsukerman, A. Plaks, Comparison of accuracy criteria for approximation of conservative fields on tetrahedra, *IEEE Trans. Magn.*, vol. 34, pp.3252-3255, 1998.
26. I.A. Tsukerman, Accurate computation of 'ripple solutions' on moving finite element meshes, *IEEE Trans. Magn.*, vol. 31, No. 3, pp.1472-1475, 1995.
27. I. Tsukerman, Flexible local approximation method for electro- and magnetostatics, *IEEE Trans. Magn.*, v. 40, No. 2, pp. 941–944, 2004.
28. H. Van der Vorst, Bi-CGSTAB: A fast and smoothly converging variant of Bi-CG for the solution of nonsymmetric linear systems. *SIAM J. Sci. Statist. Comput.*, 13:631-644, 1992.
29. Takumi Washio, NEC Europe, private communication.
30. B.B. Yellen, G. Friedman, Programmable assembly of heterogeneous colloidal particles using magnetic micro-well templates, *Langmuir* 20: 2553-2559, 2004.

Parallel Model Reduction of Large Linear Descriptor Systems via Balanced Truncation*

Peter Benner[1], Enrique S. Quintana-Ortí[2], and Gregorio Quintana-Ortí[2]

[1] Fakultät für Mathematik, Technische Universität Chemnitz, D-09107 Chemnitz, Germany
benner@mathematik.tu-chemnitz.de
[2] Depto. de Ingeniería y Ciencia de Computadores, Universidad Jaume I, 12.071–Castellón, Spain
{quintana, gquintan}@icc.uji.es

Abstract. In this paper we investigate the use of parallel computing to deal with the high computational cost of numerical algorithms for model reduction of large linear descriptor systems. The state-space truncation methods considered here are composed of iterative schemes which can be efficiently implemented on parallel architectures using existing parallel linear algebra libraries. Our experimental results on a cluster of Intel Pentium processors show the performance of the parallel algorithms.

Keywords: Model reduction, balanced truncation, linear descriptor systems, parallel algorithms.

1 Introduction

Model reduction is of fundamental importance in many modeling and control applications involving continuous-time linear descriptor systems. In generalized state-space form these systems are defined by

$$\begin{aligned} E\dot{x}(t) &= Ax(t) + Bu(t), \quad t > 0, \quad x(0) = x^0, \\ y(t) &= Cx(t) + Du(t), \quad t \geq 0, \end{aligned} \tag{1}$$

where $E, A \in \mathbb{R}^{n \times n}$, $B \in \mathbb{R}^{n \times m}$, $C \in \mathbb{R}^{p \times n}$, $D \in \mathbb{R}^{p \times m}$, $x^0 \in \mathbb{R}^n$ is the initial state of the system, and n is said to be the order of the system. The transfer function matrix (TFM) associated with (1) is then given by

$$G(s) = C(sE - A)^{-1}B + D.$$

* P. Benner was supported by the DFG Research Center "Mathematics for key technologies" (FZT 86) in Berlin. E.S. Quintana-Ortí and G. Quintana-Ortí were supported by the CICYT project No. TIC2002-004400-C03-01 and FEDER.

M. Daydé et al. (Eds.): VECPAR 2004, LNCS 3402, pp. 340–353, 2005.

Hereafter, we assume that the matrix pencil $A - \lambda E$ is regular, (i.e., $\exists \lambda \in \mathbb{C}$ with $\det(A - \lambda E) \neq 0$) and $A - \lambda E$ is stable, implying that the finite part of the spectrum of $A - \lambda E$ is contained in the open left half plane.

In model reduction, we are interested in finding a reduced-order realization

$$\begin{aligned} \hat{E}\dot{\hat{x}}(t) &= \hat{A}\hat{x}(t) + \hat{B}\hat{u}(t), \quad t > 0 \quad \hat{x}(0) = \hat{x}^0, \\ \hat{y}(t) &= \hat{C}\hat{x}(t) + \hat{D}\hat{u}(t), \quad t \geq 0, \end{aligned} \tag{2}$$

of order $r \ll n$, and TFM $\hat{G}(s) = \hat{C}(s\hat{E} - \hat{A})^{-1}\hat{B} + \hat{D}$ which approximates $G(s)$.

Model reduction of large-scale linear descriptor systems arises in control of multibody (mechanical) systems, manipulation of fluid flow, circuit simulation, VLSI chip design, in particular when modeling the interconnections via RLC networks, simulation of MEMS and NEMS, etc.; see, e.g.,[12, 13, 14, 21, 22]. State-space dimensions n of order 10^2–10^4 are common in these applications.

There is no general technique for model reduction that can be considered as optimal in an overall sense. In this paper we focus on the so-called state-space truncation approach [2, 20] and, in particular, on balanced truncation (BT) of the system [19, 26, 31, 33]. BT methods for linear descriptor systems as proposed in [27, 28] with dense state matrix pencil $A - \lambda E$ present a computational cost of $\mathcal{O}(n^3)$ floating-point arithmetic operations (flops). Large-scale applications, as those listed above, thus clearly benefit from using parallel computing techniques to obtain the reduced-order system. Parallel model reduction of standard linear systems ($E = I_n$) using the state-space approach has been investigated elsewhere; see, e.g., the review in [9].

Although there exist several other approaches for model reduction, see, e.g., [2, 15], those are specialized for certain problem classes and often lack properties such as error bounds or preservation of stability, passivity, or phase information.

Our procedure for parallel BT model reduction of linear descriptor systems is composed of two major tasks: First, it is necessary to decouple (separate) the finite poles of the system from the infinite ones, which is equivalent to computing an additive decomposition of the TFM. BT is then applied to the part of the system associated with the finite poles, which requires the solution of two generalized Lyapunov equations. These two tasks can be performed via iterative schemes which are easily parallelized using parallel linear algebra libraries and provide high efficiency on parallel distributed-memory architectures.

The rest of the paper is structured as follows. In Section 2 we review specific algorithms for the two major tasks involved in BT model reduction of linear descriptor systems. In Section 3 we describe a few parallelization and implementation details of the model reduction algorithm. Finally, the numerical and parallel performances of the algorithms are reported in Section 4, and some concluding remarks follow in Section 5.

2 Model Reduction of Descriptor Systems

In this section we briefly describe a model reduction method for linear descriptor systems proposed in [6]. It is based on the observation that the method of [27,28] is mathematically equivalent to the following procedure. First, the TFM of the system is decomposed as $G(s) = G_0(s) + G_\infty(s)$, where G_0 and G_∞ contain, respectively, the finite and infinite poles of the system. This is followed by the application of BT to approximate $G_0(s)$ by $\hat{G}_0(s)$. The reduced-order model is then given by $\hat{G}(s) = \hat{G}_0(s) + G_\infty(s)$; see [29] for further details.

2.1 Decoupling the Infinite Poles of a TFM

Consider the system defined by $(A - \lambda E, B, C, D)$. Our goal here is to find a transformation defined by nonsingular matrices $U, V \in \mathbb{R}^{n\times n}$ such that

$$\hat{A} - \lambda\hat{E} := U(A - \lambda E)V^{-1} = \begin{bmatrix} A_0 & 0 \\ 0 & A_\infty \end{bmatrix} - \lambda \begin{bmatrix} E_0 & 0 \\ 0 & E_\infty \end{bmatrix},$$

where $A_0 - \lambda E_0$ and $A_\infty - \lambda E_\infty$ contain, respectively, the finite and infinite eigenvalues of $A - \lambda E$. If this transformation is now applied to the rest of the system,

$$\hat{B} := UB =: \begin{bmatrix} B_0 \\ B_\infty \end{bmatrix}, \qquad \hat{C} := CV^{-1} =: [\, C_0 \; C_\infty \,],$$

we obtain the desired additive decomposition of the TFM into

$$\begin{aligned} G(s) &= C(sE - A)^{-1}B + D = \hat{C}(s\hat{E} - \hat{A})^{-1}\hat{B} + \hat{D} \\ &= \{C_0(sE_0 - A_0)^{-1}B_0 + D\} + \{C_\infty(sE_\infty - A_\infty)^{-1}B_\infty\} = G_0(s) + G_\infty(s). \end{aligned}$$

Thus, the problem of computing an additive decomposition is redefined in terms of finding appropriate matrices U and V that block-diagonalize the system.

In order to compute these matrices we will proceed in two steps. First, we compute orthogonal matrices $Q, Z \in \mathbb{R}^{n\times n}$ that reduce $A - \lambda E$ to block-triangular form; i.e,

$$Q^T(A - \lambda E)Z = \begin{bmatrix} A_0 & W_A \\ 0 & A_\infty \end{bmatrix} - \lambda \begin{bmatrix} E_0 & W_E \\ 0 & E_\infty \end{bmatrix}. \tag{3}$$

We propose to perform this stage using an iterative scheme for spectral division of a matrix pencil introduced by Malyshev in [18]. The technique was made truly inverse-free in [4] and was further refined in [30] where a deflating subspace extraction scheme was described that provides both Q and Z from a single inverse-free iteration, saving thus half of the computational cost. The spectral division algorithm can be used to separate the eigenvalues of a matrix pencil inside and outside the unit circle. A careful use of Möbius transformations allows to separate the spectrum along other regions. In particular, the finite eigenvalues of $\Lambda(A, E)$ can be separated from the infinite ones by applying the method to the matrix pencil $\alpha A - \lambda E$, with $\alpha > \max\{|\lambda|, \lambda \in \Lambda(A, E)\backslash\{\infty\}\}$. In many practical applications like in [29], the number of infinite eigenvalues is known. This can be used to verify that the computed decomposition (using a given α) correctly decouples only the infinite poles. However, it remains an open question how to estimate the value of α.

Next, in a second step, (3) is further reduced to the required block-diagonal form by solving the following generalized Sylvester equation

$$A_0 Y + X A_\infty + W_A = 0, \quad E_0 Y + X E_\infty + W_E = 0. \tag{4}$$

Thus,

$$\begin{aligned}\hat{A} - \lambda \hat{E} &:= \hat{U}(A - \lambda E)\hat{V}^{-1} \\ &:= \begin{bmatrix} I & X \\ 0 & I \end{bmatrix} \left(\begin{bmatrix} A_0 & W_A \\ 0 & A_\infty \end{bmatrix} - \lambda \begin{bmatrix} E_0 & W_E \\ 0 & E_\infty \end{bmatrix} \right) \begin{bmatrix} I & Y \\ 0 & I \end{bmatrix} = \begin{bmatrix} A_0 & 0 \\ 0 & A_\infty \end{bmatrix} - \lambda \begin{bmatrix} E_0 & 0 \\ 0 & E_\infty \end{bmatrix},\end{aligned}$$

and $U = \hat{U}Q^T$, $V^{-1} = Z\hat{V}^{-1}$.

In case the system is of index 1 (i.e., the infinite eigenvalues of $A - \lambda E$ are non-defective), denoted by $\operatorname{ind}(A, E) = 1$, we obtain that $E_\infty = 0$ in (3) and the Sylvester equation in (4) decouples into two linear systems of equations

$$E_0 Y = -W_E, \quad X A_\infty = -(W_A + A_0 Y).$$

As many applications with $\operatorname{ind}(A, E) > 1$ have an equivalent formulation with $\operatorname{ind}(A, E) = 1$, we assume that the index is one hereafter.

2.2 Balanced Truncation of Linear Descriptor Systems

Consider now the realization $(A_0 - \lambda E_0, B_0, C_0, D_0)$ resulting from the deflation of the infinite part of $(A - \lambda E, B, C, D)$. In the second step we will proceed to apply BT to reduce this realization further by computing the controllability Gramian W_c and the observability Gramian W_o of the system. These Gramians are given by the solutions of two coupled generalized Lyapunov equations

$$A_0 W_c E_0^T + E_0 W_c A_0^T + B_0 B_0^T = 0, \quad A_0^T \hat{W}_o E_0 + E_0^T \hat{W}_o A_0 + C_0^T C_0 = 0, \tag{5}$$

with $W_o = E_0^T \hat{W}_o E_0$. As $A - \lambda E$ is assumed to be stable, so is $A_0 - \lambda E_0$, and W_c and W_o are positive semidefinite; therefore there exist factorizations $W_c = S^T S$ and $W_o = R^T R$. The matrices S and R are often referred to as "Cholesky" factors of the Gramians (even if they are not square triangular matrices). Efficient parallel algorithms for the solution of (5), based on the matrix sign function method [25], are described in detail in [8].

Consider now the singular value decomposition (SVD)

$$SR^T = [U_1 \; U_2] \begin{bmatrix} \Sigma_1 & 0 \\ 0 & \Sigma_2 \end{bmatrix} \begin{bmatrix} V_1^T \\ V_2^T \end{bmatrix}, \tag{6}$$

where the matrices are partitioned at a given dimension r such that $\Sigma_1 = \operatorname{diag}(\sigma_1, \ldots, \sigma_r)$, $\Sigma_2 = \operatorname{diag}(\sigma_{r+1}, \ldots, \sigma_n)$, $\sigma_j \geq 0$ for all j, and $\sigma_r > \sigma_{r+1}$. Here, $\sigma_1, \ldots, \sigma_n$ are the Hankel singular values of the system.

Both square-root (SR) and balancing-free square-root (BFSR) BT algorithms use the information contained in (6) to compute a realization $\hat{G}_0$ which satisfies

$$\|\Delta_a\|_\infty = \|G_0 - \hat{G}_0\|_\infty \leq 2 \sum_{j=r+1}^{n} \sigma_j, \tag{7}$$

and allows an adaptive choice of the state-space dimension r of the reduced-order model. In particular, SR BT algorithms determine the reduced-order model as

$$\hat{E}_0 := LE_0T, \quad \hat{A}_0 := LA_0T, \quad \hat{B}_0 := LB_0, \quad \hat{C}_0 := C_0T, \quad \hat{D}_0 := D_0, \tag{8}$$

where the truncation matrices L and T are given by

$$L = \Sigma_1^{-1/2} V_1^T R E_0^{-1} \quad \text{and} \quad T = S^T U_1 \Sigma_1^{-1/2}. \tag{9}$$

Note that $\hat{E}_0 = I_r$ and needs not be computed. Even though the reduction yields a standard state-space representation, it is still beneficial not to transform the generalized system to standard form by inverting E_0 directly as rounding errors introduced by this operation are delayed as much as possible in the given procedure.

BFSR algorithms often provide more accurate reduced-order models in the presence of rounding errors [33]. These methods and further details on (parallel) model reduction of linear descriptor systems are given in [27, 10].

3 Parallelization and Implementation Details

The numerical algorithms that we describe in this section are basically composed of traditional matrix computations such as matrix factorizations, solution of triangular linear systems, and matrix products. All these operations can be efficiently performed employing parallel linear algebra libraries for distributed memory computers [11, 32]. The use of these libraries enhances the reliability and improves the portability of the model reduction algorithms. The performance will depend on the efficiencies of the underlying serial and parallel computational linear algebra kernels and the communication routines.

Here we employ the parallel kernels in the ScaLAPACK library [11]. This is a freely available library that implements parallel versions of many of the routines in LAPACK [1], using the message-passing paradigm. ScaLAPACK employs the PBLAS (a parallel version of the serial BLAS) for computation and BLACS for communication. The BLACS can be ported to any (serial and) parallel architecture with an implementation of the MPI (our case) or the PVM libraries.

3.1 Spectral Division via the Inverse-Free Iteration

Malyshev's iteration for spectral division of a matrix pencil $F - \lambda G$, $F, G \in \mathbb{R}^{n \times n}$, as refined in [4], is given by:

$F_0 \leftarrow F,\ G_0 \leftarrow G,\ R_0 \leftarrow 0,\ j \leftarrow 0$
`repeat`
Compute the QR decomposition

$$\begin{bmatrix} F_j \\ -G_j \end{bmatrix} = \begin{bmatrix} U_{11} & U_{12} \\ U_{21} & U_{22} \end{bmatrix} \begin{bmatrix} R_{j+1} \\ 0 \end{bmatrix}$$

$F_{j+1} \leftarrow U_{12}^T F_j,\ G_{j+1} \leftarrow U_{22}^T G_j,\ j \leftarrow j+1$
`until` $\|R_{j+1} - R_j\|_1 > \tau$

This iteration is easily parallelized using parallel kernels in ScaLAPACK for the QR factorization (`pdgeqrf`) and the matrix product (`pdgemm`). Instead of building the complete orthogonal matrix resulting from the QR factorization, we can reduce the computational cost of the iteration by only computing the last n columns of this matrix, corresponding to blocks U_{12} and U_{22}. For this purpose, we utilize ScaLAPACK routine `pdormqr` to apply the Householder transformations generated during the QR factorization to the matrix $\left[0_n^T, I_n^T\right]^T$ from the left.

The computation of the orthogonal matrices which decouple the matrix pencil described in [30] requires first the computation of the rank-revealing QR (RRQR) factorization [17] of the matrix $F_\infty := \lim_{j\to\infty} F_j$ at convergence

$$F_\infty^T = \bar{Z}\bar{R}\bar{\Pi}, \quad r = \operatorname{rank}(F_\infty) = \operatorname{rank}(\bar{R}).$$

Now, let $\bar{Z} = [Z_1, Z_2]$, $Z_1 \in \mathbb{R}^{n\times(n-r)}$, $Z_2 \in \mathbb{R}^{n\times r}$. The sought-after orthogonal matrices are then given by $Z = [Z_2, Z_1]$ and the orthogonal matrix in the RRQR factorization

$$[AZ_2, BZ_2] = QR\Pi.$$

These two RRQR factorizations can be obtained by means of the traditional QR factorization with column pivoting [17], which is implemented in ScaLAPACK as routine `pdgeqpf`. In order to improve the performance of our parallel model reduction routines we have designed and implemented a parallel BLAS-3 variant of the QR factorization with partial pivoting [24] that outperforms (the BLAS-2) routine `pdgeqpf` and maintains the same numerical behavior.

In our implementation of the inverse-free iteration we propose to set $\tau = n\sqrt{\varepsilon}\|R_j\|_1$ and perform two more iterations once this criterion is satisfied. Due to the ultimate quadratic convergence of the iteration this ensures in most cases the maximum possible accuracy while avoiding convergence stagnation problems.

All the parallel routines involved in the spectral division are composed of block-oriented, coarse-grain computations which are highly efficient on parallel distributed-memory architectures. We therefore expect a considerable degree of parallelism for the implementation. The parallelization of the inverse-free iteration has been previously reported in [3, 23].

3.2 Solving Generalized Lyapunov Equations via the Sign Function

We propose to employ the following variant of the Newton iteration for the sign function that solves the generalized Lyapunov equations in (5) and allows great computational savings when applied to the type of equations arising in large-scale non-minimal systems [9].

$$
\begin{array}{l}
A_0 \leftarrow A,\ S_0 \leftarrow B,\ R_0 \leftarrow C,\ j \leftarrow 0 \\
\texttt{repeat} \\
\quad c_j \leftarrow (\det(A_j)/\det(E))^{\frac{1}{n}},\ A_{j+1} \leftarrow \frac{1}{\sqrt{2c_j}}\left(A_j + c_j^2 E A_j^{-1} E\right) \\
\quad \text{Compute the RRQR decompositions} \\
\quad \frac{1}{\sqrt{2c_j}}\left[S_j \ \ c_j E A_j^{-1} S_j\right]^T = Q_s \begin{bmatrix} R_s \\ 0 \end{bmatrix} \Pi_s \\
\quad \frac{1}{\sqrt{2c_j}}\begin{bmatrix} R_j \\ c_j R_j A_j^{-1} E \end{bmatrix} = Q_r \begin{bmatrix} R_r \\ 0 \end{bmatrix} \Pi_r \\
\quad R_{j+1} \leftarrow (R_r \Pi_r)^T,\ \ S_{j+1} \leftarrow R_s \Pi_s,\ \ j \leftarrow j+1 \\
\texttt{until}\ \|A_j - E\|_1 < n\sqrt{\varepsilon}\|A_j\|_1 \\
S \leftarrow \frac{1}{\sqrt{2}} \lim_{j\to\infty} S_j E^{-T},\ \ R \leftarrow \frac{1}{\sqrt{2}} \lim_{j\to\infty} R_j
\end{array}
$$

This iteration requires parallel kernels for the LU factorization (`pdgetrf`), solution of triangular linear systems (`pdtrsm`), matrix products, and the computation of RRQR factorizations.

The convergence criterion for the iteration relies on $A_\infty := \lim_{j\to\infty} A_j = E$. Thus, taking advantage of the ultimate quadratic convergence of the iteration, we perform two more iterations once the criterion is satisfied to achieve the maximum possible accuracy.

Note that, because of the parallel performance of the matrix kernels involved in the iteration, we can expect a high efficiency of the implementation of the Lyapunov solvers on parallel distributed-memory architectures. For further details on the numerical accuracy, converge rate, and (parallel) implementation of this iteration, see [7, 8].

4 Experimental Results

All the experiments presented in this section were performed on a cluster of 32 nodes using IEEE double-precision floating-point arithmetic ($\varepsilon \approx 2.2204 \times 10^{-16}$). Each node consists of an Intel Pentium Xeon processor at 2.4 GHz with 1 GByte of RAM. We employ a BLAS library specially tuned for this processor that achieves around 3800 Mflops (millions of flops per second) for the matrix product (routine `DGEMM`). The nodes are connected via a Myrinet multistage network and the MPI communication library is specially developed and tuned for this network. The performance of the interconnection network was measured by a simple loop-back message transfer resulting in a latency of 18 μsec. and a bandwidth of 1.4 Gbit/sec. We made use of the LAPACK, BLAS, and ScaLAPACK libraries whenever possible.

As in all our experiments both SR and BFSR BT algorithms obtained closely similar results we only report data for the first algorithm.

4.1 Numerical Aspects

In order to evaluate the numerical behavior of our model reduction algorithms, in this subsection we employ a "random" linear descriptor system with $n = 500$ states, $m = 5$ inputs, and $p = 10$ outputs. This system was generated so that the number of finite/infinite poles are 470/30, and $\text{ind}(A, E) = 1$.

Our first experiment is designed to evaluate the numerical accuracy of the inverse-free iteration as a spectral division tool. For this purpose we compare the residuals (backward errors for spectral division) of $\|[A_{21}, E_{21}]\|_F$ and $\|E_{22}\|_F$ in (3) obtained both with the inverse-free iteration and the QZ algorithm. The latter is proved to be a numerically stable algorithm [17] and therefore should provide a measure of how accurate our inverse-free approach is. The results in Table 1 demonstrate that, at least for this example, the inverse-free iteration provides orthogonal bases Q and Z that are as "accurate" as those resulting from the QZ algorithm.

Once the TFM of the system has been decoupled into $G(s) = G_0(s) + G_\infty(s)$, the next stage consists in applying BT to the finite decoupled realization, $G_0(s)$,

Table 1. Accuracy of the bases for the deflating subspaces

	inverse-free iteration	QZ alg.
$\|[A_{21}, E_{21}]\|_F$	4.2×10^{-13}	1.0×10^{-12}
$\|E_{22}\|_F$	1.1×10^{-15}	1.1×10^{-15}

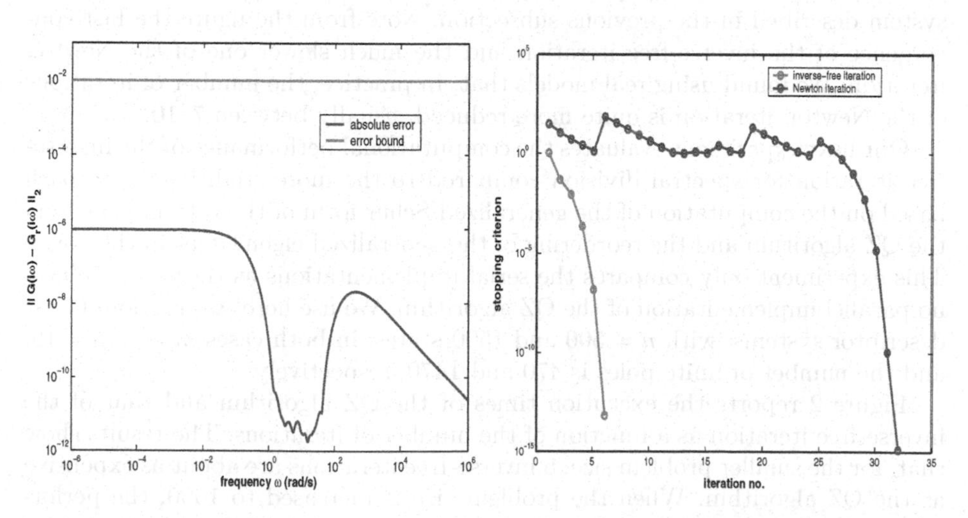

Fig. 1. Bode plot (magnitude) of error system (left plot) and convergence rate for the iterative schemes (right plot)

to obtain a reduced-order realization. In our case we forced this realization to satisfy $\|G_0 - \hat{G}_0\|_\infty \leq 0.01$. As a result, the order of the system was reduced from $n = 470$ to $r = 60$ and

$$\|\Delta_a\|_\infty = \|G_0 - \hat{G}_0\|_\infty \leq 2 \sum_{j=61}^{470} \sigma_j = 9.796 \times 10^{-3}.$$

We report the absolute error between the original and the reduced-order TFM in the left-hand plot in Figure 1. The results show that the reduced-order model is very accurate and the error bound (7) is fairly pessimistic in this example.

4.2 Convergence Rate of Iterative Schemes

The efficiency of our algorithms strongly depends on the convergence rates of the iterative schemes for the inverse-free iteration and the Lyapunov solver.

As the inverse-free iteration computes a splitting of the spectrum of a matrix pencil along a certain curve in the complex plane, convergence problems (and rounding errors) can be expected when the algorithm is applied to a matrix pencil $\alpha A - \lambda E$ with eigenvalues close to this curve. Note however that we are given the freedom to select the parameter α as long as $\alpha > \max\{|\lambda|, \lambda \in \Lambda(A, E)\backslash\{\infty\}\}$. Therefore, a slow convergence could always be detected and the algorithm could be restarted with a slightly different value of α. On the other hand, the convergence rate of the Newton iteration employed in our Lyapunov solver depends on the distance between the eigenspectrum of the stable pencil $A_0 - \lambda E_0$ and the imaginary axis.

The right-hand plot in Figure 1 reports the convergence rate of the inverse-free iteration for spectral division and the Newton iteration for the solution of the coupled generalized Lyapunov equations using the random linear descriptor system described in the previous subsection. Note from the figure the fast convergence of the inverse-free iteration and the much slower one of the Newton iteration. We found using real models that, in practice, the number of iterations of the Newton iteration is quite more reduced, usually between 7–10.

Our next experiment evaluates the computational performance of the inverse-free iteration for spectral division compared to the more traditional approach based on the computation of the generalized Schur form of the matrix pencil via the QZ algorithm and the reordering of the generalized eigenvalues in this form. This experiment only compares the serial implementations as there is currently no parallel implementation of the QZ algorithm. We use here two random linear descriptor systems, with $n = 500$ and 1500 states; in both cases $m = 5$, $p = 10$, and the number of finite poles is 470 and 1470, respectively.

Figure 2 reports the execution times of the QZ algorithm and that of the inverse-free iteration as a function of the number of iterations. The results show that, for the smaller problem size, 5 inverse-free iterations are about as expensive as the QZ algorithm. When the problem size is increased to 1500, the performance threshold between the inverse-free iteration and the QZ algorithm is also increased to 6 iterations. Although these results seem to suggest that unless the inverse-free iteration converges rapidly we should employ the QZ algorithm,

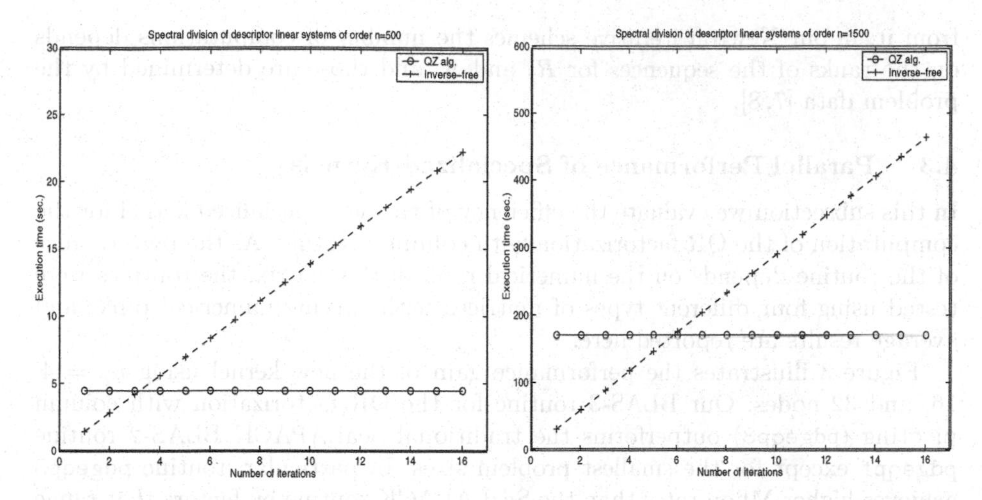

Fig. 2. Execution time of the inverse-free iterative scheme (as a function of the number of iterations) and the QZ algorithm

notice that the inverse-free iteration can be easily and efficiently parallelized while the same cannot be claimed for the QZ algorithm. Thus, a comparison between the serial QZ algorithm and a parallel inverse-free iteration would shift the performance threshold much in favor of the latter.

Although a a similar comparison between the generalized Lyapunov solver based on the Newton iteration scheme and a generalization of the Bartels-Stewart algorithm [5, 16] could also be done, it would be hard to obtain any general results

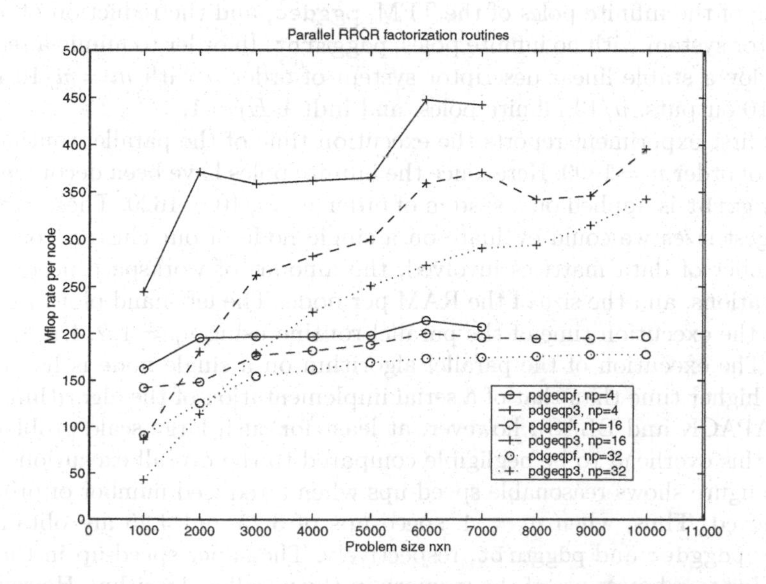

Fig. 3. Mflop rate of the parallel routines for the RRQR factorization

from it: in our Newton iterative schemes the number of computations depends on the ranks of the sequences for R_j and S_j and those are determined by the problem data [7, 8].

4.3 Parallel Performance of Specialized Kernels

In this subsection we evaluate the efficiency of the new specialized kernel for the computation of the QR factorization with column pivoting. As the performance of the routine depends on the numerical rank of the matrix, the routines were tested using four different types of matrices with varying numerical ranks and average results are reported here.

Figure 3 illustrates the performance gain of the new kernel using $n_p = 4$, 16, and 32 nodes. Our BLAS-3 routine for the QR factorization with column pivoting (`pdgeqp3`) outperforms the traditional ScaLAPACK BLAS-2 routine `pdgeqpf` except for the smallest problem sizes. In particular, routine `pdgeqp3` achieves higher Mflop rates than the ScaLAPACK routine by factors that range between 1.50–2.14, 1.22–2.01, and 1.15–1.93 on 4, 16, and 32 nodes, respectively.

Although our routine is based on BLAS-3 kernels, in order to maintain the numerical behavior of routine `pdgeqpf`, the routine applies the Householder transformations accumulated up to a certain column whenever a catastrophic cancellation in the computation of the column norms is detected. This yields a much lower performance than that of the blocked QR factorization (with no column pivoting).

4.4 Parallel Performance of the Model Reduction Algorithms

In this subsection we report the efficiency of the parallel routines for the decoupling of the infinite poles of the TFM, `pggdec`, and the reduction of a linear descriptor system with no infinite poles, `pdggmrbt`. In order to mimic a real case, we employ a stable linear descriptor system of order n with $m = n/10$ inputs, $p = n/10$ outputs, $n/10$ infinite poles, and $\text{ind}(A, E) = 1$.

Our first experiment reports the execution time of the parallel routines on a system of order $n = 1800$. Here, once the infinite poles have been decoupled, routine `pdggmrbt` is applied on a system of order $n - n/10 = 1620$. These are about the largest sizes we could evaluate on a single node of our cluster, considering the number of data matrices involved, the amount of workspace necessary for computations, and the size of the RAM per node. The left-hand plot in Figure 4 reports the execution time of the parallel routine using $n_p = 1, 2, 4, 6, 8$, and 10 nodes. The execution of the parallel algorithm on a single node is likely to require a higher time than that of a serial implementation of the algorithm (using, e.g., LAPACK and BLAS); however, at least for such large scale problems, we expect this overhead to be negligible compared to the overall execution time.

The figure shows reasonable speed-ups when a reduced number of processors is employed. Thus, when $n_p = 4$, speed-ups of 3.25 and 4.25 are obtained for routines `pdggdec` and `pdggmrbt`, respectively. The super speed-up in the latter case is due to a better use of the memory in the parallel algorithm. However, the efficiency decreases when n_p gets larger (as the system dimension is fixed, the

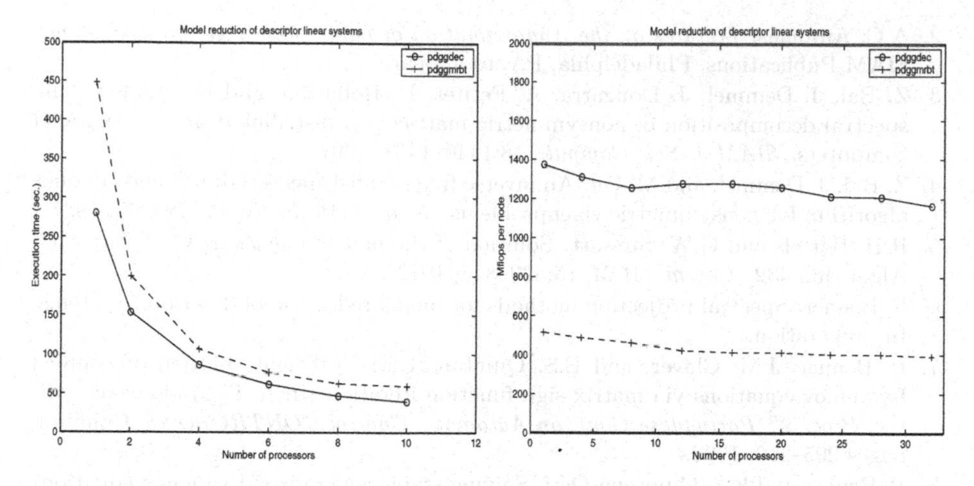

Fig. 4. Performance of the parallel model reduction algorithms

problem size per node is reduced) so that using more than a few processors does not achieve a significant reduction in the execution time for such a small problem.

We finally evaluate the scalability of the parallel routine when the problem size per node is constant. For that purpose, we fix the problem dimensions to $n/\sqrt{n_p} = 1800$ for routine `pggdec`, and $n/\sqrt{n_p} = 1620$ for routine `pdggmrbt`. The number of inputs/outputs in both cases is $m = p = n/10$. The right-hand plot in Figure 4 shows the Mflop rate per node of the parallel routine. These results demonstrate the scalability of our parallel kernels, as there is only a minor decrease in the performance of the algorithms when n_p is increased while the problem dimension per node remains fixed.

5 Concluding Remarks

We have described the design and use of parallel algorithms for model reduction of large-scale linear descriptor systems that are composed of highly parallel iterative schemes.

Experimental results report a high efficiency and scalability of the parallel algorithms on a cluster of Intel Pentium Xeon processors: Reduced-order models of moderate-scale systems, of order less than 2000, can be obtained rapidly using several processors. Also, our parallel algorithm for model reduction can be employed to reduce large systems, with tens of thousands of states, a problem dimension that is not tractable on a single processor due to memory restrictions.

References

1. E. Anderson, Z. Bai, C. Bischof, J. Demmel, J. Dongarra, J. Du Croz, A. Greenbaum, S. Hammarling, A. McKenney, and D. Sorensen. *LAPACK Users' Guide*. SIAM, Philadelphia, PA, third edition, 1999.

2. A.C. Antoulas. *Lectures on the Approximation of Large-Scale Dynamical Systems.* SIAM Publications, Philadelphia, PA, to appear.
3. Z. Bai, J. Demmel, J. Dongarra, A. Petitet, H. Robinson, and K. Stanley. The spectral decomposition of nonsymmetric matrices on distributed memory parallel computers. *SIAM J. Sci. Comput.*, 18:1446–1461, 1997.
4. Z. Bai, J. Demmel, and M. Gu. An inverse free parallel spectral divide and conquer algorithm for nonsymmetric eigenproblems. *Numer. Math.*, 76(3):279–308, 1997.
5. R.H. Bartels and G.W. Stewart. Solution of the matrix equation $AX + XB = C$: Algorithm 432. *Comm. ACM*, 15:820–826, 1972.
6. P. Benner. Spectral projection methods for model reduction of descriptor systems. In preparation.
7. P. Benner, J.M. Claver, and E.S. Quintana-Ortí. Efficient solution of coupled Lyapunov equations via matrix sign function iteration. In A. Dourado et al., editor, *Proc.* 3rd *Portuguese Conf. on Automatic Control CONTROLO'98,* Coimbra, pages 205–210, 1998.
8. P. Benner and E.S. Quintana-Ortí. Solving stable generalized Lyapunov equations with the matrix sign function. *Numer. Algorithms*, 20(1):75–100, 1999.
9. P. Benner, E.S. Quintana-Ortí, and G. Quintana-Ortí. State-space truncation methods for parallel model reduction of large-scale systems. *Parallel Comput.*, 29:1701–1722, 2003.
10. P. Benner, E.S. Quintana-Ortí, and G. Quintana-Ortí. Parallel model reduction of large-scale linear descriptor systems via Balanced Truncation. Preprint #53, DFG Research Center "Mathematics for Key Technologies", Berlin, Germany, `http://www.fzt86.de`, 2004.
11. L.S. Blackford, J. Choi, A. Cleary, E. D'Azevedo, J. Demmel, I. Dhillon, J. Dongarra, S. Hammarling, G. Henry, A. Petitet, K. Stanley, D. Walker, and R.C. Whaley. *ScaLAPACK Users' Guide.* SIAM, Philadelphia, PA, 1997.
12. J. Cheng, G. Ianculescu, C.S. Kenney, A.J. Laub, and P. M. Papadopoulos. Control-structure interaction for space station solar dynamic power module. *IEEE Control Systems*, pages 4–13, 1992.
13. P.Y. Chu, B. Wie, B. Gretz, and C. Plescia. Approach to large space structure control system design using traditional tools. *AIAA J. Guidance, Control, and Dynamics*, 13:874–880, 1990.
14. L. Fortuna, G. Nummari, and A. Gallo. *Model Order Reduction Techniques with Applications in Electrical Engineering.* Springer-Verlag, 1992.
15. R. Freund. Reduced-order modeling techniques based on Krylov subspaces and their use in circuit simulation. In B.N. Datta, editor, *Applied and Computational Control, Signals, and Circuits*, volume 1, chapter 9, pages 435–498. Birkhäuser, Boston, MA, 1999.
16. J.D. Gardiner, A.J. Laub, J.J. Amato, and C.B. Moler. Solution of the Sylvester matrix equation $AXB + CXD = E$. *ACM Trans. Math. Software*, 18:223–231, 1992.
17. G.H. Golub and C.F. Van Loan. *Matrix Computations.* Johns Hopkins University Press, Baltimore, third edition, 1996.
18. A.N. Malyshev. Parallel algorithm for solving some spectral problems of linear algebra. *Linear Algebra Appl.*, 188/189:489–520, 1993.
19. B.C. Moore. Principal component analysis in linear systems: Controllability, observability, and model reduction. *IEEE Trans. Automat. Control*, AC-26:17–32, 1981.

20. G. Obinata and B.D.O. Anderson. *Model Reduction for Control System Design*. Communications and Control Engineering Series. Springer-Verlag, London, UK, 2001.
21. P. Papadopoulos, A.J. Laub, G. Ianculescu, J. Ly, C.S. Kenney, and P. Pandey. Optimal control study for the space station solar dynamic power module. In *Proc. Conf. on Decision and Control CDC-1991*, pages 2224–2229, Brighton, December 1991.
22. C.R. Paul. *Analysis of Multiconductor Transmission Lines*. Wiley–Interscience, Singapur, 1994.
23. E.S. Quintana-Ortí. *Algoritmos Paralelos Para Resolver Ecuaciones Matriciales de Riccati en Problemas de Control*. PhD thesis, Universidad Politécnica de Valencia, 1996.
24. G. Quintana-Ortí, X. Sun, and C.H. Bischof. A BLAS-3 version of the QR factorization with column pivoting. *SIAM J. Sci. Comput.*, 19:1486–1494, 1998.
25. J.D. Roberts. Linear model reduction and solution of the algebraic Riccati equation by use of the sign function. *Internat. J. Control*, 32:677–687, 1980. (Reprint of Technical Report No. TR-13, CUED/B-Control, Cambridge University, Engineering Department, 1971).
26. M.G. Safonov and R.Y. Chiang. A Schur method for balanced-truncation model reduction. *IEEE Trans. Automat. Control*, AC–34:729–733, 1989.
27. T. Stykel. Model reduction of descriptor systems. Technical Report 720-2001, Institut für Mathematik, TU Berlin, D-10263 Berlin, Germany, 2001.
28. T. Stykel. *Analysis and Numerical Solution of Generalized Lyapunov Equations*. Dissertation, TU Berlin, 2002.
29. T. Stykel. Balanced truncation model reduction for semidiscretized Stokes equation. Technical Report 04-2003, Institut für Mathematik, TU Berlin, D-10263 Berlin, Germany, 2003.
30. X. Sun and E.S. Quintana-Ortí. Spectral division methods for block generalized Schur decompositions. *Mathematics of Computation*, 73:1827–1847, 2004.
31. M.S. Tombs and I. Postlethwaite. Truncated balanced realization of a stable non-minimal state-space system. *Internat. J. Control*, 46(4):1319–1330, 1987.
32. R.A. van de Geijn. *Using PLAPACK: Parallel Linear Algebra Package*. MIT Press, Cambridge, MA, 1997.
33. A. Varga. Efficient minimal realization procedure based on balancing. In *Prepr. of the IMACS Symp. on Modelling and Control of Technological Systems*, volume 2, pages 42–47, 1991.

A Parallel Algorithm for Automatic Particle Identification in Electron Micrographs

Vivek Singh, Yongchang Ji, and Dan C. Marinescu

School of Computer Science, University of Central Florida, Orlando, USA
{vsingh, yji, dcm}@cs.ucf.edu

Abstract. Three dimensional reconstruction of large macromolecules like viruses at resolutions below 10 Å requires a large set of projection images. Several automatic and semi-automatic particle detection algorithms have been developed along the years. We have developed a general technique designed to automatically identify the projection images of particles. The method is based on Markov random field modelling of the projected images and involves a preprocessing of electron micrographs followed by image segmentation and post processing. In this paper we discuss the basic ideas of the sequential algorithm and outline a parallel implementation of it.

1 Introduction

Over the past decade cryo-transmission electron microscopy (Cryo-TEM) has emerged as an important tool, together with X-ray crystallography, to examine the three dimensional structures and dynamic properties of macromolecules. Cryo-TEM specimen preparation procedures permit viruses and other macromolecules to be studied under a variety of conditions, which enables the functional properties of these molecules to be examined [1], [19], [21]. The three dimensional model of a specimen is normally represented as a density function sampled at points of a regular grid. The images of individual particles in the electron micrograph are approximate projections of the specimen in the direction of the electron beam. The *Projection Theorem* that connects the Fourier Transform of the object with the transforms of its projections is used to construct the spatial density distribution of the particle [4].

The initial step in three-dimensional structural studies of single particles and viruses after electron micrographs have been digitized is the *selection* (boxing) of particles images. Traditionally, this task has been accomplished by manual or semi-automatic procedures. However, the goal of greatly improving the solution of structure determinations to 8 Å or better comes with a requirement to significantly increase the number of images. Though 50 or fewer particle projections often suffice for computing reconstructions of many viruses in the 20 Å - 30 Å range [1], the number of particle projections needed for a reconstruction of a virus with unknown symmetry at ~ 5 Å may increase by three orders of magnitude. Hence, the manual or semi-automatic particle identification techniques create a burdensome bottleneck in the overall process of three dimensional

M. Daydé et al. (Eds.): VECPAR 2004, LNCS 3402, pp. 354–367, 2005.

structure determination [12]. It is simply unfeasible to manually identify tens to hundreds of thousands of particle projections in low contrast micrographs, and, even if feasible, the manual process is prone to errors. At high magnification, noise in cryo-TEM micrographs of unstained, frozen hydrated macromolecules is unavoidable and makes automatic or semi-automatic detection of particle positions a challenging task. Since biological specimens are highly sensitive to the damaging effects of the electron beam used in cryo-TEM, minimal exposure methods must be employed and this results in very noisy, low contrast images.

A difficult problem in automatic particle identification schemes is the so called *labelling* step, the process of associating a label with every pixel of the micrograph, e.g., a logical 1 if the pixel belongs to a projection of the virus particle and a logical 0 if the pixel belongs to the background. Once all pixels in the image are labelled, morphological filtering [8] followed by clustering [10] to connect together components with the same label can be used. Then, we may take advantage of additional information, if available, to increase the accuracy of the recognition process. In this paper we present a parallel algorithm for labelling based upon Hidden Markov Random Field (HMRF) modelling combined with expectation maximization.

The paper is organized as follows. In Section 2 we discuss approaches taken elsewhere for particle selection in micrographs. The preprocessing step and it's relation to segmentation is presented in Section 3. It is followed by a description of the segmentation algorithm and it's schematic representation. Boxing, a post processing step, is described in Section 5. The quality of the solution is discussed in Section 6. The parallelization of the algorithm described in Section 4 is presented in Section 7 where the results of the parallel version of the algorithm are discussed. We present the conclusions and future work in Section 8 followed by acknowledgements in Section 8.

2 Related Work

Nicholson and Glaeser [11] provide a rather comprehensive review of automatic particle selection algorithms and methods. Automatic particle selection methods were first described by van Heel [20]. The method relies on computation of the local variance of pixel intensities over a relatively small area. The variance is calculated for each point of the image field and a maximum of the local variance indicates the presence of an object.

Several methods make use of a technique called *template matching*. In such algorithms a reference image is cross-correlated with an entire micrograph to detect the particles. The method proposed by Frank [5] uses correlation functions to detect the repeats of a motif. This approach is restricted to either a particular view of an asymmetric particle, or to any view of particles with high point group symmetry, such as icosahedral viruses. Cross-correlation procedures are also performed during the post-processing phase of the Crosspoint method presented in [10].

Another class of automatic particle identification methods are feature-based. For example, the local and the global spatial statistical properties of the micrograph images are calculated in the method discussed in [16]. In order to isolate the particles and distinguish them from the noise, discriminant analysis based supervised clustering is performed in the feature space. This technique leads to many false positives.

Approaches based upon techniques developed in the computer vision field such as, edge detection [6] are also very popular. Edges are significant cues to a transition from a view of the particle projection to the view of the background. Edges are generally signalled by changes in the gradient of the pixel intensity and indicate the presence of an object boundary. Thus, sometimes edge detection is used as a preprocessing step, as in the method proposed in [7]. In this case the edges are further processed by spatial clustering. Particles are detected by symbolic processing of such a cluster of edge cues representing a particle. The approach taken by [24] is to first obtain an edge map of the micrograph consisting of the edge cues. The Hough transform is used to spatially cluster the edges to represent particles.

3 Preprocessing and Segmentation

An automatic particle identification method often involves a pre-processing step designed to improve the signal to noise ratio of a noisy micrograph. Various techniques, such as histogram equalization, and different filtering methods are commonly used. We describe briefly an anisotropic filtering technique we found very useful for enhancing the micrographs before the segmentation and labelling steps.

The aim of anisotropic diffusion is to enhance the "edges" present in the image by preferential smoothing where regions with lower gradient are smoothed more than the areas with higher gradients. This is achieved by drawing an analogy between smoothing and thermal diffusion [14]. A diffusion algorithm is used to modify iteratively the micrograph, as prescribed by a partial differential equations (PDE) [14]. Consider for example the isotropic diffusion equation

$$\frac{\partial I(x,y,t)}{\partial t} = div(\nabla I) \tag{1}$$

In this partial differential equation t specifies an artificial time and ∇I is the gradient of the image. Let $I(x,y)$ represent the original image. We solve the PDE with the initial condition $I(x,y,0) = I(x,y)$, over a range of $t \in \{0,T\}$. As a result, we obtain a sequence of Gaussian filtered images indexed by t. Unfortunately, this type of filtering produces an undesirable blurring of the edges of objects in the image. Perona and Malik [14] replaced this classic isotropic diffusion equation with

$$\frac{\partial I(x,y,t)}{\partial t} = div[g(\|\nabla I\|)\nabla I] \tag{2}$$

where $\|\nabla I\|$ is the modulus of the gradient and $g(\|\nabla I\|)$ is the "edge stopping" function chosen to satisfy the condition $g \to 0$ when $\|\nabla I\| \to \infty$.

The modified diffusion equation prevents the diffusion process across edges. As the gradient at some point increases sharply, signalling an edge, the value of the "edge stopping" function becomes very small making $\frac{\partial I(x,y,t)}{\partial t}$ effectively zero. As a result, the intensity at that point on the edge of an object is unaltered as t increases. This procedure ensures that the edges do not get blurred in the process. Clearly a higher number of iterations benefits segmentation. However increasing the number of iterations is detrimental as it causes widespread diffusion resulting in joining of nearby projections in the segmentation output. The filtered image is then segmented using the Hidden Markov Random Field based method.

The essence of the segmentation algorithm is based on the assumption that the temporal evolution of a physical process does not suffer abrupt changes and that the spatial properties of objects in a neighborhood exhibit some degree of coherence. Markov models are used to describe the evolution in time of memoryless processes, i.e., those processes whose subsequent states do not depend upon the prior history of the process. Such models are characterized by a set of parameters such as the number of states, the probability of a transition between a pair of states, and so forth.

Hidden Markov models exploit the "locality" of physical properties of a system and have been used in speech recognition applications [15] as well as object recognition in two dimensional images [2]. The term "hidden" simply signifies the lack of direct knowledge for constructing the temporal or spatial Markov model of the object to be detected, e.g., the virus particle, or the physical phenomena to be investigated, e.g., the speech.

A classical example wherein a hidden Markov model is used to describe the results of an experiment involves the tossing of a coin in a location where no outside observer can witness the experiment itself, only the sequence of head/tail outcomes is reported. In this example, we can construct three different models assuming that one, two, or three coins were used to generate the reported outcomes. The number of states and the probability of a transition between a pair of states are different for the three models. The information about the exact setup of the experiment, namely the number of coins used, is hidden: we are only told the sequence of head/tail outcomes.

Expectation maximization is a technique for selecting from all possible models, the one that best fits our knowledge, as well as the parameters of the model that best fit the identifying information available. The more information we can gather, the more reliable we can expect the choice of model to be. Also, once the model is available, the more accurate the parameters of the model are. The digitized micrographs may constitute the only data available. It is not uncommon, for example, to have additional information such as a low resolution, preliminary 3D reconstruction of the object of interest. The extent of prior knowledge will vary among different applications and hence the model construction becomes application-dependent.

Hidden Markov techniques are often used to construct models of physical systems when the information about the system is gathered using an apparatus that distorts in some fashion the physical reality being observed. For example, images of macromolecules obtained using cryo-TEM methods are influenced by the Contrast Transfer Function (CTF) characteristic of the imaging instrument and the conditions under which images are obtained, just as an opaque glass separating the observer from the individual tossing the coins prevents the former from observing the details of the experiment discussed above. This explains why hidden Markov techniques are very popular in virtually all experimental sciences.

The term "field" is used to associate the value of the property of interest with a point in a time-space coordinate system. For example, in a 2D image each pixel is described by two coordinates and other properties such as it's intensity, label, etc. In HMRF models, the field is constructed based upon a set of observed values. These values are related by an unknown stochastic function to the ones of interest to us. For example, a step in automatic particle selection is the labelling of individual pixels. Labelling means to construct the stochastic process $X(i,j)$ describing the label (0 or 1) for the pixel with coordinates (i,j) given the stochastic process $Y(i,j)$ describing the pixel intensity. The pixel intensities are observable, thus a model for $Y(i,j)$ can be constructed.

Our interest in HMRF modelling was stimulated by the successful application of this class of methods to medical imaging. The diagnosis of cancer tumors from NMR images [23] based on HMRF modelling is used to distinguish between healthy cells and various types of cancerous cells, subject to the artifacts introduced by the NMR exploratory methods. These artifacts are very different than the ones introduced by an electron microscope. While the basic idea of the method is the same, the actual modelling algorithms and the challenges required to accurately recognize the objects of interest in the two cases are dissimilar. The most difficult problem we are faced with is the non-uniformity of the layer of ice in cryo-TEM micrographs.

4 The Segmentation Algorithm

First, we give an informal description of the method illustrated in Figure 1 and then present the algorithm in detail. It is common practice to de-noise a noisy image before processing it for detection of cues such as edges. Very frequently some form of isotropic smoothing, e.g., Gaussian smoothing is used to achieve this. However, such a method of smoothing leads to blurring of the important cues such as edges in addition to smoothing of the noise. Anisotropic diffusion [14] [22] is a form of smoothing that dampens out the noise while keeping the edge cues intact. Following is a stepwise description of the particle identification procedure.

1. The image is split into rectangular blocks that are roughly twice the size of the projection of a particle. This is done to reduce the gradient of the background across each processed block. A high gradient degrades the quality of the segmentation process carried out in Step 3. The gradient of the background affects the algorithm in the following ways

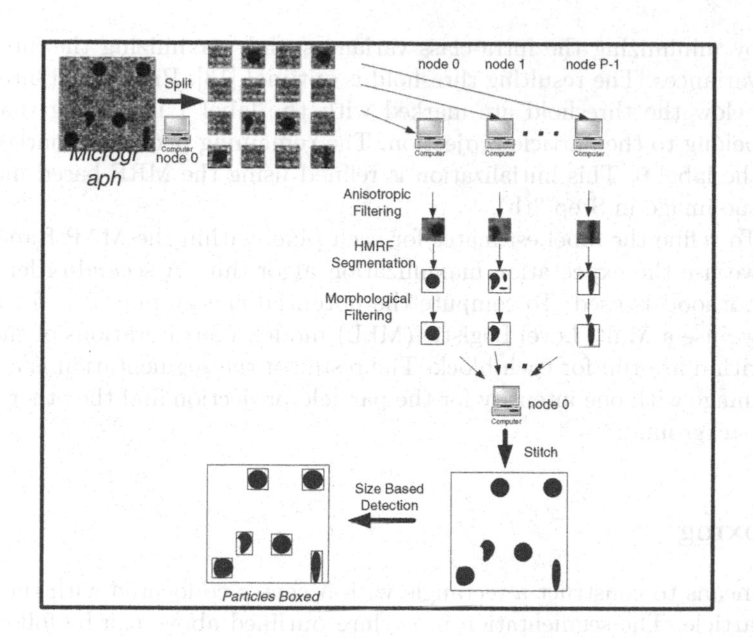

Fig. 1. The schematic representation of the steps for HMRF/EM-based automatic particle identification procedure introduced in [17]

(a) The initialization is based solely on intensity histogram which encode only the frequencies of occurrence of any pixel intensity in the image. Due to the presence of a gradient, the contribution to the count of an intensity for example, may come from the background of a darker region, as well as from the inside of a projection of a virus in a brighter region. When the initialization is done for the entire image it performs poorly.

(b) The parameters μ_0, σ_0, μ_1, σ_1 are fixed for an image. However, they are not the true parameters for intensity distribution across the whole image. The means and variances of pixel intensities are significantly different across the image due to the presence of the gradient. Cutting the image into blocks ensures a lack of drift in these parameters across each block.

2. As a pre-processing step individual blocks are filtered by means of anisotropic diffusion. Such filtering ensures that "edges" are preserved and less affected by smoothing. The edge stopping function is

$$g(\nabla I) = e^{-(\|\nabla I\|/K)^2} \quad (3)$$

For each block we run 5 iteration of the algorithm. $K = 0.2$ to ensure the stability of the diffusion process.

3. The blocks filtered through the anisotropic diffusion based filter are segmented using the HMRF method. The following steps are taken for segmentation

(a) The initialization of the model is done using a discriminant measure based thresholding method proposed by Otsu [13]. The threshold is found by minimizing the intra-class variances and maximizing the inter-class variances. The resulting threshold is optimal [13]. Pixels with intensities below the threshold are marked with the label 1 indicating that they belong to the particle projection. The remaining pixels are marked with the label 0. This initialization is refined using the MRF based model of the image in Step 3(b).

(b) To refine the label estimates for each pixel within the MAP framework, we use the expectation maximization algorithm. A second order neighborhood is used. To compute the potential energy functions for cliques we use a Multi Level Logistic (MLL) model. Four iterations of the algorithm are run for each block. The result of the segmentation is a binary image with one intensity for the particle projection and the other for the background.

5 Boxing

Boxing means to construct a rectangle with a center co-located with the center of the particle. The segmentation procedure outlined above can be followed by different boxing techniques. We now discuss briefly the problem of boxing for particles with unknown symmetry.

Boxing particles with unknown symmetry is considerably more difficult than the corresponding procedure for iscosahedral particles. First, the center of a projection is well defined in case of icosahderal particle, while the center of the projection of an arbitrary 3D shape is more difficult to define. Second, the result of pixel labelling, or segmentation, is a shape with a vague resemblance to the actual shape of the projection. Typically, it consists of one or more clusters of marked pixels, often disconnected from each other. Recall that when the shape of the projection is known we can simply run a connected component algorithm, like the one described in [10], and then determine the center of mass of the resulting cluster. In addition, we run a procedure to disconnect clusters corresponding to particle projections next to each other.

The post-processing of the segmented image to achieve boxing involves morphological filtering operations of opening and closing [18]. These two morphological filtering operations are based on the two fundamental operations called dilation and erosion. For a binary labelled image, dilation is informally described as taking each pixel with value '1' and setting all pixels with value '0' in its neighborhood to the value '1'. Correspondingly, erosion means to take each pixel with value '1' in the neighborhood of a pixel with value '0' and re-setting the pixel value to '0'. The term "neighborhood" here bears no relation to the therm "neighborhood" in the framework of Markov Random Field described earlier. Pixels marked say as "1", separated by pixels marked as "0" could be considered as belonging to the same neighborhood if they dominate a region of the image. The opening and closing operations can then be described as erosion followed by dilation and dilation followed by erosion respectively.

The decision of whether a cluster in the segmented image is due to noise, or due to the projection of a particle is made according to the size of the cluster. For an icosahedral particle, additional filtering may be performed when we know the size of the particle. Such filtering is not possible for particles of arbitrary shape. A fully automatic boxing procedure is likely to report a fair number of false hits, especially in very noisy micrographs.

New algorithms to connect small clusters together based upon their size and the "distance" between their center of mass must be developed. Without any prior knowledge, about the shape of the particle boxing poses tremendous challenges. It seems thus natural to investigate a model based boxing method, an approach we plan to consider next.

The boxing of spherical particles in Figure 2 is rather ad hoc and cannot be extended for particles of arbitrary shape. We run a connected component algorithm and then find the center of the particle and construct a box enclosing the particle.

6 The Quality of the Solution

We report on some results obtained with the method of automatic particle identification presented in this paper. The experiments were performed on micrographs of frozen-hydrated reovirus and the Chilo Iridescent virus (CIV) samples recorded in an FEI/Philips CM200 electron microscope.

The HMRF particle selection algorithm does not assume any particular shape or size for the virus projection. Even though most of our experiments were performed with micrographs of symmetric particles the results of particle detection for asymmetric particles gives us confidence in the ability of our algorithm to identify particles of any shape. However, several aspects of particle selection pro-

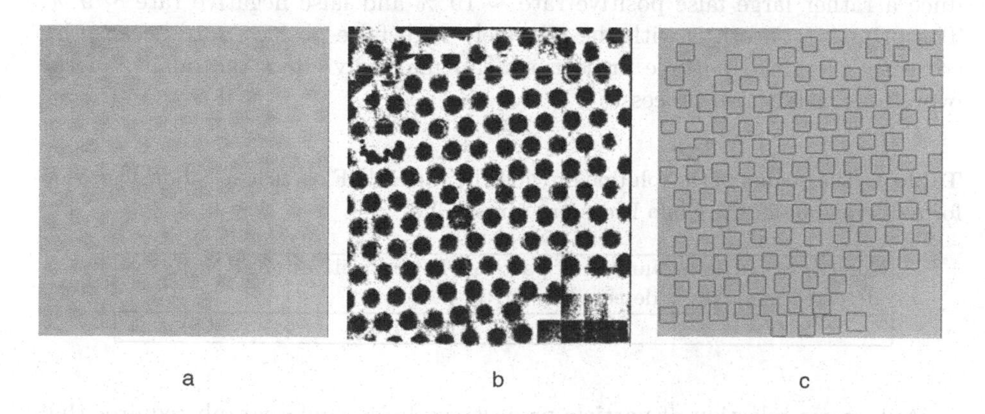

Fig. 2. a) A portion of a micrograph of frozen hydrated sample of Chilio Iridescent virus (CIV). b) The histogram of the pixel intensities for the image in (a). c) The micrograph after anisotropic diffusion filtering. d) Micrograph after the initialization step of HMRF. e) Segmented micrograph. f) Boxed particles in the micrograph

cessing demand improvement. First, the algorithm is computationally intensive. The time for automatic particle selection is hundreds of seconds for a 6000×6000 pixel micrograph with some $50 - 70$ particles.

A significant portion of the time is spent in obtaining an optimization for the MRF. This can be overcome if a multi-scale technique is adopted. With a multi-scale technique, a series of images of smaller size, with larger pixel dimensions, are constructed. The optimization starts with the smallest size image, corresponding to the largest scale. The results are propagated to the optimization for the same image but of larger size, at next scale. Multi-scale MRF techniques are already very popular within the vision community [3].

More intelligent means of distinguishing a projection of the virus from the projection due to noise could be introduced such as some discrimination method based on features of the projection more informative than mere projection size. Such features can be extracted from projections of the lower resolution 3D map of the particle.

The ICM algorithm performs a local optimization of the Markov Random Field and it is very sensitive to initialization. Global optimization methods like simulated annealing [9] are likely to produce better results, but would require a fair amount of CPU cycles and, thus, increase the processing time. High pass filtering may be used to remove the background gradient. However, this does not perform very well if the background has a steep gradient. The other possible solution is to work on smaller regions where the background gradient is expected to vary gradually.

The result of particle identification for a Chilo Iridescent virus (CIV) micrograph is presented in Figure 2. The size of the micrograph is $5,424 \times 5,968$ pixels. The total time taken is ≈ 1400 seconds on a single processor machine. Table 1 compares results obtained by manual selection and by the use of the HMRF algorithm. The present implementations of the HMRF algorithm produce a rather large false positive rate ~ 19 % and false negative rate ~ 9 %. The primary concern is with the relatively high false positive rate because incorrectly identified particle projections can markedly effect the quality of the overall reconstruction process.

Table 1. The quality of the solution provided by the HMRF particle selection algorithm for a micrograph of the Chilo Iridescent virus (CIV)

Number of particles detected manually	Number of particles detected by MRF	False Positives	False Negatives
277	306	55	26

Automatic selection of particle projections from a micrograph requires that the results be obtained reasonably fast. Hence, in addition to analysis pertinent to the quality of the solution, we report the time required by the algorithm for different size and number of particles in a micrograph. Since the preprocessing step has not yet been optimized, we are confident that significant efficiency can be realized after further algorithm development. Indeed, the rather large number of false positives (Table 1, column 3), demonstrates an inability to currently distinguish particles from junk. Clearly, the existing algorithm is unsuitable for real time electron microscopy which would allow the microscopist to record a very low magnification image at very low electron dose and from which it would be possible to assess particle concentration and distribution to aid high magnification imaging.

7 Parallelization: Algorithm and Experimental Results

The performance data reported in [17] indicate that the execution time of the segmentation process is prohibitively large. Considering the fact that even larger micrographs may need to be processed and that we may need to improve the quality of the solution and use global optimization techniques it becomes abundantly clear that we should consider parallel algorithms for automatic particle identification, rather than sequential ones.

We now describe such an algorithm designed for a cluster of PCs. We divide the image into blocks, where the size of the blocks is roughly twice the estimated size of the particles. Let the image be divided into m by n blocks. Further, let there be p processors. A pseudo code for the parallel algorithm for segmentation is shown below.

```
if master node
 then
   divide the image into m by n blocks
   for i = 0 to m-1
     for j = 0 to n-1
       send block(i,j) to node (i+j) mod p
     end for
   end for
receive the block/s form the master node
filter the block/s received using anisotropic diffusion
segment the block/s received
perform morphological filtering on the block/s
send the block/s to the master node
if  master node
 then
   while not all blocks are received
     receive the next block
   end while
   Stitch all the blocks into final segmented image
```

The computation was performed on a cluster of 44 dual processing (2.4 GHz Intel^{TM}Xeon processors) nodes, each with 3 GB of main memory. The interconnection network is a 1 Gbps Ethernet switch. To study the performance of the algorithm we performed the segmentation of three images of different size,

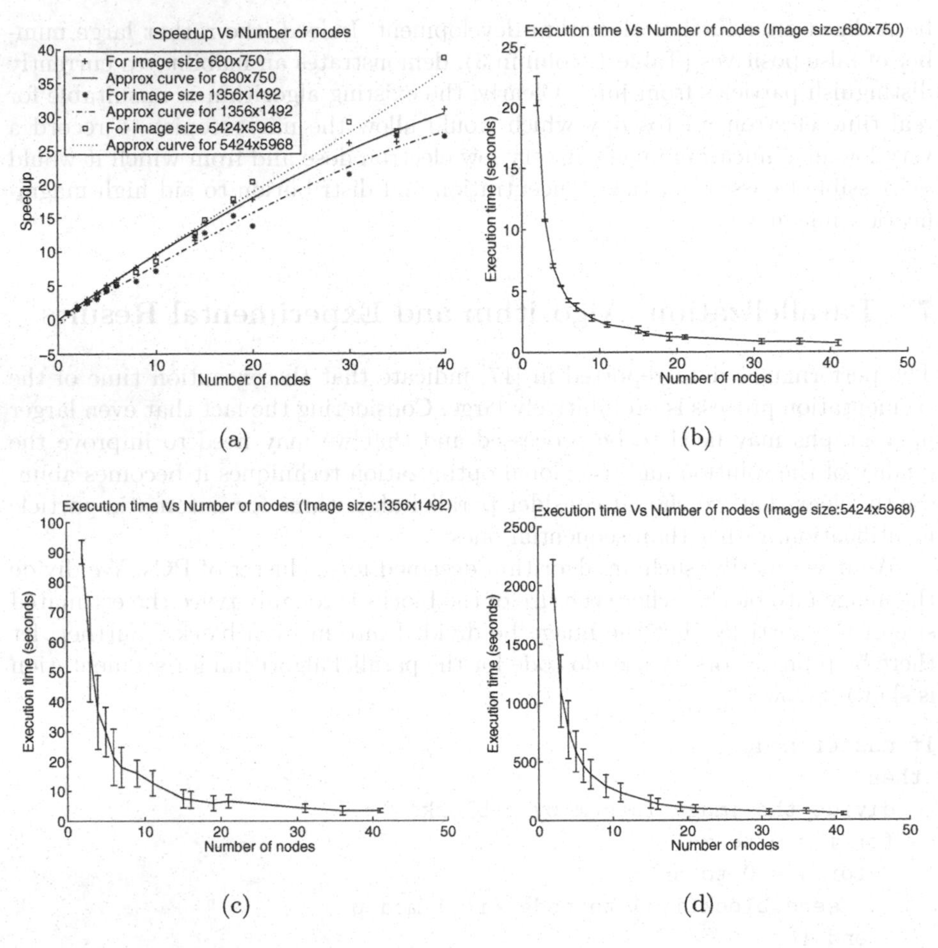

Fig. 3. (a) The speedup of the algorithm for different sizes of micrographs. (b), (c) and (d). The *total computation time* versus the *number of nodes* for three different sizes of micrograph images (in pixels) 680×750, 1356×1492 and 5424×5968, respectively. The vertical bars indicate 95% confidence intervals, computed using 30 readings for each value of *number of nodes*

680x750, 1356x1492 and 5424x5968 pixels. The speedup is nearly linear when the number of nodes is less than 20 and then levels off (see Figure 7) reflecting the fact that communication costs start playing a more significant role as the number of nodes increases. However this levelling off is not observed for the image of size 5424x5968 since communication time becomes less significant as the image size, and hence the computation time is increased. The vertical bars in Figures 3(b), 3(c) and 3(d) correspond to the 95% confidence interval. The image of size 5424x5968 corresponds to the image shown in Figure 2. The confidence intervals for the image of size 680x750 are very narrow indicating a high degree of reproducibility of the results.

The experiments were conducted on a cluster shared with other unrelated computations. We believe that the size of confidence intervals for the mean of the measured execution time for the three image sizes reflects the contention for communication bandwidth and other resources, among the processes running on behalf of our algorithm and unrelated processes.

8 Conclusions and Future Work

The HMRF particle selection algorithm does not assume any particular shape or size for the virus projection. However, several aspects of particle selection processing demand improvement. First, the algorithm is computationally intensive. The time for automatic particle selection is hundreds of seconds for a 6000×6000 pixel micrograph with some $50 - 70$ particles. A significant portion of the time is spent in obtaining an optimization for the MRF. This can be overcome if a multi-scale technique is adopted. With a multi-scale technique, a series of images of smaller size, with larger pixel dimensions, are constructed. The optimization starts with the smallest size image, corresponding to the largest scale. The results are propagated to the optimization for the same image but of larger size, at next scale. Multi-scale MRF techniques are already very popular within the vision community [3].

More intelligent means of distinguishing a projection of the virus from the projection due to noise could be introduced such as some discrimination method based on features of the projection more informative than mere projection size. Such features can be extracted from projections of the lower resolution 3D map of the particle.

The ICM algorithm performs a local optimization of the Markov Random Field and it is very sensitive to initialization. Global optimization methods like simulated annealing [9] are likely to produce better results, but would require a fair amount of CPU cycles and, thus, increase the processing time. High pass filtering may be used to remove the background gradient. However, this does not perform very well if the background has a steep gradient. The other possible solution is to work on smaller regions where the background gradient is expected to vary gradually. Current work is focussed on the following issues:

1. Construct a better “edge stopping” function for the anisotropic diffusion equation to improve noise smoothing operations.
2. Use a multiscale Markov random field to reduce the computation time and make the algorithm more robust to initialization.
3. Use low resolution 3D reconstruction maps to enhance the identification of particles and their distinction from “junk”.

Future work would involve a study of performance of various parallelization strategies.

Acknowledgments

We thank T. S. Baker for his insights in particle identification methods, W. Zhang and X. Yan for supplying image data.

The study was supported in part by the National Science Foundation grants MCB9527131, DBI0296107, ACI0296035, and EIA0296179.

References

1. T. S. Baker, N. H. Olson, and S. D. Fuller. Adding the third dimension to virus life cycles: Three-dimensional reconstruction of icosahedral viruses from cryo-electron micrographs. *Microbiology and Molecular Biology Reviews*, 63(4):862–922, 1999.
2. J. Besag. On the statistical analysis of dirty pictures. *Journal of the Royal Statistical Society*, 48(3):259–302, 1986.
3. C. Bouman and M. Shapiro. A multiscale random field model for Bayesian image segmentation. *IEEE Trans. on Image Processing*, 3(2):162–177, March 1994.
4. R. A Crowther, D. J. DeRosier, and A. Klug. The reconstruction of a three-dimensional structure from projections and its application to electron microscopy. *Proc. of the Royal Society of London*, A 317:319–340, 1970.
5. J. Frank and T. Wagenkknecht. Automatic selection of molecular images from electron micrographs. In *Ultramicroscopy*, 2(3):169–175, 1983-84.
6. R. C. Gonzales and R. E. Woods *Digital Image Processing.* Prentice Hall, 1996.
7. G. Harauz and F. A. Lochovsky. Automatic selection of macromolecules from electron micrographs. *Ultramicroscopy*, 31:333–344, 1989.
8. H. J. A. M. Heijmans Morphological image operators. *Academic Press, Boston*, 1994.
9. S. Kirkpatrick, C. D. Gelart Jr. and M. P. Vecchi Optimization by simulated annealing. *Science*, 31:671-680, 1983.
10. I. A. B. Martin, D. C. Marinescu, R. E. Lynch and T. S. Baker. Identification of spherical virus particles in digitized images of entire electron micrographs. *Journal of Structural Biology*, 120:146–157, 1997.
11. W. V. Nicholson and R. M. Glaeser Review: automatic particle detection in electron microscopy. *Journal of Structural Biology*, 133:90–101, 2001.
12. E. Nogales and N. Grigorieff. Molecular machines: putting the pieces together. *Journal of Cell Biology*, 152:F1F10, 2001.
13. N. Otsu. A threshold selection method from gray level histogram. *IEEE Trans. on Systems Man and Cybernetics.*, SMC-8:62–66, 1979.
14. P. Perona and J. Malik. Scale-space and edge detection using anisotropic diffusion. *IEEE Trans. on Pattern Analysis and Machine Intelligence*, 12(7):629–639, 1990.
15. L. R. Rabiner. A tutorial on hidden Markov models and selected applications in speech recognition. *Proc. of the IEEE*, 77(2):257–286, 1989.
16. K. Ramani Lata, P. Penczek, and J. Frank. Automatic particle picking from electron micrographs. *Ultramicroscopy*, 58:381–391, 1995.
17. V. Singh, D. C. Marinescu, and T. S. Baker. Image segmentation for automatic particle identification in electron micrographs based on hidden Markov random field models and expectation maximization, *Journal of Structural Biology*, 145(1-2):123–141, 2004.
18. P. Soille. Morphological Image Analysis: Principles and Applications. *Springer-Verlag*, 1999.

19. P. A. Thuman-Commike, W. Chiu. Reconstruction principles of icosahedral virus structure determination using electron cryomicroscopy. *Micron*, 31:687–711, 2000.
20. M. van Heel. Detection of objects in quantum-noise-limited images. *Ultramicroscopy*, 7(4):331–341, 1982.
21. M. van Heel, B. Gowen, R. Matadeen, E. V. Orlova, R. Finn, T. Pape, D. Cohen, H. Stark, R. Schmidt, M. Schatz, and A. Patwardhan. Single-particle electron cryo-microscopy: towards atomic resolution. *Quarterly Reviews of Biophysics*, 33(4):307–369, 2000.
22. J. Weickert. *Anisotropic Diffusion in Image Processing.* Teubner-Verlag, 1998.
23. Y. Zhang, S. Smith, and M. Brady. Segmentation of brain MRI images through a hidden Markov random field model and the expectation-maximization algorithm. *IEEE Trans. on Medical Imaging*, 20-1:45–57, 2001.
24. Y. Zhu, B. Carragher, D. Kriegman, R. A. Milligan, and C. S. Potter. Automated identification of filaments in cryoelectron microscopy images. *Journal of Structural Biology*, 135-3:302–312, 2001.

Parallel Resolution of the Two-Group Time Dependent Neutron Diffusion Equation with Public Domain ODE Codes

Víctor M. García[1], V. Vidal[1], G. Verdú[2], J. Garayoa[1], and R. Miró[2]

[1] Departamento de Sistemas Informáticos y Computación, Universidad Politécnica de Valencia, Valencia, 46022 Spain
[2] Departamento de Ingeniería Química y Nuclear, Universidad Politécnica de Valencia, Valencia, 46022 Spain

Abstract. In this paper it is shown how specialised codes for the resolution of ordinary differential equations (FCVODE[3], DASPK[2]) can be used to solve efficiently the time dependent neutron diffusion equation. Using these codes as basis, several new codes have been developed, combining the sequential and parallel versions of DASPK and FCVODE with different preconditioners. Their performance has been assessed using a two-dimensional benchmark (The TWIGL reactor).

1 Introduction

The time-dependent neutron diffusion equation is a partial differential equation with source terms, which describes the neutron behaviour inside a nuclear reactor. In order to carry out better and faster safety analyses, it is needed to solve this equation as fast as possible, while keeping the accuracy requirements.

The resolution method usually includes discretizing the spatial domain, obtaining a system of linear, stiff ordinary differential equations (ODEs). In order to guarantee the reliability of the safety analysis, it becomes necessary to use fine discretizations, which results in large ODE systems. In the past, these systems were solved through simple fixed time step techniques. This can result in errors (if the time step is too large) or in long computing times (if the time step is too small).

The way chosen to avoid this problem has been to use specialised public domain codes for the resolution of large systems of ODEs. We have selected the codes DASPK and FCVODE, which are powerful codes for the resolution of large systems of stiff ODEs. These codes can estimate the error after each time step, and, depending on this estimation can decide which is the new time step and, possibly, which is the integration method to be used in the next step. With these mechanisms, it is possible to keep the overall error below the chosen tolerances, and, when the system behaves smoothly, to take large time steps increasing the execution speed.

The development of fast sequential codes based on DASPK and FCVODE was reported in [8] and some results are summarized in section 4.

M. Daydé et al. (Eds.): VECPAR 2004, LNCS 3402, pp. 368–381, 2005.

The goal of the work presented here has been to develop parallel versions based on the sequential ones and on DASPK and FCVODE; in the case of FCVODE, it has a parallel version which eased the process. On the other hand, DASPK only has a sequential version, so that only the routines written by the user can be parallelized. This paper shows the techniques used to obtain the parallel versions and compares the results with both versions and with the sequential ones.

The rest of the paper has the following structure: first, the neutron diffusion equation and the spatial discretization techniques used shall be briefly described. Next, FCVODE and DASPK will be described with more detail followed by the test cases and the sequential versions, showing some results. Finally, the parallel versions, the techniques used, and the results will be compared and discussed.

2 Problem Description

2.1 The Time Dependent Neutron Diffusion Equation

To model properly the behaviour of neutrons in a reactor, it is usually needed to divide the neutron energies in two or more groups. This is so because the neutrons have a wide range of energies, properties and constants associated. In this case, a two group approximation has been chosen, so that neutrons are divided in fast neutrons (whose flux is denoted as ϕ_1) and thermal neutrons (with flux denoted as ϕ_2). With this assumption, the time dependent neutron diffusion equations can be expressed as:

$$[v^{-1}]\dot{\phi} + \mathcal{L}\phi = (1-\beta)\mathcal{M}\phi + \sum_{k=1}^{K} \lambda_k \mathcal{C}_k \chi \quad , \tag{1}$$

where:

$$\phi = \begin{bmatrix} \phi_1 \\ \phi_2 \end{bmatrix} , \quad \mathcal{L} = \begin{bmatrix} -\boldsymbol{\nabla}\cdot(D_1\boldsymbol{\nabla}) + \Sigma_{a1} + \Sigma_{12} & 0 \\ -\Sigma_{12} & \boldsymbol{\nabla}\cdot(D_2\boldsymbol{\nabla}) + \Sigma_{a2} \end{bmatrix} ,$$

and

$$\mathcal{M} = \begin{bmatrix} \nu\Sigma_{f1} & \nu\Sigma_{f2} \\ 0 & 0 \end{bmatrix} , \quad [v^{-1}] = \begin{bmatrix} \frac{1}{v_1} & 0 \\ 0 & \frac{1}{v_2} \end{bmatrix} , \quad \chi = \begin{bmatrix} 1 \\ 0 \end{bmatrix} .$$

where:

- ϕ is the neutron flux on each point of the reactor; so, it is a function of time and position.
- $\mathcal{C}_k$ is the concentration of the k-th neutron precursor on each point of the reactor (it is as well a function of time and position). $\lambda_k \mathcal{C}_k$is the decay rate of the k-th neutron precursor.
- K is the number of neutron precursors.
- β_k is the proportion of fission neutrons given by transformation of the k-th neutron precursor; $\beta = \sum_{k=1}^{K} \beta_k$.

$\mathcal{L}$ models the diffusion $(-\nabla \cdot (D_1 \nabla))$, absorption $(\Sigma_{a1}, \Sigma_{a2})$,and transfer from fast group to thermal group (Σ_{12}).
$\mathcal{M}$ models the generation of neutrons by fission.
$\nu\Sigma_{fg}$ gives the amount of neutrons obtained by fission in group g.
v^{-1} gives the time constants of each group.

In addition, the equations governing the concentration of precursors are:

$$\frac{\partial \mathcal{C}_k}{\partial t} = \beta_k[\nu\Sigma_{f1} + \nu\Sigma_{f2}]\phi - \lambda_k \mathcal{C}_k \quad k = 1, \cdots, K \tag{2}$$

The diffusion terms makes (1) a partial differential equation; in fact, it can be seen as two coupled time-dependent diffusion equations with source terms. Equation (2) models the time evolution of the K neutron precursors, which act as source in (1).

It is assumed that there is transfer of neutrons from the fast group to the thermal group (modelled by the term Σ_{12}) but not from the thermal to the fast group. Also, the precursors contribute to the fast group but not to the thermal group, and it is finally assumed that only fast neutrons are produced by fission ($\mathcal{M}$ operator). A more detailed description of the problem can be found in [4, 5, 6, 7].

It is customary to simplify the system by making some assumptions about (2), so that (2) can be replaced with a simple algebraic expression which gives $\mathcal{C}_k$ as a linear function of ϕ (See [4, 5, 6]). So, independently from the spatial discretization method, if (1) and (2) are discretized with N unknowns the resulting ODE systems (and the linear systems) will have dimension $2N$ (N for the fast neutrons, N for the thermal neutrons).

2.2 Spatial Discretization Techniques

Usually, the resolution of these equations is tackled carrying out first the spatial discretization, either by standard methods (finite differences, finite elements) or by special collocation methods. Once the spatial discretization is performed, the result is a stiff linear ODE system of large dimension. This system can be written exactly as equation (1), but substituting the flux ϕ and the operators $\mathcal{L}$ and $\mathcal{M}$ by their discretized versions: ϕ becomes a vector and the operators become the matrices L and M.

In the previous work [8], the discretization methods used were centered finite differences and a nodal collocation method, based on Legendre polynomials. However, the parallel codes based on Legendre polynomials are still under development and the results presented here shall be only those obtained with centered finite differences.

Applying centered finite differencing to the diffusion operator in the operator $\mathcal{L}$ results in the standard matrix with five non-zero diagonals in the two-dimensional case and with seven non-zero diagonals in the three-dimensional case. The discretization of $v\Sigma_{f1}$, $v\Sigma_{f2}$, Σ_{a1} , Σ_{a2} ,and Σ_{12} results in diagonal matrices, so that (for a small two-dimensional case) the structure of nonzero elements in matrices L and M is like the shown in Fig. 1.

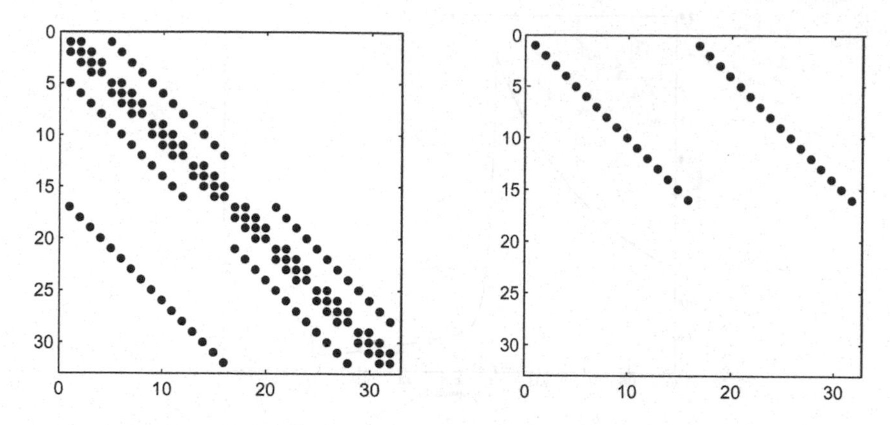

Fig. 1. Matrices L and M for Finite Diffferences dicretization

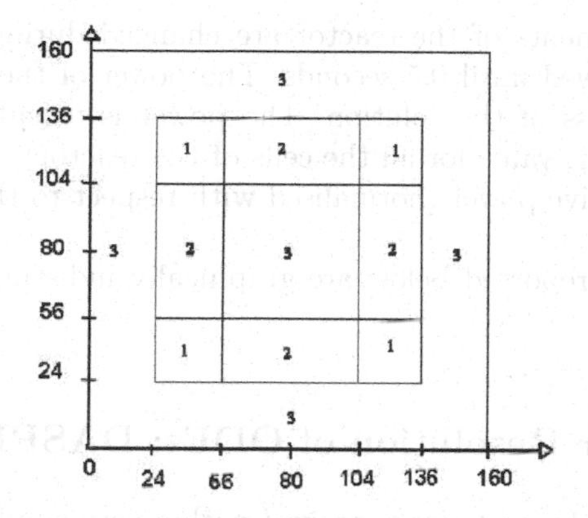

Fig. 2. Geometry of TWIGL reactor

An advantage of the finite differencing scheme is that the matrix L can be stored in diagonal format which allows very fast products matrix-vector and easy implementations of standard methods such as Gauss-Seidel, SOR or Multigrid. However, the accuracy obtained is not very good (compared with other discretization methods), so that the number of cells needed is quite large; and, therefore, the ODE system becomes quite large as well.

2.3 Test Case(s)

The problem used as test is a simplified two-dimensional reactor known as Seed-Blanket or TWIGL, which has three different materials and a single neutron precursor. Again, a detailed description can be found in [4, 5, 6].

The problem starts from a steady-state solution (which is found solving a generalised eigenvalue problem). From the steady-state solution, the properties

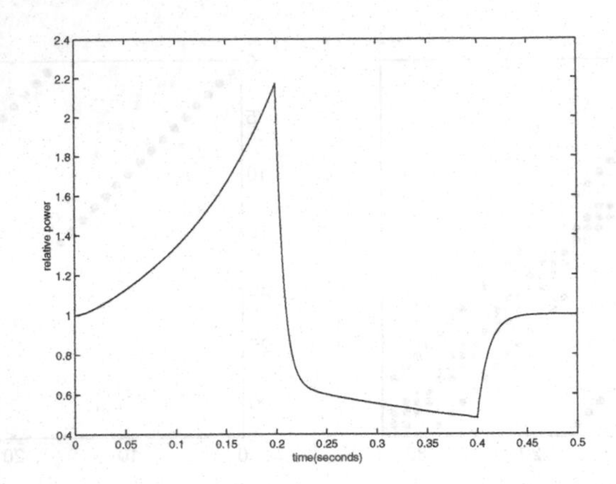

Fig. 3. Relative Power in TWIGL Transient

of one of the components of the reactor are changed during 0.4 seconds, and the transient is followed until 0.5 seconds. The power of the reactor is used to calibrate the goodness of the solution. The power is calculated as $\nu\Sigma_{f1}\phi_1 + \nu\Sigma_{f2}\phi_2$ (summing this value for all the cells of the reactor). Figure 3 shows the evolution of the relative power (normalised with respect to the initial power) in this problem.

All the solutions reported below are graphically indistinguishable from the displayed in Fig. 3.

3 Software for Resolution of ODEs: DASPK, FCVODE

DASPK and FCVODE are codes conceived with a very similar structure. First, the common characteristics shall be described, and next the differences shall be highlighted.

As mentioned above both are variable step, variable order codes, which can solve large systems of stiff ODE systems. This resolution is made combining BDF (Backward Differentiation Formulae) formulae with orders 1 to 5 with the Newton method to solve the nonlinear systems that should be solved in each time step.

Both offer similar options to solve the linear systems that appear during the resolution of the nonlinear system: LU decomposition, LU banded or an iterative method for large systems, the preconditioned GMRES (Generalised Minimum Residual). Since our problem is a large one,in this work we have chosen always the preconditioned GMRES solver. As will be described later, the routines that must provide the user are quite similar in both cases. Now follows a brief description of the details specific to each code:

FCVODE: It is a FORTRAN interface for the code CVODE, written in C. It is a part of the package SUNDIALS, written and developed in the Lawrence

Livermore National Laboratory. This package includes as well a nonlinear solver, an ODE solver computing sensibilities and a DAE (Differential-Algebraic Equations) solver.
FCVODE on its own only solves ODE systems, and can do it with the described stiff combination (BDF+Newton) or with a non-stiff combination (Adams-Moulton formulae with orders 1 to 12, combined with fixed point iteration to solve the nonlinear systems) . If (in the stiff case) the GMRES solver is used, the user can choose among several options:

1. No preconditioning.
2. Provide a single routine which computes the preconditioner and apply it.
3. Provide two routines, one for computing the preconditioner and other for applying it.
4. Provide three routines; two like in option 3, and another for computing a product matrix-vector, where the matrix is the Jacobian of the system.

FCVODE has a parallel version based on MPI (Message Passing Interface) [9], based on the Single Program-Multiple Data (SPMD) Paradigm. We will describe this in detail later.

DASPK: DASPK is a code fully written in Fortran 77, which must be credited mainly to Professor L. Petzold. This code can solve ODE systems but is designed with the more difficult goal of solving Differential-Algebraic systems of equations (DAEs). Due to this orientation, DASPK does not have a non-stiff option (useless for solving DAEs). It can also compute sensibilities (derivatives) of the solution with respect to any parameter with influence on the solution. It offers similar options for the GMRES solver to those of FCVODE. Viewed as a solver, DASPK is a purely sequential solver, although it can use MPI for computing sensitivities in parallel.

3.1 User Subroutines

The user must provide in all cases a subroutine that describes the system, which we will name RES; and optionally others for computing the preconditioner, applying the preconditioner or computing a matrix-vector product. In our case (apart from the routine RES) only one is needed, which has to compute and apply the preconditioner in the same routine; we will name this subroutine PRECON.

Both subroutines are called by FCVODE or DASPK several times in each time step; the number of calls will depend on many details; speed of convergence of linear iterations, speed of convergence of nonlinear iterations, local behaviour of the system, and many others. In an "easy" step, RES and PRECON may be called only 3 or 4 times; in a "difficult" step, they can be called 30 or more times. Clearly, the coding of these two subroutines is crucial for the overall efficiency of the code.We will describe the subroutine for FCVODE only, since the subroutines for DASPK are quite similar.

FCVODE Subroutines: FCVODE is designed to solve systems which can be expressed as:

$$\dot{Y} = \frac{dY}{dt} = F(t, Y)$$
$$Y(t_0) = Y_0$$

In our case, Y would be the vector of neutron fluxes $\begin{pmatrix} \phi_1 \\ \phi_2 \end{pmatrix}$, so that the subroutine RES must take as arguments the actual time t and the flux vector Y and return $F(t, Y)$. This means reordering (1) and computing:

$$\dot{\phi} = F(t, \phi) = \nu \left(-L\phi + (1 - \beta)M\phi + \sum_{k=1}^{K} \lambda_k \mathcal{C}_k \chi \right) \tag{3}$$

The PRECON subroutine must be used to solve (at least, approximately) a system of the form $(I - \gamma J)\,x = r$, where I is the identity matrix, γ is a real constant and J is the Jacobian of the system: $J = \frac{\partial F}{\partial Y}$. The right hand side of the system r and the constant γ are inputs of the routine, which must return the approximation to the solution x. In this case, since our ODE system is linear, the Jacobian is simply:

$$J = \nu \left(-L + (1 - \beta)M + \sum_{k=1}^{K} \lambda_k \frac{\partial \mathcal{C}_k}{\partial \phi} \chi \right)$$

which can be computed and stored easily.

4 Sequential Implementations; Results

4.1 Implementations

The sequential implementations are (compared with the parallel ones) conceptually simple. The main decision to be made is the choice of preconditioner. It must be remembered that the role of the preconditioner is to improve the efficiency of the built-in preconditioned GMRES method, not to solve the system. Therefore, some features that should be included if we were writing a full solver (such as convergence control) are not really needed.

The most straightforward choices are point Gauss-Seidel and SOR methods. These have the advantage of the ease of implementation, especially when the matrices are stored with Diagonal format. We have implemented as well a Line Gauss-Seidel preconditioned and a simplified Multigrid preconditioner(this is a standard multigrid code but only with three grids, the original one and two coarser grids). The best results were obtained with our simplified Multigrid

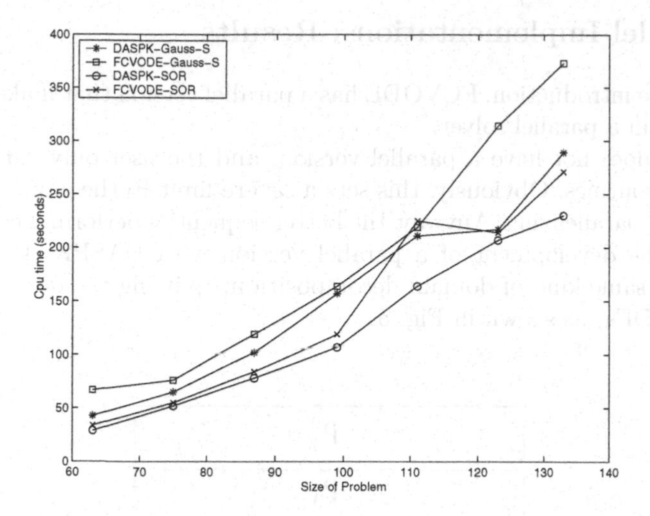

Fig. 4. CPU Time for TWIGL transient with sequential codes

preconditioner, followed by point SOR and Gauss-Seidel (Line Gauss-Seidel did not work well in this problem). However, the multigrid parallel version has not been developed yet, so that for this paper the tests have been made with point SOR and Gauss-Seidel preconditioning.

Therefore, the results below have been obtained with a DASPK-based sequential code, which shall be named SEQ-DAS, and a FCVODE-based sequential code, named SEQ-FCV. Both solve the TWIGL transient using Gauss Seidel or SOR preconditioning (10 Gauss-Seidel or SOR iterations on each preconditioning call), compiled with the same options and using the same tolerances.

4.2 Results

All the tests have been performed in a cluster of PC's, formed by twelve servers IBM xSeries 330 SMP, mounted in a Netfinity Rack. Each one of the twelve servers includes two 866 Mhz Pentium III processors with 256 KB of cache and 512 MB of 133MHz RAM memory.

Figure 4 shows the CPU time consumed by the SEQ-DAS code and by the SEQ-FCV code in our test case using several discretizations of increasing size. It is clear that the DASPK-based code obtains better results.

The problem has been discretized with different mesh sizes, starting from cases where the accuracy was considered acceptable: that is, with more than 50*50 cells. The actual sizes chosen have been 63*63, increasing until 133*133. For all these cases, the results of the relative power of the reactor are practically indistinguishable form the results in Fig. 3. In the last case, the dimension of the ODE system is 133*133*2=35378 equations.

5 Parallel Implementations; Results

As said in the introduction, FCVODE has a parallel version that makes relatively easy to obtain a parallel solver.

DASPK does not have a parallel version, and the user only can parallelize his own subroutines. Obviously, this sets a severe limit to the gains that can be obtaining by parallelizing. Anyway, the better sequential performance of DASPK motivated the development of a parallel version with DASPK. Both versions perform the same kind of domain decomposition, splitting the domain (and the system of ODEs) as shown in Fig. 5.

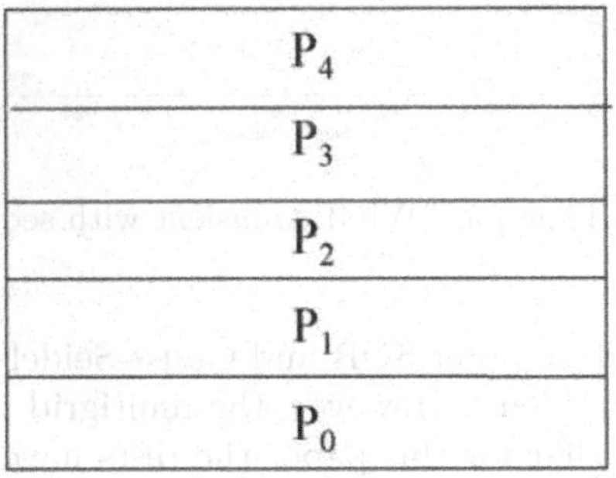

P_4
P_3
P_2
P_1
P_0

Fig. 5. Domain Decomposition of the TWIGL Reactor with 5 Processors

The reactor is split in horizontal stripes, as many as processors available. Each processor has to perform the calculations corresponding to the unknowns in its stripe, both in the RES and PRECON subroutines. This means also that the matrices L and M are split also in stripes, each containing a group of contiguous rows.

First we will describe the parallel version of FCVODE, which shall be named PAR-FCV, and afterwards we will describe the DASPK parallel version (PAR-DAS).

5.1 Parallel Implementation Based on FCVODE

A program wich calls the parallel version of FCVODE has to follow several steps. First, must initialize MPI; then, the needed data and initial conditions must be sent to each processor, and some calls to initialization of FCVODE must be performed. (These steps must be programmed by the user). Then FCVODE can be called and the numerical integration of the system starts. Each processor performs the numerical integration of its part of the system, and all the communications needed are handled by FCVODE, except those needed in RES and PRECON, which must be coded explicitly by the user.

In the RES subroutine, each processor has the values of the unknowns vector (Y) corresponding to its zone, and must evaluate f(t,Y) (see (3)) for its set of

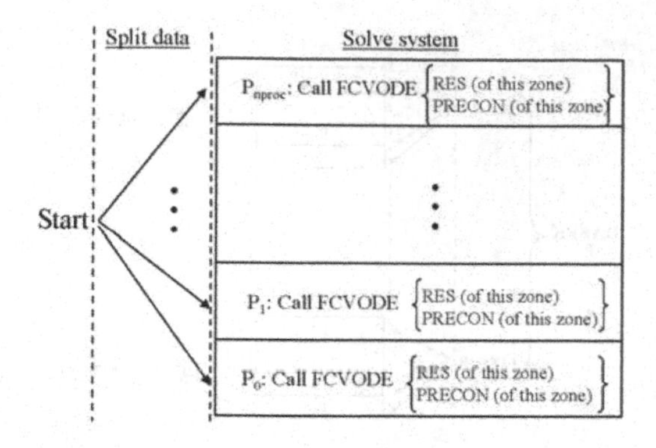

Fig. 6. Data decomposition in PAR-FCV

equations. In our case, given the domain decomposition chosen, and taking the vector Y as a matrix (`Y(1:meshx,1:meshy)`), each processor owns a group of rows of the matrix (`Y(1:meshx,InitialRow:FinalRow)`).

An update line of the RES subroutine looks like this (recall that the matrices are stored as diagonals):

```
do j=InitialRow,FinalRow
 do i=1,meshx
  res(i,j)=Y(i,j)*main_diag(i,j)+
         Y(i-1,j)*diag_left(i,j)+Y(i+1,j)*diag_right(i,j)+
         Y(i,j-1)*diag_down(i,j)+Y(i,j+1)*diag_up(i,j)+
          .../*other terms*/
 end do
end do
```

If there are N processors and the domain decomposition is like in Fig. 5, then all processors (but the 0) need to receive the row of unknowns located just below of its zone; that is, the row of unknowns `InitialRow-1`; and all processors (but the N-1) need to receive the row of unknowns inmediately above (`FinalRow+1`). These communications must be programmed by the user.

However, the calculations for all inner rows (from the row `InitialRow+1` to the row `FinalRow-1`) do not need communication. Using the non-blocking MPI calls MPI_ISend and MPI_IReceive becomes possible to overlap communications and computing; while the sends of the upper and lower row and the receives of neighbouring rows are performed, each processor can perform the calculations corresponding to the inner part of its domain.

The completion of the sends, receives and computing of the values of rows `InitialRow` and `FinalRow` is deferred until the calculations of the inner part have been carried out.

The implementation of the Gauss_Seidel or SOR preconditioner follows the same guidelines; however, for a "true" SOR or Gauss_Seidel iteration the com-

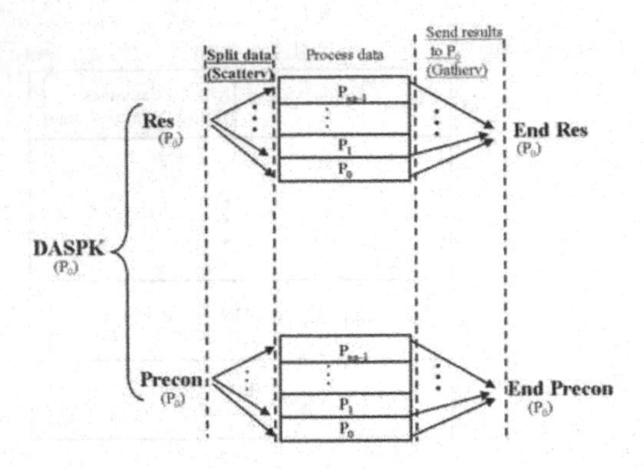

Fig. 7. Data Decomposition in PAR-DAS

munication should take place in every iteration. These communications overhead can be reduced by limiting the number of sends and receives of the data from "neighbor" processes, so that this communication takes place only in certain iterations. Through experimentation, it has been been found that a good choice is to perform the communications only in the 1st and 5th iteration.

5.2 Parallel Implementation Based on DASPK

As said above, DASPK does not have a parallel version. However, motivated by the good performance of DASPK in the sequential case, a parallel version was written in which DASPK runs in a single processor but the user subroutines RES and PRECON have been parallelized.

This means that both RES and PRECON are invoked from a single processor (the one in which DASPK is running). Then a parallel version of these subroutines must start by spreading the data needed over all the available processors (which in our case was done with the MPI_Scatterv subroutine) and must end collecting the results from all processors (which was done with the MPI_Gatherv subroutine). Once the communication is carried out, the rest of the code is nearly identical to the written for PAR-FCV.

This approach was not very sucessful, since the amount of data to be sent was much larger than in the FCVODE case, and because the sends and receives have to be necessarily blocking. Also, all the internal work in DASPK is made sequentially, where in FCVODE most of these calculations are distributed.

5.3 Results

The parallel codes were tested solving the TWIGL transient. The problem has been solved using several mesh sizes, but, due to lack of space, the results presented here are only the most relevant ones, the obtained discretizing the domain with 160*160 cells, giving a total of 160*160*2= 51200 equations. The precon-

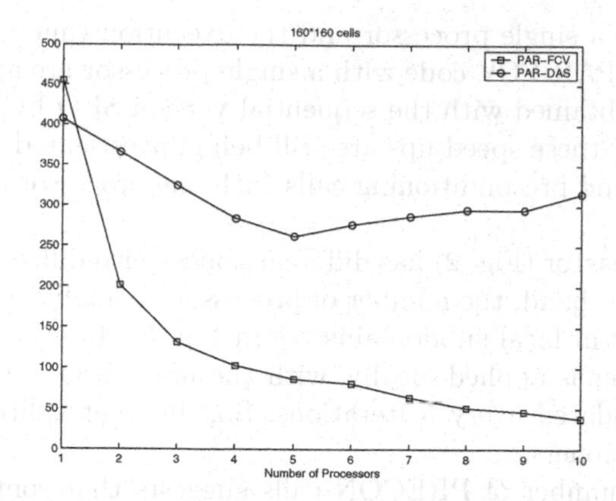

Fig. 8. CPU Time for TWIGL transient with parallel codes in a 160*160 discretization

Table 1. CPU time, number of time steps, and PRECON Calls for PAR-FCV code in the case with 160*160 cells

Processors	CPU time	time steps	PRECON Calls
1	455	302	2083
2	202.6	272	1856
3	132.2	261	1829
4	103.8	292	2030
5	86.4	317	2139
6	81.3	351	2398
7	63.8	328	2237
8	51.1	310	2130
9	46.3	309	2102
10	38.2	293	1960

ditioner chosen was SOR, and the transient was run using different number of processors, from 1 to 10. The timings for this case for PAR-FCV and PAR-DAS are shown in Fig. 8, and the timings, number of time steps and PRECON Calls with PAR-FCV are shown in table 1.

Figure 8 shows an excellent performance of the PAR-FCV code, while the PAR-DAS code, limited by the heavy communication, cannot get even close to the PAR-FCV results. These show that the approach chosen is quite promising, and can be used for three-dimensional simulations.

However, there are some surprising points about these results, which are discussed below.

Table 1 shows that the PAR-FCV code executed in a single processor performs worse than when executed in several processors (though the solution obtained is equally accurate). In the table, it can be seen that the speed-up for N processors ($2 < N < 10$) is in many cases greater than the perfect speed-up N (Recall that the speed-up for N processors is defined as the quotient between the execution time for a single processor and the execution time for N processors). The results of the PAR-FCV code with a single processor are approximately the same than those obtained with the sequential version SEQ-FCV.

The reasons for these speed-ups are still being investigated. The larger number of time steps and preconditioning calls with one processor suggests a partial explanation:

The TWIGL reactor (Fig. 2) has different zones, with different physical properties. On the other hand, the number of processors available determines a splitting of the reactor in local subdomains (as in Fig. 5). In every subdomain, the SOR preconditioner is applied locally, with the unknowns of the neighbor subdomains being updated every 5 iterations. So, different splittings give rise to different preconditioners.

The different number of PRECON calls suggests that some splittings "fit" the geometry of the TWIGL reactor better than others, and that, at least for this case, these splittings can be better that no splitting at all. It has been confirmed (through a sequential code that simulates the parallel communication between subdomains) that in this problem, this kind of splitting enables FCVODE to complete the transient in fewer steps than using a full SOR preconditioning (the used when PAR-FCV is run in a single processor). It also decreases the average CPU time to compute a time step.

However, not all the extra CPU time can be accounted for with this explanation, so that this is only a partial explanation, and other possible causes (memory effects) are being investigated.

6 Conclusions and Future Work

The results (and our experience) show that, while DASPK performs somewhat better in sequential solving, the parallel FCVODE code gives a tool for solving really large systems, since its parallel version can obtain excellent speed-ups.

In [8] it was shown that our sequential codes were faster (and more reliable) than fixed time step codes, reported in [6]. Despite the strange details commented in Sect. 5.3, the results show that the parallel version of FCVODE is a basis to speed-up further the resolution process.

The future work (already going on) will be to obtain a three-dimensional parallel version of the code, based on the nodal collocation method, and apply it to real reactors.

Acknowledgements

This work has been partially supported by the Spanish Ministerio de Ciencia y Tecnología and the projects FEDER DPI2001-2766-C02-01 and FEDER DPI2001-2766-C02-02.

References

1. W.L. Briggs: A Multigrid Tutorial. SIAM, Philadelphia, 1987.
2. P.N. Brown, A.C. Hindmarsh and L.R. Petzold: Using Krylov methods in the solution of large-scale differential-algebraic systems. SIAM J. Sci. Comput. 15(6), pp. 1467-1488, November 1994
3. A.C. Hindmarsh and Allan G. Taylor, PVODE and KINSOL: Parallel Software for Differential and Nonlinear Systems, Lawrence Livermore National Laboratory report UCRL-ID-129739, February 1998.
4. R. Bru, D. Ginestar, J. Marín, G. Verdú, and V. Vidal: A second degree method for the neutron diffusion equation. In F. J. Cobos, J. R. Gomez, and F. Mateos, editors, Encuentro de Análisis Matricial y Aplicaciones, EAMA97, pages 87-93, Sevilla, September 1997. Universidad de Sevilla.
5. D. Ginestar, G. Verdú, V. Vidal, R.Bru, J. Marín and J.L. Munoz-Cobo: High Order Backward Discretization for the Neutron Diffusion Equation. Ann. Nucl. Energy, 25(1-3):47-64, 1998.
6. J. Marín: Métodos Numéricos para la Resolución de la Ecuación de la Difusión Neutrónica Dependiente del Tiempo. PhD Thesis, Universidad Politécnica de Valencia, 2000.
7. V. Vidal: Métodos Numéricos para la Obtención de los Modos Lambda de un Reactor Nuclear. Técnicas de Aceleración y Paralelización. PhD Thesis, Universidad Politécnica de Valencia, 1997.
8. V.M. García, V. Vidal, J. Garayoa, G. Verdú and R. Gómez: Fast Resolution of the Neutron Diffusion Equation Through Public Domain Codes. International Conference on Super Computing in Nuclear Applications (SNA 2003) Paris 2003.
9. W. Gropp, E. Lusk, and A. Skjellum: Using MPI Portable Parallel Programming with the Message-Passing Interface The MIT Press, Cambridge, MA, 1994.

FPGA Implementations of the RNR Cellular Automata to Model Electrostatic Field*

Joaquín Cerdá-Boluda[1], Oscar Amoraga-Lechiguero[2], Ruben Torres-Curado[2], Rafael Gadea-Gironés[1], and Angel Sebastià-Cortés[1]

[1] Group of Digital Systems Design, Dept, Of Electronic Engineering, Universidad Politécnica de Valencia, 46022 Valencia, Spain
{joacerbo, rgadea, asebasti}@eln.upv.es
http://dsd.upv.es

[2] Telecommunication School, Universidad Politécnica de Valencia, 46022 Valencia, Spain
{osamlec, rutorcu}@teleco.upv.es

Abstract. The classic way to describe electrostatic field is using Partial Differential Equations and suitable boundary conditions. In most situations we need numerical methods mainly based on discretisation of time and space. In this paper we follow a different approach: we introduce RNR Cellular Automata that is capable of modeling the electrostatic field at macroscopic level. Iterations of the automata are averaged on time and space to get continuous variables. We implement RNR Cellular Automata with FPGA and compare the performance with the results using classic sequential programming. We get some regular architectures specially adapted to fully-parallel machines, and explore its benefits and drawbacks.

1 Introduction

Cellular Automata (CA) model massively parallel computation and physical phenomena [1]. They consist of a lattice of discrete identical sites called *cells*, each one taking a value from a finite set, usually a binary set. The value of the cells evolve in discrete time steps according to deterministic rules that specify the value of each site in terms of the values of the neighboring sites. This is a parallel, synchronous, local and uniform process [1, 2, 3].

CA are used as computing and modeling tools in biological organization, self-reproduction, image processing and chemical reactions. Also, CA have proved themselves to be useful tools for modeling physical systems such as gases, fluid dynamics, excitable media, magnetic domains and diffusion limited aggregation [2, 4, 5, 6]. CA have been also applied in VLSI design in areas such as generation of pseudorandom sequences and their use in built-in self test (BIST), error-correcting codes, private-key cryptosystem, design of associative memory and testing the finite state machine [7, 8].

* Candidate to the best student paper award.

M. Daydé et al. (Eds.): VECPAR 2004, LNCS 3402, pp. 382–395, 2005.

This paper has two main goals: in one hand, we will present the RNR Cellular Automata, that is capable of modeling electrostatic phenomena. On the other hand, we will develop some computational architectures orientated to the particular problem, and implement them with dedicated hardware. This will show the convenience of using special hardware when working in a fully parallel environment and the increase of performance that is obtained.

2 Electrostatic Field and the RNR Cellular Automata

Electrostatic field in a space free of charge [9] can be completely described by one of the oldest and most important equations in physics: the Laplace Equation

$$\nabla^2 \Phi = 0 \ , \tag{1}$$

where Φ is the *Electrostatic Potential*. This equation also appears in several physical phenomena. Its time dependant version describes the diffusion of particles and heat. It is also useful for the study of incompressible, nonviscous fluids. It corresponds, in imaginary time, to the Schrödinger equation of quantum mechanics for a free particle. Recently, it has been also considered to play a major role in fractal growth processes such as diffusion limited aggregation and the dielectric breakdown model.

Almost every problem in electrostatics can be solved applying the necessary boundary conditions to (1) and then using classical methods to obtain the analytical solution. Although this procedure is simple from a conceptual point of view, in the practice mathematical work can be difficult and, in some situations, completely impossible. This is the reason why many numerical methods have been developed to solve the problem, at least in an approximate way. These methods are, generally, based on the idea of discretisation: both time and space are discretised on a lattice. We replace the function itself by a discrete approximation in a 2D square lattice of N-by-N nodes

$$\Phi(x,y) \longrightarrow u_{i,j} = \Phi(ih, jh) \ . \tag{2}$$

$h=L/N$ is the space between nodes on the lattice. Most numerical methods are based on solving this discretised system by Gaussian reduction or by iterative methods [10]. In this way we can mention the *Liebmann Method* that, essentially, resorts to iterating the equation

$$u_{i,j} = \frac{u_{i+1,j} + u_{i-1,j} + u_{i,j+1} + u_{i,j-1}}{4} \tag{3}$$

until two successive configurations differ by less than some small δ.

In the following we will focus our attention on solving the Laplace Equation in a rectangular region described by

$$(x,y) \in \{[0, L_x], [0, L_y]\} \tag{4}$$

in which boundary conditions are

$$\begin{array}{c}\Phi(0,y)=\Phi(L_x,y)=\Phi(x,0)=0\\ \Phi(x,L_y)=\Phi_0\end{array} \tag{5}$$

It is well-known that the analytical solution for this problem is

$$\Phi(x,y)=\Phi_0\cdot\sum_{n=1}^{\infty}\frac{4}{n\pi}\sin\left(\frac{n\pi x}{L_x}\right)\frac{\sinh(n\pi y/L_x)}{\sinh(n\pi L_y/L_x)}(n\neq\dot{2})\ . \tag{6}$$

This solution consists of an infinite summation that is very difficult to operate with. Several terms of the summation are needed to obtain good numerical results, especially near the boundaries, where some disturbing oscillations appear, due to discontinuities[1]. Using (3) on the discretised lattice, or directly iterating (6), we obtain the numerical solution that is depicted in Fig. 1 (values are normalized for Φ_0=1).

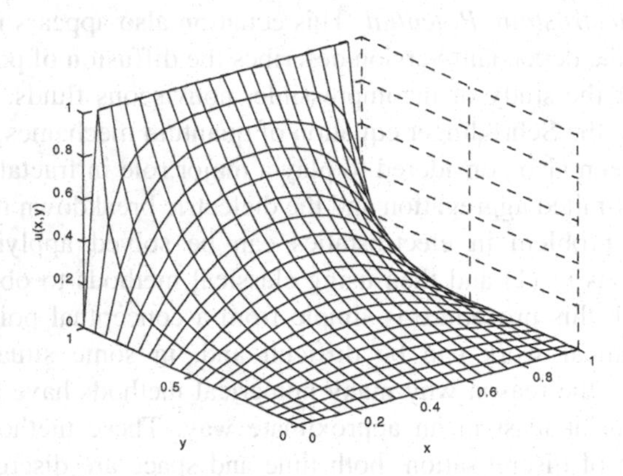

Fig. 1. Solution of the Laplace Equation on a 2D square region using the Liebmann Method

Computationally efficient methods of solution (as iterating (3)) are based on the discretisation of space and time. There is also the possibility of replacing the value of the function itself with a discrete version, so that variables at each point of the grid are only allowed to range over "0" and "1". This leads us to the field of Cellular Automata.

2.1 RNR Cellular Automata

In [11] we introduce the *Random Neighbor Rule* (RNR) Cellular Automata. This Cellular Automata is based on a probabilistic approach in the fashion of *Lattice Boltzmann Automata* [12] but in a much simpler way. It has been designed to behave as a solution of the Laplace Equation at macroscopic level. The rule uses the von Neumann Neighborhood[2] of radius 1, and its behavior can be summarized as follows:

---

[1] These oscillations are characteristic of Fourier Series based solutions.

[2] To be precise, von Neumann neighborhood includes also the central cell, that the RNR rule does not need to have.

```
for all cells in the lattice:
        randomly choose one of the four neighbors
        copy its value
```

Some steps in the evolution of this Automata are shown in Fig. 2. In this simulation we used a two-dimensional lattice of 50x50 connected cells, each cell implementing the RNR rule, with boundary conditions (3) (assuming Φ_0=1). We started from a random initial configuration with $p_{i,j}(1)$=1/4 ($p_{i,j}(1)$ is the probability for the cell (i,j) of being in the state '1').

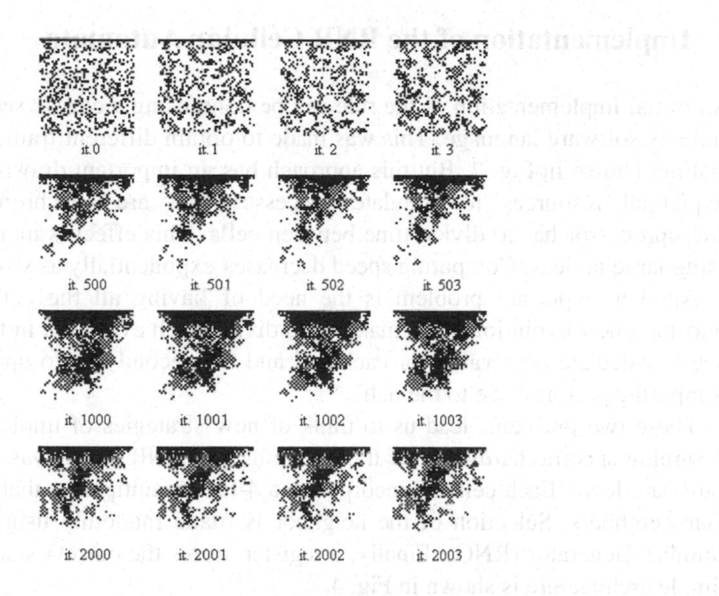

Fig. 2. Some steps in the evolution of the automata in a 50x50 matrix

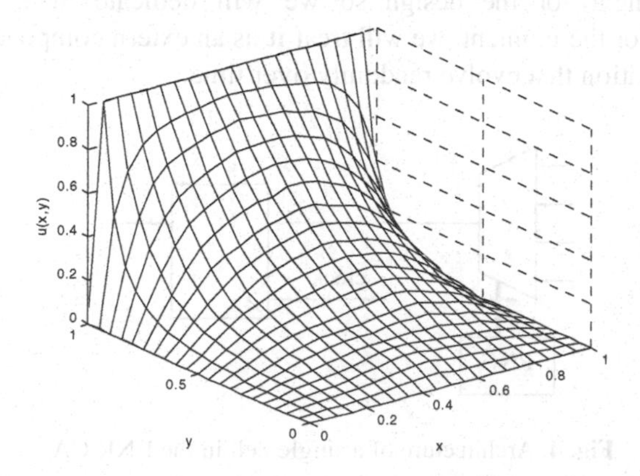

Fig. 3. Average value of a lattice of 20x20 cells after 5000 generations

The frames given in Fig. 2. show that, after some transient steps, the system evolves to a stationary situation where cells near the upper boundary have greater probability of being '1'. If we average over time, the result is very similar to Fig.1. For example, Fig. 3 is the average on a 20x20 lattice after 5000 generations. Similarity with Fig. 1 is notorious.

In [11] we show that, as the number of generations or cells averaged increases, the system follows the analytic solution more accurately.

Similar results can be found for a wider range of different boundary conditions.

3 Implementation of the RNR Cellular Automata

An initial implementation of the rule can be done using a classic sequential computer and any software language. This was made to obtain different frames in the evolution that are shown in Fig. 2. But this approach has an important drawback: we are using sequential resources to simulate processes that are synchronous in time. A microprocessor has to divide time between cells. This effect is more important when using large lattices. Computing speed decreases exponentially as size increases.

Another important problem is the need of having all the cells updated before starting a new evolution. This make us to divide each evolution in two steps: the first one to calculate next value for each cell and the second one to update. This reduces computing performance to the half.

These two problems lead us to think of new strategies of implementation, maybe designing specific hardware for the problem. The RNR rule is easily implemented at hardware level. Each cell can incorporate a 4-input multiplexer that selects one of the four neighbors. Selection of the neighbor is made randomly using a 2-bit Random Number Generator (RNG). Finally, a register stores the current state of the cell. This simple architecture is shown in Fig. 4.

In this cell, the only point that needs some explanation is the RNG. This is a very important component for the design so we will dedicate Sect. 4 to study its implementation. For the moment, we will treat it as an extern component that supplies two bits of information that evolve randomly over time.

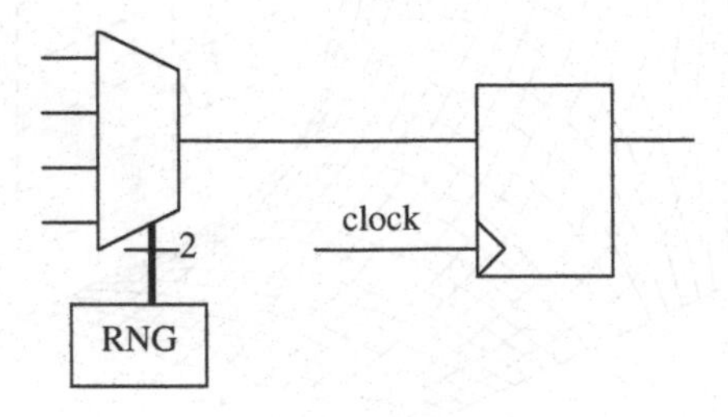

Fig. 4. Architecture of a single cell in the RNR CA

Using the cell depicted in Fig. 4, the lattice of cells implemented is shown in Fig. 5. In this picture each cell is labeled as "RNR Rule". We can see the two-dimensional arrangement of cells that defines the connectivity used. According to it, each cell is connected with its nearest neighbors, placed up, down, left and right. Note that some particular cells have been placed in the boundaries to supply the suitable Boundary Conditions.

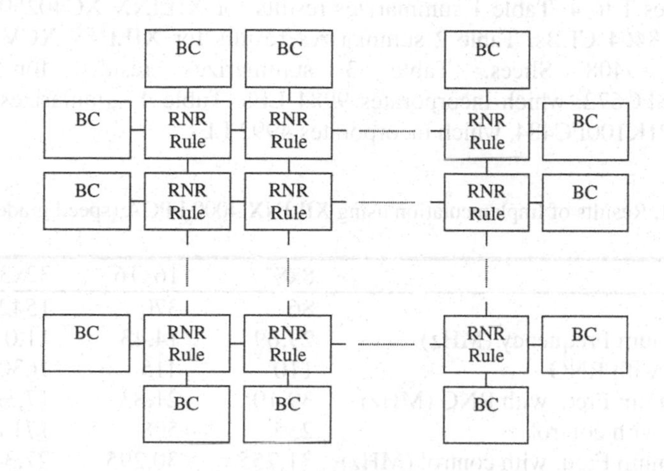

Fig. 5. Lattice of RNR Cellular Automata cells to obtain the complete cellular space

The proposed architecture was implemented on several FPGA families to compare performances. For the logic synthesis we used Leonardo Spectrum from Exemplar Logic. This allows us to synthesize on different families from several manufactures. The results shown here are those obtained for ALTERA and XILINX devices. For the ALTERA devices the implementation and simulation was performed with MAX+PLUSII. Implementation and simulation on XILINX devices was performed using FOUNDATION SERIES. Results of the implementation are shown in Tables 1 to 4, each for a different FPGA family.

As a first stage for our design we implemented only the lattice of cells (without RNG) and obtained values of Maximum Frequency reached and Logic Cells (LCs) or Configurable Logic Blocks (CLBs) needed for each device and size of the matrix[3]. After that, we included the RNG and checked how those values changed.

At this point we have a system that evolves correctly, but it is embedded inside the FPGA. There is no system for accessing to the value of the cells and managing data. To achieve this, we have designed a little control unit that selects cells sequentially and presents their states as an output by means of an standard serial transmission. Accessing to the whole matrix is performed using 10-bit frames with transmission rate of 19200 bauds. It is important to remark that each frame consists on 8 data bits and 2 control bits. This allows us to calculate transmission times using the formula

[3] These results are given in different units. CLBs are used for XILINX 4000 devices. Slices are used for XILINX VIRTEX devices. ALTERA devices use the concept of Logic Element (LE). A detailed comparison of structures is given in [11].

$$t = \frac{N}{15360}, \tag{7}$$

where t is the time for a complete transmission (given in seconds) and N is the total size of the matrix (i. e., the number of cells). Results for the complete system are given in tables 1 to 4. Table 1 summarizes results for XILINX XC40250XV, which incorporates 8464 CLBs. Table 2 summarizes results for XILINX XCV800, which incorporates 9408 Slices. Table 3 summarizes results for ALTERA EPF10K200SFC672, which incorporates 9984 LEs. Table 4 summarizes results for ALTERA EP1K100FC484, which incorporates 4992 LEs.

Table 1. Results of Implementation using XILINX 4000 FPGA (speed grade –07)

Size	8x8	16x16	32x32
CLBs	86	370	1542
Maximum Frequency (MHz)	29,69	14,43	11,03
CLBs with RNG	110	415	1630
Maximum Freq. with RNG (MHz)	39,10	21,83	17,38
CLBs with control	235	508	1714
Maximum Freq. with control (MHz)	31,755	30,295	27,345
Evolution + Transmission time (ms)	4,167	16,667	66,667

Table 2. Results of Implementation using XILINX VIRTEX FPGA (speed grade –4)

Size	8x8	16x16	32x32
Slices	95	362	1489
Maximum Frequency (MHz)	80,85	59,34	48,473
Slices with RNG	128	395	1516
Maximum Freq. with RNG (MHz)	74,538	58,05	65,28
Slices with control	281	553	1710
Maximum Freq. with control (MHz)	85,34	67,40	57,195
Evolution + Transmission time (ms)	4,167	16,667	66,667

Table 3. Results of Implementation using ALTERA FLEX10KE FPGA (speed grade –1)

Size	8x8	16x16	32x32
LEs	192	768	3072
Maximum Frequency (MHz)	105,26	84,74	94,33
LEs with RNG	224	896	3584
Maximum Freq. with RNG (MHz)	102,04	97,08	96,15
LEs with control	355	1095	4046
Maximum Freq. with control (MHz)	95,23	80,64	70,42
Evolution + Transmission time (ms)	4,167	16,667	66,667

Table 4. Results of Implementation using ALTERA ACEX1K FPGA (speed grade –1)

Size	8x8	16x16	32x32
LEs	192	768	3072
Maximum Frequency (MHz)	138,88	136,98	116,27
LEs with RNG	224	896	3584
Maximum Freq. with RNG (MHz)	140,84	117,64	107,52
LEs with control	355	1095	4046
Maximum Freq. with control (MHz)	105,26	102,98	88,49
Evolution + Transmission time (ms)	4,167	16,667	66,667

It is important to note that the evolution of the whole matrix is performed in only one clock cycle, so this time is completely negligible when compared to transmission time. This is the reason why values given as "Evolution + Transmission Time" seem to fit perfectly with formula (7). As it is evident from the given data, with the inclusion of transmission our system loose its best feature: concurrency. Though our system continues evolving synchronously in a single clock period (independently of the size), sequential transmission depends on size. This fact is the reason why total time increases exponentially with size.

To overcome this drawback, we will introduce some modifications in the basic cell. It must be clear, from the considerations given above, that we are interested in average values rather than particular ones. So, we don't want to know the state of cell (x,y) in the step i. We want to know how many times that cell (x,y) has been in a particular state in the last N iterations. Therefore we can explore the possibility of not transmit each generation, but the average value of a cell after N generations.

Fig. 6 shows the modified cell. In each cell we include a counter to store the value. Average value is performed by a simple addition. This way, the counter increases in 1 when the state of the cell is '1', and remains unchanged if the state of the cell is '0'. So, the counter tells how many times that particular cell has been in state '1' in the last N generations.

In the modified cell there is an important parameter that is not fixed: the size of the counter. Intuitively, as the size of the counter increases, the cell also increases its complexity. So, we will need more Logic Cells to implement our system. On the other hand, speed increases with the size of the counter because we need less transmissions to supply all the information (although each transmission will need more time to be performed, as the number of bits increases).

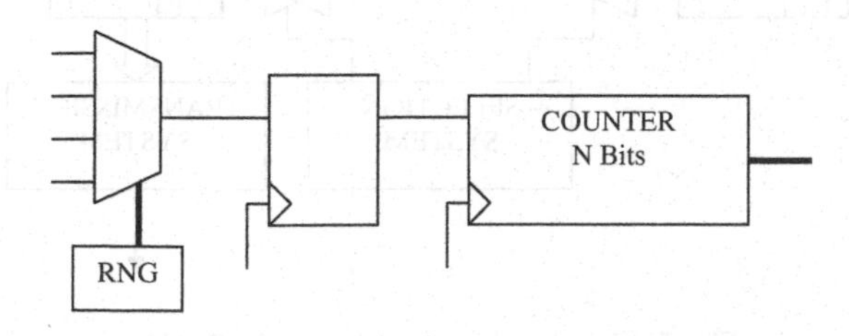

Fig. 6. Architecture of a modified cell in the RNR CA

To study these effects, some results are given in Table 5. This table shows the influence of the size of the counter in a complete design. For the values given we have used a 8x8-cell matrix implemented on a ALTERA FLEX10K. We present result for one device only because we are interested in general behavior rather than particular values.

Table 5. Results of modified cell implementation using ALTERA FLEX10K FPGA (speed grade –1)

Evolutions Averaged	1	3	7	15	31	63	127	255
Counter bits	-	2	3	4	5	6	7	8
LEs	192	570	698	891	1018	1209	1339	1531
Max. Freq. (MHz)	75,30	62,54	58,34	52,46	55,78	50,62	44,68	40,16
Evolution time (ns)	13,28	15,99	17,14	19,06	17,93	19,76	22,38	24,90
Trx. time (ms)	4,167	12,5	16,67	20,83	25,0	29,17	33,33	37,5
Total time for 255 evolutions (ms)	1065	797,9	531,2	332,0	199,2	116,3	66,67	37,5

Given data show that, with this modified cell, we are reducing time but increasing logic resources used. This leads us to think in another strategies to obtain intermediate solutions.

The limiting factor, as evident from the tables given above, is the serial transmission. Let's use the original cells (depicted in fig. 4) and introduce a new resource: memory. Some FPGA families give the possibility of implementing memory by means of embedded resources (such as the RAM blocks in VIRTEX devices). This memory will store the number of times that cell (x,y) has been in state '1' in the last N evolutions. From this point of view, each position in memory behaves just like the counter in the modified cell (fig. 6). But memory is not capable of perform additions by itself, so an external adder must be included in the design.

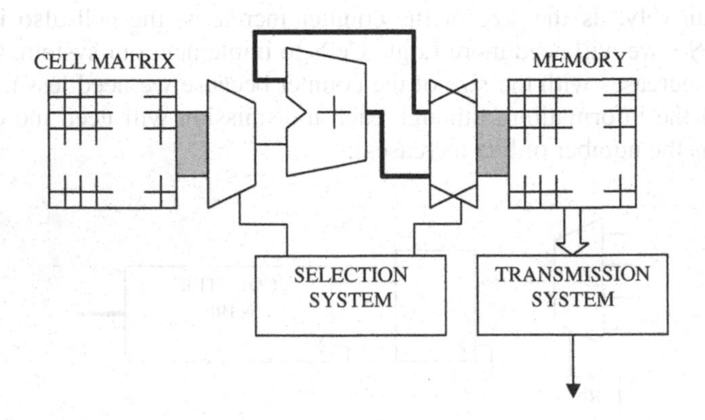

Fig. 7. Mixed architecture for the RNR CA

This leads us to think about the mixed architecture shown in fig. 7. Cell matrix evolves synchronously in only one clock period. After that, sequential control performs a loop and updates the values stored in memory. Transmission only takes place after 2^n generations, where n is the memory width.

The performances of this mixed architecture are given in Table 6. As in table 5, we can see the influence of the size of the memory and the adder in a complete design. For the values given we have used a 16x16-cell matrix implemented on a XILINX XCV800.

Table 6. Results of mixed architecture implementation using XILINX VIRTEX FPGA (speed grade –4)

Evolutions Averaged	1	3	7	15	31	63	127	255
Memory width	-	2	4	4	8	8	8	8
Slices	553	653	655	658	656	660	665	668
Max. Freq. (MHz)	67,40	49,70	52,45	52,29	45,40	45,60	44,38	49,45
Evolution time (ns)	14,84	20,12	19,07	19,12	22,03	21,93	22,53	20,22
Trx. time (ms)	16,67	33,33	66,66	66,66	133,33	133,33	133,33	133,33
Total time for 255 evolutions (ms)	4240	2130	2130	1060	1060	533,32	266,66	133,33

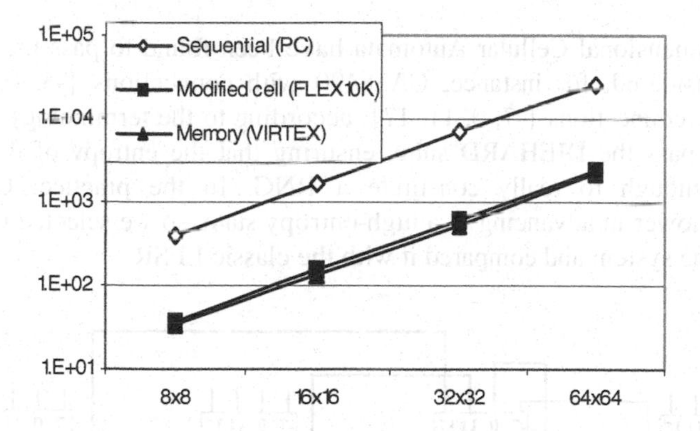

Fig. 8. Time (ms) to perform 255 evolutions using the systems described

As most FPGA families incorporate on-chip memories, this last architecture continues fitting into one single device.

Having presented all those possibilities, let's evaluate their performances. From the above tables we can see the enormous increase of logic resources used by the modified cell when compared to the mixed architecture. In order to compare velocities, times for performing 255 evolutions are shown in fig. 8. We show this time for the modified cell using 8-bit counters implemented on a ALTERA EPF10K200SFC672, for different matrix sizes. We also show times for mixed architecture implemented on XILINX XCV800. We can see that there is not much variation. To have a reference, we show the performance of a sequential program implemented in VISUAL C++ and running in a PC (PENTIUM III 800MHz) a little variation. Although all three implementations show an exponential increase of time (due to the serial transmission process in the FPGA implementations) there is an important improvement of speed (about 14 times) using dedicated parallel hardware.

4 Random Number Generation

One of the most important needs of our system is to incorporate a high-quality RNG. In VLSI, this is a very common need since random numbers appear in several areas such as Monte-Carlo simulations, chip testing, or computer-based gaming. Initially, linear feedback shift registers (LFSR) were used to implement the random pattern generator. But in 1986 Wolfram showed that CA30 1D Cellular Automata offers better performance in the generation of random numbers than LFSR. Several authors have followed Wolfram's suggestions [8]. There is no single test to determine the quality of an RNG –a battery of different tests is required. In our work, we used George Marsaglia's widely acknowledged battery of tests: DIEHARD [13].

In our work, we tested the convenience of using a Cellular Automata RNG, comparing its performance with the classical LFSR.

4.1 CA50745 RNG

Several one-dimensional Cellular Automata have been found to pass the DIEHARD suite. We considered, for instance, CA38490 with connections {-3, 0, 4, 9} and CA50745 with connections {-7, 0, 11, 17}, according to the terminology used in [8]. Both of them pass the DIEHARD suite, ensuring that the entropy of the generated sequence is enough to really constitute a RNG. In the practice, CA38490 is considerably slower in advancing to a high-entropy state, so we selected CA50745 as the RNG for the system and compared it with the classic LFSR.

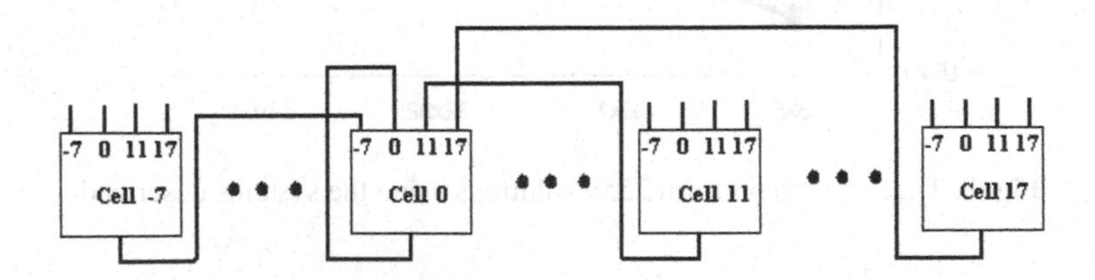

Fig. 9. Cell connectivity for the CA50745 RNG

We implemented the RNG as a one-dimensional automata with 4-input cells, connected to the lattice following the {-7, 0, 11, 17} schema. This is shown in Fig. 9.

2-bit RNG can be implemented by selecting two random bits from a greater RNG. Therefore, a N-bit RNG gives support to a number n of 2-bit RNG that is given by the formula:

$$n = \binom{N}{2} = \frac{N(N-1)}{2} . \quad (8)$$

4.2 Implementation and Results

The proposed Cellular Automata has been implemented on FPGAs. In order to make comparisons of performances, we also implemented a 32-bit Leap-Forward LFSR RNG [14], defined by a shit register with the connectivity formula:

$$q_{31} = q_0 \otimes q_4 \otimes q_6 \otimes q_{12} \otimes q_{13} \otimes q_{14} \otimes q_{15} \otimes q_{18} \otimes q_{19} \otimes q_{20} \otimes q_{21} \otimes q_{23} \otimes q_{24} \otimes q_{26} \otimes q_{27} \otimes q_{29} \otimes q_{31} . \quad (9)$$

Results of both RNG are given in table 7. Here we have used XILINX XC40250XV, but similar results have been found for all the FPGAs. CA50745 uses less logic resources and reaches higher frequency.

Table 7. Implementation of RNGs on XILINX XC40250XV (speed grade –07)

RNG	CLBs	Max. Freq. (MHz)
LF LFSR	62	42,963
CA50745	25	68,4

As both RNG are found to be good generators (having passed DIEHARD), there must be no influence on the overall behavior of the system. Fig. 10 shows the evolution of the error as the number of generations increases. Error has been calculated with respect to the analytic solution with precision 10^6. There is no significant difference between RNGs.

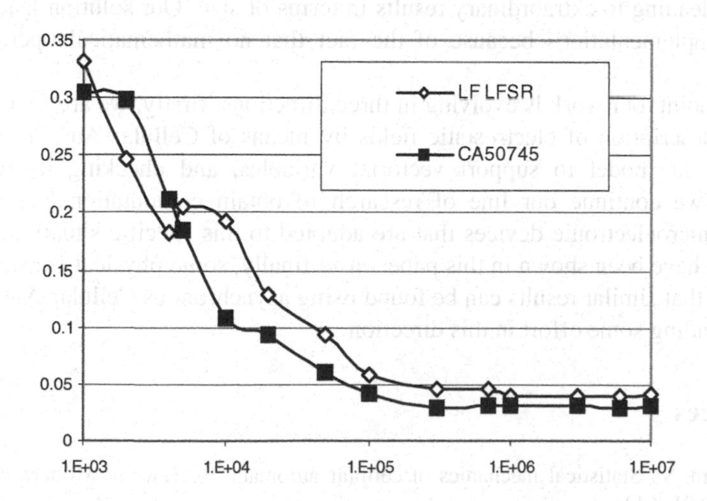

Fig. 10. Relative error of the automata with respect to the analytic solution using different RNGs

5 Conclusions and Further Work

We have introduced RNR, a Cellular Automata for numerically solving the Laplace Equation. To achieve this, it was necessary to replace continuous space and time by a discrete grid, replace continuous variables by binary values and replace continuous equations by finite differences. Scalar quantities appear by averaging the binary value of cells over time and space. RNR has been tested at macroscopic level, leading to very accurate results. The accuracy of the solution increases as the averaged data increases (over time and in space). This procedure can be generalized for other physical phenomena and equations.

Sequential machines like classical microprocessors do not adapt well to cellular automata because they have to emulate concurrent behavior, so new architectures have been introduced. First of all, we consider a fully-parallel system. Access to all the cells is made often by a serial communication. Other possibilities include more complex cells or a mixed architecture using memory to store intermediate values, thus reducing accessing time. This last architecture shows itself the best choice according to performance reached and logic resources used.

Finally, our system needs a good quality RNG. We have showed that Cellular Automata based RNG presents better features than classic LFSR RNG. This way, the system described is a fully cellular one. With all this resources, the whole systems was implemented and tested on available development platforms (UP1 board from ALTERA and XS board from XILINX). This fact explains the choice of devices in our work.

We must remark the important difference between the given method and conventional techniques for solving the same problem (a good reference is [15]). RNR CA does not use arithmetic resources but logic ones, thus reducing the overall complexity of the circuit. In the basic implementation each point is represented by a single bit, leading to extraordinary results in terms of size. Our solution leads to very efficient implementations because of the fact that no mathematical operations are needed.

At this point, our work is evolving in three directions: firstly, we are focused in the complete description of electrostatic fields by means of Cellular Automata. We are extending our model to support vectorial variables, and checking its properties. Secondly, we continue our line of research to obtain computational architectures based on microelectronic devices that are adapted to this specific situation. Some of this results have been shown in this paper. And finally, some physical constraints lead us to think that similar results can be found using asynchronous Cellular Automata, so we are spending some effort in this direction.

References

1. Wolfram, S.: Statistical mechanics of cellular automata. Reviews of Modern Physics, 55 (1983) 601–644
2. Toffoli, T.: Occam, Turing, von Neumann, Jaynes: How much can you get for how little? (A conceptual introduction to cellular automata). The Interjournal (October 1994)

3. Wolfram, S.: A New Kind of Science. Wolfram media (2002)
4. Margolus, N.: Physics-Like models of computation. Physica 10D (1984) 81-95
5. Toffoli, T.: Cellular Automata as an alternative to (rather than an approximation of) Differential Equations in Modeling Physics. Physica 10D (1984) 117-127
6. Smith, M.A.: Cellular Automata Methods in Mathematical Physics. Ph.D. Thesis. MIT Department of Physics (May 1994).
7. Sarkar, P.: A brief history of cellular automata. ACM Computing Surveys, Vol. 32, Issue 1 (2000) 80–107
8. Shackleford, B., Tanaka, M., Carter, R.J., Snider, G.: FPGA Implementation of Neighborhood-of-Four Cellular Automata Random Number Generators. Proceedings of FPGA 2002 (2002) 106–112
9. Levich, B.G.: Curso de Física teórica. Volumen I: Teoría del campo electromagnético y teoría de la relatividad. Reverté (1974)
10. Gerald, C.F., and Wheatley, P.O.: Applied Numerical Analysis, 6th ed. Addison Wesley Longman (1999).
11. Cerdá-Boluda, J.: Arquitecturas VLSI de Autómatas Celulares para Modelado Físico. Ph.D Thesis. UPV Electronic Engineering Dept. (to be published).
12. Chopard, B., Droz, M.: Cellular Automata Modeling of Physical Systems. Cambridge University Press (1999).
13. Marsaglia, G.: DIEHARD. http://stat.fsu.edu/~geo/diehard.html, 1996
14. Jaquenod, G., De Giusti. M.R.: Performance y uso de recursos de contadores basados en Linear Feedback Shift Registers (LFSRs). VIII Workshop IBERCHIP, Guadalajara, México (2002)
15. Chew, W.C., Jin, J.M., Lu, C.C., Michielssen, E. Song, J.M.: Fast Solution Methods in Electromagnetics. IEEE Transactions on Antennas and Propagation, vol. 45, no.3 (1997) 523-533

PerWiz: A What-If Prediction Tool for Tuning Message Passing Programs*

Fumihiko Ino, Yuki Kanbe, Masao Okita, and Kenichi Hagihara

Graduate School of Information Science and Technology, Osaka University
1-3 Machikaneyama, Toyonaka, Osaka 560-8531, Japan
{ino,y-kanbe,m-okita,hagihara}@ist.osaka-u.ac.jp

Abstract. This paper presents PerWiz, a performance prediction tool for improving the performance of message passing programs. PerWiz focuses on locating where a significant improvement can be achieved. To locate this, PerWiz performs a post-mortem analysis based on a realistic parallel computational model, LogGPS, so that predicts what performance will be achieved if the programs are modified according to typical tuning techniques, such as load balancing for a better workload distribution and message scheduling for a shorter waiting time. We also show two case studies where PerWiz played an important role in improving the performance of regular applications. Our results indicate that PerWiz is useful for application developers to assess the potential reduction in execution time that will be derived from program modification.

1 Introduction

Message passing paradigm [1] is a widely employed programming paradigm for distributed memory architectures such as clusters and Grids. This paradigm enables writing high performance parallel applications on these architectures. However, it assumes implicit parallelism coded by application developers. Therefore, developers have to take responsibility for detecting the bottleneck code of sequential applications and for determining which code should be parallelized, according to the performance analysis of the applications. Thus, performance tuning is an important process for developing high performance parallel applications.

One issue for tuning parallel programs is an enormous amount of performance data, which makes this process difficult and time-consuming task. Earlier research addresses this issue by visualizing the performance data collected during program execution [2–5]. Although this visual approach enables intuitive understanding of program behavior, it has a scalability problem. It is not easy for developers to detect performance bottlenecks from the complicated visualizations rendered for large-scale applications.

Therefore, some research projects address the issue by automating program instrumentation [6], performance problem search [7–10], and performance prediction [11]. These automatic approaches are attractive for developers who want to locate performance bottlenecks and adapt applications to other computing environments. However,

* This work was partly supported by JSPS Grant-in-Aid for Young Scientists (B)(15700030), for Scientific Research (C)(2)(14580374), and Network Development Laboratories, NEC.

M. Daydé et al. (Eds.): VECPAR 2004, LNCS 3402, pp. 396–409, 2005.

these projects raise another question that must be answered: how much improvement in execution time will be derived with what kind of tuning techniques?

Hollingsworth [12] gives a preliminary answer to this question by developing a runtime algorithm to compute a variant of critical path (CP), called CP zeroing. CP zeroing provides an upper bound on the reduction in the length of CP possible by tuning a specific procedure. Because CP zeroing directs developers to procedures expected to provide a significant improvement, it is useful to prevent spending time tuning procedures that will not give an improvement. A similar analysis [13] is also useful to balance the workload in a coarse-grain manner. This analysis predicts the potential improvement in execution time when changing the assignment of processes to processors.

Thus, many tools focus on helping performance tuning. However, to the best of our knowledge, there is no tool that gives an answer to the further question: what performance will be achieved if various tuning techniques are applied to the programs? This *what-if prediction* support is important for developers because it enables them to prevent wasted effort with no improvement. Note here that CP zeroing predicts the improvement if a procedure is removed, so that considers no tuning technique during prediction. Thus, addressing what-if prediction with various tuning techniques is lacking, for instance, what-if predictions for fine-grain load balancing and message scheduling, aiming to obtain a better workload distribution and a shorter waiting time, respectively.

In this paper, we present Performance Wizard (PerWiz), a what-if prediction tool for tuning Message Passing Interface (MPI) programs [1]. PerWiz reveals hidden bottleneck messages in a program by predicting the program's execution time if zeroing the waiting time of a message. Furthermore, by specifying the program's *parallel region*, where developers intend to parallelize, PerWiz presents a lower bound on the execution time if balancing the workload in a parallel region. These what-if predictions are based on a post-mortem analysis using LogGPS [14], a parallel computational model that models the computing environment by several parameters. Predicted results are presented in a timeline view rendered by logviewer [4], a widespread visualization tool.

The paper is organized as follows. We begin in Section 2 by abstracting the execution of message passing programs with LogGPS. In Section 3, we describe a method for assessing the potential improvement of message passing programs. Section 4 describes PerWiz, which implements the method. Section 5 shows two case studies and Section 6 discusses related work. Finally, Section 7 concludes this paper.

2 Modeling Message Passing Programs

2.1 Definition of Parallel Region

Let $\mathcal{A}$ denote a message passing program. In order to analyze the workload distribution of $\mathcal{A}$, the program code that developers intend to execute in parallel must be identified. We call this code a parallel region, $\mathcal{R}$, composed of a set of blocks in $\mathcal{A}$. That is, any block b such that $b \in \mathcal{R}$ can be executed in parallel (or even in sequential, against developer's intension). Note here that any $b \in \mathcal{R}$ can include a routine call for message passing. Figure 1 shows two examples of a parallel region specified by two directives: PRGN_BEGIN and PRGN_END. Developers aim to parallelize the entire N iterations in Figure 1(a) and each of N iterations in Figure 1(b).

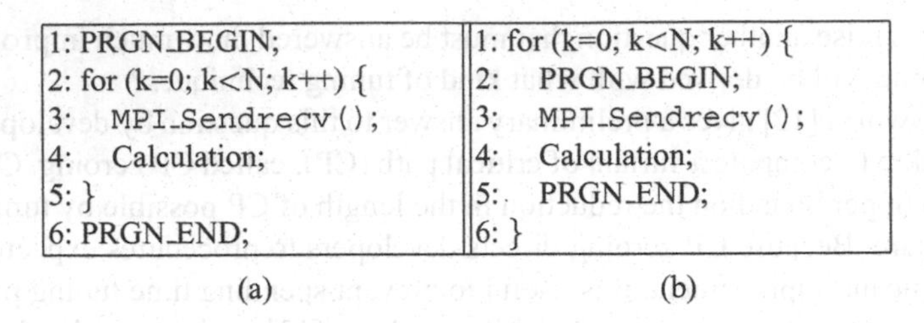

```
1: PRGN_BEGIN;              1: for (k=0; k<N; k++) {
2: for (k=0; k<N; k++) {    2:   PRGN_BEGIN;
3:   MPI_Sendrecv();        3:   MPI_Sendrecv();
4:   Calculation;           4:   Calculation;
5: }                        5:   PRGN_END;
6: PRGN_END;                6: }
          (a)                          (b)
```

Fig. 1. Parallel regions specified (a) for the entire N iterations and (b) for each of N iterations

In the following discussion, we call a logical step that corresponds to an execution of a parallel region as a *parallel step*.

2.2 Modeling Program Execution

An execution of $\mathcal{A}$ can be represented as a directed acyclic graph (DAG), $\mathcal{G} = (\mathcal{V}, \mathcal{E})$, where $\mathcal{V}$ and $\mathcal{E}$ denote a set of vertexes and that of edges, respectively. While a vertex corresponds to an event occurred during an execution, an edge corresponds to *a happened-before relation* [15], $\rightarrow$, a precedence relationship defined between events.

An event occurs when a process executes a sequence of statements in $\mathcal{A}$. Events can be classified into two groups: communication and calculation events. A communication event corresponds to an execution of a communication routine, from its call to return. On the other hand, a calculation event corresponds to an execution of other statements processed between two successive communication events. Communication events can be more classified into send and receive events according to whether their corresponding routine sends or receives a message.

An execution time for $\mathcal{A}$ can be represented as the CP length of weighted $\mathcal{G}$, which has a weight associated with each vertex and edge. One method for weighting $\mathcal{G}$ is to use a realistic parallel computational model such as the LogP [16] family of models [14, 17]. LogGPS [14] is an extension of LogGP [17] and captures both synchronous and asynchronous messages by using the following seven parameters (see Figure 2).

- L: an upper bound on the latency, incurred in sending a message from its source processor to its target processor.
- o: the overhead, defined as the length of time that a processor is engaged in the transmission or reception of each message; During this time the processor cannot perform other operations.
- g: the gap between messages, defined as the minimum time interval between consecutive message transmissions or consecutive message receptions at a processor.
- G: the Gap per byte for long messages, defined as the time per byte for a long message.
- P: the number of processor/memory modules.
- S: the threshold for message length, above which messages are sent in a synchronous mode.
- s: the threshold for message length, above which messages are sent in multiple packets.

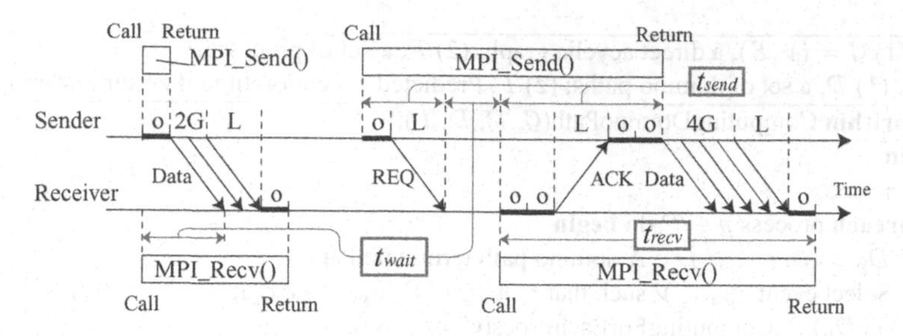

Fig. 2. Asynchronous and synchronous messages under LogGPS. An arrow from the sender to the receiver represents a 1-byte message

In a precise sense, o should be distinguished for the send and receive overheads and also be represented by a linear function on message length. However, because the following explanation requires no precise notation, this paper uses a simple notation, o.

As illustrated in Figure 2, the execution times for send and receive events include waiting time t_{wait}, in addition to sending and receiving times, t_{send} and t_{recv}, respectively. Here, the waiting time for a receive event is defined as the receiver's time from the call of a receive routine to the arrival of the message. The waiting time for a send event, which appears only in a synchronous mode, is also defined as the sender's time from the arrival of the send request REQ to the call of a matching receive routine.

In this paper, waiting time zeroing for communication event e means obtaining $t_{wait} = 0$ for e. This is achieved by making a hypothesis in which e's matching routine is scheduled so that called for t_{wait} earlier than present.

3 Potential Improvement Assessment

This section describes a method for assessing the potential improvement of message passing programs, which perform communication and calculation.

3.1 Assessing Communication Bottlenecks

Tuning methods for communication bottlenecks can be classified into two groups.

T1: Message scheduling, aiming to minimize t_{wait}.
T2: Message reduction, aiming to minimize t_{send} and t_{recv}.

CP zeroing analysis is effective for T2 because applying this analysis to communication routines shows an upper bound on the potential improvement by assuming the amount of messages be zero byte. Therefore, this paper tackles on T1.

In general, focusing on a message with maximum t_{wait} does not always result in a significant improvement. That is, even if the message is scheduled to obtain $t_{wait} = 0$, the program will not reduce its execution time unless the length of CP is shortened. By contrast, scheduling a message with small t_{wait} can trigger *domino effect*, which can significantly improve the overall performance by reducing the waiting time of its subsequent messages, like as domino toppling. Thus, to evaluate the potential improvement

Inputs: (1) $\mathcal{G} = (\mathcal{V}, \mathcal{E})$, a direct acyclic graph; (2) $\mathcal{P}$, a set of processes. Outputs: (1) $\mathcal{D}$, a set of domino paths; (2) T, Predicted execution time if zeroing waiting time.

1: **Algorithm** ComputingDominoPath($\mathcal{G}, \mathcal{P}, \mathcal{D}, T$);
2: **begin**
3: $T := \infty$;
4: **foreach** process $p \in \mathcal{P}$ **do begin**
5: $D_p := \phi$; // D_p: A domino path terminated at $e_{p,i}$
6: Select event $e_{p,i} \in \mathcal{V}$ such that $\neg \exists e_{p,j} \in \mathcal{V}\ (e_{p,i} \rightarrow e_{p,j})$;
7: $w(D_p)$:= ComputingForEachProcs($\mathcal{G}, e_{p,i}, \infty$);
8: **if** ($T > w(D_p)$) **then** $T := w(D_p)$; // Select the minimum execution time
9: **end**
10: $\mathcal{D} := \{D_p \mid w(D_p) = T\}$;
11: **end**

12: **Function** ComputingForEachProcs($\mathcal{G}, e_{p,i}, T_p$);
13: **begin**
14: **foreach** event id $j \in \{1, 2, \cdots, i\}$ **do begin**
15: $T_{p,j}$:= WaitingTimeZeroingSimulation($e_{p,j}$); // See Section 4
16: **end**
17: Select event id $m \in \{1, 2, \cdots, i\}$ such that $\forall j \in \{1, 2, \cdots, i\}\ (T_{p,j} \geq T_{p,m})$;
18: **if** (($e_{p,m} = e_{p,i}$) || ($T_{p,m} > T_p$)) **then**
19: **return** T_p;
20: **else begin**
21: Select event $e_{q,j} \in \mathcal{V}$ such that $(p \neq q) \wedge ((e_{q,j} \rightarrow e_{p,m}) \vee (e_{p,m} \rightarrow e_{q,j}))$;
22: Add events $e_{p,m}$ and $e_{q,j}$ to D_p;
23: **return** ComputingForEachProcs($\mathcal{G}, e_{q,j}, T_{p,m}$); // Recursive call
24: **end**
25: **end**

Fig. 3. Algorithm for computing the domino paths of a DAG

in terms of T1, our method searches a message that triggers the domino effect rather than a message with maximum t_{wait}.

We now propose an algorithm for locating a communication event that triggers the domino effect, aiming to achieve a significant improvement with little effort. Figure 3 describes our algorithm, which requires a DAG and a set of processes, $\mathcal{G}$ and $\mathcal{P}$, respectively, then returns a set of domino paths, $\mathcal{D}$, with the minimum of predicted execution time, T. To locate the event, our algorithm recursively backtracks happened-before relations $\rightarrow$ in $\mathcal{G}$, and repeatedly predicts the execution time with a hypothesis in which a specific communication event $u \in \mathcal{V}$ is scheduled to have $t_{wait} = 0$. As a result, it computes a *domino path* (DP), a set of events that participate in the domino effect. As presented later in Section 4, our prediction is based on LogGPS, which clearly defines t_{wait} for MPI routines so that improves the prediction accuracy for synchronous messages, compared to LogP and LogGP [14].

Let $e_{p,i}$ denote the i-th event occurred on process p. The relation backtracking is performed in three steps as follows. First, the algorithm starts with event $e_{p,i}$ last occurred on process p (line 6), then searches event number m, where $1 \leq m \leq i$, such that $e_{p,m}$ accompanies the minimum execution time $T_{p,m}$ if assuming $t_{wait} = 0$ for $e_{p,m}$

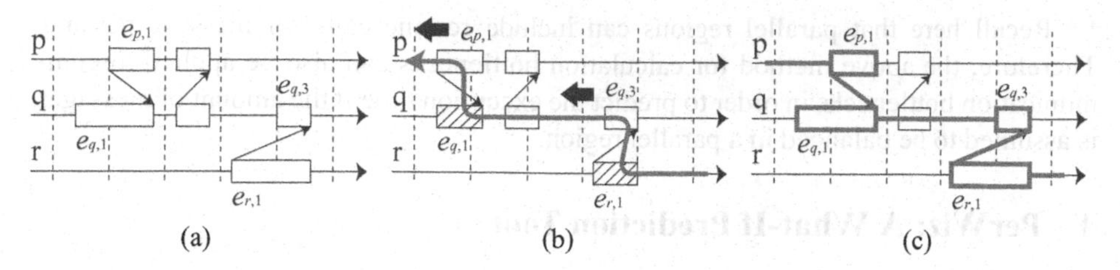

Fig. 4. Computing process for a domino path. (a) Given a DAG, the algorithm (b) recursively backtracks happened-before relations so that (c) locates a domino path

(line 14–17). Here, the algorithm in Figure 3 is simplified to obtain such m in unique, however, our actual algorithm performs round robin prediction on every m with minimum $T_{p,m}$. In the next step, it terminates this backtracking (1) if current computed DP D_p already includes $e_{p,m}$ or (2) if current predicted time $T_{p,m}$ exceeds to T_p, the minimum execution time predicted before (line 18). Otherwise, the algorithm searches $e_{p,m}$'s matching event $e_{q,j}$ (line 21), then add $e_{p,m}$ and $e_{q,j}$ to D_p (line 22). Finally, applying this backtracking recursively to $e_{q,j}$ (line 23) computes a DP terminated on p, and performing this for all processes $p \in \mathcal{P}$ (line 4) returns $\mathcal{D}$ and T (lines 8 and 10).

Figure 4 illustrates an example of computing process for a DP. Our algorithm backtracks $e_{r,1} \rightarrow e_{q,3}$, then $e_{q,1} \rightarrow e_{p,1}$, so that points out a DP containing $e_{p,1}$, $e_{q,1}$, $e_{q,3}$ and $e_{r,1}$. The waiting times of $e_{q,1}$ and $e_{r,1}$ can be reduced by calling $e_{p,1}$'s corresponding send routine earlier in order to cause the occurrence of $e_{p,1}$ forward.

3.2 Assessing Calculation Bottlenecks

As we did for communication bottlenecks, tuning methods for calculation bottlenecks can also be classified into two groups.

T3: Load balancing, aiming to obtain a better workload distribution.
T4: Time complexity reduction, aiming to minimize calculation time.

Like for T2, CP zeroing is also effective for T4 because it predicts an upper bound on the potential improvement by assuming the execution time of a specific procedure be zero second. Therefore, we focus on T3.

The concept of parallel region is useful to evaluate the potential improvement that will be derived by T3, because it enables predicting the execution time if the workload is balanced among processors. For example, the most imbalanced parallel step can be detected by computing the standard deviation $\sigma(t_k)$ of execution time t_k, where t_k represents the time that a process requires in the k-th parallel step. However, as we mentioned for T1, focusing on a parallel step with maximum $\sigma(t_k)$ does not always derive a significant improvement. Therefore, by using LogGPS simulation, our method searches a parallel step that derives the minimum execution time if its workload is assumed to be balanced ($\sigma(t_k) = 0$).

Recall here that parallel regions can include routine calls for message passing. Therefore, the above method for calculation bottlenecks can also be applied to communication bottlenecks in order to predict the execution time if the amount of messages is assumed to be balanced in a parallel region.

4 PerWiz: A What-If Prediction Tool

PerWiz has the following three functions.

- **F1: Function for modification choice search.**
  To assist developers in achieving a significant improvement with little effort, PerWiz presents modification choices sorted by their significances. For communication bottlenecks, it shows a DP computed by the algorithm presented in Section 3.1. For calculation bottlenecks, as mentioned in Section 3.2, it locates a parallel step that reduces the program's execution time by load balancing.
- **F2: Function for what-if prediction.**
  This function is a basis for function F1 and predicts the program's execution time for the following what-if questions: what performance will be derived if making a hypothesis in which
  - the waiting time or the execution time of a specific event becomes zero second;
  - all processes pass the same amount of messages or require the same execution time in a specific parallel step.

  The predicted results for the above questions can be visualized by logviewer, as shown in Figure 5(c). In addition, developers can give PerWiz a combination of the questions, allowing them to consider various tuning techniques.
- **F3: Function for lower bound prediction.**
  In order to assist developers in changing the parallel algorithm employed in the target program, this function computes a lower bound on execution time for each parallel region. Since different algorithms produce different DAGs, it predicts a lower bound by simply summing up the breakdown of execution time. That is, F3 needs only the breakdown of calculation, sending, receiving, and waiting times while F2 requires a DAG in order to simulate a program execution with proper precedence. Developers can give and combine the following what-if questions: what performance will be derived if making a hypothesis in which
  - the waiting time or the execution time spent for a specific parallel region becomes zero second;
  - all processes pass the same amount of messages or require the same execution time in a specific parallel region.

To tune MPI programs with PerWiz, developers first must specify parallel regions in their code, as presented in Figure 1. After this instrumentation, compiling the instrumented code generates an object file and linking it with an instrumentation library generates an executable binary file. By executing the binary file in parallel, PerWiz generates a trace file, $\mathcal{L}$, logged in ALOG format [18]. The instrumentation library for ALOG is widely distributed with MPICH [19], a basis of many MPI implementations. Finally, giving a what-if question and $\mathcal{L}$ generates a predicted result in a text file and

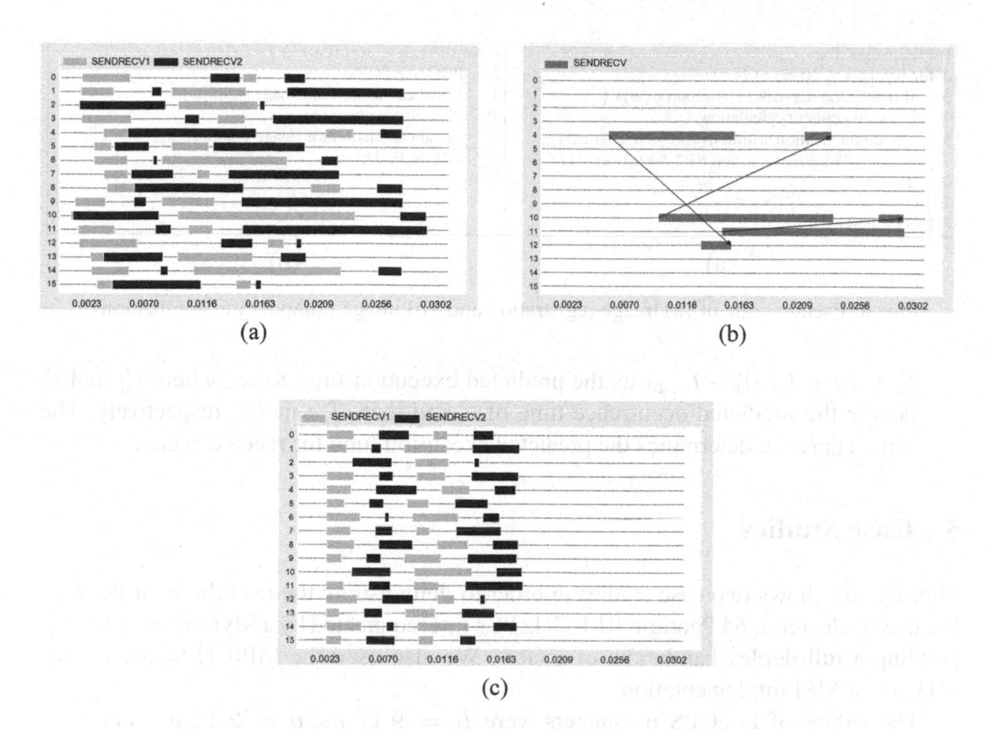

Fig. 5. Predicted results in timeline views rendered by logviewer. (a) The original behavior on 16 processes, (b) its computed domino path, and (c) a predicted behavior if the workload in parallel steps is assumed to be balanced. A colored box represents an execution of MPI_Sendrecv(). Gray and black boxes correspond to an execution in odd and even parallel steps, respectively

a reconstructed ALOG file, $\mathcal{L}'$. Note here that giving $\mathcal{L}'$ to PerWiz enables iterative prediction without additional program execution.

We now present how PerWiz reconstructs predicted trace file $\mathcal{L}'$ from recorded trace file $\mathcal{L}$. In order to realize accurate performance predictions, PerWiz utilizes the measured execution time in $\mathcal{L}$ as much as possible. For calculation events, PerWiz applies the measured time in $\mathcal{L}$ to the predicted time in $\mathcal{L}'$. For communication events, PerWiz employs two approaches according to whether the event passes a synchronous message or an asynchronous message:

- For asynchronous communication events, PerWiz uses the LogGPS execution time, computed for every length of messages [14];
- For synchronous communication events, PerWiz estimates both sending time t_{send} and receiving time t_{recv} from measured $\mathcal{L}$, then utilizes them for $\mathcal{L}'$. For example, given send event u, it first decomposes its measured time t_{all} recorded in $\mathcal{L}$ into $t_{wait} = max(t_v - t_u - (o + L), 0)$ and $t_{send} = t_{all} - t_{wait}$ (see Figure 2), where t_u and t_v denote the occurrence time of u and that of a matching receive event v recorded in $\mathcal{L}$, respectively, and $o + L$ represents the estimated time defined from the occurrence of v to the arrival of REQ. PerWiz then regards $t'_{all} = max(t'_v -$

(a)	(b)

1: PRGN_BEGIN;
2: for (ϕ=1; $\phi \leq$ N; ϕ++) {
3: if (`MPI_Comm_rank()` $\in$ group $g(\phi)$) {
4: Local gradient calculation;
5: Communication within $g(\phi)$ by `MPI_Send()`, `MPI_Recv()`, and `MPI_Sendrecv()`;
6: }
7: }
8: PRGN_END;
	1: for (k=1; k $\leq$ log P; k++) {
	2: PRGN_BEGIN;
	3: Local calculation for image splitting;
	4: `MPI_Sendrecv()`;
	5: Local calculation for image compositing;
	6: PRGN_END;
	7: }

Fig. 6. Pseudo code of (a) image registration and (b) image compositing applications

$t'_u - (o + L), 0) + t_{send}$ as the predicted execution time for u, where t'_u and t'_v denote the predicted occurrence time of u and that of v in $\mathcal{L}'$, respectively. The same approach determines the predicted execution time for receive events.

5 Case Studies

This section shows two case studies in order to demonstrate the usefulness of PerWiz. We used a cluster of 64 Pentium III 1-GHz PCs interconnected by a Myrinet switch [20], yielding a full-duplex bandwidth of 2 Gb/s. We also used the MPICH-SCore library [21], a fast MPI implementation.

The values of LogGPS parameters were $L = 9.11$ μs, $o = 2.15$ μs, and $S =$ 16 383 bytes, measured using the method presented in [14]. Note here that the remaining LogGPS parameters had no influence on the predicted results because the applications employed for the studies passed only synchronous (long) messages, which require only L and o for prediction, as presented in Section 4.

5.1 Case 1: Performance Improvement by Message Scheduling

We applied PerWiz to an image registration application [22], which aligns a pair of three-dimensional (3-D) images. We improved its performance by reducing waiting time. Figure 6(a) shows its pseudo code with its parallel region. In this application, every process holds a portion of the images and computes the gradient of a function that represents the similarity between the images. This gradient has to be calculated for each point ϕ, placed dynamically in the images. In addition, the gradient calculation at ϕ requires the neighborhood pixels of ϕ. To parallelize this application, developers dynamically organized processes into groups, aiming at concurrent processing of gradient calculations. All processes that hold the neighborhood pixels of ϕ compose group $g(\phi)$, and all processes in $g(\phi)$ participate in the calculation at ϕ.

We first detected the hotspot of this application by using gprof [23], a profiling tool. The hotspot was the gradient calculation repeated for 14 times. Therefore, we decided to instrument only one of the 14 repetitions. We executed the instrumented code on 16 dedicated processors, each with one process, so that generated 720 KB of a trace file that contains 4634 communication events. The execution time was 23.8 s including a run-time overhead of 0.1 s for trace generation.

Table 1. Lower bound prediction for image registration and image compositing applications by PerWiz (function F3). Q1, Q2, and Q3 denote questions if zeroing waiting time, balancing the workload and the amount of messages in its parallel region, respectively

What-if question	Predicted execution time: Registration T_1 (s)	Predicted execution time: Compositing T_2 (ms)	What-if question	Predicted execution time: Registration T_1 (s)	Predicted execution time: Compositing T_2 (ms)
—	23.8	30.6	$Q1 \wedge Q2$	5.3	16.5
Q1	8.2	22.3	$Q1 \wedge Q3$	8.0	16.8
Q2	23.4	34.9	$Q2 \wedge Q3$	23.6	38.2
Q3	21.4	25.7	$Q1 \wedge Q2 \wedge Q3$	5.0	11.0

Table 2. Predicted and measured times for image registration application after l-th modification, where $0 \leq l \leq 7$. Code modification is based on the events located by the domino path approach and the longest waiting time approach

l	Domino path approach by PerWiz: Execution time (s) $T_{P,l}$: Predicted	Domino path approach by PerWiz: Execution time (s) $T_{M,l}$: Measured	Domino path approach by PerWiz: Waiting time (s) $t_{wait,l}$	Longest waiting time approach: Execution time (s) $T_{M,l}$: Measured	Longest waiting time approach: Waiting time (s) $t_{wait,l}$
0	—	23.8	—	23.8	—
1	21.6	21.8	11.8	23.9	12.5
2	19.5	19.5	10.7	22.6	14.7
3	18.2	18.2	7.5	22.6	11.9
4	15.9	15.8	5.8	21.5	13.5
5	14.8	14.8	3.4	21.5	11.4
6	14.4	14.4	0.4	20.7	12.3
7	13.7	13.7	0.7	20.8	10.7

Table 1 shows the execution time, T_1, predicted by function F3, the lower bound prediction. We obtained $T_1 < 10$ s if Q1 is specified: what performance will be derived if zeroing the waiting time in the parallel region? Therefore, reducing waiting time is necessary to improve the performance of this application.

In order to clarify the usefulness of PerWiz, we now compare two approaches for reducing waiting time in MPI programs: (1) the DP approach, which locates an event that triggers the domino effect computed by PerWiz and (2) the longest waiting time (LWT) approach, which locates an event with LWT. Because the waiting time in this application was due to the processing order of points (loops at line 2), we repeatedly modified the code to obtain an appropriate order that reduces the waiting time of the event located by each approach. To compute a DP, PerWiz took approximately 10 s on a Pentium III 1-GHz system.

Table 2 compares predicted time $T_{P,l}$ and measured time $T_{M,l}$ after the l-th modification, where $0 \leq l \leq 7$. PerWiz enables developers to reduce $T_{M,l}$ from 23.8 to 13.7 s after the seventh modification. By contrast, the LWT approach results in 20.8 s. Thus, our DP approach allows developers to efficiently improve this application and to derive a performance improvement of 42% while the LWT approach gives that of 13%.

Furthermore, our DP approach successfully gives a performance improvement for every modification, whereas the LWT approach fails to reduce execution time $T_{M,l}$ at

$l = 1, 3, 5$, and 7. This is due to the LWT approach, which lacks the guarantee on the reduction of the length of CP. By contrast, our approach guarantees it by LogGPS simulation. For example, the LWT approach reduces $t_{wait,l}$, where $l = 1, 3, 5$, and 7, however, this reduction increases the waiting times of the succeeding events, resulting in a similar total performance.

The domino effect appears at $t_{wait,l}$, the waiting time of the event located at the l-th modification. In our approach, $t_{wait,l}$ decreases as l increases. This is due to the domino effect, which reduces the waiting time of many events that compose the computed DP. Actually, at every modification, our DP approach reduces the total amount of the waiting time by approximately 25 s, whereas the LWT approach reduces it by 0–10 s. Thus, our DP approach directs developers to events with a shorter waiting time but with a promised improvement, and the LWT approach directs them to events with a longer waiting time but with an uncertain improvement. Therefore, PerWiz enables developers to efficiently improve the application's performance.

5.2 Case 2: Performance Improvement by Load Balancing

We also applied PerWiz to an image compositing application [24] for 3-D volume rendering systems (see Figure 6(b)). We improved its performance by load balancing. On P processes, it merges P locally rendered images into the final image in $\log P$ stages. At each stage, all processes are paired up, and the two processes involved in a compositing split the image plane into two pieces. Each process then takes responsibility for one of the two pieces and exchanges the other piece. Repeating this splitting and exchanging with different pairs of processes for $\log P$ times produces the final image in a distributed manner. Since every process takes responsibility for $1/2^k$ image at the k-th stage, where $1 \leq k \leq \log P$, developers expected a good workload distribution. However, in practice, it had a load imbalance issue due to transparent pixels that can omit calculation for compositing. The trace file generated on 64 dedicated processors was 100 KB in size, containing 384 communication events.

The lower bound on execution time, T_2, presented in Table 1 indicates that its execution time can be reduced from 30.6 to 11.0 ms by applying all of the following three tuning techniques: reducing waiting time, and balancing the workload and the amount of messages in its parallel region. However, the most effective technique is unclear because T_2 decreases by approximately 5 s when adding these techniques successively to the question given to PerWiz (22.3, 16.5, and 11.0 ms).

In this application, because every process synchronizes at every stage, we first intended to reduce its waiting time by balancing the workload. We then used function F2 to predict the execution time, T_3, possible by balancing the workload in the k-th parallel step, where $1 \leq k \leq 6$. Table 3 shows the results, which give two hints toward a performance improvement: (1) load balancing in all parallel steps reduces T_3 from 30.6 to 11.0 ms; and (2) this load balancing is effective especially for parallel step S_2.

First, we discuss on hint (1). The minimum time of $T_3 = 11.0$ ms equals to that of $T_2 = 11.0$ ms predicted for Q1 $\wedge$ Q2 $\wedge$ Q3, as presented in Table 1. This indicates that, as we intended before, load balancing in all parallel steps successfully reduces the waiting time in this application, because it yields the same performance possible by

Table 3. What-if prediction for image compositing program by PerWiz (function F2). Q4(k) represents a question if balancing the workload in parallel step S_k, where $1 \leq k \leq 6$

What-if question	Predicted time T_3 (ms)	What-if question	Predicted time T_3 (ms)
—	30.6	Q4(4)	27.2
Q4(1)	29.1	Q4(5)	29.6
Q4(2)	21.6	Q4(6)	29.9
Q4(3)	26.4	Q4(1) $\wedge \cdots \wedge$ Q4(6)	11.0

zeroing the waiting time of every communication event. Thus, the most efficient modification was to balance the workload in all parallel steps. To realize this, we developed the BSLBR method [24], which splits images into interleaved pieces rather than block pieces. BSLBR achieved an execution time of 14.8 ms.

For hint (2), we examined the sparsity of images at each stage. We then found that exploiting the sparsity of images was inadequate immediately after the first compositing step, so that it augmented load imbalance in S_2. To address this issue, we developed the BSLMBR method [24], which exploits more sparsity of images with a low overhead. BSLMBR reached a minimum execution time of 11.8 ms, which is close to the minimum predicted time, $T_3 = 11.0$ ms.

6 Related Work

Paradyn [7] performs an automatic run-time analysis for searching performance bottlenecks based on thresholds and a set of hypotheses structured in a hierarchy. For example, it locates a synchronization bottleneck if waiting time is greater than 20% of the program's execution time. Aksum [8] is also based on a similar approach that is capable of multi-experiment performance analysis for different problems and machine sizes in order to automate performance diagnosis of parallel and distributed programs. Because these diagnoses are based on dynamic instrumentation technology, sophisticated search strategies for reducing run-time overhead [9] and for quickly finding bottlenecks [10] are developed recently.

SCALEA [25] is a performance instrumentation, measurement, and analysis system that explains the performance behavior of each program by computing a variety of performance metrics regarding data movement, synchronization, control of parallelism, additional computation, loss of parallelism, and unidentified overheads.

Prophesy [26] gives insight into how system features impact application performance. It provides a performance data repository that enables the automatic generation of performance models. To develop these models, it uses coupling parameter [27], a metric that quantifies the interaction between kernels that compose an application.

TAU performance system [28] aims to enable online trace analysis in order to overcome the inherent scalability limitations of post-mortem analysis. Although its preliminary testbed is presented in [28], scalability performance tests are left as future work.

Dimemas [29] is a simulator for analyzing the influence of sharing a multiprocessor system among several parallel applications. It enables studying the influence of different scheduling policies in the global system performance.

In contrast to these previous works, the novelty of PerWiz is the assessment of performance bottlenecks to guide developers to bottlenecks with a promised improvement.

7 Conclusion

We have presented a performance prediction tool named PerWiz, which focuses on assessing the potential improvement of message passing programs. PerWiz directs developers to where a significant improvement can be achieved, as a result of applying typical tuning techniques. To enable this, PerWiz performs a post-mortem analysis based on a parallel computational model, then locates domino paths and parallel steps in the program. Furthermore, PerWiz predicts the execution time under specific assumptions such as if zeroing wait time, balancing the workload and the amount of messages.

In case studies where PerWiz played a key role in tuning MPI programs, we confirmed that focusing on a domino path effectively improves the performance of MPI programs. Therefore, we believe that PerWiz is useful for developers to investigate the reduction in execution time that will be derived from a modification.

Future work includes incorporating dynamic instrumentation technology into PerWiz to reduce instrumentation overhead for large-scale applications. Furthermore, run-time optimization techniques are required to tune irregular applications that dynamically vary program behavior according to run-time situations.

References

1. Message Passing Interface Forum: MPI: A message-passing interface standard. Int'l J. Supercomputer Applications and High Performance Computing **8** (1994) 159–416
2. Heath, M.T., Etheridge, J.A.: Visualizing the performance of parallel programs. IEEE Software **8** (1991) 29–39
3. Nagel, W.E., Arnold, A., Weber, M., Hoppe, H.C., Solchenbach, K.: VAMPIR: Visualization and analysis of MPI resources. The J. Supercomputing **12** (1996) 69–80
4. Zaki, O., Lusk, E., Gropp, W., Swider, D.: Toward scalable performance visualization with Jumpshot. Int'l J. High Performance Computing Applications **13** (1999) 277–288
5. Rose, L.A.D., Reed, D.A.: SvPablo: A multi-language architecture-independent performance analysis system. In: Proc. 28th Int'l Conf. Parallel Processing (ICPP'99). (1999) 311–318
6. Yan, J., Sarukkai, S., Mehra, P.: Performance measurement, visualization and modeling of parallel and distributed programs using the AIMS toolkit. Software: Practice and Experience **25** (1995) 429–461
7. Miller, B.P., Callaghan, M.D., Cargille, J.M., Hollingsworth, J.K., Irvin, R.B., Karavanic, K.L., Kunchithapadam, K., Newhall, T.: The Paradyn parallel performance measurement tool. IEEE Computer **28** (1995) 37–46
8. Fahringer, T., Seragiotto, C.: Automatic search for performance problems in parallel and distributed programs by using multi-experiment analysis. In: Proc. 9th Int'l Conf. High Performance Computing (HiPC'02). (2002) 151–162
9. Cain, H.W., Miller, B.P., Wylie, B.J.N.: A callgraph-based search strategy for automated performance diagnosis. Concurrency and Computation: Practice and Experience **14** (2002) 203–217
10. Roth, P.C., Miller, B.P.: Deep Start: a hybrid strategy for automated performance problem searches. Concurrency and Computation: Practice and Experience **15** (2003) 1027–1046

11. Block, R.J., Sarukkai, S., Mehra, P.: Automated performance prediction of message-passing parallel programs. In: Proc. High Performance Networking and Computing Conf. (SC95). (1995)
12. Hollingsworth, J.K.: Critical path profiling of message passing and shared-memory programs. IEEE Trans. Parallel and Distributed Systems **9** (1998) 1029–1040
13. Eom, H., Hollingsworth, J.K.: A tool to help tune where computation is performed. IEEE Trans. Software Engineering **27** (2001) 618–629
14. Ino, F., Fujimoto, N., Hagihara, K.: LogGPS: A parallel computational model for synchronization analysis. In: Proc. 8th ACM SIGPLAN Symp. Principles and Practice of Parallel Programming (PPoPP'01). (2001) 133–142
15. Lamport, L.: Time, clocks, and the ordering of events in a distributed system. Communications of the ACM **21** (1978) 558–565
16. Culler, D., Karp, R., Patterson, D., Sahay, A., Schauser, K.E., Santos, E., Subramonian, R., von Eicken, T.: LogP: Towards a realistic model of parallel computation. In: Proc. 4th ACM SIGPLAN Symp. Principles and Practice of Parallel Programming (PPoPP'93). (1993) 1–12
17. Alexandrov, A., Ionescu, M., Schauser, K., Scheiman, C.: LogGP: Incorporating long messages into the LogP model for parallel computation. J. Parallel and Distributed Computing **44** (1997) 71–79
18. Herrarte, V., Lusk, E.: Studying parallel program behavior with `upshot`. Technical Report ANL–91/15, Argonne National Laboratory (1991)
19. Gropp, W., Lusk, E., Doss, N., Skjellum, A.: A high-performance, portable implementation of the MPI message passing interface standard. Parallel Computing **22** (1996) 789–828
20. Boden, N.J., Cohen, D., Felderman, R.E., Kulawik, A.E., Seitz, C.L., Seizovic, J.N., Su, W.K.: Myrinet: A gigabit-per-second local-area network. IEEE Micro **15** (1995) 29–36
21. O'Carroll, F., Tezuka, H., Hori, A., Ishikawa, Y.: The design and implementation of zero copy MPI using commodity hardware with a high performance network. In: Proc. 12th ACM Int'l Conf. Supercomputing (ICS'98). (1998) 243–250
22. Schnabel, J.A., Rueckert, D., Quist, M., Blackall, J.M., Castellano-Smith, A.D., Hartkens, T., Penney, G.P., Hall, W.A., Liu, H., Truwit, C.L., Gerritsen, F.A., Hill, D.L.G., Hawkes, D.J.: A generic framework for non-rigid registration based on non-uniform multi-level free-form deformations. In: Proc. 4th Int'l Conf. Medical Image Computing and Computer-Assisted Intervention (MICCAI'01). (2001) 573–581
23. Graham, S.L., Kessler, P.B., McKusick, M.K.: gprof: a call graph execution profiler. In: Proc. SIGPLAN Symp. Compiler Construction (SCC'82). (1982) 120–126
24. Takeuchi, A., Ino, F., Hagihara, K.: An improved binary-swap compositing for sort-last parallel rendering on distributed memory multiprocessors. Parallel Computing **29** (2003) 1745–1762
25. Truong, H.L., Fahringer, T.: SCALEA: a performance analysis tool for parallel programs. Concurrency and Computation: Practice and Experience **15** (2003) 1001–1025
26. Taylor, V., Wu, X., Stevens, R.: Prophesy: An infrastructure for performance analysis and modeling of parallel and grid applications. ACM SIGMETRICS Performance Evaluation Review **30** (2003) 13–18
27. Geisler, J., Taylor, V.: Performance coupling: Case studies for improving the performance of scientific applications. J. Parallel and Distributed Computing **62** (2002) 1227–1247
28. Brunst, H., Malony, A.D., Shende, S.S., Bell, R.: Online remote trace analysis of parallel applications on high-performance clusters. In: Proc. 5th Int'l Symp. High Performance Computing (ISHPC'03). (2003) 440–449
29. Labarta, J., Girona, S., Cortes, T.: Analyzing scheduling policies using Dimemas. Parallel Computing **23** (1997) 23–34

Maintaining Cache Coherency for B$^+$ Tree Indexes in a Shared Disks Cluster

Kyungoh Ohn and Haengrae Cho

Department of Computer Engineering, Yeungnam University,
Kyungsan, Kyungbuk 712-749, Republic of Korea
{ondal, hrcho}@yu.ac.kr

Abstract. A shared disks (SD) cluster couples multiple computing nodes for high performance transaction processing, and all nodes share a common database at the disk level. To reduce the number of disk access, the node may cache both data pages and index pages in its local memory buffer. In general, index pages are accessed more often and thus cached at more nodes than their corresponding data pages. Furthermore, tree-based indexes such as B$^+$ trees require complicated operations, e.g., root-to-leaf traversal and structure modification operation due to a page split or a page concatenation. This means that it is strongly required to devise a dedicated cache coherency scheme for index pages that takes advantage of the semantics of their access. In this paper, we propose a new cache coherency scheme for B$^+$ tree indexes in the SD cluster. The proposed scheme can reduce the message traffic between nodes and the number of tree re-traversals. Using a simulation model of the SD cluster, we show that the proposed scheme exhibits substantial performance improvement over the previous schemes.

Keywords: Cluster and Grid Computing, Data Processing, Parallel and Distributed Computing.

1 Introduction

A cluster is a collection of interconnected computing nodes that collaborate on executing an application and presents itself as one unified computing resource. Depending on the nature of disk access, there are two primary flavors of cluster architecture designs: *shared nothing* (SN) and *shared disks* (SD) [7,19]. In the SN cluster, each node has its own set of private disks and only the owner node can directly read and write its disks. On the other hand, the SD cluster allows each node to have direct access to all disks. The SD cluster offers a number of advantages compared to the SN cluster, such as dynamic load balancing and seamless integration, which make it attractive for high performance transaction processing. Furthermore, the rapidly emerging technology of storage area networks (SAN) makes SD cluster the preferred choice for reasons of higher system availability and flexible data access. The recent parallel database systems using

M. Daydé et al. (Eds.): VECPAR 2004, LNCS 3402, pp. 410–423, 2005.

the SD cluster include IBM DB2 Parallel Edition [6] and Oracle Real Application Cluster [15, 17].

Each node in the SD cluster has its own buffer pool and caches database pages in the buffer. Caching may substantially reduce the amount of expensive and slow disk I/O by utilizing the locality of reference. However, since a particular page may be simultaneously cached in multiple nodes, modification of the page in any buffer invalidates copies of that page in other nodes. This necessitates the use of a *cache coherency scheme* (CCS) so that the nodes always see the most recent version of database pages [3, 4, 5, 13].

The CCS incurs two overheads: *locking overhead* and *page transfer overhead*. The locking overhead comes from implementing distributed latches in the SD cluster to satisfy an inter-node consistency of a page. Specifically, each node is required to check the page validity whenever it accesses a record on the page. This involves message exchanging between the node and the locking service [3, 13]. The page transfer overhead occurs when the cached page is out of date. In this case, the node has to receive the most recent version of the page from other node via message passing or from the shared disks [13].

Most of previous CCSs in the SD cluster do not discriminate between index pages and data pages. However, index pages are accessed more frequently and cached at more nodes than their corresponding data pages. This means that the locking overhead should be significant if we just apply the conventional CCSs to index pages. Furthermore, the tree-based indexes such as B^+ trees require complicated operations, e.g., root-to-leaf traversal and structure modification operation due to a page split or a page concatenation. As a result, it is strongly required to devise a dedicated CCS for index pages.

In this paper, we propose a new CCS for B^+ tree indexes in the SD cluster. The proposed CCS is novel in the sense that it takes advantage of the semantics of complicated B^+ tree operations: hence, it can minimize the message traffic between nodes and the number of tree re-traversals.

The rest of this paper is organized as follows. Sect. 2 introduces the previous CCSs for index pages. Sect. 3 presents the proposed CCS in detail. We have evaluated the performance of the CCSs using a simulation model. Sect. 4 describes the simulation model and Sect. 5 analyzes the experiment results. Concluding remarks appear in Sect. 6.

2 Related Work

To the best of our knowledge, ARIES/IM-SD [14] is the only CCS for B^+ tree indexes in the SD cluster. ARIES/IM-SD tries to reduce the locking overhead on index pages by taking into account the frequency of updates of various types of pages. Specifically, ARIES/IM-SD uses an *invalidation* scheme [5] for non-leaf pages that are updated rarely. Hence, a node can access non-leaf pages immediately if they are not invalidated. For leaf pages that are updated frequently, ARIES/IM-SD uses a *check-on-access* scheme [5]. Each node has to request the global lock manager (GLM) to check whether its cached leaf page is valid.

To preserve the consistency of root-to-leaf traversal, ARIES/IM-SD extends the lock-coupling scheme of ARIES/IM [12]. After a child page is latched but before searched, ARIES/IM-SD checks if the cached parent page is still valid. If the parent page is invalidated, its current version is delivered from other node or disks. Then the B^+ tree has to be traversed again. The drawbacks of ARIES/IM-SD are two-fold: (a) it does not exploit any idea to reduce the page transfer overhead, and (b) the B^+ tree could be re-traversed unnecessarily, which will be illustrated in the next section.

We have reviewed CCSs for B^+ tree indexes proposed in different architectures such as client-server database systems or SN clusters. Most of the CCSs try to improve the performance by caching part of a B^+ tree to reduce the locking overhead [10, 18, 20]. However, this should not be applied to the SD cluster because the SD cluster allows every node to access the entire database. Each node has to be able to cache the whole B^+ tree.

RIC [8] is a CCS proposed in client-server database systems, which allows each client to cache the whole B^+ tree. Unlike ARIES/IM-SD, RIC adopts an optimistic approach where a client does not check the validity of cached pages during the root-to-leaf traversal. When the client requests a data lock to the server, it sends cache coherency information together. The cache coherency information consists of an identifier of index segment and its version number. Using the information, the server validates the currency of the index segment. If the index segment is not recent, the server returns a PageLSN[1] vector of all pages in the index segment. Then the client invalidates changed pages according to the PageLSN vector. If any pages in the traversal path are invalidated, the B^+ tree has to be traversed again. The main drawback of RIC is heavy message traffic due to PageLSN vectors. RIC also does not try to reduce the page transfer overhead and could suffer from unnecessary tree re-traversals.

3 Cache Coherency Scheme for B^+ Trees

In this section, we propose a new CCS for B^+ trees, named path cognitive CCS (PCCS). We first describe our assumption and basic idea of PCCS. Then we describe the details of PCCS for each index operation. We consider four basic operations on the B^+ tree [12].

- **Fetch:** Given a key value, check if it is in the index by traversing the B^+ tree with root-to-leaf manner.
- **Fetch Next:** Having opened a range scan with a Fetch call, fetch the next key within a specified range by traversing the chain of leaf pages.

[1] Every page in the database is assumed to have a PageLSN field that contains the log sequence number (LSN) of the log record that describes the latest update to the page. Since LSN is monotonically increasing, we can determine which page is more recent by comparing PageLSNs of two pages [13].

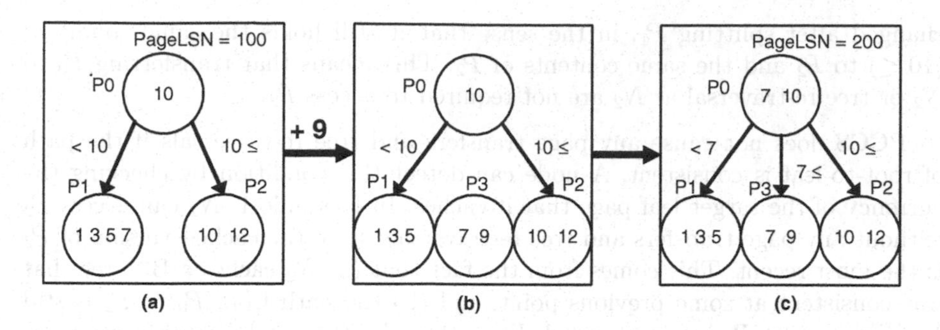

Fig. 1. Example of a structure modification by insertion

- **Insert:** Insert the given key to the B^+ tree. It begins with a root-to-leaf traversal to locate a leaf page, and then the work of insertion, splitting, and promotion proceeds upward from the leaf.
- **Delete:** Delete the given key from the leaf page of the B^+ tree.

We assume that a delete operation does not cause page concatenations even though the leaf page could be underflow. Due to the advantages of keeping delete operations cheap and reducing the probability of page split for insert operations, most practical implementations of B^+ tree adopt this approach [9]. Sect. 3.4 will discuss issues to be considered in case of page concatenation.

3.1 Assumption and Basic Idea

Similar to ARIES/IM-SD, we assume that there is a global lock manager (GLM), which not only provides global locking functions for every node, but also tracks which node has a valid copy of a page. The GLM maintains two tables: (1) lock table, which registers record locks held by active transactions, and (2) cache table, which registers (page identifier, PageLSN, owner) for each cached page. The PageLSN field of a page P_i is set to the maximum PageLSN among the cached version of P_i. An owner is a node that caches the most recent version of P_i and can update P_i. By allowing at most one owner per page, the physical consistency of cached pages can be satisfied.

As we have described in the previous section, both ARIES/IM-SD and RIC cause unnecessary root-to-leaf traversals and page transfers. Example 1 illustrates this problem.

Example 1: Suppose two nodes N_1 and N_2 in the SD cluster cache a B^+ tree of Fig. 1(a). Suppose also that N_1 is an owner of pages P_0 and P_1, and tries to insert a new key 9 to the B^+ tree. N_1 updates the B^+ tree into Fig. 1(b) and then into Fig. 1(c). In ARIES/IM-SD, this invalidates P_0 cached at N_2, and N_2 has to receive the recent version of P_0 from N_1. Furthermore, if N_2 was at the stage of accessing P_2, it has to traverse the B^+ tree again since the parent page (P_0) of P_2 is invalidated. The same scenario also occurs in RIC when N_2 requests any data locks on keys at P_2. However, note that the path from P_0 to P_2 is not changed after splitting P_1, in the sense that it still holds the same condition ($10 \leq$) to P_2 and the same contents of P_2. This means that transferring P_0 to N_2 or tree re-traversal at N_2 are not required to access P_2. □

PCCS does not cause any page transfers and tree re-traversals if the path of root-to-leaf is consistent. A node can detect this condition by checking the currency of the target leaf page that it caches. In Example 1, N_2 can access P_2 without any page transfers and tree re-traversals since the cached version of P_2 is the most recent. This comes from the fact that (a) N_2 caches a B^+ tree that was consistent at some previous point, and (b) the path from P_0 to P_2 is still consistent until P_2 is not changed. Even though other nodes may insert more keys to P_1 or P_3, it does not affect the consistency of the path from P_0 to P_2.

3.2 Fetch

A node may have many transactions accessing a cached B^+ tree concurrently. To synchronize them, PCCS traverses the B^+ tree with a lock-coupling manner like ARIES/IM [12]. In the lock-coupling, a latch to a parent page is released after acquiring a latch to the child page. The latch means a local latch to support an intra-node consistency of pages accessed by transactions in the node [2, 11]. A distributed latch to support an inter-node cache coherency is implemented by a page lock that causes message exchanging with the GLM [3, 13].

Step 0: Suppose that a transaction T_i of a node N_i traverses a B^+ tree cached at N_i. T_i performs the following steps in order.
Step 1: Acquire a shared latch to a root page of the B^+ tree.
Step 2: Find a child page P_c in N_i's buffer.
1. If P_c is cached in the buffer and it is a leaf page, then Goto Step 5.
2. If P_c is cached in the buffer and it is a non-leaf index page, then Goto Step 4 after requesting a shared latch on P_c.

Step 3: If P_c is not cached in N_i's buffer, then N_i sends the GLM a cache check message $\mathcal{C}$, which is a list of (page identifier, PageLSN) of P_c and its ancestor pages. P_c's PageLSN in $\mathcal{C}$ is set to 0.
1. The GLM sends N_i a cache validation message $\mathcal{V}$, generated by comparing each (P_i, PageLSN$_i$) $\in \mathcal{C}$ with that in the cache table (CTable) as follows.
   – If PageLSN$_i$ < CTable[P_i].PageLSN, then includes (P_i, CTable[P_i].owner) to $\mathcal{V}$ and requests CTable[P_i].owner node to transfer P_i to N_i.
   – If CTable[P_i] is empty, then include (P_i, 0) to $\mathcal{V}$.
2. For each (P_i, owner$_i$) $\in \mathcal{V}$, N_i receives the current version of P_i from owner$_i$ if owner$_i \neq 0$. Otherwise N_i accesses P_i from the disk.
3. If any parent pages of P_c are included in $\mathcal{V}$, then Goto Step 1.

Step 4: After acquiring a shared latch to P_c, release the shared latch on parent page. If P_c is not a leaf page, then Goto Step 2.
Step 5: Validate the root-to-leaf traversal as Fig. 3 describes.

Fig. 2. B^+ tree traversal with lock-coupling

Fig. 2 describes the procedure of B+ tree traversal at a node N_i. PCCS does not check the currency of non-leaf pages during the traversal. So additional communications between N_i and the GLM are not required if the pages are cached in the N_i's buffer. When a page P_c is not cached, N_i sends the GLM a cache check message to request P_c (step 3). The cache check message is a list of (page identifier, PageLSN) of P_c and its ancestor pages. This is required to guarantee the validity of a path from the root page to P_c. Suppose N_i sends only the information of P_c to the GLM, and P_c has been split due to insert operations at other node. Then receiving the current version of P_c could cause an inconsistent traversal, since N_i caches an old version of P_c's parent page that does not include the result of split.

The GLM returns a list of invalidated pages to N_i by comparing the PageLSNs of cache check message and its cache table. If any ancestor pages of P_c are invalidated, N_i has to receive the current versions from the corresponding page owners and re-traverse the B+ tree from the root page. This way the consistency of the path from the root page to P_c can be preserved.

After finding a key at the leaf page, N_i has to check the validity of the root-to-leaf traversal by sending a cache check message piggybacked with the data lock request of the key. The cache check message is a list of (page identifier, PageLSN) for all pages from the root to the leaf in the traversal path.

Fig. 3 shows the procedure of validating the consistency of the root-to-leaf traversal. The idea is similar to the procedure of handling cache misses at the B+ tree traversal. The difference is that page transfers or tree re-traversals are required only when leaf pages are out of date. Releasing a data lock at step 2.2 is required since N_i traversed an old version of B+ tree and thus it might access a wrong data record. Example 2 illustrates the validation procedure.

Step 1: A node N_i sends the GLM a data lock request piggybacked with a cache check message $\mathcal{C}$, which is a list of (page identifier, PageLSN) for all pages from the root to the leaf in the traversal path.

Step 2: If the data lock is granted, the GLM checks whether the leaf page P_{leaf} is valid. If $\text{PageLSN}_{leaf} = \text{CTable}[P_{leaf}].\text{PageLSN}$, then P_{leaf} is valid.

1. If P_{leaf} is valid, the GLM sends N_i a lock grant message.
2. If P_{leaf} is invalid, the GLM first releases the data lock. Then the GLM sends N_i the cache validation message $\mathcal{V}$ as step 3.1 of Fig. 2 describes.

Step 3: N_i can execute the operation if it receives a lock grant message.

Step 4: If N_i receives a cache validation message, it performs the following steps.

1. Acquire exclusive latches for all pages in $\mathcal{V}$.
2. Receive the page from the owner node if the owner field is not 0.
3. Access the page from the disk if the owner field is 0.
4. Release latches after the pages are cached.
5. Re-traverse the B+ tree from the root again. Goto Step 1.

Fig. 3. Validation of root-to-leaf traversal

Example 2: Suppose that a node N_1 caches a B$^+$ tree of Fig. 1(c) and N_2 caches a tree of Fig. 1(a). Suppose also that the P_1.PageLSN is 80 and 180 in tree (a) and (c) respectively, and the P_2.PageLSN is 50 in both trees. If N_2 accesses a record whose key is 10, then after traversing from P_0 to P_2 it sends the GLM a lock request message including $\{(P_0, 100), (P_2, 50)\}$. The GLM returns a lock grant message to node N_2 because P_2.PageLSN is the highest. On the other hand, if N_2 accesses a record whose key is 7, the lock request message includes $\{(P_0, 100), (P_1, 80)\}$. Then the GLM returns a cache validation message including a list of $\{(P_0, N_1), (P_1, N_1)\}$ to node N_2 because P_1.PageLSN is not the highest. The GLM also requests N_1 to transfer both P_0 and P_1 to node N_2. After N_2 receives the recent version of P_0 and P_1, it traverses the B$^+$ tree again. Finally, N_2 can access P_3 according to the new version of P_0. □

The potential advantages of PCCS compared to ARIES/IM-SD and RIC are as follows. First, PCCS does not cause any message traffic during the root-to-leaf traversal as long as the non-leaf pages in the traversal path are cached. Next, even though a node updates a non-leaf page, PCCS does not invalidate immediately the page cached in other nodes. This increases the scalability of PCCS since the invalidation based CCSs suffer from heavy message traffic when there are a lot of nodes [5]. Last, PCCS can prevent unnecessary page transfers and tree re-traversals since it does not require the most recent versions of non-leaf pages as long as the traversal path is consistent.

It is worthy to discuss the overhead of PCCS. PCCS checks the validity of root-to-leaf traversal when a data lock is requested to the GLM. This must increase the length of lock request message. The response time of the data lock will also be delayed due to the validation processing if the leaf page is out of date. However, we believe that the communication overhead due to the slightly longer message is not significant, since in the SD cluster the fixed protocol processing (i.e., CPU overhead) dominates the on-the-wire time for messages in network. The delay of response time depends on the height of B$^+$ tree, which might be reduced effectively with prefix and/or suffix compression [9]. Furthermore, the validation has to be performed only once per root-to-leaf traversal.

3.3 Other Index Operations

In this section, we sketch roughly how the remaining index operations can be extended from the conventional index management scheme, such as ARIES/IM [12]. The basic idea is to replace the latch into the page lock that involves message exchanging with the GLM similar to ARIES/IM-SD.

After locating the target leaf page by the fetch operation, a fetch next operation accesses the next key within a specified range by traversing the chain of leaf pages. The fetch next operation does not cause validation of root-to-leaf traversal since it accesses leaf pages only. Instead a node has to check the validity of a cached leaf page by sending the GLM a PageLSN of the page together with a data lock request. If the page is out of date, the node has to receive the current version and then search the next key again. In a special case where the page was split, the node also has to invalidate its cached parent page if any. This is because the parent page does not include the result of split. The node will cache the recent version of the invalidated page in the procedure of root-to-leaf traversal of other leaf page.

PCCS executes insert and delete operations on a leaf page sequentially so that there is no missing updates. Specifically, at both operations, a node N_i sends the GLM an exclusive page lock on a leaf page P_{leaf} together with a data lock request. If N_i is not an owner of P_{leaf}, the GLM requests the owner node of P_{leaf} to transfer P_{leaf} to N_i. Note that only an owner can update a page. After that, the GLM registers N_i as a new owner of P_{leaf}.

An insert operation may cause a structure modification operation (SMO) that splits a page. Suppose N_i is an owner of P_{leaf} and a transaction T_i at N_i tries to insert a new key into P_{leaf}. If P_{leaf} is full, T_i has to execute an SMO to split P_{leaf} to make a space for the new key. T_i first acquires a tree lock and a tree latch to prevent simultaneous SMOs by other transactions in different node or N_i, respectively. Then T_i sets the SM_bit of P_{leaf} to 1. The SM_bit is used to indicate that a page is currently under SMO. If another transaction tries to access a page with SM_bit = 1, then it has to wait until the SMO completes and traverse the B+ tree again [12]. After choosing a new page P_{new}, T_i distributes the keys in P_{leaf} evenly at P_{leaf} and P_{new}. Then T_i promotes the result of split to the parent page P_{parent}. In this case, N_i has to be an owner of P_{parent}. T_i executes the insertion into P_{parent} with the same manner. When the SMO is complete, SM_bits are set to 0 and T_i releases the tree latch and tree lock.

It is important to note that PCCS can reduce the duration of SMO compared to ARIES/IM-SD. This is because the SMO does not include any message exchanging to invalidate pages cached in other nodes.

3.4 Page Concatenation

PCCS validates the result of root-to-leaf traversal by checking the currency of the target leaf page only. If the page concatenation due to a delete operation is not allowed, this policy is correct since the separator in a non-leaf page will not be removed. Separators might be moved to different non-leaf page as a result of page split, but PCCS updates the non-leaf pages only when there are no transactions accessing the pages by requiring the exclusive latches on the pages (see step 4 of Fig. 3). Then the succeeding transactions can find the correct path again. If the page concatenation is allowed, this policy would cause incorrect search results as Example 3 illustrates.

Example 3: Suppose that two nodes N_1 and N_2 are caching a tree of Fig. 4(a). N_1 first deletes a key 21 and the tree is changed to Fig. 4(b) as a result of the page concatenation. PageLSNs of page P_{10}, P_3, and P_1 increase. N_1 deletes again a key 5 and the tree is changed to Fig. 4(c). A PageLSN of page P_6 increases as a result. After that, suppose that N_2 caching a tree of Fig. 4(a) tries to search a key 2. Since N_2's P_6 is out of date, N_2 receives P_6 and P_1 from N_1. Now N_2 caches a tree of Fig. 4(d). If N_2 searches a key 22 after accessing P_9, it makes

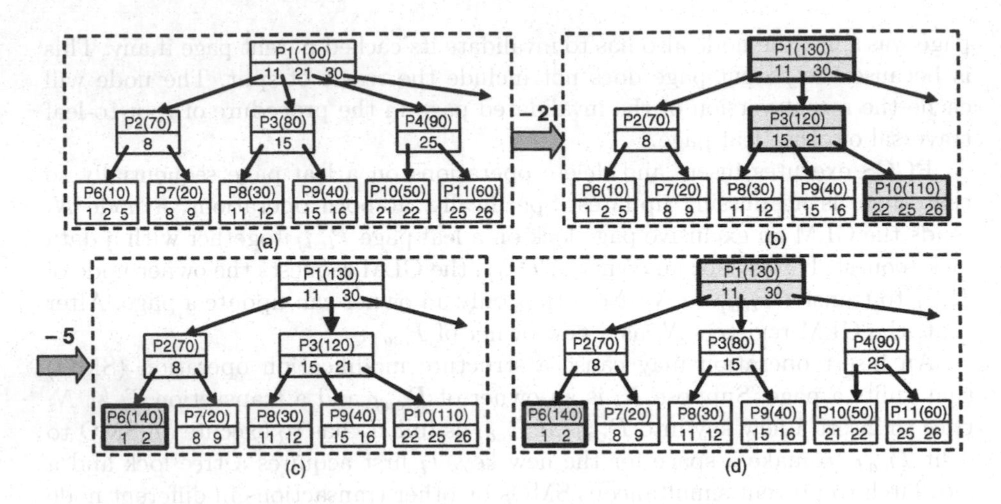

Fig. 4. A scenario of incorrect searching due to page concatenation

an incorrect result that there is not a key 22. However, the GLM cannot detect this error because P_9 cached in N_2 is the most recent! □

This problem comes from the fact that (a) the page concatenation procedure deletes a separator 25 at P_4, and (b) the new version of P_3 is not delivered to N_2 since P_3 is not in the traversal path from P_1 to P_6. Note that if the page concatenation is not allowed, the problem does not occur since separators on non-leaf pages are not deleted.

We can resolve the problem by broadcasting invalidation messages on concatenated pages as ARIES/IM-SD does. Specifically, after N_1 concatenates the tree as Fig. 4(b), it informs the updates of P_1, P_3, and P_4 to the GLM. Then the GLM broadcasts invalidation messages on the pages to other nodes. N_2 invalidates the pages after receiving the messages. Thereafter N_2 can cache the current versions of them with the B$^+$ tree traversal procedure of Fig. 2.

4 Simulation Model

We evaluate the performance of PCCS using a simulation model. Fig. 5 shows the simulation model of the SD cluster. The model was implemented using the CSIM discrete-event simulation package [16].

The SD cluster consists of a GLM plus varying number of nodes, all of which are connected via high speed local area network. Every node shares disks. A node has a *buffer manager* that manages its LRU buffer and a *latch manager* that manages local latches for cached pages. A *transaction generator* generates transactions, each of which is modelled as a sequence of database operations. The GLM has a role to perform the concurrency control and cache coherency control. A *transaction scheduler* supports two phase locking with record gran-

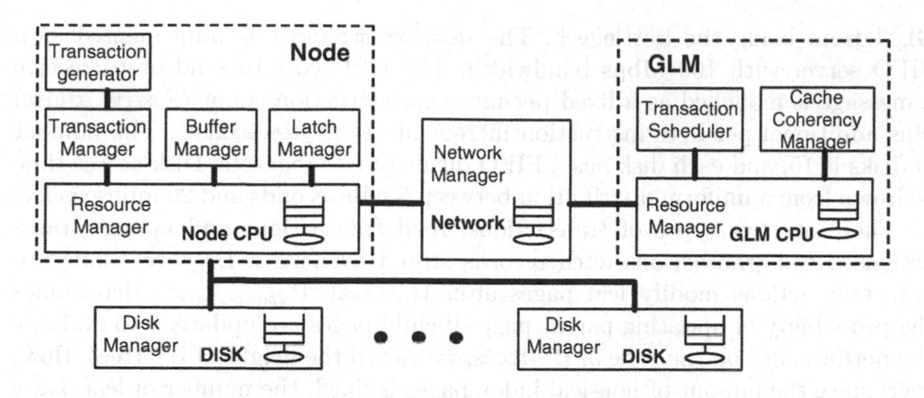

Fig. 5. A simulation model of SD cluster

ularity and deadlock resolution. The cache coherency manager implements one of three CCSs, including PCCS, ARIES/IM-SD, and RIC.

Table 1 shows the simulation parameters used in experiments. Many of system parameter values extend those of [3, 4] to reflect the current status of cluster technology. The values of index parameters are adopted from [8, 20]. GLM's CPU is assumed to perform much better than each node's CPU to prevent the

Table 1. Simulation parameters

System Parameters		
NodeCPUSpeed	Instruction rate of a node CPU	1 GIPS
GLMSpeed	Instruction rate of a GLM CPU	2 GIPS
NetBandwidth	Network Bandwidth	100 Mbps
NumNode	Number of computing nodes	$1 \sim 40$
NumDisk	Number of shared disks	10
MinDiskTime	Minimum disk access time	5 milliseconds
MaxDiskTime	Maximum disk access time	20 milliseconds
BufSize	Per-node buffer size	500 pages
FixedMsgInst	Fixed number of instructions per message	20,000
PerByteMsgInst	Additional instructions per-message byte	10,000 per page
PerIOInst	Number of instructions per disk I/O	5000
TRSize	Number of leaf pages accessed by transaction	10
TRSizeDev	Deviation of transaction size	10%
P_{lookup}	Probability of read transactions (%)	$20 \sim 100$
P_{update}	Probability of update transactions (%)	$0 \sim 80$
Index Parameters		
IdxDepth	Height of B+ tree index	3, 4, 5
AccessedLeaf	Number of accessed leaf pages	1000
Degree	Degree (fan-out) of index pages	100
PerPageInst	Number of instructions per index page	5000
P_{update_parent}	Probability of updating parent pages	10%

GLM from being the bottleneck. The `network manager` is implemented as a FIFO server with 100 Mbps bandwidth. The CPU cost to send or to receive a message is modelled as a fixed per-message instruction count (`FixedMsgInst`) plus additional per-byte instruction increment (`PerByteMsgInst`). The number of disks is 10, and each disk has a FIFO queue of I/O requests. Disk access time is drawn from a uniform distribution between 5 milliseconds and 20 milliseconds.

There are two types of transactions, read transactions and update transactions. Read transactions fetch records after traversing a B^+ tree, while update transactions modify leaf pages after traversal. P_{update_parent} determines the probability of updating parent pages if child pages are updated. To evaluate the performance on the type of B^+ trees, we varied the height of B^+ trees. However, since the fan-out of non-leaf index pages is fixed, the number of leaf pages increases at deep B^+ trees and the cache hit ratio may be distorted as a result. So we assume to access only a fixed number of leaf pages (`AccessedLeaf`).

5 Experiment Results

We compare the performance of PCCS with ARIES/IM-SD and RIC using the simulation model. The performance metric used in the experiments is the average response time of transactions. The communication time and the tree traversal time are also used in analyzing the experiment results. The communication time gives the total time for transferring messages and pages per transaction commit. The tree traversal time gives the aggregation of time spent in traversing the B^+ tree per transaction commit.

5.1 Experiment 1: Varying the Probability of Update Transactions

We first compare the performance of CCSs by varying the probability of update transactions (P_{update}). Fig. 6 shows the experiment results when there are 10 nodes and the height of B^+ tree is set to 3.

All the CCSs perform worse as P_{update} increases. At high P_{update}, a lot of update transactions deteriorate the cache hit ratio of each node due to the frequent buffer invalidations. PCCS outperforms other CCSs significantly at high P_{update}. This is because PCCS can reduce the message traffic as Fig. 6(b) shows. When P_{update} is 80%, the communication time of PCCS is only about half of ARIES/IM-SD and under 10% of RIC. Compared to PCCS, ARIES/IM-SD causes considerable message traffic due to broadcasting the buffer invalidation messages. ARIES/IM-SD also suffers from frequent tree re-traversals as Fig. 6(c) shows, even though its effect to the total response time is not significant. RIC performs worst because of the heavy message traffic for transferring PageLSN vectors from the GLM. Note that the size of PageLSN vector is larger at least two orders of magnitude compared to control messages (e.g., cache check message or buffer invalidation message) used in other CCSs. Furthermore, the tree traversal time of RIC is the longest since a node has to re-traverse the tree if any page in the traversal path has been updated.

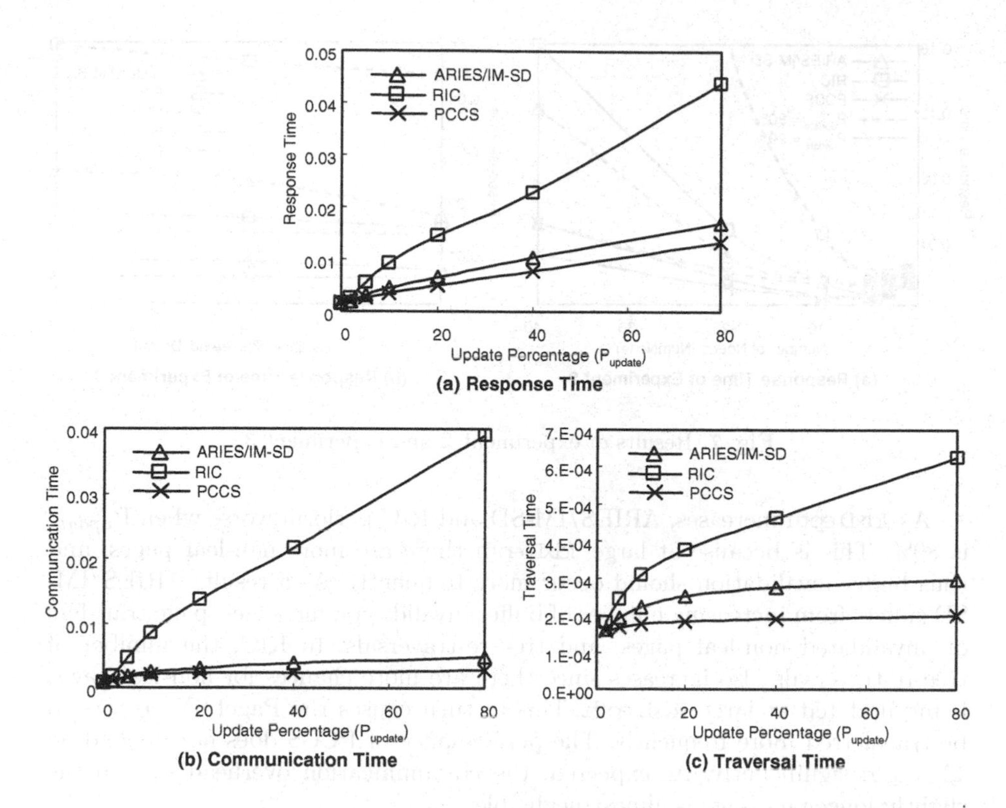

Fig. 6. Results of experiment 1

5.2 Experiment 2: Varying the Number of Nodes

We also compare the performances of CCSs by varying the number of nodes (NumNode). Fig. 7(a) shows the experiment result when the height of B^+ tree is set to 3 and P_{update} is set to 20% and 80%, respectively.

As NumNode increases, PCCS outperforms other CCSs especially when P_{update} is 80%. The reasons are as follows. First, when NumNode is large, many update transactions will be executed at the same time and the probability of buffer invalidation should increase as a result. This is why RIC performs worst again. Second, many nodes would cache same index pages when NumNode is large. Then if any non-leaf index page is updated, ARIES/IM-SD has to send large number of buffer invalidation messages. It is well-known that the invalidation based CCSs suffer from heavy message traffic as NumNode increases [5].

5.3 Experiment 3: Varying the Height of B^+ Tree

We finally compare the performances of CCSs by varying the height of B^+ tree (IdxDepth). Fig. 7(b) shows the experiment result when NumNode is set to 10 and P_{update} is set to 20% and 80%, respectively.

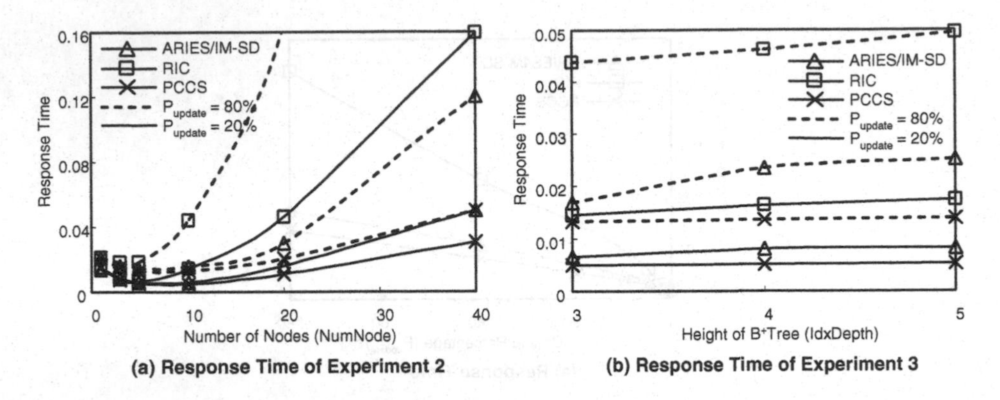

Fig. 7. Results of experiment 2 and experiment 3

As IdxDepth increases, ARIES/IM-SD and RIC perform worse when P_{update} is 80%. This is because at large IdxDepth there are more non-leaf pages, and thus buffer invalidation should occur more frequently. As a result, ARIES/IM-SD suffers from increasing number of buffer invalidation messages, page transfers on invalidated non-leaf pages, and tree re-traversals. In RIC, the number of tree re-traversals also increases since there are more chances for non-leaf pages being updated at large IdxDepth. This in turn causes the PageLSN vectors to be transferred more frequently. The performance of PCCS does not depend on IdxDepth significantly. As expected, the communication overhead due to the slightly longer message is almost negligible.

6 Conclusions

In this paper, we proposed a new CCS for B^+ tree indexes, named path cognitive CCS (PCCS), in the SD cluster. The novel features of PCCS are summarized as follows. First, PCCS does not cause any message traffic during the root-to-leaf traversal as long as the non-leaf pages in the traversal path are cached. Next, even though a node updates a non-leaf page, PCCS does not invalidate the page cached in other nodes. This increases the scalability of PCCS since the invalidation based CCSs such as ARIES/IM-SD suffer from heavy message traffic when there are a lot of nodes. Last, PCCS can prevent unnecessary page transfers and tree re-traversals since it does not require non-leaf pages to be the most recent as long as the traversal path is consistent.

We have explored the performance of PCCS under a variety of system configurations. PCCS outperforms previous CCSs significantly (1) when there are a lot of nodes, (2) when the number of update transactions is large, or (3) when the height of B^+ tree is large. This result is encouraging because the main applications of SD cluster are large-scale OLTP applications where data volumes are huge and a lot of update transactions execute concurrently. Furthermore, considering the ever increasing scalability of the SD cluster, it is very important to achieve good performance at large number of nodes.

References

1. Carey, M., Franklin, M., Zahrioudakis, M.: Adaptive, Fine-Grained Sharing in a Client-Server OODBMS: A Callback-Based Approach. ACM Trans. on Database Syst. **22** (1997) 570-627
2. Cha, S., Hwang, S., Kim, K., Kwon, K.: Cache-Conscious Concurrency Control of Main-Memory Indexes on Shared-Memory Multiprocessor Systems. In: Proc. 27th Int. Conf. VLDB (2001) 181-190
3. Cho, H.: Cache Coherency and Concurrency Control in a Multisystem Data Sharing Environment. IEICE Trans. Information and Syst. **E82-D** (1999) 1042-1050
4. Cho, H., Park, J.: Maintaining Cache Coherency in a Multisystem Data Sharing Environment. J. Syst. Architecture **45** (1998) 285-303
5. Dan, A., Yu, P.: Performance Analysis of Buffer Coherency Policies in a Multisystem Data Sharing Environment. IEEE Trans. Parallel and Distributed Syst. **4** (1993) 289-305
6. DB2 Universal Database for OS/390 and z/OS - Data Sharing: Planning and Administration. IBM SC26-9935-01 (2001)
7. DeWitt, D., Gray, J.: Parallel Database Systems: The Future of High Performance Database Systems. Comm. ACM **35** (1992) 85-98
8. Gottemukkala, V., Omiecinski, E., Ramachandran, U.: Relaxed Index Consistency for a Client-Server Database. In: Proc. 12th Int. Conf. Data Eng. (1996) 352-361
9. Gray, J., Reuter, A.: Transaction Processing: Concepts and Techniques. Morgan Kaufmann Publishers (1993)
10. Miyazaki, J., Yokota, H.: Concurrency Control and Performance Evaluation of Parallel B-tree Structures. IEICE Trans. Information and Syst. **E85-D** (2002) 1269-1283
11. Mohan, C., Haderle, D., Linsay, B., Pirahesh, H., Schwarz, P.: ARIES: A Transaction Recovery Method Supporting Fine-Granularity Locking and Partial Rollbacks Using Write-Ahead Logging. ACM Trans. on Database Syst. **17** (1992) 94-162
12. Mohan, C., Levine, F.: ARIES/IM: An Efficient and High Concurrency Index Management Method Using Write-Ahead Logging. In: Proc. ACM SIGMOD (1992) 371-380
13. Mohan, C., Narang, I.: Recovery and Coherency Control Protocols for Fast Intersystem Page Transfer and Fine-Granularity Locking in a Shared Disks Transaction Environment. In: Proc. 17th Int. Conf. VLDB (1991) 193-207
14. Mohan, C., Narang, I.: Locking and Latching Techniques for Transaction Processing Systems Supporting the Shared Disks Architecture. IBM Research Report (1995)
15. Oracle Application Server 10g - High Availability. An Oracle White Paper (2004)
16. Schwetman, H.: User's Guide of CSIM18 Simulation Engine. Mesquite Software, Inc. (1996)
17. Vallath, M.: Oracle Real Application Clusters. Elsevier Digital Press (2004)
18. Yokota, H., Kanemasa, Y., Miyazaki, J.: Fat-Btrees: An Update-Conscious Parallel Directory Structure. In: Proc. 15th Int. Conf. Data Eng. (1999) 448-457
19. Yousif, M.: Shared-Storage Clusters. Cluster Comp. **2** (1999) 249-257
20. Zaharioudakis, M., Carey, M.: Highly Concurrent Cache Consistency for Indices in Client-Server Database Systems. In: Proc. ACM SIGMOD (1997) 50-61

Message Strip-Mining Heuristics for High Speed Networks

Costin Iancu[1], Parry Husbands[1], and Wei Chen[2]

[1] Computational Research Division, Lawrence Berkeley National Laboratory
{cciancu, pjrhusbands}@lbl.gov
[2] Computer Science Division, University of California at Berkeley
wychen@cs.berkeley.edu

Abstract. In this work we investigate how the compiler technique of message strip-mining performs in practice on contemporary high performance networks. Message strip-mining attempts to reduce the overall cost of communication in parallel programs by breaking up large message transfers into smaller ones that can be overlapped with computation. In practice, however, network resource constraints may negate the expected performance gains. By deriving a performance model and synthetic benchmarks we determine how network and application characteristics influence the applicability of this optimization. We use these findings to determine heuristics to follow when performing this optimization on parallel programs. We propose strip-mining with variable block size as an alternative strategy that performs almost as well as a highly tuned fixed block strategy and has the advantage of being performance portable across systems and application input sets. We evaluate both techniques using synthetic benchmarks and an application from the NAS Parallel Benchmark suite.

1 Introduction

Reducing the overhead of message transfers is the goal of many optimization techniques for parallel programs. One such strategy, message vectorization [20], reduces communication time by hoisting fine-grained remote reads and writes outside loops and coalescing them into bulk transfers. This has the benefit of speeding up communication both by amortizing startup costs and by taking advantage of the higher bandwidths realized for larger messages. Implemented in optimizing compilers for a variety of parallel programming languages [9, 13, 12], vectorization is widely recognized as an extremely useful optimization for message-passing programs.

While message vectorization is a common and effective optimization, it alone is not enough to minimize total run time, as the processor remains idle while the network is busy performing the bulk message transfer. In this paper we focus on a technique called *message strip-mining* [19] that has the potential to further enhance the effectiveness of message vectorization. While a vectorized loop waits for the remote memory access to complete before it proceeds with local

M. Daydé et al. (Eds.): VECPAR 2004, LNCS 3402, pp. 424–437, 2005.

computation, message strip-mining divides communication and computation into phases and pipelines their executions by skewing the loop. This increases the number of messages sent and thus message startup costs, but has the potential to reduce communication overhead through the overlapping of non-blocking send and receive operations with independent computation. Loop unrolling can in addition be applied to increase the number of operations available for overlap.

Applying message strip-mining carelessly may result in performance degradation due to both increased startup costs and network contention. Furthermore, performance is directly influenced by application characteristics; the data transfer size and the ratio between communication and computation affect the amount of available overlap. A systematic scheme is therefore required to evaluate, for a vectorizable loop, whether it is worthwhile to perform message strip-mining and loop unrolling based on network parameters and application communication patterns. Our contributions in this paper are: 1) An enumeration of application and machine characteristics that influence the applicability and performance of message strip-mining; and 2) A determination of how these characteristics guide us in choosing good message decompositions for message strip-mining based on analytical models and experimental results.

Our work differs from previous investigations [18, 19] in that we consider both network and application characteristics, use a performance model based on LogGP [6], and use both synthetic and application benchmarks that capture the characteristics of a wide class of programs. While previous approaches use a static decomposition, we introduce a variable block size message decomposition that performs in practice almost as well as the fixed size strategy and it is less sensitive to variations in the performance parameters. Experimental results suggest that message strip-mining is a highly effective optimization for any vectorizable loop whose total data transfer size exceeds a minimal threshold that has a typical value of 2-4 KBytes for today's high performance networks. Furthermore, the optimization does not require a large amount of local computation to be effective, provided the fetched remote data is loaded at least once; the overhead of the associated cache misses is usually enough to hide most of the communication time. Therefore, we believe message strip-mining holds great potential as an automatic compiler optimization for reducing application communication costs.

The rest of this paper is organized as follows. Section 2 introduces message strip-mining and identifies network and application characteristics that affect the success of this optimization. The analytical model for the performance of strip-mining is described in Section 2.2, and our experimental results are presented in Section 3. Related work is surveyed in Section 4 and we conclude in Section 5.

2 Message Strip-Mining

Strip-mining is a loop transformation commonly associated with vectorizing compilers. In this context a single loop is first transformed into a doubly nested loop and the inner loop is then replaced by a sequence of vector instructions. This transformation is illustrated in Figures 1 and 2 (the generation of vector in-

```
for(i=0; i<N; i++)
    a[i] = b[i]+r[i];
```

Fig. 1. Unoptimized loop

```
for(i=0; i<N; i+=S)
    for(ii=i; ii<min(i+S-1,N), ii++)
        a[ii] = b[ii]+r[ii];
```

Fig. 2. Strip-mined loop

```
get(lr, r, N);
for(i=0; i<N; i++)
    a[i] = b[i]+lr[i];
```

Fig. 3. Vectorized loop

structions is omitted). The parameter S, often dependent on the size of a vector register, is usually referred to as the strip size.

In this paper we explore a less well-known application of strip-mining: its ability to reduce the communication overhead of parallel programs. Following the standard "owner computes" rule that most parallel paradigms adhere to, we consider only remote read operations as candidates for message strip-mining, though the analytical model and optimization techniques in this paper can easily be adapted to support remote writes.

Assuming that arrays `a` and `b` in Figure 1 are local and `r` is remote, we now demonstrate message strip-mining. Figure 3 displays the results of applying message vectorization to the unoptimized loop. Performance is significantly improved by copying all remote value in one bulk transfer instead of performing a read in every iteration. One disadvantage with such a transformation, however, is that the processor must wait for the completion of the remote transfer (denoted by $T(N)$) before proceeding with the local computation. Figure 4 demonstrates the process of message strip-mining, where the single bulk transfer[1] from Figure 3 is divided into several blocks based on the strip size, and the loop is then skewed so that the communication and computation code can be performed in a pipelined manner. The transformation is similar in spirit to software pipelining, as it exploits the parallelism within the loop body to allow the communication and computation phases of several iterations of the loop to be processed simultaneously. Moreover, the computation is not the only source of overlap; as Figure 5 shows, loop unrolling can be additionally applied to increase the amount of overlap available in the unrolled loop body, by issuing operations for different strips at the same time. In an ideal scenario (ignoring the overheads of initiating and completing remote operations), the communication time of each strip is completely overlapped with independent computation or communication from previous iterations, leaving only the overhead $T(S)$ of transferring the very first strip. Because S is typically much smaller than N, the maximum performance gain $T(N) - T(S)$ can be significant. A less obvious benefit of this optimization is that it can help reduce the amount of shadow memory used to store remote elements. While a vectorized loop requires an N element buffer, the strip-mined loop can reuse its buffers and therefore only requires $U * S$ elements, where U is the unroll depth of the loop.

[1] We denote a blocking memory read operation by `get(dest, src, nbytes)` and a non-blocking read operation by `h = nbget(dest, src, nbytes)`. The completion test for a non-blocking remote memory operation is denoted by `sync(h)`.

```
h0 = nbget(lr, r, S);
for(i=0; i<N; i+=S)
  h1 = nbget(lr+S*(i+1), r+S*(i+1), S);
  sync(h0);
  for(ii=i; ii<min(...); ii++)
    a[ii] = b[ii]+lr[ii];
  h0=h1;
```

Fig. 4. Message strip-mining

```
h[0] = nbget(lr, r, S);
for(i=0; i<N; i+=S*U)
  h[1] = nbget(..., S);
  ...
  h[U] = nbget(..., S);
  sync(h[0])
  for(ii=i; ii<S; ii++)
    a[ii] = b[ii]+lr[ii];
  ...
  sync(h[U-1]);
  for(ii=i+S*(U-1); ii<min(...); ii++)
    a[ii] = b[ii]+lr[ii];
  h[0] = h[U];
```

Fig. 5. Unrolling a strip-mined loop

2.1 Practical Considerations for Message Strip-Mining

In this section, we discuss the impact of both machine and application characteristics on the effectiveness of message strip-mining.

Machine Characteristics. The LogGP [6, 10] (Figure 6) network performance model approximates the cost of a data transfer as the sum of the costs incurred by the transfer in all system components. Parameters of special relevance to us are o_s and o_r, the send and receive overhead of a message; G, the inverse network bandwidth; and g, the minimal gap required between the transmission of two consecutive messages.

According to the model, the cost of a single message transfer can be divided into two components, the software overhead on both the send and receive side, and the time the message actually spends in the network. The total communication cost of the vectorized loop in Figure 3 is thus $T_{vect}(N) = o + G * N$, where o is the sum of o_s and o_r. The total communication cost of the strip-mined loop (in Figure 4), on the other hand, must be expressed as the sum of the time spent issuing the non-blocking gets and the time spent waiting for the corresponding syncs. It is now easy to see how message strip-mining could have a negative impact on performance. Although the transformation does not increase message volume, it incurs more software overhead due to more messages being issued. Furthermore, since smaller messages are transferred, this overhead becomes a greater fraction of communication time. While picking a small strip size allows for more overlapping of the waiting times, it is also accompanied by an increase in issue time. Loop unrolling introduces further complications, as the effective time to issue a non-blocking get operation can depend not only on the transfer size S but also on the number of these operations issued in sequence U. Hence large unroll depths, though theoretically attractive, may not be desirable in practice due to this network limitation. In Section 2.2 we analyze these tradeoffs in greater detail.

Another potential source of performance degradation comes from the network itself. The LogGP model assumes an "ideal" network with no resource constraints and considers the parameters to be constant. In practice, combining loop unrolling with message strip-mining increases the number of outstanding messages on the Network Interface Card at a given time and can have a large performance impact on networks where resources are scarce.

In addition to NIC resources, remote DMA message transfers also compete with the processor for access to a node's memory system. This "interference" may adversely affect the speed of local computation by increasing the time to access memory. These effects are considered further in Section 3.4.

Application Characteristics. The data transfer size is perhaps the most important factor in determining the effectiveness of message strip-mining for a vectorized loop. An intuitive rule of thumb is that each strip transfer should be a bandwidth-bound ($G*N > o$) message, so that the increase in message startup costs can be compensated for by the performance gain from hiding the message transfer time. We further explore this concept in Section 3.1.

Even with a large transfer size, message strip-mining is still not useful unless we can discover enough computation to overlap; the computation cost of a strip, $C(S)$, should not be significantly smaller than its communication cost $T(S)$. While loop unrolling mitigates this problem by increasing the number of iterations that execute simultaneously, it also increases message initiation time. Since network performance is generally orders of magnitude worse than CPU performance, it may appear that sufficient overlap could not be attained without a large amount of computation. This assumption neglects the cost of memory access, however. Because remote data is typically transferred (by DMA) directly to main memory (and not the cache), they must then be brought into the cache for further processing. Cache miss penalties, which are also incurred by the vectorized loop, provide more time for the overlap of communication with computation. Thus, when estimating the "effective" local computation cost, one must also take into account the effect of local memory bandwidth. We study the implications further in Section 3.2.

Finally, the application's communication pattern determines the degree of contention in the network system and so influences the effectiveness of our transformation. We consider the following communication patterns, ranging from least to most contention: `one-to-one` and `all-to-one`.

2.2 Estimating the Impact of Strip-Mining

We begin our analysis by using the LogGP model we derive the communication costs for the loops shown in Figures 1, 3, 4, and 5.

Individual Small Messages (Figure 1). $T_{\text{small}}(N) = N * \text{EEL}$. One is forced to pay the full network cost for a small message on every access.

With Message Vectorization (Figure 3). $T_{\text{vect}}(N) = o + G * N$. This is the cost of one large message transfer.

Message Strip-Mining and Unrolling (Figures 4 and 5). In this case each `sync` waits if the transfer is not completed by the time the execution encounters the `sync`. The total communication cost is the sum of the overheads of initiating all the transfers and the waiting time for the `sync` call of each transfer. If the message is ultimately divided into m blocks $S_1, ...S_m$, the waiting time $W(S_i)$

for each block can be expressed as an equation based on the network parameters and computation costs:

$$\begin{aligned} W(S_1) &= G * S_1 - \mathrm{issue(S_2)} \\ W(S_2) &= G * S_2 - C(S_1) - W(S_1) - \mathrm{issue(S_3)} \\ \cdots & \quad \cdots \\ W(S_m) &= G * S_m - C(S_{m-1}) - W(S_{m-1}) \end{aligned} \tag{1}$$

We also have the constraint that $W(S_i) \geq 0$. The above represents the case where the unroll depth is one; equations for different depths can be derived in a similar fashion. $C(S_i)$ refers to the costs of computation performed on a block, and for simplicity we assume it can be expressed as a linear function $K * S_i$, without much loss of generality[2]. As the goal of message strip-mining is to minimize the communication cost, it is equivalent to solving the above system to find m and the sequence $S_1, \ldots, S_m$ that minimizes the objective function

$$T_{strip+unroll} = \sum_{i=1}^{m} \mathrm{issue(S_i)} + \mathrm{W(S_i)} \tag{2}$$

Finding the optimal message decomposition involves tradeoffs on m, the total number of transfers. A small m may minimize the waiting time but cause an excessive startup overhead, while a large m may exhibit the opposite problem. A similar tradeoff is encountered when varying the unroll depth. In the next section we analytically develop message decomposition heuristics that are straightforward to implement yet can achieve significant performance gains over a vectorized loop.

2.3 Message Decomposition

A simple yet effective decomposition for determining S_i in the above equation is to make each block equal-sized, so that the total communication cost for a given loop only depends on one variable[3]. To find the optimal block size S under this scheme, we simply search sample decompositions with a representative amount of computation. Our synthetic experiments in the next section detail our findings.

While the fixed block size scheme is straightforward, it may not achieve optimal message overlap. In particular, the fixed-size decomposition may cause the waiting times for the blocks to oscillate between two extremes. It can be best explained by examining Equation (1): the second sync can be completely overlapped with the waiting and computation time of the first block, making $W(S_2)$ zero. This, however, means the next sync does not benefit from $W(S_2)$ at all (though it can still be overlapped with computation), resulting in a large

[2] We are interested in balanced algorithms where communication and computation are of the same order of magnitude.

[3] We assume that U is also fixed; certainly the optimal block size may vary with the unroll depth.

$W(S_3)$, which in turn makes $W(S_4)$ small. It thus appears that for message strip mining, a sequence of varying block sizes that better captures the nature of the waiting time as recurrence relations may outperform the fixed-size scheme. As a heuristic, we pick the block sizes to be a geometric series $S_i = (1+f)S_{i-1}$, where f is between 0 and 1. For the starting block size, we choose the size of the smallest message that still benefits from strip mining as described in Section 3.1. A good value of f is determined mostly by application characteristics, and is clearly critical to the performance of our varying block size heuristic. We discuss in Section 3.5 heuristics for determining a lower bound for the values of the parameter f.

3 Experiments

In this section we provide quantitative data for the issues discussed in previous sections. The data is collected on the systems detailed in Table 1. All benchmarks are implemented in Unified Parallel C [5] and run over a customized communication layer called GASNet [4, 8], which provides a one-sided communication API with performance very close to that of each system's native communication API. Thus, our results apply for other one-sided communication layers such as MPI-2.

First, we discuss the minimal total transfer size required for message strip-mining to be effective. We then use synthetic benchmarks to study the impact of the application's computation on the message decomposition strategy. For each benchmark, we provide implementations corresponding to the strip-mining strategies outlined in Figures 4 and 5. We vary the following parameters: N - the total problem size, S - the strip size, U - the unroll depth, P - the number of processors, and the communication pattern. In evaluating the various message strip-mining techniques we seek to determine how much each technique improves performance upon the basic vectorization strategy. As such, we present performance results showing the ratio of time taken by the "optimized" strategy to the vectorized strategy. We conclude with a discussion of the influence of the application's communication pattern on the message decomposition strategy.

3.1 Minimum Message Size

From the model established in Section 2, we note that any strip-mining strategy must pay a minimum cost of $o+\max(o, g)+\epsilon$, because it decomposes the transfer

Table 1. Systems Used for Benchmarks

System	Network	CPU type
IBM Netfinity cluster [2]	Myrinet 2000	866 MHz Pentium III
IBM RS/6000 SP [1]	SP Switch 2	375 MHz Power 3+
Compaq Alphaserver ES45 [14]	Quadrics	1 GHz Alpha

into at least two messages. We include an extra ϵ to account for any syncs with a non-zero waiting time. For message strip-mining to be worthwhile, we must thus have $o + G * N > o + \max(o, g) + \epsilon$, which implies $N > \frac{\max(o,g)+\epsilon}{G}$; a lower bound for N is $\frac{\max(o,g)}{G}$. Using values for the ratio $\frac{g}{G}$ obtained in [7] this lower bound is between 1KB and 3KB for all platforms. This value is verified by the experiments described in the next section, where message strip-mining performs well only for transfers larger than 2KB.

3.2 Effects of Memory Accesses on the Available Overlap

The model described in Section 2.2 indicates that the performance of message strip-mining heavily depends on whether the vectorized loop contains a sufficient amount of computation that can be used to hide the communication latency. In most of today's high performance computing systems the network is not tightly integrated within the memory hierarchy; the three platforms examined in this paper have their NICs attached either to a PCI bus (Myrinet and Quadrics) or to a proprietary bus (the SP), and two of them (Myrinet and Quadrics) use remote DMA operations that bypass the processor's cache. Accordingly, the cost of the computation that operates on the transferred data is composed of two parts: 1) the cache miss penalties incurred by accessing the transfered data, and 2) the execution time required by the computation itself. While the second component obviously varies from application to application, the cache miss penalty is an inherent part of the computation overhead and does not depend on the type of computation performed.

To measure the available overlap contributed by memory accesses, we compare the cost of moving data across the network with the overhead of moving the same data through the memory hierarchy. The cost of moving data over the network is captured by the inverse network bandwidth, the G parameter of LogGP. We measure this by timing the end-to-end latency of a large get The transfer size is intentionally chosen to be large (1MB) so that the communication cost is dominated by the wire latency rather than the overhead of initiating the transfer.

To determine the latency of moving data across the memory system, we measure the execution time of an integer vector reduction on a 1MB (the size used to determine G) array. With an unit stride access and one integer operation per element, the code snippet represents, in some sense, the minimal amount of computation possible on the transferred data, provided that each data element is referenced at least once.

Table 2. Comparison on the network and memory data transfer rate

System	Inverse Network Bandwidth (μsec/Kb)	Inverse Memory Bandwidth (μsec/Kb)	Ratio (memory/network)
Myrinet	6.089	4.06	67%
Quadrics	4.117	0.46	11%
SPSwitch	3.35	1.85	55%

Table 2 presents the results for our test platforms. Memory access time proves to be an important source of computation overhead. With the exception of the Quadrics system, the cost of fetching data from main memory is at least half the cost of fetching the same data over the network. This provides an important insight, that essentially no limitation exists on the minimal amount of computation a vectorized loop must perform for it to benefit from message strip-mining. As the performance gap between the processor and memory subsystem grows, sustained memory bandwidth will likely become the bottleneck in determining the amount of computation available that can be overlapped with communication using message strip-mining.

To confirm our hypothesis that memory traffic in applications alone can be a sufficient source for computation overlap, we manually applied message-strip mining (with fixed-size decomposition) on the ghost region exchange portion of the NAS-MG [3] benchmark. The results show speedups ranging from 3% (Quadrics) to 64% (IBM SP), illustrating message strip-mining's effectiveness even for code that performs a relatively small amount of computation.

3.3 Fixed-Size Versus Variable-Size Decomposition

To compare the effectiveness of fixed-size and variable-size decomposition for message strip-mining, we construct two benchmarks that perform computation on a double precision floating point remote array. The `Reduction` benchmark conducts a multiply accumulate reduction operation, which is often encountered in scientific programming and has a computation overhead comparable to the latency of communication. For the `Vector` benchmark we consider a $vector-to-vector$ operation that also involves function calls, and thus has a cost that is several times higher than the latency of communication. For each benchmark, the number of elements per processor is varied between 2^8 and 2^{20}. For the fixed-size decomposition, we experimented with strip size in powers of two, ranging from $2KB$ to the total size. The starting block size for the variable-size decomposition is picked to be $2KB$, and we experimented with different value of f between 0 and 1.

We present selected results in Figure 8. As a first observation, we note that a tuned strip-mining implementation, using either a fixed or variable size decomposition, always improves performance compared to a vectorized-only implementation. For fixed-size decomposition, there exists an optimal range of the block size that is influenced both by the platform and by the total transfer size; a large total transfer size generally will also have a correspondingly larger block size. Examining the results of the `Reduction` and `Vector` benchmark, we also note that the former, with a smaller amount of computation for overlap, produces a narrower range of optimal block size for the fixed-size decomposition. Equivalently, increasing the amount of local computation can simplify the task of tuning message strip-mining, by exposing more values for block size that can deliver good performance.

Comparing the performance of the variable-size strategy with the fixed-size strategy, we find that the best block size for the latter generally outperforms

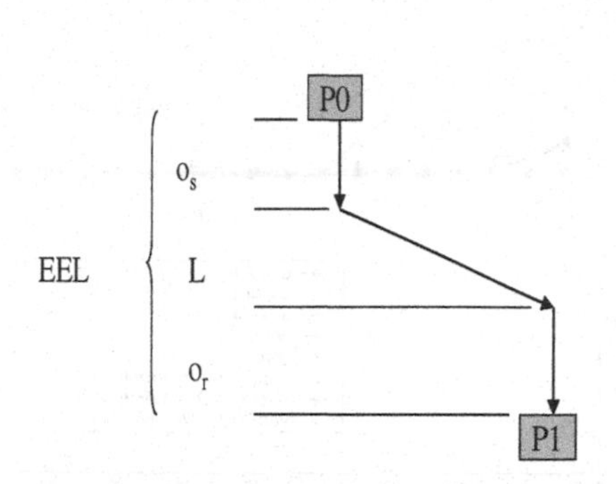

Fig. 6. Traditional LogP model for sending a message from processor P0 to processor P1

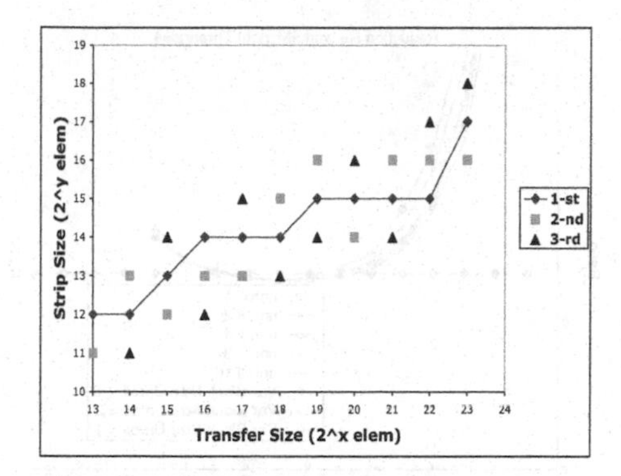

Fig. 7. Variation of the optimal fixed strip size with the total transfer size for Myrinet

the best f for the former, with a difference of less than 10%. The optimal block size for the fixed-size decomposition, however, heavily depends on the platform and problem size, and a good block size for one configuration may be outside the the optimal range for another, causing a performance degradation of more than 50%. Variable-size decomposition, on the other hand, is more performance-portable; the same starting block size can be used in all tests, and even a badly chosen f will incur no more than a 15% performance degradation compared to the optimal value.

Finally, the performance of combining strip-mining with unrolling is influenced by the gap network parameter g. On Myrinet and the IBM SP network, where $o \approx g$, unrolling with small factors (2 or 4) improves performance. On the Quadrics network, where $g >> o$, unrolling causes a performance degradation.

3.4 Influence of Communication Pattern

So far we have examined only applications with a one-to-one communication pattern. We now examine the impact of resource contention, which include both contention on the memory system of a node and on the network interface card.

To determine the effect of network transfers on local computation, we time computation loops on a node where the networking subsystem is serving remote read requests from a variable number of processors. As local computation we use the loop from the Reduction benchmark, and another loop performing the same operation but using indirect accesses with a uniform index distribution. The latter exhibits a much higher processor-to-memory traffic. Both loops are large enough so data do not fit in the cache. For all systems, our results show a slowdown of the computation loop ranging from 3% to 6%. The slowdown increases with the increase in local memory traffic and is not affected by the number of nodes reading from the memory of the computation node. We conclude that interference from the network DMAs does not substantially affect the cost of memory access, and thus the effectiveness of strip-mining.

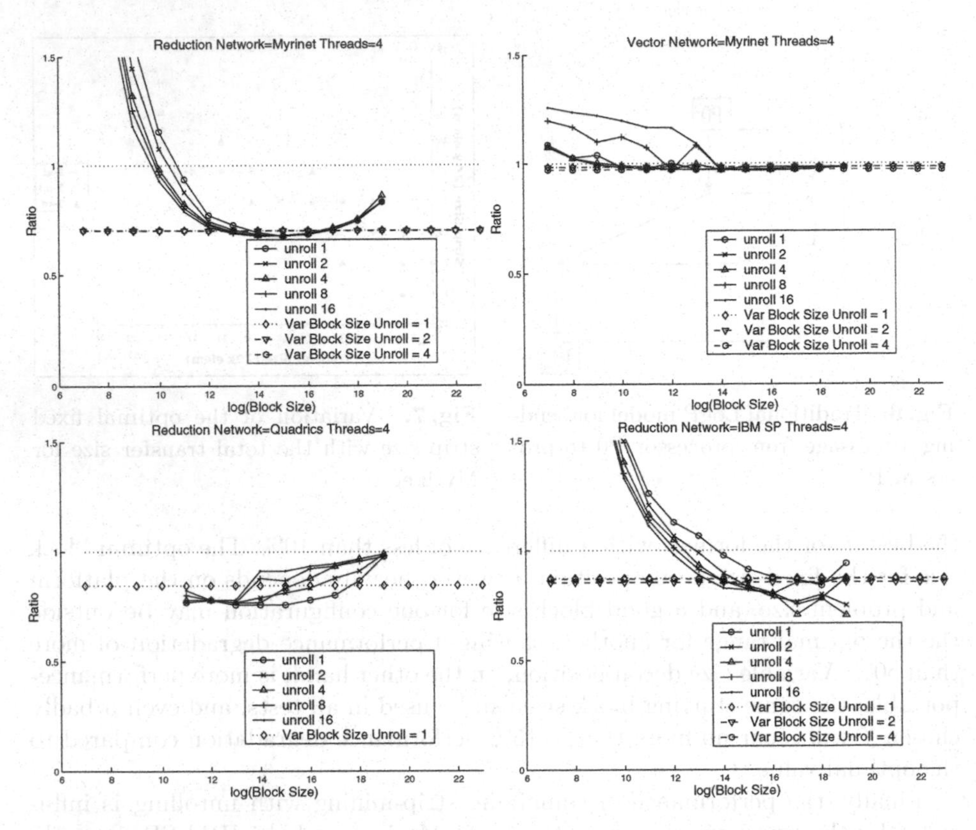

Fig. 8. Performance results for *one-to-one* communication. Problem size = 2^{20} doubles

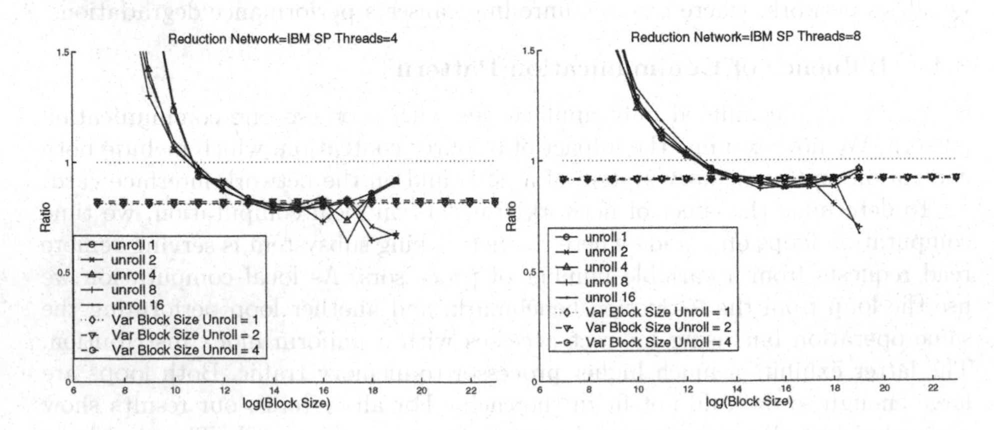

Fig. 9. IBM SP performance results for *Reduction* benchmark with an *all-to-one* communication pattern

To determine the impact of network interface card contention on the performance of strip-mining we use the same Reduction and Vector benchmarks modified to have all the processors read from the memory space of a single processor. In this case, the congested NIC serializes the remote read requests, resulting in an increase in the average message latency and large variability in the observed latency at each endpoint. Thus, we expect message decompositions that issue fewer requests to perform better. Figure 9 presents selected performance results for the IBM SP platform; similar trends are observed on the other platforms. As the figure suggests strip-mining remains an effective strategy in the presence of contention, though speedup is decreased due to poorer overlap. The fixed-size strategy outperforms the variable-size strategy by 2%-4%. Increasing the degree of contention causes an increase in the size of the optimal decomposition compared to the one-to-one communication pattern.

3.5 Message Decomposition in Practice – A Summary

We now present some guidelines for choosing an "optimal" message decomposition for a given problem.

For fixed-size decomposition, the determining parameter is the ratio $\frac{N}{S}$, the total number of blocks transferred. Figure 7 presents the best three decomposition values based on the total transfer size for the Reduction benchmark on the Myrinet platform. The "optimal" decomposition for any given application can be found by employing a search strategy using the findings in this paper to prune the parameter search space. We note that increasing the total size increases the block size, and increasing the degree of contention also increases the block size. Increasing the amount of computation would decrease the strip size, but as indicated by Amdahl's Law, the benefits of strip-mining will be less and less pronounced with the growing cost of computation.

While employing a fixed-size decomposition, any variation in the total problem size, computation and communication pattern parameters requires a complete "re-tuning" of the application. On the other hand, the performance of the variable strip size strategy requires only an approximation of the computation cost. The performance of this approach is highly dependent on the value of the multiplication coefficient $1+f$. This value determines the number of messages issued and the size of each message. For small values of f, the algorithm issues a large number of transfer requests and thus incurs a high message initiation and control overhead. Increasing the value of f increases the probability of poor overlap since transfers might have to wait for completion. For best performance, a search based strategy is also required to determine the value of this parameter. For the benchmarks in this paper, we obtained an optimal value of 0.45 by tuning the Myrinet implementation. This value performed equally well on the IBM-SP platform and a little worse on the Quadrics platform, where the value of f in the range of 0.1 to 0.4 performs best.

4 Related Work

Wakatani and Wolfe [19, 18] introduce message strip-mining and analyze its impact for array redistribution in HPF and a code that implements a simple inspector-executor. They consider only strip-mining with fixed block size and use a simple model tailored for the array redistribution problem to predict performance benefits.

The idea of decomposing message traffic also appears in [16, 17] where Prylli et al. implement transparent message decomposition inside the BIP networking layer. They model the steps involved in message transmission and derive formulas for predicting the optimal decomposition, and overall message passing performance. Unlike our variable strategy, their message decomposition strategy chooses a fixed strip size based on the total message size. Furthermore, they do not expose this decomposition to the application layer.

Gupta and Banerjee [11] present a methodology for estimating the communication costs in HPF programs. They use the analysis to guide the data partitioning decisions of their compiler and also to select communication primitives. Their work is directly applicable to the analysis of UPC programs, but it does not take into account contention as generated by the program communication pattern or the interference caused by the communication primitives on the memory performance of the communication peers. Evidence presented by Prieto et.al.[15] suggests that memory interference on the communication peers can severely affect performance.

5 Conclusion

In this paper we investigated message strip-mining as a communication optimization technique, introduced strip-mining with a variable size, demonstrated the effectiveness of both techniques, and analyzed the factors that influence the performance of these program transformations.

We find the potential performance gain to be heavily dependent on both network characteristics and application characteristics. We empirically determine a lower bound on the total problem size after which our optimizations are effective. This value is 2KB for all networks studied. While we find that a well tuned fixed size strategy usually outperforms the variable size strategy by 0%-4%, its performance is very sensitive to all optimization parameters. On the other hand, the performance of the variable size strategy is determined only by the computation pattern and we recommend this approach when developing performance portable applications.

Besides application development, the heuristics that we examine in this paper are also of interest when developing collective communication libraries and vectorizing compilers. We plan on using the variable strip size strategy in our UPC implementation work in both areas.

Acknowledgment

This work was supported in part by the U.S. Department of Energy under contracts DE-FC03-01ER25509 and DE-AC03-76SF00098, by the National Science Foundation under ACI-9619020 and EIA-9802069, and by the Department of Defense. The authors would like to thank Paul Hargrove for the GASNet program that determines network parameters.

References

1. IBM SP – Seaborg. http://hpcf.nersc.gov/computers/SP/.
2. NERSC Alvarez Cluster. http://www.nersc.gov/alvarez.
3. The NAS Parallel Benchmarks. http://www.nas.nasa.gov/Software/NPB.
4. The UPC Runtime Specification, v1.0. http:/upc.lbl.gov/docs/system.
5. UPC Language Specification, Version 1.0. Available at http://upc.gwu.edu.
6. A. Alexandrov, M. F. Ionescu, K. E. Schauser, and C. Scheiman. LogGP: Incorporating Long Messages into the LogP Model for Parallel Computation. *Journal of Parallel and Distributed Computing*, 44(1):71–79, 1997.
7. C. Bell, D. Bonachea, Y. Cote, J. Duell, P. Husbands, P. Hargrove, C. Iancu, M. Welcome, and K. Yelick. An Evaluation of Current High-Performance Networks. In *Proceedings of 17th International Parallel and Distributed Processing Symposium (IPDPS)*, 2003.
8. D. Bonachea. GASNet Specification, v1.1. Technical Report CSD-02-1207, University of California at Berkeley, October 2002.
9. Z. Bozkus, A. Choudhary, G. Fox, T. Haupt, and S. Ranka. A Compilation Approach for Fortran 90D/HPF Compilers on Distributed Memory MIMD Computers. In *Proceedings of the Sixth Workshop on Languages and Compilers for Parallel Computing*, Portland, OR, 1993.
10. D. E. Culler, R. M. Karp, D. A. Patterson, A. Sahay, K. E. Schauser, E. Santos, R. Subramonian, and T. von Eicken. LogP: Towards a Realistic Model of Parallel Computation. In *Principles Practice of Parallel Programming*, pages 1–12, 1993.
11. M. Gupta and P. Banerjee. Compile-Time Estimation of Communication Costs on Multicomputers. In *Proceedings of the 6th International Parallel Processing Symposium*, Beverly Hills, CA, 1992.
12. M. Gupta, S. Midkiff, E. Schonberg, V. Seshadri, D. Shields, K.-Y. Wang, W.-M. Ching, and T. Ngo. An HPF compiler for the IBM SP2. In *Proceedings of the 1995 ACM/IEEE conference on Supercomputing (CDROM)*, page 71. ACM Press, 1995.
13. S. Hiranandani, K. Kennedy, and C.-W. Tseng. Compiling Fortran D for MIMD Distributed-Memory Machines. *Communications of the ACM*, 35(8):66–80, 1992.
14. Lemieux. http://www.psc.edu/machines/tcs/lemieux.html.
15. M. Prieto and I. M. Llorente and F.Tirado. Data Locality Exploitation in the Descomposition of regular Domain Problems. In *IEEE Trans. on Parallel and Distributed Systens*, volume Vol 11, pages 1141–1150, 2000.
16. L. Prylli, B. Tourancheau, and R. Westrelin. Modeling of a high speed network to maximize throughput performance: the experience of BIP over myrinet. In *Parallel and Distributed Processing Techniques and Applications (PDPTA'98)*, 1998.
17. L. Prylli, B. Tourancheau, and R. Westrelin. The Design for a High-Performance MPI Implementation on the Myrinet Network. In *PVM/MPI*, pages 223–230, 1999.
18. A. Wakatani and M. Wolfe. A New Approach to Array Redistribution: Strip Mining Redistribution. In *Proceedings of PARLE'94 (Athen, Greece)*, Jul 1994.
19. A. Wakatani and M. Wolfe. Effectiveness of Message Strip-Mining for Regular and Irregular Communication. In *PDCS (Las Vegas)*, Oct 1994.
20. H. Zima and B. Chapman. Compiling for Distributed-Memory Systems. In *Proceedings of the IEEE*, 1993.

Analysis of the Abortion Rate on Lazy Replication Protocols*

Luis Irún-Briz, Francesc D. Muñoz-Escoí, and Josep M. Bernabéu-Aubán

Instituto Tecnológico de Informática,
Universidad Politécnica de Valencia, 46071 Valencia, Spain
{lirun, fmunyoz, josep, iti}@iti.upv.es

Abstract. Shared memory applications are principal to solve a big number of problems in distributed systems. From high performance applications, where the different computational units use this technique to simplify its designs (and often improve the performance) to database applications, where a replicated database can also be considered as a flavor of shared memory for the different involved nodes.

Any of these applications use replication as the basis for the implementation of shared memory, and they frequently share common characteristics in respect to access locality to particular portions of the global state. Replication is also a technique commonly used in distributed systems in order to provide fault tolerance.

Many techniques have been designed to perform the necessary consistency management for the different views on a replicated memory system. Some of these techniques try to take advantadge of the access locality, by propagating the changes performed by any node in a lazy style (i.e. as late as possible). Nevertheless, lazy update protocols have proven to have an undesirable behavior due to their high abortion rate in scenarios with high degree of access conflicts.

In this paper, we present the problem of the abortion rate in such protocols from a statistical point of view, in order to provide an expression capable to predict the probability for an object to be out of date during the execution of a transaction in a contextual environment.

It is also suggested a pseudo-optimistic technique that makes use of this expression to reduce the abortion rate caused by accesses to out of date objects.

The proposal is validated by means of an empirical study of the behavior of the expression, including measurements of a real implementation. Finally, we discuss the application of these results to improve lazy update protocols, providing a technique to determine the theoretical boundaries of the improvement.

1 Introduction

Nowadays, many distributed (i.e., networked) applications have to manage large amounts of data. Despite the increasing ubiquitousness of information, the access patterns to distributed data often feature a noticeable degree of geographical locality. Moreover, many applications require a high degree of availability, in order to satisfy the need of

* This work has been partially supported by the EU grant IST-1999-20997 and the Spanish grant TIC2003-09420-C02-01.

M. Daydé et al. (Eds.): VECPAR 2004, LNCS 3402, pp. 438–453, 2005.

offering services at any time, to clients that are either internal or external to the networked application. A predominance of locality of access patterns usually suggests a partitioning of the data repository [1, 2]. In many scenarios, it may be convenient or even necessary to replicate the information in a set of servers, each one attending its local clients. The different replicas must then be interconnected.

Other scenarios, as shared memory systems, also make use of the principle of locality. In these environments, each node performs intensive accesses to a particular region of the shared data, and the rest of the information is only marginally accessed.

In general, any application making use of shared memory, and requiring an adequate management of the consistency of the replicated state is suitable to benefit from the locality.

Particular examples of these applications can be found in distributed database applications (as telecommunication providers, travel businesses, logistics, customer relationship management, e-banking, virtually all kinds of e-business as well as most e-government applications).

Moreover, also high performance applications often make intensive use of shared memory, in order to interchange information along the system. Another family of applications that make use of replication as the basis of its services is fault-tolerant applications, where replicating the state is the most suitable way to increase the availability of the provided services.

In summary, we are centering our attention to any application that make use of replication to provide shared memory. Thus, replication makes it necessary to manage consistently the different views (respectively observed by each node in the system) of the global state.

Common to all of these applications is that potentially huge amounts of data are maintained and replicated on distributed sites, while access patterns (or at least significant contingents thereof) are highly local. Efficiency and high availability of such services is key to their acceptance and success.

The GlobData project [3, 4, 5] strives to provide a solution for the kinds of distributed database applications just outlined. It does so by defining a specific architecture for replicated databases, together with an API and a choice of consistency modes for data access.

In the GlobData project, a number of consistency protocols have been proposed, that are capable of meeting the consistency requirements of a wide range of database applications, including the ones described here. Although these protocols differ in the priorities of their particular goals, all of them share a common characteristic: They are (more or less) optimistic (or, at least, cannot be classified as pessimistic), since transactions are allowed to proceed locally and are checked for consistency violation only at commit time. When a consistency violation is encountered, the transaction is aborted.

During the development of the project, we discovered that the protocol needs of our target scenario (i.e. geographical distributed applications with a high degree of locality) could be well-fitted with a family of replication protocols classified as *lazy update protocols*. However, we encountered that this kind of protocols introduces a high number of abortions in the system, because they propagate the updates beyond the commit phase (in contrast to *eager update protocols*, where the whole system is updated inside the commit phase), making it possible for a transaction to perform undesirable accesses to outdated data, thus appearing consistency violations.

But, an additional inconvenience was found for the *optimistic lazy update protocols* when they are used in contextual (e.g. transactional) environments like distributed databases: the use of such techniques makes it impossible to guarantee[6] one-copy serializability, unless the protocol is redesigned, in which case it becomes unusable due to the huge number of aborted transactions it introduces.

Nevertheless, the consistency and isolation guarantees provided by optimistic lazy update protocols can still be useful to a wide range of applications, where there is no need to include different operations inside a transactional context.

In [3, 7], we presented the implementations of two consistency protocols, using different approaches for the update propagation.

As expected, lazy update protocols proved to have a critical inconvenience in contrast to eager approaches: the dramatical increase of the abortion rate in scenarios with a high degree of access conflicts. This inconvenience makes unusable the traditional lazy update protocols in such scenarios, because an unacceptable number of started transactions terminate with an undesirable abort.

To understand the problem, this work presents a set of expressions describing the abortion rate. In the presentation, we model a complete system including nodes, the sessions executed, and the objects accessed by the sessions. Moreover, we use our Lazy Optimistic Multicast Protocol (LOMP) algorithm as an example of a lazy update protocol where all these expressions may be directly applied. A full description of this protocol may be found in Section 3.2 of [6].

The rest of the paper is organized as follows: section 2 includes the description of such modeled system, in order to formalize an analysis of the abortion rate in section 3. In sections 4 and 5, an empirical validation of the model is presented, and section 6 will provide a theoretical analysis of an improvement over a lazy approach. Finally, section 7 describes some related work, and section 8 includes some conclusions about the applicability of the expressions.

2 The Modeled System

In the GlobData project, a software platform is used to support database replication. This platform, is called COPLA, and provides an object-oriented view of a network of relational DBMSs. The use of COPLA allows multiple applications (even running in different nodes) to gain access the same database replica using its local "*COPLA Manager*". The COPLA Manager of each node propagates all updates locally applied to other database replicas using its replication management components.

One of the problems solved by COPLA[3, 8] is to ensure the consistency among transactions being executed in different database replicas. The solution to this problem depends mainly on the update propagation technique. COPLA is flexible enough to allow multiple update approaches, each one being managed by a dedicated consistency protocol. A particular COPLA component implements the consistency protocol. Multiple implementations of its interfaces are possible, thus allowing a GlobData system to change its consistency protocol at will. So, we use the COPLA architecture as a platform to experiment with different consistency algorithms working on the top of a transactional system (in fact, a database) and replicated over a distributed system.

The target system of the following analysis takes a number of considerations, designed to configure a scenario as close as possible to a general environment that, although simplified, is able to fit the requirements of the kind of environment we are centered in. This environment was described in the introduction, and has considerations about client applications, system load, pattern of accesses, interconnection network, etc.

In summary, these adopted assumptions are the following:

- There are K COPLA managers running in the system. Each one can be considered as a "node" $N_{k=1..K}$.
- Each node in the system manages a complete replica of the database. This database contains N objects.
- A session S can be written as a tuple $S = [R(S), W(S)]$ where:
  - $R(S)$ is the set of objects read by the session S. It is also named "readset of S". $R = \{r_i\}_{i=1..|R|}$
  - $W(S)$ is the set of objects written by the session S (or "writeset of S"). $W = \{w_i\}_{i=1..|W|}$
- We assume that $W(S) \subseteq R(S)$. And the objects contained in $R(S)$ and $W(S)$ can be expressed as tuples: $o_i = [id(o_i), ver(o_i), val(o_i), t(o_i), ut(o_i)]$ where:
  - $id(o_i)$ is a unique identifier for the object. The identifier includes the owner node (the node where the object was originated), a sequential number established within the context of each node, and other information used to calculate conflicts.
  - $ver(o_i)$ is the version number of the accessed object.
  - $val(o_i)$ is the value read (or written) by the session for the object.
  - $t(o_i)$ is the local time the object was accessed at.
  - $ut(o_i)$ is the local time the object was more recently updated at.

2.1 Calculable Parameters

For each node N_k, it is possible to calculate a number of variables:

- nw_k The average number of objects written per write-session committed in the system at N_k.
- $wtps_k$ The number of write-sessions committed per second at N_k.

3 Probability of Abortion

We can define the probability for a session S to be aborted as: $PA(S) = 1-(PC_{conc}(S) \cdot PC_{outd}(S))$ where:

- $PC_{conc}(S)$ is the probability that the session concludes without concurrency conflicts.
- $PC_{outd}(S)$ is the probability that the session concludes without accessing to outdated objects.

The goal of this section is to determine the value of $PC_{outd}(S)$, in order to predict the influence our LOMP has on the abortion rate in the system. To this end, we can calculate this probability in terms of the probability of a session to conclude with conflicts produced by the access to outdated objects (i.e. $PA_{outd}(S)$):

$$PC_{outd}(S) = PC_{outd}(r_i)^{nr}$$

taking nr as the number (in mean) of objects read by a session,

$$PC_{outd}(S) = PC_{outd}(r_i)^{\frac{\sum_k nr_k}{K}} \tag{1}$$

moreover, $PC_{outd}(r_i)$ is the probability for an object r_i to have an updated version in the instant the session accesses it. This probability can be expressed in terms of the probability for an object to be accessed in an outdated version ($PA_{outd}(r_i)$) as:

$$PC_{outd}(r_i) = 1 - PA_{outd}(r_i)$$

now, let's see the causes of these conflicts: we took r_i as an asynchronous object in the active node that has not been updated since $ut(r_i)$; the outdated time for r_i satisfies $\delta(r_i) = t(r_i) - ut(r_i)$; it can be seen that $PA_{outd}(r_i)$ depends on the number of sessions that write r_i having the chance to commit during $\delta(r_i)$. Let PC_{T,r_i} be the probability for another concurrent session T (that has success in its commit) to finalize with $r_i \notin W(T)$. Then,

$$PA_{outd}(r_i) = 1 - (PC_{T,r_i})^C$$

where C depends on the number of write-sessions that can be committed in the system during $\delta(r_i)$ ...

$$PA_{outd}(r_i) = 1 - (PC_{T,r_i})^{\sum_k wtps_k \times \delta(r_i)} \tag{2}$$

now, we can reformulate PC_{T,r_i} as

$$PC_{T,r_i} = P[r_i \notin W(T)]$$

and, considering $W(T) = \{w_1, w_2, \ldots w_{nw(T)}\}$, then in mean, it will be satisfied that:

$$PC_{T,r_i} = \left(P[r_i \neq w_{j\in\{1..nw\}}]\right)^{nw}$$

taking nw as the mean of $|W(T)|$ for every write-session in the system.

$$PC_{T,r_i} = \left(P[r_i \neq w_{j\in\{1..nw\}}]\right)^{\frac{\sum_k nw_k}{K}} \tag{3}$$

The next step consists of the calculation of $P[r_i \neq w_{j\in\{1..nw\}}]$. To do this, we must observe the number of objects in the database (N). The probability that an accessed object is a given one is $\frac{1}{N}$, thus: $P[r_i \neq w_{j\in\{1..nw\}}] = 1 - \frac{1}{N}$

Finally, the complete expression can be rewritten as follows:

$$PA_{outd}(r_i) = 1 - \left(1 - \frac{1}{N}\right)^{\frac{\sum_k nw_k}{K} \times \sum_k wtps_k \times \delta(r_i)} \tag{4}$$

This expression provides a basic calculation of the probability for an object access to cause the abortion of the session by an out-of-date access.

The expression can be calculated with a few parameters. Only nw_k and $wtps_k$ must be collected in the nodes of the system in order to obtain the expression. Thus, it becomes possible for a node to estimate the convenience for an object to be locally updated before being accessed by a session. This estimation will be performed with a certain degree of accuracy, depending on the "freshness" of the values of nw_k and $wtps_k$ the node has. The way the expression can be used, and an adequate mechanism for the propagation of these parameters has been presented in [9].

3.1 Quality of the Expression

In order to validate the expressed expression, we designed a simulation of a simplified system, in order to measure the confidence level of the predictions done with the expression. In addition, a real implementation of a basic algorithm was also performed. This basic algorithm makes use of the expression to decide, for each access request (made from the user application), whether the requested object should be updated from the owner node of such object. This section shows the obtained results of these validations.

The simulated system -detailed in section 4- was implemented as a quick starting point for the basic validation. Once the obtained results were analyzed, we encountered the quality of the provided expression satisfactory. As a result, an algorithm [9] including our analytical results was implemented, exploiting the benefits discussed in this paper.

Section 5 will discuss the measurements performed in the real implementation to study the quality of the expression shown in this section.

4 Experimental Validation of the Model

We have validated the algorithm presented above by implementing a simulation of the system. In this simulation, we have implemented nodes that concurrently serve sessions, accessing to different objects of a distributed database. We have also modeled the concurrency control, and the lazy update propagation used by LOMP.

4.1 Assumptions

The assumptions for the implementation of the simulation [10, 11, 12] are compatible with the ones taken for the model calculation, and the values have been established to increase the number of conflicts produced by the transactions executed in the system (i.e., this configuration shows a "worst-case" scenario for our system):

- There are 4 nodes in the system, each holding a full replica of the database, that contains 20 objects. Each node executes transactions, accessing the database.
- For every object, a local replica holds the value of the object, and the version corresponding to the last modification of the object. That is, the only synchronous replica for each object is its owner node.
- There are three kinds of transactions, with a probability to appear of 0.2, 0.4, and 0.4 respectively:

  - "Type 0", or read-only transactions: reads three objects.
  - "Type 1", or read-write transactions: reads three objects, then writes these three objects.
  - "Type 2", or read&read-write transactions: reads six objects, then writes three of the objects read.
- The model supports the locality of the access by means of the probability for an accessed object to be owned by that node (i.e. the node where the transaction is started).
  - For read-only transactions, this is 1/4, (as the system contains 4 nodes, this models no locality for read-only transactions).
  - For read-write transactions, and read&read-write transactions, the probability is 3/4 (i.e. the number of local accessed objects should be 3 times higher than the number of accessed objects owned by other nodes).
- The cost of each operation are shown in time units (t.u.):
  - Read operation in a local database: $LR = 0.01\ t.u.$
  - Write operation in a local database: $LW = 0.02\ t.u.$
  - Cost of a local-update request: $LUR = K_{LUR} \times LR, K_{LUR} = 5$
  - Cost of a confirmation request: $CR = 0.6 \times LUR$
- The simulation time has been set at 1,422 t.u., discarding the first 2 t.u. as stabilization time for the simulation. This allows to start up to 60,000 transactions.

4.2 Accuracy of the Prediction

As described above, we use the expression as a measure of the convenience for a requested object to be updated from a synchronous (i.e. updated) node. This *convenience* is determined using a threshold. When the calculation of the expression for an object exceeds the established threshold, an update request is sent. Then, it is possible for the response of this request to contain the same version for the requested object (e.g. when the object was, in fact, up to date). We name this situation "Inaccurate prediction".

The more accurate the predictions are, the less overhead an algorithm using the expression introduces in the system. This accuracy of the predictions will be given by the set threshold: higher values for the threshold should provide more accurate predictions.

Figure 1 shows the evolution of the inaccuracy of the prediction, for different values of the threshold. For lower values of the threshold, the number of update requests is very high, and many of them are unnecessary. In contrast, a higher threshold produces a lower number of update requests, and only the most likely stale objects will be asked for update.

In general, it can be observed that higher values for the threshold increase the accuracy of the prediction. In addition, relatively low values for the threshold (e.g. 0.15) produce a high degree of accuracy (e.g. 75%).

The evolution of the inaccuracy in respect to the amount of *update requests* is shown in figure 2. The optimum line is also shown, and corresponds with the diagonal. The more accurate the prediction is, the closest the curves are. The studied implementation differs from the ideal line with a lower bound pattern, and it is shown that is quite proximal to the ideal.

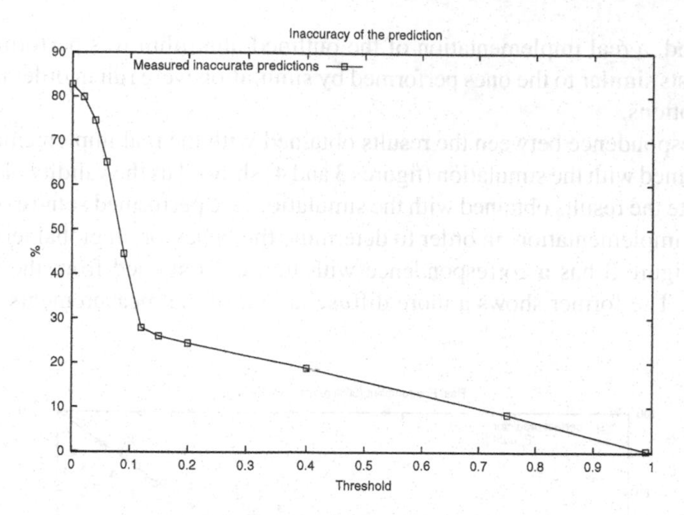

Fig. 1. Evolution of the inaccuracy of the prediction for different thresholds

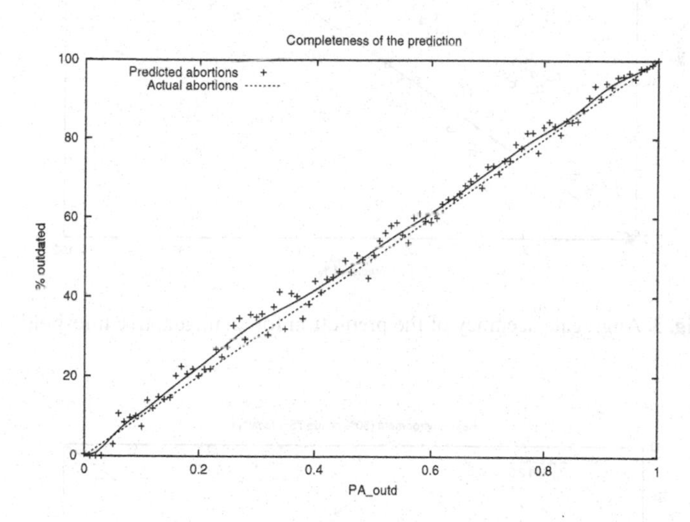

Fig. 2. Evolution of the inaccuracy for different P[update]

5 Validation of the Model with a Real Implementation

The performed simulation shown in section 4 provided a number of conclusions in respect to the viability of using the expression for P_{out}.

However, further validations should be necessary to ensure that the assumptions considered by both the simulation and the analysis have a correspondence with real environments.

To this end, a real implementation of the outlined algorithm was performed, and a number of tests similar to the ones performed by simulation were run in order to validate these assumptions.

The correspondence between the results obtained with the real implementation, and the ones obtained with the simulation (figures 3 and 4) showed us the validity of our work.

To validate the results obtained with the simulation, we performed some experiments with the real implementation, in order to determine the behavior, in global terms, of the prediction. Figure 3 has a correspondence with figure 2 obtained from the simulated environment. The former shows a more diffuse layout of the measurements in respect

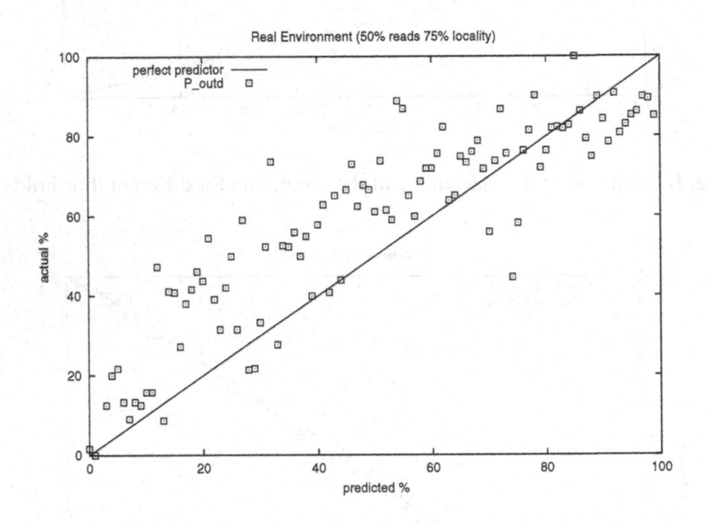

Fig. 3. Aggregate accuracy of the prediction for an unadaptive threshold

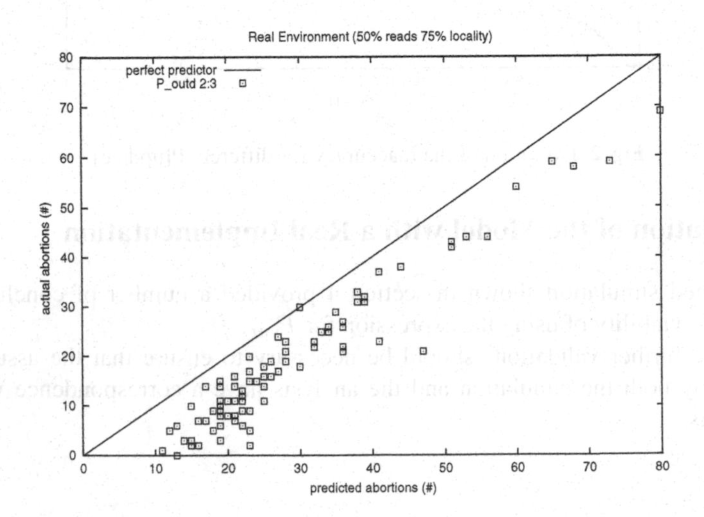

Fig. 4. Accuracy of the prediction for an unadaptive threshold

to the diagonal than the simulated results offered. Nevertheless, these differences are smaller enough to keep the validity of our conclusions. We find the cause of the differences in the fact that the simulated networked system did not consider the variability of the transport layer behavior, thus providing a more stable results for the predicted stale time of each object.

Figure 4 compares the number of objects predicted to be outdated, in contrast to the number of objects actually outdated. The prediction, as it can be seen, evidences an upper bound -in global terms- of the probability for an object to be stale. In the figure, the behavior of a perfect predictor (for which the same number of predictions should terminate with an abortion) is represented with the diagonal line. The points shown in the figure are, although close to this ideal line, below it, evidencing the pessimistic behavior of the prediction.

6 Theoretical Boundary of an Improvement

The first sections of this paper have been dedicated to the study of a statistical expression determining the probability for a particular object access, to obtain an out to date value (written by a previously committed transaction, but not yet propagated to the requesting node), thus causing the abortion of the requesting transaction.

The expression can be used by an improved lazy algorithm having the performance benefits provided by laziness, but relaxing the disadvantages in respect to abortion rate these protocols hamper. In this protocol, the expression discussed in section 3 can be used as an *oracle*, in order to determine whether a transaction is requesting an object that has not been updated in the local node. If such situation is predicted, the protocol will request the last modification of such object to the *owner* node of the object (we assume that there exist, for each object, an *owner* node to which every modification is always notified during the commit phase, thus guaranteeing that the held value in the local database is the more updated version of the object).

We can make use of such expression, in order to determine the achievable improvement for a transaction, in terms of abortion rate, when the average stale time of the objects is decreased.

Unfortunately, this decrement in the stale time will cause a degradation in the service time of the executed transactions, and this must also be taken into account.

In this section, we present a theoretical boundary for the reduction of the abortion rate that an adequate exploit of the expression can provide.

6.1 Preliminaries

The used expression for the probability of an access to be a stale-access was presented in 4 as:

$$PA_{outd}(r_i) = 1 - \left(1 - \frac{1}{N}\right)^{\frac{\sum_k nw_k}{K} \times \sum_k wtps_k \times \delta(r_i)} \quad (5)$$

In the expression, the elapsed time between two consecutive updates of the object r_i is expressed as $\delta(r_i)$.

We can perform a serial analysis in order to determine the mean value for $PA_{outd}(r_i)$. The analysis can be easily performed applying differentiate calculus to the expression. The obtained expression showed to be:

$$PA_{outd} = 1 - \left(1 - \frac{1}{N}\right)^{\frac{\sum_k nw_k}{K} \times \sum_k wtps_k \times \delta} \quad (6)$$

where δ is the mean value for the $\delta(r_i)$ in a system execution, and PA_{outd} is the mean value for the $PA_{outd}(r_i)$ in the same system execution.

Now, considering *nwwt* as the mean in the entire system of objects written per write-transaction, and *wtps* as the global amount of write-transactions per second served in the system, we can reformulate PC_{outd} as $1 - PA_{outd}$, being:

$$PC_{outd} = PC = \left(1 - \frac{1}{N}\right)^{nwwt \times wtps \times \delta} \quad (7)$$

6.2 Average Outdate (Stale) Time

To simplify, let's suppose that transactions are distributed homogeneously along the system history.

Imagine the system execution history as a line, where a number of transactions are sequentially executed. For a certain object o_i, the probability for a executed transaction to read o_i will be $\frac{nr}{N}$, where *nr* is the number, in mean, of objects read by any executed transaction (either read or write transactions are included here), and N is the number of objects in the database.

The probability of the object o_i to be updated by a lazy replication protocol depends on the probability for a transaction that read o_i to be aborted by a stale-access (i.e. $PA_T = PA^{nr}$). Thus the probability for an object to be updated by a generic transaction will be:

$$PA^{nr} \times \frac{nr}{N}$$

Now, let *tps* be the number of transactions executed in the system per second. Thus, there will be $up(o_i) = tps \times PA^{nr} \times \frac{nr}{N}$ updates of o_i per second. Finally, we can express $\delta(o_i)$ as $\frac{1}{up(o_i)}$, and, in mean:

$$\delta = \frac{1}{tps \times PA^{nr} \times \frac{nr}{N}} \quad (8)$$

If the accessed objects are updated along the transaction, the value for δ will be decreased proportionally to the amount of updates performed during the transaction execution.

To model this, a simple approach can be expressed with the following expression:

$$\delta' = PC \times d'_T + (1 - PC) \times \delta \quad (9)$$

where d'_T is the duration of a transaction when the updates are performed along its execution. For the aborted transactions, (i.e. $(1 - PC)$), the mean stale time is unchanged (δ). In contrast, for committed transactions, the new stale time is decreased to d'_T (i.e. the duration of the transaction).

Now, the duration of a transaction when the updates are performed will depend on the number of requested accesses that are actually updated along the transaction execution ($nr \times P_{UPD}$), and the cost of each of these updates (K_{UPD}). Note that P_{UPD} is the probability for a requested object to be updated. Thus, if $P_{UPD} = \frac{1}{2}$, there will be forced to be updated the half of the objects requested by a transaction.

In summary, the expression for d'_T can be built as:

$$d'_T = d_T + nr \times P_{UPD} \times K_{UPD} \quad (10)$$

Replacing 10 in 9, the new stale time will follow the expression:

$$\delta' = \delta \times (1 - PC) + PC \times (d_T + nr \times P_{UPD} \times K_{UPD}) \quad (11)$$

This result will be useful in section 6.3, where the achievable abortion rate is specified in terms of δ'.

6.3 Abortion Rate

In mean, we can say that the achievable commit rate will be, observing equation 7:

$$PC' = \left(1 - \tfrac{1}{N}\right)^{nwwt \times wtps \times \delta} \quad (12)$$

Replacing δ' in the expression, we obtain:

$$PC' = \begin{cases} \left(1 - \frac{1}{N}\right)^{nwwt \times wtps \times \delta \times (1-PC)} \\ \times \\ \left(1 - \frac{1}{N}\right)^{nwwt \times wtps \times PC \times (d_T + nr \times P_{UPD} \times K_{UPD})} \end{cases} \quad (13)$$

That can be rewritten as:

$$PC' = \begin{cases} PC^{(1-PC)} \\ \times \\ PC^{PC \times \frac{d_T + nr \times P_{UPD} \times K_{UPD}}{\delta}} \end{cases} \quad (14)$$

and simplified as:

$$PC' = PC^{1 + PC\left(\frac{d_T + nr \times P_{UPD} \times K_{UPD}}{\delta} - 1\right)} \quad (15)$$

Now, we can replace δ with the expression obtained in 8, the resulting expression is:

$$\frac{PC'}{PC} = PC^{PC\left((d_T + nr \times P_{UPD} \times K_{UPD}) \times (tps \times (1-PC)^{nr} \times \frac{nr}{N}) - 1\right)} \quad (16)$$

From the equation 16 we obtain that the improvement of the probability for an object to be accessed in an adequate way (i.e. not a stale access), is determined by $\frac{PC'}{PC}$, and it will be benefited from the decrease of the established value or any of the following expressions:

- d_T, the duration of the transactions.
- $nr \times P_{UPD} \times K_{UPD}$, the number of updated objects, and the computational cost of each of these updates.

- tps, the amount of committed transactions per second (including both read-only and read-write transactions).
- $\frac{nr}{N}$, the relation between the amount of objects accessed per transaction (read-only or read-write transactions) and the total number of objects contained in the database.

To simplify the expression 15, we can denote as Δ to the existing relation between d'_T and δ (i.e. $\Delta = \frac{d_T + nr \times P_{UPD} \times K_{UPD}}{\delta}$). The resulting expression is:

$$PC' = PC^{1+PC \times (\Delta - 1)} \Longrightarrow \frac{PC'}{PC} = PC^{PC \times (\Delta - 1)} \tag{17}$$

Let's see an example for the improvement achievable in an extremely simple system, where $nr = 1$. In such system, we can establish as a parameter the probability for a requested object to be previously updated (i.e. P_{UPD}), and then study the achieved improvement for different values of Δ (and, consequently, different computational overheads).

Figure 5 shows the commit probability for a session accessing **only one object**. There is shown how the commit probability can be improved, when the update-time is decreased to the half, up to the 120% of the original commit time. When the update-time is decreased ten times, the improvement reaches the 140%. Lower values of Δ provides marginal improvements, at a higher computational costs.

When more than just an object is accessed in a session, the results show higher differences between the commit promise, due to the factorization of such probabilities. For instance, considering more complex transactions accessing to four objects in mean, we obtain an improvement of 207% when $\Delta = 0.5$ (i.e. the system will abort a half of transactions if the update time is decreased to the half). Moreover, for such transactions, the obtained improvement for a $\Delta = 0.1$ is 380% (almost four times less abortions when the update time is decreased to 1/10 of the original time).

These results point to the convenience, in the scenarios fitting the parameters described above, to apply the techniques postulated by the presented discussion.

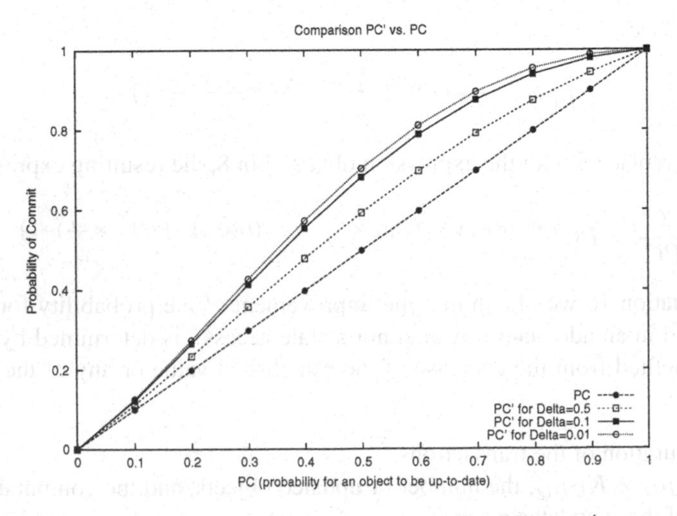

Fig. 5. Evolution of the improvement for different Δ

7 Related Work

Consistency is a key to distributed applications. Many approaches to provide consistency guarantees when replication is used have been proposed and discussed in the literature. From the use of synchronized clocks [13] to the management of replication by means of Fault-Tolerant Broadcasts [14].

Current work in consistency protocols can be found using either eager [15, 16] or lazy approaches [17, 18] to manage replication of the state.

Each one has its pros and cons, as described in [2]. Eager protocols usually hamper the update performance and increase transaction response times but, on the positive side, they can yield serializable execution of multiple transactions without requiring too much effort. On the other hand, lazy protocols may allow a transaction to read outdated versions of objects, hamper the abortion rate, but they can improve transaction response times.

Lazy update protocols, introduced in [19, 17, 18], presented a new approach for the propagation of the updates, in contrast to the traditionally used "eager replication". Lazy update protocols take advantage of the fact that an object can be written several times in a particular node before another node tries to read it.

In respect to the consistency control, both pessimistic and optimistic approaches are commonly used. Pessimistic consistency control [20] is commonly based on the principle of "locks" in order to avoid concurrent operations to access to the same object in an inadequate mode. The use of "locks" minimizes the number of abortions, but degrades the performance of the system, because the complexity introduced by the management of the locks, as well as the temporal inactivity these techniques introduce in the transaction execution.

On the other hand, the traditional approach for optimistic consistency control was presented in [21], and its main advantage consists on the reduction of the blocking time of the operations, using "versions" (or "timestamps" [22]) as the basis for its implementation. The main disadvantage of optimistic consistency protocols consists in the increase of the abortion rate.

In the context of Distributed Shared Memory (DSM) systems, pessimistic consistency control techniques are commonly used, providing a number of relaxations of the consistency models in order to reduce the latency that remote memory accesses introduces. In these systems, lazy techniques have been also used (mainly in the context of Release Consistency (RC) [23]) to reduce the number of messages needed to propagate the updates performed in the shared objects. Systems like TreadMarks[24] make use of this technique to improve the performance of DSM applications.

8 Conclusions

Shared memory systems often make use of replication as the main technique to improve fault-tolerance, and performance to distributed applications. Access locality often is a key in these kind of applications, that often is managed by implementing lazy update protocols to propagate the changes along the system.

However, lazy update protocols have not been widely exploited due to its excessive abortion rate on scenarios with high probability of access conflicts. Nevertheless, such protocols can provide important improvements in the performance of a distributed system, when the abortion rate can be kept low, and the locality of the accesses is appreciable.

We have presented a statistical study of the abortion rate (as disadvantage of lazy protocols), in order to provide an expression for the probability for an accessed object to be out of date ($PA_{outd}(o_i)$), and cause a further abortion of the accessing transaction.

A particular application of the study has been used to validate the assumptions and conclusions provided by the analysis. The presented validation includes both a simulation in a general environment, and a real implementation in the area of distributed databases.

The application of the expression has been also discussed, in order to determine the convenience, using a general algorithm, to update along the execution of a transaction the objects predicted to be stale. This discussion has provided a set of conditions, in base to a number of parameters, where these generic algorithms can improve the abortion rate of a lazy update protocol.

Thus, the improvement has also discussed, in base to the established decrement of the update-time of the accessed objects, giving as conclusion that such reductions may considerably improve the probability for an object to be updated, thus reducing dramatically the abortion rate of lazy update protocols.

This work theoretically validates the implementation of such *"not-so-lazy update protocols"* based on the statistical prediction of stale accesses to reduce the abortion rate. These protocols conform possible implementations of the mentioned generic algorithm, as an update algorithm implementable in a real environment.

Acknowledgements

We want to thank the development consortium of the GlobData Project the performed work and support provided, and specially to the development team of the GlobData Project in the Instituto Tecnológico de Informática. This specially includes Antonio Calero, Paco Castro, Javier Esparza and Félix García.

References

1. Rahm, E.: Empirical performance evaluation of concurrency and coherency control protocols for database sharing systems. ACM Trans. on Database Sys. **18** (1993) 333–377
2. Gray, J., Helland, P., O'Neil, P., Shasha, D.: The dangers of replication and a solution. In: Proc. of the 1996 ACM SIGMOD International Conference on Management of Data, Montreal, Quebec, Canada (1996) 173–182
3. Muñoz, F., Irún, L., Galdámez, P., Bernabéu, J., Bataller, J., Bañuls, M.C.: Globdata: A platform for supporting multiple consistency modes. In Press, A., ed.: Information Systems and Databases (ISDB'02), Tokyo (2002) 244–249
4. Instituto Tecnológico de Informática: GlobData Web Site (2002) Accessible in URL: *http://globdata.iti.es*.
5. Rodrigues, L., Miranda, H., Almeida, R., Martins, J., Vicente, P.: The globdata fault-tolerant replicated distributed object database. In: Proceedings of the First Eurasian Conference on Advances in Information and Communication Technology, Teheran, Iran (2002) 426–433
6. Irún-Briz, L.: Implementable Models for Replicated and Fault-Tolerant Geographically Distributed Databases. Consistency Management for GlobData. PhD thesis, Departamento de Sistemas Informáticos y Computación, Universidad Politécnica de Valencia, Valencia, Spain (2003) Available at *http://www.iti.upv.es/˜lirun*.

7. Irún, L., Muñoz, F., Decker, H., Bernabéu-Aubán, J.M.: Copla: A platform for eager and lazy replication in networked databases. In: 5th Int. Conf. Enterprise Information Systems (ICEIS'03). Volume 1., Angers, France (2003) 273–278
8. Decker, H., Muñoz, F., Irún, L., Calero, A., Castro, F., Esparza, J., Bataller, J., Galdámez, P., Bernabéu, J.: Enhancing the availability of networked databases services by replication and consistency maintenance. In: 14th International Conference on Database and Expert Systems Applications - DEXA 2003. Lecture Notes in Computer Science, Prague, Czech Republic, Springer-Verlag (2003) 531–535
9. Irún-Briz, L., Muñoz-Escoí, F.D., Bernabéu-Aubán, J.M.: An improved optimistic and fault-tolerant replication protocol. In: Proceedings of 3rd. Workshop on Databases in Networked Information Systems (DNIS). Lecture Notes in Comp. Sci., Springer-Verlag (2003)
10. Chandy, K.M., Misra, J.: Distributed simulation: A case study in design and verification of distributed programs. IEEE Transactions on Software Engineering **SE-5** (1979) 440–452
11. Bagrodia, R.L., Chandy, K.M., Misra, J.: A message-based approach to discrete-event simulation. IEEE Transactions on Software Engineering **SE-13** (1987) 654–665
12. Bagrodia, R.L.: An integrated approach to the design and performance evaluation of distributed systems. In: Proceedings of the First International Conference on Systems Integration, Morristown, NJ, USA, IEEE Computer Society (1990) 662–671
13. Liskov, B.: Practical uses of synchronized clocks in distributed systems. In Logrippo, L., ed.: Proceedings of the 9th Annual ACM Symposium on Principles of Distributed Computing (PODC'90), Montéal, Québec, Canada, ACM Press (1991) 1–10
14. Hadzilacos, V., Toueg, S.: A modular approach to fault-tolerant broadcasts and related problems. Technical Report TR94-1425, Cornell University, Computer Science Department (1994)
15. Agrawal, D., Alonso, G., El Abbadi, A., Stanoi, I.: Exploiting atomic broadcast in replicated databases. Lecture Notes in Computer Science **1300** (1997) 496–503
16. Wiesmann, M., Schiper, A., Pedone, F., Kemme, B., Alonso, G.: Database replication techniques: A three parameter classification. In: Proc. of the 19th IEEE Symposium on Reliable Distributed Systems (SRDS'00), Nuernberg, Germany (2000) 206–217
17. Breitbart, Y., Korth, H.F.: Replication and consistency: being lazy helps sometimes. In: Proceedings of the sixteenth ACM SIGACT-SIGMOD-SIGART symposium on Principles of Database Systems, Tucson, Arizona, EE.UU., ACM Press (1997) 173–184
18. Holliday, J., Agrawal, D., Abbadi, A.E.: Database replication: If you must be lazy, be consistent. In: Proceedings of 18th Symposium on Reliable Distributed Systems SRDS'99, Lausanne, Switzerland, IEEE Computer Society Press (1999) 304–305
19. Chundi, P., Rosenkrantz, D.J., Ravi, S.S.: Deferred updates and data placement in distributed databases. In: Proceedings of the 12th International Conference on Data Engineering, New Orleans, Louisiana, IEEE Computer Society (1996) 469–476
20. Bernstein, P.A., Shipman, D.W., Rothnie, J.B.: Concurrency control in a system for distributed databases (SDD-1). ACM Transactions on Database Systems **5** (1980) 18–51
21. Kung, H.T., Robinson, J.T.: On optimistic methods for concurrency control. ACM Transactions on Database Systems **6** (1981) 213–226
22. Bernstein, P.A., Hadzilacos, V., Goodman, N.: Concurrency Control and Recovery in Database Systems. Addison Wesley, Reading, MA, EE.UU. (1987)
23. Keleher, P., Cox, A.L., Zwaenepoel, W.: Lazy release consistency for software distributed shared memory. In: Proc. of the 19th Annual Int'l Symp. on Computer Architecture (ISCA'92), Queensland, Australia (1992) 13–21
24. Amza, C., Cox, A.L., Dwarkadas, S., Keleher, P., Lu, H., Rajamony, R., Yu, W., Zwaenepoel, W.: Treadmarks: Shared memory computing on networks of workstations. IEEE Computer **29** (1996) 18–28

protoRAID: A User-Level RAID Emulator for Fast Prototyping in Fibre Channel SAN Environment

Dohun Kim and Chanik Park

Department of Computer Science and Engineering/PIRL,
Pohang University of Science and Technology,
790-784 San31, Hyoja-dong, Nam-gu, Pohang, Kyungbuk, Republic of Korea
Tel: +82-54-279-5668, +82-54-279-2248, Fax: +82-54-279-5699
{hunkim, cipark}@postech.ac.kr

Abstract. To measure accurate performance of the new RAID storage system, simulators and emulators must consider the behavior of system components, such as the host connection interface, internal system bus, internal I/O bus, and physical disk. Moreover, the behavior of RAID functional operations with the buffer cache is also considered important. Although many RAID simulators exist, they do not consider all the components of RAID storage systems, resulting in inaccurate performance measurements. As far as we know, there exists no RAID emulator that considers all system components. Therefore, we present a RAID emulator to correct this defect. It is implemented as a user-level process in Linux. It interacts with a host via a physical I/O interface like FibreChannel; that is, the effect of interactions between the host and the RAID controller is physically considered. It interacts with physical disks via SCSI or FibreChannel; that is, the effect of interaction between RAID controller and disks is considered. Moreover, RAID functional operations such as the buffer cache and block mapping function are also considered in overall timing measurements.

1 Introduction

Careful investigation of the operational behavior of a RAID storage system leads us to reveal the most important steps that affect the resulting performance: host connection interface, RAID functional software, internal system bus, internal I/O bus, and physical disk. Therefore, to develop a new RAID storage system, we must consider these steps in addition to validating the RAID functions.

Previous studies showed several RAID simulators and emulators. However, these studies were limited to validating the correctness of RAID functions and measuring performance while only a few steps of the operational behavior were considered. All simulators simulate a few steps by modelling, resulting in inaccurate performance and RAID function validation. As a result, they are incomplete and inaccurate. In contrast to simulators, emulators use actual system compo-

M. Daydé et al. (Eds.): VECPAR 2004, LNCS 3402, pp. 454–467, 2005.

nents to emulate each step and measure more accurate performance. However, existing emulators also fail to consider all steps.

Therefore, we present a RAID emulator called protoRAID which can validate the correctness of new RAID functions and measure more accurate performance than the other emulators, because it tries to consider all the important operational steps using as many hardware components as possible. The protoRAID considers operations involved in the host connection interface, internal system bus, internal I/O bus, and physical disks as a hardware environment and operations involved in RAID functional software with buffer cache as a software environment. Moreover, it is implemented under the Linux Operating System as a user-level process.

This paper shows the design of the protoRAID and its two implementation variations to implement data transfer between user-level RAID software and the host connection interface. Next, comparison with Linux Software RAID's performance is described. Lastly, we discuss the overhead of the protoRAID.

This paper is organized as follows. Section 2 describes related works. In Section 3, our design model is described. Section 4 describes the implementation of the protoRAID. In section 5, evaluations of a prototyped RAID, comparison between the protoRAID and Linux Software RAID, and overhead of protoRAID are presented. Conclusion and future works are given in Section 6.

2 Related Work

DiskSim[8] and Memulator[10] are simulators that can be configured for testing a RAID storage system. DiskSim[8] and Memulator[10] were not originally made for validating RAID functions but measuring performance. They simulate actual system components by modeling them. Therefore, they depends on only small number of parameters. Note that Memulator[10] uses the host connection interface of the actual system, while still using simulation for the other components. As a result, they are insufficient for validating the correctness of RAID functions and measuring accurate performance.

RAIDsim[9] is a simulator that validates RAID functions. Although it can validate RAID functions, it considers only two system components, such as RAID software and simulated disks. Therefore, it is not sufficient for measuring accurate performance.

RAIDframe[6] shows both simulator and emulator features. It can validate RAID functions and use such simulated system components as RAID software and disks. Moreover, it can run as a kernel device driver under Digital Unix OS. Therefore, it emulates actual system components and accepts the actual application's workload. Currently, it is being ported to NetBSD[7]. However, RAIDframe still shows a limit as a storage emulator. It is dedicated to only one host system and does not consider the host connection interface. Therefore, it cannot be used as a storage node that can be shared by several hosts.

Software RAID[4] known as MD is an emulator. It runs as a block device driver under Linux. Therefore, it emulates actual system components. However, it has many limits to configure RAID system. Under heavy workload, it uses up whole system memory, because it is implemented as a Linux block device driver. This incurs unnecessary swap in/out if applications want to use memory. In spite of such problem, there is no method to control its memory usage. Although it supports RAID level 0, 1, and 5, it cannot configure RAID buffer cache except RAID level 5. Moreover, it cannot control the speed of dirty block flushing by itself and does not consider the host connection interface.

Our work differs from previous studies. The protoRAID chooses an emulator approach. Therefore, it does not use a simulated system component. The protoRAID is different from RAIDframe[6] and Software RAID as an emulator in that it operates as a user-level process in Linux OS and contains a host connection interface that supports Fibre Channel. It contains interfaces to easily insert new features and can control more system parameters such as read/write cache size, start time of dirty block flush, and etc. Therefore, our approach simplifies debugging and provides persons studying the Storage Area Network with a true storage node emulator.

3 Emulator Design

Real hosts can use the protoRAID as a RAID storage system, because it emulates actual system. Therefore, actual applications such as file system and database can store/retrieve their data into/from the protoRAID. This section describes how the protoRAID can target a new RAID system and how it is designed to emulate the RAID storage system.

3.1 Hardware Part

The protoRAID can emulate various H/W environments that do not exist in real RAID systems, because it runs under Linux OS. For example, these days, the more many hosts share RAID systems, the more RAID systems needs powerful CPUs with high-bandwidth internal I/O BUS. However, there would not be a real H/W environment which considers this. The protoRAID can configure such an ideal environment using many H/W components such as Intel Pentium 4 CPU and PowerPC CPU with SCSI BUS or VME BUS.

To configure a new RAID system, the protoRAID needs many H/W components such as a CPU, Memory, a host connection interface, and a disk I/O interface. However, this part does not describe adopting CPU and Memory, because it is simple to adopt. Figure 1 shows the relationship between the hardware part and the software part.

Host Connection Interface. The host connection interface exchanges host I/O requests/responses between the host and RAID system. It consists of a target device driver, an internal system bus, a target host adapter and the Fibre Channel SAN. Because all messages are exchanged through the host connection interface, the host connection interface affects the performance of the RAID

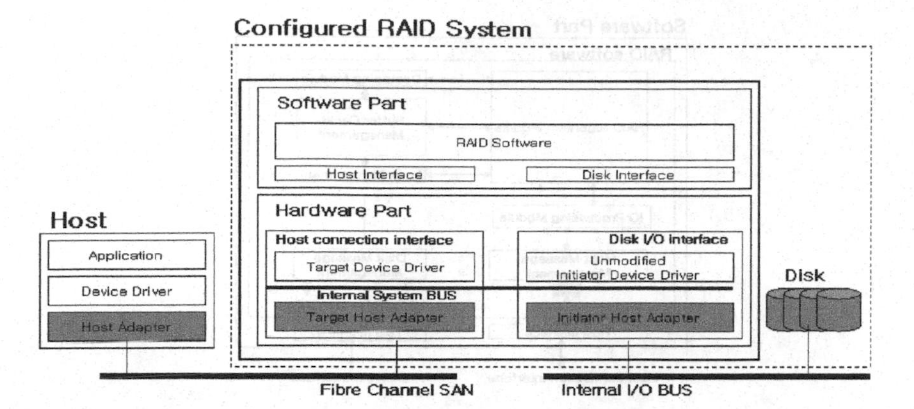

Fig. 1. Relationship between the hardware part and the software part. The hardware part consists of the host connection interface and the disk I/O interface. The software part consists of the RAID software and two interfaces between the hardware part and the software part

storage system. There are many variations for configuring the host connection interface such as selecting 66MHz PCI as an internal system bus and Fibre Channel host adapter as a host adapter.

Selecting a target device driver depends on its host adapter. We made two device drivers composing a target device driver: a middle layer SCSI target driver and a lower layer FC target driver. We separated the target device driver, because we will provide another lower layer target driver.

Disk I/O Interface. The disk I/O interface exchanges disk I/O requests/ responses between the RAID software and disks. It consists of an unmodified initiator device driver, an internal system bus, an initiator host adapter and an internal I/O bus. The disk I/O interface affects the performance of the RAID system, because all disk I/O requests/responses are exchanged through it. There are also many variations for configuring the disk I/O interface such as selecting SCSI host adapters or FC host adapters as initiator host adapters.

We adopted Linux device drivers as an initiator device driver. Therefore, we marked it as unmodified in figure 1. The initiator device driver is composed of SCSI Generic driver, a middle layer SCSI driver, and a lower layer SCSI driver. SCSI generic driver is a type of character device driver and supports raw device I/O mode. Moreover, it can access SCSI disks using SCSI CDB.

3.2 Software Part

The protoRAID supports a software part composed of a host interface, a disk interface, and RAID software. Because the RAID software is modularized, newly designed features such as new RAID mapping algorithm, buffer cache management, and host/disk queue management can easily be inserted. Their relationship is described in figure 2.

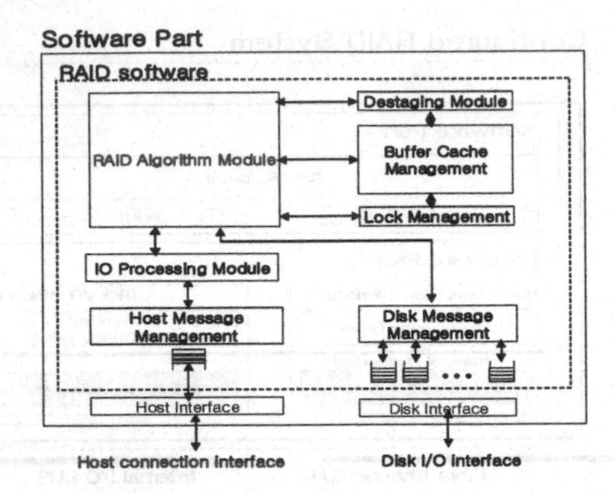

Fig. 2. Relationship between components of the protoRAID's software part

The protoRAID needs two interfaces between RAID software and the hardware part to exchange host I/O requests/responses and disk I/O requests/responses. To implement the host interface, we designed two versions: TCP/IP socket version and device file system version. Designing the disk interface depends on the Linux SCSI generic driver. It is described in the **disk interface**.

RAID Software. RAID software is organized into several modules such as host message management, IO processing, RAID algorithm, buffer cache management, destaging, lock management, and disk message management.

The host interface transforms host I/O requests into host messages and queues them into the host queue. After a while, the host message management module get one host message from the queue, and it is transferred to IO processing module. After each host I/O message is processed by IO processing module, the host message management module sends its responses to the host connection interface.

The IO processing module makes a thread for each host message transferred from the host message management module. Each thread makes stripe I/O requests by using the RAID algorithm module.

The RAID algorithm module includes RAID level specific operations such as block mapping operation for normal/degraded status and rebuild operation to reconstruct a disk image. Each stripe I/O request tries to find blocks from buffer cache. If any block is missed, missed blocks are transformed into disk I/O requests according to the RAID level. These disk I/O requests are sent to the disk message management module. If the current RAID status is disk failure, it tries to continue disk I/O requests using a degraded operation. A rebuild operation can be activated by the user's request. Currently, the protoRAID supports RAID level 0, 1, 5, 0+1 and RM2[5].

The buffer cache management module manages buffer cache operations such as block replacement and block sharing. Because the protoRAID supports a background write I/O for fast response, a destaging module is needed.

The destaging module flushes dirty buffer blocks from the buffer cache to disks. Before starting flush, it makes destaging threads per each stripe to increase I/O concurrency. Each destage thread makes disk I/Os by using the RAID algorithm module. According to the ratio of dirty blocks to the total buffer blocks, the destaging module starts destaging at high water mark and stops at low water mark.

The lock management module preserves data consistency using a hashing function. There are two kinds of lock in the protoRAID. The first is a stripe lock, preserving data consistency across disks by stripe. The second is a buffer cache lock, preserving data consistency for each buffer cache block.

The disk message management module controls disk I/O requests and their responses. It transforms these disk I/O requests into disk messages such as SCSI CDB and data blocks, and queues them into each proper disk queue. After a while, the disk interface gets a disk message from the disk queue and transfers it to disk I/O interface. After all the disk I/O messages of the same host I/O request are completed, the disk message management relays this information to the IO processing module.

Host Interface. To poll whether host I/O requests arrives from the middle layer SCSI target driver, we designed a Request Poller. First, we designed TCP/IP socket version because of simplicity of implementation. However, this version has a limit in that only one kernel thread of the middle layer SCSI target driver processes all host I/Os with socket overhead, resulting in showing low performance. Therefore, we made another kernel thread. The original thread(sending thread) sends host I/O requests to the host interface and the new thread(receiving thread) receives host I/O responses from the host message management module. There are five steps to communicate between the RAID software and hosts.

Step 1: the lower layer FC(LLFC) target driver receives FC messages from the host through FC SAN.
Step 2: the kernel thread in the LLFC target driver extracts SCSI CDB and data blocks from the FC message. Next, it transfers them to middle layer SCSI(MLS) target driver.
Step 3: the sending thread in MLS target driver transfers SCSI CDB and data blocks to Request Poller in the host interface using TCP/IP socket.
Step 4: after processing host I/O requests, the host message management module transfers replies to the receiving thread in the MLS target driver.
Step 5: the receiving thread in the MLS target driver transfers replies and data blocks to the host using a transfer function in LLFC target driver. Design and communication path of TCP/IP socket version are described in figure 3(a).

After several experiments, we found out that the TCP/IP socket version still shows low performance because of socket processing, although it gives us implementing simplicity. Therefore, we added device file system features into the MLS target driver. This version uses only one kernel thread in the MLS target driver. There are five steps for communicating between the RAID software and

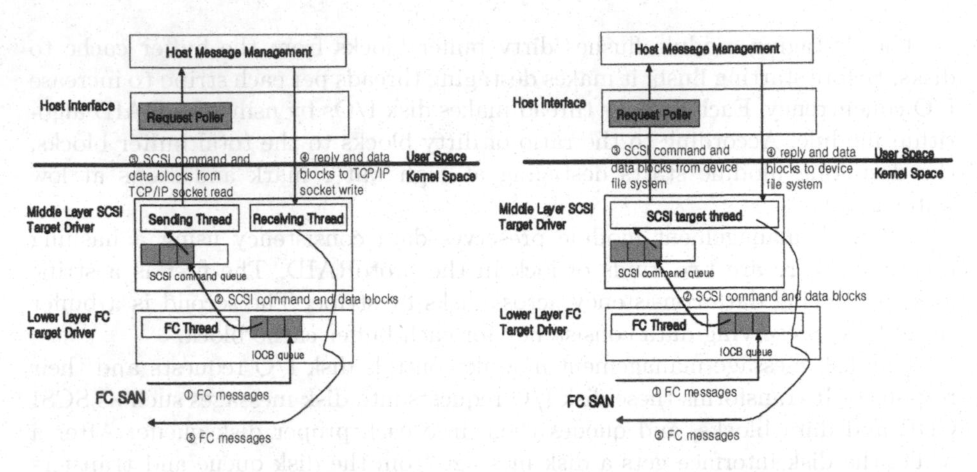

(a) TCP/IP socket version of the host interface and its communication path

(b) Device file system version of the host interface and its communication path

Fig. 3. Diagrams of each host interface implementation and its communication path. Each version shows five communication steps. Each step is described in the host interface item

hosts. Five steps of this version are not described, because they are similar to the TCP/IP socket version. The design and communication path of device file system version are described in figure 3(b).

Disk Interface. The protoRAID uses SCSI generic driver[2] with asynchronous I/O mode, because it needs the capability of accessing raw disks and high performance I/O. To provide asynchronous I/O, we made an I/O Response Poller for each disk, which determines whether a disk I/O response arrives from each disk. There are two steps to communicate between the RAID software and disks. Step 1: disk I/O messages are transferred to the disk interface. The disk interface merges SCSI CDB and data blocks into an sg header. Next, it transfers the sg header to the SCSI generic device driver by using system call.

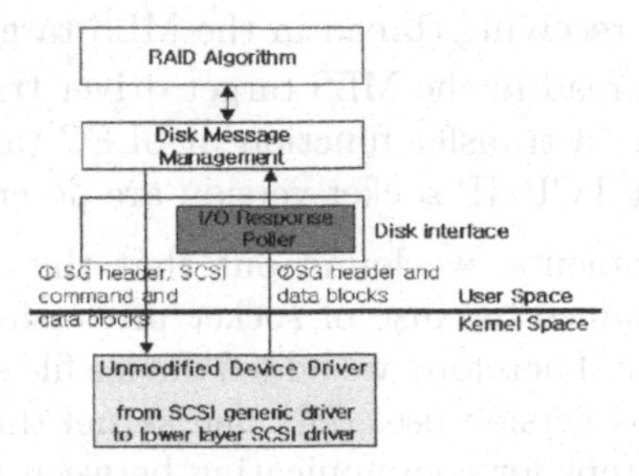

Fig. 4. Design and communication path of the disk interface

Step 2: after each disk processes disk I/O requests, it informs the I/O Response Poller that a disk I/O request is done. Next, the I/O Request Poller informs the disk message management module that a disk I/O request is completed. Figure 4 shows the design and communication path of the disk interface.

4 Implementation of the protoRAID

The protoRAID uses POSIX-compliant thread, mutual exclusion and semaphore. The host message management module and the disk message management module use threads to process each messages. The IO processing module and destaging module use threads for each striped I/O request. The lock management module uses semaphores and mutual exclusions.

We implemented a Request Poller using a thread. The Request Poller of the TCP/IP socket version uses select() system call for polling whether a host I/O request arrives while the Request Poller of the device file system uses a poll() system call which we added in the MLS target driver. We also implemented read(), write() system call in the MLS target driver. To use the TCP/IP socket version, we also added a kernel thread in the MLS target driver. It always polls whether a host I/O response arrives from the RAID software.

We implemented the I/O Response Poller using a thread. It always polls whether a disk I/O response arrives from disks using poll() system call which SCSI generic driver supports. It transfers disk I/O response to the disk message management module by using an sg header which contains SCSI CDB and a pointer of buffer for data blocks.

To act as a true RAID storage system, the protoRAID supports SCSI commands such as 10 byte READ/WRITE, READ CAPACITY, INQUIRY, and TEST UNIT READY. These SCSI commands are sufficient for using the protoRAID as a basic storage node in the Storage Area Network.

To provide fast prototyping, the protoRAID currently supports mapping and RAID algorithm function interface which depends on RAID status and RAID level. There are already built-in mapping and RAID algorithm functions for RAID 0, 1, 0+1, 5, and RM2. The rest interfaces for host/disk message management, buffer cache, and destaing remain as a future work. Current interface of the protoRAID is described in the **Appendix section.**

5 Evaluation

Two testbeds are used to evaluate a prototyped RAID. One of them is used as a host composed of one P4-1.4G CPU, 128MB main memory, 33Mhz PCI, and one Qlogic ISP2200 FC HBA for Fibre Channel SAN. Another testbed is used as a RAID controller which is similar to the host, except that it has another FC HBA for the internal I/O bus. Testbeds use a Linux kernel 2.4.2 and SCSI generic driver 3.1.23 of which command queue size is 64 to increase disk I/O concurrency. The RAID controller is connected to an enclosure that contains four

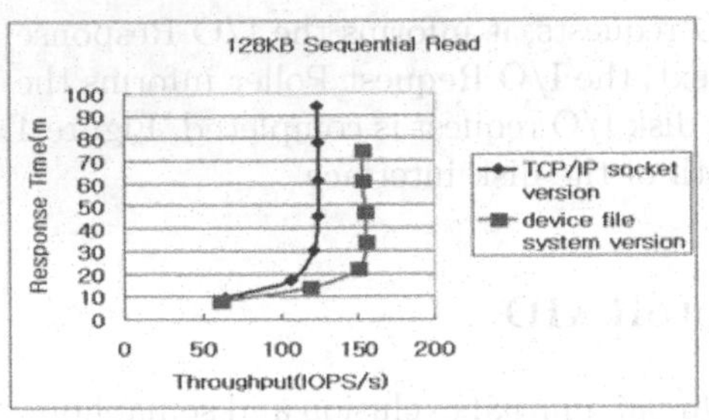

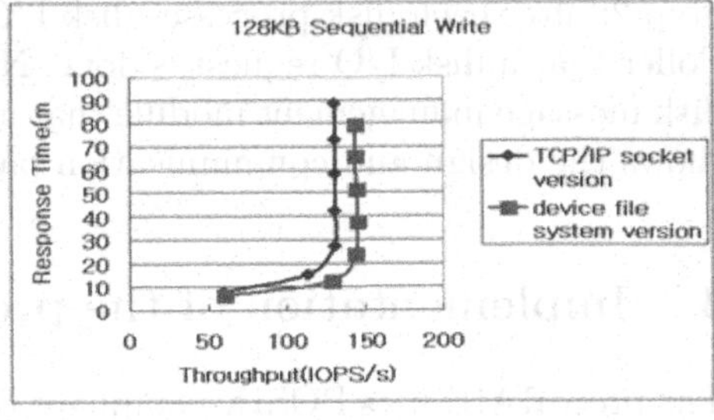

(a) IOps vs. response time graph (b) IOps vs. response time graph

Fig. 5. Performance comparison between two versions of the host interface. The TCP/IP socket version has higher additional overhead, such as socket processing than the device file system version

Segate Barracuda FC disks(ST373307FC). The protoRAID uses a write-back cache policy, that is, it starts flushing dirty blocks according to LRU. Throughout the experiment, we use a I/O benchmark utility used in [1]. It produces some I/O processes and each I/O process invokes blocked I/O workloads after some thinking time(we used exponential distribution with 10ms mean).

5.1 Comparison Between Variations of the Host Interface

We implemented two variations of the host interface: the TCP/IP socket version and the device file system version. Because the actual RAID storage system does not consider additional overhead such as socket processing, we made a device file system version where data is copied to and from user space directly. Figure 5 shows the effect of socket processing overhead by comparing performance between two versions. Performance is measured using a 128KB sequential workload where the protoRAID operates at RAID level 0, has 32MB buffer cache and uses 32KB stripe unit. Because the device file system version has lower additional overhead, it is used for measuring performance throughout the rest experiments.

5.2 Performance According to Simulation Parameters

The protoRAID can show various performance according to several parameters, such as the RAID level, stripe unit size, buffer cache size, the number of destaging threads, the ratio of dirty blocks of starting/ending destaging, and etc. This section shows the performance of the prototyped RAID system affected by the number of destaging threads that other simulators and emulators do not consider. Figure 6(a) shows a performance which the protoRAID operates at RAID level 5 and has 32MB buffer cache. Its stripe unit size is 32KB and the high water mark of destage is 70 percent and the low water mark of destage is 30 percent. The number of destaging threads is limited to 21 threads. Performance is measured by varying the I/O concurrency from 1 to 14 enough for saturating the RAID system. 32KB random workload is used for this experiment. Figure 6(b) shows the performance where the protoRAID has

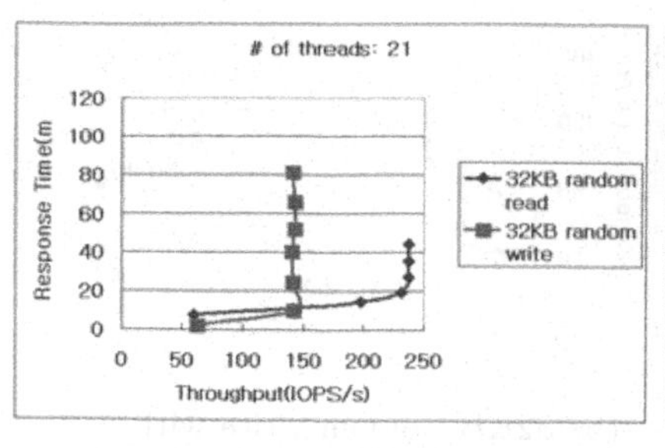

(a) IOps vs. response time graph with 21 destaging threads

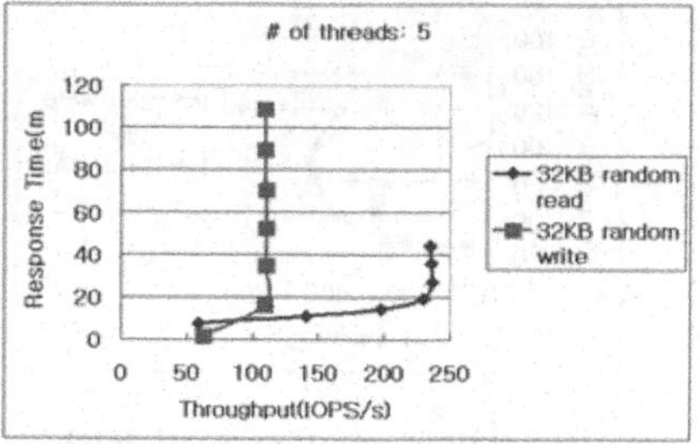

(b) IOps vs. response time graph with 5 destaging threads

Fig. 6. An example of configuring parameters that affect performance(device file system version). Because the number of destaging threads affects the speed of making clean buffer blocks, a lower number of destaging threads results in degraded write performance

changed the number of destaging threads from 21 to 5. Because all read operations are not related to destage threads which are for write operations, they show the same performance. However, random write performance is degraded, because insufficient cache blocks are allocated to write requests for fast writing. This comes from shortage of destaging threads that makes clean buffer blocks available.

5.3 Comparison with Other Emulator

The protoRAID is compared with Software RAID, because RAIDframe runs under BSD OS. We configured Software RAID as RAID level 5 with 32MB buffer cache and measured its performance by increasing I/O concurrency until it is fully saturated. In spite of having its own 32MB buffer cache, Software RAID used up all system buffer cache, because it is a block device driver in Linux. We configured the protoRAID and measured its performance with the same method as Software RAID except that it can have maximum 30 destaging threads. However, the protoRAID did not use additional memory unlikely Software RAID. We measured each performance by using two types of FC disk such as ST373307FC and ST39175FC. ST373307FC has higher performance than ST39175FC. Figure 7 shows each emulator's performance measured by using 32KB random workload which is the same size to stripe unit size.

In figure 7, Software RAID shows 1.8 times higher performance. There are three reasons for it. First, the protoRAID runs as a user process. Second, Software RAID uses up all system memory in addition to its own buffer cache. Effectively, it has two-level buffer cache. Third, Software RAID does not consider host connection interface. To validate first reason, we took two experiments: 1) send host I/O response as soon as the MLS receives host I/O request 2) send host I/O response as soon as the host interface receives host I/O message. We find out that 1) shows two times higher performance. Because second and third rea-

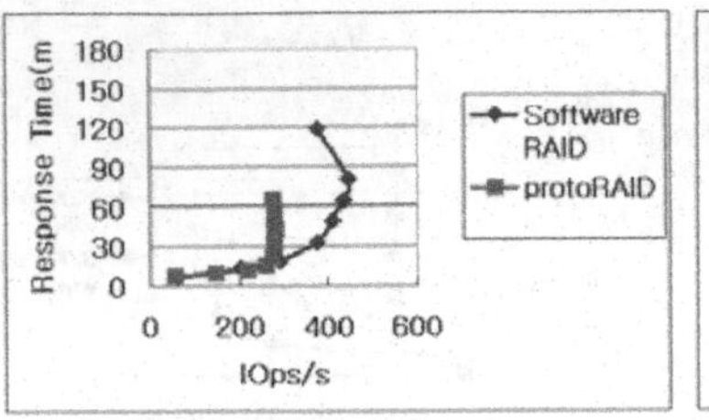

(a) 32KB random read with high performance disk

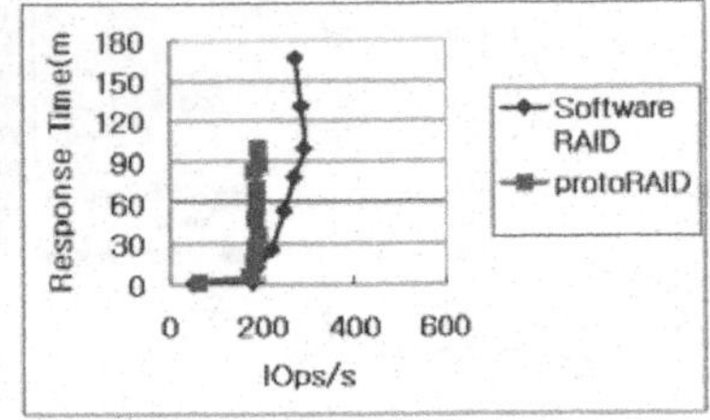

(b) 32KB random write with high performance disk

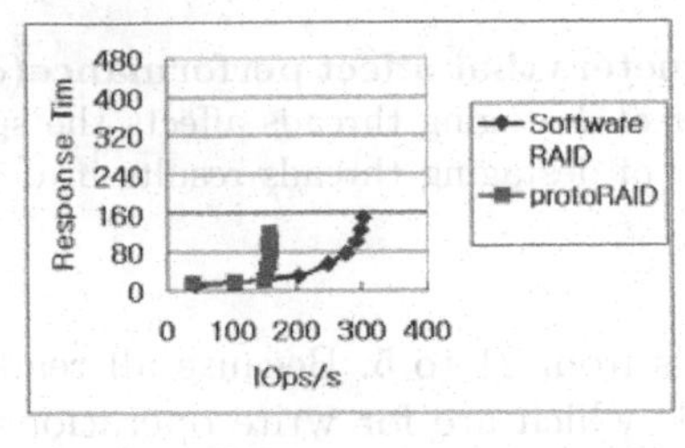

(c) 32KB random read with low performance disk

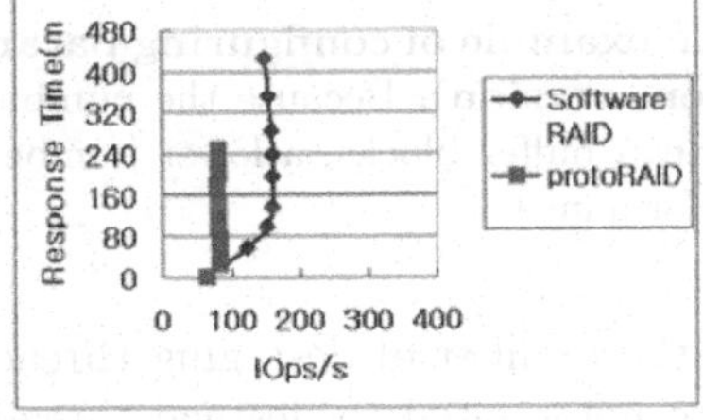

(d) 32KB random write with low performance disk

Fig. 7. Comparison between the protoRAID and Software RAID(device file system version). Performance of Software RAID shows 1.8 times higher, because the protoRAID runs as user level process

sons are straightforward, we will not discuss it. Although Software RAID shows higher performance, the protoRAID has many advantages such as easy debugging, no system crash, and developing directly on the target system, because it runs as user level process.

5.4 Overhead of the protoRAID

Because the protoRAID runs at user-level, the measured performance may include unnecessary overheads such as kernel/user space data transferring and system calls. Therefore, we measured this overhead where various I/O workloads are issued. We measured the time portions of all system components, such as the host connection interface, the host interface, RAID software, disk interface, and disk I/O interface. During the experiment, the buffer cache size is fixed to only 2MB. Therefore, it affects the performance little. Moreover, the protoRAID operates at RAID level 0 with only one disk, because many disks may affect the entire performance by the effect of parallelism. We used 32KB, 64KB, 96KB, 128KB random reads by varying I/O concurrency from 1 to 14, because random reads always run in the foreground and go through the entire I/O path. For these reasons, we can accurately measure each time portion of each component. Figure 8 shows each system component's portion of the response time. We did

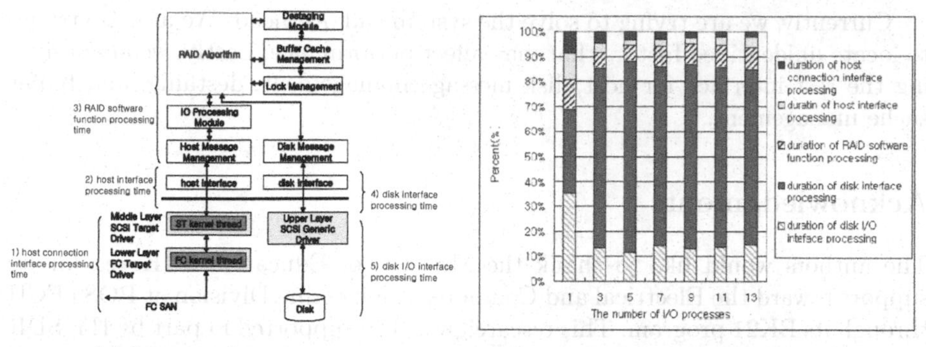

(a) Measured components

(b) processing time portion of each component in response time(96KB random read)

Fig. 8. Each component's portion of the response time(device file system version). This figure shows only one case of 96KB random read, because other request sizes show the similar results. The host interface of the device file system version takes almost no portion of response time. However, the disk interface takes the largest portion. This comes from system calls that user-level threads have called for processing disk I/O requests and their responses

not displayed the other workload cases, because they show similar results. The host interface takes almost no portion of the response time(average 4%). The graph also shows that the processing time of the disk interface takes the largest portion. This comes from many system calls that user-level I/O threads have called. The system call is an additional overhead that an actual RAID storage system may not process. Currently, we are trying to solve this problem.

6 Conclusion and Future Works

The protoRAID can measure more accurate performance than other simulators and emulators, because it emulates all actual system components, such as the host connection interface, RAID software, disk I/O interface and disk. The protoRAID needs two interfaces to use a host connection interface and disk I/O interface: the host interface and the disk interface. There are two variations for implementing the host interface: the TCP/IP socket version and the device file system version. Because the TCP/IP socket shows more additional overhead than device file system version, we used the device file system version. The protoRAID has many configuration parameters that affect performance. For example, changing the number of destaging threads shows that the write performance is affected by the number of destaging threads. The protoRAID's performance shows about a half times lower performance than RAID software, because it runs as a user level process. However, the protoRAID has more advantages such as easy debugging, no system crash, direct development on testbed, and etc.

Currently, we are trying to solve the system call overhead. We are also trying to locate unidentified factors that may affect performance. Lastly, we are designing the rest interface for host/disk message management, destaing, and buffer cache management.

Acknowledgments

The authors would like to thank the Ministry of Education of Korea for its support toward the Electrical and Computer Engineering Division at POSTECH through its BK21 program. This research was also supported in part by HY-SDR IT Research Center, and in part by grant No. R01-2003-000-10739-0 from the Basic Research Program of the Korea Science and Engineering Foundation.

References

1. Young Jin Nam, Dae-Woong Kim, Tae-Young Choe and Chanik Park, "Enhancing Write I/O Performance of Disk Array RM2 Tolerating Double Disk Failures," *Proceedings of the 31st International Conference on Parallel Processing (ICPP2002)*, Vancouver, Canada, Aug. 2002
2. "The Linux SCSI Generic (sg) Driver," http://www.torque.net/sg/
3. "The Linux 2.4 SCSI subsystem HOWTO," http://www.tldp.org/HWOTO/SCSI-2.4-HOWWO/
4. "The Software-RAID HOWTO," http://www.tldp.org/HWOTO/Software-RAID-HWOTO.html/
5. Chan-Ik Park, "Efficient Placement of Parity and Data To Tolerance Two Disk Failures In Disk Array Systems," *IEEE Trans. Parallel and Distributed Systems, Vol. 6, No. 11*, Nov. 1995
6. Garth Gibson, William V. Courtright II, Mark Holland, and Jim Zelenka, "RAIDframe: Rapid prototyping for disk arrays," Proc. of the 1996 Conference on Measurement and Modeling of Computer Systems (SIGMETRICS), May 1996, Vol. 24 No. 1, pp. 268-269.
7. "NetBSD and RAIDframe," http://www.cs.usask.ca/staff/oster/raid.html
8. G. Ganger, Bruce L. Worthington, and Yale N. Patt, "The DiskSim Simulation Environment Version 3.0," Jan. 2003. http://www.pdl.cmu.edu/DiskSim/index.html
9. Edward K. Lee, and Randy H. Katz, "The Performance of Parity Placements in Disk Arrays," IEEE Trans. Computers, Vol. 42, No. 6, June 1993, pp. 651-664.
10. John Linwood Griffin, Jiri Schindler, Steven W. Schlosser, John C. Bucy, and Gregory R. Ganger, "Timing-accurate Storage Emulation," Conference on File and Storage Technologies(FAST) Jan. 2002

Appendix: Interface for Fast Prototyping

The protoRAID has a function table to support new mapping and RAID algorithm functions. At each RAID status, each RAID algorithm can have its own mapping and RAID algorithm functions. In figure 9(a), column means RAID status such as normal, 1 disk fault, 2 disk fault, and etc. Row means RAID

Level \ Status	RAID Status 1 (Normal)	RAID Status 2 (1 disk fault)	RAID Status 2 (2 disk fault)	...
RAID 0	Normal RAID 0	None	None	NULL
RAID 1	Normal RAID 1	1-degraded RAID 1	1-fault RAID 1	NULL
RAID 0+1	Normal RAID 0+1	1-degraded RAID 0+1	1-fault RAID 0+1	NULL
RAID 5	Normal RAID 5	1-degraded RAID 5	1-fault RAID 5	NULL
RAID RM2	Normal RM2	1-degraded RAID RM2	2-degraded RAID RM2	NULL
...	NULL	NULL	NULL	NULL

(a) Function table for each RAID status and RAID level

	Parameters	Description
doReadIO	block address block length	start block address of stripe I/O request the size of stripe I/O request
doWriteIO	block address block length destaging link	start address of stripe I/O request the size of stripe I/O request destaging thread maintains this request using this link
doCloseIO	information of stripe I/O request	information on stripe lock, buffer cache lock
mapBlock	block address physical disk number physical block address	logical block address to be mapped disk number of the mapped block mapped block address in mapped disk

(b) Function interface for each cell of the function table(a)

Fig. 9. Fast prototyping interface of the protoRAID. Currently, the protoRAID supports mapping and RAID algorithm interface

level. Each cell means function interface for each RAID level and RAID status. Figure 9(b) shows function interface format for each cell in (a).

Parallel Computational Model with Dynamic Load Balancing in PC Clusters

Ricardo V. Dorneles[1], Rogério L. Rizzi[2], André L. Martinotto[1], Delcino Picinin Jr.[1], Philippe O.A. Navaux[3], and Tiarajú A. Diverio[3]

[1] Departamento de Informática, CCET, UCS, Caxias do Sul, RS, Brasil
[2] CCET, UNIOESTE, Campus de Cascavel, Cascavel, PR, Brasil
[3] PPGC, Instituto de Informática, UFRGS, Porto Alegre, RS, Brasil

Abstract. This work describes the use of dynamic load balancing in a PC cluster, applied to a multi-physics model that combines the parallel solution for three-dimensional (3D) PDEs of shallow water bodies flow and the parallel solution for the three-dimensional PDEs of scalar transportation of substances. The dynamic load balancing is obtained via diffusion algorithms. Three approaches for dynamic load balancing were implemented and are described here.

1 Introduction

Computational modeling has been used for environmental studies for a long time. In its scope, simulation of hydrodynamics and mass transportation in water bodies such as oceans, lakes and rivers is an area of very active research and very demanding of computing power. Usually this computing power is obtained parallelizing the models and running it on parallel machines.

The parallelization of models used in environmental studies is a very complex task and several issues must be considered to get good results. Two important issues are the way the partitioning of the domain is done and, in some applications, the mechanisms used to keep the load balanced between the processors along the simulation. In a previous paper [1] we have presented a parallel model for hydrodynamics with static partitioning called HIDRA. After that, a mass transportation model with a dynamic mesh and dynamic load balancing mechanisms have been incorporated to HIDRA. Presenting the mass transportation model and the dynamic load balancing mechanisms in the HIDRA model is the object of this paper.

This paper is structured as follows: in section 2 a short description of the mass transportation model is provided, as it is necessary to contextualize the mechanisms described; in section 3 the multi-physics problem of hydrodynamics and mass transportation is presented; in section 4 an overview of dynamic load balancing is presented; in sections 5, 6 and 7 the three different approaches for dynamic load balancing implemented in the model are presented; in section 8 the metrics used to load evaluation are shown; in section 9 the algorithms for calculation and selection of cells to be transferred are presented; in section

M. Daydé et al. (Eds.): VECPAR 2004, LNCS 3402, pp. 468–479, 2005.

10 some results obtained are presented and in section 11 some conclusions and future works are presented.

2 Mass Transportation Model and Numerical Scheme

The mass transportation equation (MTE) is the mathematical model for the transportation (advection and diffusion) of substances dispersed in water. It is a hyperbolic non-linear PDE. The MTE solution also requires the specification of Dirichlet and Neumannz boundary conditions(BC). The Dirichlet BC is defined specifying a concentration at the inflow frontiers for a specific time period, so that it acts like a source, supplying mass to the domain, or as a drain, taking away mass from the domain. In this case, the BC is given by $S = S(x, y, z, t)$, where $(x, y, z, t) \in \partial\omega \times [0, T]$, in which $\partial\omega$ denotes the border of the domain and $S = S(x, y, z, t)$ is the concentration on this border.

The Neumann BC occurs when the concentration gradient is normal at the border, so that a disperse flux is specified there. This BC is expressed by $\partial S/\partial n$, where $n = n(x, y, z, t)$ is the unitary vector normal at the point $(x, y, z, t) \in \partial\omega$. In this work the null gradient, $\nabla S_{\partial\omega} = 0$, was taken at the outflow border, indicating that the space variation rate is zero, so that the substances can go out of the domain without accumulating at this border. Figure 1 shows the behavior of a contaminant plume being eliminated from hydrodynamics domain.

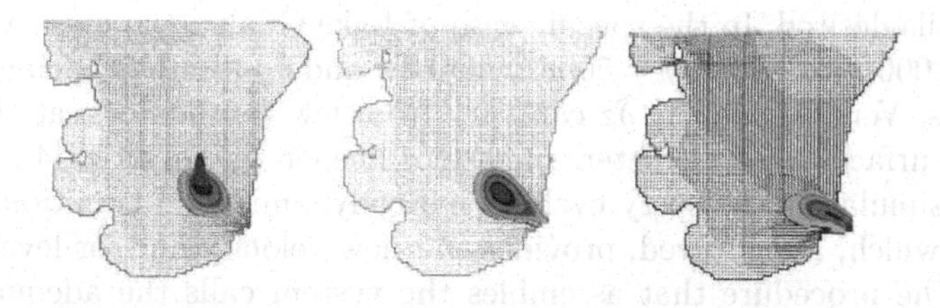

Fig. 1. Sequence of images showing substances being eliminated from hydrodynamics domain with BC of null gradient

The finite difference method was the discretization method used in this work to obtain the PDE in discrete form both in hydrodynamics and mass transportation. The hydrodynamics model has been described in a previous paper [1] therefore we focus this section on the mass transportation model.

In the discretization of 3D MTE, the approach used was that of building a semi-implicit numerical system for vertical transportation and an explicit scheme for the remaining terms, except for horizontal diffusion, which is semi-implicit. Horizontal advection is approximated considering the TVD (Total Variation Diminishing) approach, in which the horizontal fluxes are obtained by the Sweby's method of flux restriction [2]. Vertical advection is approximated through the numerical flux approach, by Gross's beta method [3]. Vertical diffusion is calculated through the Crank-Nicolson differences method. The source and the dumping are obtained in terms of concentration , and as a dumping matrix respectively.

Upper and lower boundary conditions for the MTE are obtained considering that there is no mass going through the solid surfaces or the free surface. That is, the BC are of the non-flux type at the free surface, at the water-sediment interface and at solid walls. With these choices, MTE 3D discretization results in 3-diagonal equation systems, one for each water column, each one with a number of equations equal to the number of layers in that column. The global numerical scheme is semi-implicit, and only vertical transportation is implicit. This is a particularly useful approach to obtain the parallelization of the solution in 2D decompositions. Solutions of the 3-diagonal independent, SPD equation systems are obtained via Thomas' algorithm.

3 Local Refinement and Multi-physics Problem

The class of multi-physics problems includes the problem addressed in this work, the parallel solution in multiprocessed PC clusters for the 3D hydrodynamics and for the 3D scalar transportation of substances in water bodies.

This characterization is due to the different physical natures of Shallow Water and Mass Transport PDEs. In the case of the SWE, an efficient computational modeling must consider the magnitude of the horizontal dimensions involved; the average depth of the water bodies; the water type and flow and the degree of details desired. In the specific case of Lake Guaba, the mesh varies from $\delta x = \delta y = 1000m$ to $\delta x = \delta y = 50m$, where δx and δy are the spacing in X and Y directions. Vertical spacing δz can vary from few centimeters, at the bottom and at the surface, to some meters in case of deeper waters as in [4].

During simulation, at every cycle of hydrodynamics, an equation system is assembled, which, after solved, provides the new velocity and/or level values of each cell. The procedure that assembles the system calls the adequate solver, and, after the system is solved, the values of the variables in the domain mesh are updated. Before starting the calculation of substance transport, the values of velocity and level in the cells of the refined mesh are obtained through interpolation from the corresponding cells in the hydrodynamics domain. After the values of velocity and level are updated, an equation system is generated to solve the concentration. The system is solved and the resulting concentration values are updated in the cells. After this updating, the concentration level near to some of the borders is checked to see if it has reached some predetermined value (10 pan/l). If this is true, the refined domain is expanded near this border by the allocation of new cells.

The Mass Transport Equation requires a well refined local mesh to capture the details of transport. A local refinement allows the mesh to be concentrated in subregions where more activities occur. Accurate solutions can be obtained, in these regions without the computational cost of a globally refined mesh.

For regular meshes, as the ones presented in this work, the local refinement can be obtained through interpolations, obtaining a mesh that is nested to the global one. As the global mesh and the local mesh are of the Arakawa C type, which is equally spaced, formulas of the Newton-Gregory type have been used where, consistent with the strategy developed to obtain SPD matrices, the algorithms calculate water levels. Velocities are recovered from the values of the levels. After that, the concentrations values are calculated in each cell of the local mesh.

The local solutions also require temporal interpolation, as the numerical schemes of hydrodynamic flow and MTE have distinct timesteps. Therefore, schemes for temporal interpolation have been built, where each timestep depends on the limitation imposed by the particular scheme of approximation used to solve each PDE. Thus, during the simulation, at each cycle of hydrodynamics, an equation system is assembled and solved, resulting in the new values of velocity or water level in each cell.

The procedure that assembles the system calls the appropriate solver and, after its solution, the values of the variables in the domain mesh are updated. Before initiating the calculation of the substance transport, the values of velocity and level of the refined mesh cells are obtained by interpolation from the corresponding cells in the hydrodynamic domain. After the values of velocity and level are updated, the equations system for the solution of the transport is assembled. The system is solved and the values of the resulting concentrations are updated in the cells. After the update, it is verified if the concentration level close to some frontier reached a predefined value (100 nmp/l). If it happened, the refined domain is expanded close to this frontier, by the allocation of more cells.

4 Dynamic Load Balancing

For the parallel solution of the hydrodynamics a domain decomposition method was used, as described in [1]. In the beginning of the simulation the hydrodynamics domain is divided between the processors using METIS, a library for graph partitioning.

This static partitioning of the computational domain used in the hydrodynamics cannot be used in the refined mesh used to model the substances transport because the substance plume can take other directions as time goes by, and frequently a load unbalance, that can be quite significant, occurs.

A non-balanced distribution of computational load leads to idleness of some processors, by the waiting time introduced by synchronization. So, the distribution needs to bear in mind that the processors may have different capacities and/or receive heavier tasks. For problems where the load is known at as initial time and stays the same, load balancing is static. If the computational load allocated for each processor is changed during execution time, this process is called dynamic balancing [5]. An algorithmic definition of dynamic balancing must consider [6]:

1. load evaluation: necessary to detect the degree of unbalancing and to identify the processors in which load transfer must occur;

2. cost/benefit evaluation: As redistribution cost is related to interrupting processing and recalculating the local or global mesh, repartitioning and remapping it again, it must be checked if the cost/benefit relation is compensatory.

Two solutions for repartitioning the domain are scratch-remap approaches and diffusion algorithms. In scratch-remap approaches a new partitioning is generated and it is mapped over the previous partition. This approach usually performs well in graphs with a high level of unbalance concentrated in some parts of the graph. In domains with a low unbalance or with the load well distributed over the entire domain, it can result in an excessive load transference.

A second approach for repartitioning is the use of diffusion algorithms that, usually, assure a new load balancing with little data movement [7]. With this scheme, the load of processor is evaluated and compared to the load of its neighbors. If the difference is above a limit, part of the load of one processor is transferred to another one, as it happens in heat diffusion [8].

Several decisions must be made when using diffusion algorithms. Some of these decisions are how frequently the load in each processor is evaluated and exchanged with the neighbors, which difference of load between neighbors can be tolerated before a load redistribution is done, what number of cells must be exchanged with each neighbor (usually it is a fraction of the difference, but which fraction?) and which cells will be chosen to be exchanged.

A solution for the dynamic load balancing of multi-physics problems like this can be addressed in several ways:

1. partitioning the hydrodynamics mesh without considering the mass transportation. The partitioning of the refined grid is done based on a coarser grid so each cell of the refined grid is kept in the same processor as the hydrodynamics cell to which it is associated. This approach reduces communication to a minimum, but it can cause a strong unbalance in the mass transportation stage;
2. partitioning both grids independently and moving, in the beginning of each stage of mass transportation, the hydrodynamics data needed by each cell of the finer grid, from the processor who owns it. This approach assures a perfect balance, but the migration of hydrodynamics data can be high;
3. partitioning the mass transportation grid and keeping in the same processors the cells of transportation and the cells of hydrodynamics associated to them and, after that, partitioning the hydrodynamics cell where transportation does not occur. This is the most complex approach and has a high cost but assures a good balancing in both stages and reduces communication to the frontiers.

We have implemented the three approaches in our model and in the next sections we will describe the mechanisms implemented and present some results obtained.

5 Transport Partitining Based On Hydrodynamics

In the three approaches described in this work, the load unbalance starts in the mesh of mass transportation, when the concentration of some cells near the frontier grows beyond a determined level. When it happens, the mesh is expanded allocating new cells. These new cells can be allocated in the processor where the high concentration occurred or in a neighbor processor.

In this approach, which is not a dynamic load balancing approach, but only an heuristic for partitioning the mass transport, each new cell is allocated on the procesor that keeps the part of the hydrodynamic mesh corresponding to it.

The advantages of this approach are its simplicity, an optimum load balance and a reduction in the communication cost in the hydrodynamics phase, as all the partitioning is based on hydrodynamics. The disadvantage is that a strong unbalance can occur in the transport phase. In an extreme case, all the refined mesh can be associated to the same subdomain and consequently its processor will process the entire transport mesh.

6 Independent Balance of Both Meshes

In this approach, the hydrodynamics partitioning is the initial partitioning, and the load balancing of transport mesh is done independently. When a transport cycle begins, the velocity field in the refined mesh is obtained from the hydrodynamic mesh through interpolation. Thus, each processor must keep allocated the cells of the hydrodynamics associated to the transport cells kept by it. If the processor that calculates the hydrodynamics of these cells is another one, they must be transferred from the processor that processes them.

The number of cells to be transferred can be very high, if the transport subdomain and the hydrodynamics subdomain kept by a processor have no overlapping. This transfer can have a high cost, and it is a problem of this approach.

The method used for partitioning the refined mesh in this approach and in the next one, was a stripwise partitioning as shown in figure 2.

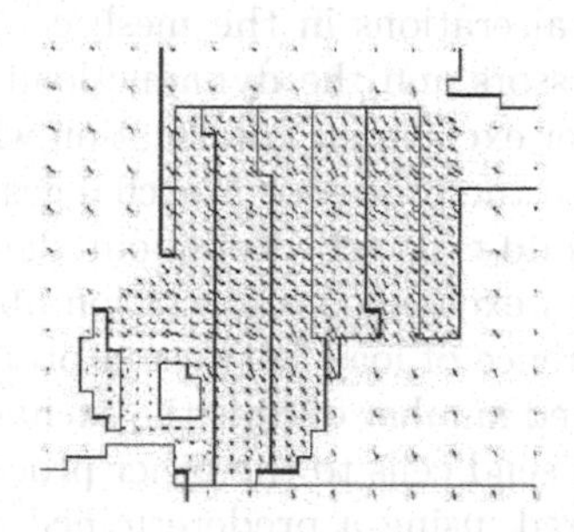

Fig. 2. Stripwise partitioning of refined mesh

The choice of this partitioning method is due to the dynamic characteristic of the refined mesh, where cells are allocated during the simulation. If an irregular partitioning as the partitioning implemented by Metis was used, before allocating a cell a processor should verify if this cell did not already belong to another subdomain. To verify it, the processor should have information about all the cells that belong to the neighbor subdomains. In the stripwise partitioning used in the refined mesh, each processor n knows, from the coordinates of the new cell to be allocated, if it belongs to its own subdomain or to the subdomain of processor $n-1$ or processor $n+1$, as contiguous subdomains belong to processors with contiguous numbers.

As in the previous approach, the unbalance starts with the allocation of new cells to the refined mesh. At predefined intervals, the load of each processor n is compared to the load of processors $n-1$ and $n+1$, and if the difference of load is higher than a predefined value, some cells are transferred from the higher loaded processor to the less loaded processor.

7 Two-Phase Balancing

In this approach, the dynamic load balancing mechanism must assure that after its execution the process responsible for a part of the refined mesh is also responsible for the processing of the part of the hydrodynamic mesh associated to it. To obtain this result, the dynamic load balancing is executed in two phases.

The first phase is similar to the mechanism described in the previous section. When a new cell is allocated to the mass transport mesh, the hydrodynamics cell corresponding to that position in domain must reside in the same processor. If it belongs to another processor, it migrates, after some message exchanges, to the processor who owns the cell just allocated. This migration can cause an unbalance in the hydrodynamics mesh.

This mechanism for verifying if the mesh must be expanded is activated at every 20 cycles. The optimal number of cycles or even if there is an optimal number is still an open issue in our work and other measures must be taken.

After some cycles, the alterations in the mesh can cause a strong load unbalance between the processors and the dynamic load mechanism is activated. It starts with each processor exchanging information with its neighbors (considering a strip partitioning, so the processor n exchanges messages with processor $n-1$ and $n+1$, using an odd-even scheme) about the load of each one.

After each processor has exchanged information about load with the neighbor processors, if the difference of load between a processor and its neighbor is greater than a predetermined number of cells the exchange of cells is effectuated.

The processor that will send cells to the other processor calculates the number of cells to be transferred, using a predetermined coefficient and select the cells that will be transferred. This selection is straight, as the strip partitioning imposes that the cells to be transferred will be the cells near to the frontier between the two subdomains. After that, the cells are sent to the other processor and all the neighbor processors that own subdomains that depend on the transferred cells, are notified about the transference, so they can update their structures.

When cells of the refined mesh are transferred, the corresponding cells of the hydrodynamics mesh must also be transferred so they reside in the same processor. This can cause an unbalance in the hydrodynamics mesh, which is solved in the next phase of the load balancing mechanism.

For the balancing of the hydrodynamics, each processor exchange information with all the neighbors (considering the irregular partitioning) and in a similar way to the refined grid, and if the difference is above a predetermined value, the load is transferred. An additional problem that does not exist in the balancing of mass transport is the selection of cells to be transferred.

8 Metrics Used For Load Evaluation

Two metrics were used for load evaluation along this work. During the development of the model, the load in each processor was based in the number of cells of its subdomain. This choice was based on the fact that the cost of assembling the matrix corresponding to a subdomain is linear for the number of cells of the subdomain, as the number of floating points operations is nearly the same for every cell (with a slight difference for the cells in the boundaries). Furthermore, as the matrices generated for hydrodynamics are sparse and solved by Conjugate Gradient, which has a linear time per iteration (for sparse matrices) and the matrices generated for mass transport are tridiagonal and solved by Thomas algorithm, which has linear cost, it is reasonable to suppose that the time for processing a subdomain is linear for the number of cells. Measures were taken to confirm this supposition and the time was shown to be linear, although some differences in the time for processing different subdomains appeared. These differences can be explained by the irregular characteristic of the subdomains, which results in areas of overlapping with different sizes and different processing cost.

The difficulty of modeling the irregularities of the subdomains led to the choice of another metric to represent the load of each processor. This metric was the processing time of the subdomain, considering the time for the assembly of the matrices and execution of the solvers. One advantage of this metric is that it considers aspects hard to consider analytically as the irregularities of the frontiers or the influence of other processes running in the same node where the model is being executed. The execution of other processes in the same node where the model is being executed reduces the computing power that the node can provide to the model, increasing the time to process the subdomain associated to that processor. This increase in the processor load is perceived by the mechanism of load balancing that reacts to it sending cells from the most loaded processor to the neighbor ones, until both processors use the same time to process their subdomains. It must be noted that this mechanism works only in the two-phase balancing, described in section VI, as in the independent balance of the meshes, the partitioning of the hydrodynamics mesh is static, causing an unbalance during the phase of hydrodynamics.

9 Calculation and Selection of Cells to be Migrated

After the load in each processor is evaluated and a significant difference of load between two neighbor processors is detected, the next step is to proceed to the calculation of how many cells will be migrated. In the diffusion mechanism used in this model, the difference of load between the two processors is calculated and a percentage of this difference is transferred. If the percentage of the difference to be transferred is too high, oscillations in the load can occur, where cells keep migrating from one processor to another without converging to a stable distribution. If the percentage of the difference of load is too low, the load will converge to a stable distribution, but it will take too many cycles of load movement. After the mechanism of diffusion was implemented, several tests were performed to study the effect of this percentage.

The tests performed with different percentages showed that, the increase in the percentage reduced the number of iterations of the load balancing mechanism until the load was balanced. The best results were obtained with a percentage of about 70%. Higher percentages caused the mechanism to fail in the convergence.

After the number of cells to be migrated is calculated, it is necessary to choose which cells will be migrated between the subdomains. In the refined mesh, where a stripwise partitioning is used, this choice is trivial as, to keep the stripwise partitioning, the cells to be transferred must be the cells close to the frontier.

Although the hydrodynamics mesh is static, in the third approach described its partitioning is modified during the simulation as result of the partitioning of the transport mesh. As an irregular partitioning is used in the hydrodynamics mesh, the problem of selecting which cells will be moved must be considered carefully. The choice of the cells to be transferred minimizing the edge cut is as complex as the problem of graph partitioning and, again, heuristics must be used to get a good result.

The algorithm used for the selection of the cells to be transferred is based in the same idea of cost of a cell that is used in the Kernighan-Lin and Fidduccia-Mattheyses algorithms for partition refinement. In these algorithms the cost of a cell is the difference between the number of edges connecting the cell to other cells in the same subdomain (internal cost) and the number of edges connecting the cell to other cells in the other domain (external cost). Thus, if a cell has an internal cost of one and an external cost of three, it has a cost of two and the migration of the cell to the other subdomain will reduce the edge-cut by two.

10 Results

The three approaches here described were implemented in the model and several tests were performed. As expected, the partitioning of the transport mesh based on hydrodynamics presented a very low and inconstant performance, and the speedup obtained in hydrodynamics was good. Table 1 presents the total time, the time used in the hydrodynamics and the time used in transport, for a total of 10.000 cycles. It can be observed that the speedups obtained in hydrodynamics are reasonable, but the low performance of transport compromises the performance of the entire model.

Table 2 presents the execution times of the cycles from 10.000 to 20.000, after the initial load distribution, of the model with independent partitioning. The total time is presented, as well as the time used in the processing of hydrodynamics and the time used in the processing of transport. The speedups obtained are shown in parentheses and, in the last column, the processors that received parts of the refined mesh to process are shown (processors 0 to 10). A limiting factor to the speedup in the processing of hydrodynamics is the processing of the regions of overlapping needed for the domain decomposition method used. As the subdomains become smaller, the contribution of this region of overlapping to the processing cost grows. The low speedup obtained in the transport processing is due to the restraints imposed to the minimum size of the subdomain, limiting the number of subdomains to 11, although 20 processors were available. Due to limitations of computational resources, we have not performed tests with larger meshes.

Table 1. 10.000 cycles without dynamic load balancing

Processors	Time in seconds		
	Total (speedup)	**Hydrodynamics (speedup)**	**Transport (speedup)**
1	6794	1847	4659
2	5306 (1.28)	1016 (1.81)	4050 (1.15)
4	4798 (1.41)	533 (3.46)	4047 (1.15)
6	3681 (1.84)	375 (4.92)	3130 (1.48)
7	3432 (1.97)	343 (5.38)	2934 (1.58)

Table 2. Cycles from 10.000 to 20.000 with independent partitioning

Processors	Time in seconds		
	Total (speedup)	**Hydrodynamics (speedup)**	**Transport (speedup)**
1	10162	2648	7512
20	965 (10.53)	188 (14.08)	776 (9.68)(0-10)

Table 3. Cycles from 10.000 to 20.000 in two-phase partitioning

Processors	Time in seconds		
	Total (speedup)	**Hydrodynamics (speedup)**	**Transport (speedup)**
1	10704	2695	8008
20	935 (11.44)	289 (9.32)	645 (12.41)(0-15)

Table 3 presents the execution times of the cycles from 10.000 to 20.000, after the initial load distribution, for the model using two-phase partitioning. In this execution the restraints on the size of the subdomain in the transport mesh were reduced, and the same mesh of the previous situation could be partitioned in 16 subdomains, improving the speedup obtained in transport and, consequently, in the total execution time of the model. However, a significant reduction in the speedup obtained in the hydrodynamic processing, resulting of the fragmentation of the hydrodynamics mesh and consequent increase in the size of the frontiers, can be observed.

The tests performed with both approaches of dynamic load balancing implemented have presented very close speedups in total execution time. The processing times for the transport mesh were practically the same in both approaches, occurring some differences in the hydrodynamics mesh. This difference in the speedup of the hydrodynamics processing can be explained by the growth of the frontiers that occurs in the hydrodynamics mesh when the two-phase partitioning is used, not occurring when the independent partitioning is used. The results obtained show that the cost to transfer the hydrodynamics cells in each cycle in the independent partitioning is nearly equivalent to the cost to transfer the cells in the larger frontiers and to the additional processing that occurs due to the larger area of overlapping, resulting of larger frontiers.

11 Conclusions and Future Works

Domain partitioning and load balancing are fundamental issues for the good performance of models of physical phenomena like the one described in this paper. The implementation of efficient mechanisms for dynamic load balancing in multi-physics models with irregular subdomains has shown to be very complex, specially due to the dependency between the cells in the irregular frontiers, which requires a careful and expensive management of frontiers so that every processor keep its information about its subdomain and its neighbors updated.

The main contribution of this work is the definition, implementation and evaluation of the mechanisms of dynamic load balancing for the HIDRA model. We have investigated 35 hydrodynamic models (the complete list can be found in [9]), and none of them presents mechanisms as the ones described here.

Due to the dependency between the hydrodynamics mesh and transport mesh, several messages must be exchanged between the processors to perform the load transfer. The definition of a correct sequence of messages who assures the correction of the model is a complex issue. The mechanisms implemented were tested on up to 20 processors, but no formal demonstration of the correction of this sequence was done.

The mechanisms presented in this paper are described in more details in [10].Although they are already operational, there are some improvements that can be done in order to get a better performance of the model. Some of these improvements are: 1) Consider the number of layers in the partitioning and in the dynamic load balancing mechanisms; 2) Use of other schemes of diffusion, as second-order schemes; 3) Disallocation of regions of the mesh where the concentration is not significant anymore; 4) Use of an adaptive scheme of frequency of balancing.

Acknowledgments

This article was partially funded by CNPq, CAPES, DELL and FAPERGS.

References

1. Rizzi, R.L., Dorneles, R.V., Navaux, P.O.A., Diverio, T.A.: Parallel solution in pc clusters by the schwarz domain decomposition for three-dimensional hydrodynamics models. In: Proceedings in International Meeting on High Performance Computing for Computational Science, VECPAR, 5., Porto, Portugal, Porto:Faculdade de Engenharia da Universidade do Porto, Porto:Faculdade de Engenharia da Universidade do Porto (2002) 655–667
2. Sweby, R.: High resolution schemes using flux limiters for hyperbolic conservation laws. SIAM Journal of Numerical Analysis (1984) 995–1011
3. Gross, E.S., Casulli, V., Bonaventura, L., Koseff, J.: A semi-implicit method for vertical transport in multidimensional models. International Journal for Numerical Methods in Fluids **28** (1998) 157–186
4. Rosatti, G., Bonaventura, L., Villa, D.A.: Three-dimensional numerical modelling of pollutant transport in the lagoon of venice. In: Proceedings in XXVII Convegno di Idraulica e Construzioni Idrauliche, IDRA 2000. (2000)
5. Quinn, M.J.: Parallel Computing : Theory and Practice. McGraw-Hill, New York (1994)
6. Watts, J.: A practical approach to dynamic load balancing. Master's thesis, Scalable Concurrent Programming Laboratory, California Institute of Technology, California, USA (1995)
7. Diekmann, R., Monien, B., Preis, R.: Load Balancing Strategies for Distributed Memory Machines. In: Multi-Scale Phenomena and their Simulation. World Scientific (1997) 255–266
8. Ghosh, G., Muthukrishnan, U., Schultz, M.: First and second order diffusive methods for rapid, coarse, distributed load balancing. In: Proceedings of the 8th. Annual ACM Symposium on Parallel Algorithms and Architectures, Padua, Italia (1996) 72–81
9. Rizzi, R.L.: Modelo Computacional Paralelo para a Hidrodinâmica e para o Transporte de Massa Bidimensional e Tridimensional. PhD thesis, Instituto de Informática, UFRGS, Porto Alegre (2002)
10. Dorneles, R.V.: Particionamento de Domínio e Balanceamento Dinâmico de Carga no Modelo HIDRA. PhD thesis, Instituto de Informática, UFRGS, Porto Alegre (2003)

Dynamically Adaptive Binomial Trees for Broadcasting in Heterogeneous Networks of Workstations

Silvia M. Figueira[1] and Christine Mendes[2]

[1] Department of Computer Engineering, Santa Clara University
Santa Clara, CA 95053-0566, USA
sfigueira@scu.edu

[2] Applied Signal Technology, 400 West California Ave.
Sunnyvale, CA 94086, USA
christine_mendes@appsig.com

Abstract. Binomial trees have been used extensively for broadcasting in clusters of workstations. In the case of heterogeneous non-dedicated clusters and grid environments, the broadcasting occurs over a heterogeneous network, and the performance obtained by the broadcast algorithm will depend on the organization of the nodes onto the binomial tree. The organization of the nodes should take into account the network topology, i.e., the communication cost between each pair of nodes. Since the network traffic, and consequently the latency and bandwidth available between each pair of nodes, is constantly changing, it is important to update the binomial tree so that it always reflects the most current traffic condition of the network. This paper presents and compares strategies to dynamically adapt a binomial tree used for broadcasting to the ever-changing traffic condition of the network, including accommodating nodes joining the network at any time and nodes suddenly leaving.

1 Introduction

Broadcasting is an important communication strategy, which is used in a variety of parallel and distributed linear algebra algorithms [7]. In fact, algorithms for broadcasting in different kinds of environments have been discussed extensively in the literature [1, 2, 3, 5, 6, 8, 9, 10, 11]. Different structures, such as binomial trees, hypercubes, and rings are used for broadcasting. Binomial trees are particularly useful for broadcasting in clusters [12] and grid environments, in which they are used in each local group of machines [8, 9]. When used in heterogeneous networks, the binomial tree needs to reflect the network topology by taking into account the communication cost between each pair of nodes [4]. It also needs to evolve according to the dynamic changes in the network traffic. In particular, the binomial tree needs to dynamically adapt to changes in latency and bandwidth, to nodes going down, and to nodes wishing to join the tree. Dynamically adapting the binomial tree is particularly important for applications that execute for a long time, facing many changes in the traffic condition of the network.

In this paper, we propose strategies that can be used to update a topology-based binomial tree as communication costs increase between nodes. We also propose

M. Daydé et al. (Eds.): VECPAR 2004, LNCS 3402, pp. 480–495, 2005.

strategies to deal with nodes joining the network at any time and nodes suddenly leaving the system. While rebuilding the tree from scratch would provide the best organization of nodes on the tree, allowing the broadcast operation to be completed in the smallest amount of time, the overhead associated with it is significant. Dynamically adapting the tree consistently by changing the placement of nodes on the tree according to the changes in the network traffic is far less expensive and still helps decrease the cost of broadcasting.

The strategies presented have not been implemented in connection with any specific application. They are general strategies to enable binomial trees to adapt to changes in the topology of the underlying network and can be implemented as part of any message passing library, such as MPICH or PVM.

This paper is organized as follows. Section 2 explains how to build a topology-based binomial tree. Section 3 presents strategies to update topology-based binomial trees dynamically according to changes in communication costs. Section 4 deals with nodes entering and leaving the binomial tree. Section 5 presents a representative set of experiments. Section 6 summarizes the results obtained.

2 Topology-Based Binomial Trees

In [4], we have presented the Balanced-Path algorithm to provide a performance-effi-cient topology-based binomial tree. This algorithm traverses the tree, choosing children for the nodes and serving the nodes with the highest number of undefined children first. When choosing a child, the algorithm picks the node, from those that are left, that is closest to the parent. This strategy balances the paths from the root to the leaves by giving priority to the parents that have the largest number of children, since parents with more children lie along longer paths. In a tie, the algorithm gives priority to the longer path. At the end, the distance in the paths is balanced, and the communication costs are distributed roughly evenly. For N nodes, the Balanced-Path algorithm takes $O(N^2)$ steps to generate an improved binomial tree.

In Figure 1, the tree on the left was obtained using the Balanced-Path algorithm. The number on each edge represents the distance between the two nodes, or machines, connected via the edge. The distances were obtained from Table 1. Distances are

Table 1. Distances between the nodes

	0	1	2	3	4	5	6	7
0	0	2	2	0	3	3	0	0
1	2	0	0	2	5	5	2	2
2	2	0	0	2	5	5	2	2
3	0	2	2	0	3	3	0	0
4	3	5	5	3	0	0	3	3
5	3	5	5	3	0	0	3	3
6	0	2	2	0	3	3	0	0
7	0	2	2	0	3	3	0	0

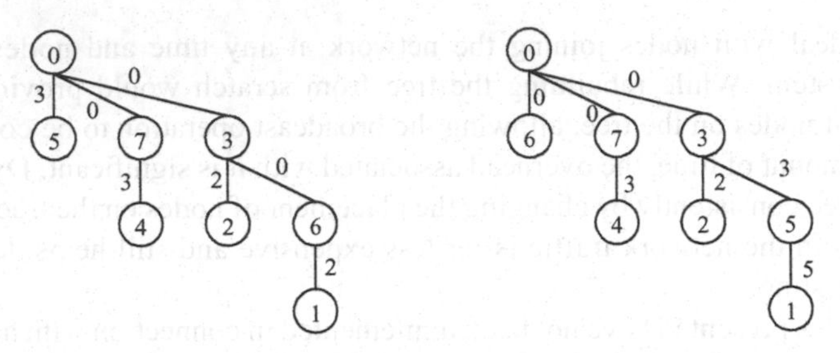

Fig. 1. Left: Balanced-Path tree formed according to the distances in Table 1. Right: Paths affected if nodes 5 and 6 are swapped

specified in number of hops, and a distance of zero between two nodes means that the two nodes are directly connected.

3 Dynamically Adapting the Binomial Tree

After building the tree, it is necessary to evolve the tree according to the changes in the network traffic. The strategies used to evolve the tree need to be cost effective, otherwise it would be more efficient to rebuild the tree instead of simply updating it. Four strategies were developed to update the tree as the network traffic changes. The cost of the tree is used as a measure to compare the different strategies. The cost is defined as the cost of the path with the highest cost, and the cost of each path is calculated as the sum of all the edges (i.e., distances) from the root to the respective leaf. This measuring strategy assumes that all paths are covered roughly in parallel, and it uses the costliest path for comparison, since this is the path that determines the time to execute a broadcast operation.

When the tree is first built, the cost of all the paths, from each leaf to the root, needs to be computed in order to determine the path with the highest cost. The cost obtained for each path in the tree shown in Figure 1 (left) is listed in Table 2. Since the highest cost is 3, the cost of the tree is 3.

Table 2. Cost of each path for the tree in Figure 1 (left)

Leaf	5	4	2	1
Cost	3	3	2	2

Fluctuations in the network traffic may cause the communication cost of a link in the tree to increase and the four strategies described below may be used to swap nodes in the tree in order to obtain a configuration that uses communication links that are of lower costs. When we swap two nodes in the tree, the cost of the tree needs to be recomputed in order to determine if the swap reduces the cost of the tree. Rather than compute the cost of all the paths in the tree to determine the cost of the tree as we have to do initially, we only compute the cost of those paths affected by the swap. This is more efficient since previously computed values for many paths in the tree are still valid. For example, if we swap node 5 and node 6 in Figure 1 (left), then node 6 is now a leaf and both the paths from the root to node 6 and from the root to node 1 have been affected by the swap. Figure 1 (right) shows the tree after the swap. It is evident that the paths leading to leaves 4 and 2 have not been impacted by the swap, and the costs for these two paths do not need to be re-computed. The cost of the affected paths to nodes 6 and 1 are re-computed as shown in Table 3, and we are able to determine that the new cost of the tree is 8.

Table 3. Cost of each path for the tree in Figure 1 (right)

Leaf	6	4	2	1
Cost	0	3	2	8

Since a tree with *N* nodes has *N*/2 leaves, the maximum number of leaves, whose path will have to be traversed to calculate its new cost, is *N*/2. The cost associated with updating the cost data structure for the tree therefore has an upper bound of *O*(*N*). The actual number of leaves whose cost needs to be recomputed depends on how high a node is located in the tree. The higher the node is in the tree, the greater the number of leaves affected by a swap with that node. On average, however, when nodes are swapped, many of the nodes affected will be located lower in the tree thus reducing the number of leaves whose cost needs to be re-computed. For this reason, in many cases, the cost for fewer than *N*/2 leaves will have to be recalculated.

Fluctuations in the network traffic may also cause the communication cost between two neighboring nodes in the tree to go down. Although, at this point, we do not take any action to modify the tree when link costs decrease, the cost of the binomial tree needs to be re-computed to reflect the change. Since a tree with *N* nodes has *N*/2 leaves, the maximum number of leaves, whose path will have to be traversed to calculate its new cost, is again *N*/2. The cost associated with updating the cost data structure for the tree therefore has an upper bound of *O*(*N*) as well but, in many cases, the cost for fewer than *N*/2 leaves will have to be recalculated.

The main goal of the proposed swapping strategies is to provide a tree with a cost that is at least as low as the cost of the original tree was, before the cost between two nodes increased. Suppose that the cost (i.e., latency and/or bandwidth) between two nodes that are neighbors in the binomial tree increases, then each of the strategies will try to exchange one of these nodes with some other node in the tree in order to keep the cost of the tree at least as low as it was before the change in the communication cost. As it tries each swap, the algorithms remember the lowest cost of the tree obtained so far, as well as the nodes whose swap produced that cost. This information is needed when the algorithms cannot reduce the cost to at least what it was when the communication cost increased, in which case the algorithms perform the swap that leads to the lowest possible cost. If the algorithms are able to find a swap that results in a tree with a cost that is at least as low as the cost was when the communication cost increased, the algorithms stop searching.

3.1 Family_Swapping Algorithm

This approach involves swapping one of the nodes connected by the link whose cost increased with another node in the family. If the link with an increased cost is the link that connects nodes a and b, where b is the child, node b will be the one to be swapped. In this case, the algorithm searches for a better position for node b by trying to swap it with its children, with its parent (node a), and with its siblings, i.e., the other children of node a. Note that the algorithm does not swap node zero, which acts as the root of the tree. The upper bound for the number of steps in this algorithm is $O(logN)$, where N is the number of nodes in the tree.

Figure 2 (left) illustrates the algorithm's behavior. The number on the left side of each node represents the order in which that node is checked to see if it can be swapped with the node of the affected link. Suppose the cost of the link between nodes 8 and 12 has increased. Node 8 is the parent, and node 12 is the child. The Family_Swapping algorithm will try to first exchange node 12 with its children, nodes 13 and 14. Then it will try to exchange node 12 with node 8, its parent, and then it will try to exchange node 12 with its siblings, which are nodes 9 and 10.

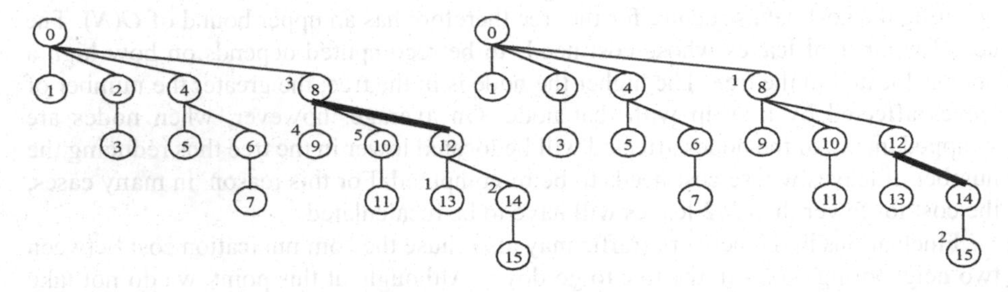

Fig. 2. Left: Using the Family_Swapping strategy. Right: Using the Path_Swapping strategy

In the example above, if swapping node 12 with its children does not reduce the cost of the tree, then the algorithm tries to swap node 12 (the child) with node 8 (the parent). Note that the link with the increased cost is still retained but now node 12 becomes the parent of node 8 and the child of node 0, and node 8 becomes the parent of nodes 13 and 14. This swap can lead to a lower-cost tree, if the communicating cost between node 12 and node 0 is less than that between nodes 8 and 0, or if the cost of communication between node 8 and its new children, nodes 13 and 14 are lower than that between node 12 and nodes 13 and 14.

The algorithm may not be able to lower the cost of the tree to a value at least as low as the cost was when the communication cost increased. In this case, the algorithm will swap node 12 with the child, parent, or sibling that results in the lowest possible cost. Each time the algorithm tries a swap, it has to re-compute the cost of the tree by examining the cost of each path to the leaves affected by the swap. The cost to perform each swap is then $O(N)$, where N is the number of nodes in the tree, since the number of leaves present in the tree is $N/2$. The swap between nodes closer to the root node will have more paths leading to leaves than the swap between nodes closer to the edge of the tree. In the tree shown in Figure 2 (left), if nodes 8 and 12 are swapped, then the cost to the leaves 9, 11, 13 and 15 need to be computed. However, had the link cost increased between nodes 12 and 14 and had the two nodes been swapped, then the leaves affected are only 13 and 15. Since the link cost randomly increases between nodes in different locations in the tree, in the average case fewer than $N/2$ leaves will be affected by a swap. Also, when a parent and a child nodes are swapped, the cost to compute the new cost of the binomial tree is less because all the leaves with paths leading to the child are also connected by a path to the parent. After swapping the parent and the child, the child will be in the parent's position and the parent will be in the child's position. The cost of only those leaves with paths to the new parent need to be calculated, because the leaves impacted by the new child are a subset of the leaves impacted by the new parent. However, when a node is swapped with its sibling there is no overlap in the leaves affected by the swap, so the cost of the leaves impacted by the new position of the node and its sibling has to be calculated independently, which is costlier.

3.2 Path_Swapping Algorithm

This approach involves swapping one of the two nodes, connected by the link with an increased cost, with another node in the same path. If the link with an increased cost is the link between nodes a and b, where a is the parent and b is the child, the algorithm alternates between trying to find a better position for node a by following its path up to the root and trying to find a better position for node b by following its longest path down to a leaf. Note that the algorithm does not try to swap with the root. The upper bound for the number of steps in this algorithm is $O(logN)$, where N is the number of nodes in the tree.

Figure 2 (right) illustrates the algorithm's behavior. The number on the left side of each node represents the order in which that node will be checked. Suppose the cost of the link between nodes 12 and 14 has increased. Node 12 is the parent, and node 14 is the child. The Path_Swapping algorithm will try to exchange node 12 with node 8, then it will try to exchange node 14 with node 15. Since the algorithm does not swap with node 0, it would not try to swap node 12 with any other node. Since node 15 does not have any children with which node 12 could be swapped, the algorithm stops. If additional nodes were still available for swapping along the longest path from node 14 down to a leaf, then the algorithm would continue in that direction. Similarly, if additional nodes were available along the path up the tree leading to the root, the algorithm would continue trying to swap the parent node with one of those nodes. The algorithm stops as soon as a swap results in a tree whose cost is at least as low as the cost of the tree before the communication cost increased. If the algorithm cannot reduce the cost to what it was before the change in the communication cost, it performs a swap that leads to a tree with the lowest possible cost.

Since the Path_Swapping algorithm swaps nodes along the same path in the tree, it takes the least amount of steps to compute the cost of the tree on average compared to the other algorithms. Each time the algorithm performs a swap between two nodes, one of the nodes will be higher in the tree or closer to the root, and the other node will be located lower in the tree or closer to a leaf. The leaves affected by the new position of the node lower in the tree will be a subset of the leaves affected by the new position of the node higher in the tree. Hence, when the cost of the tree is re-computed, only calculating the cost of the paths to the leaves forming the larger set is sufficient. So, although the cost associated with re-computing the cost of the tree after a swap is still $O(N)$, where N is the number of nodes, the cost for only one set of leaves needs to be recalculated, even though two nodes have changed positions.

3.3 Leaf_Swapping Algorithm

This approach involves swapping one of the nodes, connected by the link with an increased cost, with one of the leaves in the tree. If the link with an increased cost is the link between nodes a and b, where a is the parent and b is the child, the algorithm alternates between looking for a better position for node a and looking for a better position for node b by trying to replace each of them by a leaf in the tree. The upper bound for the number of steps in this algorithm is $O(N)$, where N is the number of nodes in the tree.

Figure 3 (left) illustrates how the algorithm works. The number on the left hand side of each node represents the order in which that node will be checked. Suppose the cost of the link between nodes 8 and 12 increases. Node 8 is the parent of node 12. The Leaf_Swapping algorithm will try to exchange nodes 8 and 12 with node 1, the first leaf. Then, it will try to exchange nodes 8 and 12 with node 3, the second leaf. The algorithm will follow the leaves as shown in Figure 3 (left), trying to exchange each of them with nodes 8 and 12. The algorithm will stop when it finds a swap that leads to a tree with a cost at least as low as the cost of the tree before the communication cost increased. If the cost cannot be reduced to at least what it was before the increase, either node 8 or node 12 is swapped with a leaf that leads to the smallest cost possible.

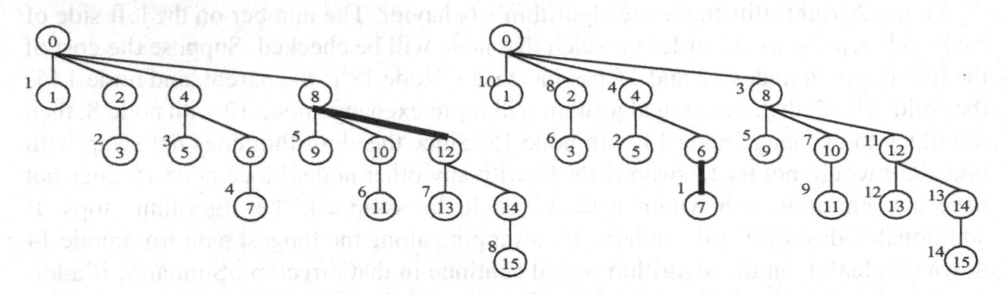

Fig. 3. Left: Using the Leaf_Swapping strategy. Right: Using the Position_Swapping strategy

The advantage of exchanging a node with a leaf node is that fewer links are affected by the exchange since the leaves are at the edge of the tree and do not have any children. However, since nodes closer to the root have paths leading to more leaves, swapping one of these nodes with a leaf still takes $O(N)$ steps to recalculate the cost of the tree. In the worst case scenario, we have to re-compute the cost for $N/4$ leaves when we swap a child of the root node with a leaf. However, since most of the time we will be swapping nodes from different positions in the tree, the number of leaves affected by a swap will be on average less than $N/4$.

3.4 Position_Swapping Algorithm

This approach involves trying to swap one of the nodes connected by the link with an increased cost with a nearby node. If the link with an increased cost is the link between nodes a and b, where a is the parent and b is the child, node a will be the one to be swapped. In this case, the algorithm searches for a better position for node a by trying to swap it with nodes in nearby positions first. This is accomplished by choosing the nodes with increasing and decreasing positions in the tree. For example, if the node to be swapped is in position x, then the nodes will be checked according to their position in the following order: node in position x+1, x-1, x+2, x-2, and so on. Note that the algorithm does not swap the root node. The upper bound for the number of steps in this algorithm is $O(N)$, where N is the number of nodes in the tree.

Figure 3 (right) illustrates the algorithm's behavior. The number on the left side of each node represents the order in which that node will be checked. Suppose that the cost between nodes 6 and 7 has increased. Node 6 is the parent, and node 7 is the child. The Position_Swapping algorithm will try to exchange node 6 with nodes 7 and 5 (distance 1 from position 6). Then it will try to exchange node 6 with nodes 8 and 4 (distance 2 from 6). Then it will continue increasing the distance and checking the nodes as shown in Figure 3 (right). Note that the node in each position x may or may not be node x, since the nodes were rearranged to reflect the network topology. In this example, there are fewer nodes whose position is less than 6 than nodes whose position is greater than 6. When the algorithm exhausts the nodes available for swapping on one side of the tree, it will continue trying to swap with nodes on the other side of the tree, in this case, with nodes whose position is greater than 6.

The algorithm will try to swap until it finds a swap, which results in a tree whose cost is at least as low as the cost of the tree before the communication cost increased, or until it runs out of nodes with which it can swap the parent. If the algorithm runs out of nodes without finding a swap that reduces the cost of the tree to at least what it was before the network change, then it will swap with the node that leads to the tree with the smallest possible cost. The Position_Swapping algorithm may run for a longer time than the Family_Swapping, Leaf_Swapping, or Path_Swapping algorithm, because it can potentially attempt a swap with all the nodes in the tree until it realizes that the cost of the tree cannot be reduced to its original value. However, because it has a greater number of potential candidates for a swap, it is more successful in reducing the cost of the tree than the other three algorithms.

The number of steps taken to compute the cost of the tree after each swap has an upper bound of $O(N)$, where N is the number of nodes in the tree and the maximum number of leaves in the tree is $N/2$. This algorithm is the most comprehensive, and it generally does more work to compute the cost of the tree, because the nodes used to swap could come from any position in the tree and may not lie in the same path.

4 Nodes Entering and Leaving the Binomial Tree

4.1 Nodes Entering the Tree

When a new node needs to be added, we need to add a new position to the binomial tree. This new position will be the next position in the binomial order. For example, in Figure 4 (left) we have an 8-node binomial tree composed of nodes 0 through 7. We add node 8 to the tree by creating position 8, which is a new sub-tree, with just one node, under node 0. The sub-tree rooted at node 8 is incomplete and can accommodate 7 more nodes.

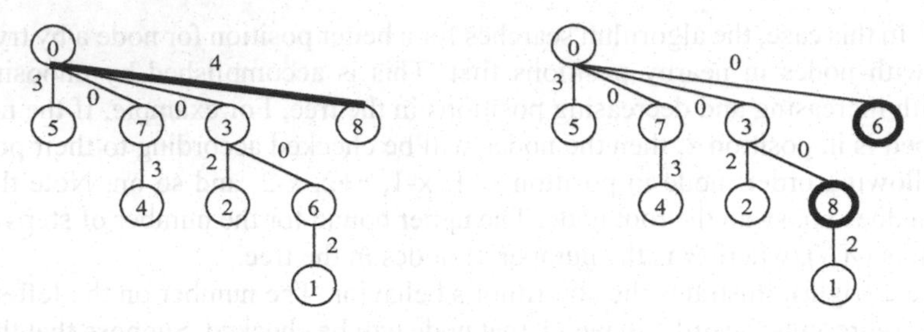

Fig. 4. Left: Adding a node to a complete 8-node tree. Right: Result of running Position_Swapping algorithm on the tree

Once a new node is added to the tree, the cost of the tree may be affected if the communication cost between the newly added node and its parent is expensive. In fact, even if adding the new node did not impact the tree's cost, swapping this node with another node in the tree may help reduce the communication cost of the tree. In order to determine if there is a better position for the new node, we first re-compute the cost of the tree. Using the example in Figure 4 (left), since we added a new branch to the tree and since node 8 is a leaf, we update our cost data structure by adding the cost of the path from the root to the new leaf, i.e., node 8, as shown in Table 4. We compute the cost by obtaining the latency from the latency matrix shown in Table 5, which reflects the distance in number of hops between node 8 and the other nodes. The latency matrix is updated with values each time a new node needs to be inserted into the tree.

Table 4. Updated costs using the latencies from Table 5

Leaf	5	4	2	1	8
Cost	3	3	2	2	4

Table 5. Distance between the nodes (in number of hops) after adding node 8

	0	1	2	3	4	5	6	7	8
0	0	2	2	0	3	3	0	0	4
1	2	0	0	2	5	5	2	2	2
2	2	0	0	2	5	5	2	2	5
3	0	2	2	0	3	3	0	0	0
4	3	5	5	3	0	0	3	3	0
5	3	5	5	3	0	0	3	3	4
6	0	2	2	0	3	3	0	0	4
7	0	2	2	0	3	3	0	0	3
8	4	2	5	0	0	4	4	3	0

After computing the cost of the tree, we then try to find a better position for node 8. A better position for node 8 would be one that reduces the cost of the binomial bymaking it equal or less than the cost of the tree before the new node was added. We do so by using the Position_Swapping algorithm, which will swap node 8 with nodes in different positions located around position 8. If we cannot reduce the cost of the tree to what it was before the node was added to the tree, the algorithm swaps node 8 with a node that produces the minimal cost possible.

In the example presented in Figure 4 (left), the cost of the tree, before adding node 8, was 3. After adding node 8, the cost of the tree increases to 4, because the communication cost between node 0 and node 8 is 4, the highest-cost path in the tree. Since there are no nodes in positions greater than 8, the algorithm can only look at positions whose values are less than 8. A swap with the node in position 7, i.e., node 1, is first attempted. If the swap does not decrease the cost of the tree, the algorithm subtracts 2 from 8 to get position 6 and attempts a swap with the node in position 6, and so on. Given the latency table in Table 5, we can reduce the cost of the tree from 4 to 3 by swapping node 8 with node 6 in position 6. The new tree formed is shown in Figure 4 (right). The algorithm stops at this point since it has performed a swap that reduced the cost of the tree to what it was before node 8 was added.

After running the Position_Swapping algorithm on the tree, the cost data structure is updated to reflect the cost of the path from node 0 to node 6, which is now a leaf node, and the cost of the path from node 0 to node 1. The cost of the paths unaffected by the swap is not re-computed. Table 6 shows the cost of all the paths in the tree after the swap.

Table 6. Updated costs after running the Position_Swapping algorithm

Leaf	5	4	2	1	6
Cost	3	3	2	2	0

4.2 Nodes Leaving the Tree

Machines attached to a network sometimes go down, and machines that are part of a binomial tree structure are no exception. If a machine that is a node in the tree is no longer running, all of its children, and their children in turn, are disconnected from the tree. Even if the node that goes down does not have any children, the tree is still impacted. In order to keep the structure of the tree consistent with that of a binomial tree and in order to keep all nodes connected to the tree, we replace a node that fails with one that is functioning. The replacement node is always the one at the last position in the tree. In the special case when the last node in the tree goes down, there is no need to find another node to replace it.

To illustrate, suppose node 7, a child of node 0, goes down as shown in the tree in Figure 5 (left). As a result of node 7 going down, node 4 is now no longer connected to the tree. Any information broadcast by node 0 will not reach node 4. We fill in the empty position, caused by node 7's failing, with node 1, the node in the last position in the tree. We pick the last node in the tree, because by doing so we avoid creating additional holes in the tree and we preserve the binomial tree structure in all the paths except for the last one. After replacing node 7 with node 1, the binomial tree will now have the structure shown in Figure 5 (right). It is clear that node 6 loses its child, and the tree is no longer complete. However, by replacing node 7, the tree remains connected and any broadcast operation will reach all the functional nodes.

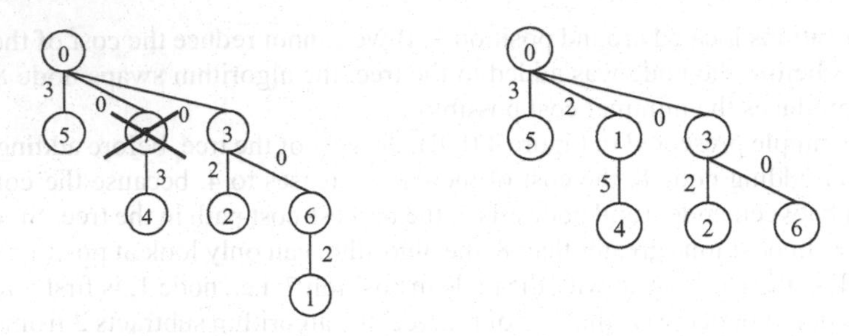

Fig. 5. Left: Binomial tree after node 7 goes down. Right: Binomial tree after node 7 is replaced by node 1

Since the position of two nodes in the tree was altered, the cost of some paths in the tree and, consequently, the cost of the tree need to be updated. In this example, replacing node 7 with node 1, increases the cost of the tree from 3 to 7, since node 1 is not close to either node 0 or node 4. The cost data structure is updated to reflect the new costs for all the paths in the tree, as shown in Table 7. Since node 6 lost its child, node 6 is now a leaf, and the cost of the path from node 0 to node 6 is 0.

Table 7. Updated costs after replacing node 7 with node 1

Leaf	5	4	2	6
Cost	3	7	2	0

After updating the cost, we try to reduce the cost of the tree by using either the Path_Swapping algorithm or the Position_Swapping algorithm. If the Path_Swapping algorithm is used in this example, we are unable to improve the cost of the tree by swapping node 1 with a node along its path. We do not swap node 1 with node 0, since node 0 is the root of the tree, and the only other available option is performing a swap with node 4, as illustrated in Figure 6 (left). The expensive link between nodes 1 and 4 is still in the tree, and the communication cost between node 0 and 4 is 3 hops, increasing the cost of the tree even further to 8 hops. In this scenario, the Path_Swapping algorithm was unable to reduce the cost of the tree.

The Position_Swapping algorithm can swap with a greater number of nodes since it is more flexible than the Path_Swapping algorithm. Hence, it has a better chance of improving the cost of the tree after a node that goes down is replaced with the last node in the tree. In the example above, the Position_Swapping algorithm determines that node 1 needs to be swapped with another node in the tree in order to have a tree with a reduced cost. Node 1 is at position 2 in the tree and the algorithm tries to swap node 1 with the nodes that are one position away from it. The algorithm swaps node 1 with node 4, which is at position 3. As demonstrated above, swapping nodes 1 and 4 does not reduce the tree's cost. So the Position_Swapping algorithm tries swapping node 1 with node 5, which is at position 1. This swap does succeed in reducing the cost of the tree down to 3, its cost before node 7 was replaced by node 1. Figure 6 (right) shows the tree after swapping nodes 1 and 5.

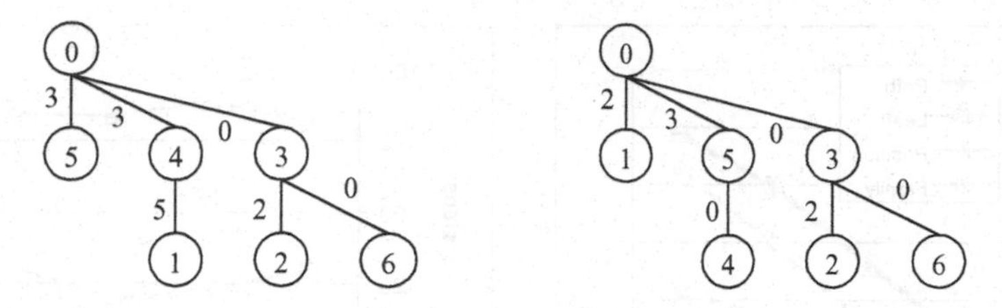

Fig. 6. Left: Binomial tree after swapping nodes 1 and 4 using the Path_Swapping algorithm. Right: Binomial tree after swapping nodes 1 and 5 using the Position_Swapping algorithm

The improved cost of the tree is reflected in the cost data structure, which maintains the cost of the paths from the root to each leaf. Since a node went down and was replaced with a node from the edge of the tree, there is one less node in the tree and the last leaf of the tree has also changed as Table 8 shows.

Table 8. Updated costs after running Position_Swapping algorithm

Leaf	1	4	2	6
Cost	2	3	2	0

5 Experiments

5.1 Comparison of the Swapping Algorithms

We compare the efficiency of the four strategies proposed to adapt the binomial tree by simulating them on randomly generated networks. We have executed experiments on networks with different number of nodes, and we show below a representative subset of these experiments. In each set of experiments, we vary the cost factor, which is the amount by which a communication cost is increased. For each cost factor, we report the average gain, the average number of steps to achieve the gain, and the average benefit obtained for 1,000 different randomly-generated topologies, determined by the distance between each pair of nodes. For each topology, the cost of a randomly chosen link is increased by the cost factor specified.

Note that each algorithm stops when it finds a swap leading to a tree with a cost that is less than or equal to the cost of the original tree. Therefore, the average number of steps reported indicates how long it takes the respective algorithm to reach that level of optimization. The average gain reported is calculated as ((increased_cost – new_cost) / increased_cost), where increased_cost is the cost of the tree after the link had its cost increased, and new_cost is the cost of the tree after being updated by the respective algorithm. The average benefit is given by the ratio (average_gain / average_steps), which will help determine which approach is more cost-effective.

It is important to note that the gain may be as low as zero, if no swap leads to an improved tree. In this case, the number of steps will be the maximum allowed by the algorithm, and the tree remains the same.

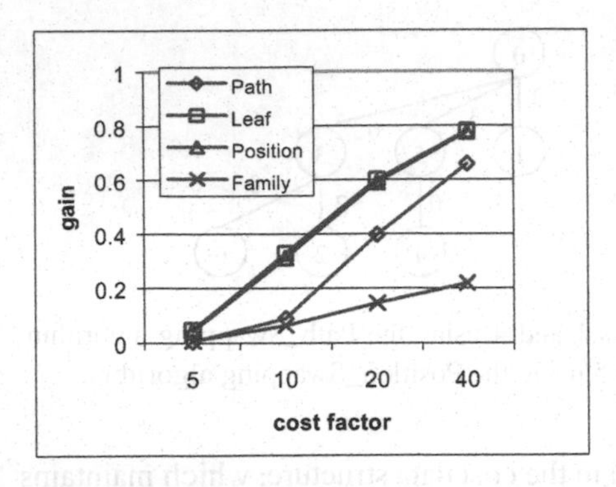

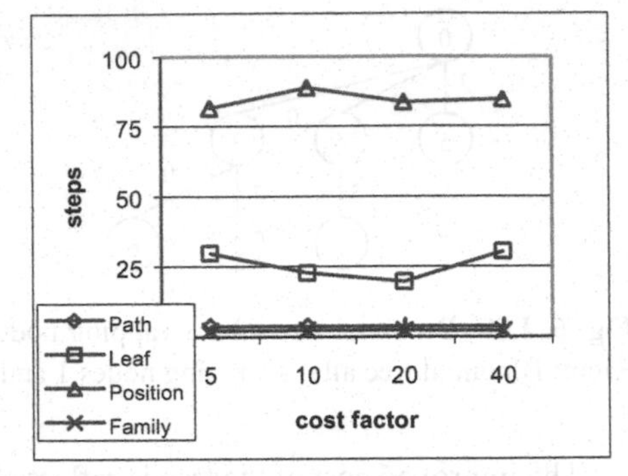

Fig. 7. Sets of 1,024 nodes, maximum distance 10 hops

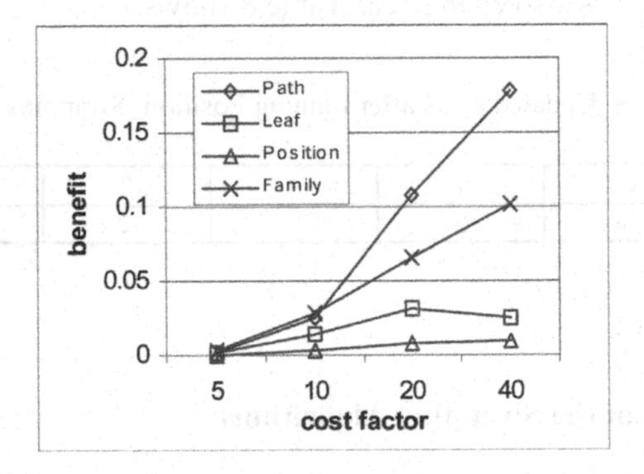

Fig. 8. Sets of 1,024 nodes, maximum distance 10 hops

Figures 7 and 8 show the results obtained for 1,024 nodes, which are apart by at most 10 hops. This is a large network, in which the nodes are close together. The graphs show that the Family_Swapping algorithm provides the smallest gain, but the Leaf_Swapping and Position_Swapping algorithms take more steps. The success of the four algorithms depends on the ratio of benefit to the cost, and the Path_Swapping algorithm has the best results when the gain and steps are examined together.

These results lead to the conclusion that the Path_Swapping algorithm is the best approach. Even though the gain obtained by this approach may not be the largest, the number of steps is always low and the benefit-to-cost ratio obtained is always the largest. In addition, the cost associated with computing the new cost of the tree after performing a swap with the Path_Swapping algorithm is the lowest since the leaves affected by the swapped node located lower in the tree is a subset of the leaves affected by the swapped node higher in the tree, i.e., closer to the root. Therefore, when two nodes are swapped with the Path_Swapping algorithm, the cost of the tree can be computed by calculating the cost of the leaves linked by a path to the swapped node located closer to the root.

It is important to note that the gain grows proportionally to the cost factor, and can be as high as 80% when the cost factor is 40. This shows the importance of keeping the binomial tree updated, especially because the overhead for the update with the Path_Swapping algorithm is quite low.

5.2 Analysis of Nodes Entering and Leaving Tree

A set of experiments were conducted to explore the effects of dynamically adding and removing nodes from a binomial tree in order to simulate nodes entering the cluster and nodes going down. The experiment presented was conducted on 100 trees, starting with 1,024 nodes each. The initial maximum distance for each set of 100 trees was set to 10, 30, and 50 hops. 1,000 nodes were then randomly added to the tree or removed from the tree. We first obtain the average cost of the 100 trees after adding and removing the nodes without using any algorithm to improve the tree's cost after each addition or deletion operation. Then we perform the experiment of adding and removing nodes, but we use the Position_Swapping algorithm to improve the tree's cost after a node is added and use the Path_Swapping algorithm to improve the tree's cost when a node is deleted. The same experiment is run a third time for each set of trees, but this time the Position_Swapping algorithm is used both to improve the tree's cost after adding a node and to reduce the tree's cost after removing a node.

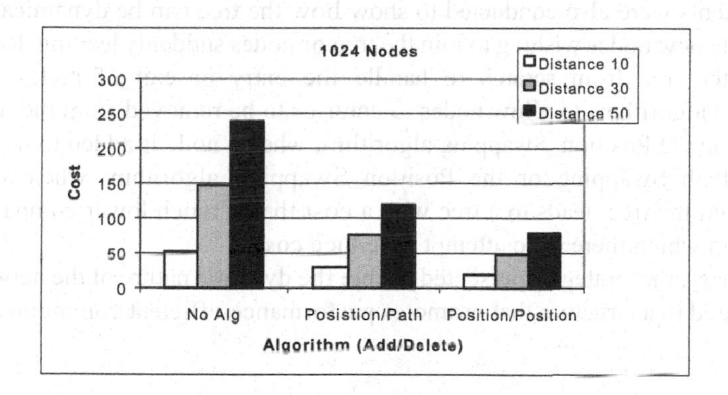

Fig. 9. 1,024 nodes, 100 trees, 1,000 nodes added or deleted

Figure 9 illustrates the outcome of adding and deleting 1,000 nodes randomly from trees with 1,024 nodes each. The Position/Path combination and the Position/ Position combination of algorithms are both able to cut the cost of the tree to nearly half of what it is when no algorithms are used. In addition, the Position/Position combination is able to reduce the cost of the tree even more than the Position/Path combination. This is because the Position_Swapping algorithm is able to attempt a swap with each node in the tree and is not restricted to nodes along the same path as the Path_Swapping algorithm is. As a result, it is able to improve the cost of the tree in more cases than the Path_Swapping algorithm.

These sets of experiments indicate that, when nodes are entering and leaving the tree, it is most beneficial to use the Position_Swapping algorithm to reduce the cost of the tree, if it is not important to perform the addition and deletion of nodes in the minimal number of steps. However, if the goal is to perform the operations as quickly as possible, then the Position_Swapping algorithm can be used when a node is added and the Path_Swapping algorithm can be used to improve the tree when a node is deleted since, together, the two algorithms take fewer steps and still provide performance-effi-cient trees.

6 Conclusion

The four strategies, Path_Swapping, Position_Swapping, Family_Swapping and Leaf_Swapping, described above provide ways to update a topology-based binomial tree in response to the changing conditions of the network traffic. Both the gain obtained and the number of steps used to reach an improved tree have been compared in order to determine the best approach. The efficiency of the strategies proposed was evaluated by comparing the benefit, i.e., the ratio of gain to number of steps, achieved by each approach. The experiments performed indicate that it is worthwhile updating the tree, since the overhead involved in the operation is outweighed by the gain obtained. The experiments also show that the Path_Swapping algorithm achieves the best results by providing improved trees in a short number of steps.

Experiments were also conducted to show how the tree can be dynamically modified to handle new nodes wishing to join the tree or nodes suddenly leaving. Rather than rebuilding the tree from scratch to handle the entry or exit of nodes, we have implemented algorithms to allow nodes to enter or to be removed from the tree at any time. Running the Position_Swapping algorithm, when a node is added to the tree, and either the Path_Swapping or the Position_Swapping algorithm, when a node is removed from the tree, leads to a tree with a cost that is much lower compared to the cost of one in which there is no attempt to reduce costs.

In summary, the strategies presented enable the dynamic nature of the network to be accommodated in a structure that promotes performance-efficient communication.

References

1. M. Banikazemi, V. Moorthy, and D. K. Panda, "Efficient Collective Communication on Heterogeneous Networks of Workstations," *Proceedings of the International Conference on Parallel Processing*, August 1998.
2. M. Banikazemi, et al., "Communication Modeling of Heterogeneous Networks of Workstations for Performance Characterization of Collective Operations," *Proceedings of the Heterogeneous Computing Workshop*, April 1999.
3. M. Bernaschi and G. Iannello, "Collective Communication Operations: Experimental Results vs. Theory," *Concurrency: Practice and Experience*, vol. 10, no. 5, pp. 359-386, 1998.
4. S. M. Figueira, "Improving Binomial Trees for Broadcasting in Local Networks of Workstations," *Proceedings of the VECPAR*, Portugal, June 2002.
5. I. Foster, J. Geisler, W. Gropp, N. Karonis, E. Lusk, G. Thiruvathukal, and S. Tuecke, "Wide-Area Implementation of the Message Passing Interface," *Parallel Computing*, vol. 24, no. 12, pp. 1735-1749, 1998.

6. I. Foster and N. Karonis, "A Grid-Enabled MPI: Message Passing in Heterogeneous Distributed Computing Systems," *Electronic Proceedings of the IEEE/ACM Supercomputing Conference*, November 1998.
7. D. Gannon and J. Van Rosendale, "On the Impact of Communication Complexity in the Design of Parallel Numerical Algorithms," *IEEE Transactions on Computers*, vol. C-33, pp. 1180-1194, December 1984.
8. N. Karonis, B. de Supinski, I. Foster, W. Gropp, E. Lusk, and J. Bresnahan, "Exploiting Hierarchy in Parallel Computer Networks to Optimize Collective Operation Performance," *Proceedings of the 14th International Parallel Distributed Processing Symposium*, pp. 377-384, May 2000.
9. T. Kielmann, H. E. Bal, and S. Gorlatch, "Bandwidth-Efficient Collective Communication for Clustered Wide Area Systems," *Proceedings of the International Parallel and Distributed Processing Symposium*, May 2000.
10. T. Kielmann, R. F. H. Hofman, H. E. Bal, A. Plaat, and R. A. F. Bhoedjang, "MagPIe: MPI's Collective Communication Operations for Clustered Wide Area Systems," *Proceedings of the Symposium on Principles and Practice of Parallel Programming*, May 1999.
11. B. B. Lowekamp and A. Beguelin, "ECO: Efficient Collective Operations for Communication on Heterogeneous Networks," *Proceedings of the 10th International Parallel Processing Symposium*, April 1996.
12. C. Martin and O. Richard, "Parallel Launcher for Cluster of PC," *Proceedings of the ParCo*, 2001.

Parallel Simulation of Multicomponent Systems
Invited Talk

Michael T. Heath and Xiangmin Jiao

Computational Science and Engineering,
University of Illinois, Urbana, IL 61801, USA
{heath, jiao}@cse.uiuc.edu

Abstract. Simulation of multicomponent systems poses many critical challenges in science and engineering. We overview some software and algorithmic issues in developing high-performance simulation tools for such systems, based on our experience in developing a large-scale, fully-coupled code for detailed simulation of solid propellant rockets. We briefly sketch some of our solutions to these issues, with focus on parallel and performance aspects. We present some recent progress and results of our coupled simulation code, and outline some remaining roadblocks to advancing the state of the art in high-fidelity, high-performance multicomponent simulations.

1 Introduction

Many real-world systems involve complex interactions between multiple physical components. Examples include natural systems, such as climate models, as well as engineered systems, such as automobile, aircraft, or rocket engines. Simulation of such systems helps improve our understanding of their function or design, and potentially leads to substantial savings in time, money, and energy. However, simulation of multicomponent systems poses significant challenges in the physical disciplines involved, as well as computational mathematics and software systems. In addition, such simulations require enormous computational power and data storage that usually are available only through massively parallel computer systems or clusters. Therefore, the algorithms and software systems must function efficiently in parallel in order to have significant impact on real-world applications of complex multicomponent systems.

At the Center for Simulation of Advanced Rockets (CSAR) at the University of Illinois (http://www.csar.uiuc.edu), we have been developing a software system for detailed simulation of solid rocket motors. A rocket motor is a complex system of interactions among various parts—propellant, case, insulation, nozzle, igniter, core flow, etc.—involving a variety of mechanical, thermal, and chemical processes, materials, and phases [23]. Compared to the simulation codes currently employed in the rocket industry, which typically use one- or two-dimensional empirical models, our software system is fully-coupled, three-dimensional, and based on first principles, involving a wide range of temporal and spatial scales [9].

M. Daydé et al. (Eds.): VECPAR 2004, LNCS 3402, pp. 496–513, 2005.

Many issues must be addressed in developing such a high-performance system, including

- easy-to-use software environment for exploiting parallelism within as well as between components,
- reusable parallel software components and high-performance physics modules,
- system tools for assessing and tuning performance of individual modules as well as the integrated system,
- management of distributed data objects for inter-module interactions,
- parallel computational methods, such as data transfer between disparate, distributed meshes.

In this paper we overview some of our recent progress in addressing these issues and outline some remaining challenges in large-scale multicomponent simulations. The remainder of the paper is organized as follows. Section 2 overviews the parallel, multiphysics software system developed at CSAR. Section 3 describes the key characteristics of the physics modules and their parallel implementations. Section 4 presents our novel integration framework for multicomponent systems and some middleware services built on top of it. Section 5 surveys a few key computational problems arising from the integration of such systems and our solutions to them. Section 6 presents sample simulation results obtained with our code, followed by performance results in Section 7. Section 8 concludes the paper with a discussion of some remaining challenges.

2 System Overview

Rocstar is our high-performance, integrated software system for detailed, whole-system simulation of solid rocket motors, currently under development at CSAR. We briefly overview the methodology and the software components of this system.

2.1 Coupling Methodology

Simulation of a rocket motor involves many disciplines, including three broad physical disciplines—fluid dynamics, solid mechanics, and combustion—that interact with each other at the primary system level, with additional subsystem level interactions, such as particles and turbulence within fluids. Because of its complex and cross-disciplinary nature, the development of Rocstar has been intrinsically demanding, requiring diverse backgrounds within the research team. In addition, the capabilities required from the individual physical disciplines are at the frontier of their respective research agendas, which entails rapid and independent evolution of their software implementations.

To accommodate the diverse and dynamically changing needs of individual physics disciplines, we have adopted a partitioned approach, to enable coupling of individual software components that solve problems in their own physical and geometrical domains. With this approach, the physical components of the

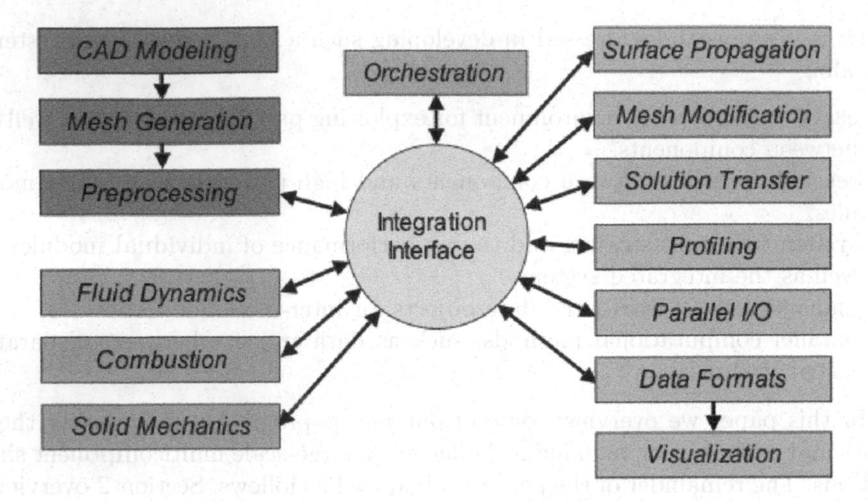

Fig. 1. Overview of *Rocstar* software components

system are naturally mapped onto various software components (or modules), which can then be developed and parallelized independently. These modules are then integrated into a coherent system through an integration framework, which, among other responsibilities, manages the distributed data objects, and performs inter-module communications on parallel machines.

2.2 System Components

To enable parallel simulations of rockets, we have developed a large number of software modules. Fig. 1 shows an overview of the components of the current generation of Rocstar. These modules serve very diverse purposes and have diverse needs in their development and integration. We group these modules into the following four categories.

Physics modules solve physical problems in their respective geometric domains. In general, they are similar to stand-alone applications, are typically written in Fortran 90, and use array based data structures encapsulated in derived types.

Service modules provide specific service utilities, such as I/O, communication, and data transfer. They are typically developed by computer scientists but driven by the needs of applications, and are usually written in C++.

Integration interface provides data management and function invocation mechanisms for inter-module interactions.

Control (orchestration) modules specify overall coupling schemes. They contain high-level, domain-specific constructs built on top of service modules, provide callback routines for physics modules to obtain boundary conditions, and mediate the initialization, execution, finalization, and I/O for physics and service modules through the integration interface.

In Rocstar, the above categories correspond to the components at the lower-left, right, center, and top, respectively, of Fig. 1. In the following sections, we describe various parallel aspects associated with these modules. In addition, our system uses some o -line tools, such as those in the upper-left corner of Fig. 1, which provide specific pre- or post-processing utilities for physics modules.

3 High-Performance Physics Modules

Fig. 2 shows the modeling capabilities of the current generation of Rocstar. These capabilities are embedded in physics modules in three broad areas:

- uid dynamics, modeling core flow and associated multiphysics phenomena,
- solid mechanics, modeling propellant, insulation, case, nozzle,
- combustion, modeling ignition and burning of propellant.

We summarize the key features and parallel implementations of these modules, focusing on their interactions and the characteristics that are most critical to the coupling of Rocstar.

3.1 Fluid Dynamics Solvers

Roc o [3] and Roc u [8] are advanced CFD solvers developed at CSAR. Both modules solve the integral form of the 3-D time-dependent compressible Euler or Navier-Stokes equations using finite-volume methods. In addition, they are both designed to interact with other physics sub-modules, including turbulence, particles, chemical reactions, and radiation [18], through the Roc uid multiphysics framework illustrated in Figure 3. Roc o uses a multiblock structured

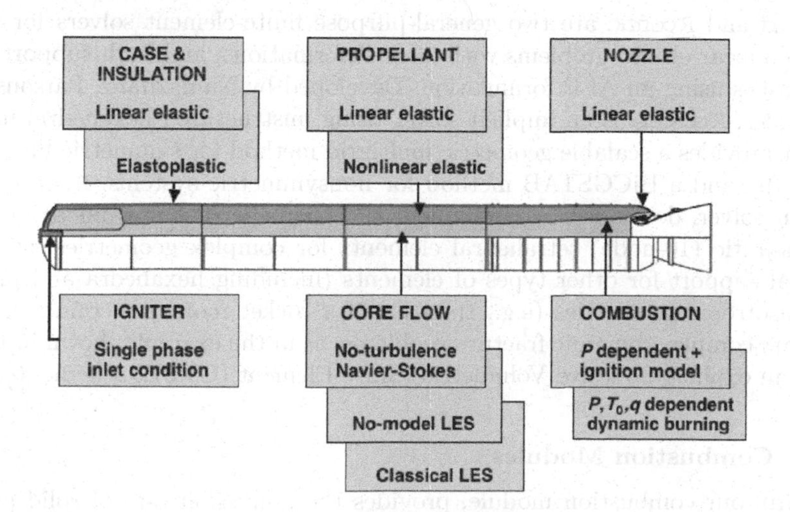

Fig. 2. Current modeling capabilities of *Rocstar*

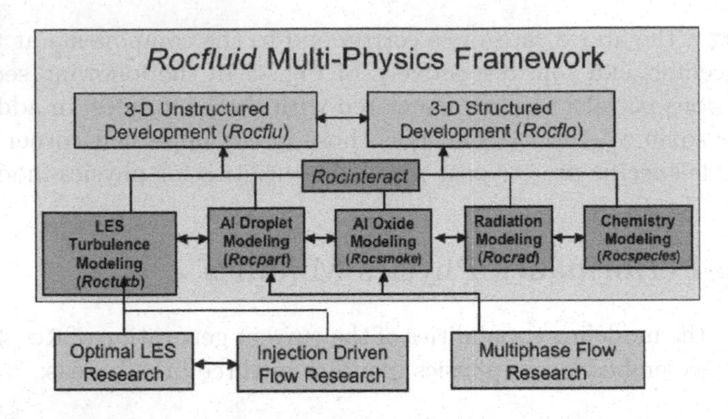

Fig. 3. Multiphysics framework of *Rocflo* and *Rocflu*

mesh, developed by Blazek, Wasistho, and colleagues. Roc u, developed primarily by Haselbacher, uses a mixed unstructured mesh consisting of arbitrary combinations of tetrahedra, hexahedra, prisms, and pyramids. Both solvers provide second- or higher-order accuracy in both space and time, notably with a novel WENO reconstruction method for unstructured grids in Roc u [8]. To accommodate dynamically changing fluid domains, these solvers allow for moving meshes using an Arbitrary Lagrangian-Eulerian (ALE) formulation. The Geometric Conservation Law (GCL) [26] is satisfied in a discrete sense to within machine-precision to avoid the introduction of spurious sources of mass, momentum, and energy due to grid motion.

3.2 Solid Mechanics Solvers

Rocsolid and Rocfrac are two general-purpose finite-element solvers for linear and nonlinear elastic problems with large deformations, and both support moving meshes using an ALE formulation. Developed by Namazifard, Parsons, and colleagues, Rocsolid is an implicit solver using unstructured hexahedral meshes [20]. It provides a scalable geometric multigrid method for symmetric linear systems [19], and a BiCGSTAB method for nonsymmetric systems. Rocfrac is an explicit solver, developed by Breitenfeld and Geubelle that uses linear (4-node) or quadratic (10-node) tetrahedral elements for complex geometries, with additional support for other types of elements (including hexahedra and prisms) for anisotropic geometries (e.g., the case of a rocket motor). Rocfrac can also simulate complex dynamic fracture problems, as in the example shown in Fig. 4, using an explicit Cohesive Volumetric Finite Element (CVFE) scheme [6].

3.3 Combustion Modules

Rocburn, our combustion module, provides the regression rate of solid propellant at the burning propellant surface, which is critical for driving the motion of the interface and achieving conservation of mass and energy at the solid-fluid

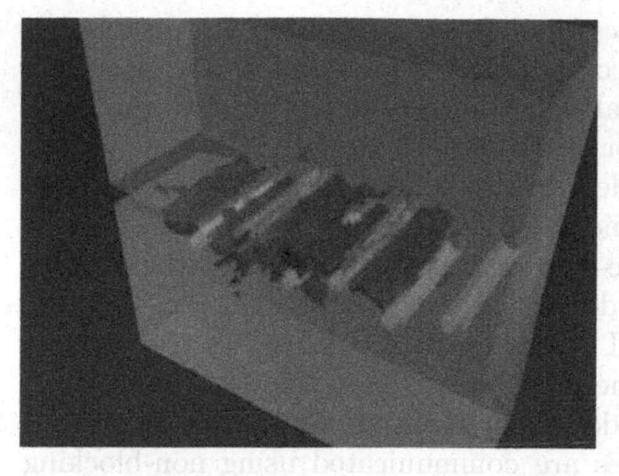

Fig. 4. Crack propagation in medium with holes

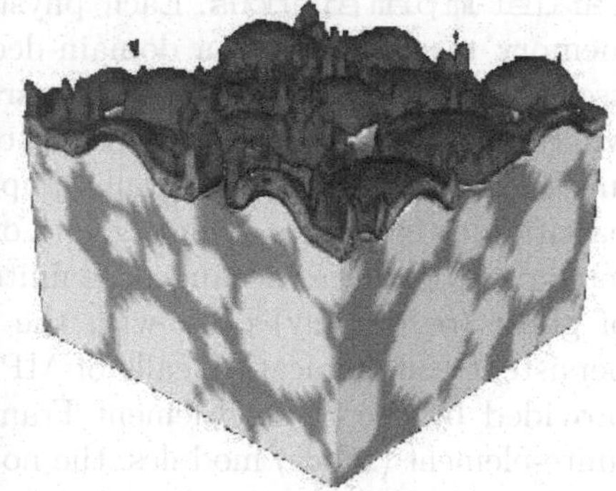

Fig. 5. Three-dimensional simulation of heterogeneous solid propellant

interface. Developed by Tang, Massa, and colleagues, Rocburn is composed of three interchangeable one-dimensional combustion models and an adaptor module. The combustion models calculate the burning rate at any given point on the burning surface using either a quasi-steady combustion model using Saint Robert's (a.k.a. Vieille's) law [11], or one of two unsteady combustion models that can capture ignition pressure spikes [24, 25]. The adaptor module manages data and interact with other modules at the level of the burning surface. Additional one-dimensional combustion models can also be plugged into the adaptor module conveniently. In addition, Rocburn simulates the ignition of the propellant under convective heat flux, when a thermal boundary layer is present in the fluids modules. The models in Rocburn are in turn informed by more detailed results from Roc re, developed by Jackson and Massa [17], which models combustion of heterogeneous solid propellant at micron-scale resolution. These scales are small enough to capture the interaction between gas phase fields and solid phase morphology, which plays a fundamental role in the combustion of heterogeneous propellant, as shown by the simulation depicted in Fig. 5. Because it is at the micron scale, Roc re is not directly integrated into Rocstar, but it is critical for calibrating the homogenized combustion models used in Rocburn.

3.4 High-Performance Implementations

All of our physics modules are implemented in Fortran 90 and employ derived data types and pointers extensively to achieve data encapsulation and modularity. For example, each block in Roc o or partition or Roc u is regarded as an object and contains all data relevant to that block, including nodal coordinates, face vectors, cell volumes, and state vectors. To achieve high serial performance, computationally intensive parts of the codes are written in Fortran 77 style (e.g., without using pointers) or utilize LAPACK routines [1] whenever possible.

Parallel Implementations. Each physics module is parallelized for distributed memory machines using a domain-decomposition approach. For Roc o, which uses multiblock structured meshes, parallelization is based on the block topology, where each processor may own one or more blocks. For other modules using unstructured meshes, the parallel implementations are based on element-based partitioning of meshes, typically with one partition per process. For inter-block or inter-partition data exchange, the finite-volume (fluids) codes utilize the concept of ghost (or dummy) cells, with the difference that Roc o uses non-blocking persistent communication calls of MPI [7], whereas Roc u utilizes the services provided by the Finite Element Framework built on top of MPI [2]. For the finite-element (solids) modules, the nodes along partition boundaries are shared by multiple processes, and their values are communicated using non-blocking MPI calls. Rocburn is solved on the burning patches of either the fluid or solid surface meshes, and its decomposition is derived from that of the volume mesh of the corresponding parent module.

Inter- and Intra-Parallelism Through Adaptive MPI. Adaptive MPI (AMPI) is a multi-threaded version of MPI that supports virtual processes and dynamic migration [10]. Developed in Kalé's group at Illinois, AMPI is based on Charm++, provides virtually all the standard MPI calls, and can be linked with any MPI code with only minor changes to driver routines. Its virtual processes allow emulating larger machines with smaller ones and overlapping communication with computation, as demonstrated by the example in Fig. 6. To take full advantage of virtual processes, we have have eliminated global variables in nearly all modules in Rocstar. For each physics module, whose data are well encapsulated to begin with, a context object (of a derived data type) is introduced to contain pointers to all the data used by the module. Each virtual process has a private copy of the context object for each physics module, and this private copy is passed to all the top-level interface subroutines of the physics module at runtime, utilizing the data registration and function invocation mechanisms of Roccom, which we will describe shortly. Utilizing AMPI, our physics code developers and users can, for example, run a 480-process data set by using 480 virtual AMPI processes on just 32 physical processors of a Linux cluster, and to take advantage of parallelism within a process, without having to use a complex hybrid MPI plus OpenMP programming paradigm.

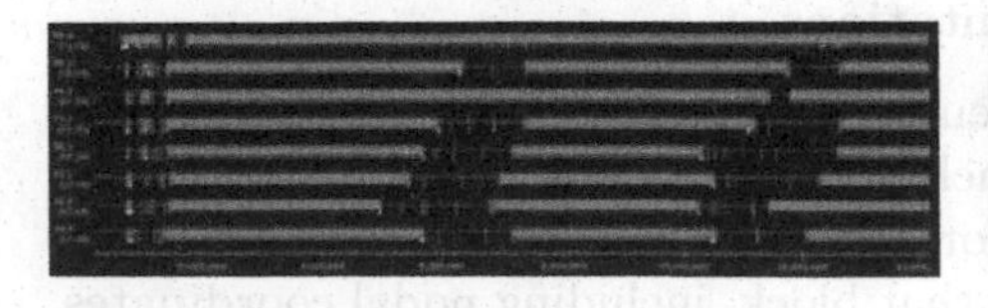

(a) 8 virtual processes on 8 physical processors

(b) 64 virtual processes on 8 physical processors

Fig. 6. AMPI minimizes processor idle times (gaps in plots) by overlapping communication with computation on different virtual processes. Courtesy of Charm group

4 Integration Framework and Middleware Services

To accommodate rapidly changing requirements of physics modules, we have developed a software framework that allows the individual components to be developed as independently as possible, and integrate them subsequently with little or no changes. It provides maximum flexibility for physics codes and can be adapted to fit the diverse needs of the components, instead of requiring the opposite. This framework is different from many traditional software architectures and frameworks, which typically assume that the framework is fully in control, and are designed for extension instead of integration.

4.1 Management of Distributed Objects

To facilitate interactions between modules, we have developed an object-oriented, data-centric integration framework called Roccom. Its design is based on an important observation and assumption of *persistent objects*. An object is said to be *persistent* if it lasts beyond a major coupled simulation step. In a typical physics module, especially in the high-performance regime, data objects are allocated during an initialization stage, reused for multiple iterations of calculations, and deallocated during a finalization stage. Therefore, most objects are naturally persistent in multipcomponent simulations.

Based on the assumption of persistency, Roccom defines a registration mechanism for data objects and organizes data into distributed objects called *windows*. A window encapsulates a number of *data attributes*, such as the mesh (coordinates and connectivities) and some associated field variables. A window can be partitioned into multiple *panes* for exploiting parallelism or for distinguishing different material or boundary-condition types. In a parallel setting, a pane belongs to a single process, while a process may own any number of panes. A module constructs windows at runtime by creating attributes and registering their addresses. Different modules can communicate with each other only through windows, as illustrated in Figure 7.

The window-and-pane data abstraction of Roccom drastically simplifies intermodule interaction: data objects of physics modules are registered and reorganized into windows, so that their implementation details are hidden from the framework and need not be altered extensively to fit the framework. Service utilities can now also be developed independently, by interacting only with window objects. Window objects are self descriptive, and in turn the interface functions

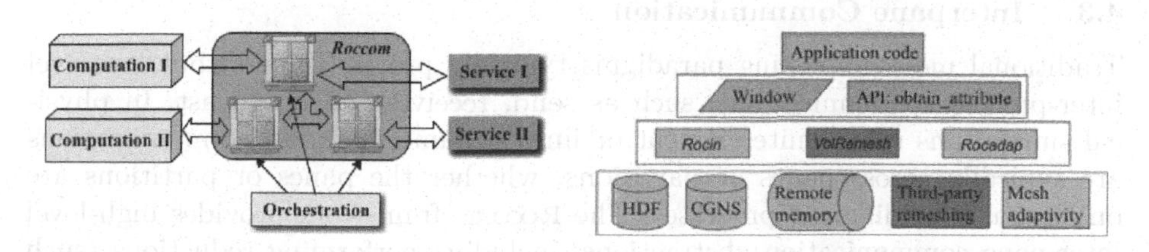

Fig. 7. Schematic of windows and panes

Fig. 8. Abstraction of data input

can be simplified substantially, frequently reducing the number of functions or the number of arguments per function by an order of magnitude. Roccom also introduces the novel concept of *partial inheritance* of windows to construct a sub-window by *using* or *cloning* a subset of the mesh or attributes of another window. In addition, the registered attributes in Roccom can be referenced as an aggregate, such as using "mesh" to refer to the collection of nodal coordinates and element connectivity. These advanced features enable performing complex tasks, such as reading or writing data for a whole window, with only one or two function calls. For more information on the novel features of Roccom, see [12].

On top of Roccom, we have developed a number of reusable service modules, including *middleware* services, such as I/O, communication, and performance tools, which we describe in the following subsections, as well mathematical service, which we discuss in the next section.

4.2 Data Input/Output

In scientific simulations, data exchange between a module and the outside world can be very complex. For file I/O alone, a developer must already face many issues, including various file formats, parallel efficiency, platform compatibility, and interoperability with off-line tools. In a dynamic simulation, the situation is even more complex, as the code may need to exchange its mesh and data attributes with mesh repair or remeshing services, or receive data from remote processes.

To meet these challenges, we use the *window* abstraction of Roccom as the medium or "virtual file" for all data exchanges for a module, regardless whether the other side is a service utility, files of various formats, or remote machines, and let middleware services take care of the mapping between the window and the other side. For example, file I/O services map Roccom windows with scientific file formats (such as HDF and CGNS), so that the details of file formats and optimization techniques become transparent to application modules. Furthermore, as illustrated in Fig. 8, all application modules obtain data from an input window through a generic function interface, obtain_attribute(), which is supported by a number of services, including file readers and remeshing tools. This novel design allows physics modules to use the same initialization routine to obtain data under different circumstances, including initial startup, restart, restart after remeshing, and reinitialization after mesh repair.

4.3 Interpane Communication

Traditional message-passing paradigms typically provide general but low-level inter-process communications, such as send, receive, and broadcast. In physical simulations using finite element or finite volume methods, communications are typically across panes or partitions, whether the panes or partitions are on the same or different processes. The Roccom framework provides high-level inter-pane communication abstractions, including performing reductions (such as sum, max, and min operations) on shared nodes, and updating values for ghost (i.e., locally cached copies of remote values of) nodes or elements. Communication patterns between these nodes and elements are encapsulated in the pane connectivity of a window, which can be provided by application modules or constructed automatically in parallel using geometric algorithms. These inter-pane communication abstractions simplify parallelization of a large number of modules, including surface propagation and mesh smoothing, which we will discuss in the next section.

4.4 Performance Monitoring

There is a large variety of tools for the collection and analysis of performance data for parallel applications. However, the use of external tools for performance tuning is too laborious a process for many applications, especially for a complex integrated system such as Rocstar. Many tools are not available on all the platforms for which we wish to collect data, and using a disparate collection of tools introduces complications as well. To obviate the need for external tools, we have extended the Roccom framework's "service-based" design philosophy with the development of a performance profiling module, Rocprof.

Rocprof provides two modes of profiling: module-level profiling and submodule-level profiling. The module-level profiling is fully automatic, embedded in Roccom's function invocation mechanism, and hence requires no user intervention. For submodule-level profiling, Rocprof offers profiling services through the standard MPI_Pcontrol interface, as well as a native interface for non-MPI based codes. By utilizing the MPI_Pcontrol interface, applications developers can collect profiling information for arbitrary, user-defined sections of source code without breaking their stand-alone codes. Internally, Rocprof uses PAPI [4] or HPM (http://www.alphaworks.ibm.com/tech/hpmtoolkit) to collect hardware performance statistics.

5 Parallel Computational Methods

In Rocstar, a physical domain is decomposed into a volume mesh, which can be either block-structured or unstructured, and the numerical discretization is based on either a finite element or finite volume method. The interface between fluid and solid moves due to both chemical burning and mechanical deformation. In such a context, we must address a large number of mathematical issues, three of which we discuss here.

5.1 Intermodule Data Transfer

In multiphysics simulations, the computational domains for each physical component are frequently meshed independently, which in turn requires geometric algorithms to correlate the surface meshes at the common interface between each pair of interacting domains to exchange boundary conditions. In a parallel setting, these meshes are typically also partitioned differently. These surface meshes in general differ both geometrically and combinatorially, and are also partitioned differently for parallel computation. To correlate such interface meshes, we have developed novel algorithms to constructs a *common refinement* of two triangular or quadrilateral meshes modeling the same surface, that is, a finer mesh whose polygons subdivide the polygons of the input surface meshes [14]. To resolve geometric mismatch, the algorithm defines a *conforming homeomorphism* and utilizes locality and duality to achieve optimal linear time complexity. Due to the nonlinear nature of the problem, our algorithm uses floating-point arithmetic, but nevertheless achieves provable robustness by identifying a set of consistency rules and an intersection principle to resolve any inconsistencies due to numerical errors.

After constructing the common refinement, we must transfer data between the nonmatching meshes in a numerically accurate and physically conservative manner. Traditional methods, including pointwise interpolation and some weighted residual methods, can achieve either accuracy or conservation, but none could achieve both simultaneously. Leveraging the common refinement, we developed more advanced formulations and optimal discretizations that minimize errors in a certain norm while achieving strict conservation, yielding significant advantages over traditional methods, especially for repeated transfers in multiphysics simulations [13]. For parallel runs, the common refinement also contains the correlation of elements across partitions of different meshes, and hence provides the communication structure needed for inter-module, inter-process data exchange.

5.2 Surface Propagation

In *Rocstar*, the interface must be tracked as it regresses due to burning. In recent years, Eulerian methods, especially level set methods, have made significant advancements and become the dominant methods for moving interfaces [21, 22]. In our context Lagrangian representation of the interface is crucial to describe the boundary of volume meshes of physical regions, but there was no known stable numerical methods for Lagrangian surface propagation.

To meet this challenge, we have developed a novel class of methods, called *face-offsetting methods*, based on a new *entropy-satisfying Lagrangian (ESL) formulation* [15]. Our face-offsetting methods exploit some fundamental ideas used by level set methods, together with well-established numerical techniques to provide accurate and stable entropy-satisfying solutions, without requiring Eulerian volume meshes. A fundamental difference between face-offsetting and traditional Lagrangian methods is that our methods solve the Lagrangian formulation face by face, and then reconstruct vertices by constrained minimization and curvature-aware averaging, instead of directly moving vertices along some approximate normal directions. Fig. 9 shows a sample result of the initial burn of a star grain section of a rocket motor, which exhibits rapid expansion at slots and contraction at fins.

Our algorithm includes an integrated node redistribution scheme that suffices to control mesh quality for moderately moving interfaces without perturbing the

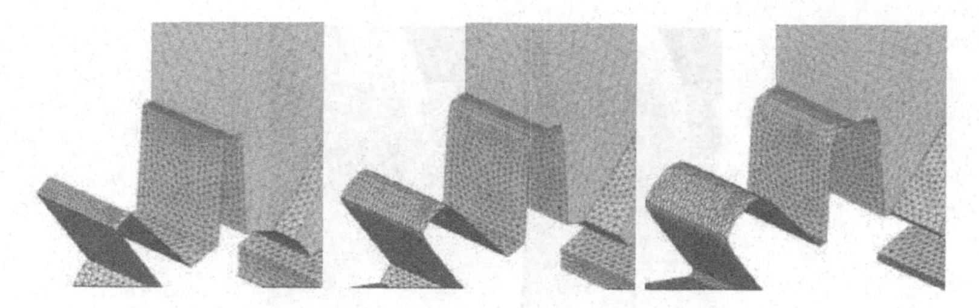

Fig. 9. Initial burn of star slice exhibits rapid expansion at slots and contraction at fins. Three subfigures correspond to 0%, 6%, and 12% burns, respectively

geometry. Currently, we are coupling it with more sophisticated geometric and topological algorithms for mesh adaptivity and topological control to provide a more complete solution for a broader range of applications.

5.3 Mesh Optimization

In Rocstar, each physics module operates on some type of mesh. An outstanding issue in integrated rocket simulations is the degradation of mesh quality due to the changing geometry resulting from consumption of propellant by burning, which causes the solid region to shrink and the fluid region to expand, and compresses or inflates their respective meshes. This degradation can lead to excessively small time steps when an element becomes poorly shaped, or even outright failure when an element becomes inverted. Some simple mesh motion algorithms are built into our physics modules. For example, simple Laplacian smoothing [27] is used for unstructured meshes, and a combination of linear transfinite interpolation (TFI) with Laplacian smoothing is used for structured meshes in Roc o. These simple schemes are insufficient when the meshes undergo major deformation or distortion. To address the issue, we take a three-tiered approach, in increasing order of aggressiveness: mesh smoothing, mesh repair, and global remeshing.

Mesh smoothing copes with gradual changes in the mesh. We provide a combination of in-house tools and integration of external packages. Our in-house effort focuses on parallel, feature-aware surface mesh optimization, and provides novel parallel algorithms for mixed meshes with both triangles and quadrilaterals. To smooth volume meshes, we utilize the serial MESQUITE package [5] from Sandia National Laboratories, which also works for mixed meshes, and we parallelized it by leveraging our across-pane communication abstractions.

If the mesh deforms more substantially, then mesh smoothing becomes inadequate and more aggressive mesh repair or even global remeshing may be required, although the latter is too expensive to perform very frequently. For these more drastic measures, we currently focus on only tetrahedral meshes, and leverage third-party tools off-line, including Yams and TetMesh from Simulog and MeshSim from Simmetrix, but we have work in progress to integrate MeshSim into our framework for on-line use. Remeshing requires that data be mapped

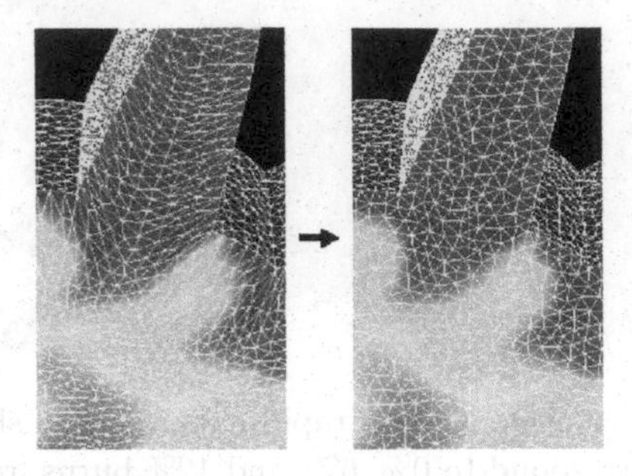

Fig. 10. Example of remeshing and data transfer for deformed star grain

from the old mesh onto the new mesh, for which we have developed parallel algorithms to transfer both node- and cell-centered data accurately, built on top of the parallel collision detection package developed by Lawlor and Kalé [16]. Fig. 10 shows an example where the deformed star grain is remeshed with the temperature field of the fluids volume transferred from the old to the new mesh.

6 Example Results

The Rocstar code suite has been tested using a number of 3-D problems, some of which provide qualitative or quantitative comparison of simulation results with empirical data (validation), while others have known analytical solutions that should be matched (verification). Most of these simulation runs are performed on five hundred or more processors and may run for days or even weeks. We give a sample of our test problems and their results in this section.

6.1 Titan IV Propellant Slumping Accident

The Titan IV SRMU Prequalification Motor motor exploded during a static test firing on April 1, 1991, caused by excessive deformation of the aft propellant segment just below the aft joint slot [29]. Fig. 11 shows a cutaway view of the fluid domain and the propellant deformation nearly 1 second after ignition, obtained from 3-D simulations with Rocstar. The deformation is due to an aeroelastic effect and is particularly noticeable near joint slots. The propellant slumping and resulting flow restriction that led to catastrophic failure on the test stand are captured by our simulations, at least qualitatively (sufficient data for detailed quantitative comparison are not available).

6.2 RSRM Ignition

To obtain insight into the ignition process in the Space Shuttle Reusable Solid Rocket Motor (RSRM), we used our ignition and dynamic burn rate models in a fully coupled 3-D simulation of the entire RSRM (Fig. 12). Our results demonstrate that hot gas from the igniter (located at upper left in each of the

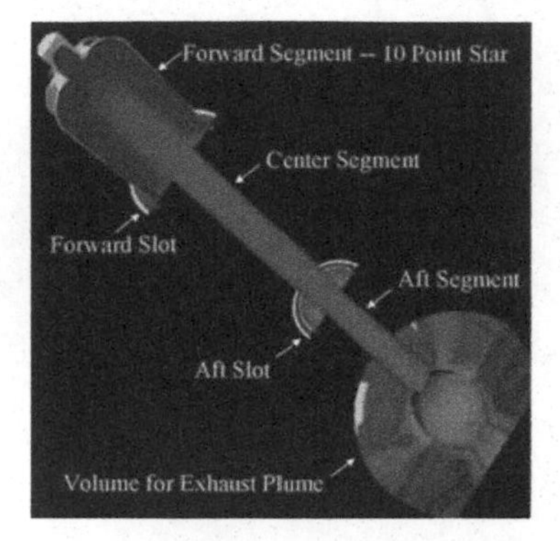

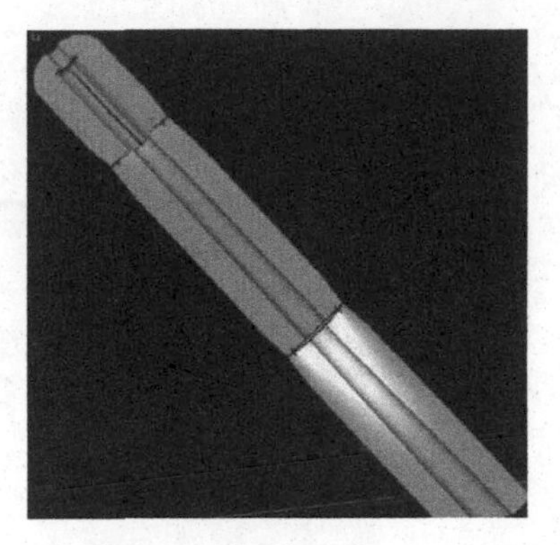

Fig. 11. Titan IV propellant slumping. Left: Cutaway view of fluids domain. Right: Propellant deformation after 1 second

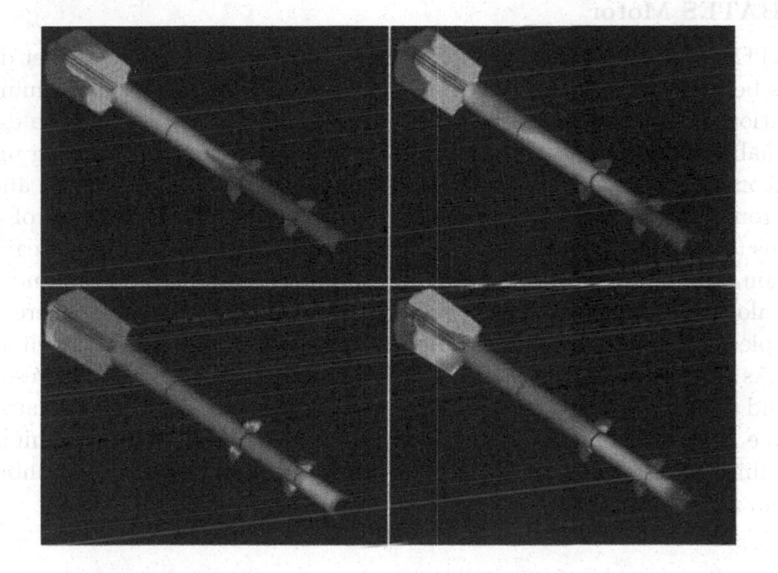

Fig. 12. RSRM propellant temperature at 100, 125, 150, and 175 ms (clockwise from upper left)

four panels of Fig. 12) reaches deep into the star-grain slots at the head end and ignites the propellant there without a significant delay, despite our omission of radiation in this calculation. Hot gas takes a long time to penetrate deeply into the two uninhibited (aft) joint slots and ignite the propellant because, unlike the star grain slots, cool gas trapped in joint slots has no easy escape route. These results agree well with experimental data obtained from NASA.

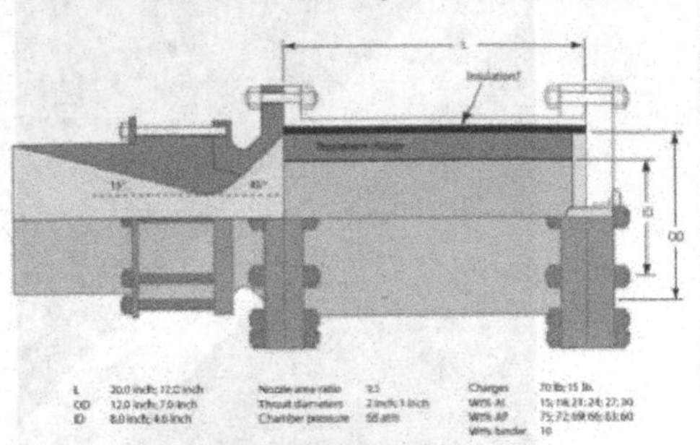

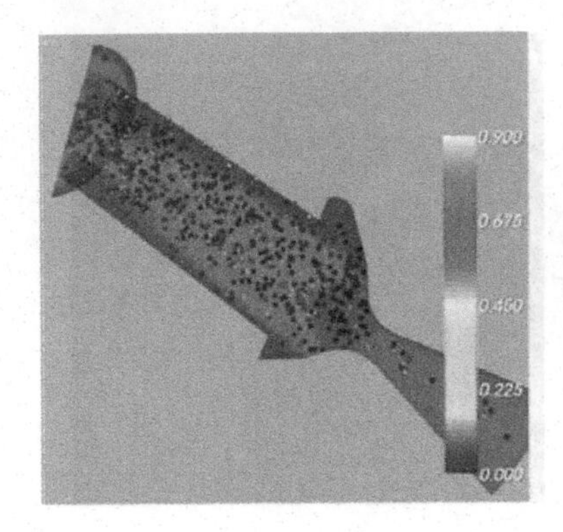

Fig. 13. BATES motor design (left) and aluminum fraction in at 15 ms (right)

6.3 BATES Motor

The BATES motor is a U.S. Air Force-sponsored, laboratory-scale rocket design that has been used to investigate the effect of various factors, such as aluminum composition of the propellant, on motor performance. Results of these experiments enable validation of the particle transport and aluminum burning models [28] in Rocstar. Fig. 13 shows the basic geometry of the BATES motor and the results from a Rocstar simulation in which burning aluminum droplets of a single radius (30 micrometers) and aluminum fraction (90%) are injected beginning at ignition. Gas and droplets injected in the narrow slot at the head end move rapidly along the axis of the rocket and out through the nozzle, whereas gas and droplets injected elsewhere on the propellant surface move at much slower speeds. As a result, droplets near the axis of the rocket have shorter residence times, and therefore a larger final aluminum fraction, than do droplets that spend more time at larger radii. The smoke concentration is correspondingly higher at larger radii. Droplets entering the nozzle with a substantial amount of unburned aluminum reduce the efficiency of the motor.

7 Performance Results

Single-processor performance is often critical to good overall parallel performance. With the aid of Rocprof, code developers can tune their modules selectively. Fig. 14 shows a sample of statistics from hardware counters for Rocfrac obtained by Rocprof on the ALC Linux Cluster at LLNL, which is based on Pentium IV technology. Fig. 15 shows the absolute performance of Rocstar and the computationally intensive Rocfrac and Rocflo on a number of platforms. Overall, Rocstar achieves a respectable 16% of peak performance on ALC, and 10% or higher on other platforms.

To measure the scalability of Rocstar we use a scaled problem, for which the problem size is proportional to the number of processors, so that the amount of work per process remain constant. Ideally, the wall-clock time should remain

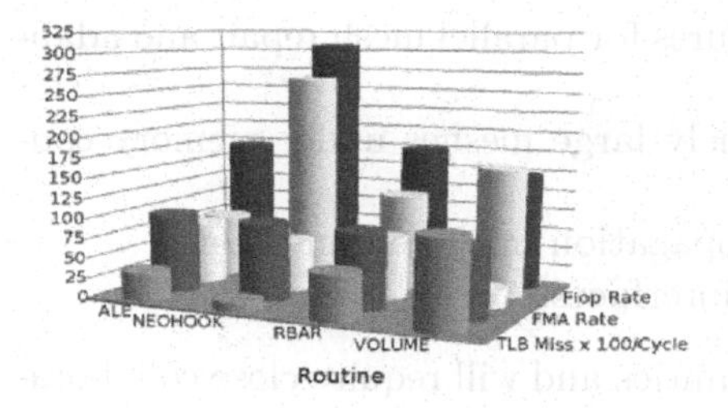

Fig. 14. Hardware counters for *Rocfrac* obtained by *Rocprof* on Linux cluster

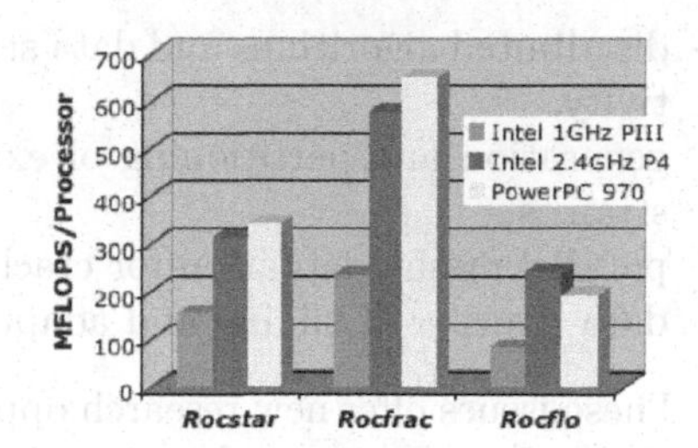

Fig. 15. Absolute performance of *Rocstar* on Linux and Mac clusters

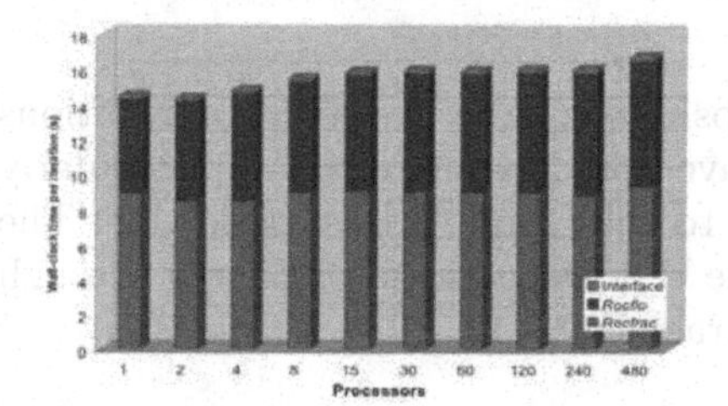

Fig. 16. Scalability of *Rocstar* with *Rocflo* and *Rocsolid* on IBM SP

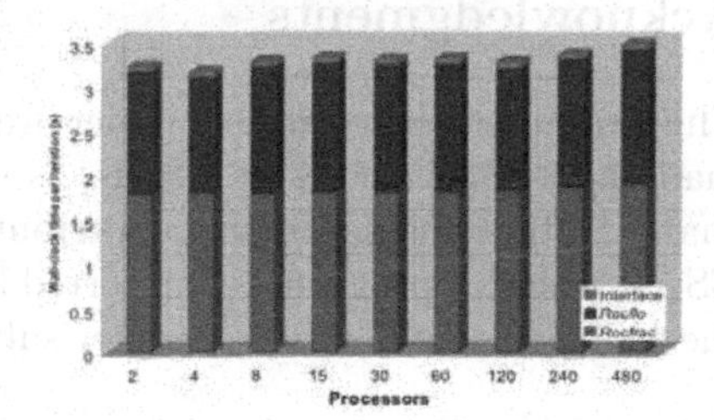

Fig. 17. Scalability of *Rocstar* with *Rocflu* and *Rocfrac* on Linux cluster

constant if scalability is perfect. Fig. 16 shows wall-clock time per iteration using an explicit-implicit coupling between Roc u and Rocsolid, with a five to one ratio (i.e., five explicit fluid time steps for each implicit solid time step), up to 480 processors on ASC White (Frost) at LLNL, which is based on IBM's POWER3 SP technology. Fig. 17 shows wall-clock time for an explicit-explicit coupling between Roc o and Rocfrac up to 480 processors on ALC. In both cases, the scalability is excellent, even for very large numbers of processors. The interface code, predominantly data transfer between fluid and solid interfaces, takes less than 2% of the overall time. Times for other modules are negligible and hence are not shown.

8 Conclusion

In this paper, we have given an overview some software and algorithmic issues in developing high-performance simulation tools for multicomponent systems, based on our experience in detailed simulation of solid propellant rockets. Working with an interdisciplinary team, we have made substantial progress in advancing the state-of-the-art in such simulations. Many challenges remain, however, including outstanding research issues such as

- distributed algorithms and data structures for parallel mesh repair and adaptivity,
- generation and partitioning of extremely large meshes under memory constraints,
- parallel mesh adaptation for crack propagation in three dimensions,
- data transfer at sliding and adaptive interfaces.

These issues offer new research opportunities and will require close collaboration between computer and engineering scientists to devise effective and practical solutions.

Acknowledgments

The results presented in this paper are snapshots of many research contributions made by our colleagues at CSAR, as we have noted in the text. We particularly thank Robert Fiedler for his contributions to the Example Results section. The CSAR research program is supported by the U.S. Department of Energy through the University of California under subcontract B523819.

References

1. E. Anderson et al. *LAPACK Users' Guide.* SIAM, Philadelphia, 2nd edition, 1995.
2. M. Bhandarkar and L. V. Kalé. A parallel framework for explicit FEM. In M. Valero et al., editors, *Proc. Internat. Conf. High Performance Computing (HiPC 2000)*, pages 385–395. Springer, 2000.
3. J. Blazek. Flow simulation in solid rocket motors using advanced CFD. In *39th AIAA/ASME/SAE/ASEE Joint Propulsion Conf. and Exhibit*, Huntsville Alabama, 2003. AIAA Paper 2003-5111.
4. S. Browne, J. Dongarra, N. Garner, G. Ho, and P. Mucci. A portable programming interface for performance evaluation on modern processors. *Int. J. High Perf. Comput. Appl.*, 14:189–204, 2000.
5. L. Freitag, T. Leurent, P. Knupp, and D. Melander. MESQUITE design: Issues in the development of a mesh quality improvement toolkit. In *8th Intl. Conf. Numer. Grid Gener. Comput. Field Sim.*, pages 159–168, 2002.
6. P. H. Geubelle and J. Baylor. Impact-induced delamination of composites: a 2D simulation. *Composites B*, 29B:589–602, 1998.
7. W. Gropp, E. Lusk, and A. Skjellum. *Using MPI: Portable Parallel Programming with the Message-Passing Interface.* MIT Press, Cambridge, MA, 2nd edition, 1999.
8. A. Haselbacher. A WENO reconstruction method for unstructured grids based on explicit stencil construction. In *43rd AIAA Aerospace Sciences Meeting and Exhibit*, Reno, NV, Jan. 2005. AIAA Paper 2005-0879, to appear.
9. M. T. Heath and W. A. Dick. Virtual prototyping of solid propellant rockets. *Computing in Science & Engineering*, 2:21–32, 2000.
10. C. Huang, O. S. Lawlor, and L. V. Kalé. Adaptive MPI. In *Proc. 16th Internat. Workshop on Languages and Compilers for Parallel Computing (LCPC 03)*, October 2003.

11. C. Huggett, C. E. Bartley, and M. M. Mills. *Solid Propellant Rockets*. Princeton University Press, Princeton, NJ, 1960.
12. X. Jiao, M. T. Campbell, and M. T. Heath. Roccom: An object-oriented, data-centric software integration framework for multiphysics simulations. In *17th Ann. ACM Int. Conf. Supercomputing*, pages 358–368, 2003.
13. X. Jiao and M. T. Heath. Common-refinement based data transfer between nonmatching meshes in multiphysics simulations. *Int. J. Numer. Meth. Engrg.*, 61:2401–2427, 2004.
14. X. Jiao and M. T. Heath. Overlaying surface meshes, part I: Algorithms. *Int. J. Comput. Geom. Appl.*, 14:379–402, 2004.
15. X. Jiao, M. T. Heath, and O. S. Lawlor. Face-offsetting methods for entropy-satisfying Lagrangian surface propagation. In preparation, 2004.
16. O. S. Lawlor and L. V. Kalé. A voxel-based parallel collision detection algorithm. In *Proc. Internat. Conf. Supercomputing*, pages 285–293, June 2002.
17. L. Massa, T. L. Jackson, and M. Short. Numerical simulation of three-dimensional heterogeneous propellant. *Combustion Theory and Modelling*, 7:579–602, 2003.
18. F. M. Najjar, A. Haselbacher, J. P. Ferry, B. Wasistho, S. Balachandar, and R. D. Moser. Large-scale multiphase large-eddy simulation of flows in solid-rocket motors. In *16th AIAA Computational Fluid Dynamics Conf.*, Orlando, FL, June 2003. AIAA Paper 2003-3700.
19. A. Namazifard and I. D. Parsons. A distributed memory parallel implementation of the multigrid method for solving three-dimensional implicit solid mechanics problems. *Int. J. Numer. Meth. Engrg.*, 61:1173–1208, 2004.
20. A. Namazifard, I. D. Parsons, A. Acharya, E. Taciroglu, and J. Hales. Parallel structural analysis of solid rocket motors. In *AIAA/ASME/SAE/ASEE Joint Propulsion Conf.*, 2000. AIAA Paper 2000-3457.
21. S. Osher and R. Fedkiw. *Level Set Methods and Dynamic Implicit Surfaces*. Springer, 2003.
22. J. A. Sethian. *Level Set Methods and Fast Marching Methods*. Cambridge University Press, 1999.
23. G. P. Sutton and O. Biblarz. *Rocket Propulsion Elements*. John Wiley & Sons, New York, 7th edition, 2001.
24. K. C. Tang and M. Q. Brewster. Dynamic combustion of AP composite propellants: Ignition pressure spike, 2001. AIAA Paper 2001-4502.
25. K. C. Tang and M. Q. Brewster. Nonlinear dynamic combustion in solid rockets: L*-effects. *J. Propulsion and Power*, 14:909–918, 2001.
26. P. D. Thomas and C. K. Lombard. Geometric conservation law and its application to flow computations on moving grids. *AIAA J.*, 17:1030–1037, 1979.
27. J. F. Thompson, B. K. Soni, and N. P. Weatherill, editors. *Handbook of Grid Generation*. CRC Press, Boca Raton, 1999.
28. J. F. Widener and M. W. Beckstead. Aluminum combustion modeling in solid propellant combustion products, 1998. AIAA Paper 98-3824.
29. W. G. Wilson, J. M. Anderson, and M. Vander Meyden. Titan IV SRMU PQM-1 overview, 1992. AIAA Paper 92-3819.

Parallel Boundary Elements: A Portable 3-D Elastostatic Implementation for Shared Memory Systems

Manoel T.F. Cunha, J.C.F. Telles, and Alvaro L.G.A. Coutinho

Universidade Federal do Rio de Janeiro - COPPE / PEC,
Caixa Postal 68506 - CEP 21945-970 - Rio de Janeiro - RJ - Brasil
manoel@ufpr.br, telles@coc.ufrj.br
alvaro@nacad.ufrj.br

Abstract. This paper presents the parallel implementation of a computer program for the solution of three-dimensional elastostatic problems using the Boundary Element Method (BEM). The Fortran code is written for shared memory systems using standard and portable libraries: OpenMP and LAPACK. The implementation process provides guidelines to develop highly portable parallel BEM programs, applicable to many engineering problems. Numerical experiments performed in the solution of a real-life problem on a SGI Origin 2000, a Cray SV1ex and a NEC SX-6 show the effectiveness of the proposed approach.

1 Introduction

Since there is always a limit to the performance of a single CPU, computer programs have evolved to increase their performance by using multiple CPUs. The great increase in the availability and affordability of shared memory systems and the development of high performance parallel libraries justify the efforts to port existing serial codes to such parallel environments.

Over almost two decades, since the first implementations of parallel applications [1, 2] using the Boundary Element Method (BEM) [3, 4], a large number of papers have been published on the subject [5, 6]. During these years hardware has evolved at an amazing rate [7]. Programming tools and techniques have changed or been created to profit from this hardware evolution [8]. In addition, scientific research also adds complexity to applications and demands considerable improvement in software resources and techniques.

In this paper, a portable approach for the development of parallel boundary element programs for shared memory systems is presented. The paper emphasizes the benefits of using available standard libraries for writing parallel codes, ensuring high performance in the simulations and increasing the lifetime and increasing the lifetime of the respective algorithm.

OpenMP [10] is a standard and portable Application Programming Interface (API) for writing shared memory parallel programs. It allows to parallelize

M. Daydé et al. (Eds.): VECPAR 2004, LNCS 3402, pp. 514–526, 2005.

an application in a stepwise process, without concerns about data distribution layouts or how the workload is distributed between multiple processes.

LAPACK[9], or Linear Algebra PACKage, is a parallel high performance library for solving linear algebra problems on workstations, vector computers and shared-memory parallel systems. This library contains routines for solving systems of linear equations and related linear algebra problems.

Both libraries are available for a variety of platforms, from PCs to high performance computers and are supported by many Fortran and C/C++ compilers. LAPACK can be freely downloaded from the Internet, including precompiled implementations.

By using high performance parallel libraries, BEM programmers can benefit from an existing large knowledge-base, continuously updated, to focus on the actual engineering problem to be solved. In the same way, efficient parallel applications can be created in a short time, to be implemented on most available platforms used nowadays, from dual and quad-processors personal computers to supercomputers such as SGI, Cray, NEC, among others.

The text is organized as follows: next section presents an outline of the BEM theory and the following section presents the selected application. Section 4 details the shared memory implementation of the program while experimental results are shown and discussed in section 5. The paper ends with a summary of the main conclusions.

2 Outline of BEM Theory

The Boundary Element Method (BEM) is a technique for the numerical solution of partial differential equations subjected to initial and boundary conditions [3, 4].

By using a weighted residual formulation, Green's third identity, Betty's reciprocal theorem or some other procedure, an equivalent integral equation can be obtained and then converted to a form that involves only surface integrals performed over the boundary. The bounding surface is divided into elements and the original integrals over the boundary are simply the sum of the integrations over each element, resulting in a reduced dense and non-symmetric system of linear equations. The discretization process involves selecting nodes, on the boundary, where unknown values are considered. Interpolation functions relate such nodes to the approximated displacements and tractions on the respective boundary elements. The simplest case places a node in the center of each element and defines an interpolation function that is constant across the entire element. For linear 2-D elements, nodes are placed at, or near, the end of each element and the interpolation function is a linear combination of the two node values. High-order elements, quadratic or cubic, can be used to better represent curved boundaries using three and four nodes, respectively.

Once the boundary solution has been obtained, interior point results can be computed directly from the basic integral equation in a post-processing routine.

2.1 Governing Equations

Elastostatic problems are governed by the well-known Navier equilibrium equation which, using commas to indicate space derivatives and the so-called Cartesian tensor notation, may be written for a domain Ω in the form:

$$G u_{j,kk} + \frac{G}{1-2\,\nu} u_{k,kj} + b_j = 0 \quad \text{in } \Omega \tag{1}$$

subject to the boundary conditions:

$$\begin{aligned} u = \bar{u} \text{ on } \Gamma_1 \quad &\text{and} \\ p = \bar{p} \text{ on } \Gamma_2 & \end{aligned} \tag{2}$$

where u are displacements, p surface distributed forces or tractions, $\bar{u}$ and $\bar{p}$ are prescribed values and the total boundary of the body is $\Gamma = \Gamma_1 + \Gamma_2$. G is the shear module, ν is the Poisson's ratio and b_j is the body force component.

Notice that the subdivision of Γ into two parts is just conceptual, i.e., the same physical point on Γ can have the two types of boundary conditions in different directions.

2.2 Integral Equation

An integral equation, equivalent to (1) and (2), can be obtained through a weighted residual formulation or Betty's reciprocal theorem [3].

This equation, also known as Somigliana's identity for displacements, can be written as:

$$u(\xi_i) = \int_\Gamma u^*_{ij}(\xi, x)\, p_j(x)\, d\Gamma(x) - \int_\Gamma p^*_{ij}(\xi, x)\, u_j(x)\, d\Gamma(x) \tag{3}$$

where $b_i = 0$ was assumed for simplicity and the fundamental solution tensors, u^*_{ij} and p^*_{ij}, represent the component in j direction of displacement and traction at the field point x due a unit point load applied at the source point ξ in i direction. Typically, these kernels depend on the Cartesian distance r between ξ and x.

In order to obtain an integral equation involving only variables on the boundary, one can take the limit of Eq. (3) as the point ξ tends to the boundary Γ. This limit has to be carefully taken since the starred u^*_{ij} and p^*_{ij} kernels present a $log(1/r)$ and $1/r$ behaviour, respectively, producing integrands at ξ.

The resulting equation is:

$$c(\xi_i)\, u(\xi_i) + ⨍_\Gamma p^*_{ij}(\xi, x)\, u_j(x)\, d\Gamma(x) = \int_\Gamma u^*_{ij}(\xi, x)\, p_j(x)\, d\Gamma(x) \tag{4}$$

where the coefficient c is a function of the geometry of Γ at the point ξ and the integral on the left is to be computed in a Cauchy principal value sense.

2.3 Discretization

Assuming that the boundary Γ is discretized into N elements, Eq. (4) can be written in the form:

$$c_i\, u_i + \sum_{j=1}^{N} \int_{\Gamma_j} p_{ij}^*\, u_j\, d\Gamma = \sum_{j=1}^{N} \int_{\Gamma_j} u_{ij}^*\, p_j\, d\Gamma \tag{5}$$

The substitution of displacements and tractions by element approximated interpolation functions leads to:

$$\mathbf{c}_i\, \mathbf{u}_i + \sum_{j=1}^{N} \mathbf{h}\, \mathbf{u}_j = \sum_{j=1}^{N} \mathbf{g}\, \mathbf{p}_j \tag{6}$$

which can be rearranged in a simpler matrix form as,

$$\mathbf{H}\, \mathbf{u} = \mathbf{G}\, \mathbf{p} \tag{7}$$

By applying the prescribed boundary conditions, the problem unknowns can be grouped on the left-hand side of Eq. (7) to obtain a system of equations ready to be solved by standard methods.

This system of linear equations can be written as:

$$\mathbf{A}\, \mathbf{x} = \mathbf{f} \tag{8}$$

where $\mathbf{A}$ is a dense square matrix, vector $\mathbf{x}$ contains the unknown tractions and displacements nodal values and vector $\mathbf{f}$ is formed by the product of the prescribed boundary conditions by the corresponding columns of matrices $\mathbf{H}$ and $\mathbf{G}$.

Note that Eq. (8) can be assembled directly from the elements $\mathbf{h}$ and $\mathbf{g}$ without need to generate first Eq. (7).

2.4 Internal Points

Since Somigliana's identity provides a continuous representation of displacements at any point $\xi \in \Omega$, it can also be used to generate the internal stresses. The discretization process, described above, can also be applied here, now in a post-process procedure.

3 The Application Program

The program presented here is a Fortran code for linear elastic analysis of three-dimensional problems with the Boundary Element Method. It allows the selection of triangular and quadrilateral elements, with constant, linear and quadratic interpolations, as shown in Figures 1 and 2. In addition, the elements can be continuous, discontinuous and of transition type [3].

According to Telles [3], one of the most interesting properties of BEM is to accept discontinuous functions and motivated the development of a family of elements for which the approximated functions is defined by Lagrangian polynomials but with all the nodes inside the elements. The method also allows transition elements with nodes inside the element and at its border, as well. The elements can be isoparametric since their geometries are still defined by corner nodes coordinates.

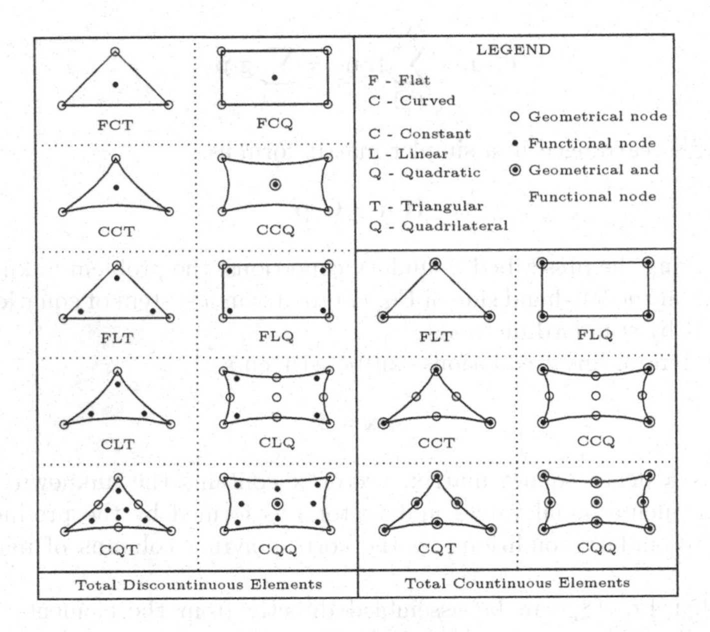

Fig. 1. Elements Types

The program defines general variables and arrays and a set of Fortran routines [11] read the input data. Another set of routines determines the position, numbering and global coordinates of functional nodes. Among the main routines in the code, `SYSTEM` computes matrix **A** and the right hand side vector **f**, while `TENCON` computes stresses at the boundary. Internal displacements and stresses are calculated by two routines: `DESINT` and `TENINT`.

Numerical integrals are performed over non-singular elements by using Gauss integration. A selective process was adopted to define the number of Gauss integration points, based on the relative distance between the source point and the field element. The shorter this distance, the greater will be the number of points. For elements with the singularity at one of its functional nodes the required integrals are computed semi-analytically to obtain more accurate results. In order

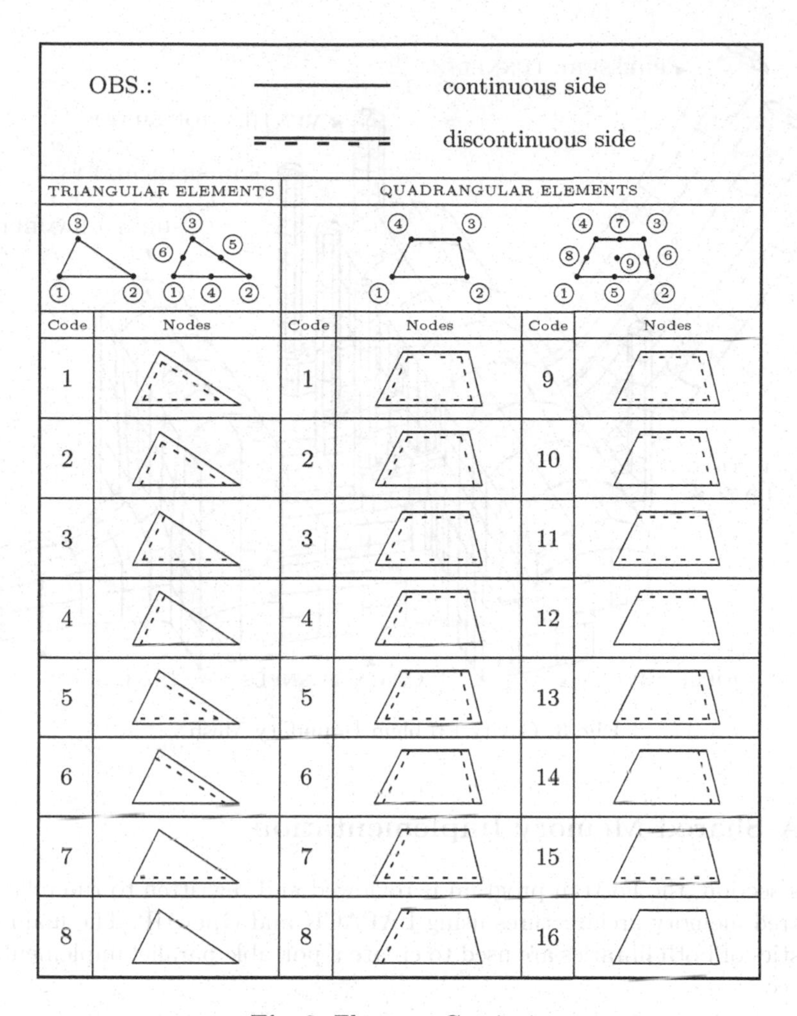

Fig. 2. Elements Continuity

to improve the precision of quasi-singular integrals a coordinate transformation over the integration path proposed by Telles [12, 13] has been adopted.

The system of equations is solved using LAPACKs solver `DGESV` [9].

Here, a real life engineering test case is used to compare the performance of the serial and parallel versions of the code. The 3-D modeling of a real cavity problem, taking in account the in-situ initial stresses is analyzed. A mesh of 6827 geometric points is used to define 1513 discontinuous rectangular elements and the input data includes the spatial coordinates of 1618 internal points. The mesh of boundary elements is shown in Figure 3, it represents the underground rock excavations for the hydraulic circuit of one of the largest hydroelectric power plants recently built in Brazil [14].

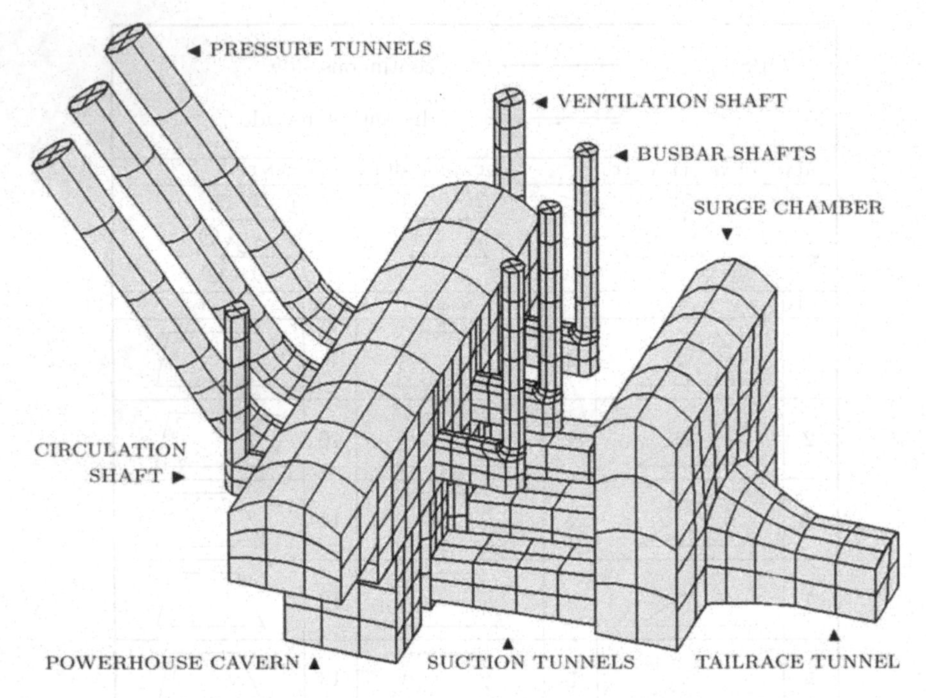

Fig. 3. Cavity Problem Boundary Mesh

4 A Shared Memory Implementation

In this section, the Fortran program is reviewed and rewritten to run efficiently on shared memory architectures using LAPACK and OpenMP. The main characteristics of both libraries are used to create a portable parallel implementation of the code.

4.1 The LAPACK Library

The Linear Algebra Package, LAPACK, is a public domain library of subroutines for solving linear equations systems and related linear algebra problems.

LAPACK routines are written so that as much as possible of the computation is performed by calls to the Basic Linear Algebra Subprograms, BLAS. Highly efficient machine-specific implementations of the BLAS are available for many high-performance computers. BLAS enables LAPACK routines to achieve high performance with a portable code [9].

LAPACK is designed to be efficient on a wide range of modern high performance computers, such as vector processors, superscalar workstations and shared memory multiprocessors. LAPACK routines are written as a single thread of execution. Nevertheless, LAPACK can accommodate shared-memory machines, provided parallel BLAS routines, already available, are used [15].

Table 1. Serial Performance on a SGI Origin 2000 - Constant Elements

```
-------------------------------------------------------------------------
Summary of ideal time data (ideal)--
            144956926278: Total number of instructions executed
             87863050700: Total computed cycles
                 351.452: Total computed execution time (secs.)
                   0.606: Average cycles / instruction
-------------------------------------------------------------------------
Function list, in descending order by exclusive ideal time
-------------------------------------------------------------------------
function             excl.secs  excl.%  cum.%  incl.secs  incl.%    calls
__scsl_dgemm_hoistc    128.477   36.6%  36.6%    128.477   36.6%    33250
tenint                  51.072   14.5%  51.1%     90.292   25.7%        1
intnsi                  41.070   11.7%  62.8%    118.758   33.8%  4735690
vetorn                  27.498    7.8%  70.6%     37.528   10.7% 36340636
subele                  24.216    6.9%  77.5%     25.251    7.2%  7183724
septin                  13.467    3.8%  81.3%     38.719   11.0%  7183724
calcd2                  11.092    3.2%  84.5%     14.770    4.2%  7183724
dforma                  10.031    2.9%  87.3%     10.031    2.9% 36343662
forma                    9.363    2.7%  90.0%      9.363    2.7% 73151494
tciqsi                   8.678    2.5%  92.5%     33.291    9.5%  7183724
__fastm_pow              7.429    2.1%  94.6%      7.429    2.1% 28734896
system                   3.614    1.0%  95.6%     61.733   17.6%        1

-------------------------------------------------------------------------
Butterfly function list, in descending order by inclusive ideal time
-------------------------------------------------------------------------
100.0%   351.424(0000001)                        351.424  main [2]
100.0%   351.424      0.0%      0.000         cav3d [3]
                     38.5%   135.461(0000001)    135.461  DGESV [4]
                     25.7%    90.292(0000001)     90.292  tenint [11]
                     17.7%    62.174(0000001)     62.174  desint [14]
                     17.6%    61.733(0000001)     61.733  system [15]
-------------------------------------------------------------------------
```

The solver is usually the most time consuming routine of BEM codes. Indeed, as shown in Table 1, the routines called by LAPACK's DGESV solver, such as _dgemm_, though optimized still take the greatest portion of the execution time. Hence, the parallelization of the solver should bring the most effective performance gain. A very simple and effective approach is to use an available parallel version of the solver.

The Cray Scientific Library, SCSL, implemented in the SGI Origin 2000 is an optimized library containing LAPACK and BLAS routines, many of them parallelized and linked into the program using an additional compiling option [16]. LAPACK and BLAS libraries are also found on Cray and NEC computers. There is no need for any change in the code.

4.2 The OpenMP Implementation

In shared memory architectures all processors have direct and equal access to all the memory in the system. As a consequence, a program can be parallelized without concern about how the work is distributed across multiple processors and how data is accessed by each CPU. Another advantage of this type of parallelization is that it can be done incrementally, that is, the user can identify and parallelize just the most time consuming parts of the code.

OpenMP is a set of compiler directives that may be embedded within a program written with a standard programming language such as Fortran or C/C++. In Fortran these directives take the form of source code comments identified by the !$OMP prefix and are simply ignored by a non-OpenMP compiler. Thus, the same source code can be used to compile a serial or parallel version of the application. The number of CPUs used by the program is defined externally by the OMP_NUM_THREADS environment variable [10].

Loops that can be executed in parallel on multiple CPUs are then surrounded by OpenMP directives. Wrapping a loop definition with OpenMP directives will distribute the workload between processors. The iterations are divided among processors by the compiler in a process hidden from the user. The programmer must pay special attention to the definition of shared and private variables. As implicit in the definition, concurrent iterations have common access to shared variables. Private elements are usually loop counters and intermediate calculations which cannot be modified by concurrent iterations.

The profile presented in Table 1 shows the routine `TENINT` as the second most consuming part of the program. This subroutine computes stresses at internal points using Gauss integration over all the boundary elements. In the code, an outer loop controls the internal points variation while internal nested loops control symmetry and boundary elements, respectively.

The boundary element method has a parallel nature since the computing of internal points are totally independent and loop iterations can safely be performed concurrently. Hence, a straightforward approach is to use OpenMP directives to parallelize the external loop, which controls the node iterations.

The third most time consuming routine is `INTNSI`, which assembles the **H** and **G** matrices for each element using numerical integration, when there is no singularity. Thus, a simple and effective approach is to parallelize the routines from where `INTNSI` is called: `SYSTEM` and `DESINT`. The first one, `SYSTEM`, generates the system of equations while `DESINT` computes displacements at internal points.

The structure of `DESINT` is quite similar to `TENINT`, described previously. It also presents the same nested loops, where an outer loop controls internal points and encloses the symmetry and boundary elements loops. Hence, the same parallelization scheme of the external loop is repeated.

Additional considerations, however, are needed to the parallelization of the routine `SYSTEM`. As already stated, this subroutine generates the system of equations and calls `INTNSI` to compute integrals over non-singular elements by using

Gauss integration. For elements with the singularity at one of its functional nodes the required integrals are computed calling INTSIN.

In routine SYSTEM, the same three main loops are presented. However, now the outer loop controls symmetry. Inside this loop, a nest of loops control the variation of source elements, their corresponding functional nodes and finally the field elements. It is easy to show that parallelization of the external loop would not bring any advantage in the case of no symmetry. Again, the parallelization of the loop controlling source elements should result in an efficient approach.

The parallelization of an internal loop demands special care due to the high cost of invoking a parallel loop at each iteration of the external loop. According to Chandra et al. [10], the overhead to invoke an empty parallel do in a SGI Origin 2000 can arise from 2400 cycles for two processors to 8000 cycles for sixteen processors.

The same authors [10] compare the performance of fine-grained (loop) versus coarse-grained (domain) decomposition and found better results performing an empty loop (!$OMP DO) within an outer parallel region (!$OMP PARALLEL). However, in the code studied here, this option cannot be implemented since an outer parallel region would include the external symmetry loop. In this case, each thread would concurrently execute a copy of all the code enclosed in the external loop.

The parallelization approach adopted in SYSTEM greatly benefits from the already noticed parallel nature of boundary elements in the sense that a set of lines in the equations system can be computed totally independent and, as at internal points, loop iterations can safely be performed concurrently. Other features of boundary elements present in the code [3] also lead to this approach.

5 Parallel Results Summary

Several discretizations, using constant, linear and quadratic elements, were run on a SGI Origin 2000 16 R10000 250 MHz processors, a Cray SV1ex with 32 processors and a NEC SX-6 with 8 processors.

The execution times on the SGI Origin 2000 are shown in Table 2[1]. Execution times obtained from running the program on a CRAY SV1ex and a NEC SX-6 are presented in Tables 3 and 4.

The size of the system of equations being solved changes as follows. In the case of constant elements, the 1513 elements generate the same number of functional nodes which have three degrees of freedom each. Thus, a 4539x4539 system of equations must be solved. Linear elements held 4 functional nodes per element and generate a 18156x18156 system of equations and in the quadratic elements case, each element contains 9 functional nodes and a 40851x40851 system of equations is created.

[1] In spite of having 8 Gb of memory, the present configuration of the SGI Origin 2000 did not allow to run larger problems without using virtual memory.

Table 2. SGI Origin 2000 - Constant Elements

Procs.	Seconds	Speedup
1	393.38	-
2	297.89	1.89
4	108.55	3.62
8	61.48	6.40
16	39.36	9.99

Table 3. Cray SV1ex Elapsed Times

	Constant		Linear		Quadratic	
Procs.	Seconds	Speedup	Seconds	Speedup	Seconds	Speedup
1	826.47	-	5041.70	-	41224.34	-
2	424.12	1.95	2547.96	1.98	20683.41	1.99
4	217.78	3.79	1362.21	3.70	10643.13	3.87
8	116.59	7.09	787.93	6.40	5562.44	7.41
12	81.60	10.13	485.52	10.38	3623.71	11.37
16	65.03	12.71	405.63	12.43	2697.24	15.29
20	57.30	14.42	400.19	12.60	2503.04	16.47
24	50.67	16.31	361.10	13.96	2315.38	17.80
28	45.68	18.09	356.31	14.31	1951.47	21.12
32	42.72	19.34	352.41	14.31	1880.40	21.92

Table 4. NEC SX-6 Elapsed Times

	Constant		Linear		Quadratic	
Procs.	Seconds	Speedup	Seconds	Speedup	Seconds	Speedup
1	1705.22	-	4459.37	-	17946.17	-
2	1170.53	1.46	2911.27	1.53	10282.61	1.75
4	589.26	2.89	1432.40	3.11	5176.77	3.47
6	389.71	4.36	974.97	4.57	3540.79	5.07
8	294.49	5.79	742.57	6.01	2690.13	6.67

6 Conclusions

The development of portable parallel applications for shared memory systems can be achieved in a simple and effortless way by using the high performance parallel libraries OpenMP and LAPACK. The results presented here justify porting existing BEM applications to parallel environments.

The code did parallelize well on Origin 2000 and on the vector machines, the Cray SV1ex and NEC SX-6. The results obtained from running the program in the SGI Origin 2000 show good parallel performance up to 8 CPUs. Adding more CPUs will not reduce much the wall clock time. In the Cray SV1ex, the program scales well up to 16 CPUs for constant and linear elements and up to

20 CPUs for the case of quadratic elements. Scalability is almost constant in the NEC and the best efficiency is achieved with quadratic elements.

The application of parallel programming techniques has proven to be very effective. For the sake of comparison, the parallel version of the code, generated with the auto parallelizing compiling option on the SGI Origin 2000 takes several times the execution time than the proposed implementation. Thus, this paper emphasizes the need for a broader approach, joining the simplicity of OpenMP with the intrinsic characteristics of the boundary element method, in order to produce efficient BEM parallel programs.

A word on the solution of the system of equations is perhaps now due. Even for dense systems, iterative solvers are faster than direct methods in BEM codes [17, 18], the present option for a direct solver is due to its scalability and portability, making this alternative very attractive. At the moment, standard, portable and public domain implementations of iterative solvers for shared and distributed memory systems are not yet available on the same level as LAPACK's.

The choice of the element type greatly affects the performance of the application on these different platforms and demands some considerations. In the constant elements case, the most time consuming routines are the generation of the system of equations and the internal point computations. On the other end, quadratic elements generate a large system of equations and the solver takes the most significant part of the execution time. Thus, the overhead of OpenMP parallelization of such routines can produce a performance gap between different architectures.

Besides the work distribution, the results from Cray SV1ex and NEC SX-6 also show a great influence of the vector register length, 64 on the former and 256 on the latter. In the Cray SV1ex, the vector registers are full almost all the time. However, on the NEC SX-6 the vector registers are only 25% busy when using constant elements (average vector length of 64.8). For linear elements the average vector length was 192.5 and for quadratic elements it was 240.8, which translated in a better utilization of the registers on the SX-6.

The portability concerns embedded in this implementation were also fulfilled since the same code could be moved effortlessly to these different computers. OpenMP and LAPACK are standard and portable libraries available for a variety of platforms, from multiprocessor PCs to high performance supercomputers. Both libraries are supported by the majority of computers vendors. LAPACK can be freely downloaded from the Internet, including its precompiled version for some machines. Thus, highly portable parallel BEM applications [19, 20] can be created to run on most platforms in a reduced development time.

Acknowledgments

The authors are indebted to Dr. Nicolas Chepurniy from Cray Inc. by his invaluable help, support and unlimited availability. Computational resources for this research were provided by SIMEPAR - Sistema de Meteorologia do Paraná, the Center for Parallel Computing of the Federal University of Rio de Janeiro and Cray Inc. M.T.F. Cunha is supported by a CAPES grant from the Ministry of Education, Brazil.

References

1. Symm, G.T.: Boundary Elements on a Distributed Array Processor. Engineering Analysis with Boundary Elements. **1/3** (1984) 162–165
2. Davies, A.J.: The Boundary Element Method on the ICL DAP. Parallel Computing. **8/1-3** (1988) 335-343
3. Brebbia, C.A., Telles, J.C.F., Wrobel L.C.: Boundary Elements Techniques: Theory and Applications in Engineering. Springer-Verlag, Berlin Heidelberg New York (1984)
4. Telles, J.C.F.: The Boundary Element Method Applied to Inelastic Problems. Springer-Verlag, Berlin Heidelberg New York (1983)
5. Davies, A.J.: Fine-grained Parallel Boundary Elements. Engineering Analysis with Boundary Elements. **19** (1997) 13-16
6. Gomez, J.E., Power, H.: A Multipole Direct and Indirect BEM for 2D Cavity Flow at Low Reynolds Number. Engineering Analysis with Boundary Elements, **19** (1997) 17-31
7. Dongarra, J.J.: High Performance Computing Trends, the Grid, and Numerical Algorithms. In: 14th Symposium on Computer Architecture and High Performance Computing. IEEE Computer Society, Washington (2002)
8. Cunha, M.T.F.: A Practical Approach in the Development of Engineering Applications to the Internet Using Object Oriented Programming. Universidade Federal do Paran, Curitiba-Brazil. (1999)
9. Anderson, E. et al.: LAPACK Users Guide. 3rd. edn. SIAM, Philadelphia. (1999)
10. Chandra R. et al.: Parallel Programming in OpenMP. Academic Press, London (2001)
11. Tang, W., Telles, J.F.C., Brebbia, C.A.: Some Subroutines for Simple and Efficient Input Data Reading for Multipurpose Pre-processors. Microsoftware for Engineers. Computational Mechanics Publications, **2/4** (1986) 209-214
12. Telles, J.F.C.: A Self-adaptative Coordinate Transformation for Efficient Numerical Evaluation of General Boundary Integrals. International Journal for Numerical Methods in Engineering. **24** (1987) 959-973
13. Telles, J.F.C.: A Self-adaptative Coordinate Transformation for Efficient Numerical Evaluation of General Boundary Integrals, Discussion. International Journal for Numerical Methods in Engineering. **26** (1988) 1683-1684
14. Franco, J.A.M. et al.: Design Aspects of the Underground Structures of the Serra da Mesa Hydroeletric Power Plant. Int. J. Rock Mech. and Min. Sci. **34/3-4** (1997) 763-771
15. Blackford, L.S. et al.: ScaLAPACK Users Guide. SIAM, Philadelphia. (1987)
16. SGI: Origin 2000 and Onyx2 Performance Tuning and Optimization Guide. Silicon Graphics Inc., Document 007-3430-002. (1998)
17. Barra, L.P.S. et al.: Iterative Solution of BEM Equations by GMRES Algorithm. Computer and Structures. **6/24** (1992) 1249–1253
18. Kreienmeyer, M., Stein, E.: Efficient Parallel Solvers for Boundary Element Equations Using Data Decomposition. Engineering Analysis with Boundary Elements. **19** (1997) 33-39
19. Cunha, M.T.F., Telles, J.C.F., Coutinho, A. L. G. A.: On the parallelization of boundary element codes using standard and portable libraries. Engineering Analysis with Boundary Elements. **28/7** (2004) 893-902
20. Cunha, M.T.F., Telles, J.C.F., Coutinho, A. L. G. A.: A Portable Implementation of a Boundary Element Elastostatic Code for Shared and Distributed Memory Systems. Advances in Engineering Software. **37/7** (2004) 453-460

On Dependence Analysis for SIMD Enhanced Processors

Patricio Bulić and Veselko Guštin

Faculty of Computer and Information Science, University of Ljubljana,
Tržaška c. 25, 1000 Ljubljana, Slovenia
{Patricio.Bulic, Veselko.Gustin}@fri.uni-lj.si

Abstract. There are a number of data dependence tests that have been proposed in the literature. In each test there is a different trade-off between accuracy and efficiency. The most widely used approximate data dependence tests are the Banerjee inequality and the GCD test. In this paper we consider parallelization for microprocessors with a multimedia extension (the short SIMD execution model). For the short SIMD parallelism extraction it is essential that, if dependency exists, then the distance between memory references is greater than or equal to the number of data processed in the SIMD register. This implies that some loops that could not be vectorized on traditional vector processors can still be parallelized for the short SIMD execution. In all of these tests the parallelization would be prohibited when actually there is no parallelism restriction relating to the short SIMD execution model.

In this paper we present a new, fast and accurate data dependence test (called D-test) for array references with linear subscripts, which is used in a vectorizing compiler for microprocessors with a multimedia extension. The presented test is suitable for use in a dependence analyzer that is organized as a series of tests, progressively increasing in accuracy, as a replacement for the GCD or Banerjee tests.

1 Introduction

Vectorizing and parallelizing compilers are common in the commercial supercomputer and mini-supercomputer market. These compilers inspect the patterns of data usage in programs, especially array usage in loops, often representing these patterns as data dependence graphs. Using this information compilers can often detect parallelism in the loops or report to the user specific reasons why a particular loop cannot be executed in parallel.

In order to determine which restructuring transformations are legal, data dependence tests are devised to detect those programs or loops whose semantics will be violated by the transformation. Most of the theory behind data dependence testing for array references in loops can be reduced to solving simultaneous Diophantine equations.

In the automatic parallelization of sequential programs, parallelizing compilers rely on subscript analysis to detect data dependencies between pairs of array

M. Daydé et al. (Eds.): VECPAR 2004, LNCS 3402, pp. 527–540, 2005.

references inside loop nests. The data dependence problem is equivalent to the integer programming problem, a well-known NP-complete problem, and so it cannot normally be solved efficiently. A number of data dependence tests have been proposed in the literature [1, 2, 4, 5, 11, 12, 13, 14, 16, 17, 18, 19]. In each test there is a different trade-off between accuracy and efficiency.

The most widely used approximate data dependence tests are the Banerjee inequality test and the GCD test [2, 18, 19]. The major advantage of these tests is their low computational cost, which is linear in terms of the number of variables in the dependence equation. This is the primary reason why they have been adopted by most parallelizing compilers. However, both the Banerjee inequality and GCD tests are necessary but not sufficient conditions for data dependence. When independence cannot be proved, both tests approximate on the conservative side by assuming dependence, so that their use never results in unsafe parallelization.

There are several well-known, exact data dependence analysis algorithms in practice (i.e. the Omega test [11] and the Power test [17]) with varying power and computational complexity. These algorithms have worst-case exponential cost, and they are all based on integer programming techniques. They determine whether an integer solution exists for a system of linear equations.

In this paper we will use the term 'short SIMD' to refer to the SIMD (Single Instruction Multiple Data) parallelism within a register. This type of data parallelism is exhibited by instructions that act on a small set of data packed into a word. Since the same instruction applies to all data in the word, this is form of a short SIMD execution.

This paper is organized as follows. In the Section 2 we give the motivation for the presented work. In the Section 3 we give some well-known definitions and theorems about the dependency problem for subscripted variable which should help the reader to better understand the presented method in the Section 4. In the Section 4 we define the dependency problem for the parallelization regarding to the short SIMD execution model and propose the effective approximate dependence test (D-test) for the defined dependency problem. In the Section 5 we present some preliminary experimental results. In the Section 6 we give comparisons with other approaches. And finally in the Section 7 we outline some conclusions and give the proposal for the future works.

2 Motivation

The increasing demand for multimedia applications has resulted in the addition of short SIMD instructions to the microprocessor's scalar instruction set. These extensions, popularly referred to as the multimedia extensions, have been added to most existing general-purpose microprocessors [7, 8, 9, 10].

A number of automatic parallelization algorithms for the short SIMD execution model have been proposed in the literature [3, 6, 15].

In [3] the authors present a vectorizer for the Intel Architecture. The authors claim that the data dependence analyzer is organized as a series of tests, progressively increasing in terms of accuracy as well as time and space cost. First, the compiler tries to prove the memory references are independent by means of simple tests. If the simple tests fail, more expensive tests are used. But, as we will show later, simple tests such as the GCD and Banerjee tests will often fail to prove independence.

A number of good dependency tests have, however, been proposed. Practically all of these tests are based on checking the existence of the integer solution to the dependency problem, which is represented by the set of linear equalities and inequalities. But as we will see in Section 4, when parallelizing loops for the short SIMD execution model we allow the existence of the integer solution to the dependency problem unless the distance between memory references is 'too short'. Thus the Banerjee and GCD tests may fail to prove independence. This could happen very often, as we allow relatively short distances between memory references in very large number of iterations, thus allowing many integer solutions to fall into the iteration space.

As a consequence of the above, we propose a fast and accurate dependency test that is based on checking whether the distance between two dependent memory references is 'large enough' within the iteration space, which will indicate whether parallelization is allowed or not.

3 Dependency Problem

In this section we will give some well-known definitions and theorems about the dependency problem for the subscripted variable, which should help the reader to better understand the method presented in Section 4.

Whether a loop can be parallelized/vectorized or not depends upon the resolution of array aliases. The problem is one of determining whether two references to the same array within a nest of loops can be references to the same element of this array. The main dependence problem is as in Definition 1. The potential parallelization of the loops in the nest depends upon the data dependence problem.

Definition 1. Let $S1$ and $S2$ be the two statements shown in Figure 1. Let $A[f(i_1, i_2, \ldots, i_p)]$ and $A[g(i_1, i_2, \ldots, i_p)]$ be two potentially con icting references to an array A, embedded in a nest of p for bops as in Figure 1. If there are two integer vectors, $(i_1, i_2, \ldots, i_p)$ and $(i_1, i_2, \ldots, i_p)$, satisfying the equation:

$$f(i_1, i_2, \ldots, i_p) = g(i_1, i_2, \ldots, i_p) \tag{1}$$

with

$$l_k \leq i_k \leq u_k \;\; and \;\; l_k \leq i_k \leq u_k \;\; for \;\; 1 \leq k \leq p$$

where l_k and u_k are the bwer and upper bounds for the k-th iteration index i_k, then we say that the statements $S1$ and $S2$ are data dependent.

```
L1:  for i1 = l1 to u1 do
L2:      for i2 = l2 to u2 do
                 .
                 .
Lp:              for ip = lp to up do
S1:                  A(f(i1, i2, ..., ip)) =  ...
S2:                  ...                   =  A(g(i1, i2, ..., ip))
                 endfor
                 .
                 .
         endfor
     endfor
```

Fig. 1. A p-level nested loop

3.1 Linearization

So far we have supposed that the subscripts are linearized, which means that each element of a multidimensional array is denoted by a single linear function of subscripts. Here we will show how to linearize subscripts if we have access to the elements of a multidimensional array. Furthermore, we have assumed a row-major order of data layout in the memory.

Lemma 1. (Zima, 1990)
Let A denote an array of dimension n where the bounds for each dimension are $L_j, U_j : L_j \leq U_j$ for all $1 \leq j \leq n$. Then the element $A(i_1, \ldots, i_n)$ can be accessed by a linearized subscript expression $A(f(i_1, \ldots, i_n))$ where:

$$f(i_1, i_2, \ldots, i_n) = \sum_{j=1}^{n} D_j \cdot i_j, \quad D_j = \begin{cases} 1 & ; j = n \\ \prod_{k=j+1}^{n} (U_k - L_k + 1) & ; 2 \leq j \leq (n-1) \end{cases}$$

3.2 Extreme Values of a Linear Function

Before describing a method to check the dependency in a vectorizing compiler for multimedia-enhanced processors, let us write some basic definitions on linear affine functions and show how to find the maximum and minimum values that an integer affine function $f : \mathbb{R}^n \rightarrow \mathbb{R}$ can take for integer-valued arguments in a bounded region of R^n.

Definition 2. Let $r \in \mathbb{R}$. Then:

$$r^+ = \begin{cases} r & r > 0 \\ 0 & r \leq 0 \end{cases}$$

$$r^- = \begin{cases} 0 & r \geq 0 \\ -r & r < 0 \end{cases}$$

Theorem 1. (Banerjee, 1997)
Let $n \geq 1$. Let L_k, U_k and ϵ_k be integer numbers such that $L_k \leq U_k$ and $0 \leq \epsilon_k \leq U_k - L_k$ for $1 \leq k \leq n$. The minimum and maximum values of the integer linear function $f(\mathbf{x}) = \sum_{k=1}^{n}(a_k x_k + a_k y_k)$, where a_k, b_k are integers for $1 \leq k \leq n$, on the bounded region $\mathcal{P}$, de ned by inequalities $L_k \leq x_k, y_k \leq U_k,\ x_k \leq y_k - \epsilon_k,\ 1 \leq k \leq n$, are:

$$\min_{\mathcal{P}} f(\mathbf{x}) = \sum_{k=1}^{n}\{(a_k + b_k)L_k + b_k\epsilon_k - (U_k - L_k - \epsilon_k)(a_k^- - b_k)^+\}$$

$$\max_{\mathcal{P}} f(\mathbf{x}) = \sum_{k=1}^{n}\{(a_k + b_k)L_k + b_k\epsilon_k + (U_k - L_k - \epsilon_k)(a_k^+ - b_k)^+\}$$

Proof. The proof can be found in [2]. □

3.3 The Banerjee Test

The Banerjee test [2, 19] is based on the Intermediate Value Theorem. It is widely used test that calculates the possible minimum and maximum values that a linear affine function can achieve under given bounds on each variables involved. If the zero does not fall between these extreme values than no dependence exists. If it does than we know only that a real solution to the given linear equation exists. However, we cannot conclude that dependence exists, since there may not be an integer solution to the linear equation.

The Banerjee test is commonly used test because of its efficiency and usefulness at disproving dependencies. But, as we will show in the next few sections, the Banerjee tests would prohibit short SIMD vectorization when there is no vectorization restriction as we very often have the case with many integer solutions within the loop limits although the loop may be parallelized. This is becasue the Banerjee test tends to ignore dependence distance constraints in the dependece problem.

4 D-Test

Assume again the general loop from Figure 1. In the parallelization for the short SIMD execution model we can only vectorize a loop where all the references have stride 1 or stride -1 and the same data type. This is almost always the case for the innermost loop, as this loop rapidly can access to the successive elements in the memory. In the short SIMD processing we can process VL (Vector Length) data simultaneously and these data must be successively stored in the memory. All data must be of the same bit length, b, and the SIMD registers must be $VL \cdot b$ bits long. Thus the problem of SIMD vectorization becomes a problem of whether we can unroll the innermost loop VL times and substitute the simultaneous statements with a single SIMD statement that performs the same operation at a time over VL b-bits data in the $(VL \cdot b)$-bits SIMD register.

For the short SIMD parallelism extraction it is essential that, if dependency exists, then the distance between memory references in the innermost loop is greater than or equal to the number of data processed in the SIMD register (VL). This implies that some loops that cannot be vectorized on traditional vector processors can still be parallelized for the short SIMD execution and thus converted into SIMD code (i.e. Intel MMX or SSE/SSE2 code). It is also essential that elements of an array are accessed sequentially with unit stride within the innermost loop.

With this constraint, the general dependency Diophantine equation becomes:

$$a_p i_p - a_p i_p = a_1 i_1 + \ldots + a_{p-1} i_{p-1} + A - \\ -a_1 i_1 - \ldots - a_{p-1} i_{p-1} - A \tag{2}$$

If we assume that:

$$a_0 = A - A$$

and:

$$\zeta(\mathbf{i}) = a_1 i_1 + \ldots + a_{p-1} i_{p-1} - a_1 i_1 - \ldots - a_{p-1} i_{p-1}$$

then we can rewrite the above equation as:

$$a_p i_p - a_p i_p = a_0 + \zeta(\mathbf{i}) \tag{3}$$

While the innermost loop is running, the integer linear function $\zeta(\mathbf{i})$ is an integer constant, which is denoted by the particular values of the loop indexes $i_1, i_2 \ldots i_{p-1}$ and where the following directions hold:

$$i_1 = i_1, i_2 = i_2, i_3 = i_3, \ldots, i_{p-1} = i_{p-1} \tag{4}$$

The distance between two (possibly) dependent memory references (d) is defined as:

$$d = i_p - i_p \tag{5}$$

and can be expressed from the Equation 3 as follows:

$$d = i_p - i_p = -(a_0 + \zeta(\mathbf{i})) + (a_p - 1)i_p - (a_p - 1)i_p \tag{6}$$

As for the short SIMD parallelism extraction it is essential that, if dependency exists, then the distance between memory references is greater than or equal to the number of data processed in the SIMD register (VL), the following condition must be satisfied:

Criterion 1. The innermost loop can be SIMD vectorized if:

$$\max_{\mathcal{P}}(d) \leq 0 \;\; \vee \;\; \min_{\mathcal{P}}(d) \geq VL \tag{7}$$

where d is the distance between memory references and $\mathcal{P}$ is a given bounded region.

If the Criterion 1 is not satisfied, there could be such a distance between two dependent memory references that would prohibit vectorization.

Also, note that the loop index in the innermost normalized loop can be changed only by ± 1 (as loops are normalized) and we allow only unit stride ($|a_p| = |a_p| = 1$). As a consequence there are four possible cases in Equation 6:

1. $a_p = a_p = 1$:

Theorem 2. Let the equation $a_p i_p - a_p i_p = a_0 + \zeta(\mathbf{i})$ represents the dependency problem in a p-level nested loop with innermost loop bounds $l_p \leq i_p, i_p \leq u_p$, and let $a_p = a_p = 1$. Then the innermost loop can be SIMD vectorized if:

$$\max(-(a_0 + \zeta(\mathbf{i}))) \leq 0$$

or

$$\min(-(a_0 + \zeta(\mathbf{i}))) \geq VL$$

Proof. Since the distance between memory references (d) is in this case:

$$d = i_p - i_p = -(a_0 + \zeta(\mathbf{i})) \tag{8}$$

the Criterion 1 is satisfied when the inequalities from the Theorem 2 hold. □

2. $a_p = a_p = -1$:

Theorem 3. Let the equation $a_p i_p - a_p i_p = a_0 + \zeta(\mathbf{i})$ represents the dependency problem in a p-level nested loop with innermost loop bounds $l_p \leq i_p, i_p \leq u_p$, and let $a_p = a_p = -1$. Then the innermost loop can be SIMD vectorized if:

$$\max(a_0 + \zeta(\mathbf{i})) \leq 0$$

or

$$\min((a_0 + \zeta(\mathbf{i})) \geq VL$$

Proof. Since the distance between memory references (d) is in this case:

$$d = i_p - i_p = (a_0 + \zeta(\mathbf{i})) \tag{9}$$

the Criterion 1 is satisfied when the inequalities from the Theorem 3 hold. □

3. $a_p = -a_p = 1$:

Theorem 4. Let the equation $a_p i_p - a_p i_p = a_0 + \zeta(\mathbf{i})$ represents the dependency problem in a p-level nested loop with innermost loop bounds $l_p \leq i_p, i_p \leq u_p$, and let $a_p = -a_p = 1$. Then the innermost loop can be SIMD vectorized if:

$$\max(a_0 + \zeta(\mathbf{i}) - 2i_p) \leq 0$$

or

$$\min((a_0 + \zeta(\mathbf{i}) - 2i_p) \geq VL$$

Proof. Since the distance between memory references (d) is in this case:

$$d = i_p - i_p = (a_0 + \zeta(\mathbf{i})) - 2i_p \tag{10}$$

the Criterion 1 is satisfied when the inequalities from the Theorem 4 hold. □

4. $a_p = -a_p = -1$:

Theorem 5. *Let the equation $a_p i_p - a_p i_p = a_0 + \zeta(\mathbf{i})$ represents the dependency problem in a p-level nested loop with innermost loop bounds $l_p \leq i_p, i_p \leq u_p$, and let $a_p = -a_p = -1$. Then the innermost loop can be SIMD vectorized if:*

$$\max(-(a_0 + \zeta(\mathbf{i}) + 2i_p)) \leq 0$$

or

$$\min(-(a_0 + \zeta(\mathbf{i}) + 2i_p) \geq VL$$

Proof. Since the distance between memory references (d) is in this case:

$$d = i_p - i_p = -(a_0 + \zeta(\mathbf{i})) - 2i_p \tag{11}$$

the Criterion 1 is satisfied when the inequalities from the Theorem 5 hold. □

5 Experimental Results

We have implemented a simple vectorizing compiler that uses a dependence analyzer in which the D-test is the first test used. If the D-test fails to disprove data dependence then a more powerful test is used. Actually, the second test is based on Fourier-Motzkin elimination (FM) and besides real shadow it calculates dark shadow also. This second test is also used for comparison purposes. We have compared the results obtained with the D-test to the results obtained from the Banerjee test and the exact test based on Fourier-Motzkin (FM) elimination. The results obtained from several example loops showed that the D-test is definitely more accurate than the Banerjee test and that there are some rare cases in which the D-test failed to prove the independence and where some powerful test should be used. The results also proved that the D-test is a good substitution for the Banerjee test in a dependence analyzer that consists of a series of tests, progressively increasing in accuracy.

To illustrate how the presented method is used in our vectorizing compiler, consider the following examples:

Example 1

```
int a[1024][512];

  for (i= 0; i<1024; i+ + )
  {
    for (j= 4; j<512; j+ + )
    {
        a[i][j] = a[i][j- 4*i - 4];
    }
  }
```

The elements of array a are 32-bit integer values and the SIMD register has a length of 128 bits. Thus the vector length, VL, is 4. We wish to vectorize this loop for a SIMD extended processor. To do this we should rst check if such a data dependence exists that prohibits the desired parallelization. We should rst linearize the subscripted variable a to obtain the dependence equation and then express the distance between memory references (d) (note that $i = i$, according to the Equation 4):

$$512 \cdot i + j = 512 \cdot i + j - 4 \cdot i - 4$$
$$512 \cdot i + j = 508 \cdot i + j - 4$$
$$d = j - j = 4 \cdot i + 4$$

Now, we should check whether $\min(d)$ and $\max(d)$ satisfy the conditions given by Theorem 2 on the bounded region given by $2 \leq i \leq 1024$:

$$\min_{2\leq i\leq 1024}(d) = \min_{2\leq i\leq 1024}(4 \cdot i + 4) = 12$$

$$\max_{2\leq i\leq 1024}(d) = \max_{2\leq i\leq 1024}(4 \cdot i + 4) = 4100$$

As the $\min(d) = 12 > VL$ holds, the Criterion 1 is satis ed. When we applied the Banerjee, D- and FM-based tests we got the following answers:

```
****************BANERJEE TEST RESULTS****************
MESSAGE: TRUE DEPENDENCE ASSUMED

*******************d-TEST RESULTS******************
D-test duration: 0.000000065900000
MESSAGE: NO DEPENDENCE

*******************FM TEST RESULTS*****************
REAL SHADOW: EMPTY,  EXACT PROJECTION
DARK SHADOW: EMPTY
INDEPENDENCE HAS BEEN PROVEN
Duration: 0.000011120000
```

Note: In the above loop Banerjee test fails to prove the independence. D-test proves the independence with the distance between memory references that is greater than VL (VL = 4). FM-based test also disproves the dependence (the dark and the real shadows are empty) but D-test is in this case approximately 150x faster (in our implementation of both tests). Thus the above loop can be safely vectorized with the SIMD instruction set.

Example 2

```
int a[1024][512][1024];

  for (i= 0; i<512; i++)
  {
    for (j= 128; j<512; j++)
    {
      for (k= 512; k<1024; k++)
      {
        a[i][j][k] = a[i][j-8*i][j+ 4*i+ k-10];
      }
    }
  }
```

The elements of array a are 32-bit integer values and the SIMD register has a length of 128 bits. Thus the vector length, VL, is 4. We wish to vectorize this loop for a SIMD extended processor. To do this we should rst check if such a data dependence exists that prohibits the desired parallelization. We should rst linearize the subscripted variable a to obtain the dependence equation and then express the distance between memory references (d) (note that $i = i$, $j = j$, according to the Equation 4):

$$1024 \cdot 512 \cdot i + 1024 \cdot j + k = 1024 \cdot 512 \cdot i \; + \\ +1024 \cdot (j \; - 8 \cdot i \;) + \\ +j \; + 4 \cdot i \; + k \; - 10$$

$$d = k \; - k = 8188 \cdot i - j + 10$$

Now, we should check whether $\min(d)$ and $\max(d)$ satisfy the conditions given by Theorem 2 on the bounded region given by $0 \leq i \leq 512, 128 \leq j \leq 512$:

$$\min_{0 \leq i \leq 512, 128 \leq j \leq 512}(d) = \min_{0 \leq i \leq 512, 128 \leq j \leq 512}(8188 \cdot i - j + 10) = -502$$

$$\max_{0 \leq i \leq 512, 128 \leq j \leq 512}(d) = \max_{0 \leq i \leq 512, 128 \leq j \leq 512}(8188 \cdot i - j + 10) = 4192138$$

As $\min(d) < VL < \max(d)$ holds, the Criterion 1 is not satisfied. In this case the D-test failed to disprove data dependence, so we should assume data dependence unless we disprove it with some more powerful test.

Example 3

```
int a[128][64][128];
int i, j ,k;

   for (i= 0; i<64; i++)
   {
     for (j= 32; j<128; j++)
     {
       for (k= 64; k<128; k++)
       {
          a[i][j][k] =
               a[i+1][2*i - 4*j - 32][- 2*i + 4*j + k - 32];
       }
     }
   }
```

The elements of array a are 32-bit integer values and the SIMD register has a length of 128 bits. Thus the vector length, VL, is 4. We wish to vectorize this loop for a SIMD extended processor. To do this we should first check if such a data dependence exists that prohibits the desired parallelization. We should first linearize the subscripted variable a to obtain the dependence equation and then express the distance between memory references (d) (note that $i = i$, $j = j$, according to the Equation 4):

$$64 \cdot 128 \cdot i + 128 \cdot j + k = 64 \cdot 128 \cdot (i + 1) + \\ +128 \cdot (2 \cdot i - 4b \cdot j - 32) + \\ +(-2 \cdot i + 4 \cdot j + k - 32)$$

$$d = k - k = -254 \cdot i + 636 \cdot j - 4064$$

Now we should check whether $\min(d)$ and $\max(d)$ satisfy the conditions given by Theorem 2 on the bounded region given by $0 \leq i \leq 64, 32 \leq j \leq 128$:

$$\min_{0 \leq i \leq 64, 32 \leq j \leq 128}(d) = \min_{0 \leq i \leq 64, 32 \leq j \leq 128}(-254 \cdot i + 636 \cdot j - 4064) = 32$$

$$\max_{0 \leq i \leq 64, 32 \leq j \leq 128}(d) = \max_{0 \leq i \leq 64, 32 \leq j \leq 128}(-254 \cdot i + 636 \cdot j - 4064) = 77344$$

As the $\min(d) = 32 > VL$ holds, the second condition in Theorem 2 is satisfied. When we applied the Banerjee, D- and FM-based tests we got the following answers:

```
****************BANERJEE TEST RESULTS****************
MESSAGE: TRUE DEPENDENCE ASSUMED

********************d-TEST RESULTS******************
D-test duration: 0.00000012320000
MESSAGE: NO DEPENDENCE

********************FM TEST RESULTS******************
REAL SHADOW: EMPTY,  EXACT PROJECTION
DARK SHADOW: EMPTY
INDEPENDENCE HAS BEEN PROVEN
Duration: 0.0005230100000
```

Note: In the above loop Banerjee test fails to disprove the existence of dependence. D-test proves the independence with the distance between memory references that is greater than VL (VL = 4). FM-based test also proves the independence but D-test is in this case approximately 400x faster (in our implementation of both tests). Thus the above loop can be safely vectorized with the SIMD instruction set. instruction set.

6 Comparisons with Other Tests

Before describing other tests we would like to emphasize an important property that arises from the short SIMD processing model. In the vectorizer for the multimedia extension we allow the existence of the integer solutions in the bounded region as long as the distance between memory references is greater than the SIMD register length or the distance between memory references causes only anti-dependence. The following methods are based on finding an integer solution to the dependence equation. They all assume the dependence if there are solutions to the dependence equation. Because SIMD registers are relatively short, some loops that could not be vectorized on traditional vector processors can still be parallelized for the short SIMD execution. This is the reason why these methods are not so practical, or even unusable, in our problem.

The GCD [19] test is based upon a theorem of elementary number theory which says that Equation 1 has an integer solution if an only if $gcd(a_1, a_2, \ldots, a_n)$ divides a_0. As coefficients a_p and a_p are ± 1 then the $gcd(a_1, a_2, \ldots, a_n) = 1$ which always divides a_0. The GCD test would thus always prohibit short SIMD parallelization in our case. Note also that the GCD test ignores the loop limits entirely and determines whether Equation 1 has the integer solution for any integer value of all variables.

The Banerjee test [2, 19] takes the loop limits into consideration but determines whether Equation 1 has a real solution within the loop limits. The Banerjee test will prohibit parallelization if Equation 1 has a real solution within the loop limits but no integer solution within the loop limits. In our problem we very often have the case with many integer solutions within the loop limits although the loop may be parallelized. In such a case the Banerjee test would prohibit short SIMD parallelization.

The problem is that neither the GCD test, nor the Banerjee test take the distance between memory references into consideration. In these tests the parallelization would be prohibited when actually there is no parallelism restriction relating to the short SIMD execution model (again, integer solutions may exist as long as the distance between memory references is "large enough"). The distance between memory references can be taken into account only in exact tests as an additional inequality when solving the dependence system. The D-test, in the other hand, is an approximate test that takes the distance between memory references into account and thus it can be a good substitution for the Banerjee test. It could prove the independence much faster than an exact test. Thus, we need only to use an exact test after the D-test fails.

7 Conclusions

In this paper we have presented an accurate and simple method for a dependency check for linear array references within a nested loop that is used in a vectorizing compiler for microprocessors with a multimedia extension. The presented method uses only integer calculations.

Whereas other approximate tests turn out to be inadequate for this type of problem, the D-test is accurate. The D-test is based on checking the minimum and maximum distances between two dependent memory references within the iteration space rather than searching for the existence of an integer solution to the dependency equation. This results in the reduction of the time cost as it computes only the bounds of an integer affine function.

The test assumes that the loop bounds are known in the compilation time. As the test computes the minimum acceptable distance between memory references for which the loop is still parallelizable, we can produce two different code segments in the parallelized program in the case of unknown loop bounds during the compilation time. In the execution time, when the loop bounds are known, we can calculate the actual values of the minimum and maximum distances between memory references that may occur in the iteration space and decide which code should be executed. As the calculation of these values is straightforward it would not affect the execution time of a program.

D-test is suitable for use in a dependence analyzer that is organized as a series of tests, progressively increasing in accuracy, as a replacement for the GCD or Banerjee tests.

References

1. Banerjee U., Eigenman R., Nicolau A., Padua D.A. Automatic Program Parallelization. Proceedings of the IEEE. 81 (2), 211-243, 1993.
2. Banerjee U. Dependence Analysis: A Book Series on Loop Transformations for Restructuring Compilers. Kluwer Academic Publishers, Dordrecht, 1997.

3. Bik A.J.C., Girkar M., Grey P.M., Tian X.M. Automatic Intra-Register Vectorization for the Intel (R) Architecture. International Journal of Parallel Programming. 30 (2), 65-98, 2002.
4. Chang W.L., Chu C.P. The Generalized Direction Vector I Test. Parallel Computing. 27 (8), 1117-1144, 2001.
5. Kong X., Klappholz D., Psarris K. The I Test: An Improved Dependence Test for Automatic Parallelization and Vectorization. IEEE Transactions on Parallel and Distributed Systems. 2 (3), 342-349, 1991.
6. Krall A., Lelait S. Compilation Techniques for Multimedia Processors. International Journal of Parallel Programming. 28 (4), 347-361, 2000.
7. Lee R. Accelerating Multimedia with Enhanced Processors. IEEE Micro. 15 (2), 22-32, 1995.
8. Lee R., Smith M.D. Media Processing: A New Design Target. IEEE Micro, 1996. 16 (4), 6-9.
9. Oberman S., Favor G., Weber F. AMD 3DNow! Technology: Architecture and Implementation. IEEE Micro. 19 (2), 37-48, 1999.
10. Peleg A., Weiser U. MMX Technology Extension to the Intel Architecture. IEEE Micro. 16 (4), 42-50, 1996.
11. Pough W. A Practical Algorithm for Exact Array Dependence Analysis. *Communications of the ACM*. Vol. 35, No. 8, pp. 102-114, 1992.
12. Psarris K., Klappholz D., Kong X. On the Accuracy of the Banerjee Test, Shared Memory Multiprocessors (special issue). Journal of Parallel and Distributed Computing. 12, 152-157, 1991.
13. Psarris K., Klappholz D., Kong X. The Direction Vector I Test. IEEE Transactions on Parallel and Distributed Systems. 4 (11), 1280-1290, 1993.
14. Psarris K. The Banerjee-Wolfe and GCD Tests on Exact Data Dependence Information. Journal of Parallel and Distributed Computing. 32, 119-138, 1996.
15. Sreraman N., Govindarajan R. A Vectorizing Compiler for Multimedia Extensions. International Journal of Parallel Programming. 28 (4), 363-400, 2000.
16. Wolfe M.J., Banerjee U. Data Dependence and its Application to Parallel Processing. International Journal of Parallel Programming. 16 (2), 137-178, 1987.
17. Wolfe M.J., Tseng C.W. The Power Test for Data Dependence. IEEE Transactions on Parallel and Distributed Systems. 3 (5), 591-601, 1992.
18. Wolfe M.J. High Performance Compilers for Parallel Computing. Addison-Wesley Publishing Company, 1996.
19. Zima H.P., Chapman B.M. Supercompilers for Parallel and Vector Computers. Addison-Wesley Publishing Company, 1990.

A Preliminary Nested-Parallel Framework to Efficiently Implement Scientific Applications

Arturo González-Escribano[1], Arjan J.C. van Gemund[2], Valentín Cardeñoso-Payo[1], Raúl Portales-Fernández[1], and Jose A. Caminero-Granja[1]

[1] Dept. de Informática, Universidad de Valladolid,
E.T.I.T. Campus Miguel Delibes, 47011 - Valladolid, Spain
Phone: +34 983 423270
arturo@infor.uva.es

[2] Parallel and Distributed Systems groep (PGS, prof.dr.ir. Henk J. Sips),
Faculty of Information Technology and Systems (ITS),
P.O.Box 5031, NL-2600 GA Delft, The Netherlands
Phone: +31 15 2786168
a.vgemund@et.tudelft.nl

Abstract. Nested-parallel programming models, where the task graph associated to a computation is series-parallel, present good analysis properties that can be exploited for scheduling, cost estimation or automatic mapping to different architectures.

In this work we present a preliminary framework approach to exploit some of these advantages. In our framework we reconstruct an application task graph from a high-level specification, where no scheduling or communication details are yet expressed. The obtained synchronization structure determines which mapping modules or back-ends are used to port the application to an specific platform.

The first results obtained with our prototype show that even simple balancing techniques for irregular scientific applications may be easily integrated in this nested-parallel framework, to obtain efficient implementations from high-level and portable specifications.

Topic: Parallel and Distributed Computing.

1 Introduction

A common practice in high-performance computing is to program applications in terms of the low-level concurrent programming model provided by the target machine, trying to exploit the maximum possible performance. Portable APIs, such as message-passing interfaces (e.g. MPI, PVM) propose an abstraction of the machine architecture, still obtaining good performance. However, programming in terms of these unrestricted coordination models can be extremely error-prone and inefficient, as the synchronization dependencies that a program can generate are complex and difficult to analyze by humans or compilers [12]. Important

M. Daydé et al. (Eds.): VECPAR 2004, LNCS 3402, pp. 541–555, 2005.

decisions in the implementation trajectory, as scheduling or data-layout become extremely difficult to optimize. Considering these problems, more abstract parallel programming models, which restrict the possible synchronization and communication structures available to the programmer, are proposed and studied (see e.g. [20]). These models, due to the restricted synchronization structures, are easier to understand and program, and can provide tools and techniques that help in mapping decisions.

Nested-parallelism models present a middle point between expressiveness, complexity and easy of programming [20]. They restrict the coordination structures and dependencies to those that can be represented by series-parallel (SP) task-graphs (DAGs). Thus, nested-parallelism is also called SP programming. Due to the inherent properties of SP structures [21], they provide clear semantics and analyzability characteristics [14], a simple compositional cost model [9,16, 19] and efficient scheduling [1,7]. These properties can lead to automatic compilation techniques that increase portability and performance. Examples of parallel programming models based on nested-parallelism include BSP [22], nested BSP (e.g. NestStep [13], PUB [2]), BMF [19], skeleton based (e.g. SCL [4], Frame [3], OTOSP/LLC [5]), SPC [9], and Cilk [1].

In previous work (see e.g. [11]) we have shown that many applications classes, including some typical irregular scientific applications, may be efficiently mapped to nested-parallelism due to some inherent load or synchronization regularities, or using simple balancing techniques. Algorithmic graph transformations to map non-SP task graphs into nested-parallel models were also introduced. See also [17] for a theoretical approach.

In this work we present a preliminary and flexible framework approach to exploit the advantages of nested-parallelism. Our framework may integrate different scheduling and mapping techniques spread across other languages and compilers into the same tool. It may automatically apply different compilation techniques driven by the synchronization structure of the application. Different back-ends may also be selected to obtain the maximum efficiency and portability across different distributed-memory or shared-memory platforms. We also propose an structured XML intermediate representation to describe programs at high level, which simplifies the application graph reconstruction and opens the possibility of developing front-ends to translate legacy code written in any other nested-parallel language. We are using a preliminary prototype to implement examples of representative scientific applications; including macro-pipeline, cellular-automata, FFT, LU reduction, and a structural engineering application based on an PDE solver using an iterative method on sparse-matrices. The first results point out that our system may map simple high-level and nested-parallel specifications of these applications to low-level implementations, with efficiency similar as MPI manually developed codes. In particular, we present an example of using a graph partitioning library to balance the irregular sparse-matrix PDE solver, and make it suitable for our nested-parallel programming framework. We also discuss performance effects in a heterogeneous Beowulf cluster.

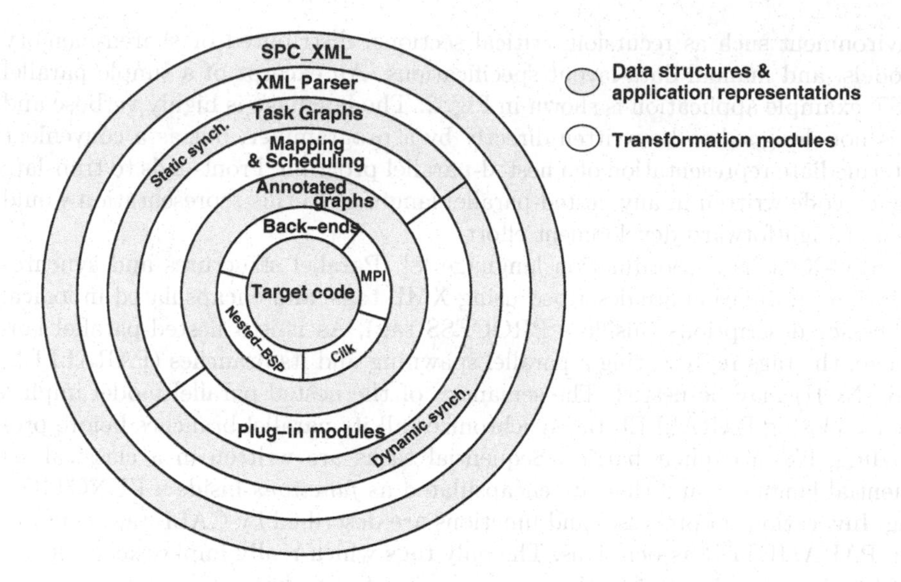

Fig. 1. Layered levels of representation and transformation tools

The paper is organized as follows: Section 2 introduces our prelimminary framework proposal. In section 3 we present some implementation details of a simple prototype of this framework. The section 4 discusses the methodology to implement an irregular application example using the framework. Description of the experiments and results obtained with this application are presented in section 5. Finally, section 6 draws our conclusion.

2 Preliminary Framework Proposal

We propose an approach to a framework with the layered structure shown in Fig. 1. A different representation language is used at each level of abstraction, along the mapping trajectory. We introduce a new nested-parallel coordination language at the highest level. Graphs, representing the program synchronization structure are used at the intermediate layers. Finally, at the lower level, a target code will be generated in a common programming language with parallel support (e.g. using message-passing libraries). Between levels, different transformation tools may be activated accordingly to the mapping decisions. Such decisions may be guided by the application structure, details about the target architecture and cost models.

2.1 SPC-XML Program Representations

In our proposal we introduce a new intermediate highly-structured coordination language based on XML, called SPC-XML. This full parallel programming language is designed to support any feature to be found in a nested-parallel environment such as recursion, critical sections, distributed or shared-memory models, and manual data-layout specifications. An excerpt of a simple parallel FFT example application is shown in Fig. 2. The language is highly verbose and it is not designed to be written directly by a programmer, but as a convenient intermediate representation of a nested-parallel program. Front-ends to translate legacy code written in any nested-parallel language to this representation would be a straightforward development effort.

SPC-XML is a coordination language [8]. Parallel structures and synchronization information are described using XML tags, and encapsulated in logical *processes* descriptions (inside a PROCESS tag). As it is a nested-parallel language, the tags representing a parallel spawning and its branches (PARALLEL, BRANCH) may be nested. The semantics of the nested-parallel model implies that a closing PARALLEL tag synchronizes all its parallel branches before proceeding, like a explicit barrier. Sequential tasks are written in a classical sequential language, and they are encapsulated as *functions* inside a FUNCTION tag. Invocations of processes and functions are described by CALL tags containing PARAMETER associations. The only tags which really imply execution are CALL tags referring to function names (sequential tasks).

SPC-XML variables are generic arrays of a base type available on the sequential language used. The actual dimensions of a variable are readable attributes. Variables are declared inside the LOCAL tag of a process. Special variables called *overlaps* are supported to define logical data partitions over other SPC-XML array variables. Thus, simple data-layouts may be devised.

The input/output interface to processes or functions is described in an unified way. The formal parameters are defined by the tags IN,OUT,INOUT declaring explicitly their interface behavior as *input, output* or *input and output* parameters. This information may be used to automatically determine dependencies at the mapping stage. Communication details are then hidden to the programmer; they are implicit on the task decomposition and calling interfaces.

SPC-XML provides a generic distributed-memory model. A process or function works with local copies of the parameters obtained from the caller. However, when the programmer knows it is safe to work with shared-memory instead of creating local copies of a variable, she may give a hint to the compiler, using the special *shared* attribute on that variable. Nevertheless, the programmer should not rely on shared-memory for communication as the underlying architecture or back-end may not support it. The only communications between parallel tasks should be driven through processes/functions interfaces.

The language includes some more features to help in the mapping trajectory. For instance, the programmer, or a profiling tool, may provide sequential functions with an optional *workload* estimation attribute as a hint to help the scheduling modules.

Although its main functionalities and semantics are clearly defined, SPC-XML is still syntactically evolving to find a more mature level. Even if SPC-XML is intended as an intermediate representation of nested-parallel applications, classical XML tools may help in the visualization, editing and consistency check of

```
<SPC-XML> <!-- *** FFT example *** -->
  <TITLE>fft</TITLE>
  <INCLUDE name="fftLowLevelDefinitions.h"/>

  <FUNCTION name="seqFFTstage" workload="dataArray_dim">
    <IN name="dataArray" baseType="double" />
    <CODE> <!-- C code to compute an FFT stage on size elements -->
      ...
    </CODE>
  </FUNCTION>

  <MAIN>
    <LOCAL> <VAR name="data" baseType="double" dim="65536" /> </LOCAL>
    <BODY>
      <CALL name="initData"> <PARAMETER name="data" /> </CALL>
      <CALL name="parallelFFT"> <PARAMETER name="data" /> </CALL>
    </BODY>
  </MAIN>

  <!-- Parallel-Recursive FFT computation -->
  <PROCESS name="parallelFFT">
    <INOUT name="data" baseType="double" />
    <LOCAL> <VAR_OVERLAP name="splitted" baseType="double" parts="2" /> </LOCAL>
    <BODY>
      <IF> <CONDITION> (data_dim \&gt; 2) </CONDITION> <THEN>
        <OVERLAP source="data" target="splitted" cutPoints="(data_dim/2)" />
        <PARALLEL>
          <BRANCH>
            <CALL name="parallelFFT"> <PARAMETER name="splitted[0]" dim="(data_dim/2)" />
            </CALL>
          </BRANCH>
          <BRANCH>
            <CALL name="parallelFFT"> <PARAMETER name="splitted[1]" dim="(data_dim/2)" />
            </CALL>
          </BRANCH>
        </PARALLEL>
      </THEN> </IF>
      <!-- Sequential stage computation -->
      <CALL name="seqFFTstage"> <PARAMETER name="data" dim="data_dim" /> </CALL>
    </BODY>
  </PROCESS>
</SPC-XML>
```

Fig. 2. SPC-XML example: Excerpt of a parallel FFT program representation

SPC-XML documents. Moreover, using structured XML tags to express coordination simplifies the identification of parallelism. The syntaxis and semantics are easy to understand and to teach to new developers.

2.2 Parsing and Graph Reconstruction

The lexical/syntactical parsing of an SPC-XML document may be completely done by generic and portable XML parsers. Due to the clear semantics and highly structured form of the documents, an application graph is easily reconstructed from its high-level specification. Task decomposition is simple, and in the high-level of abstraction the dependencies are determined directly from the tags structure. We present here the main lines of an unified technique to reconstruct and represent applications by task-graphs.

In this first stage, we find two different main types of application specifications, which lead to different graph representations:

Static: The data partition and layout is statically specified in coarse-grain.
Dynamic: Dynamic or recursive data partitions are specified, to allow simple fine-grain parallel specifications.

The type of a given program is easily determined by the presence or not of flow-control tags with data-dependent conditions. For instance, the FFT program in Fig. 2 implements a recursive partition of the data, controled by an IF conditional tag. The recursion level and grain must be chosen by the compiler, or dynamically in run-time, depending on the target machine.

When considering programs in the static class, the number of tasks and the the exact shape of the graph is completely deterministic. In this task-graph class we only consider one type of *nodes* (tasks which may contain computation functions) and *edges* (precedence dependencies which may derive in data-flow at lower implementation levels). The structure of a static application is then reconstructed as a *DAG (Direct Acyclic Graph)* along the following guidelines. An example of their application is shown in Fig. 3.

1. A process invocation is always expanded, inlining its content where the CALL tag is found, doing parameter substitution.
2. A *task-node* contains the function calls found inside the piece of code it represents. One task-node is created for each of the following cases: (a) from the program start-point until an opening PARALLEL tag; (b) between a closing PARALLEL tag, and the next opening one; (c) from the last closing PARALLEL tag to the end of the program.
3. The content of a BRANCH tag is processed as a subgraph. The predecessor of the subgraph root is the task ending at the opening PARALLEL tag which contains the branch. The successor of the subgraph leaf is the task starting at the closing PARALLEL tag which contains the branch.
4. Loops with a deterministic number of iterations are expanded. The closing tag of the loop is in the same task as the opening tag of the same loop in the next iteration.

For dynamic or recursive applications, which are detected by the presence of data-dependent control-flow structures, the exact number of tasks cannot be determined. It depends on the number of available resources (processors), the number of recursive invocations and/or the exact values of the input data. The graphs generated for these cases must contain indications of the potentially generated structure, but cannot yet be fully expanded. Different types of edges are introduced which describe: (a) data-dependent conditional execution of the target subgraph; and (b) recursive or iterative calls –which create graph cycles–. The resulting graph is *not acyclic* in the presence of recursion or data-dependent iterations.

2.3 Mapping and Scheduling Modules

A first subdivision in the compilation chain is then determined by the type of graph obtained: *static* or *dynamic*, depending on the presence or not of dynamic

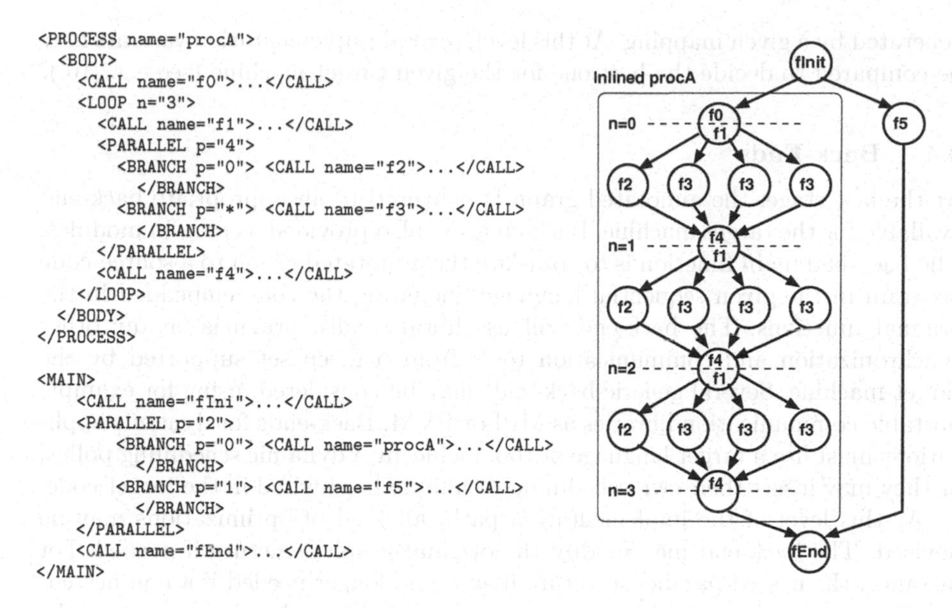
<PROCESS name="procA">
<BODY>
<CALL name="f0">...</CALL>
<LOOP n="3">
<CALL name="f1">...</CALL>
<PARALLEL p="4">
<BRANCH p="0"> <CALL name="f2">...</CALL>
</BRANCH>
<BRANCH p="*"> <CALL name="f3">...</CALL>
</BRANCH>
</PARALLEL>
<CALL name="f4">...</CALL>
</LOOP>
</BODY>
</PROCESS>
<MAIN>
<CALL name="fIni">...</CALL>
<PARALLEL p="2">
<BRANCH p="0"> <CALL name="procA">...</CALL>
</BRANCH>
<BRANCH p="1"> <CALL name="f5">...</CALL>
</BRANCH>
</PARALLEL>
<CALL name="fEnd">...</CALL>
</MAIN>
fInit
Inlined procA
n=0
f0
f1
f5
f2
f3
f3
f3
n=1
f4
f1
f2
f3
f3
f3
n=2
f4
f1
f2
f3
f3
f3
n=3
f4
fEnd

Fig. 3. Example of graph reconstruction

synchronization structures. A dynamic mapping strategy is always possible for static applications. However, better compilation techniques may be available for them. In both cases, static and dynamic, further structural details of the graph may be inspected to determine the best suitable mapping strategy The nested-parallel structure of the graphs opens the possibility of using specific cost-models and mapping techniques not available for more generic synchronization forms.

Our framework introduces a *mapping-decision engine*, which allows to *register* different mapping plug-in modules, indicating the structural characteristics of the graph (a) needed, and (b) recommended; to use it. The best matching module will receive the graph as input to modify it or annote its elements with mapping details; mainly resource allocation. The mapping module typically uses a *machine-model* (which will be also specified in another XML application). The simplest machine-model may contain only one data-item: the number of processors; while more complex models may contain other resources information or allocating costs (e.g. communication parameters, different performances on heterogeneous clusters).

Static-application graphs are fully expanded and then perfectly suitable for classical graph-based scheduling techniques. The scheduling algorithm selected will supply the graph with annotations about task-to-processor bindings. Workload annotations introduced in the specification level through the appropriate attributes will be a key-point to obtain efficient schedules and accurate cost models. Mapping decisions are also guided by communication costs. Thus, mapping modules should use the information about the input/output interfaces of functions contained on the nodes to compute the communication-volume generated by a given mapping. At this level, several implementation variants may be compared to decide the best one for the given target machine (see e.g. [10]).

2.4 Back-Ends

At the last stage, the annotated graph is delivered to an appropriate back-end available for the target machine. Back-ends are also provided as plug-in modules. The back-end main function is to translate the annotated graph to a source-code program in the given sequential language, including the code embedded in the original functions. The back-end will use library calls, pragmas or any other synchronization and communication tools from a given set supported by the target machine. Several generic back-ends may be considered, using for example portable communication libraries as MPI or PVM. Back-ends for dynamic applications must use a target language or tool including a dynamic scheduling policy, or they may insert their own scheduling technique hard-wired in the target code.

At this level of the implementation path, all kind of optimizations may be devised. The back-end may modify the original graph structure if needed. For instance, the nested-parallel structure may be no longer needed if a non nested-parallel programming tool (as a message-passing library) will be used at the lower level. Thus, empty synchronization tasks may be eliminated and communications redirected for performance improvement. However, these decisions may be taken with the costs of new data-movements in mind, to avoid degenerating performance due to unbalanced or extra communications. A back-end may provide aggressive as well as conservative levels of optimization. The final target code can be compiled with the appropriate machine-specific tools to generate the executable.

3 Prototype Implementation

In this section we describe the simple prototype we have developed as a first testing tool of our framework structure. The first stage, parsing SPC-XML, is done using a free XML parser using SAX mode (*expat* [6]). Graph reconstruction is guided by the tags structure. A full mapping-decision engine has not yet been implemented. In our prototype, the main class of graph (static or dynamic) directly selects one of the two main mapping modules. The dynamic module is still under development. An alpha version appropriate for a nested-BSP library back-end exists, and a porting to Cilk language is being proposed and studied.

The static mapping module implements a trivial scheduling policy, based on simulating the behavior of a dynamic *Work-Stealing Technique* (see e.g. [1]), during the compilation phase. The module uses the *workload* attribute values to determine task-costs. The communications which will be produced through graph edges are computed using: (a) the information contained in the nodes about actual parameter substitutions; and (b) their role (input/output) in the interfaces. We modify the original scheduling technique to penalize the displacement of locally precomputed or modified data across processors, using a basic communication-load cost-model. The final policy creates reasonable schedules for regular applications such as cellular-automata, or pipelines. The compiler also offers to the developer an estimated cost of the final program.

The only available back-end in our prototype targets to C code with added MPI library calls for communication and synchronization. The back-end generates one different program for each processor, including its allocated tasks and communications in terms of an SMPD model. Replicated tasks, as the ones generated by a loop, are compacted in C loops when possible to avoid inconvenient huge source programs.

Some experimental optimizations may be devised for this back-end. In the prototype they are still fired manually by developers with knowledge of the graph structure and back-end details. These optimizations include simple communications interleave to diminish reception bottle-necks, and elimination of empty synchronization nodes, which in many cases derive on codes with a structure and number of communications very similar to cleverly programmed MPI codes. More complex techniques have been previously used to reorganize communication in nested-parallel programming environments, obtaining further performance improvements (see e.g. [2]).

4 Study Case: Sparse-Matrix PDE Solver Example

To test our framework prototype we have successfully implemented and executed some efficient real and simulated scientific applications (e.g. cellular-automata, FFT, LU reductions). In this paper we present an interesting study-case with more complex implementation problems; an irregular PDE solver for structural engineering. The input of this program is a finite-element model of a real 3D structure, stored as a sparse-matrix. The matrix contains information about which nodes are adjacent in the graph-model. The main routine executes a given number of iterations, updating node values as a function of the previous values from adjacent nodes. Let *oldV* be the array of values computed in a previous iteration, V the new computed values, and M the symmetric adjacency matrix of the input graph-model. The main routine algorithm may be written as follows:

```
DO iteration=1,numIterations
   DO i=1,matrix_dim
      V(i) = f(oldV(i),oldV(j_0),oldV(j_1),...,oldV(j_k)) : M(i,j_i)=1
   ENDDO
ENDDO
```

A node value may be a structured type containing several data, as 3D position, velocity, energy or whatever. In our test implementations we use a simple double precision number. To simulate the load of a real scientific application, we write a dummy loaded function, which issues 8 mathematical library operations per function-input.

When the data are distributed across processors, adjacent nodes in different processors must communicate their values on each iteration. Determining a good data-partition is a key point for load-balancing and minimizing the

ALGORITHM 1

```
Compute partition
CREATE numProc In,Out arrays

DO iteration=1,numIterations
  PARALLEL p=numProcs
    BRANCH default
      compute( In(p), Out(p) )
    END-BRANCH
  END-PARALLEL
  COPY needed values
    from any Out(i) to any In(j)
END-DO
```

ALGORITHM 2

```
Compute partition
CREATE numProc In,Out arrays

DO iteration=1,numIterations
  PARALLEL p=numProcs
    BRANCH default
      compute( In(p), Out(p) )
      COPY needed values
        from Out(p) to any In(j)
    END-BRANCH
  END-PARALLEL
END-DO
```

Fig. 4. Two nested-parallel algorithms for the application example

communication-volume. Several graph-partitioning tools are available. We use a free and state-of-the-art multi-level partitioning software for unstructured graphs: METIS [18]. This software may be applied to a given matrix, and a given number of parts, creating specific SPC-XML code for that graph model. On the other hand, in classical manually-developed MPI code, the partition process is done at run-time. Thus, the same code may be applied to different matrices and number of processors. The consequence of giving up such flexibility is that specific optimizations can be done by the compiler for the communication structure generated by a given input matrix. Moreover, the partitioning stage is taking a relevant initialization time for big real-case matrices. With our approach, this initialization time is moved from run-time to compile-time.

We have tested two front-ends, which generate different SPC-XML implementations. The main algorithms are shown in Fig. 4. In both cases data to be sent is aggregated to avoid multiple messages sent between the same processors. In the first version, we create one `In` and one `Out` array for each processor (array specific sizes are computed by a post-process of the partitioning algorithm output). The `In` array have space for all the values needed on a given partition, neighbors from other partitions included. On each iteration each task receives its initialized In array, and computes new values on its `Out` array. All these results are collected out of the parallel region, and locally copied to the appropriate places of any other `In` arrays, for the next iteration. The second version avoids the copy tasks at the synchronization point, doing the copies of the local results to any `In` subarray immediately, inside the branch. At the low-level, all these copies work like aliases; real data-duplication will be skipped because the In arrays are not used anymore inside the tasks which modify them. In both versions, the synchronization task only receives and redirects the updated subarrays.

The second version is also chosen as a basis for a manually-coded MPI version, which is used as a reference. However, the MPI version is not nested-parallel. Sends and receives are issued for each non-empty output subarray directly to the target processor, without the need of a explicit synchronization between iterations. To obtain a similar solution, we also test our proposed back-end optimization technique which eliminates empty synchronization nodes from the graph, redirecting communications. We give the program version the following identification names:

v1. SPC-XML with copies in the synchronization point
v2. SPC-XML with aliased copies inside branches
v3. SPC-XML with eliminated empty synchronization task
vMPI. Manually developed MPI version (used as reference)

5 Results

In this section we describe our results. The inputs for our programs are real 3D structural engineering models, widely used in benchmarks, obtained from the Harwell-Boeing collection of sparse-matrices (they are available on the Matrix Market web page [15]). An example model, and the structural representation of its sparse-matrix, are shown in Fig. 5. We have selected inputs with up to more than 4,800 nodes and 140,000 non-zero entries in the sparse-matrix. We execute the programs for 50, 100, 150 and 200 iterations. The machine used to measure is a heterogeneous Beowulf cluster built with a 100Mbits/sec Ethernet network, and the following workstations: 1 Intel-Xeon 1.4GHz processor; 3 AMD-800Mhz, 1 AMD-750Mhz, and 5 AMD-500Mhz. When experimenting, we add machines in this order to maximize performance instead of productivity. METIS library functions allow us to include *weight* factors to adapt the number of nodes allocated to each part to the different processor speeds. This balancing is highly important to achieve proper scalability. See the obtained execution times of the MPI reference program, for one example matrix, on the left of Fig. 6.

Let v be a program version, and τ_v the execution time (eliminating start-up) of that program with given input parameters We define the relative performance-loss comparing with the MPI reference program Γ, as:

$$\Gamma_v = \frac{\tau_v}{\tau_{MPI}}$$

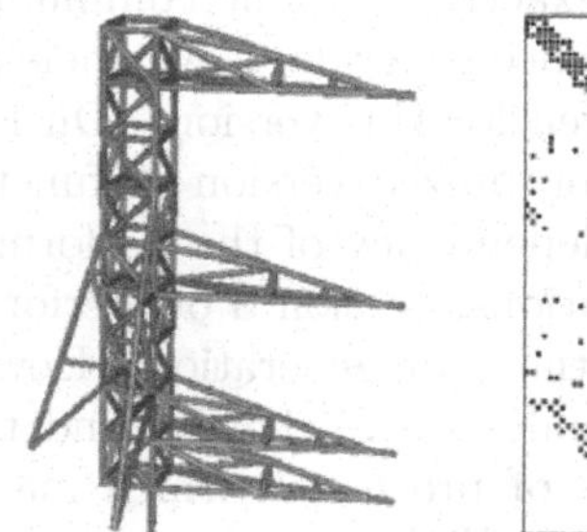

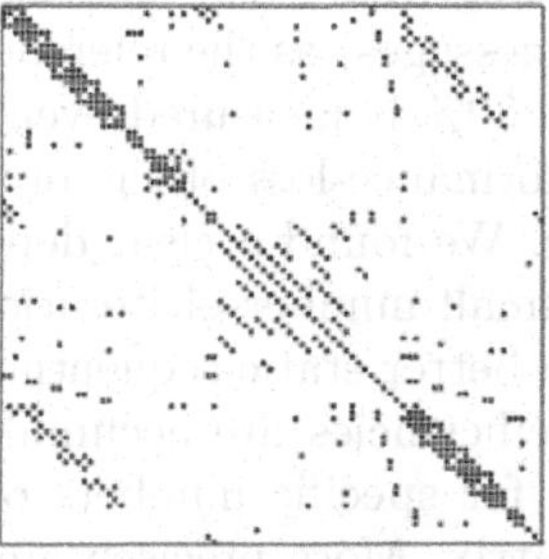

Fig. 5. Example of an input model and its associated sparse-matrix

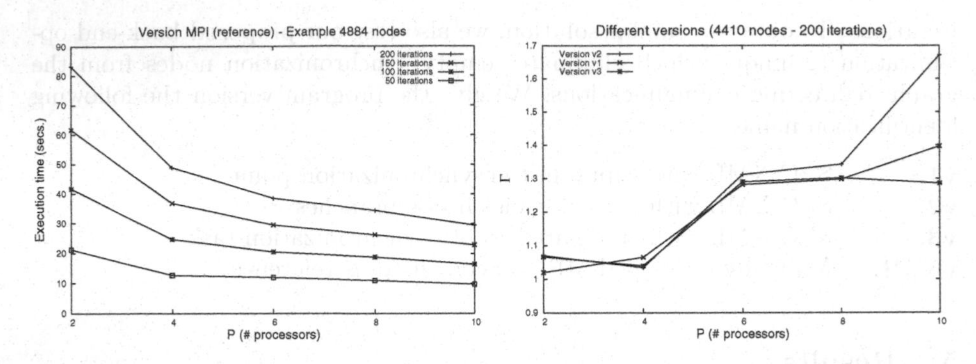

Fig. 6. Scalability and relative performance of versions

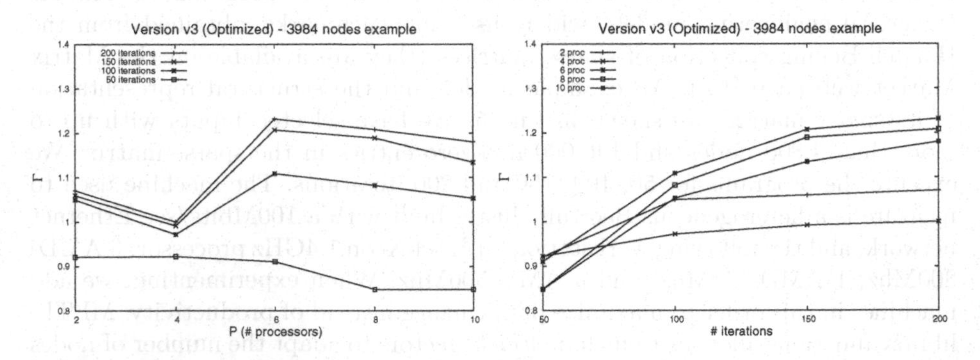

Fig. 7. Relative performance-loss of the optimized version

On the right of Fig. 6 we compare typical Γ values for the three program versions generated with our prototype. Their behavior is similar except when the number of processors approximates the maximum. Version 1 and 2 scale bad, because of the network bottle-neck produced at the synchronization task. This effect is more noticeable on version 2 due to the higher amount of smaller messages used to transfer the same communication-volume. On the other hand, our optimized version 3, which directs the messages to the final target processors, continues on scaling. Indeed, it has exactly the same communication structure (interchange of messages) as the reference program. Nevertheless, a performance penalty of up to 30% is measured even for this version. On Fig. 7 we present the relative performance-loss of the optimized version as function of the number of iterations. We found a clear dependency of the performance-loss on the loop limit. For small number of iterations, version 3 outperforms the reference program (due to better static sequential-code generation). However, on each iteration some inefficiencies are accumulated. We have found that the effect is more noticeable for specific numbers of processors which many times depend on the input matrix. More precisely, in Fig. 8 we show how this effect is correlated with the communication-volume produced by the partition for a given matrix structure. This indicates that our versions include some inefficiencies on

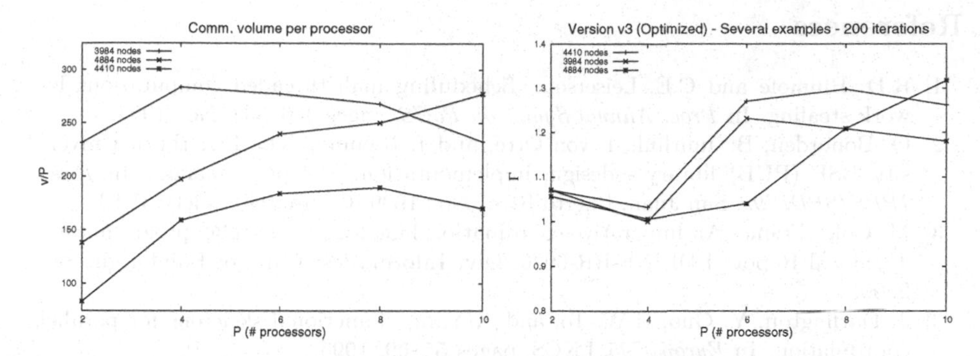

Fig. 8. Performance effects of the communication-volume

the sequential code used for marshaling and/or buffering the lists of data to be communicated. Similar or improved results are also found for more regular applications tested, where the communication-volume is only dependent on the input size. Our framework may automatically produce optimized implementations with the same communication structure as manually programmed MPI codes. A careful framework implementation should be able to avoid the performance losses previously discussed.

6 Conclusion

The preliminary framework proposed in this paper is the base for a flexible tool, which will be able to produce efficient implementations of many important classes of scientific applications from high-level, portable, and nested-parallel structured specifications. Different mapping and scheduling techniques found in other tools may be integrated in this framework. The selection of the best suitable ones for a specific case, may be driven by the structure of the application, as detected on the reconstructed task-graph. Moreover, different back-ends may be provided for maximum efficiency and portability across different platforms. This framework may also derive on a profitable tool for experimental testing of new scheduling and mapping techniques oriented to nested-parallel programming environments.

Future work will include a consistent analysis technique of graph and application characteristics to judiciously select new and more sophisticated mapping and scheduling modules. In particular, modules to map applications to hierarchically distributed clusters or Grids will be of high interest.

Acknowledgements

This work has been partially supported by: JCyL under contract number VA-083/03, the EC(FEDER) and the Spanish MCyT (Plan Nacional de I+D+I, TIC2002-04498-C05-05 and TIC2002-04400-C03). We want to thank the Parallel Computing Group at La Laguna University (Spain) for their support. We also want to thank the anonymous referees for their useful comments.

References

[1] R.D. Blumofe and C.E. Leiserson. Scheduling multithreaded computations by work stealing. In *Proc. Annual Symp. on FoCS*, pages 356–368, Nov 1994.

[2] O. Bonorden, B. Juurlink, I. von Otte, and I. Rieping. The Paderborn University BSP (PUB) library - design, implementation, and performance. In *Proc. IPPS/SPDP'99*, San Juan, Puerto Rico, Apr 1999. Computer Society, IEEE.

[3] M. Cole. Frame: An imperative coordination language for parallel programming. Technical Report EDI-INF-RR-0026, Div. Informatics, Univ. of Edinburgh, Sep 2000.

[4] J. Darlington, Y. Guo, H.W. To, and J. Yang. Functional skeletons for parallel coordination. In *Europar'95*, LNCS, pages 55–69, 1995.

[5] A.J. Dorta, J.A. Gonzlez, C. Rodrguez, and F. de Sande. Llc: A parallel skeletal language. *Parallel Processing Letters*, 13(3):437–448, Sept 2003.

[6] <eXpat/>. The Expat XML parser. WWW. http://expat.sourceforge.net/.

[7] L. Finta, Z. Liu, I. Milis, and E. Bampis. Scheduling UET–UCT series–parallel graphs on two processors. *Theoretical Computer Science*, 162:323–340, Aug 1996.

[8] D. Gelernter and N. Carriero. Coordination languages and their significance. *Communications of the ACM*, 35(2):97–107, Feb 1992.

[9] A.J.C. van Gemund. The importance of synchronization structure in parallel program optimization. In *Proc.* 11th *ACM ICS*, pages 164–171, Vienna, Jul 1997.

[10] A. González-Escribano, A.J.C. van Gemund, and V. Cardeñoso. Predicting the impact of implementation level aspects on parallel application performance. In *CPC'2001 Ninth Int. Workshop on Compilers for Parallel Computing*, pages 367–374, Edinburgh, Scotland UK, Jun 2001.

[11] A. González-Escribano, A.J.C. van Gemund, and V. Cardeñoso-Payo. Mapping unstructured applications into nested parallelism. In J.M.L.M. Palma, J. Dongarra, V. Hernández, and A.A. Sousa, editors, *High Performance Computing for Computational Science - VECPAR 2002*, number 2565 in LNCS, Porto (Portugal), 2003. Springer.

[12] S. Gorlatch. Send-Recv considered harmful? myths and truths about parallel programming. In V. Malyshkin, editor, *PaCT'2001*, volume 2127 of *LNCS*, pages 243–257. Springer-Verlag, 2001.

[13] C.W. Kessler. NestStep: nested parallelism and virtual shared memory for the BSP model. In *Int. Conf. on Parallel and Distributed Processing Techniques and Applications (PDPTA'99)*, Las Vegas (USA), Jun-Jul 1999.

[14] K. Lodaya and P. Weil. Series-parallel posets: Algebra, automata, and languages. In *Proc. STACS'98*, volume 1373 of *LNCS*, pages 555–565, Paris, France, 1998. Springer.

[15] National Institute of Standards and Technology (NIST). Matrix Market. WWW, 2002. On http://math.nist.gov/MatrixMarket/.

[16] R.A. Sahner and K.S. Trivedi. Performance and reliability analysis using directed acyclic graphs. *IEEE Trans. on Software Eng.*, 13(10):1105–1114, Oct 1987.

[17] A.Z. Salamon. *Task Graph Performance Bounds Through Comparison Methods*. MSc.Thesis, Fac. of Science, Univ. of the Witwatersrand, Johannesburg, Jan 2001.

[18] K. Schloegel, G. Karypis, and V. Kumar. *CRPC Parallel Computing Handbook*, chapter Graph Partitioning for High Performance Scientific Simulations. Morgan Kaufmann, 2000.

[19] D.B. Skillicorn. A cost calculus for parallel functional programming. *Journal of Parallel and Distributed Computing*, 28:65–83, 1995.

[20] D.B. Skillicorn and D. Talia. Models and languages for parallel computation. *ACM Computing Surveys*, 30(2):123–169, Jun 1998.
[21] J. Valdés, R.E. Tarjan, and E.L. Lawler. The recognition of series parallel digraphs. *SIAM Journal of Computing*, 11(2):298–313, May 1982.
[22] L.G. Valiant. A bridging model for parallel computation. *Comm.ACM*, 33(8):103–111, Aug 1990.

Exploiting Multilevel Parallelism Within Modern Microprocessors: DWT as a Case Study[1]

C. Tenllado, C. Garcia, M. Prieto, L. Piñuel, and F. Tirado

Departamento de Arquitectura de Computadores y Automática,
Universidad Complutense, 28040 Madrid, Spain
Phone: +34-91-3944625 Fax: +34-91-3944687
{tenllado, garsanca, mpmatias, lpinuel, ptirado}@dacya.ucm.es

Abstract. Simultaneous multithreading (SMT) is being incorporated into modern superscalar microprocessors, allowing several independent threads to issue instructions to the functional units in a single cycle. Effective use of the SMT can hide the inefficiencies caused by long operation latencies, thereby yielding a better utilization of the processor's resources. In this paper we explore techniques to efficiently exploit this capability and its interaction with short-vector processing. We put special emphasis on the differences in algorithm tuning between SMT architectures and shared memory symmetric multiprocessors. As a case study we have chosen the well known Discrete Wavelet Transform (DWT), a central-piece in some image and video coding standards such as MPEG-4 or JPEG-2000.

1 Introduction

As a consequence of the growth in transistor density and some technological constraints [1], there is a clear trend in general-purpose microprocessor toward incorporating additional levels of parallelism. Unlike instruction level parallelism (ILP), the exploitation of such levels is not transparent to the software. Either the programmer or the compiler needs to figure out how to efficiently exploit such levels. In this paper we have mainly focused on simultaneous multithreading (SMT), although we have also analyzed its interaction with short-vector processing.

The availability of short-vector processing within current microprocessors is a consequence of the increasing importance of multimedia, which in recent years has become one of the key drivers of new computer architectures. Whereas media codes typically process small precision data types, general-purpose systems are optimized for processing wider size data. Hence, rather than wasting wider ALUs and datapaths, instruction-set architectures (ISA) have been extended with new instructions (Multimedia ISA Extensions) that can operate on several narrower data items at the same time. In some cases, this level of parallelism can be hidden from the code developer by using a compiler with automatic vectorization capabilities. However, even if the target platform provides such software infrastructure (which is not always

[1] This work has been supported by the Spanish research grant TIC 2002-750.

M. Daydé et al. (Eds.): VECPAR 2004, LNCS 3402, pp. 556–568, 2005.

the case), real applications cannot take advantage of automatic vectorization without some programmer tuning. Successful and well-known examples are the Intel SSE/SSE2 ISA extensions of the Pentium family [2] and the IBM-Motorala PowerPC Altivec [3].

In the high-performance commercial sphere, the traditional focus on ILP is being expanded to include thread-level parallelism (TLP) [1]. Single-chip multiprocessors (CMP) are becoming ubiquitous: IBM's Power4 and Power5 integrate two processors [4], Compaq WRL's proposed Piranha processor had eight [1], and Intel has announced plans to build CMP-based IA-64 processors [1]. On the other hand, most superscalar-style cores likely will have some form of simultaneous multithreading (SMT) capabilities [1]. SMT, as its name suggests, allows several independent threads (with separate architectural states) to execute instructions simultaneously in a single cycle [5]. Its main goal is to yield better utilization of the processor's resources, hiding the inefficiencies caused by long operational latencies. SMT has already been incorporated into the Intel's Pentium 4 and Xeon [6] families and has been announced for use in other microprocessors such as Sun's Ultrasparc V [7] and IBM's Power5 [4].

Despite these design trends, there have been few studies about algorithm tuning in such architectures, and there is even less analysis about the interaction between different levels of parallelism.

Our main concern in this paper is to evaluate its potential benefits of SMT and its synergies with short-vector processing. We also put special emphasis on the differences in algorithm tuning between SMT architectures and shared memory symmetric multiprocessors. As a case study we have chosen the well known 2-D Discrete Wavelet Transform (DWT), a central-piece in some image and video coding standards such as MPEG-4 or JPEG-2000, which presents enough inherent data and functional parallelism to allow for the exploration of different kinds of parallelization strategies.

The rest of this paper is organized as follows. Sections 2 and 3 describe the target algorithm and the computing platform respectively. Section 4 describes the proposed optimizations. The paper ends with experiment results and conclusions.

2 Discrete Wavelet Transform. The Lifting Scheme

The Lifting Scheme [6] is one of the most popular algorithms used to compute the Discrete Wavelet Transform (DWT). As figure 1 shows, its 1-D version consists of four different stages: lazy wavelet transform, prediction, update and normalization. The first stage splits the input signal into even and odd coefficients. It is followed by a set of prediction and update stages that depend on the particular filter used. In the prediction stage, even samples are used to predict the odd ones, which are then replaced by the prediction error. Every prediction stage is followed by an update one in which even elements are modified to maintain certain signal features. Finally, after normalization even coefficients represent a coarse approximation of the original signal, and odd elements store the high frequency information lost when reducing the resolution level (descending a subband). The resulting signals are known as approximation and detail. The full decomposition is obtained by iterating this process on the coarse version of the signal.

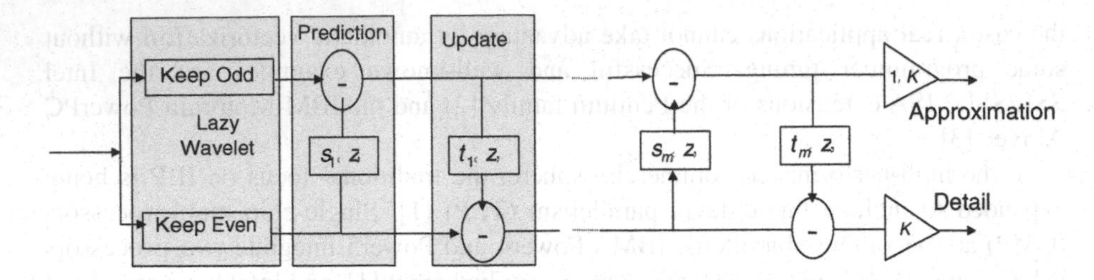

Fig. 1. 1-D Wavelet decomposition using lifting

The DWT is usually expanded to higher dimensions performing separate 1-D transforms along each dimension. In the most common approach, known as the square decomposition, the computation on the image rows and columns is interleaved. One level of the DWT is applied first horizontally (i.e. treating image rows as 1-D signals) and then vertically (on the columns). This process is applied recursively to the quadrant containing the coarse scale approximation in both directions.

For the sake of conciseness, the measurements reported in this paper have been obtained using the popular Daubechies (9,7) biorthogonal filter. Among others important applications, it is employed in the lossy compression version of the JPEG-2000 standard [9]. As equation 1 shows, in this case lifting consists of two prediction (alpha and gamma) and update (beta and delta) stages. As mentioned above, other filters involve a different set of prediction and update stages. Nevertheless, the proposed optimizations can also be applied to other kinds of sets.

$$
\begin{array}{ll}
s_l^{(0)} = x_{2l} & \\
d_l^{(0)} = x_{2l+1} & \\
\hline
d_l^{(1)} = d_l^{(0)} + \alpha(s_l^{(0)} + s_{l+1}^{(0)}) & \textit{prediction} \\
s_l^{(1)} = s_l^{(0)} + \beta(d_l^{(1)} + d_{l-1}^{(1)}) & \textit{update} \\
d_l^{(2)} = d_l^{(1)} + \gamma(s_l^{(1)} + s_{l+1}^{(1)}) & \textit{prediction} \\
s_l^{(2)} = s_l^{(1)} + \delta(d_l^{(2)} + d_{l-1}^{(2)}) & \textit{update} \\
\hline
s_l = \zeta\, s_l^{(2)} & \text{approximation} \\
d_l = s_l^{(2)} / \zeta & \text{detail}
\end{array}
\quad (1)
$$

3 Computing Platform

Our study has been performed on a HP Proliant ML570 G2 server equipped with four Xeon-MP microprocessors. It offers state-of-the-art SIMD extensions [2], symmetric multiprocessing (SMP), and some sort of simultaneous multithreading (commercially named Hyper-Threading by Intel) [5]. Its main characteristics are summarized in Table 1.

Intel's Hyper-Threading allows a single physical processor to act as two logical processors. They share nearly all the resources on the physical processor (caches, execution units, branch predictors, buses…) but maintain a separate set architecture state.

Table 1. Main features of the target computing platform

<table>
<tr><td rowspan="4">Processor</td><td colspan="2">Intel® Xeon-MP 2.8 GHz
FSB: 400MHz</td></tr>
<tr><td>L1 Data Cache</td><td>8KB ; 4-way ;
64 bytes/line ;
1 line/sector</td></tr>
<tr><td>L2 Unified Cache</td><td>512KB ; 8-way ;
64 bytes/line ;
2 lines/sector</td></tr>
<tr><td>L3 Unified Cache</td><td>2MB ; 8-way ;
64 bytes/line ;
2 lines/sector</td></tr>
<tr><td>Main Memory</td><td colspan="2">8GB (PC1600 DDR SDRAM)</td></tr>
<tr><td rowspan="2">Operating system</td><td colspan="2">GNU Debian</td></tr>
<tr><td colspan="2">Linux kernel 2.6.5 (+ perfctr patch)</td></tr>
<tr><td>Intel's ICC v8.0 Compiler Switches</td><td colspan="2">-O3 –tpp7 –xW -ip –restrict
Automatic Vectorization: -vec
Parallelization with OpenMP: -openmp</td></tr>
</table>

From the operating system perspective, our 4-way Xeon sever appears as an 8-way SMP, which allows multiprocessor capable software to run unmodified on twice as many logical processors. This virtual view makes a bit difficult to assess the performance differences between exploiting SMT or SMP. To overcome this problem we have explicitly bound our application threads into specific logical processors using the *sched_setaffinity syscall* [10].

The exploitation of both SMT and SMP has been performed by means of OpenMP, which is directly supported by the Intel ICC compiler [11].We have also made use of this compiler, namely its automatic vectorization capabilities, some intrinsic functions and its disambiguation directives, in order to take advantage of the Intel's SSE/SSE2 ISA extensions [2].

4 Implementation Details: SMT-Aware DWT

The baseline code employed in this research already includes several optimizations inspired by our previous work; namely a memory management optimization denoted as *inplace-mallat* [14], vectorization [12][14] and an alternative way to perform the transformation, denoted as *pipeline computation*, that improves temporal locality and functional parallelism [12].

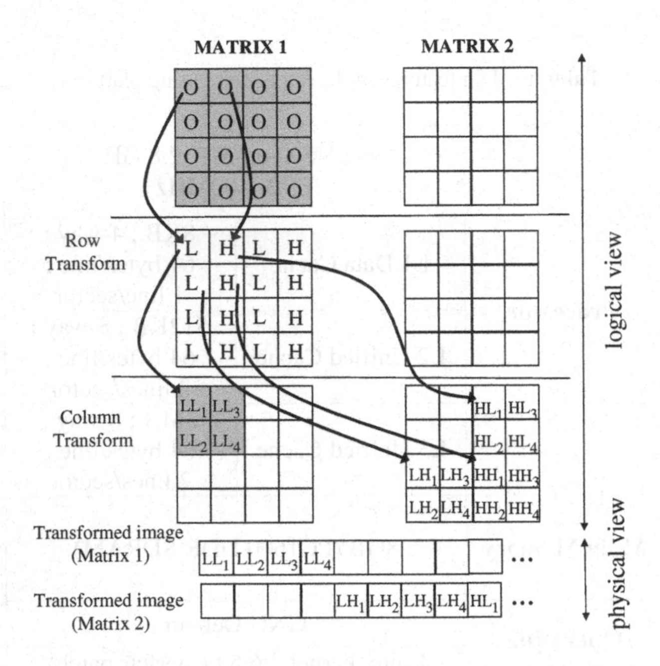

Fig. 2. *Inplace-mallat* strategy (logical view on the top; physical data layout on the bottom)

Lifting allows the computation of the DWT *inplace*, i.e. without allocating auxiliary memory [8]. However, this memory saving is at the cost of scattering the wavelet coefficients throughout the original matrix, which in the context of many applications, such us JPEG 2000 coding [9], may involve a post-processing step where the coefficients are rearranged. In *inplace-mallat* (see figure 2), the horizontal filtering is performed *inplace* but uses an auxiliary matrix to store the final wavelet coefficients. In this way, at the end of the calculations, the different subbands are stored contiguously in memory. Only the low frequency components in each direction (denoted by LL) are stored in the original matrix, whereas the other components (denoted by LH, HL and HH) are moved into the auxiliary matrix in their correct final positions. In particular, we have opted to store each wavelet subband following a column-major layout. This decision allows an efficient semi-automatic vectorization of the horizontal filtering [12], i.e. vectorization is performed introducing some rearrangements at source code level and guiding the compilation process with some directives and compiler's switches [12]. Although, it is also possible to vectorize the vertical filtering using the compiler's intrinsic functions [11], our baseline algorithm does not include it since it does not provide any performance improvement in our target platform. Although we have not addressed these optimizations here, the interested reader can refer to [12][14] for a detailed description.

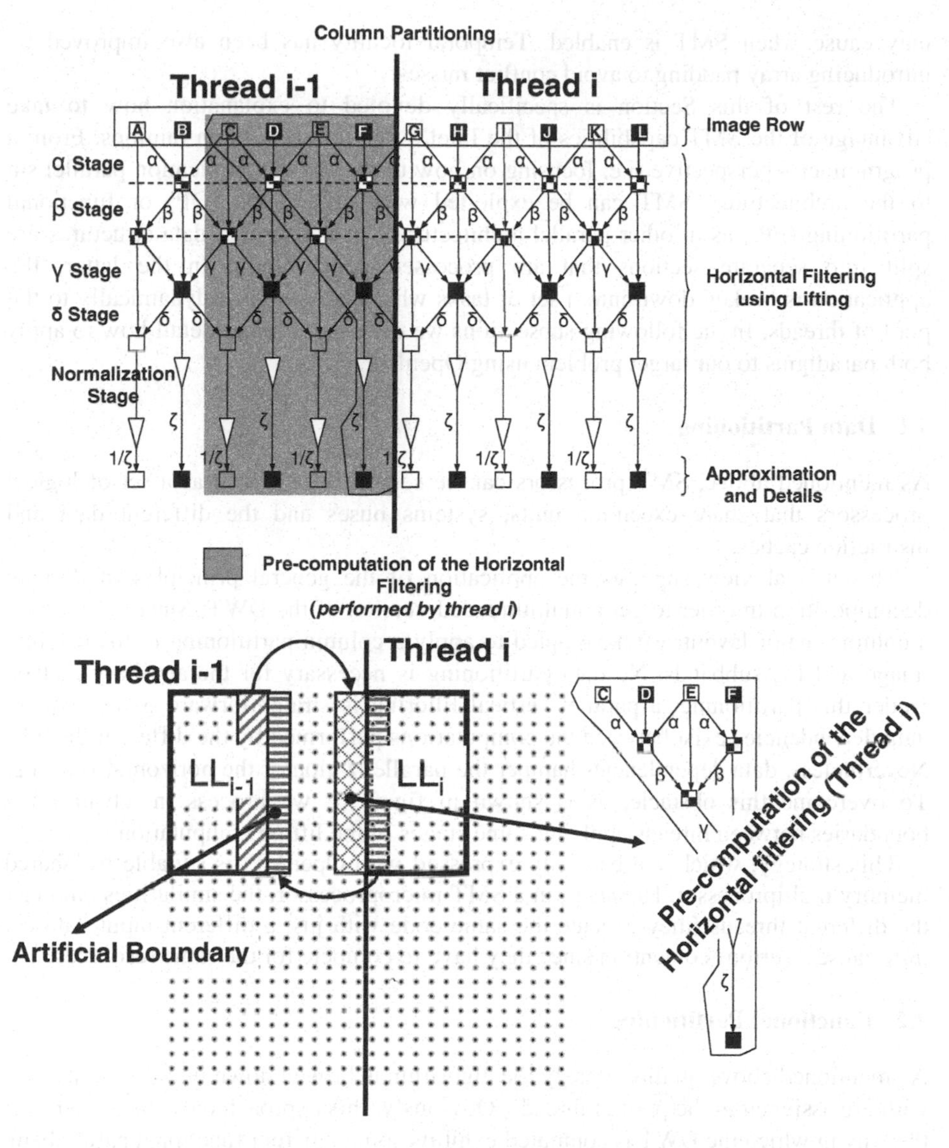

Fig. 3. Column partitioning of the original image (on the top) and LL sub-bands (on the bottom)

In the traditional implementation of the 2-D DWT, which some authors denote as *row-column* scheme [13], the horizontal filtering on a certain decomposition level must be completed before starting the corresponding vertical filtering. Instead of applying this ordering, we have interleaved the horizontal and the vertical filtering. This transformation, denoted as *pipeline computation* [12], enhances the temporal locality of the transform. Nevertheless, we should anticipate that our main interest in filter interleaving is not its memory access improvements but the advantages that it may cause when SMT is enabled. Temporal locality has been also improved by introducing array padding to avoid conflict misses.

The rest of this Section is specifically devoted to explanation how to take advantage of the SMT capabilities of the Intel's Pentium and Xeon families. From a programmer's perspective (i.e. focusing on how to expose the application parallelism to the architecture), SMT can be exploited with either data (DP) or functional partitioning (FP), as in other parallel architectures. In the former, data structures are split into separate sections that are processed concurrently. In the latter, the application is broken down into a set of tasks which are assigned dynamically to the pool of threads. In the following subsections we describe in more detail how to apply both paradigms to our target problem using OpenMP.

4.1 Data Partitioning

As mentioned above, SMT processors can be seen at first glance as a set of logical processors that share execution units, systems buses and the different data and instruction caches.

This logical view suggests the application of the general principles of domain decomposition in order to get a multithreaded version of the DWT. Since we employ a column-major layout, we have opted to apply a column partitioning to the original image and LL subbands. No data partitioning is necessary for the auxiliary matrix. Under this partitioning, a parallel vertical filtering is straightforward given that no data dependencies exist between the computations performed by the different threads. Nevertheless, data dependencies hamper the parallelization of the horizontal filtering. To overcome this obstacle, as is shown in figure 3, we process in advance the boundaries between threads applying some stages of the lifting computation.

This strategy, which can be easily expressed with OpenMP, is suitable for shared memory multiprocessor. However, in a SMT microprocessor, the similarities amongst the different threads (they execute the same code with just a different input dataset) may cause a serious contention since they have to compete for the same resources.

4.2 Functional Partitioning

As mentioned above, in this strategy the algorithm is broken down into different tasks with are assigned to the pool of threads. Obviously, this approach only makes sense if the way in which the DWT is computed exhibits some inherent functional parallelism. This means that only those schemes that allow some sort of filtering interleaving, such as the *pipeline computation* [12], can take advantage of this model. Otherwise, no speedup can be expected.

Figure 4 shows a pseudo-code of this strategy based on the OpenMP's *parallel* and *critical* directives. Firstly, the horizontal filtering is applied to a certain number of columns. This number, which we have denoted as *thread_distance*, specifies how many columns are horizontally filtered before the vertical filtering starts. After this initial stage, each thread acts as scheduler and decides the next task to be performed.

```
// Initial Stage
for( column=0; column < thread_distance; column++)
     Horizontal_Filtering(matrix,column);

#pragma omp parallel private(task, more_tasks) shared(control variables)
{
     more_tasks=true;
     while(more_tasks)
     {
               #pragma omp critical
               Scheduler.next_task(&task);

               if(task.type == HTask)
                         Horizontal_Filtering(matrix,task.column);
               if(task.type == VTask)
                         Vertical_Filtering(matrix,task.column);

               #pragma omp critical
               more_tasks = Scheduler.commit_task(task);
     }
}
```

Fig. 4. Functional partitioning of the pipeline computation using OpenMP

Two kinds of tasks are possible, either the horizontal or the vertical filtering of a certain column (denoted as *HTask* and *VTask* respectively). In the scheduling, it is necessary to guarantee that the vertical filtering of a given column is always performed after its horizontal counterpart. The employment of critical sections is necessary to preserve this correct interleaving.

Although this strategy cannot scale beyond two processors, this drawback is not so critical in this context given that :

1. Current trends suggest that the number of threads per SMT core (the target of this scheme is SMT architectures) is unlikely to increase much beyond the small number currently appearing [1] (two threads in both the Intel's Pentiun4 [6] and IBM's Power5 [4]).
2. It can scale beyond this limit in combination with data partitioning.

Intuitively, this model may present some advantages over the data parallel counterpart. In the former, threads perform the whole DWT on their share of the image and hence, they compete for the memory bandwidth and data caches. Under this model, there is lower competition for the memory bandwidth and data caches since there is no such data partitioning among threads. In fact, when a certain thread (*thread_h*) performs the horizontal filtering of a given column it may act indirectly as a helper thread if the corresponding *VTask* is assigned to a different thread (*thread_v*), i.e. apart from doing its task, *thread_h* also prefetches data for *thread_v*. In other words, whereas in the DP model threads compete for memory resources, in the FP threads they may cooperate. Consequently, TP can improve DP whenever its scheduling overheads are compensated by its better data cache exploitation.

5 Experiment Results

In this section we analyze the performance of the proposed parallelization methods for different configurations.

5.1 SMT-Aware DWT: Data Versus Funtional Partitioning

In this sub-section we evaluate how well data and funtional partitionings exploit SMT (the corresponding configuration has been denoted as 1P-2T: 2 threads in one physical processor). Figure 5 shows the speedup achieved by both strategies over the baseline code (with vectorization enabled) for different image sizes.

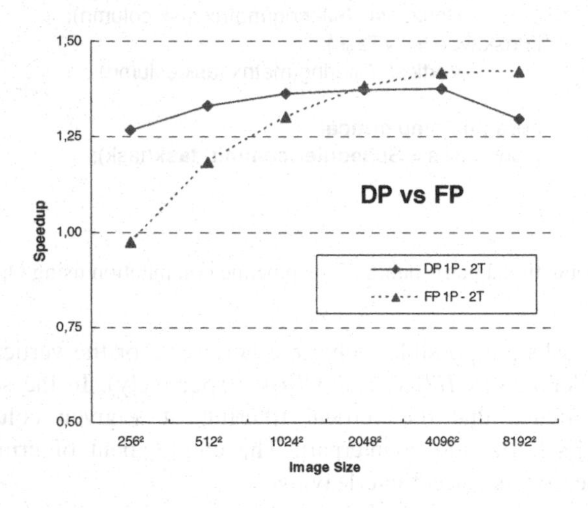

Fig. 5. Data versus functional partitioning when vectorization is enabled

Firstly, we should remark that speedups are quite satisfactory. On average, speedup is close to 40%, which matches other performance analysis carry out about Intel's Hyperthreading technology [6].

As can be noticed, the election of the most suitable strategy depends on the size of the target image. For small and medium size images DP outperforms TP. This is the expected behavior given that for small and medium working sets, memory bandwidth and data cache exploitation are not a key issue and DP beats TP on performance due to the overheads involve in FP's task scheduling. However, for large images we observe the opposite behavior given that the overheads involved in task scheduling become negligible whereas the competition for memory resources becomes critical in the DP version (under FP threads cooperate instead of compete) .

In our target platform, the *breakeven point* between both strategies is a relative large image due to its large L3 cache (2 MB). However, as figure 6 shows, in other platforms with smaller caches (the measurements shown in figure 6 have been obtained on an Intel Pentium 4 processor running at 3.06 GHz equipped with 512 KB L2 Cache), this point is reached with a smaller image size. In addition, DP becomes clearly inefficient for large images.

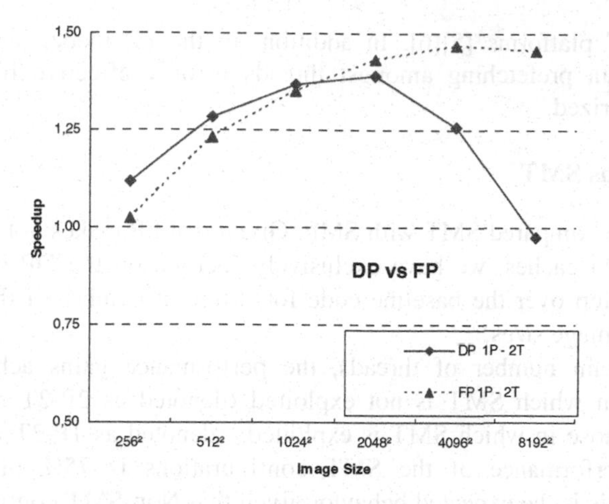

Fig. 6. Data versus functional partitioning when vectorization is enabled in an Intel Pentium 4 processor without third level cache

5.2 Interaction Between Vectorization and SMT

Figure 7 shows the speedup achieved with SMT with and without enabling vectorization. As can be noticed, a certain synergy between both capabilities exists, given that the performance gain achieved with SMT is higher when vectorization is enabled.

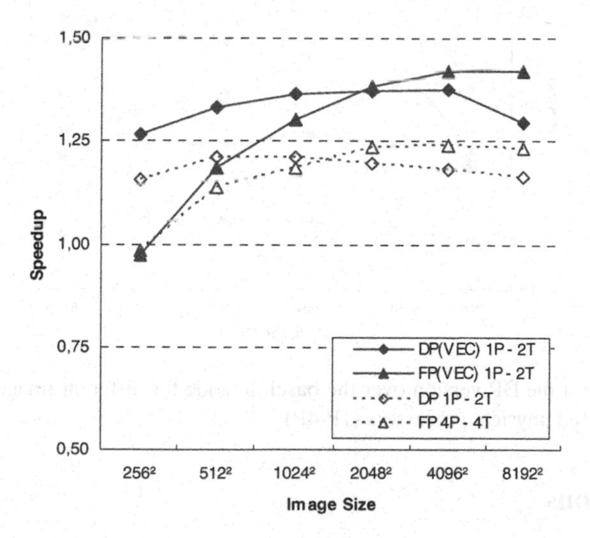

Fig. 7. SMT exploitation with and without enabling vectorization

One of the reason for this behavior is that vectorization reduces the number of load instructions [12], enhancing the use of the physical registers which are shared among threads in SMT platforms [5][6]. In addition, in the FP model, experiments also indicate that data prefetching amongst threads is only effective if the horizontal filtering is vectorized.

5.3 SMP Versus SMT

Finally, we have compared SMT with SMP. Given that FP makes not sense if threads do not share data caches, we have exclusively focused on the DP model. Figure 8 shows the speedup over the baseline code for different number of threads, physical processors and image sizes.

Given a certain number of threads, the performance gains achieved in those configurations in which SMT is not exploited (denoted as 2P-2T and 4P-4T), are higher that in those in which SMT is exploited (denoted as 1P-2T and 2P-4T). On average, the performance of the SMT configurations is 75% of the non-SMT counterparts. This is the expected behavior given that Non-SMT configurations suffer a lower contention for shared resources.

Focusing on configurations with 4 threads, the speedup drops significantly for large images due to the memory contention caused by a higher bandwidth demand. In fact, for large images the performance gap between the SMT and the Non-SMT configuration is reduced given that the memory bus becomes the main bottleneck.

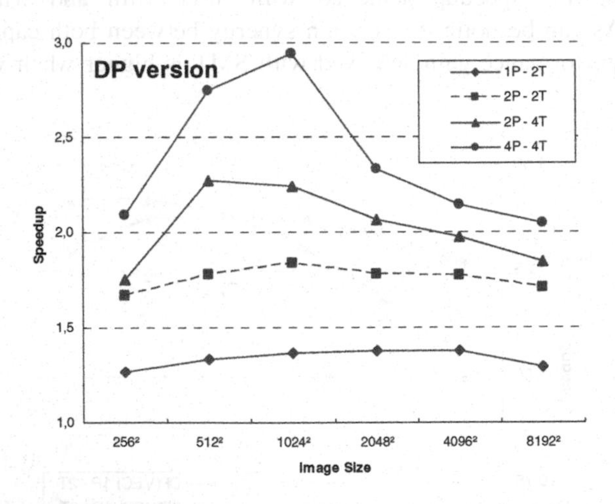

Fig. 8. Speedup of the DP version over the baseline code for different image sizes, number of threads (1T-4T) and physical processors (1P-4P)

6 Conclusions

In this paper we analyze two strategies (denoted as FP and DP) to parallelize the DWT on both SMT and SMP platforms. From the results presented above, we can draw the following conclusions:

1. FP outperforms DP on SMT platforms when large images are processed. Competition among threads for memory bandwidth and data cache works against the DP strategy for large image sizes. Whereas for small images, the scheduling overheads involved in FP does not compensate its better exploitation of the temporal locality due to cooperation amongst threads.
2. A certain synergy between SMT and short-vector processing exists. Vectorization allows a better exploitation of some of the shared resources and allows a better cooperation amongst thread in the FP model.
3. Under SMP, DP scales properly with the number of processors although its performance is limited for large images due to system bus contention.

These encouraging results suggest that the algorithm could take advantage of both FP and DP strategies. Our future work will be focused on hybrid approaches in which DP will be used to distribute data amongst the physical processors and FP to exploit thread level parallelism inside each SMT core.

References

[1] D. Burger, J.R. Goodman. "Billion-Transistor Architectures: There and Back Again". IEEE Computer Magazine, Vol. 37, No. 3; March 2004, pp. 22-28.

[2] S. Thakkar, T. Huff. "The Internet Streaming SIMD Extensions". Intel Technology Journal, Q2, 1999.

[3] K. Diefendorff, P. Dubey , R. Hochsprung, H. Scales "AltiVec Extension to PowerPC Accelerates Media Processing".IEEE Micro, pp 85-96, Apr. 2000.

[4] R. kalla, B. Sinharoy, J. M. Tendler, "IBM Power5 Chip: A Dual-Core Multithreaded Processor". IEEE Micro, Vol. 24, No. 2; March-April 2004, pp. 40-47.

[5] D. M. Tullsen, S. J. Eggers, and H.M. Levy. "Simultaneous Multithreading: Maximizing On-Chip Parallelism". Proceedings of the 22nd Annual International Symposium on Computer Architecture (ISCA 22), 1995, pages 392-403.

[6] T. Marr, F. Binns, D. L. Hill, G. Hinton, D. A. Koufaty, J. A. Miller, M. Upton. "Hyper-Threading Technology Architecture and Microarchitecture", Intel Technology Journal Vol. 6 (1), 2002.

[7] T. Ungerer , B. Robič , J. Šilc. "A survey of processors with explicit multithreading". ACM Computing Surveys. Vol. 35, Issue 1 (March 2003). Pages: 29 – 63, 2003.

[8] W. Sweldens. The lifting Scheme: A construction of second generation wavelets. Technical Report 1995:6, Department of Mathematics, University of South Carolina, 1995.

[9] D. S. Taubman and M. W. Marcellin. "Jpeg2000: Image Compression Fundamentals, Standards, and Practice" Kluwer International Series in Engineering and Computer Science. 2002.

[10] Robert Love. Kernel Corner: Cpu Affinity. Linux Journal Issue 111, July 2003. Available at http://www.linuxjournal.com/

[11] X. Tian, A. Bik, M Girkar, P. Grey, H. Saito, E. Su. "Intel OpenMP C++/Fortran Compiler for Hyper-Threading Technology: Implementation and Performance. Intel Technology Journal Vol. 6 (1), 2002

[12] D. Chaver, C. Tenllado, L. Piñuel, M. Prieto and F. Tirado. "2-D Wavelet Transform Enhancement on General-Purpose Microprocessors: Memory Hierarchy and SIMD Parallelism Exploitation". Proceeding of the 2002 International Conference on High Performance Computing, pp. 9-21. Bangalore, India, December 2002.
[13] G. Lafruit, L. Nachtergaele, J. Bormans, M. Engels and I. Bolsens, "Optimal Memory Organization for Scalable Texture Codecs in MPEG-4," IEEE Trans. on Circuits and Systems for Video Technology, vol.9, No.2, pp. 218-243, March 1999.
[14] D. Chaver, C. Tenllado, L. Piñuel, M. Prieto and F. Tirado. "Vectorization of the 2D Wavelet Lifting Transform using SIMD extensions". Workshop on Parallel and Distributed Image Processing, Video Processing, and Multimedia (PDIVM'03). Nize, France, April 2003.

Domain Decomposition Methods for PDE Constrained Optimization Problems*

Ernesto Prudencio, Richard Byrd, and Xiao-Chuan Cai

Department of Computer Science,
University of Colorado, Boulder, CO80309, USA
{prudenci, richard, cai}@cs.colorado.edu

Abstract. Optimization problems constrained by nonlinear partial differential equations have been the focus of intense research in scientific computing lately. Current methods for the parallel numerical solution of such problems involve sequential quadratic programming (SQP), with either reduced or full space approaches. In this paper we propose and investigate a class of parallel full space SQP Lagrange-Newton-Krylov-Schwarz (LNKSz) algorithms. In LNKSz, a Lagrangian functional is formed and differentiated to obtain a Karush-Kuhn-Tucker (KKT) system of nonlinear equations. Inexact Newton method with line search is then applied. At each Newton iteration the linearized KKT system is solved with a Schwarz preconditioned Krylov subspace method. We apply LNKSz to the parallel numerical solution of some boundary control problems of two-dimensional incompressible Navier-Stokes equations. Numerical results are reported for different combinations of Reynolds number, mesh size and number of parallel processors.

1 Introduction

In this paper we describe a general framework for solving optimization problems in interaction with nonlinear partial differential equations (PDEs). The focus is on how to adapt state-of-the-art PDE solvers to the requirements of optimization methods, while allowing for an efficient parallel implementation. Our method treats the differential equations as equality constraints. In order to demonstrate its effectiveness, the problem of optimizing fluid flows modeled by incompressible Navier-Stokes equations on two-dimensional domains is considered. In optimal control problems one usually searches for the best feasible values of the control variables, such as boundary values or external forces that minimize or maximize a certain system behavior, such as turbulence. In this paper, we only consider

* The work of Prudencio and Cai was partially supported by the Department of Energy, DE-FC02-01ER25479, and by the National Science Foundation, CCR-0219190, ACI-0072089 and ACI-0305666. The work of Byrd was supported in part by the National Science Foundation, CCR-0219190, and by Army Research Office contract DAAD19-02-1-0407.

M. Daydé et al. (Eds.): VECPAR 2004, LNCS 3402, pp. 569–582, 2005.

boundary control problems, which refer to the control of the system through boundary conditions.

Reduced space SQP methods have been the most widely used SQP approaches for PDE constrained problems until recently since they require much less memory, even though many sub-iterations are needed to converge the outer-iterations and the parallel scalability is less ideal. As more powerful computer systems with lots of memory and many processors become available, full space methods seem to be more appropriate due to their increased scalability. One such method, Lagrange-Newton-Krylov-Schur (LNKSr), was introduced in [2, 3], where four block factorization based preconditioners, as well as globalization techniques and heuristics, are proposed and tested. In this paper we replace the Schur type preconditioner with an overlapping Schwarz method which has a better asymptotical convergence rate and is easier to use as a nonlinear preconditioner [6, 7] for highly nonlinear problems.

The rest of the paper is organized as follows. Section 2 discusses the numerical solution of optimal control problems with equality constraints. Section 3, the core of this paper, presents the full space SQP *Lagrange-Newton-Krylov-Schwarz* (LNKSz) method for the parallel numerical solution of such problems. Section 4 presents a boundary flow control problem, which is then solved in Section 5 with LNKSz. Numerical experiments are performed and analyzed for different combinations of Reynolds number, mesh size and number of parallel processors. Final conclusions are given in Section 6.

2 Optimal Control with Equality Constraints

In this paper we focus on optimal control problems with equality constraints:

$$\begin{cases} \min\limits_{(\mathbf{s},\mathbf{u})\in\mathbf{S}\times\mathbf{U}} & \mathcal{F}(\mathbf{s},\mathbf{u}) \\ \quad\text{s.t.} & \mathbf{C}(\mathbf{s},\mathbf{u}) = \mathbf{0} \in \mathbf{Y}. \end{cases} \tag{2.1}$$

Here $\mathbf{S}$ and $\mathbf{U}$ are called state and control spaces respectively; the state variables $\mathbf{s}$ represent the state of the system being controlled; the control variables $\mathbf{u}$ represent the means one has to control the system; the objective or cost functional $\mathcal{F} : \mathbf{S} \times \mathbf{U} \to \mathbb{R}$ to be minimized (or maximized) represents the reason why one wants to control the system; constraints $\mathbf{C}(\mathbf{s},\mathbf{u}) = \mathbf{0}$ represent the system behavior and other constraints imposed to state and control variables.

The Lagrangian functional $\mathcal{L} : \mathbf{S} \times \mathbf{U} \times \mathbf{Y}^* \to \mathbb{R}$ associated with (2.1) is defined by

$$\mathcal{L}(\mathbf{s},\mathbf{u},\boldsymbol{\lambda}) \equiv \mathcal{F}(\mathbf{s},\mathbf{u}) + \langle \boldsymbol{\lambda}, \mathbf{C}(\mathbf{s},\mathbf{u}) \rangle_{\mathbf{Y}}, \quad \forall\, (\mathbf{s},\mathbf{u},\boldsymbol{\lambda}) \in \mathbf{S} \times \mathbf{U} \times \mathbf{Y}^*, \tag{2.2}$$

where $\mathbf{Y}^*$ is the adjoint space of $\mathbf{Y}$, $\langle\cdot,\cdot\rangle_{\mathbf{Y}}$ denotes the duality pairing and variables $\boldsymbol{\lambda}$ are called Lagrange multipliers or adjoint variables.

When the constraints are PDEs over a domain Ω, the discretization necessary for the solution of (2.1) can occur at two different points of the logical development of an algorithm. In the first case one demonstrates that, if $(\hat{\mathbf{s}}, \hat{\mathbf{u}})$

is a (local) solution of (2.1) then there exist Lagrange multipliers $\hat{\boldsymbol{\lambda}}$ such that $(\hat{\mathbf{s}}, \hat{\mathbf{u}}, \hat{\boldsymbol{\lambda}})$ is a critical point of $\mathcal{L}$. So, under sufficient smoothness assumptions, one obtains, as necessary condition for a solution, a system of equations, called Karush-Kuhn-Tucker (KKT) or optimality system, which is then discretized, generating a finite dimensional system of nonlinear equations [14, 15, 18, 24].

In the second case one begins by creating a mesh Ω_h of characteristic size $h > 0$ and then discretizing problem (2.1), obtaining a finite dimensional equality constrained optimization problem with $\mathbf{S} = \mathbb{R}^{n_{s,h}}$, $\mathbf{U} = \mathbb{R}^{n_{u,h}}$ and $\mathbf{Y} = \mathbb{R}^{m_h} = \mathbf{Y}^*$. Under sufficient smoothness conditions the KKT system becomes $\nabla \mathcal{L}_h(\hat{\mathbf{s}}, \hat{\mathbf{u}}, \hat{\boldsymbol{\lambda}}) = \mathbf{0} \in \mathbb{R}^{n_{s,h}+n_{u,h}+m_h}$. The theory for finite dimensional constrained optimization problems guarantees, under appropriate assumptions, the existence of such Lagrange multipliers $\hat{\boldsymbol{\lambda}}$. It should be pointed out that the discrete KKT systems from both approaches are *not* necessarily the same.

From now on, we only work with the second approach. For simplicity, let us omit the symbols "h" and "$\hat{}$", and use the notations $N \equiv n_s + n_u + m$ and $\mathbf{X} \equiv (\mathbf{x}, \boldsymbol{\lambda}) \equiv (\mathbf{s}, \mathbf{u}, \boldsymbol{\lambda}) \in \mathbb{R}^N$. The KKT system becomes

$$\mathbf{F}(\mathbf{X}) \equiv \begin{pmatrix} \nabla_{\mathbf{x}} \mathcal{L} \\ \nabla_{\boldsymbol{\lambda}} \mathcal{L} \end{pmatrix} = \begin{pmatrix} \nabla \mathcal{F} + [\boldsymbol{\nabla} \mathbf{C}]^T \boldsymbol{\lambda} \\ \mathbf{C}(\mathbf{s}, \mathbf{u}) \end{pmatrix} = \begin{pmatrix} \nabla_{\mathbf{s}} \mathcal{F} + [\boldsymbol{\nabla}_{\mathbf{s}} \mathbf{C}]^T \boldsymbol{\lambda} \\ \nabla_{\mathbf{u}} \mathcal{F} + [\boldsymbol{\nabla}_{\mathbf{u}} \mathbf{C}]^T \boldsymbol{\lambda} \\ \mathbf{C}(\mathbf{s}, \mathbf{u}) \end{pmatrix} = \mathbf{0}, \quad (2.3)$$

where $\mathbf{F} : \mathbb{R}^N \to \mathbb{R}^N$, $\nabla_{\mathbf{x}} \mathcal{L}$ denotes the gradient of $\mathcal{L}$ w.r.t. state and control variables, with similar meaning holding for $\nabla_{\boldsymbol{\lambda}} \mathcal{L}$, $\nabla_{\mathbf{s}} \mathcal{F}$ and $\nabla_{\mathbf{u}} \mathcal{F}$, $\boldsymbol{\nabla} \mathbf{C}$ denotes the Jacobian of $\mathbf{C}$ and $\boldsymbol{\nabla}_{\mathbf{s}} \mathbf{C}$ and $\boldsymbol{\nabla}_{\mathbf{u}} \mathbf{C}$ denote the Jacobian of $\mathbf{C}$ w.r.t. state and control variables, respectively. In (2.3), we refer to the first equation as the adjoint equation, the second as the control equation, the third as the state equation or the forward problem, and $\boldsymbol{\nabla}_{\mathbf{s}} \mathbf{C}$ as the linearized forward operator.

System (2.3) can be solved with an inexact Newton method [12, 13]. Given an initial guess $\mathbf{X}^{(0)}$, at each iteration $k = 0, 1, 2, \ldots$ an approximate solution

$$\mathbf{p}^{(k)} \equiv \left(\mathbf{p}_{\mathbf{x}}^{(k)}, \mathbf{p}_{\boldsymbol{\lambda}}^{(k)} \right) \equiv \left(\mathbf{p}_{\mathbf{s}}^{(k)}, \mathbf{p}_{\mathbf{u}}^{(k)}, \mathbf{p}_{\boldsymbol{\lambda}}^{(k)} \right)$$

of the linearized KKT system

$$\left[\mathbf{K}^{(k)} \right] \mathbf{p}^{(k)} = -\mathbf{F}^{(k)} \quad (2.4)$$

is computed, where $\mathbf{K}^{(k)} = \boldsymbol{\nabla} \mathbf{F}(\mathbf{X}^{(k)})$ and $\mathbf{F}^{(k)} = \mathbf{F}(\mathbf{X}^{(k)})$. The KKT matrix $\mathbf{K}^{(k)}$ is the transpose of the Hessian of the Lagrangian $\mathcal{L}$, is symmetric indefinite under sufficient smoothness assumptions and can be computed by a finite difference approximation. If $\boldsymbol{\nabla} \mathbf{C}^{(k)}$ has full rank and $\boldsymbol{\nabla}_{\mathbf{xx}} \mathcal{L}^{(k)}$ is positive definite in the tangent space of the constraints (i.e., $\mathbf{d}^T [\boldsymbol{\nabla}_{\mathbf{xx}} \mathcal{L}^{(k)}] \mathbf{d} > 0$ for all $\mathbf{d} \neq \mathbf{0}$ such that $[\boldsymbol{\nabla} \mathbf{C}^{(k)}] \mathbf{d} = \mathbf{0}$), then we can interpret the solution $(\mathbf{p}_{\mathbf{x}}^{(k)}, \mathbf{p}_{\boldsymbol{\lambda}}^{(k)})$ of (2.4) as being the *unique* solution and respective Lagrange multipliers of

$$\begin{cases} \min\limits_{\mathbf{p}_{\mathbf{x}} \in \mathbf{S} \times \mathbf{U}} \frac{1}{2} \mathbf{p}_{\mathbf{x}}^T [\boldsymbol{\nabla}_{\mathbf{xx}} \mathcal{L}^{(k)}] \mathbf{p}_{\mathbf{x}} + \left(\nabla_{\mathbf{x}} \mathcal{L}^{(k)} \right)^T \mathbf{p}_{\mathbf{x}} \\ \text{s.t. } [\boldsymbol{\nabla} \mathbf{C}^{(k)}] \mathbf{p}_{\mathbf{x}} + \mathbf{C}^{(k)} = \mathbf{0} \in \mathbf{Y}. \end{cases}$$

This interpretation justifies the use of terminology *sequential quadratic programming* (SQP) for methods involving (2.4), [19].

After approximately solving (2.4), one may use a globalization method like *line search* or *trust region*. In this study we focus on a line search approach, where the next iterate is $\mathbf{X}^{(k+1)} = \mathbf{X}^{(k)} + \alpha^{(k)}\mathbf{p}^{(k)}$ and the step length $\alpha^{(k)}$ is selected by backtracking until the sufficient decrease condition

$$\phi(\mathbf{X}^{(k)} + \alpha^{(k)}\mathbf{p}^{(k)}) \leqslant \phi(\mathbf{X}^{(k)}) + \alpha^{(k)} c_1 \left(\nabla\phi(\mathbf{X}^{(k)})\right)^T \mathbf{p}^{(k)} \tag{2.5}$$

is satisfied. Here ϕ is a merit function and c_1 is a constant satisfying $0 < c_1 < 1/2$. For constrained optimization, merit functions such as l_1 or augmented Lagrangian are most commonly used. In contrast to the standard merit function $\|\mathbf{F}(\mathbf{X})\|_2^2/2$, which is commonly used for systems of nonlinear equations, these merit functions try to balance the sometimes conflicting goals of reducing the objective function and satisfying the constraints [19]. We use the augmented Lagrangian $\phi_{AL} : \mathbb{R}^N \to \mathbb{R}$ in our experiments in this paper. Given a penalty parameter $\rho > 0$, it is defined by

$$\phi_{AL}(\mathbf{X};\rho) = \mathcal{L}(\mathbf{s},\mathbf{u},\boldsymbol{\lambda}) + \frac{\rho}{2}\,\|\mathbf{C}(\mathbf{s},\mathbf{u})\|_2^2 \quad \forall\, \mathbf{X} = (\mathbf{s},\mathbf{u},\boldsymbol{\lambda}) \in \mathbf{S}\times\mathbf{U}\times\mathbf{Y}.$$

At iteration k, $\rho = \rho^{(k)}$ must be such that we obtain descent directions $\mathbf{p}^{(k)}$, i.e.,

$$(\nabla\phi_{AL}(\mathbf{X}^{(k)};\rho^{(k)}))^T\mathbf{p}^{(k)} = (\nabla\mathcal{L}^{(k)})^T\mathbf{p}^{(k)} + \rho^{(k)}\ {\mathbf{C}^{(k)}}^{\mathbf{T}}[\boldsymbol{\nabla}\mathbf{C}^{(k)}]\mathbf{p}_{\mathbf{x}}^{(k)} < 0. \tag{2.6}$$

Since ${\mathbf{C}^{(k)}}^{\mathbf{T}}[\boldsymbol{\nabla}\mathbf{C}^{(k)}]\mathbf{p}_{\mathbf{x}}^{(k)} = -\|\mathbf{C}^{(k)}\|_2^2$ for an exact step, it is reasonable to expect

$${\mathbf{C}^{(k)}}^{\mathbf{T}}[\boldsymbol{\nabla}\mathbf{C}^{(k)}]\mathbf{p}_{\mathbf{x}}^{(k)} < 0 \tag{2.7}$$

for approximate steps, if $\mathbf{C}^{(k)} \neq \mathbf{0}$ and the tolerances for the Krylov subspace method are small enough and the preconditioner is good enough to guarantee that the Krylov subspace method does not stop by achieving the maximum allowed number of iterations. If (2.7) does not hold we can continue the Krylov iterations, with eventual smaller tolerances, until it does. Once (2.7) holds, we then use a fairly common strategy where we demand $\rho^{(k)}$ to satisfy

$$(\nabla\phi_{AL}(\mathbf{X}^{(k)};\rho^{(k)}))^T\mathbf{p}^{(k)} \leqslant \frac{\rho^{(k)}}{2}\ {\mathbf{C}^{(k)}}^{\mathbf{T}}[\boldsymbol{\nabla}\mathbf{C}^{(k)}]\mathbf{p}_{\mathbf{x}}^{(k)},$$

that is,

$$\rho^{(k)} \geqslant \overline{\rho}^{(k)} = -2\frac{(\nabla\mathcal{L}^{(k)})^T\mathbf{p}^{(k)}}{{\mathbf{C}^{(k)}}^{\mathbf{T}}[\boldsymbol{\nabla}\mathbf{C}^{(k)}]\mathbf{p}_{\mathbf{x}}^{(k)}}.$$

We then choose

$$\rho^{(k)} = \max\{\overline{\rho}^{(k)}, \rho^{(k-1)}\},$$

where $\rho^{(-1)} > 0$ is a given positive value.

However, if $\mathbf{C}^{(k)} = \mathbf{0}$ then there is no way to guarantee a descent direction. This is a fundamental issue with line search methods. Some algorithms handle this by modifying the Hessian to make it positive definite on the null space of $\nabla \mathbf{C}^{(k)}$, but for problems of the size we are considering there is no efficient way to check positive definiteness.

In all tests described in this paper, (2.7) held for every step generated, as long as we made the absolute Krylov tolerance small enough that at least one Krylov iteration was performed. In addition, it is worth noting that, in each run, the value of $\rho^{(k)}$ became fixed before 60% of the iterations had been made. Thus the heuristic merit parameter updating strategy described above appeared to work well for the examples of this paper.

3 Parallel Full Space SQP Lagrange-Newton-Krylov-Schwarz

A key element of a successful full space approach is the preconditioning of the Jacobian of the KKT system, which is indefinite and extremely ill-conditioned. A good preconditioner has to be able to substantially reduce the condition number and, at the same time, to provide good scalability, so that the potential of massively parallel computers can be realized. The Schur complement preconditioner used in LNKSr [2, 3] is an *operator-splitting* type technique, in which a sequential block elimination step is needed to form the Schur complement w.r.t. the control variable. In contrast to operator-splitting, Schwarz type preconditioners are *fully coupled* in the sense that all variables are treated equally and the partition is based completely on the physical domain Ω. Because there is no need to eliminate any variables from the system, there is one less sequential step in the preconditioning process. Another advantage of LNKSz method is that it does not demand $m = n_s$. With LNKSz we can, for instance, deal directly with full [17] boundary control problems, where an equation like (4.3) is explicitly added to the constraints.

Schwarz preconditioners can be used in one-level or multi-level approaches and, at each case, with a combination of additive and/or multiplicative algorithms [23]. In this paper we deal with one-level additive algorithms only. Let $\Omega \subset \mathbb{R}^2$ be a bounded open domain on which the control problem is defined. We only consider simple box domains with uniform mesh of characteristic size h here. To obtain the overlapping partition, we first partition the domain into non-overlapping subdomains Ω_j^0, $j = 1, \cdots, N_s$. Then we extend each subdomain Ω_j^0 to a larger subdomain Ω_j^δ, i.e., $\Omega_j^0 \subset \Omega_j^\delta$. Here $\delta > 0$ indicates the size of the overlap. Only simple box decomposition is considered in this paper; i.e., all the subdomains Ω_j^0 and Ω_j^δ are rectangular and made up of integral number of fine mesh cells. For boundary subdomains, we simply cut off the part that is outside Ω. Let $H > 0$ denote the characteristic diameter of subdomains Ω_j. Let N, N_j^0 and N_j^δ denote the number of degrees of freedom associated to Ω, Ω_j^0 and Ω_j^δ, respectively. Let $\mathbf{K}$ be a $N \times N$ matrix of a linear system that needs to be solved during the numerical solution process of the differential problem. For each subdomain Ω_j^0, we define $\mathbf{R}_j^0$ as an $N_j^0 \times N_j^0$ block sub-identity matrix whose diagonal element, $(\mathbf{R}_j^0)_{k,k}$, is either an $d \times d$ identity matrix if the mesh point $x_k \in \Omega_j^0$ or a $d \times d$ zero matrix if x_k is outside of Ω_j^0, where d indicates the degree of freedom per mesh point. Similarly we introduce a block sub-identity matrix $(\mathbf{R}_j^\delta)_{k,k}$ for each Ω_j^δ. The multiplication of $\mathbf{R}_j^\delta$ with a vector will zero out all of its components outside of Ω_j^δ. We denote by $\mathbf{K}_j$ the subdomain matrix given by

$$\mathbf{K}_j = \mathbf{R}_j^\delta \, \mathbf{K} \, (\mathbf{R}_j^\delta)^T.$$

Let $\mathbf{B}_j^{-1}$ be either a subspace inverse of $\mathbf{K}_j$ or a preconditioner for $\mathbf{K}_j$. The classical one-level additive Schwarz preconditioner $\mathbf{B}_{asm}^{-1}$ for $\mathbf{K}$ is defined as

$$\mathbf{B}_{asm}^{-1} = \sum_{j=1}^{N_s} \mathbf{R}_j^\delta \mathbf{B}_j{}^{-1} \mathbf{R}_j^\delta.$$

In addition to this standard additive Schwarz method (ASM) described above, we also consider the newly introduced restricted version (RAS) of the method [4, 8]. For some applications, the restricted version requires less communication time since one of the restriction or extension operations does not involve any overlap. The RAS preconditioner is defined as

$$\mathbf{B}_{ras}^{-1} = \sum_{j=1}^{N_s} \mathbf{R}_j^\delta \mathbf{B}_j{}^{-1} \mathbf{R}_j^0.$$

Some numerical comparisons of the ASM and RAS are presented later in the paper.

When the Schwarz preconditioner is applied to symmetric positive definite systems arising from the discretization of elliptical problems in $H_0^1(\Omega)$, the condition number κ of the preconditioned system satisfies $\kappa \leqslant C\,(1 + H/\delta)\,/H^2$, where C is independent of h, H, δ and the shapes of Ω and Ω_j^δ [23], that is, a Schwarz preconditioned Krylov subspace method is expected to have the following properties:

The number of iterations grows approximately proportional to $1/H$; (3.1)

If δ is maintained proportional to H, the number of iterations is bounded independently of h and H/h (a parameter related to the number of degrees of freedom of each subproblem); (3.2)

The convergence gets better as δ is increased. (3.3)

Theoretically, results (3.1)-(3.3) may not be applied immediately to Krylov subspace methods, e.g. GMRES [22], for the solution of indefinite linearized KKT systems. Nonetheless, we carefully examine all the properties in our numerical experiments. In particular, let $\bar{l}$ be the average number of Schwarz preconditioned GMRES iterations per linearized KKT system. We look for the following scalability properties:

For fixed h and δ, $\bar{l}$ increases as H decreases; (3.4)

For fixed H and δ, $\bar{l}$ is not very sensitive to the mesh refinement; (3.5)

For fixed h and H, $\bar{l}$ decreases as δ increases. (3.6)

4 Boundary Control of Incompressible Navier-Stokes Flows

In this section we discuss the boundary control of the two-dimensional steady-state incompressible Navier-Stokes equations in the velocity-vorticity formulation [21]. The velocity is denoted by $\mathbf{v} = (v_1, v_2)$ and the vorticity by ω. Let Ω be an open and bounded polygonal domain in the plane, $\Gamma = \partial\Omega$ its boundary and $\boldsymbol{\nu}$ the unit outward normal vector along Γ. Let $\mathbf{f}$ be a given external force defined in Ω. *Re* is the Reynolds number and curl $\mathbf{f} = -\frac{\partial f_1}{\partial x_2} + \frac{\partial f_2}{\partial x_1}$. A boundary control problem consists on finding $(v_1, v_2, \omega, u_1, u_2)$ such that the minimization

$$\min_{(\mathbf{s},\mathbf{u})\in \mathbf{S}\times\mathbf{U}} \mathcal{F}(\mathbf{s},\mathbf{u}) = \frac{1}{2}\int_{\Omega} \omega^2 \, d\Omega + \frac{c}{2}\int_{\Gamma_{u,1}} u_1^2 \, d\Gamma + \frac{c}{2}\int_{\Gamma_{u,2}} u_2^2 \, d\Gamma \tag{4.1}$$

is achieved subject to the constraints

$$\begin{cases} -\Delta v_1 - \frac{\partial \omega}{\partial x_2} & = 0 & \text{in } \Omega, \\ -\Delta v_2 + \frac{\partial \omega}{\partial x_1} & = 0 & \text{in } \Omega, \\ -\Delta \omega + Re\ v_1 \frac{\partial \omega}{\partial x_1} + Re\ v_2 \frac{\partial \omega}{\partial x_2} - Re \text{ curl } \mathbf{f} & = 0 & \text{in } \Omega, \\ v_i - v_{D,i} & = 0 & \text{on } \Gamma_{D,i}, \quad i = 1,2, \\ v_i - u_i & = 0 & \text{on } \Gamma_{u,i}, \quad i = 1,2, \\ \omega + \frac{\partial v_1}{\partial x_2} - \frac{\partial v_2}{\partial x_1} & = 0 & \text{on } \Gamma, \\ \int_{\Gamma} \mathbf{v}\cdot\boldsymbol{\nu}\, d\Gamma & = 0, & \end{cases} \tag{4.2}$$

where, for $i = 1, 2$, $\Gamma = \Gamma_{D,i} \cup \Gamma_{u,i}$, $\Gamma_{D,i}$ is the part of the boundary where the v_i velocity component is specified through a Dirichlet condition with a prescribed velocity $v_{D,i}$, and $\Gamma_{u,i}$ is the part of the boundary where the v_i velocity component is specified through a Dirichlet condition with a control velocity u_i. The positive constant parameter c is used to adjust the relative importance of the control norms on achieving the minimization, so indirectly constraining the size of those norms. The physical objective behind problem (4.1)-(4.2) is the minimization of turbulence [16, 17]. The last constraint, given by

$$\int_{\Gamma} \mathbf{v}\cdot\boldsymbol{\nu}\, d\Gamma = 0, \tag{4.3}$$

is necessary for the consistency with the physical law of mass conservation. So the control $\mathbf{u} = (u_1, u_2)$ cannot be any control: it must belong to the space of functions satisfying (4.3). We denote problems like (4.1) – (4.2), where controls are allowed to assume nonzero normal values at the boundary, as *full* boundary control problems (BCPs), [17]. In these kind of problems one has $m \neq n_s$ due to the extra constraint (4.3). This fact also complicates the parallel finite differences approximation of Jacobian matrices, since one does not have only PDEs (i.e., equations with local behavior) anymore. One can also deal with *tangential* BCPs, where the control is allowed to be just tangential to the boundary and the velocity $\mathbf{v}_b = (v_{b,1}, v_{b,2})$, defined as, for $i = 1, 2$,

$$v_{b,i} = \begin{cases} v_{D,i} \text{ on } \Gamma_{D,i}, \\ 0 \quad\ \ \text{on } \Gamma \setminus \Gamma_{D,i}, \end{cases}$$

is assumed to satisfy $\int_\Gamma \mathbf{v}_b \cdot \boldsymbol{\nu}\, d\Gamma = 0$, and so, $m = n_s$. Since tangential BCPs restrict even more the space where the control $\mathbf{u} = (u_1, u_2)$ can exist, one naturally expects better objective function values with full BCPs. In this paper we only study tangential boundary control problems.

5 Numerical Experiments

We consider both a simulation and a tangential BCP over the cavity $\Omega = (0,1) \times (0,1)$, with

$$\mathbf{f} = \begin{pmatrix} -\sin^2(\pi x_1)\cos(\pi x_2)\sin^2(\pi x_2) \\ \sin^2(\pi x_2)\cos(\pi x_1)\sin^2(\pi x_1) \end{pmatrix}.$$

The simulation problem has slip boundary conditions. In the case of the tangential BCP, the objective function has $c = 10^{-2}$ and $\Gamma_{u,1} = \Gamma_{u,2} = \Gamma$. We run flow problems for $Re = 200$ and $Re = 250$.

To discretize the flow problems, we use a five-point second order finite difference method on a uniform mesh. All derivative terms of interior PDEs are discretized with a second order central difference scheme. The boundary condition $\omega + \partial v_1/\partial x_2 - \partial v_2/\partial x_1 = 0$ on Γ is also discretized with a second order approximation [20].

In all experiments, the Jacobian matrix is constructed approximately using a multi-colored central finite difference method with step size 10^{-5}, [9]. For control problems, central finite differences provide KKT matrices closer to be symmetric than the ones computed by forward finite differences. To solve the Jacobian systems we use restarted GMRES with an absolute tolerance equal to 10^{-6}, a relative tolerance equal to 10^{-4}, a restart parameter equal to 90 and a maximum number of iterations equal to 5000. The GMRES tolerances are checked over preconditioned residuals. Regarding the one-level additive Schwarz preconditioner, the number of subdomains is equal to the number of processors and the extended subdomain problems have zero Dirichlet interior boundary conditions and are solved with sparse LU. Line search with the merit function defined in Section 2 is performed with cubic backtracking, with $c_1 = 10^{-4}$ in (2.5) and a minimum allowed step length $\alpha^{(k)}$ equal to 10^{-6}. For augmented Lagrangian merit functions we follow the strategy explained in Section 2 with $\rho^{(-1)}$=10. For Newton iterations we use an absolute stopping tolerance equal to 10^{-6} and a relative tolerance equal to 10^{-10} times the initial residual. The maximum allowed number of Newton iterations is 100.

All tests were performed on a cluster of Linux PCs and our parallel software was developed using the Portable, Extensible Toolkit for Scientific Computing (PETSc) library [1], from Argonne National Laboratory. Our main concern is the scalability of the algorithms in terms of the linear and nonlinear iteration numbers. CPU times are also reported, but they should not be taken as a reliable measure of the scalability of the algorithms because our network is relatively slow and is shared with other processes.

Results are grouped into tables according to a unique combination of problem type (simulation or control), Reynolds number Re, ASM type (standard or restricted), overlap δ and merit function (standard or augmented Lagrangian). Each table presents results for nine situations, related to three different sizes of meshes (parameter h) and three different numbers of processors (parameter H). For each case we report:

- the total number of Newton iterations: n,
- the average number of GMRES iterations per Newton iteration: $\bar{l}$,
- the total CPU time in seconds spent on all Newton iterations: t_n,
- the average CPU time, in seconds, per Newton iteration, spent on solving for the Newton steps: $\bar{t}_l$.

For each table we compare the behavior of $\bar{l}$ against (3.4)-(3.6). Predictions (3.4) and (3.5) can be checked by observing the values of $\bar{l}$ in a column (fixed h and δ) and in a row (fixed H and δ), respectively. Prediction (3.6) can be checked by observing the values of $\bar{l}$ at the same situation (fixed problem type, Re, ASM type, merit function, H and h) in different tables (different δ). We also compare approximate values of

$$\|\omega\|_h^2 = \int_{\Omega_h} \omega^2 \, d\Omega_h. \tag{5.1}$$

Table 1 presents results for the simulation problem with $Re = 200$. The preconditioner is ASM with $\delta = 1/64$ and the standard merit function is used in the line search. The total number of Newton iterations does not change with the mesh size or the number of processors. The average number of Krylov iterations per Newton iteration changes as expected in predictions (3.4) and (3.5).

The next three tables present results for the tangential BCP with $Re = 200$. An augmented Lagrangian merit function is used in the line search. Several different overlap values are used in the ASM preconditioner and the results are summarized as follows: Table 2 for $\delta = 1/64$, Table 3 for $\delta = 1/32$ and Table 4 for $\delta = 1/16$. Changes on the total number of Newton iterations w.r.t. the mesh

Table 1. Results for the cavity flow simulation problem with $Re = 200$, standard ASM with overlap $\delta = 1/64$ and standard merit function $\|\mathbf{F}\|_2^2/2$. n is the total number of Newton iterations, $\bar{l}$ is the average number of Krylov iterations per Newton iteration, t_n is the total time in seconds spent on all Newton iterations and $\bar{t}_l$ is the average time in seconds, per Newton iteration, spent on solving for Newton steps. For the case of finest mesh, the number of variables is 198, 147 and $\|\omega\|_h^2 \approx 55.4$. See (5.1)

# Procs.	Mesh 64 × 64		128 × 128		256 × 256	
16	$n = 5$	$t_n \approx 0.83$	$n = 5$	$t_n \approx 3.6$	$n = 5$	$t_n \approx 18.3$
	$\bar{l} \approx 46$	$\bar{t}_l \approx 0.13$	$\bar{l} \approx 47$	$\bar{t}_l \approx 0.62$	$\bar{l} \approx 47$	$\bar{t}_l \approx 3.29$
32	$n = 5$	$t_n \approx 0.71$	$n = 5$	$t_n \approx 2.7$	$n = 5$	$t_n \approx 12.1$
	$\bar{l} \approx 60$	$\bar{t}_l \approx 0.091$	$\bar{l} \approx 63$	$\bar{t}_l \approx 0.46$	$\bar{l} \approx 63$	$\bar{t}_l \approx 2.20$
64	$n = 5$	$t_n \approx 0.70$	$n = 5$	$t_n \approx 1.99$	$n = 5$	$t_n \approx 7.50$
	$\bar{l} \approx 69$	$\bar{t}_l \approx 0.077$	$\bar{l} \approx 80$	$\bar{t}_l \approx 0.32$	$\bar{l} \approx 79$	$\bar{t}_l \approx 1.35$

Table 2. Results for the cavity tangential boundary flow control problem with $Re = 200$, standard ASM with overlap $\delta = 1/64$ and augmented Lagrangian merit function. n is the total number of Newton iterations, $\bar{l}$ is the average number of Krylov iterations per Newton iteration, t_n is the total time in seconds spent on all Newton iterations and $\bar{t}_l$ is the average time in seconds, per Newton iteration, spent on solving for Newton steps. For the case of finest mesh, the number of variables is 528,392 and $\|\omega\|_h^2 \approx 32.5$. See (5.1)

# Procs.	Mesh					
	64 × 64		128 × 128		256 × 256	
16	$n = 7$ $\bar{l} \approx 92$	$t_n \approx 9.92$ $\bar{t}_l \approx 1.21$	$n = 8$ $\bar{l} \approx 85$	$t_n \approx 52.1$ $\bar{t}_l \approx 5.78$	$n = 6$ $\bar{l} \approx 100$	$t_n \approx 238$ $\bar{t}_l \approx 36.4$
32	$n = 7$ $\bar{l} \approx 208$	$t_n \approx 10.3$ $\bar{t}_l \approx 1.34$	$n = 8$ $\bar{l} \approx 204$	$t_n \approx 53.3$ $\bar{t}_l \approx 6.26$	$n = 6$ $\bar{l} \approx 272$	$t_n \approx 264$ $\bar{t}_l \approx 42.5$
64	$n = 7$ $\bar{l} \approx 187$	$t_n \approx 5.39$ $\bar{t}_l \approx 0.67$	$n = 8$ $\bar{l} \approx 182$	$t_n \approx 25.6$ $\bar{t}_l \approx 2.96$	$n = 6$ $\bar{l} \approx 216$	$t_n \approx 108$ $\bar{t}_l \approx 17.1$

Table 3. Results for the cavity tangential boundary flow control problem with $Re = 200$, standard ASM with overlap $\delta = 1/32$ and augmented Lagrangian merit function. For the case of finest mesh, the number of variables is 528,392 and $\|\omega\|_h^2 \approx 32.5$. Case "(*)" is discussed in Section 5

# Procs.	Mesh					
	64 × 64		128 × 128		256 × 256	
16	$n = 7$ $\bar{l} \approx 58$	$t_n \approx 8.22$ $\bar{t}_l \approx 0.96$	$n = 8$ $\bar{l} \approx 59$	$t_n \approx 49.1$ $\bar{t}_l \approx 5.38$	$n = 7$ $\bar{l} \approx 62$	$t_n \approx 254$ $\bar{t}_l \approx 33.1$
32	$n = 7$ $\bar{l} \approx 119$	$t_n \approx 8.32$ $\bar{t}_l \approx 1.05$	$n = 8$ $\bar{l} \approx 130$	$t_n \approx 46.6$ $\bar{t}_l \approx 5.41$	$n = 6$ $\bar{l} \approx 140$	$t_n \approx 199$ $\bar{t}_l \approx 31.6$
64	$n = 7$ (*) $\bar{l} \approx 155$	$t_n \approx 6.21$ $\bar{t}_l \approx 0.78$	$n = 8$ $\bar{l} \approx 132$	$t_n \approx 27.6$ $\bar{t}_l \approx 3.18$	$n = 6$ $\bar{l} \approx 143$	$t_n \approx 110$ $\bar{t}_l \approx 17.4$

Table 4. Results for the cavity tangential boundary flow control problem with $Re = 200$, standard ASM with overlap $\delta = 1/16$ and augmented Lagrangian merit function. For the case of finest mesh, the number of variables is 528,392 and $\|\omega\|_h^2 \approx 32.5$

# Procs.	Mesh					
	64 × 64		128 × 128		256 × 256	
16	$n = 7$ $\bar{l} \approx 47$	$t_n \approx 9.65$ $\bar{t}_l \approx 1.16$	$n = 8$ $\bar{l} \approx 44$	$t_n \approx 59.2$ $\bar{t}_l \approx 6.62$	$n = 7$ $\bar{l} \approx 50$	$t_n \approx 744$ $\bar{t}_l \approx 103$
32	$n = 7$ $\bar{l} \approx 100$	$t_n \approx 10.8$ $\bar{t}_l \approx 1.39$	$n = 8$ $\bar{l} \approx 108$	$t_n \approx 75.1$ $\bar{t}_l \approx 8.96$	$n = 7$ $\bar{l} \approx 149$	$t_n \approx 616$ $\bar{t}_l \approx 86.3$
64	$n = 6$ $\bar{l} \approx 104$	$t_n \approx 6.36$ $\bar{t}_l \approx 0.94$	$n = 8$ $\bar{l} \approx 115$	$t_n \approx 56.5$ $\bar{t}_l \approx 6.79$	$n = 7$ $\bar{l} \approx 128$	$t_n \approx 331$ $\bar{t}_l \approx 46.4$

size and the number of processors are not pronounced. We observe that $\bar{l}$ follows (3.6). With the same δ used on the simulation problem, we can see in Table 2 that the average number of GMRES iterations is now more sensitive to both h and, especially, H. Table 3 is the one where $\bar{l}$ best follows both (3.4) and (3.5) and the CPU times t_n and $\bar{t}_l$ for the finest mesh decrease with the increase on the number of processors. Table 4 shows that if δ gets too big then the consequent decrease on $\bar{l}$ might not compensate the increased time taken by sparse LU on the larger extended subdomains; that is, $\bar{t}_l$ increases. Comparing values of $\bar{t}_l$ in Tables 2 and 3 with the values in Table 1, we see that the average time spent on computing $\mathbf{p}^{(k)}$ can be more than 10 times bigger in control problems than in simulation problems on the same mesh, instead of being around $8/3 \approx 3$ times bigger, in accordance to the ratio between the number of variables per mesh point on control problems and on simulation problems.

As reported before we use a GMRES relative tolerance of 1.0×10^{-4}, a GMRES absolute tolerance of 1.0×10^{-6} and a Newton absolute tolerance of 1.0×10^{-6}. Case "(*)" in Table 3, however, gives results for a Newton absolute tolerance of 1.2×10^{-6}. When a Newton absolute tolerance of 1.0×10^{-6} is used, although full steps are accepted in some iterations, the line search stalls once $||F||_2 \approx 1.14 \times 10^{-6}$. If we change the GMRES relative tolerance to 1.0×10^{-6} and the GMRES absolute tolerance to 1.0×10^{-13} (in order to obtain a more accurate Newton step) then we achieve $||F||_2 < 1.00 \times 10^{-6}$ with $n = 6$, $\bar{l} \approx 272$, $t_n \approx 9.38$ and $\bar{t}_l \approx 1.42$. Although we performed our tests with fixed GMRES tolerances, this experiment suggests that for more demanding Newton tolerances one might need to use decreasing GMRES tolerances as the outer loop proceeds, as expected by the theory for superlinear convergence of the inexact Newton method [10, 19].

In the next two tables, we change the preconditioner to RAS and everything else stays the same; i.e., these results are for the tangential BCP with Re = 200 and we use an augmented Lagrangian merit function in the line search. We increase the overlap size in the RAS preconditioner as follows: Table 5 for

Table 5. Results for the cavity tangential boundary flow control problem with Re = 200, restricted ASM with overlap $\delta = 1/32$ and augmented Lagrangian merit function. n is the total number of Newton iterations, $\bar{l}$ is the average number of Krylov iterations per Newton iteration, t_n is the total time in seconds spent on all Newton iterations and $\bar{t}_l$ is the average time in seconds, per Newton iteration, spent on solving for Newton steps. For the case of finest mesh, the number of variables is 528,392 and $||\omega||_h^2 \approx 32.5$. See (5.1)

# Procs.	Mesh 64 × 64		128 × 128		256 × 256	
16	$n = 7$	$t_n \approx 8.33$	$n = 9$	$t_n \approx 53.1$	$n = 7$	$t_n \approx 253$
	$\bar{l} \approx 59$	$\bar{t}_l \approx 0.98$	$\bar{l} \approx 57$	$\bar{t}_l \approx 5.15$	$\bar{l} \approx 62$	$\bar{t}_l \approx 32.9$
32	$n = 7$	$t_n \approx 8.47$	$n = 8$	$t_n \approx 46.9$	$n = 6$	$t_n \approx 211$
	$\bar{l} \approx 131$	$\bar{t}_l \approx 1.07$	$\bar{l} \approx 134$	$\bar{t}_l \approx 5.45$	$\bar{l} \approx 154$	$\bar{t}_l \approx 33.6$
64	$n = 7$	$t_n \approx 6.38$	$n = 8$	$t_n \approx 30.1$	$n = 6$	$t_n \approx 132$
	$\bar{l} \approx 175$	$\bar{t}_l \approx 0.81$	$\bar{l} \approx 162$	$\bar{t}_l \approx 3.50$	$\bar{l} \approx 184$	$\bar{t}_l \approx 21.1$

Table 6. Results for the cavity tangential boundary flow control problem with $Re = 200$, restricted ASM with overlap $\delta = 1/16$ and augmented Lagrangian merit function. For the case of finest mesh, the number of variables is 528,392 and $\|\omega\|_h^2 \approx 32.5$

# Procs.	Mesh					
	64 × 64		128 × 128		256 × 256	
16	$n = 7$	$t_n \approx 9.66$	$n = 8$	$t_n \approx 60.7$	$n = 7$	$t_n \approx 743$
	$\bar{l} \approx 47$	$\bar{t}_l \approx 1.15$	$\bar{l} \approx 46$	$\bar{t}_l \approx 6.80$	$\bar{l} \approx 51$	$\bar{t}_l \approx 103$
32	$n = 7$	$t_n \approx 9.49$	$n = 8$	$t_n \approx 65.5$	$n = 6$	$t_n \approx 436$
	$\bar{l} \approx 90$	$\bar{t}_l \approx 1.21$	$\bar{l} \approx 86$	$\bar{t}_l \approx 7.7$	$\bar{l} \approx 98$	$\bar{t}_l \approx 71.0$
64	$n = 7$	$t_n \approx 7.18$	$n = 9$	$t_n \approx 56.9$	$n = 7$	$t_n \approx 309$
	$\bar{l} \approx 105$	$\bar{t}_l \approx 0.91$	$\bar{l} \approx 97$	$\bar{t}_l \approx 6.1$	$\bar{l} \approx 114$	$\bar{t}_l \approx 43.3$

Table 7. Results for the cavity tangential boundary flow control problem with $Re = 250$, restricted ASM with overlap $\delta = 1/16$ and augmented Lagrangian merit function. For the case of finest mesh, the number of variables is 528,392 and $\|\omega\|_h^2 \approx 50.2$

# Procs.	Mesh					
	64 × 64		128 × 128		256 × 256	
16	$n = 11$	$t_n \approx 15.8$	$n = 13$	$t_n \approx 110$	$n = 10$	$t_n \approx 1090$
	$\bar{l} \approx 51$	$\bar{t}_l \approx 1.22$	$\bar{l} \approx 55$	$\bar{t}_l \approx 7.67$	$\bar{l} \approx 55$	$\bar{t}_l \approx 106$
32	$n = 11$	$t_n \approx 17.2$	$n = 13$	$t_n \approx 126$	$n = 9$	$t_n \approx 726$
	$\bar{l} \approx 107$	$\bar{t}_l \approx 1.42$	$\bar{l} \approx 112$	$\bar{t}_l \approx 9.23$	$\bar{l} \approx 123$	$\bar{t}_l \approx 79.0$
64	$n = 10$	$t_n \approx 12.6$	$n = 13$	$t_n \approx 102$	$n = 9$	$t_n \approx 504$
	$\bar{l} \approx 135$	$\bar{t}_l \approx 1.16$	$\bar{l} \approx 139$	$\bar{t}_l \approx 7.62$	$\bar{l} \approx 180$	$\bar{t}_l \approx 55.1$

$\delta = 1/32$ and Table 6 for $\delta = 1/16$. The average number of GMRES iterations continues to follow (3.6) but now it better follows (3.4) and (3.5). The computing times t_n and $\bar{t}_l$ for the finest mesh in Table 6 decrease with the increase on the number of processors. The average number of GMRES iterations is larger in Table 5 than in Table 3, but the saving in communications of RAS compensates this increase so that $\bar{t}_l$ does not increase proportionally. By comparing finest mesh results on Tables 6 and 4 we see that RAS performs better than the standard ASM in terms of both $\bar{l}$ and $\bar{t}_l$, resulting on a smaller t_n.

Table 7 presents results for the tangential BCP with $Re = 250$, restricted ASM with $\delta = 1/16$ and augmented Lagrangian merit function. We can see that, with a Reynolds number greater than that in the previous table, both nonlinear (n) and average linear ($\bar{l}$) complexities increase.

6 Conclusions

We have developed a general LNKSz algorithm for PDE constrained optimization problems and applied it to some tangential boundary control problems involving two-dimensional incompressible Navier-Stokes equations. In our numerical experiments the LNKSz algorithm, together with an augmented Lagrangian merit function, provides a fully parallel and robust solution method. The one-level additive Schwarz preconditioned GMRES, with a proper overlap, works well for the indefinite linearized KKT systems. A proper overlap for a control problem seems to be greater than a proper overlap for a simulation problem. More precisely, in our experiments the proper overlaps are two to four times greater in the control problems. For larger overlaps the restricted version of ASM seems to perform better than the standard ASM as a preconditioner for linearized KKT systems. Theoretically, as a full space SQP method, LNKSz does not guarantee descent directions, and the solution of the KKT system through proper steps is guaranteed only to be a local minimum. However, in our numerical tests, the computed steps are always descent directions and $\|\omega\|_h^2$ decreases with the computed boundary control.

References

1. S. Balay, K. Buschelman, W. D. Gropp, D. Kaushik, M. Knepley, L. C. McInnes, B. F. Smith, and H. Zhang, *Portable, Extensible Toolkit for Scientific computation (PETSc) home page.* http://www.mcs.anl.gov/petsc, 2003.
2. G. Biros and O. Ghattas, *Parallel Lagrange-Newton-Krylov-Schur methods for PDE-constrained optimization, part I: the Krylov-Schur solver*, SIAM J. Sci. Comput., to appear.
3. ——, *Parallel Lagrange-Newton-Krylov-Schur methods for PDE constrained optimization, part II: the Lagrange-Newton solver and its application to optimal control of steady viscous flows*, SIAM J. Sci. Comput., to appear.
4. X.-C. Cai, M. Dryja, and M. Sarkis, *Restricted additive Schwarz preconditioners with harmonic overlap for symmetric positive definite linear systems*, SIAM J. Numer. Anal., 41 (2003), pp. 1209-1231.
5. X.-C. Cai, W. D. Gropp, D. E. Keyes, R. G. Melvin, and D. P. Young, *Parallel Newton-Krylov-Schwarz algorithms for the transonic full potential equation*, SIAM J. Sci. Comput., 19 (1998), pp. 246-265.
6. X.-C. Cai and D. E. Keyes, *Nonlinearly preconditioned inexact Newton algorithms,* SIAM J. Sci. Comput., 24 (2002), pp. 183-200.
7. X.-C. Cai, D. E. Keyes, and L. Marcinkowski, *Nonlinear additive Schwarz preconditioners and applications in computational fluid dynamics,* Int. J. Numer. Meth. Fluids, 40 (2002), pp. 1463-1470.
8. X.-C. Cai and M. Sarkis, *A restricted additive schwarz preconditioner for general sparse linear systems*, SIAM J. Sci. Comput., 21 (1999), pp. 792–797.
9. T. F. Coleman and J. J. Moré, *Estimation of sparse Jacobian matrices and graph coloring problem*, SIAM J. Numer. Anal., 20 (1983), pp. 243-209.
10. J. E. Dennis Jr. and R. B. Schnabel, *Numerical Methods for Unconstrained Optimization and Nonlinear Equations*, SIAM, second ed., 1996.
11. M. Dryja and O. Widlund, *Domain decomposition algorithms with small overlap*, SIAM J. Sci. Comp., 15 (1994), pp. 604–620.
12. S. C. Eisenstat and H. F. Walker, *Globally convergent inexact Newton method,* SIAM J. Optim., 4 (1994), pp. 393-422.
13. ——, *Choosing the forcing terms in an inexact Newton method,* SIAM J. Sci. Comput., 17 (1996), pp. 16-32.

14. M. D. Gunzburger and L. S. Hou, *Finite-dimensional approximation of a class of constrained nonlinear optimal control problems*, SIAM J. on Control and Optimization, 34 (1996), pp. 1001–1043.
15. M. D. Gunzburger, L. S. Hou, and T. P. Svobodny, *Optimal control and optimization systems arising in optimal control of viscous incompressible flows*, Incompressible Computational Fluid Dynamics: Trends and Advances, Max D. Gunzburger and Roy A. Nicolaides, eds., Cambridge University Press, (1993), pp. 109–150.
16. L. S. Hou and S. S. Ravindran, *A penalized Neumann control approach for solving an optimal Dirichlet control problem for the Navier-Stokes equations*, SIAM J. Sci. Comput., 20 (1998), pp. 1795–1814.
17. ——, *Numerical approximation of optimal flow control problems by a penalty method: Error estimates and numerical results*, SIAM J. Sci. Comput., 20 (1999), pp. 1753–1777.
18. A. Ioffe and V. Tihomirov, *Theory of Extremal Problems*, North-Holland Publishing Company, 1979. Translation from russian edition ©NAUKA, Moscow, 1974.
19. J. Nocedal and S. J. Wright, *Numerical Optimization*, Springer-Verlag, New York, first ed., 2000.
20. E. Prudencio, R. Byrd, and X.-C. Cai, *Parallel full space SQP Lagrange-Newton-Krylov-Schwarz algorithms for PDE-constrained optimization problems*, 2004, submitted.
21. L. Quartapelle, *Numerical Solution of the Incompressible Navier-Stokes Equations*, International Series of Numerical Mathematics, Vol. 113, Birkhäuser Verlag, 1996.
22. Y. Saad, *Iterative Methods for Sparse Linear Systems*, SIAM, second ed., 2003.
23. B. Smith, P. Bjørstad, and W. Gropp, *Domain Decomposition: Parallel Multilevel Methods for Elliptic Partial Differential Equations*, Cambridge University Press, 1996.
24. V. M. Tihomirov, *Fundamental Principles of the Theory of Extremal Problems*, John-Wiley & Sons, 1986.

Parallelism in Bioinformatics Workflows*

Luiz A.V.C. Meyer[1], Shaila C. Rössle[2], Paulo M. Bisch[2], and Marta Mattoso[1]

[1] Federal University of Rio de Janeiro - Computer Science Department – COPPE
{vivacqua, marta}@cos.ufrj.br,
[2] Federal University of Rio de Janeiro - Institute of Biophysics – IBCCF
{shaila, pmbisch}@biof.ufrj.br

Abstract. Parallel processing is frequently used in bioinformatics programs and in Database Management Systems to improve their performance. Parallelism can be also used to improve performance of a combination of programs in bioinformatics workflows. This work presents a characterization of parallel processing in scientific workflows and shows real experimental results with different configurations for data and programs distribution within bioinformatics workflow execution. The implementation was done with real structural genomic and automatic comparative annotation workflows and the experiments run on a cluster of PCs.

1 Introduction

Scientific and particularly biological resources like programs, data and workflows present characteristics of distribution, heterogeneity and in some cases huge volume of information. In the bioinformatics community and also in others scientific areas it is very common to find sites [7, 8, 16, 24, 29] allowing scientists to remotely access their specific data and run programs through the use of html pages or Perl scripts. One alternative to this remote use is downloading these resources, i.e. data and programs, and installing them in their own site. Often, scientists need to combine these resources through the use of a workflow orchestration [18]. In this case, programs and data are composed along an execution chain such that the output of a program execution can be used as the input of another. As observed by Deelman et al [13], scientific communities like physicists, astronomers and biologists are no longer developing applications as monolithic codes. Instead, standalone components are being combined to process data. In this scenario, bioinformatics applications can be viewed as bioinformatics workflows.

In general, executing a bioinformatics workflow is time consuming. For instance, processing large databases can be slow if the operation needs to scan all of their records. PDB [29], Genbank [16] and SwissProt [33] are examples of large databases widely used within bioinformatics applications, causing slow execution of common programs like Blast[3] and Fasta[28], which perform operations such as protein and genome sequence alignment or sequence comparison over these databases. Other

* This work is partially funded by CNPQ and CAPES Brazilian Agencies.

M. Daydé et al. (Eds.): VECPAR 2004, LNCS 3402, pp. 583–597, 2005.

kinds of operations like comparative protein structure modeling performed by programs like Modeller [31], although not data intensive are CPU intensive and also a time consuming task. The large size of biological data sets and the inherent complexity of biological problems result in large run-time and memory requirements. Thus, parallel processing should be used to enhance the performance of bioinformatics workflows.

One typical scenario in bioinformatics is having a set of different input sequences to be processed by a workflow. Thus, one question that can be made is how to process this bioinformatics workflow in parallel. The adoption of the traditional parallel techniques to bioinformatics workflows may not apply. Many bioinformatics workflow components are legacy programs. Therefore, it is not possible to parallelize their code and they have to be treated like black box components. Although applying data parallelism for each program individually is straightforward, this may not lead to the best solution. There are many alternatives to execute a bioinformatics workflow in parallel because programs and data can be distributed among the nodes in many ways. Choosing the best strategy for parallel execution is difficult because this choice must consider: (i) the dependencies among the components, (ii) the unbalanced execution time of the programs, (iii) the different size of datasets, and (iv) the available computational resources.

In addition to defining a parallel strategy for a workflow execution, there is also a need to manage the parallel execution of the workflow, to perform the initial distribution of the work among the nodes and also to collect and re-distribute partials results to have the workflow properly processed. These complex aspects involved in a workflow parallel processing indicates the need to develop tools to provide users with a workflow parallel design and a workflow parallel execution. The development of such tools allow scientists to be independent from parallel processing specialists in order to have their scientific workflows executed with better performance.

There are several studies [9, 12, 26, 34] addressing the parallel execution of individual bioinformatics programs in PC cluster environments, using techniques like data and program distribution. However, these works deal with the parallelism of isolated programs. There are also many projects [2, 4, 21, 32] that enable the design and execution of bioinformatics workflows, but with no exploitation of parallel processing. Bioinformatics workflows can benefit from both data and program parallelism. The combined use of data and program parallelism can improve the execution of: (i) multiple instances of a workflow, (ii) one instance of a workflow, and (iii) a single component of a workflow. However, it is mandatory to have an adequate workflow execution strategy.

Database Management Systems (DBMS) have been successfully using parallel processing to achieve better performance in their operations. The parallel hardware allows DBMS to use inter-query parallelism, intra-query parallelism and intra-operation parallelism [25]. In the former case, many queries can be executed at the same time, each by one respective processor. In the intra query parallelism, also called "pipelined", the output of one operator is streamed into the input of another operator so both operators can work in parallel within the same query or if two operators from the same query are independent they can also execute in parallel.

Finally, by partitioning the table among multiple processors, one single operation from a query can execute in parallel with each process running the same operation working on a specific part of the table. This partitioned data with parallel execution characterizes the intra operation or partitioned parallelism.

Similarly, parallel hardware can be used to gain better performance to execute scientific workflows. Like a relational query, formed by a set of operators that can communicate their results, a workflow is composed by a chain of programs, in general processing the antecessor results. Thus, an analogy to parallel processing in DBMS can be done, and three ways of parallelism when executing workflows can be expected to be accomplished: "inter-workflow", "intra-workflow" and "intra-program" parallelism.

Aiming to provide non-specialists in parallel processing a parallel solution to execute their bioinformatics workflow, we are experimenting with different forms of parallel execution, to propose heuristics to automate the execution design for the workflow. The main goal of our work is to provide a software layer that can propose a workflow plan for parallel execution. We contribute by making a characterization of parallel processing in bioinformatics workflows and presenting guidelines to execute workflow components in parallel. We also present performance results for different distribution configurations on a cluster of PCs. We evaluated three strategies using structural genomic and automatic comparative annotation workflows. We show speedup results for inter-workflow, intra-workflow and intra-program parallelism.

The rest of this work is organized as follows. Section 2 discusses the related works that deal with definition and execution of bioinformatics applications in Cluster, Grid and Web environments. The third section characterizes workflow parallel processing. The fourth section presents the structured genomic and the automatic comparative annotation workflows used in our experiments. In section 5 we describe the architecture implemented in this work, while in section 6 the experimental results are analyzed. Finally, section 7 concludes this work and points to future directions.

2 Related Work

Cluster of PCs, Grids and the Web are being used as platforms for the execution of scientific programs and workflows. Many initiatives can be found in the literature exploiting these distributed environments to enhance the performance of scientific applications [5, 9, 12, 14, 26, 35]. In the bioinformatics field, we can find several works exploiting cluster of PCs to improve the response time in sequence comparison and alignment operations [9, 12, 26, 35]. Braun et al [9] explore in their work, the use of Blast in batch mode, processing multiple query requests against a database replicated at all nodes. However, details of the implementation are not shown nor are practical results. Pappas [26] used a network of Dec Alpha workstations to set up a service for Blast requests. The service was implemented using PVM for parallel interface. The sequence database was fragmented and accessed via NFS simulating a shared-disk configuration. Costa and Lifschitz [12] present a detailed work where different approaches to distribute the query requests are examined. The services were implemented using MPI and were executed in a 32 PC cluster with several input databases and with different replication and fragmentation polices. Their parallel algorithms show significant improvements for the sequential implementation of BLASTP, which compares protein queries to protein databases. However, none of these works exploit parallelism to improve the performance of a combination of programs. They are focused on one single isolated program execution. Already widely used in projects by physicists, astronomers and engineers, the grids are beginning to be also used by the bioinformatics community. The Grid Application Development Software (GrADS) project [20] is developing a framework to simplify the preparation and execution of programs on a computational Grid. Yarkhan and Dongarra [35] describe how to enable the biological sequence alignment application FastA to run on the GrADS framework. Their work adopts database replication strategies to distribute data with a master-slave approach to process the query sequences. However, results are shown for a single application.

Meteor [19, 21], SDM [2], gRNA [6], BioOpera [4, 5] and SRMW [10, 11] are projects addressing the definition and execution of bioinformatics workflows in Web and PC cluster environments. These projects allow scientists to design their bioinformatics workflows using graphical interfaces or abstract workflow definition languages, which hide the low-level details and intricacies of program interactions and invocations. However, parallelism seems to be achieved only in restricted and explicit situations.

Chimera [14, 15], Pegasus [14] and myGrid [17] are projects being developed to address the definition and the execution of scientific workflows in computational grids. Chimera defines an architecture to integrate data, programs and the computations performed to produce data. Pegasus addresses the problem of generating abstract and concrete workflows for execution in Grids and can be integrated to Chimera. Chimera and Pegasus have been used in physics workflows. myGrid aims to develop middleware to support *in silico* experiments in biology. The architecture of the project is based on grid services. Like Chimera, myGrid also addresses the provenance of datasets. However, there are no reports describing performances results. Also, issues like strategies to perform data and program distribution and replication are not mentioned.

The simultaneous execution of the same workflow, i.e., inter-workflow parallelism, is only showed in Chimera work. But in this case, the input data to be processed by the workflow was previously fragmented over the several machines. None of the other works perform the distribution of the workflow input data during the workflow execution. Therefore, in these works the first program of the workflow always processes the input data with no data parallelism. Also, we did not find systems allowing the specification for data fragmentation and/or data replication for single components. Therefore, intra-program parallelism, which is very used to improve the performance of isolated scientific programs, appears not to be exploited during the execution of a workflow component in these systems. The graphical user interfaces and languages used to design the bioinformatics workflows do not allow users to specify this kind of feature. Thus, in order to have intra-program parallelism exploited during the workflow processing, it is necessary to have this information provided by the workflow designer. Our proposal is to develop tools, based in parallel database techniques like data distribution and replication, and also in heuristics, according to the different workflow patterns, in order to help scientists to execute their bioinformatics workflows in parallel.

3 Workflow Parallelism

A workflow can be characterized by a collection of programs organized to accomplish some scientific experiment. Therefore, a workflow *Wf* is represented by a triple (*P*, *I*, *O*) where: *P* is a set $\{p_1, p_2, \ldots, p_p\}$ of programs of *Wf*; *I* is a set $\{i_1, i_2, \ldots, i_m\}$ of input data elements; and O_j is a set of output data elements for each *j* program.

We propose three ways to achieve workflow parallelism: inter-workflow, intra-workflow and intra-program. **Inter-workflow** parallelism can be characterized by simultaneous execution of the same workflow *Wf*, each one processing a subset of the entire input data. This parallelism can be reached by allocating all the *P* programs of a workflow *Wf* at each node of the system but it can only be adopted if the set of input data elements *I* can be processed independently. The set of output data O_k generated in each *k* node must be merged in order to produce the final result. The distribution of the input data can be done basically in two ways: by groups (subsets of input data elements) or individually (each element at a time). Figure 1 illustrates this kind of parallelism, where the same workflow is replicated in the *n* nodes. The elements of *I* set are distributed individually to the nodes in a round-robin way.

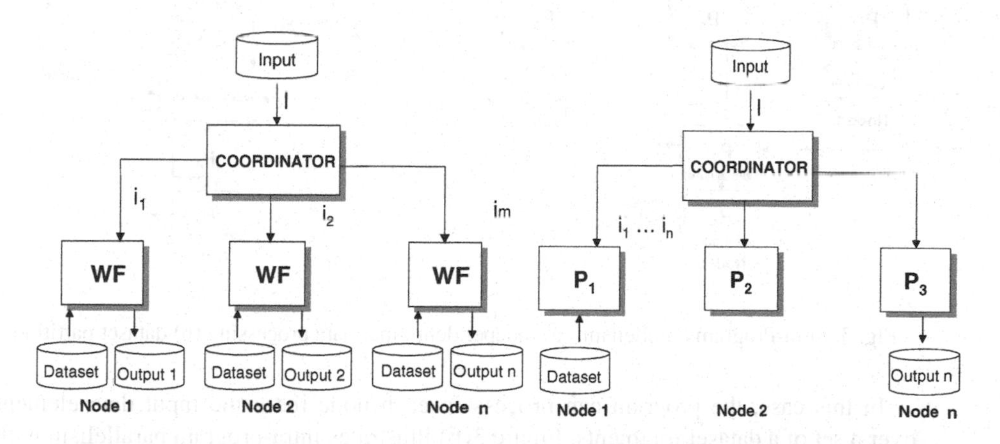

Fig. 1. Inter-Workflow Parallelism

Fig. 2. Intra-Workflow Parallelism

Intra-workflow parallelism can be characterized by simultaneous execution of more than one program p_j of the same workflow *Wf*. This parallelism can be reached by allocating the programs p_j of a *Wf* workflow at different nodes of the system. Intra-query parallelism is achieved in DBMS when an operator continually streams its result to another operator allowing both operators to work in parallel. This pipeline parallelism can also be achieved in a workflow execution. For example, supposing a workflow with three programs in a sequential pattern. **Intra-workflow** parallelism could be achieved by having program p_3 in node 3 processing an output O_2 previously generated by program p_2, program p_2 processing the output O_1 previously generated by program p_1, and program p_1 processing an input data i_3 provided by the coordinator. There are other execution chains of programs where intra-workflow parallelism can be naturally exploited. Whenever two or more programs are composed in a workflow in such way that their executions are independent, intra-workflow parallelism can be applied. Figure 2 illustrates **intra-workflow** parallelism, where three programs in a pipeline workflow are distributed to three nodes.

Intra-Program parallelism can be characterized by simultaneous execution of the same program p_j of a workflow Wf in different nodes. If the set of input data elements I for a program can be processed independently then intra-program parallelism can be achieved by allocating the same program p_j at different nodes and by distributing the set of input data between the nodes, like in the inter-workflow strategy. Figure 3(a) illustrates intra-program parallelism for independent input data processing. However, if the set of input data cannot be processed independently, **intra-program** parallelism can still be achieved if a program p_j processes a dataset that can be partitioned.

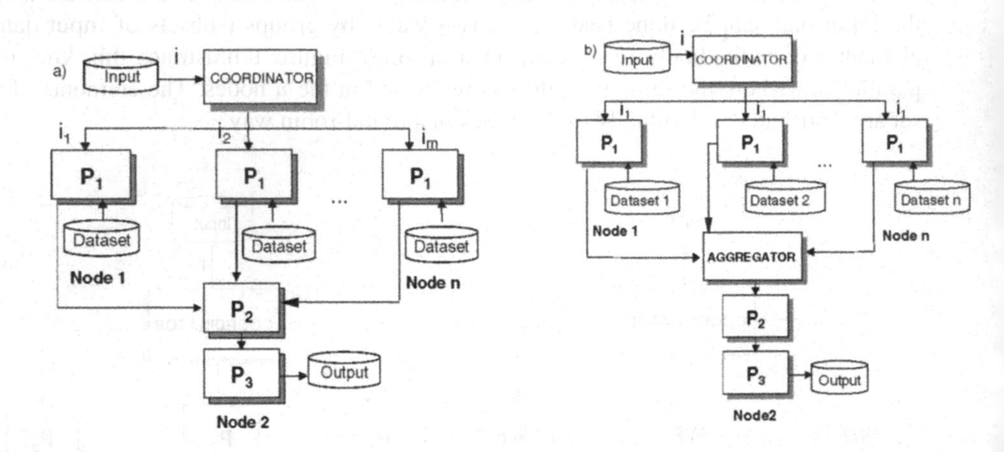

Fig. 3. Intra-Program Parallelism: (a) independent input data processing (b) dataset partition

In this case, the program can process in each node the same input data element over a set of a dataset fragments. Figure 3(b) illustrates intra-program parallelism with dataset partition where dataset 1, dataset 2 and dataset n represent fragments of a dataset processed by program p_1. This kind of data parallelism has been widely used by the database community and offers great opportunities to increase performance in shared-nothing DBMS [25]. In a parallel DBMS architecture, the database is fragmented according to some criteria and the generated data fragments are allocated to the different nodes. Consequently, slow operations like a full table scan in a large table, can be performed in much less time since each node have to scan a subset of the original table. In workflows, such strategy could be adapted for programs that process large datasets if this data partitioning does not interfere with the final result.

Aalst et al [1] define a number of workflow patterns addressing the execution of workflow components. However, these patterns do not address the parallel execution of the workflow components. Thus, based on these patterns, we propose in table 1 a set of initial heuristics that can allow the parallel execution of the workflow programs. Depending on the kind of the heuristic, additional actions must be performed in order to have the parallel execution achieved.

Table 1. Heuristics for workflow programs parallel execution

Pattern	Description	Heuristics	Additional Action
A → B → C Sequential	The execution of a program can be done after the execution of the predecessor	Execution of the programs in the same site	None
A → AND; AND → B, AND → C Parallel Split	Programs B and C have to be executed after program A	Programs B and C executes in different sites. Program A executes along with either B or C	The output of program A must be sent to the sites of programs B and C
A → XOR; XOR → B, XOR → C Exclusive Choice	Program B or program C must be executed after program A	Programs A, B and C execute in the same site	None
A, B → AND → C Synchronization	Program C can only executes when programs A and B finished	Program C execute in the same site of program A or program B	The output of programs A or B must be sent to the site of program C
A, B → XOR → C Simple Merge	Program C can execute if program A or program B finish	Program C executes in the same site of program A or program B, preferably along the faster one	The output of program A or the output of program B must be sent to the site of program C

4 The Experimental Workflows

This section describes the two bioinformatics workflows used to test our experiments. We chose these applications because they are real bioinformatics workflows and they represent the scenario which we are working: MholLine has many legacy programs, the workflow is executed for many entry sequences, the execution time of the programs is very distinct and the execution time can vary depending on the entry sequence. The automatic comparative annotation workflow has a time consuming component that processes a very large dataset that can be partitioned for parallel data processing.

Genome sequencing projects are producing a vast amount of protein sequences, emerging the need to use high throughput methods to predict structures and assign functions of these proteins. Analysis of several genome sequences indicates that the function cannot be inferred to a significant fraction of the gene products. In fact, isolate sequence homology searches do not always provide all of the answers, since some proteins may not keep sequence homology throughout evolution. On the contrary, the molecular (biochemical and biophysical) function of a protein is tightly coupled to its three-dimensional structure.

One of the methods that contributes to the prediction of protein three-dimensional structures is the comparative modeling, a computational procedure that predict the most reliable structure for a sequence using related protein structure as template. This approach consists of four steps: finding known structures (templates) related to the sequence to be modeled; alignment of the sequence with the templates, building a model, and the validation of the structure. Actually, other type of structure information can also be generated by threading approach, detecting structural similarities that are not accompanied by any detectable sequence similarity, becoming possible its use for fold recognition, pointing for a possible protein function. There are several programs addressing each of these steps.

MholLine is a biological workflow that combines a specific set of programs for the comparative modeling approach. For template structure identification it uses the Blast algorithm [3] for the search against the Protein Data Bank. A refinement in the template search step was implemented with the development of a program called Bats (Blast Automatic Targeting for Structures). Bats identifies the sequences that comparative modeling technique can be applied, choose the templates sequences from the Blast output file depending on the given scores for expectation values, identity and sequence coverage, and also construct the input files for the automated alignment and the model building carried out by Modeller [31], of the selected sequences. When a 3D model cannot be built using the comparative modeling approach, Bats identifies these sequences and construct the input file for the execution of Threader3 [22] that will aggregate structural information through threading, for sequences that did not generate any 3D model. The MholLine workflow is illustrated in figure 4.

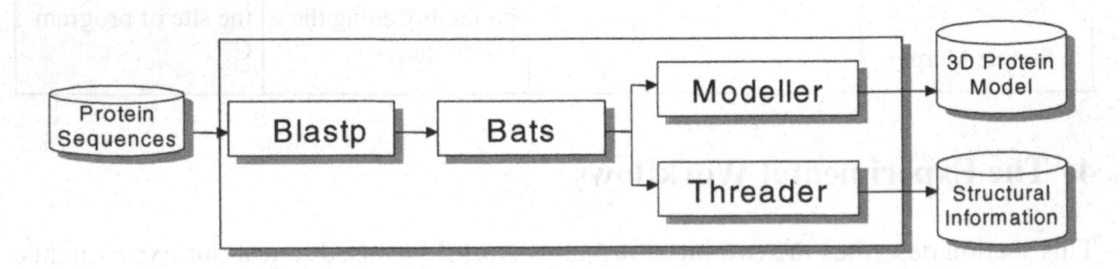

Fig. 4. The MholLine Workflow components

Functional annotation is a major objective in genome projects. The purpose of computerized sequence analysis is to discover functional information encoded in the nucleotide or amino acid sequence data. With an increasing number of complete genome sequences becoming available and specialized tools for genome comparison being developed, the comparative approach is becoming the most powerful strategy for genome analysis. The General approach involves the use of a set of algorithms such as the Blast programs to compare a query sequence to all the sequences in a specified database. Comparisons are made in a pair wise fashion. Each comparison is given a score reflecting the degree of similarity between the query and the sequence being compared. The higher is the score, the greater is the degree of similarity. Sequence alignments provide a powerful way to compare novel sequences with previously characterized genes. Many tasks in genome analysis can be automated, and, given the rapidly growing amount of data, automation is critical for the progress of genomics. The ultimate success of comparative genome analysis and annotation is based on complex decisions. Therefore, human expertise is necessary for avoiding errors and extracting the maximum possible information from the genome sequences. The Bats program can be used for selection of the best results obtained by Blast, overcoming human intervention in automatic sequence comparison. Bats calculates new scores from the Blast output file depending on the given scores for expectation values, identity, similarity, gaps and sequence coverage, merging human expertise for the automatic analysis. Figure 5 illustrates the automatic comparative annotation workflow.

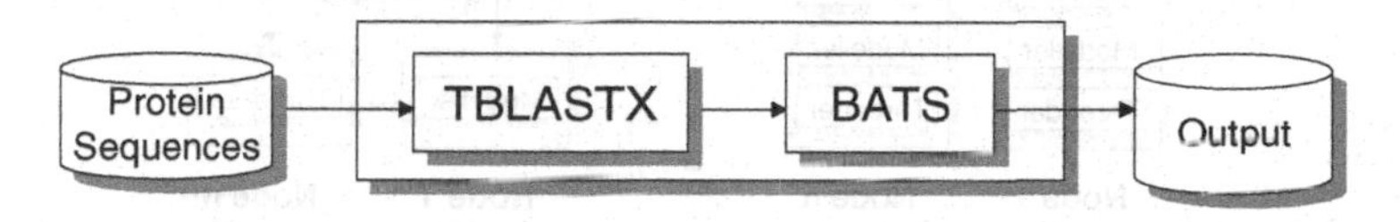

Fig. 5. The automatic comparative annotation workflow components

5 The Parallel Workflows Designs

The distribution of the programs and the data among the nodes can be done in different ways. For the MholLine workflow we adopted two scenarios. In the first one, all the workflow programs and respective files were placed in each cluster node. This strategy follows the heuristics for sequential and exclusive choice presented in section 3 and allowed inter-workflow parallelism to be exploited. On a second approach, the Modeller and the Threader modules, which are the slowest ones, were replicated, while Blast and Bats run on only one node. This strategy allowed intra-workflow parallelism to be exploited. Figure 6 illustrates these approaches. The distribution of the entries to the nodes plays an important role, due to the number of messages exchanged between the nodes. This distribution can be done basically in two ways: grouped or individually. In the first case, the number of messages exchanged by the coordinator and the nodes is proportional to the number of the nodes in the system, while in the second case, this number is proportional to the number of entries to be processed. Supposing that the number of nodes is less than the number of entries to be processed, the first alternative apparently seems to be better. However, there are no guarantees that the time needed to process each entry will be even. So if the coordinator performs the distribution in groups of entries, there are many possibilities to have one node finishing its job and stay idle while another one still has a lot of work to process. This situation leads to a load balance problem. The second approach, the individual distribution of the entries to the nodes, minimizes this problem since the coordinator can distribute the entries as soon as a node becomes able to process it. However the number of messages exchanged between the nodes is proportional to the number of entries to be processed.

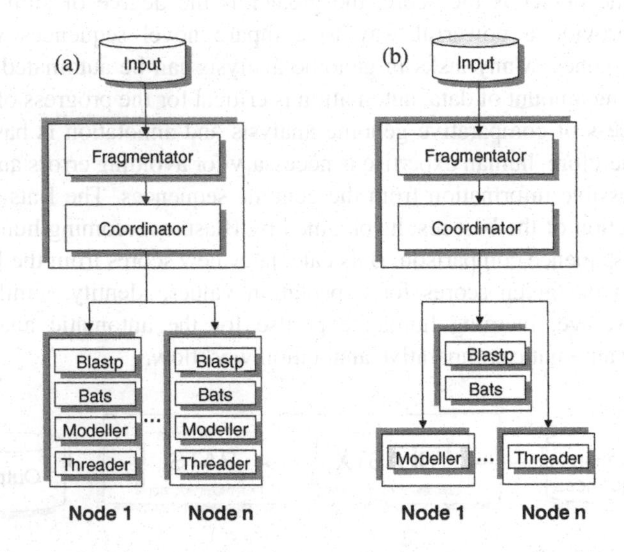

Fig. 6. MholLine implemented architectures: (a) Inter-workflow Parallelism (b) Intra-workflow Parallelism

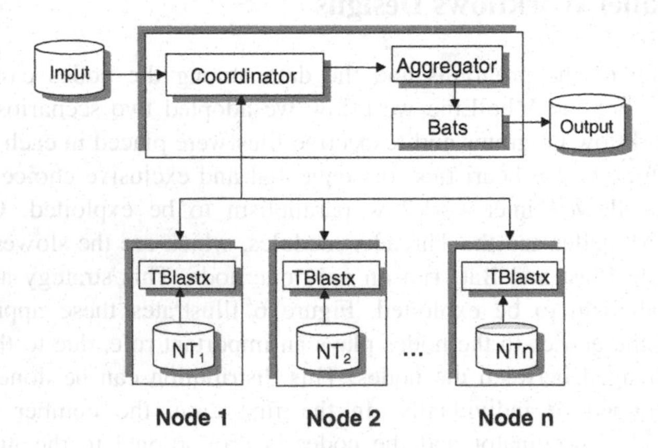

Fig. 7. The implemented architecture for automatic comparative annotation workflow

The parallel design for the automatic comparative annotation workflow exploited a different parallel strategy. Two programs compose this workflow and the Blast execution over the nt dataset [24] is very slow. Therefore, we adopted the intra-program parallelism strategy for Blast execution. In such way, in every node the Blast program run over a fragment of the nt dataset. In order to have the Blast output processed by Bats, all Blast partial results had to be aggregated. The aggregator module was responsible to merge the partial Blast results. Figure 7 illustrates the parallel design for the Automatic Comparative Workflow.

6 Experimental Results

This section presents the results obtained during the execution of the two workflows according to the architectures described in section 5. The experiments took place in an Itautec Cluster Mercury running Linux Red Hat 7.3, with 16 nodes. Each node has 512Mb of RAM memory, 18 Gb of disk storage and two Pentium III processors with 1 Ghz. They are linked with a Fast-Ethernet (100 MB/s) network. Perl programs wrapped the workflow components and we used MPI to implement the execution manager. The entry file used to process the MholLine workflow had 66 sequences and the data sources used in our experiments were: pdbaa [27] with nearly 30 MB used by Blastp, a set of pdb files extracted from Protein Data Bank with approximately 5 MB, used by Modeller, and the tdb files with almost 300 MB used by Threader. The automatic comparative annotation workflow processed one sequence over the nt dataset, which has almost 11Gb.

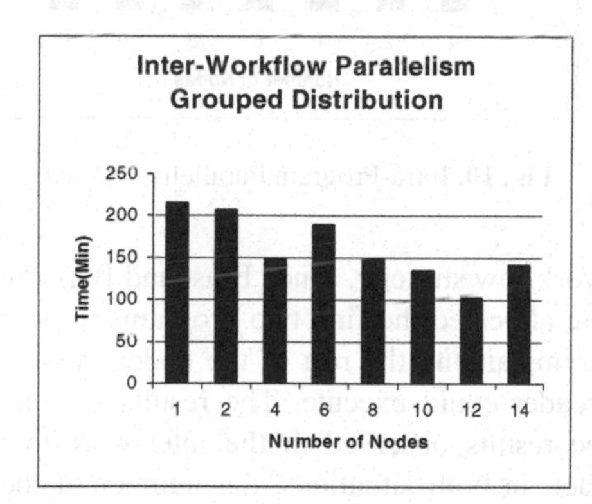

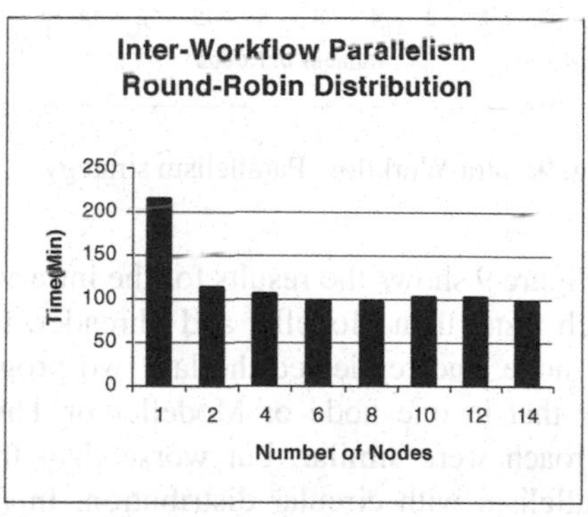

Fig. 8. The results for inter-workflow parallelism strategy. On the left the results for grouped sequences distribution. On the right the results for circular sequences distribution

The processing of the MholLine workflow began with the Blast program. The distribution of the 66 sequences to Blast was done in two ways: by groups of sequences or individually. In the first case, the sequences were grouped in files, according to the number of nodes used in the processing, and these files were distributed to the nodes. In the second case, the coordinator distributed the sequences to the nodes in a round-robin way as soon as a node became able to process it. Bats selected seven entry sequences to Modeller (sequences 4, 6, 7, 8, 11, 64 and 65) and three entry sequences to Threader (sequences 12, 14, and 22). Figure 8 shows the results for the executions using the inter-workflow parallelism strategy with a grouped and a circular distribution of the sequences. As can be observed, the circular sequences distribution acquired better performance than the grouped sequences distribution. Since Threader is the slowest program of the workflow, the best results were obtained when Threader could be executed in parallel. The fully parallel execution of Threader only happened in the grouped distribution, when running with 12 nodes. Running with 4, 8, 10 and 14 nodes implied having two sequences submitted for threading in the same node. When running with 2 and 6 nodes, the three sequences were submitted for threading in the same node. The second distribution strategy provided better load balance among the nodes and consequently best results.

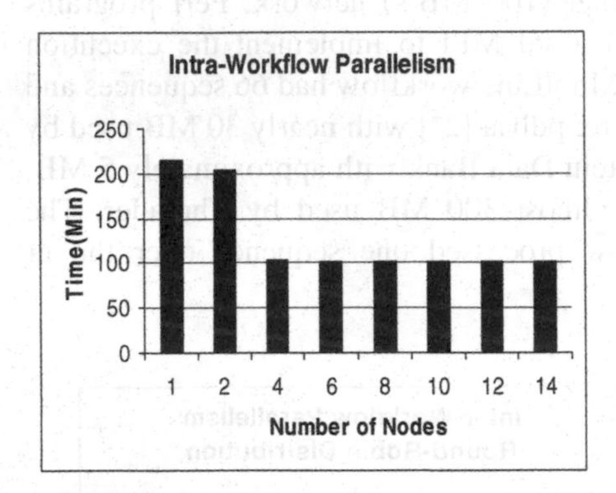

Fig. 9. Intra-Workflow Parallelism strategy

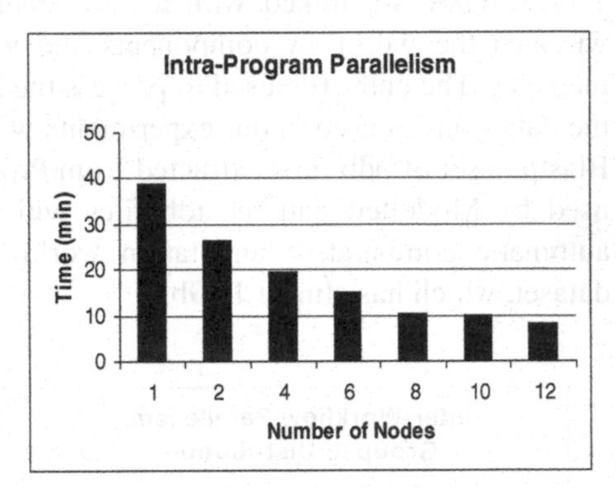

Fig. 10. Intra-Program Parallelism strategy

Figure 9 shows the results for the intra-workflow strategy. Since Blast and Bats run much faster than Modeller and Threader, we allocated the first two programs in only one node, and replicated the last two programs among the rest of the nodes in such way that in one node or Modeller or Threader could execute. The results for this approach were similar, but worse than the results observed in the inter-workflow parallelism with circular distribution. In fact, in both situations, the number of the sequences processed by Threader in parallel was the same, except when running with two nodes. The results for MholLine parallel execution show that the heuristics adopted in the second strategy were correct. As can be observed, using more than 6 nodes did not interfere in the speedup. This happened due to the few number of sequences selected by Bats to be submitted to Threader and Modeller in our experiments. However, the results show linear speedup when executing with two nodes in the inter-workflow parallelism strategy and when executing with four nodes in the intra-workflow parallelism strategy.

Figure 10 shows the results for automatic comparative annotation workflow execution. The bottleneck in this workflow execution is the time that Blast requires to compare a query sequence to all the sequences in the nt dataset. Since Bats and Aggregator are very fast, we allocated them along the Coordinator and replicated Blast among the rest of the nodes with a respective fragment of the nt dataset. The intra-program parallelism strategy adopted for Blast execution presented good results, but did not show improvements when using more than 8 nodes. Therefore, the other nodes could be used to process another sequence allowing inter-workflow parallelism to be employed simultaneously with intra-program parallelism.

7 Conclusion

We presented in this work several alternatives to explore parallel processing in bioinformatics workflows. Three approaches were defined with different entry sequence distribution and program allocation strategies. The results show that parallelism can be used to increase the performance of MholLine and the automatic comparative annotation workflows. We believe that the techniques showed here can also be applied to other bioinformatics workflows, in situations where many entries have to be processed, or different distribution strategies for programs and data can be exploited. However, the choice of the best parallel strategy is not trivial and the development of tools to help workflow designers to exploit parallelism in their applications is essential. We are still working on identifying heuristics to choose automatically the best parallel strategy. We are also working with web services alternatives [11] together with grid computing to replace Perl scripts.

References

1. Aalst, V., Hofstede, A., Kiepuszewski, B., Barros, A., 2003, "*Workflow patterns*", Distributed and Parallel Databases, Vol. 14, num 3, pp. 5-51
2. Altintas, I., Bhagwanani, S., Buttler, D., Chandra, S., Cheng, Z., Coleman, M., Critchlow, T., Gupta, A., Han, W., Liu, L., Ludascher, B., Pu, C., Shoshani, A., Vouk, M., 2003,"*A Modelling and Execution Environment for Distributed Scientific Workflows*", Proceedings of the 15th International Conference on Scientific and Statistical Database Management, SSDBM 2003, pp. 247-250, Massachusetts, USA
3. Altschul S.F., Gish W., Miller W., Myers E.W., Lipman D.J., 1990, "*Basic local alignment search tool.*" J. Mol. Biol., pp. 215-403
4. Bausch, W., Pautasso, C., Schaeppi, R., Alonso, G., 2002, "*BioOpera: Cluster-aware Computing*", Proceedings of the 4th IEEE International Conference on Cluster Computing, Chicago, USA
5. Bausch, W., Pautasso, C., Schaeppi, R., Alonso, G., 2003, "*Programming for Dependability in a Service-Based Grid*", 3rd International Symposium on Cluster Computing and the Grid, Tokyo, Japan
6. Bhowmick,S., Singh,D., Laud, A., 2003, "*Data Management in Metaboloinformatics: Issues and Challenges*", DEXA 2003, pp. 392-402
7. Bio Grid, http://www.eurogrid.org/wp1.html

8. BLAST, www.ncbi.nlm.nih.gov/BLAST
9. Braun R., Pedretti K., Casavant T., Scheetz T., Birkett C., Roberts C., 1999, "*Three Complementary Approaches to Parallelization of Local BLAST Service on Workstation Cluster*", 5th International Conference on Parallel Computing Technologies (PaCT), LNCS. Vol. 1662, pp. 271-282
10. Cavalcanti, M., Baião, F., Rössle, S., Bisch, P., Targino, R., Pires, P., Campos, M, Mattoso, M., 2003, "*Structural Genomic Workflows Supported by Web Services*", DEXA 2003, International Workshop on Biological Data Management (BIDM'03), IEEE CS Press, ISBN 0-7695-1993-8, pp. 45-50, Prague, Czech Republic
11. Cavalcanti, M., Targino, R., Baião, F., Rössle, S., Bisch, P., Pires, P., Campos, M, Mattoso, M., 2004, "*Managing Structural Genomic Workflows Using Web Services*", submitted to Data and Knowledge Engineering Journal special issue on Bioinformatics
12. Costa R., Lifschitz S., 2003, "*Database Allocation Strategies for Parallel BLAST Evaluation on Clusters*", Distributed and Parallel Databases, Vol. 13, num. 1, Jan 2003, pp. 99-127
13. Deelman E., Blythe J., Gil Y., Kesselman C., Mehta G., Vahi K., Blackburn K., Lazzarini A., Arbree A., Cavanaugh R., Koranda S., 2003, "*Mapping Abstract Complex Workflows onto Grid Environments*", Journal of Grid Computing, Vol.1, num. 1, pp.25-39
14. Foster,I., Voeckler, J., Wilde, M., Zhao, Y., 2002, "*Chimera: A Virtual Data System for Representing, Querying and Automating Data Derivation*", Proceedings of the 14th Conference on Scientific and Statistical Database Management, Edinburgh, Scotland
15. Foster,I., Voeckler, J., Wilde, M., Zhao, Y., 2003, "*The Virtual Data Grid: A New Model and Architecture for Data-Intensive Collaboration*", Proceedings of the First Bienal Conference on Innovative Data System Research – CIDR-2003, Asilomar, CA, USA
16. GenBank, www.ncbi.nlm.nih.gov/Genbank
17. Goble, C., De Roure, D., 2002, "*The Grid: An Application of the Semantic Web*", SIGMOD Record Vol. 31, num. 4, pp. 65-70
18. Greenwood, M., Wroe, C., Stevens, R., Goble, C., Addis, M., 2002, "*Are bioinformaticians doing e-Business?*" Proceedings Euroweb 2002: The Web and the GRID - from e-science to e-business, Oxford, UK
19. Hall,D., Miller,J., Arnold,J., Kochut, K., Sheth,A., Weise,M., 2003, "*Using Workflow to Build an Information Management System for a Geographically Distributed Genome Sequence Initiative,*" Genomics of Plants and Fungi, Eds, Marcel Dekker, Inc., New York, NY, 2003, pp. 359-371
20. Kennedy, K., Mazina, M., Crummey, J., Cooper, K., Torczon, L., Berman, F., Chien, A., Dail, H., Sievert, O., Angulo, D., Foster, I., Gannon, D., Johnson, L., Kesselman, C., Aydt, R., Reed, D., Dongarra, J., Vadhiyar, S., Wolski, R., 2002, "*Toward a Framework for preparing and Executing Adaptative Grid Programs*", Proceedings of Next Generation Systems Program Workshop, International Parallel and Distributed Processing Symposium, Fort Lauderdale, Florida, USA
21. Kochut, K., Arnold, J., Seth, A., Miller, J. Kraemer, E., Arpinar, B., Cardoso, J., 2003, "*IntelliGEN: A Distributed Workflow System for Discovering Protein-Protein Interactions*", Distributed and Parallel Databases, Vol. 13, num. 1, pp. 43-72
22. Miller R.T., Jones D.T. and Thornton J.M., 1996, "*Protein fold recognition by sequence threading tools and assessment techniques*". FASEB Journal, num. 10, pp. 171-178
23. ModBase, http://salilab.org/modbase/
24. NT, ftp://www.ncbi.nih.gov/blast/db/FASTA/nt.gz
25. Özsu, T., Valduriez, P., 1999, "*Principles of Distributed Database Systems*", Prentice Hall, 2nd Edition

26. Pappas, A., "*Parallelizing the Blast applications on a network of Dec Alpha Workstations*", available at http://www.cslab.ece.ntua.gr/~pappas
27. PDBAA, ftp://www.ncbi.nih.gov/blast/db
28. Pearson. W, Lipman.D.,"Improved tools for biological sequence comparison", Proceedings of National Academy of Sciences of United States of America,1988
29. Protein Data Bank, www.rcsb.org/pdb/
30. Rössle S., Carvalho, P., Dardenne, L., Bisch, P., 2003, "*Development of a Computational Environment for Protein Structure Prediction and Functional Analysis*", Second Brazilian Workshop on Informatics, Macaé, Brazil
31. Sali A. and Blundell T.L., 1993, "*Comparative protein modeling by satisfaction of spatial restraints.*" J. Mol. Biol., 234, pp. 779-815
32. Stevens, R., Robinson,A., Goble, C., 2003, "*myGrid: Personalised Bioinformatics on the Information Grid*", Bioinformatics, Vol. 19, suppl. 1, 2003, Eleventh International Conference on Intelligent Systems for Molecular Biology
33. SwissProt, http://www.ebi.ac.uk/swissprot
34. Waugh A., Willians G., Wei L., Altman R., 2001, "*Using Metacomputing Tools To Facilitate Large-Scale Analyses of Biological Databases*", Proceedings of Pacific Symposium of Biocomputing, 2001, pp. 360-371
35. Yarkhan, A., Dongarra, J., 2003, "*Biological Sequence Alignment On The Computational Grid Using The Grads Framework*", submitted to Journal on Grid Computing.

Complete Pattern Matching: Recursivity Versus Multi-threading

Nadia Nedjah and Luiza de Macedo Mourelle

Department of Systems Engineering and Computation,
Faculty of Engineering, State University of Rio de Janeiro,
Rua São Francisco Xavier, 524, Maracanã,
Rio de Janeiro, RJ, Brazil
{nadia, ldmm}@eng.uerj.br
http://www.eng.uerj.br/ldmm/index.html

Abstract. Pattern matching is essential in many applications such as information retrieval, logic programming, theorem proving, term rewriting and DNA-computing. It usually breaks down into two categories: root and complete pattern matching. Root matching determines whether a subject term is an instance of a pattern in a pattern set while complete matching determines whether a subject term contains a subterm that is an instance of a pattern in a pattern set. For the sake of efficiency, root pattern matching need to be deterministic and lazy. Furthermore, complete pattern matching needs also to be parallel. Unlike root pattern matching, complete matching received little interest from the researchers of the field. In this paper, we present a novel deterministic multi-threaded complete matching method. This method subsumes a deterministic lazy root matching technique that was developped by the authors in previous work. We evaluate the performance of proposed method using theorem-proving and DNA-computing applictions.

1 Introduction

Pattern matching [6], [8] is a fundamental operation in most applications such as information retrieval, functional, equational and logic programming [3], [4], [5], [19], theorem proving [4] and DNA-computing. Given a pattern set and a subject term, root pattern matching consists of discovering whether the subject term is an instance of one of the prescribed patterns. Moreover, pattern matching usually falls into two categories: *Root* matching techniques determine whether a given subject term is an instance of a pattern in a given set of patterns; *Complete* matching techniques determine whether the subject term contains a subterm (including the term itself) that is an instance of a pattern in the pattern set [17]. With ambiguous patterns, a subject term may be an instance of more than one pattern. Usually, patterns are partially ordered using priorities.

For instance, let $\Pi = \{faxy, fxby, fgxay, gx\}$ be the set of patterns wherein f and g are two functions of arity 3 and 2 respectively, a and b are two constants and x and y are two variables. Let $T = fagba$ be the subject term. Then T

M. Daydé et al. (Eds.): VECPAR 2004, LNCS 3402, pp. 598–609, 2005.

is an instance of pattern $faxy$ and subterm gb is an instance of pattern gx. Consequently, root matching would announce a match of the first pattern while besides this, the complete matching would also declare a match of the last pattern at the second argument of f, viewing the subject term as a parse tree.

Thus, complete matching subsumes root matching and root matching may be used to implement complete matching. This can be done by a recursive descent into the subject term. In this paper, we will show that using multi-threading instead of recursivity in monoprocessing systems improves matching time.

A simple-minded way of root pattern matching is to try each pattern sequentially until it is matched or the whole pattern set is exhausted. However, this method may consume considerable effort unnecessarily. Usually, patterns are pre-processed to produce an intermediate representation allowing the matching to be performed more efficiently. One such representation consists of a matching automaton [2], [7], [17], [20]. Moreover, the matching automaton must be deterministic so to avoid backtracking.

Pattern matching of terms does not need to be performed left-to-right. It may be performed according to a prescribed traversal order. Hence, one way to improve matching efficiency consists of adapting the traversal order to suit the input patterns. Lazy pattern matching [1], [9], [10], [18] declares a match or failure after a minimal number of symbol comparisons.

In the remainder of this paper, we first introduce a deterministic lazy root matching method [11], [12], [13]. We then generalise the root matching technique using a straightforward recursive descent then using multi-threading. We subsequently evaluate both ways of performing complete pattern matching by comparing space and time requirements necessary for an abstract rewriting machine to reduce some specific subject terms. A brief description of the machine will be given. (For further details on this issue, interested readers can refer to [14], [15], [16].) We also show the practicallity and evaluate the performance of the proposed method using two kind of applications: theorem-proving and DNA-computing.

2 Deterministic Lazy Root Matching

In this section, we briefly describe a practical method to construct a deterministic lazy tree matching automaton for a pattern set [11], [12]. The pattern set L is extended to its closure $\overline{L}$ while generating the matching automaton. First, we introduce some usefull definitions:

Definition 1. A *position* in a term is a path specification, which identifies a node in the parse tree of the term. Positions are specified here using a list of positive integers. The empty list Λ denotes the position of the root of the parse tree and the position $p.k (k \leq 1)$ denotes the root of the $k^{th\cdot}$ argument of the function symbol at position p.

Definition 2. A *matching item* is a pattern in which all the symbols already matched are now ticked i.e., they have the check-mark $\surd$. Moreover, it contains the *matching dot* $\circ$ that only designates the matching symbol i.e., the symbol to be accepted next. The position of the matching symbol is called *matching position.* A final matching item, namely one of the form $\pi\circ$, has the final matching position, which we write ∞. Final matching item may contain unchecked positions. These positions are irrelevant for announcing a match and so must be labelled with the symbol ω.

Definition 3. The term obtained from a given item by replacing all the terms with an unticked root by the placeholder _ is called the *context* of the items.

For instance, the context of the item $f^{\surd} a \circ g\omega a a^{\surd}$ is the term $f(\_, \_, a)$ where the arities of f, g and a are 3, 2 and 0. The arity of function f, will be denoted by $\#f$. In fact, no symbol will be checked until all its parents are checked. So, the positions of the placeholders in the context of an item are the positions of the subterms that have not been checked yet. The set of such positions for an item i is denoted by $up(i)$ (short for *u*nchecked *p*ositions).

Definition 4. A *matching set* is a set of matching items that have the same context and a common matching position. The initial matching set contains items of the form $\circ\pi$ because we recognise the root symbol first whereas final matching sets contain items of the form $\pi\circ$, where π is a pattern. For initial matching sets, no symbol is ticked. A final matching set must contain at least one final matching item i.e., in which all the unticked symbols are ωs.

Since the items in a matching set M have a common context, they all share a common list of unchecked positions and we can safely write $up(M)$. The only unchecked position for an initial matching set is clearly the empty position Λ.

The automaton is represented by the 4-tuple $\langle S_0, S, Q, \delta\rangle$ where S is the state set, $S_0 \in S$ is the initial state, $Q \subseteq S$ is the final state set and δ is the state transition function. The states are labelled by matching sets, which consist of original patterns whose prefixes match the current input prefix, together with extra instances of the patterns, which are added to avoid backtracking in reading the input. In particular, the matching set for S_0 contains the initial matching items formed from the original patterns and labelled by the rules associated with them. Transitions are considered according to the symbol at the matching position, i.e. that immediately after the matching dot. For each symbol $s \in F \cup \{\omega\}$ and state with matching set M, a new state with matching set $\delta(M, s)$ is derived using the composition of the functions *accept*, *choose* and *close* defined in the equations below:

$$accept(M, s, p) = \{(\alpha s^{\surd}\beta)_{\circ p} \mid \alpha \circ s\beta \in M\} \tag{1}$$

$$choose(M, s) = \begin{cases} \{p | p \in up(M) \backslash \{p_s\} \cup \{p_s.1, p_s.2, \ldots, p_s.\#s\} & \#s > 0 \\ \{p | p \in up(M) \backslash \{p_s\} & \#s = 0 \end{cases} \tag{2}$$

$$close(M) = M \cup \{\alpha \circ f\omega^{\#f}\mu \mid \alpha \circ \omega\mu \in M \; and \; \exists \alpha \circ f\lambda \in M, \; f \in F\} \tag{3}$$

$$\delta(M, s) = close\,(accept(M, s, choose(M, s))) \tag{4}$$

The items obtained by recognising the symbols in those patterns of M, where s is the next symbol, form the set *accept*(M, s), which is called the *kernel* of $\delta(M, s)$. However, the set $\delta(M, s)$ may contain more items. The presence of two items $\alpha \circ \omega\mu$ and $\alpha \circ f\lambda$ in M creates a non-deterministic situation since the variable ω could be matched by a term having f as head symbol. The item $\alpha \circ f\omega^{\#f}\mu$ is added to remove this non-determinism and avoid backtracking. The transition function thus implements simply the main step in the closure operation described by Grǎf [6] and set out in the previous section. Hence the pattern set resulting from the automaton construction using the transition function of Equation 4 coincides with the closure operation of Equation 3. The item labels simply keep account of the originating pattern for when a successful match is achieved. As we deal here with root matching, every failure transition ends up in a single global failure state. Non-determinism is worst where the input can end up matching the whole of two different patterns. Then we need a priority rule to determine which pattern to select.

Notice that the presence of the items $\alpha \circ \omega\beta$ together with the items $\alpha \circ f\mu$ in the same matching state creates a non-deterministic situation for a pattern-matcher since ω can be substituted with a term having f as head symbol. The items $\alpha \circ f\omega^{\#f}\beta$ are added to remove such non-determinism and avoid backtracking. For instance, let $L = \{f\omega\omega a, fc\omega c\}$ and let s be the matching state obtained after accepting the root symbol f so $s = \{f \circ \omega\omega a, f \circ c\omega c\} \cup \{f \circ c\omega a\}$. The item $f \circ c\omega a$ is added because a target term with the prefix fc could match the pattern $f\omega\omega a$ too if the last argument of f were a rather than c. So supplying the instance $fc\omega a$ would allow the pattern-matcher to decide deterministically which option to take. Without this new item, the pattern-matcher would need to backtrack to the first argument of f if the option offered by $f\omega\omega a$ were taken and a symbol c encountered as the last argument of f in the target term. (For more details and formal proofs see [9].)

Function *choose* selects the position that needs to be inspected next. It should choose those positions that need to be inspected first. Such positions are called *indexes*. Inspecting indexes first improves the size of the matching automaton, the matching time as well as the termination properties [15]. Function *choose* returns ∞ when a match can already be declared.

Example 1. Let $L = \{f\omega a\omega,\ f\omega\omega a,\ f\omega g\omega\omega g\omega\omega\}$ be the pattern set where $\#f = 3$, $\#g = 2$ and $\#a = \#b = 0$. Assuming a textual priority rule, the matching automaton for L is given in Fig. 1. Transitions corresponding to failures are omitted. Each state is labelled with its matching set. In the construction process, each new item is associated with the pattern from which it is directly derived. So, an added item $a \circ f\omega^{\#f}\beta$ is associated with the same pattern as is its parent $a \circ \omega\beta$. At the final nodes, whatever item is matched, the matched pattern of highest priority is chosen. This rule may be different from the one inherited by the item at that node.

During pattern matching, an ω-transition is only taken when there is no other available transition, which accepts the current symbol. The automaton can be used to drive pattern matching with any chosen term rewriting strategy.

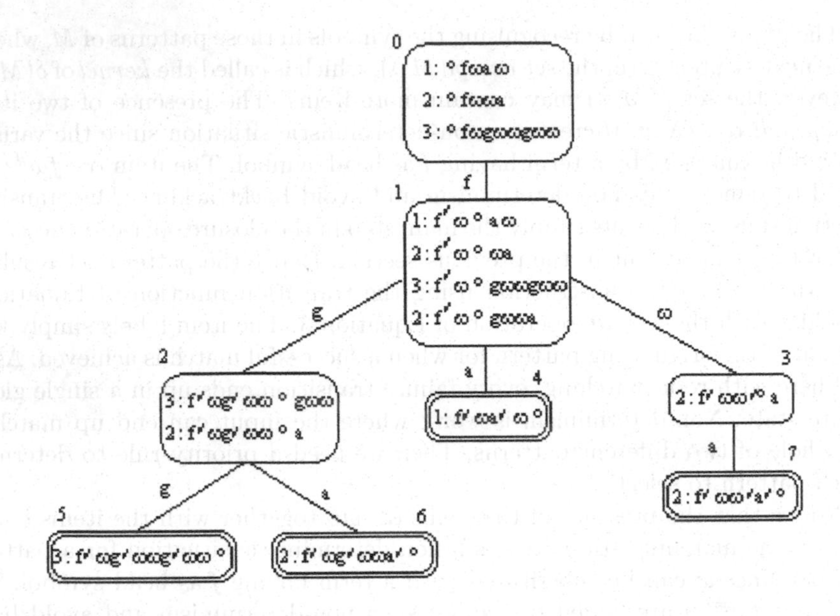

Fig. 1. Matching automaton for $\{f\omega a\omega,\ f\omega\omega a,\ f\omega g\omega\omega g\omega\omega\}$

The finite matching automaton corresponding to a given pattern set is represented using a matching table. Matching tables are simple, compact and expressive. Also, they allow a direct access to a given matching state. A matching table is an $N \times (L+1)$ matrix of transitions where N is the number of non-final states in the automaton (4 in the automaton of Fig. 1) and L is the number of function symbols in F_π. The extra column of N entries is used for variable occurrences in the patterns, all of which are denoted by ω. The distinction between ω and the other symbols permits a concise representation of the matching table. This also enables the use of matching tables for equational programs with an infinite alphabet.

For a matching state I and a symbol $s \in F_\pi \cup \{\omega\}$, let $\delta(I, s) = J$. Then the matching table entry $MT[I, s]$ is defined as follows:

$$MT[I,s] = \begin{cases} accept\_symbol_J & if\ J \ni Q\ and\ s \in F_\pi \\ accept\_term_J & if\ J \ni Q\ and\ s = \omega \\ match_\pi & if\ J \in Q\ and\ \pi\ is\ the\ matched\ pattern \\ fail & otherwise,\ i.e.\ J = \emptyset \end{cases} \quad (5)$$

In Equation 5, the J indicates the next state to be entered. Matching actions *accept_symbol* and *accept_term* mean that the current input symbol, respectively the current input term, is matched and the state $\delta(I, s)$ is to be entered. Matching

Table 1. Matching table for $\{f\omega a\omega,\ f\omega\omega a,\ f\omega g\omega\omega g\omega\omega\}$

state	f	a	g	ω
$\langle 0, \Lambda\rangle$	accept_symbol$_1$	fail	fail	fail
$\langle 1, 2\rangle$	fail	match$_{\pi_1}$	accept_symbol$_2$	accept_term$_3$
$\langle 2, 3\rangle$	fail	match$_{\pi_2}$	match$_{\pi_3}$	fail
$\langle 3, 3\rangle$	fail	match$_{\pi_2}$	fail	fail

action $reduce_\pi$ means that pattern π has been matched and should be applied whereas *fail* means matching has failed at symbol s.

The matching table corresponding to the matching automaton of Fig. 1 is shown in Table 1. It corresponds to the finite automaton of that figure. For convenience, the states/matching sets have been numbered as before. Notice that rows representing final states do not exist in the matching table. The decision whether the target expression matchs the pattern is anticipated in the state leading to the corresponding final state. This allows for the reduction of the matching table by R rows, where the number R of final states is also the number of patterns in the closure of the original pattern set. States are paired with the appropriate matching positions i.e., the position of the next symbol to be examined. For instance, in Fig. 1, state 1 is associated with position 2 while states 2 and 3 are both associated with position 3 as it is shown in Table 1.

With a matching table, whenever a state is entered, the current position in the target term is updated according to the matching position provided for that state. For instance, when state 2 is entered, the current matching position is updated as position 3. The state transition rule for the abstract matching machine is given below, whereby MP_J denotes the matching position that should be used when state J is entered. When the machine succeeds, it returns the pattern matched together with the position of the redex root R, while when it fails, it returns position λ paired with the empty pattern ϵ. In the remainder, we shall call $PatternMatch(T, P, L)$ the repetitive application of the state transition rule.

⟨I, P, MT, R, T⟩ ←
  case MatchingAction Of
    acceptsymbol$_J$: ⟨J, MP$_J$, MT, R, T⟩;
    match$_\pi$: return ⟨R, π⟩;
    fail: Case MT[I,ω] Of
        accept-term$_J$: ⟨J, MP$_J$, MT, R, T⟩;
        match$_\pi$: return ⟨R, π⟩;
        fail: return $\langle\Lambda, \epsilon\rangle$;
      End;
  where MatchingAction = if T[R.P] $\in$ F$_L$ then MT[I,T[R.P]] else fail.

3 Recursive Versus Multi-threaded Complete Matching

Complete matching subsumes root matching. One way of implementing complete matching consists of applying a recursive descent into the subject term. If the subject term T is not an empty term then we try to find a match between T and one of the proposed patterns. Subsequently, we try to find a match between the arguments of the root symbol of T and one of these patterns. The arguments are considered one by one. Using a java-like code, we codify the recusive complete matching as follows:

```
void RecursiveCM(T:Term, P:Position, L:PatternSet) {
   if (T != null) {
      PatternMatch(T, P, L);
      for (i = 0; i < #T[P]; i++) RecursiveCM(Arg(T,P,i), P.i, L);
   }
}
```

Observe that, it is only after the pattern matching of subject term $T = f(T_1, T_2, \ldots, T_{\#f})$ is completed that the matching of the first argument T_1, if any, takes place. Subsequently, only when the matching of T_1 finishes that the matching of the second argument T_2, if any, starts, and so on until all the arguments are exhausted. Hence, the matching work is performed sequentially.

Another way of implementing complete pattern matching is using multi-threading. This applies in mono-processing as well as multi-processing systems. However, multi-threading should have more effect in the latter. We demonstrate this later in this paper. For now, let us understand the multi-threaded complete matching that is coded hereafter. This is represented by a java-like class. The class implements the built-in java interface *Runnable* so that we can instantiate new threads of it. Note that the matching work can be done in any order and hence can reduce matching time by better schedule of the matching work carried by the matching threads. In a multi-processor environment, each thread can be run by an available processor. As in the recursive solution, for every arguments, a new thread is created and its execution started. Starting the execution of a thread, i.e. *start*() invokes automatically the method *run*. In the code below, function $Args(T, P, i)$ provides access to the $i^{th.}$ argument of function symbol at position p in term T, i.e. $T[p]$.

```
class MultiThreadedCM implements Runnable {
   Term T; Position P; PatternSet L;
   public void MultiThreadedCM(T:Term, P:Position, L:PatternSet) {
      this.T = T; this.P = P; this.L = L;
   }
   public void run() {
      if (T != null) {
         for(i = 0; i < #T[P]; i++) {
            Runnable R = new MultiThreadedCM(Args(T,P,i),P.i,L));
```

```
            new Thread(R).start();
          PatternMatch(T, P, L);
        }
    }
}
```

Observe that the actual pattern matching work performed by *PatternMatch* on term T is performed after the yileding of the required matching threads that deal with the subterms of the root of T. This gives a free hand to the thread scheduler of the run system so that no processor time is wasted. It is perfectly clear that the multi-threaded matching is more suitable for multi-processor systems or hyper-threaded processors. In the next section, we investigate these options.

4 Experimental Results

In this section we demonstrate the applicability and practicallity of the proposed method. We also acess its performance in real-world applications. Two kind of appliactions are explored to do so, namely theorem-proving related problems in Section 4.1 as well as DNA-computing in Section 4.2.

4.1 Theorem-Proving Related Applications

Some well-known therorem proving problems were used as benchmarks to evaluate the effect of recursivity vs. multi-threading. These problems were first used by Christian [2]. The *Kbl* benchmark is the ordinary three-axiom group completion problem. The *Comm* benchmark is the commutator theorem for groups. The *Ring* problem is to show that if $x^2 = x$ is a ring then the ring is commutative. Finally, the *Groupl* problem is to derive a complete set of reductions for Highman's single-law axiomatisation of groups using division.

4.2 DNA-Computing

In DNA-computing, information is not numerically encoded. Instead, it is encoded by genetically engineered DNA single strands. DNA strands are lists of *nucleotides*, which are chemical coumpounds such as *Adenine*(A), *Guanine*(G), *Cytosine*(C) and *Thymine*(T). For instance, the information of Table 2 will be coded as in Table 3 if attribute $X = 0, 1$, $Y = 0, 1, 2$, $Z = 0, 1, 2$ are encoded as $X =$ ATCAA, ATGGG, $Y =$ CACGG, CACTT, CACAA and $Z =$ AAAGG, AAAGT, AATAG respectively. Note that the record identification is also genetically encoded.

The advantage of using DNA-computing is that DNA strands allow parallel computation and very fast complemetary operations [11]. DNA-computing presents elegant and efficient sollutions to combinatorial problems [1].

DNA-computing needs several operations on strands. Among these operations, one finds operation *extract*. Given a tube Γ and DNA strand σ, this oper-

Table 2. Ealuation times: recursivity vs. multi-threading

Record	X	Y	Z
R_1	0	2	0
R_2	0	1	1
R_3	1	0	1
R_4	1	2	2

Table 3. Ealuation times: recursivity vs. multi-threading

Record	X	Y	Z
AAAGTC	ATCAA	CACAA	AAAGG
AAAGTA	ATCAA	CACTT	AAAGT
AAATGC	ATGGG	CACGG	AAAGT
AAATGA	ATGGG	CACAA	AATAG

ation returns two tubes: the first one consists of all strands in Γ which contain σ and the second consists of all the strands in Γ which do not contain σ. A *tube* is simply the concatenation of all the attributes including the record identification. We implemented operation *extract* using the complete pattern matching method described throughout this paper.

4.3 Performance Figures

For each of the benchmarks, we obtained the evaluation times. For the theorem proving-related applications, we used very large subject terms under our rewriting machine (for details see [11]) using both *RecursiveCM* and *MultiThreadedCM*. For the DNA-computing, we used a hundered of records each of which containing ten distinct attributes. Evaluation times were obtained for Intel^{TM}Pentium^{TM}III processor and Intel^{TM}Pentium^{TM}IV processor supporting hyper-threading technology. The time requirements(ms) are as in Table 4, wherein *RCM* and *MCM* stand for Recursive and Multi-threaded Complete Matching method respectively.

Table 4. Ealuation times: recursivity vs. multi-threading

Benchmark	Time - *RCM* PentiumIII	Time - *RCM* PentiumIV-HT	Time - *MCM* PentiumIII	Time - *MCM* PentiumIV-HT
Kbl	95	92	52	11
Comm	1630	1571	509	94
Ring	2351	2012	870	142
Groupl	2048	1962	988	229
Extract	1578	1014	776	112

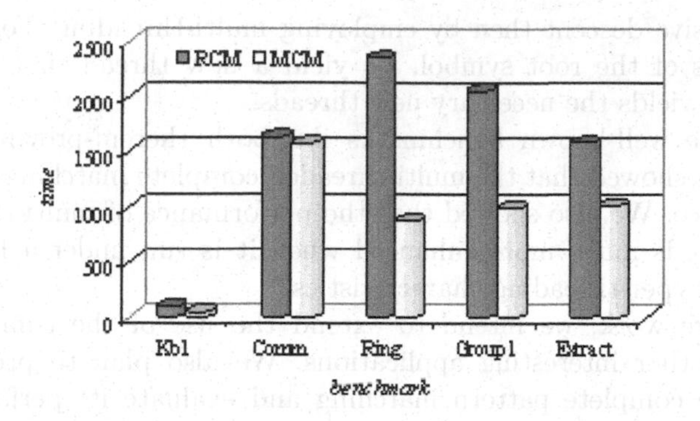

Fig. 2. Performance comparison: recusivity vs. multi-threading under Pentium^{TM}III

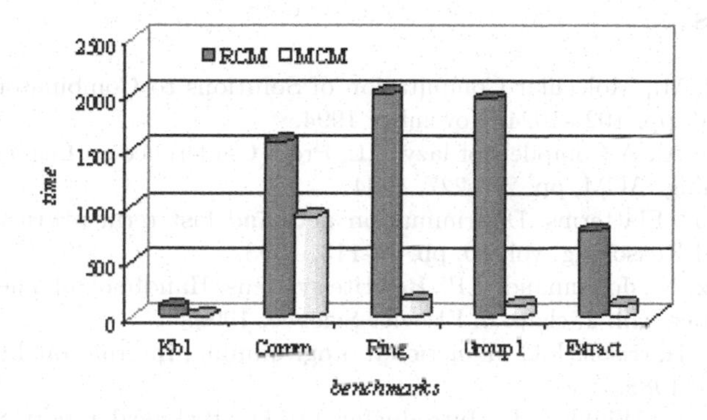

Fig. 3. Performance comparison: recusivity vs. multi-threading Pentium^{TM}IV-HT

In Fig. 2 and Fig. 3, we compare graphically the performance of the recursive and multi-threaded complete pattern matching under the Pentium^{TM}III and Pentium^{TM}IV-HT respectively. The performance figures of Table 4 show that using the multi-threaded complete matching yields about 50% of average time reduction over the recursive one on a Pentium^{TM}III processor. This time improvement is much more considerable in the case of using a hyper-threaded Pentium^{TM}IV processor. The gain is of about 80%.

5 Conclusion

In this paper, we first described a practical technique to compile pattern sets into deterministic lazy automata for root matching. We then showed that a matching automaton can be efficiently represented by a matching table. Subsequently, we illustrated how such matching table can be interpreted as to perform term pattern matching.

The paper goes on describing how the root matching can be generalised in order to allow complete pattern matching. This was done first using a straightforward recursive descent then by employing multi-threading. For each one of the arguments of the root symbol, we yield a new thread that perfoms root matching and yields the necessary new threads.

Using some well-known benchmarks ibn both therem-proving and DNA-computing, we showed that the multi-threaded complete matching provides better performance. We also showed that the performance of multi-threaded complete matching is much more enhanced when it is run under a Pentium^{TM}IV with enabled hyper-threading characteristics.

As a future work, we intend to extend the use of the complete pattern matching to other interesting applications. We also plan to provide parallel version of the complete pattern matching and evaluate its performance on a high-performance parallel computer.

References

1. Adleman L.M., Molecular Computation of Solutions to Combinatorial Problems, Science 266, pp. 1021-1024, November 1994.
2. Augustsson A., A Compiler for lazy ML, Proc. Conference on Lisp and Functional Programming, ACM, pp. 218-227, 1984.
3. Christian J., Flatterms, Discrimination nets and fast term rewriting, Journal of Automated Reasoning, vol. 10, pp. 95-113, 1993.
4. Dershowitz, N., Jouannaud, J.P., Rewrite systems, Handbook of Theoretical Computer Science, vol. 2, chap. 6, Elsevier Science, 1990.
5. Field, A.J., Harrison, P.G., Functional programming, International Computer Science Series, 1988.
6. Goguen, J.A., Winkler, T., Introducing OBJ3, Technical report SRI-CSL-88-9, Computer Science Laboratory, SRI International, 1998.
7. Gräf, A., Left-to-right tree pattern matching, Proc. Conference on Rewriting Techniques and Applications, Lecture Notes in Computer Science, Springer-Verlag, vol. 488, pp. 323-334, 1991.
8. Hoffman, C.M., ODonnell, M.J., Pattern matching in trees, Journal of ACM, vol. 29, no. 1, pp. 68-95, 1982.
9. Hudak, P., al., Report on the programming language Haskell: a Non-Strict, Purely Functional Language, Sigplan Notices, Section S, May 1992.
10. Laville, A., Comparison of priority rules in pattern matching and term rewriting, Journal of Symbolic Computation, no. 11, pp. 321-347, 1991.
11. Lipton J.L., DNA Solution to Hard Computational Problems, Science 268, pp. 542-545, April 1985.
12. Nedjah, N., Walter and C.D., Eldridge, S.E., Optimal left-to-right pattern matching automata, Proc. Conference on Algebraic and Logic Programming, Southampton, UK, Lecture Notes in Computer Science, M. Hanus, J. Heering and K. Meinke (Eds.), Springer-Verlag, vol. 1298, pp. 273-285, 1997.
13. Nedjah, N., Postponing redex contractions in equational programs, Proc. Symposium on Functional and Logic Programming, Kyoto, Japan, M. Sato and Y. Toyama (Eds.), World Scientific, pp. 40-60, 1998.
14. Nedjah, N., Walter, C.D., Eldridge, S.E., Efficient automata-driven pattern matching for equational programs, Software-Practice and Experience, vol. 29, n 9, pp. 793-813, John Wiley, 1999.

15. Nedjah, N. and Mourelle, L.M, Dynamic deterministic pattern matching, Proc. Computing: the Australasian Theory Symposium, Canberra, Australia, Elsevier Science, D.A. Wolfram (Ed.), vol. 31, 2000.
16. Nedjah, N. and Mourelle, L.M, Improving time, space and termination in term rewriting-based programming, Proc. International Conference on Industrial & Engineering Applications of Artificial Intelligence & Expert Systems, Budapest, Hungary, Lecture Notes in Computer Science, L. Monostori, J. Váncsa and A. M. Ali (Eds.), Springer-Verlag, vol. 2070, pp. 880-890, June 2001.
17. Nedjah, N. and Mourelle, L.M, Optimal Adaptive Pattern Matching. IEA/AIE 2002: 768-779, Proc. International Conference on Industrial & Engineering Applications of Artificial Intelligence & Expert Systems, Cairns, Australia, Lecture Notes in Computer Science, T. Hendtlass and A. M. Ali (Eds.), Springer-Verlag, vol. 2358, pp. 768-779, June 2002.
18. O'Donnell, M.J, Equational logic as programming language, MIT Press, 1985.
19. Paun G., Rozenberg, G. and Salomaa, A., DNA-computing - New Computing Paradigms, Springer-Verlag, Berlin Heidelberg New York, 1998.
20. Sekar, R.C., Ramesh, R. and Ramakrishnan, I.V., Adaptive pattern matching, SIAM Journal, vol. 24, no. 5, pp. 1207-1234, 1995.
21. Turner, D.A., Miranda: a Non strict functional language with polymorphic Types, Proc. Conference on Lisp and Functional Languages, ACM, pp. 1-16, 1985.
22. Wadler, P., Efficient compilation of pattern matching, In "The Implementation of Functional Programming Languages", In S. L. Peyton-Jones, Prentice-Hall International, pp. 78-103, 1987.

Probabilistic Program Analysis for Parallelizing Compilers

I.M. Forsythe*, P. Milligan, and P.P. Sage

School of Computer Science, Queen's University of Belfast,
Malone Road BT9 5HN Belfast Northern Ireland
(I.Forsythe, P.Milligan, P.Sage)@qub.ac.uk

Abstract. Parallelizing compilers have difficulty analysing and optimising complex code. To address this, some analysis may be delayed until run-time, and techniques such as speculative execution used. Furthermore, to enhance performance, a feedback loop may be setup between the compile time and run-time analysis systems, as in iterative compilation. To extend this, it is proposed that the run-time analysis collects information about the values of variables not already determined, and estimates a probability measure for the sampled values. These measures may be used to guide optimisations in further analyses of the program. To address the problem of variables with measures as values, this paper also presents an outline of a novel combination of previous probabilistic denotational semantics models, applied to a simple imperative language.

1 Introduction

Parallelizing compilers are a long investigated method for transforming legacy code into efficient parallel code. In addition, the techniques used in imperative code compilation (e.g. Fortran) may be used in new distributed systems that use object oriented languages with an imperative base (e.g. Java). Examples of these include dataflow analysis and dependence analysis. Compilers commonly have to analyse code that contains control flows involving non-local branches, exceptions or complex conditions. Alternatively, there is the problem of subscripted subscripts and input dependent variables. Various approaches are used, for instance speculative run-time analysis and iterative compilation, which search for information that is not given statically in the code being analysed.

Run-time analysis may be used to enhance the performance of the final program. Furthermore, information may be passed between the compile time and run-time analysis systems, which enables the compiler to tune the code to the target hardware. However, such analysis is perhaps best confined to frequently executed code hotspots that contain input dependent or complex constructs due to the expense of the analysis. Reducing the number of iterations required is also a useful goal. To deal with these issues, it is proposed that a sampling analysis

* Funded by QUB/ESF Studentship.

M. Daydé et al. (Eds.): VECPAR 2004, LNCS 3402, pp. 610–622, 2005.

collects information from programs of the values of variables not already determined by compile time analysis, and then hypothesises probability measures for these variables. These may be input to further iterations of the compilation process. The probability measures may be propagated by an adapted dataflow analysis algorithm.

It may also be possible to determine characteristics of codes affecting the probabilities, and as the accuracy of the estimated probabilities increases through sampling of typical applications, the probabilities can be estimated from the characteristics of a given code for use in compile-time analysis.

As an example of the utility of a probabilistic approach, regard the following code taken from the QCD Perfect Benchmark, where an array in a condition may hinder a precise code summary.

```
C----------------------------------------------------------------C
      SUBROUTINE MD(ID)
C----------------------------------------------------------------C
C PERFORMS MOLECULAR DYNAMICS OVER ONE TRAJECTORY
      INCLUDE 'PARAMS'
      INCLUDE 'VARBLES'
      CALL MDINIT(ID)
      IF (CONTROL1(ID) .GT. .5) THEN
         CALL STEPP(.5)
         IF (FERMNS) CALL STEPPF(.5)
         DO 100 ISTEP = 1, NSTEP
            CALL STEPU(1.)
            CALL STEPP(1.)
            IF (FERMNS) CALL STEPPF(1.)
  100    CONTINUE
         CALL STEPU(.5)
         ELSE
         CALL STEPU(.5)
         DO 200 ISTEP = 1, NSTEP
            CALL STEPP(1.)
            IF (FERMNS) CALL STEPPF(1.)
            CALL STEPU(1.)
  200    CONTINUE
         CALL STEPP(.5)
         IF (FERMNS) CALL STEPPF(.5)
      END IF
      RETURN
      END
```

This is a particularly acute problem if the values of the $CONTROL1$ array are input dependent, so requiring a run-time analysis to determine them for each run of the program. However, if an analysis collects information about the values of $CONTROL1$ in a series of program runs, it is possible to hypothesise probable values for the array for future runs, and hence to determine the likely effects of the code. This may be used in parallelization to determine whether to apply an optimisation to the program fragment. For instance, there are multiple loops and procedure calls within the code being analysed. If there is a high probability of the $CONTROL1$ branch being executed, and its run-time takes up a large fraction of the execution of the program, then optimisations may be profitably applied. Alternatively, if there is a low probability of the loop being executed, then it may be best to save the optimisations for elsewhere.

For a code analysis system to be able to determine the probable effect of a piece of code requires firstly that the system can work with variables with measures as values, and secondly, that it is given or can determine those values. To achieve the first goal requires sound theory to be developed for probabilistic programs. A unified model for a general imperative program will therefore be developed, using a probabilistic denotational semantics.

Section 2 will deal with some mathematical preliminaries. Section 3 will outline a probabilistic denotational semantics for imperative programs, and will address the question of fixed points for the semantics. Section 4 will summarise relevant related work.

2 Mathematical Preliminaries

The mathematics found in this section is based on that described in many texts in real analysis, functional analysis and measure theory, for instance [1], [2], [3], [4], [5]. This section and the next are a unification of the models of [6], [7], [8] and [9] for probabilistic imperative programs. The model for the semantics of the programs is based on a Hilbert space of probability measures, and bounded, linear operators acting on the space, which form a C^*-algebra [10].

The first requirement is the concept of a σ-algebra. A collection of of subsets of a set X forms a σ-algebra if it is closed under countable unions and complements of sets and contains X [4] Sect. 1.3. The closed, open and semi-open intervals on the real line form a σ-algebra, as do sets of integers. Closely related to this is the idea of a measurable space, which is a structure (X, $\mathcal{S}$), where X is a set and the $\mathcal{S}$ form a σ-algebra covering X. The members of $\mathcal{S}$ are known as measurable sets.

The next important concept is that of a bounded signed measure. Signed measures are maps from a measurable space to the real numbers. More formally, a signed measure μ is a map that is [5] Sect. 28:

1. Defined on a domain of measurable sets
2. Extended real valued
3. Zero on the empty set
4. Countably additive

Countably additive means that:

$$\mu(E) = \sum_{i=1}^{\infty} \mu(E_i). \tag{1}$$

for every disjoint sequence $\{E_i\}$ of sets in E (see [5] Sect 7). Note that only *bounded* measures are considered here, rather than those capable of taking extended real values. This will be useful later, when operators are considered to act on the measures. The definition of measure allows the definition of a measure space, which is an ordered triple $(X, \mathcal{S}, \mu)$, where X is a set, S is a σ-algebra of measurable sets and μ is a measure.

It is useful to generalise to products of measure spaces containing a finite or infinite number of components. It is probably clearer to state the definition of a countable Cartesian product of measure spaces $(\mathbb{R}_i, \mathcal{S}_i, \mu_i)$, and then explain the meaning of the components of that structure.

$$(\prod_{i=1}^{\infty} \mathbb{R}_i, \prod_{i=1}^{\infty} \mathcal{S}_i, \prod_{i=1}^{\infty} \mu_i) \tag{2}$$

The first term in (2) is the Cartesian product of a sequence of real number fields. The topology may be taken to be that of the real Hilbert space l^2. The second term in (2) is the smallest σ-algebra containing all measurable rectangles of the form $\prod_{i=1}^{\infty} \mathcal{B}_i$, where each $\mathcal{B}_i$ is a measurable set in the corresponding $\mathbb{R}_i$. To explain the third term in (2) is a little trickier. First, note that $X^{(n)}$ is defined to be:

$$X^{(n)} = \prod_{i=n+1}^{\infty} X_i. \tag{3}$$

Next, define the measure μ, where $\mu = \prod_{i=1}^{\infty} \mu_i$ by:

$$\mu(A_1 \times \cdots A_n \times X^{(n)}) = \prod_{i=1}^{n} \mu_i(A_i). \tag{4}$$

Now we can apply Th.(B) in Sect. 38 of [5].

Theorem 1. *If $\{(X_i, \mathcal{S}_i, \mu_i)\}$ is a sequence of totally finite measure spaces with $\mu_i(X_i) = 1$, then there exists a unique measure μ on the σ-algebra $\prod_{i=1}^{\infty} \mathcal{S}_i$ with the property that, for every measurable set E of the form $A \times X^{(n)}$,*

$$\mu(E) = (\mu_1 \times \cdots \times \mu_n)(A). \tag{5}$$

Therefore, given totally finite measures $\{\mu_i\}$ such that $\mu_i(\mathbb{R}_i) = 1$, a measure for a countably infinite product measure space can be defined, which may be used to represent the variables of an imperative program.

The signed measures can have the structure of a commutative algebra $\mathcal{U}$ defined on them, which obeys the axioms:

$$(\mu + \nu)(M) = \mu(M) + \nu(M) \tag{6}$$

$$(\mu \times \nu)(M) = \mu(M) \times \nu(M) \tag{7}$$

$$(a \cdot \mu)(M) = a(\mu(M)) \tag{8}$$

Where μ and ν are measures, M is a measurable set, and a is a scalar in the real field. It is also possible to define an inner product on this space, to represent how *close* two measures are. Define as an inner product on the signed measures μ, ν:

$$< \mu, \nu >= |\mu|(X) \cdot |\nu|(X) \tag{9}$$

The vector space of measures can be shown to obey the requisite axioms, making them into an inner product space. Note that it gives as norm the following (10).

$$\|\mu\| = |\mu|(X) \tag{10}$$

This coincides with the total variation norm for signed measures. The space of measures can also be shown to be complete in the metric derived from this norm, and therefore forms a Hilbert space.

The environment of the program is represented by the set of subprobability measures $\mathcal{V}$, which form a positive subset of the space of signed measures, with norms not greater than one. The variables of the program form a product probability measure defined on the measurable sets of the real part of a Hilbert space l^2, with each measurable set representing possible values for the variables. The measure of each set represents the probability that the variables have values lying in the set.

A set of linear operators $\mathcal{A}$ acting on the algebra $\mathcal{U}$ may be given an algebraic structure as follows.

$$(A + B)(\mu) = A(\mu) + B(\mu) \tag{11}$$

$$(A \times B)(\mu) = A(\mu) \times B(\mu) \tag{12}$$

$$(c \cdot A)(\mu) = c \cdot A(\mu) \tag{13}$$

Together with the uniform norm:

$$\|A\| = sup\{\|A(\mu)\| : \|\mu\| \leq 1\} \tag{14}$$

Due to the fact that the measures are bounded, the measure obtained by applying the operator A to μ in (14) will be bounded, and hence A will be a bounded operator. It is possible to define an adjoint on the Hilbert space, and hence together with the facts that the operators are linear and bounded, implies that the operators $\mathcal{A}$ form a C*-algebra [10] p3. Define another set of linear, bounded operators $\mathcal{B}$ that act as $\mathcal{A}$, but which preserve positive, real measures whose norm is always less than or equal to one; then these operators $\mathcal{B}$ preserve the subprobability measures.

It is possible to consider both discrete and continuous probability measures in this model. This flexibility may be useful for considering approximations of discrete measures - it may be quicker to consider continuous models described by a few parameters, rather than the whole set of values of the measure separately. Finally, using C*-algebras for the semantics raises the possibility of the model for the imperative language using the considerable body of work on these spaces as found in the theory of operator algebras [1, 10].

3 A Probabilistic Denotational Semantics for an Imperative Program

3.1 Syntax for an Imperative Language

The syntax for the model program can be represented in BNF as in Tables 1, 2, 3. The term *rand* represents a random number generator of unspecified distribution, and s a statement. The construct a represents an arithmetic expression, composed of variables v, addition and multiplication. The bs are boolean expressions, while *atom* is an atomic Boolean expression (representing for instance $x \leq y$).

3.2 Semantics: Arithmetic and Boolean Expressions

Arithmetic expressions map measurable sets in l^2 to other sets in l^2. To show that they form measurable functions, it is sufficient to prove that they form

Table 1. Syntax for Arithmetic Variables

a ::=	v	Variable
	v + v	Addition
	v × v	Multiplication

Table 2. Syntax for Boolean Variables

b :: =	atom	Atomic Expression
	¬b	Negation
	b ∧ b	Conjunction
	b ∨ b	Disjunction

Table 3. Syntax for Program Statements

s :: =	v:=a	Assignment
	v:=rand	Random Assignment
	s;s	Sequence
	if b then s else s	Selection
	while b do s	Iteration

Table 4. Semantics of Logical Expressions

$B[\![\neg b]\!]$	=	$B[\![b]\!]^C$	Negation
$B[\![b0 \wedge b1]\!]$	=	$B[\![b0]\!] \cap B[\![b1]\!]$	Conjunction
$B[\![b0 \vee b1]\!]$	=	$B[\![b0]\!] \cup B[\![b1]\!]$	Disjunction

continuous functions on the open sets of l^2 considered as a topological space. This is because the open sets belong to the σ-algebra of Borel sets [4] Sect. 1.11. It is important to note that the topology on infinite dimensional vector spaces is not unique, so one has to be chosen beforehand. Here Hilbert space is used, as it may be considered as a generalisation of $\mathbb{R}^n$. It is easy to prove that addition and multiplication are continuous functions, which will be omitted here.

The semantics of a logical expression are as in Table 4. Atomic logical expressions represent measurable sets in l^2, while the constructs in Table 4 describe how different logical expressions represent combinations of the measurable sets. For example, the third equation implies that the measurable set corresponding to matching $b0$ or $b1$ is the union of the measurable sets for matching $b0$ or matching $b1$.

An alternative viewpoint on Boolean expressions is from the point of view of linear operators on the set of measures $\mathcal{V}$. Let A and B be measurable sets, and μ a measure. Define:

$$\mu_B(A) = \mu(A \cap B) \tag{15}$$

and

$$e_B(\mu) = \mu_B \tag{16}$$

From this it is possible to prove that e_B is a linear operator on the set of measures $\mathcal{V}$. Following Kozen [6], we use this operator to act on the measures for the program variables following a branch statement. Inotherwords, it acts as a *weighting* operator.

3.3 Semantics: Statements

The effect of assignment is to combine the variables in the RHS of the expression, and assign the result to the variable on the LHS. Therefore, this statement replaces the variable x with the expression a and does nothing to other variables. Extending the approach of [6] to include an infinite number of variables, assignment can be represented as an operator A:

$$A(\mu) = \mu \circ F^{-1} \tag{17}$$

A(μ) will act on measurable sets to produce probabilities provided that F is a measurable function. F is defined by:

$$F(x_0, \ldots) = (x_0, \ldots, f(x_0, \ldots), \ldots) \tag{18}$$

Noting the earlier remark that continuous functions on l^2 are measurable, the measurability of F depends on the continuity of f. Since f is a continuous function on the variables that occur in the RHS of the assignment, and the other maps $x_i \to x_i$ are identity maps, F is continuous and hence measurable.

The random assignment generates a probability measure, and assigns it to x. Extending [6] to the infinite dimensional case gives:

$$R(\mu)(B_0 \times \cdots) = \mu(B_0 \times \cdots X_i \times \cdots) \cdot \rho(B_i) \tag{19}$$

Table 5. Semantics of Statements

S⟦x:=a⟧σ	= σ[x → A⟦a⟧σ]	Assignment
S⟦x:=rand⟧	= σ[x → ρ]	Random Assignment
S⟦s1; s2⟧	= S⟦s2⟧ ∘ S⟦s1⟧	Composition
S⟦if b then s1 else s2⟧	= S⟦s1⟧ ∘ B⟦b⟧ + S⟦s2⟧ ∘ B⟦¬b⟧	Selection
S⟦while b do s⟧	= fix(λg.g ∘ S⟦s⟧ ∘ B⟦b⟧ + B⟦¬b⟧)	Iteration

Table 6. Fixed Point Operator for the Language Constructs - adapted from [9]

$\Phi[\![x := a]\!]$	$= \mathcal{A}$	Assignment
$\Phi[\![x := rand]\!]$	$= \mathcal{R}$	Random Assignment
$\Phi[\![b]\!]$	$= \mathcal{B}$	Weighting Statement
$\Phi[\![S;T]\!]$	$= [\![T]\!] \circ [\![S]\!]$	Composition
$\Phi[\![$if b then S else T$]\!]$	$= [\![S]\!] \circ [\![b]\!] + [\![T]\!] \circ [\![\neg b]\!]$	Selection
$\Phi[\![$while B do S$]\!]$	$= [\![$while b do S$]\!] \circ [\![S]\!] \circ [\![b]\!] + [\![\neg b]\!]$	Iteration

Note that every measurable set is generated by the Borel sets B_i, so that the measure R(μ) is well-defined by (19) above. The analysis in Kozen applies to the sequence and selection statements. Iteration is different, and the semantics of the *while* statement are given by a fixed point construction, which will be discussed below.

3.4 Fixed Points for the Semantics

To determine the semantics of a program, the standard method is to start with a least element, and then iteratively apply a fixed point operator. For this to work a fixed point operator must exist, and so a definition follows in Table 6.

The operators $\mathcal{B}$, $\mathcal{A}$ and $\mathcal{R}$ are weighting, assignment and measure generating operators respectively. Following Di Pierro and Wiklicky [9], the Brouwer-Schauder fixed point theorem can be used to prove the existence of a fixpoint for the above semantics. However, they do not consider the case of realistic variables, so their approach will be extended to deal with this case. They also do not consider random assignments. To see what needs to be proved for the existence of a fixed point, consider the following theorem:

Theorem 2. Brouwer-Schauder. *Let F: A → A be a continuous function from a non-empty, closed convex set in a Banach Space to itself, with the closure of F(A) compact. Then F has a fixed point.*

To show that adding random assignment and realistic variables does not affect the existence of a fixed point, the approach taken in [9] may be followed. The process goes through as in that paper, except for two important points. First, it must be shown that the basic operators are compact. Second, it must be shown that the fixed point operator still maps the simplex (defined below) into itself.

Now it must be demonstrated that the basic operators $\{\mathcal{A}_i\}$ and $\{\mathcal{R}_i\}$ are bounded and compact. The norm of the assignment operator is defined by (20):

$$\|\mathcal{A}\| = sup\{\|\mathcal{A}(\mu)\| : \|\mu\| \leq 1\}. \tag{20}$$

This is certainly less than one, due to the assignment operator mapping measures to measures, which all have norm less than one. Therefore, the $\{\mathcal{A}_i\}$ are bounded, and because the measures are compact, the assignment operators are also compact. The definition of the other basic operator, random assignment, is as follows:

$$R(\mu)(B_0 \times \cdots) = \mu(B_0 \times \cdots X_i \times \cdots) \cdot \rho(B_i) \tag{21}$$

The norm for random assignment is:

$$\|\mathcal{R}\| = sup\{\|\mathcal{R}(\mu)\| : \|\mu\| \leq 1\}. \tag{22}$$

Since $\|R(\mu)\| = |R(\mu)|(X_0 \times \cdots) = \mu(X_0 \times \cdots) \cdot \rho(X) \leq 1$, $\|\mathcal{R}\|$ is also bounded. So the basic operators $\{\mathcal{A}_i\}$ and $\{\mathcal{R}_i\}$ are all bounded and compact.

Before showing that the fixed point operator maps the simplex into itself, a definition of the simplex is required.

Definition 1. *Given a C*-algebra $\mathcal{A}$ and a set of basic operators $\mathcal{O}_i$, the simplex $\mathcal{S} = \mathcal{S}(\mathcal{O})$ generated by the $\{\mathcal{O}_i\}$ is the smallest closed subset of $\mathcal{A}$ containing all the convex combinations of elements in $\mathcal{O}_i$.*

Take the simplex generated by each of the $\{\mathcal{A}_i\}$ and $\{\mathcal{R}_i\}$: it is closed and convex, by definition of the simplex. It must now be proved that $\mathcal{S}(\mathcal{O})$ is invariant under the operator Φ. Since the operators $\{\mathcal{A}_i\}$ and $\{\mathcal{R}_i\}$ are all bounded operators with a norm of one or less, we have the base case of the structural induction proved. Clearly for any S, T, of norms less than or equal to one, $\|T \circ S\| \leq \|T\| \cdot \|S\| \leq 1$. Therefore, sequencing maps the simplex into itself. For selection, the norm of:

$$S \circ b + T \circ \neg b \tag{23}$$

must be determined, and for iteration, the norm of:

$$g \circ S \circ b + \neg b \tag{24}$$

must be determined. Assuming that by the induction hypothesis, the norms of S, T and g are all less than one, the following equation is obtained for selection,

$$\|S \circ b + T \circ \neg b\| \leq \|S\| \cdot \|b\| + \|T\| \cdot \|\neg b\| \leq \|b\| + \|\neg b\|, \tag{25}$$

and this one for iteration:

$$\|g \circ S \circ b + \neg b\| \leq \|g\| \cdot \|S\| \cdot \|b\| + \|\neg b\| \leq \|b\| + \|\neg b\|. \tag{26}$$

By the definition of the norms of operators,

$$\|b\| + \|\neg b\| = sup\{\|b(\mu)\| + \|\neg b(\mu)\| : \|\mu\| \leq 1\}, \tag{27}$$

and by the properties of the norm and the definition of b,

$$\|b(\mu)\| + \|\neg b(\mu)\| \geq \|b(\mu) + (\neg b)(\mu)\| = \|\mu\|. \quad (28)$$

Therefore,

$$\|b\| + \|\neg b\| = sup\{\|b(\mu)\| + \|\neg b(\mu)\| : \|\mu\| \leq 1\} \geq \|\mu\|. \quad (29)$$

Therefore because the norm of the measures is less than or equal to one,

$$\|b\| + \|\neg b\| \leq 1, \quad (30)$$

so the operators for the selection and iteration statements are bounded. Therefore selection and iteration map the simplex to itself, so Φ maps the simplex to itself.

It can be seen that adding the operators corresponding to random assignment, or generalised variables, does not destroy the linearity of the operator Φ. Due to the proof of boundedness above, Φ is continuous. The $\{\mathcal{A}_i\}$ and $\{\mathcal{R}_i\}$ are operators forming a non-empty simplex in a C^*-algebra. Since a C*-algebra is also a Banach space, these premises of Th. (2) are also satisfied. Next, it has also been shown above that the operator Φ maps the simplex to itself, and that the basic operators are compact. Corollary (4.17) in [9] implies that the operator Φ will be compact if we assume that all the basic operations are compact, which has been proved above. Therefore by Th. (2), a fixpoint of the operator Φ exists.

4 Related Work

There are two themes in this paper - probabilistic semantics and program analysis. Since the aim of the probabilistic approach is ultimately to provide an enhancement of run-time program analysis, a brief analysis of how the semantics herein relates to work in run-time analysis follows.

4.1 Run-Time Analysis

The main approaches to run-time analysis are inspector-executor (IE) and speculative execution (SE). IE runs an analysis phase (the inspector) to determine wavefronts that can be executed in parallel, and then it runs the wavefronts with appropriate synchronization (the executor) [11] [12]. The main problem with this is that a lot of time may be spent analysing code that is unimportant or probably not potentially parallel [13]. The proposed probabilistic approach would instead determine critical points in the program that probably have a large effect on the amenability of the program to parallelization. The run-time analyses could then focus on these points. If, for instance, the probability of the existence of parallelization preventing dependences is found to be high at these points, it may be wise to treat them as actually existing, thus saving analysis time.

Speculative execution [13] [14] speculatively executes a loop as a DOALL, with run-time checks for dependences impeding parallelism. However if the dependences do exist, the attempt to run in parallel fails. Hardware is generally fast but financially expensive, software is cheaper but slower. Cintra and Ferraris [15] have produced an implementation in software that achieves 71% of the speedup of manual parallelized code on some typical benchmarks, if there are few data dependence violations. It would be useful to know the probability with which such violations occur, rather than just treating their existence as unknown. If their likelihood of occurrence was high, but the time taken to determine at run-time whether the violations actually existed was high, speculative execution could be omitted.

The Hybrid Analysis [16] approach to compilation bridges the gap between compile-time and run-time analysis, with analyses ordered from quick compile time to expensive reference by reference run-time tests. However, no attempt is made to determine the precise nature of dependences - only array privatisation and reduction recognitions are sought, and there is no attempt made to quantify the imprecision of the analysis. The efficiency is about < 1 to < 5 minutes on the four Perfect Benchmarks codes tested. This work may be furthered by extending the run-time system to hypothesize probability measures for the variables, and passing the measures to the compile-time analysis system, after a sample run of program executions. It remains to be seen whether a probability based analysis can provide a measure of the imprecision within a reasonable time.

4.2 Probabilistic Semantics

The model of Morgan et al [17] successfully links the predicate transformers of Kozen and general operational semantics of He to provide a probabilistic semantics of imperative programs. However, it is only really suitable for small programs. No attempt is made at a denotational semantics - only operational semantics are considered. Abstract Interpretation [18] is not considered either.

In Probabilistic Concurrent Constraint Programming [8], Di Pierro and Wiklicky develop a denotational Semantics of PCCP. Their successes include the development of a denotational semantics for probabilistic domains, and extending the notion of fixpoints to those domains by extending the concept of ordering to linear spaces.

They also develop a probabilistic abstract interpretation of PCCP in [19], where they show a way to develop consistent abstractions for vector spaces using the Moore-Penrose inverse. However, in neither case do they consider realistic imperative programs. A justification of the space used for the semantics is not given.

Monniaux also develops a probabilistic semantics and abstract interpretation in [7]. He deals only with simple imperative programs however, and it would be interesting to see extensions such as procedures and parallelism, and also to unbounded numbers of variables.

5 Conclusion and Future Work

The Hilbert space of measures defined on l^2 and the C^*-algebra of bounded operators on it provide a suitable space for the semantics of probabilistic imperative programs. Assignment, sequence, selection, and iteration can all be modelled within the same framework. The models of Kozen and Di Pierro et al., have been applied to imperative programs with general probability measures representing the program variables.

The next step is building a generalised dataflow analyser that can handle probabilities, instead of just treating uncertain values as unknown. Once this is done, a probabilistic dataflow analysis can propagate probabilities so that the system can provide better estimates of probable values of the other variables.

Overall, this simple model for the denotational semantics of imperative languages provides a good starting point for both further theoretical work (adding further code constructs, unifying functional analytic and order theoretic approaches), and practical (building a probabilistic dataflow analyzer), as well as offering the prospect of enhancing runtime analysis and iterative compilation. Through the quantitative measures proposed in this paper, there is also the prospect of a deeper understanding of the structure of typical HPC programs.

Acknowledgements

The authors would like to thank J. Yaqub and A. González-Beltrán for their useful comments on this paper.

References

1. Richard V. Kadison and John R. Ringrose. Fundamentals of the Theory of Operator Algebras Volume I, Academic Press, 1997
2. Marek Capiński and Ekkehard Kopp. Measure, Integral and Probability, Springer, 1999
3. Walter Rudin. Principles of Mathematical Analysis, McGraw-Hill, 1976
4. Walter Rudin. Real and Complex Analysis, McGraw-Hill, 1974
5. Paul R. Halmos. Measure Theory, Springer, 1974
6. D Kozen. Semantics of Probabilistic Programs. Journal of Computer and System Science, 22(3), pages 328-350, 1981
7. David Monniaux. Abstract Interpretation of Probabilistic Semantics, SAS 2000, pages 322-339
8. Alessandra Di Pierro, Herbert Wiklicky. Probabilistic Concurrent Constraint Programming. Towards a Fully Abstract Model, MFCS 1998, pages 446-455
9. Alessandra Di Pierro and Herbert Wiklicky. Linear Structures for Concurrency in Probabilistic Programming Languages, Electronic Notes in Theoretical Computer Science (40), 2001
10. Masamichi Takesaki. Theory of Operator Algebras I, Springer, 2002
11. J. Saltz, R. Mirchandaney, and K. Crowley. Run-time Parallelization and Scheduling of Loops, IEEE Trans. on Computers, Vol. 40, No. 5, pages 603-611, May 1991
12. S.-T. Leung and J. Zahorjan. Improving the performance of run-time parallelization, in 4th ACM SIGPLAN Symp. on Principles and Practice of Parallel Programming, pages 83-91, May 1993

13. L. Rauchwerger and D. Padua. The LRPD Test: Speculative Run-Time Parallelization of Loops with Privatization and Reduction Parallelization. In Proceedings of the SIGPLAN 1995 Conference on Programming Language Design and Implementation, pages 218–232, June 1995
14. F. Dang, H. Yu, and L. Rauchwerger. The R-LRPD Test: Speculative Parallelization of Partially Parallel Loops, Intl. Parallel and Distributed Processing Symp., pages 20-29, April 2002.
15. Marcelo H. Cintra, Diego R. Llanos Ferraris. Toward efficient and robust software speculative parallelization on multiprocessors, PPOPP 2003, pages 13-24
16. Silvius Rus, Lawrence Rauchwerger, Jay Hoeflinger. Hybrid analysis: static & dynamic memory reference analysis, ICS 2002, pages 274-284
17. Carroll Morgan, Annabelle McIver, Karen Seidel. Probabilistic Predicate Transformers. TOPLAS 18(3), pages 325-353 (1996)
18. Patrick Cousot, Radhia Cousot. Abstract Interpretation: A Unified Lattice Model for Static Analysis of Programs by Construction or Approximation of Fixpoints, POPL 1977, pages 238-252
19. Alessandra Di Pierro, Herbert Wiklicky. Concurrent constraint programming: towards probabilistic abstract interpretation, PPDP 2000, pages 127-138

Parallel Acceleration of Krylov Solvers by Factorized Approximate Inverse Preconditioners

Luca Bergamaschi and Ángeles Martínez

Dipartimento di Metodi e Modelli Matematici per le Scienze Applicate,
Università di Padova, via Belzoni 7, Padova (Italy)
{berga, acalomar}@dmsa.unipd.it

Abstract. This paper describes and tests a parallel implementation of a factorized approximate inverse preconditioner (FSAI) to accelerate iterative linear system solvers. Such a preconditioner reveals an efficient accelerator of both Conjugate gradient and BiCGstab iterative methods in the parallel solution of large linear systems arising from the discretization of the advection-diffusion equation. The resulting message passing code allows the solution of large problems leading to a very cost-effective algorithm for the solution of large and sparse linear systems.

1 Introduction

The solution of large and sparse linear systems $Ax = b$ is one of the central problems in scientific computing. Krylov subspace methods are widely used in combination with preconditioners that accelerate or even ensure convergence. The preconditioned conjugate gradient (PCG) method for symmetric positive definite (SPD) linear systems; GMRES and BiCGstab and their variants (see among the others [16, 18]) for general linear systems are the best known methods. The incomplete Cholesky (IC) and incomplete LU (ILU) factorizations prove to be the most robust accelerator for the symmetric and the general case, respectively. However, on modern high performance computers the effectiveness of Krylov subspace methods is more limited and the search for efficient parallel preconditioners is still an open question. The reason is that the application of the ILU type preconditioners to a vector is intrinsically sequential, and the attempts to parallelize their construction and their use have not been completely successful. Results in this direction can be found for example in [10] or [14] where the multilevel ILU preconditioner is blended with the algebraic multigrid approach.

In the last decade, a class of preconditioners with more natural parallelism, known as *approximate inverse preconditioners*, have been extensively studied by many authors. We quote among the others the SPAI preconditioner [9], the AINV preconditioner described in [3, 2] and the FSAI (Factorized Sparse Approximate Inverse) preconditioner proposed in [13]. These preconditioners explicitly compute an approximation to A^{-1} and their application needs only matrix vector products, which are more effectively parallelized than solving two triangular systems, as in the ILU preconditioner. Factorized sparse approximate inverses

M. Daydé et al. (Eds.): VECPAR 2004, LNCS 3402, pp. 623–636, 2005.

offer a number of advantages over their non factorized counterparts. They preserve the positive definiteness of the problem and provide better approximations to A^{-1} for the same amount of storage (than non-factorized ones). For an exhaustive comparative study of sparse approximate inverse preconditioners the reader is referred to [4]. Both FSAI and AINV compute the approximation to A^{-1} in factorized form. While the construction of FSAI is inherently parallel, in its current formulation AINV offers limited opportunity for parallelization of the preconditioner construction phase. In [5] a fully parallel AINV algorithm is devised by using a graph partitioning technique. The AINV preconditioner is generally more efficient than FSAI as accelerator of Krylov solvers, due to its flexibility in the generation of the pattern of the approximate inverse factor. On the contrary, the FSAI preconditioner requires the *a priori* specification of the sparsity pattern of the triangular factors. FSAI preconditioners have been successfully used in the acceleration of Jacobi-Davidson and CG-type methods in the partial solution of symmetric eigenvalue problems [6].

In this paper we propose an efficient parallelization of the construction of the FSAI factors both in the symmetric and in the nonsymmetric case. The parallelization is accomplished by assuming that the rows of the system matrix and of the preconditioning factors are partitioned among the processors, which exchange only the smallest amount of information needed for the computation of their local part of the preconditioner. Our code allows the *a priori* pattern to be that of A or that of A^2 (see [11]). However, in most cases using the pattern of A^2 results in a too dense preconditioning factor, which slows the iterative process. A partial cure to this inefficiency, is represented by postfiltration [12], which briefly consists in dropping out all the elements whose absolute value is below a fixed threshold. As a consequence, careful choice of the dropping parameter leads to a slight increase of the iteration number, which is counterbalanced by a large reduction of the solve timing. This is so because the CPU time per iteration is reduced since the preconditioner is sparser. All the variants of the FSAI preconditioner (with and without postfiltration) have been implemented in parallel to accelerate the CG and BiCGstab methods, to solve symmetric and nonsymmetric linear systems, respectively. These implementations use an efficient parallel matrix-vector product which minimizes communication among processors [7].

Experimental results upon large matrices arising from various discretizations of the advection diffusion equation show that the FSAI preconditioners are in several of the test cases almost as efficient as IC(0)/ILU(0) in sequential environments. Parallel results obtained on an SP4 supercomputer reveal that the code achieves good efficiency in both phases: construction of the preconditioner as well as iterative solution of linear systems. In all the problems considered, the parallel efficiency is comparable to the one obtained using a diagonal preconditioner (when applicable). The parallel code has been compared with ParaSails, a code written in C which implements the least square-based sparse approximate inverse preconditioner studied by Chow in [8].

2 The FSAI Preconditioner

Let A be a symmetric positive definite matrix and $A = L_A L_A^T$ be its Cholesky factorization. The FSAI method gives an approximate inverse of A in the factorized form $H = G_L^T G_L$, where G_L is a sparse nonsingular lower triangular matrix approximating L_A^{-1}. To construct G_L one must first prescribe a selected sparsity pattern $S_L \subseteq \{(i,j) : 1 \leq i \neq j \leq n\}$, such that $\{(i,j) : i < j\} \subseteq S_L$, then a lower triangular matrix $\hat{G}_L$ is computed by solving the equations

$$(\hat{G}_L A)_{ij} = \delta_{ij}, \qquad (i,j) \notin S_L. \tag{1}$$

The diagonal entries of $\hat{G}_L$ are all positive. Defining $D = [\text{diag}(\hat{G}_L)]^{-1/2}$ and setting $G_L = D\hat{G}_L$, the preconditioned matrix $G_L A G_L^T$ is SPD and has diagonal entries all equal to 1. As it has been described above the matrix $\hat{G}_L$ is computed by rows: each row requires the solution of a small SPD dense linear system of size equal to the number of nonzeros allowed in that row. Each row of $\hat{G}_L$ can be computed independently of each other.

In the context of a parallel environment and assuming a row-wise distribution of matrix A (that is, with complete rows assigned to different processors), each processor can independently compute its part of the preconditioner factor G_L provided it has access to a (small) number of non local rows of A.

A common choice for the sparsity pattern is to allow nonzeros in G_L only in positions corresponding to nonzeros in the lower triangular part of A. A slightly more sophisticated and more costly choice is to consider the sparsity pattern of the lower triangle of A^k where k is a small positive integer, e.g., $k = 2$ or $k = 3$; see [11]. While the approximate inverses corresponding to higher powers of A are often better than the one corresponding to $k = 1$, they may be too expensive to compute and apply. It is possible to considerably reduce storage and computational costs by working with sparsified matrices. Dropping entries below a prescribed threshold in A produces a new matrix $\tilde{A} \approx A$; using the structure of the lower triangular part of $\tilde{A}^k$ often results in a good pattern for the preconditioner factor. In [15] a prefiltration technique via vector aggregation is presented. This approach allows to minimize (sometimes significantly) the construction costs of low density high quality FSAI preconditioners.

The extension of FSAI to the nonsymmetric case is straightforward; however the solvability of the local linear systems and the non singularity of the approximate inverse is only guaranteed if all the principal submatrix of A are non singular (which is the case, for instance, if A is positive definite, i.e., $A + A^T$ is SPD).

2.1 Postfiltration of FSAI Preconditioners

In [12] a simple approach, called *postfiltration*, was proposed to improve the quality of FSAI preconditioners in the SPD case. This method is based on a posteriori sparsification of an already constructed FSAI preconditioner by using a small drop–tolerance parameter. The aim is to reduce the number of nonzero elements of the preconditioner factors to decrease the arithmetic complexity of the iteration phase. Also, in a parallel environment, a substantial reduction of the communication complexity of multiplying the preconditioner by a vector can be achieved. The drop–tolerance parameter can not be too large because the less deterioration of the preconditioner quality is desired.

In the nonsymmetric case both preconditioner factors, G_L and G_U, must be sparsified. We limit ourselves to nonsymmetric matrices with a symmetric nonzero pattern (which is the common situation in matrices arising from i.e. discretization of PDEs) and assume $S_L = S_U^T$. We perform what can be described as a symmetric filtration of factors G_L and G_U by filtering out the same number of small entries (in absolute value) of row i and column i of both preconditioner factors, respectively.

3 Parallel Implementation of the FSAI Preconditioner

We implemented the construction of the FSAI preconditioner both for SPD and nonsymmetric matrices. Our code allows the specification of both A or A^2 as sparsity patterns. No prefiltration of A has at the moment been implemented, but only postfiltration of the already constructed preconditioner to reduce the cost of applying the preconditioner during the iterative phase.

The code is written in FORTRAN 90 and exploits the MPI library for exchanging data among the processors. Even if the computation of an FSAI preconditioner is an inherently parallel computation there are some issues to be faced in order to create an efficient implementation on a distributed memory computer. These mainly regard the data distribution and the information exchanges among the processors. We used a block row distribution of matrices A, G_L (and also G_U in the nonsymmetric case), that is, with complete rows assigned to different processors. All these matrices are stored in static data structures in CSR format.

In the SPD case any row i of matrix G_L can be computed independently of each other, by solving a small SPD dense linear system of size n_i equal to the number of nonzeros allowed in that row of G_L. However, to do so the processor that computes row i must be able to access n_i rows of A. On a shared memory computer this is straightforward, but on distributed memory systems several of the needed rows may be non local (they are stored in the local memory of another processor). Therefore a communication algorithm must be devised to provide every processor with the non local rows of A needed to compute its local part of G_L. Since the number of non local rows needed by each processor is relatively small we chose to temporarily replicate the non local rows on auxiliary data structures. We implemented a routine called *get_extra_row* to carry out all the row exchanges among the processors, before starting the computation of G_L, which proceed afterwards entirely in parallel. Any processor computes its local rows of G_L by solving, for any row, a (small) dense linear system. To form the dense submatrix the local part of A and the previously received auxiliary non local rows are needed. The dense factorizations needed are carried out using

BLAS3 routines from LAPACK. Once G_L is obtained a parallel transposition routine provides every processor with its part of G_L^T.

In the nonsymmetric case we assumed a symmetric non zero pattern for matrix A and set $S_L = S_U^T$. The preconditioner factor G_L is computed as described before while G_U is computed by columns. In this way no added row exchanges is needed respect to the SPD case. Every processor computes a set of rows of G_L and a set of columns of G_U, completely in parallel. We notice that G_U is stored in CSC format only during the computation phase and it is transposed in parallel to CSR format before the start of the iterative solver. Finally we mention that the implementation of the FSAI preconditioner in the nonsymmetric case is roughly twice time and memory demanding than in the SPD case.

4 Numerical Results

The runs are carried out on a set of test matrices of dimensions varying up to $n = 10^6$. These matrices arise in the discretization of the steady state or transient advection-diffusion equation in two and three spatial dimensions

$$\frac{\partial u}{\partial t} + \mathrm{div}\,(K\nabla u) + \mathrm{div}\,(\boldsymbol{v}u) = 0 \tag{2}$$

subject to Dirichlet and Neumann boundary conditions. Different numerical integration schemes are employed, namely Finite Elements (FE), Finite Differences (FD), and Mixed Hybrid Finite Elements (MFE). The test problems arise from the discretization of regular and irregular domains Ω, with uniform or nonuniform computational meshes, and homogeneous or heterogeneous diffusion coefficients K.

Both the CG and the BiCGstab are left preconditioned. The starting vector is $x_0 = (0, \cdots, 0)^T$ and the right hand side b is computed by imposing the exact solution to be $x = (1, \ldots, 1)^T$. The diagonal (Jacobi) and the IC(0)/ILU(0) preconditioners are used for a comparison with FSAI with different sparsity patterns. The iteration is stopped when the residual r_k satisfies:

$$\|r_k\| \leq \alpha + tol\|b\| \tag{3}$$

where we set $\alpha = 10^{-8}$, $tol = 10^{-7}$. This choice guarantees that the initial residual norm is reduced by a factor 10^{-7} if the 2-norm of the right hand side is not too small. However, if $\|b\| \ll 1$, a common situation in linear systems arising from discretization of PDEs with no source terms, the iteration is stopped when the 2-norm of the residual is less than 10^{-8}.

All the sequential numerical experiments are performed on a Compaq DS20 equipped with an alpha-processor "ev6" at 500Mhz, 1.5 Gb of core memory, and 8 Mb of secondary cache. The parallel tests were performed on a SP4 Supercomputer with up to 512POWER 4 1.3 GHz CPUs with 1088 GB RAM. The actual configuration is 48 (virtual) nodes: 32 nodes with 8 processors and 16 GB RAM each, 14 nodes with 16 processors and 32 GB RAM and 2 nodes with 16

Table 1. Main characteristics of the symmetric matrices

#	matrix	n	nz	domain	K	discret.	λ_n/λ_1
1	`monte-carlo`	77 231	384 209	2d	heter	MFE	4.29×10^6
2	`fe_trans`	154 401	1 077 601	2d	heter	FE	3.82×10^3
3	`fe_ss`	154 401	1 077 601	2d	heter	FE	1.10×10^7
4	`mixed`	460 000	2 296 804	2d	homo	MFE	2.14×10^7
5	`heter`	268 515	3 926 823	3d	heter	FE	1.82×10^5
6	`fd_100`	1 000 000	6 940 000	3d	homo	FD	3.84×10^3

processors and 64 GB of RAM each. Each node is connected with 2 interfaces to a dual plane switch. The codes are written in Fortran 90. The CPU times are measured in seconds. In the SP4 Supercomputer, to obtain meaningful wall-clock times (with up 10% fluctuation) we reserved to our own use one 8-processor node at each run with up to 8 processors. Timings concerning 16 processors were obtained using a 16–processor node in dedicated mode.

4.1 Numerical Results on SPD Matrices

The main properties of the SPD matrices under consideration are summarized in Table 1, where we include the dimension n, the number nz of nonzero elements, the type of discretization employed, the characteristic of the coefficient K and the spectral condition number λ_n/λ_1 of each matrix. These matrices are the result of discretization of eq. (2) where $\boldsymbol{v}$ is set equal to 0. Matrix `fd_100` is the result of a 3d FD discretization of (2) on a unitary cube with the spatial discretization parameter $h = 0.01$. Matrices `fe_trans, fe_ss` arise from a 2d FE discretization of the same equation. The first one is obtained by setting the time step $\Delta t = 0.1$ while the second one refers to the corresponding steady-state problem. Problems `monte-carlo` and `mixed` arise from MFE discretization using a stochastic and a constant coefficient K, respectively. The last matrix `heter` comes from a FE discretization of the diffusion equation which models saltwater intrusion in porous media with a variable diffusion coefficient.

Sequential Runs. We report in Table 2 the results of the CG iteration with left preconditioning, applied to the test problems previously described. We show the sparsity ratio ρ, computed as the ratio of the number of nonzeros in the preconditioner to the number of nonzeros in the original matrix, the CPU time required for the computation of the preconditioner, T_p, the time for the CG solver iteration, T_{sol}, and the number of iterations, *iter*. Both A and A^2 are used as sparsity patterns of the FSAI preconditioner factors. The more accurate preconditioner obtained using the pattern of A^2 will be referred to as "enlarged" FSAI preconditioner. In this case three filter values are considered ($\varepsilon = 0$ (no filtration), $\varepsilon = 0.05, \varepsilon = 0.1$). We note from Table 2 that in all the test cases reported, IC(0) is always the best (sequential) preconditioner. The optimal FSAI preconditioner requires an iteration time which is usually between that needed by IC(0) and Jacobi. In particular when using the sparsity pattern

Table 2. Results of the sequential runs of the CG preconditioned with FSAI, Jacobi and IC(0) preconditioners

Matrix monte-carlo

pat	ϵ	ρ	T_p	T_{sol}	iter
A^2	0.00	1.96	0.93	17.39	469
A^2	0.05	1.77	0.93	15.81	469
A^2	0.10	1.77	0.93	16.17	468
A		1.00	0.34	26.36	701
Jacobi			0.01	15.64	1283
IC(0)		1.00	0.02	8.77	330

Matrix fe_trans

pat	ϵ	ρ	T_p	T_{sol}	iter
A^2	0.00	2.49	2.57	4.61	49
A^2	0.05	0.75	2.64	4.09	50
A^2	0.10	0.75	2.64	4.09	50
A		1.00	0.73	6.61	72
Jacobi			0.04	5.52	134
IC(0)		1.00	0.13	3.96	38

Matrix fe_ss

pat	ϵ	ρ	T_p	T_{sol}	iter
A^2	0.00	2.49	2.80	626.93	5677
A^2	0.10	0.75	2.92	542.81	5690
A		1.00	0.83	829.08	8146
Jacobi			0.04	742.49	15186
IC(0)		1.00	0.10	128.63	1080

Matrix mixed

pat	ϵ	ρ	T_p	T_{sol}	iter
A^2	0.05	2.11	5.51	319.71	1447
A^2	0.10	1.67	5.56	329.06	1511
A		1.00	1.97	526.99	2251
Jacobi			0.12	479.75	4281
IC(0)		1.00	0.18	267.83	1050

Matrix heter

pat	ϵ	ρ	T_p	T_{sol}	iter
A^2	0.00	3.98	21.91	58.98	179
A^2	0.05	1.03	23.34	33.81	201
A^2	0.10	0.71	23.52	29.38	180
A		1.00	2.38	42.64	253
Jacobi			0.12	46.43	478
IC(0)		1.00	0.50	17.51	56

Matrix fd_100

pat	ϵ	ρ	T_p	T_{sol}	iter
A^2	0.00	3.22	21.33	65.25	96
A^2	0.05	1.74	22.86	55.58	103
A^2	0.10	1.00	22.20	79.63	125
A		1.00	5.44	79.62	127
Jacobi			0.28	68.83	218
IC(0)		1.00	0.63	65.32	83

of A, the resulting FSAI preconditioner does not yield any substantial improvement with respect to diagonal preconditioning. However, especially in the most difficult problems, such as fe_ss, mixed, the A^2 pattern with filtration produces a significant reduction in both the iteration number and the CPU time. In these (2d) examples, the time required for the preconditioner computation is negligible as compared with the CG solution time. On the contrary, the results concerning the 3d problems (see the last two matrices in table 2) point out that the construction of the FSAI preconditioner with pattern A^2 may take a substantial percentage of the total CPU time.

Concerning the total CPU time ($T_p + T_{sol}$), FSAI with A^2-pattern and the "optimal" ε does not substantially improve the diagonal preconditioner. The situation may change if a sequence of linear systems with the same coefficient matrix (or a slightly modified one) has to be solved.

Parallel Runs. The parallel results which will be analyzed here, refer to a subset of the problems presented in the previous section. We choose only those tests for which the sequential performance of FSAI with optimal parameters

Table 3. Iteration number, CPU timings and speedups S_p for the solution of symmetric problems with CG and different preconditioners

Matrix `monte-carlo`

p	Jacobi (iter=1283)				FSAI(A) (iter= 700 ÷ 702)				FSAI(A^2, 0.05) (iter=469)			
	T_p	T_{sol}	T_{tot}	S_p	T_p	T_{sol}	T_{tot}	S_p	T_p	T_{sol}	T_{tot}	S_p
1	0.01	8.00	8.01		0.36	8.69	9.05		0.81	6.67	7.48	
2	0.01	4.60	4.60	1.7	0.19	5.03	5.22	1.7	0.43	3.47	3.90	1.9
4	0.00	2.92	2.92	2.7	0.12	2.74	2.86	3.2	0.26	2.18	2.44	3.1
8	0.00	1.84	1.84	4.4	0.09	1.78	1.87	4.8	0.16	1.38	1.54	4.9
16	0.00	1.02	1.02	7.9	0.09	0.92	1.01	9.0	0.14	0.71	0.85	8.8

Matrix `mixed`

p	Jacobi (iter=4281)				FSAI(A) (iter=2251)				FSAI(A^2, 0.05) (iter=1447)			
	T_p	T_{sol}	T_{tot}	S_p	T_p	T_{sol}	T_{tot}	S_p	T_p	T_{sol}	T_{tot}	S_p
1	0.03	174.24	174.27		2.07	183.51	185.58		4.72	169.23	173.95	
2	0.02	97.15	97.17	1.8	1.12	99.63	100.75	1.8	2.55	89.96	92.51	1.9
4	0.02	65.92	65.94	2.6	0.68	61.75	62.43	3.0	1.45	58.39	59.84	2.9
8	0.01	50.06	50.07	3.5	0.51	43.38	43.89	4.2	0.91	38.73	39.64	4.4
16	0.01	24.17	24.18	7.2	0.55	25.78	26.33	7.1	0.78	17.58	18.36	9.5

Matrix `heter`

p	Jacobi (iter=478)				FSAI(A) (iter=253)				FSAI(A^2, 0.1) (iter=180)			
	T_p	T_{sol}	T_{tot}	S_p	T_p	T_{sol}	T_{tot}	S_p	T_p	T_{sol}	T_{tot}	S_p
1	0.03	20.83	20.86		1.93	26.41	28.34		16.47	15.32	31.79	
2	0.02	12.25	12.27	1.7	1.04	13.85	14.89	1.9	8.45	8.39	16.84	1.9
4	0.02	7.50	7.52	2.8	0.65	8.50	9.15	3.1	4.37	5.14	9.51	3.3
8	0.01	4.96	4.97	4.2	0.50	5.73	6.23	4.6	2.45	3.75	6.20	5.1
16	0.01	3.30	3.31	6.3	0.53	3.26	3.79	7.5	1.60	2.37	3.97	8.0

improves that of the diagonal preconditioner (at least taking into account only the CG iteration time) being at the same time not too far from that of IC(0).

We report in Table 3 the number of iteration and the elapsed time using $p = 1, 2, 4, 8, 16$ processors. Jacobi, FSAI(A) and FSAI(A^2) with the optimal value of ε experimentally found, were used as preconditioners.

From Table 3 we see that the speedups obtained with the FSAI preconditioner are comparable to that obtained with Jacobi, thus showing the good degree of parallelism of the proposed preconditioner. Using A as the sparsity pattern, the scalability of the FSAI construction phase is not so good (we note for example that this time increases from 8 to 16 processors). This is due to the fact that the data to be exchanged among the processors remain roughly constant (owing to the band structure of the matrices involved) when the number of processors p increases whereas the complexity of the local computation decreases. This obviously worsens the parallel performance. The situation changes when the pattern of A^2 is used. In this case the local amount of work to construct the

Table 4. Main characteristics of the nonsymmetric matrices

#	matrix	n	nz	domain	discret.	porous medium
7	mphase1	94 482	376 782	2d	FE	heterogeneous
8	mphase2	94 482	412 582	2d	FE	heterogeneous
9	struct	27 306	566 352	3d	FE	heterogeneous
10	tran3d_small	78 246	1 058 736	3d	FE	homogeneous
11	tran3d_homo	534 681	7 837 641	3d	FE	homogeneous
12	tran3d_steady	534 681	7 837 641	3d	FE	heterogeneous

preconditioner is far larger with respect to the communication time. See for example the CPU times to construct the FSAI(A^2, 0.1) preconditioner in Table 3, matrix `heter`: from $p = 1$ to $p = 16$ time decreases of about 10.3.

Concerning the total elapsed time (T_{tot}), we found that FSAI is more efficient than Jacobi on the `mixed` and the `monte-carlo` problems. In the `mixed` problem, for instance, the CG preconditioned by FSAI(A^2, 0.1) is 25% more efficient than Jacobi-CG on 16 processors.

4.2 Numerical Results on Nonsymmetric Matrices

The main properties of the nonsymmetric matrices are summarized in Table 4. In detail, matrices `mphase1` and `mphase2` arise from 2D FE discretizations of the advection-diffusion equation which models multiphase flow in porous media by coupling together the mass conservation law with the Darcy velocity [1]. They are the results of the discretization at two successive time steps. The number of nonzero entries, and the condition number in the matrices grow as the time increases. Matrix `struct` is a 3D Finite Element discretization of a steady state structural transport problem using a variable velocity field.

As a last set of examples we consider the FE 3D discretization of the advection diffusion equation which models transport of contaminant in groundwater [1]. Matrix `tran3d_small` refers to a 2d mesh replicated in the z-plane to get 6 layers. The velocity field $\boldsymbol{v}$ is constant and the time-step relatively small. The other two problems deal with 41 layers (as a consequence the number of unknowns is roughly 7 times larger); they are characterized by a constant velocity field (`tran3d_homo`), and by a variable velocity field together with a heterogeneous porous medium and a very high Courant number (matrix `tran3d_steady`).

Sequential Runs. We report in table 5 the sequential results of the BiCGstab iteration preconditioned with Jacobi, FSAI(A), FSAI(A^2), with various choices of the filtration parameter ε, and ILU(0).

Here the situation is somewhat different with respect to the SPD case. For example, the diagonal preconditioner may not converge (matrices `mphase2`, `struct`) or converge very slowly (example `tran3d_steady`). As for problems `mphase1` and `mphase2` we notice that the performance of FSAI(A^2) with optimal ε is very close to that of ILU(0). This accounts for the efficiency of the

Table 5. Results of the sequential BiCGstab iteration with various preconditioners. Symbol † stands for convergence not achieved

Matrix `mphase1`

pat	ϵ	ρ	T_p	T_{sol}	iter
A^2	0.00	2.00	2.16	22.49	280
A^2	0.05	1.99	2.16	23.16	299
A		1.00	0.84	35.08	408
Jacobi			0.01	28.07	766
ILU(0)		1.00	0.08	18.38	261

Matrix `mphase2`

pat	ϵ	ρ	T_p	T_{sol}	iter
A^2	0.002	2.24	2.42	31.01	335
A^2	0.01	2.20	2.42	47.17	528
A		1.00	0.88	35.65	400
Jacobi			†	†	†
ILU(0)		1.00	0.09	27.67	379

Matrix `struct`

pat	ϵ	ρ	T_p	T_{sol}	iter
A^2	0.00	2.61	10.24	668.69	7460
A		1.00	†	†	†
Jacobi			†	†	†
ILU(0)		1.00	†	†	†
ILUT	0.001	2.50	2.34	45.27	407

Matrix `tran3d_small`

pat	ϵ	ρ	T_p	T_{sol}	iter
A^2	0.00	3.74	12.69	43.61	236
A^2	0.05	1.36	13.02	31.05	264
A^2	0.10	1.22	13.04	29.99	267
A		1.00	1.83	38.10	346
Jacobi			0.03	45.44	791
ILU(0)		1.00	0.13	32.46	245

Matrix `tran3d_homo`

pat	ϵ	ρ	T_p	T_{sol}	iter
A^2	0.01	2.01	108.42	33.89	38
A^2	0.05	1.13	108.75	27.58	39
A^2	0.10	0.64	109.30	32.83	52
A		1.00	13.67	36.62	52
Jacobi			0.25	44.28	112
ILU(0)		1.00	0.97	37.87	34

Matrix `tran3d_steady`

pat	ϵ	ρ	T_p	T_{sol}	iter
A^2	0.01	2.17	107.88	408.70	438
A^2	0.05	1.17	109.12	378.31	522
A^2	0.10	0.66	110.15	447.47	657
A		1.00	15.32	872.65	989
Jacobi			0.22	1182.90	2974
ILU(0)		1.00	0.94	220.74	212

proposed preconditioner also in sequential computations. For matrix `struct` neither Jacobi nor ILU(0) yields a converging scheme. The BiCGstab find a solution, though very slowly, only with FSAI(A^2, 0.0) as the preconditioner. Trying ILUT(50,0.001) (see [17]) produces a preconditioner with a sparsity ratio, ρ, similar to that of FSAI(A^2, 0.0), while the solution is reached after 407 iteration vs 7460.

In the last three examples the preconditioners display different behaviors as a function of the varying parameters of the discretization. FSAI(A^2, 0.05) is always more efficient than the diagonal preconditioner in terms of number of iterations and CPU time of the iterative phase. Here matrices `tran3d_small` and `tran3d_homo` are the result of transient discretization. In a complete time simulation, hundreds of systems with a coefficient matrix with the same sparsity pattern need to be solved; therefore a large part of the preconditioner computation is done only once at the beginning of the simulation. In the example `tran3d_steady` (steady-state computation) the FSAI preconditioner is much more efficient than the diagonal preconditioner even if one considers the total time (preconditioner computation + BiCGstab).

Table 6. Iteration number, CPU timings and speedups S_p for the solution of non-symmetric problems with BiCGstab and different preconditioners

Matrix `mphase2`

p	FSAI (A)					FSAI (A^2, 0.002)				
	iter	T_p	T_{sol}	T_{tot}	S_p	iter	T_p	T_{sol}	T_{tot}	S_p
1	461	0.37	13.69	14.06		298	1.19	12.16	13.35	
2	405	0.22	6.38	6.60	2.1	302	0.64	6.68	7.32	1.8
4	462	0.13	3.80	3.93	3.6	331	0.35	4.40	4.75	2.8
8	490	0.10	2.75	2.85	4.9	270	0.28	2.24	2.52	5.3
16	450	0.13	1.32	1.45	9.7	332	0.20	1.37	1.57	8.5

Matrix `tran3d_homo`

p	Jacobi					FSAI (A)					FSAI (A^2, 0.05)				
	iter	T_p	T_{sol}	T_{tot}	S_p	iter	T_p	T_{sol}	T_{tot}	S_p	iter	T_p	T_{sol}	T_{tot}	S_p
1	126	0.07	28.36	28.43		52	5.29	25.04	30.33		39	49.63	19.89	69.52	
2	106	0.05	12.25	12.30	2.3	52	2.91	13.48	16.39	1.9	39	25.79	11.51	37.30	1.9
4	124	0.03	9.15	9.18	3.1	52	1.72	8.53	10.25	3.0	39	13.11	6.72	19.83	3.5
8	108	0.02	5.27	5.29	5.4	52	1.28	5.82	7.10	4.3	39	7.10	4.70	11.80	5.9
16	108	0.01	2.41	2.42	11.7	52	1.13	2.66	3.79	8.0	39	4.42	2.37	6.79	10.2

Matrix `tran3d_steady`

p	Jacobi					FSAI (A^2, 0.05)				
	iter	T_p	T_{sol}	T_{tot}	S_p	iter	T_p	T_{sol}	T_{tot}	S_p
1	2745	0.07	605.66	605.73		470	49.54	244.80	294.34	
2	2841	0.05	351.60	351.65	1.7	505	25.67	142.59	168.26	1.8
4	2774	0.03	193.90	193.93	3.1	533	13.42	91.98	105.40	2.8
8	2775	0.02	139.31	139.33	4.4	537	7.12	70.74	77.86	3.8
16	2931	0.02	64.01	64.03	9.5	515	4.29	30.77	35.06	8.4

Parallel Runs. We report in Table 6 the parallel results concerning the matrices `mphase2`, `tran3d_homo` and `tran3d_steady`. The results of the Jacobi preconditioner are reported in case of convergence of the BiCGstab. Otherwise we only show the behavior of FSAI(A) and FSAI(A^2, ε) with optimal choice of the filtration parameter.

Differently from the symmetric case, here the speedup values are affected by the varying number of iterations when changing the number of processors. This happens because some of the floating point operations are computed in different orders depending on the number of processors used. In the test case `mphase2` the diagonal preconditioner fails to converge. FSAI-BiCGstab with $p = 16$ obtains satisfactory speedup values of 9.7 and 8.5.

The last two test cases reported are representative of opposite situations. In matrix `tran3d_homo` the convergence of BiCGstab is very fast even using a diagonal preconditioner, so that it is not worth spending so much time (more than 70% of the total CPU time) in the computation of the enlarged FSAI preconditioner. In the more difficult test case `tran3d_steady`, due to the com-

Table 7. Comparison between "best" FSAI and "best" ParaSails running on a single processor of an SP4 node

#	FSAI (CG/BiCGstab)							ParaSails (CG/GMRES)							
	lev	ε	ρ	iter	T_p	T_{sol}	T_{tot}	*lev*	*thr*	ε	ρ	iter	T_p	T_{sol}	T_{tot}
1	1	0.05	1.77	469	0.8	6.7	7.5	1	0.00	0.05	1.73	470	1.6	7.8	9.3
4	1	0.05	2.11	1447	4.7	169.2	173.9	2	0.05	0.09	2.22	1339	10.0	169.5	179.6
5	0	–	1.00	253	1.9	26.4	28.3	2	0.05	0.05	0.80	167	7.8	18.1	25.9
8	1	0.002	2.42	298	1.2	12.2	13.4	1	0.00	0.002	1.45	2354	5.5	59.7	65.2
11	0	–	1.00	52	5.3	25.0	30.3	0	0.05	0.00	0.34	98	14.9	32.1	47.9
12	1	0.05	1.17	470	49.5	244.8	294.3	1	0.07	0.05	0.44	1392	29.7	366.9	396.6

parable speedup obtained by the Jacobi and FSAI preconditioner, and due to the scalability of the FSAI construction phase, the FSAI-BiCGstab algorithm is roughly twice as efficient than Jacobi-BiCGstab even with $p = 16$ processors (35.06 seconds vs 64.03 seconds, cf. Table 6).

4.3 Comparisons with ParaSails

In this section we summarize the results obtained with the ParaSails package to solve the same symmetric and nonsymmetric linear systems described in the previous sections. ParaSails is a parallel implementation of a sparse approximate inverse preconditioner, using *a priori* sparsity patterns and least-squares (Frobenius norm) minimization due to Chow [8]. ParaSails handles SPD problems by using a factorized SPD sparse approximate inverse (FSAI) while general problems are handled with an unfactored sparse approximate inverse (SPAI). ParaSails also uses the postfiltration technique [12], to reduce the cost of applying the preconditioner. The package includes parallel CG and FGMRES solvers. The software is available from `http://www.llnl.gov/CASC/parasails`.

In Table 7 we report the comparison between FSAI and ParaSails on a single SP4 processor. The ParaSails iteration were stopped when the initial residual norm is reduced by a factor $tol = 10^{-7}$. ParaSails uses two parameters to construct the sparsity pattern: *lev*, the nonnegative integer used to specify the sparsity pattern of the preconditioner to be that of A^{lev+1}; and *thr* which stands for the threshold value used to sparsify the original matrix. The remaining entries of Table 7 have been already explained in the previous sections. The results reported are those obtained with the choice of level, threshold and filter value (ϵ) parameters which led to the smallest total CPU time, experimentally found. Table 8 reports results of the experimental study carried on with matrix `tran3d_steady` using different *a priori* patterns. A similar study was performed for each test matrix.

The results summarized in Table 7 show that our parallel implementation of FSAI acceleration of iterative solvers reveals as efficient as that of ParaSails both in terms of iteration number and total CPU time for the symmetric problems. In the nonsymmetric case, the two codes give very different results since both the preconditioner and the iterative method used are different. We found that naive ParaSails was not able to accurately solve the three nonsymmetric test

Table 8. Results for matrix `tran3d_steady` with the ParaSails package using different *a priori* patterns

lev	thr	ε	ρ	T_p	T_{sol}	T_{tot}	iter
0	0.05	0.00	0.56	24.3	543.6	567.9	1547
1	0.00	0.05	0.78	457.7	363.4	821.1	999
1	0.05	0.03	0.99	171.9	400.2	572.1	1059
1	0.07	0.05	0.44	29.7	366.9	396.6	1392
1	0.10	0.05	0.41	21.2	382.2	403.4	1484

Table 9. Timings and speedups for matrix `mixed` with the ParaSails package, using $lev = 2, thr = 0.05, \varepsilon = 0.09$

p	iter	T_p	T_{sol}	T_{tot}	S_p
1	1339	10.0	169.5	179.6	
2	1339	5.1	82.8	87.8	2.1
4	1339	3.1	45.9	49.0	3.7
8	1339	2.9	28.4	31.3	5.7
16	1339	2.8	15.4	18.2	9.9

problems. More in detail, when solving problem `mphase2` the computed relative error norm was of the order of units, after fulfilling the exit test. To reach a relative error $\approx 10^{-3}$ (still larger than the one produced with our code) we had to set $tol = 10^{-14}$. The results obtained are reported in Table 7. For the `tran3d_` matrices, since their coefficients may vary by many orders of magnitude, a preliminary diagonal scaling had to be performed to allow ParaSails to converge to the right solution.

Regarding the parallel performance ParaSails obtained, in all the test cases, speedup values similar to those achieved by our parallel FSAI implementation. As an example, we show in Table 9 the number of iteration and the elapsed time using $p = 1, 2, 4, 8, 16$ processors for matrix `mixed`.

5 Conclusions

We have proposed a parallel version of the FSAI preconditioner with postfiltration, in the acceleration of iterative methods for linear systems solvers. Such a preconditioner, with optimal choice of the filtration parameter yields convergence of the iterative schemes on a wide class of matrices arising from various discretizations of advection-diffusion equations. In the SPD instances the FSAI acceleration is only as effective than that of Jacobi. In the nonsymmetric case, however, FSAI is more robust and may be twice as efficient than Jacobi. The parallel performance of the iterative solvers preconditioned with FSAI, obtained on an SP4 supercomputer, is satisfactory, yielding speedup values comparable to those obtained with the diagonal preconditioner.

Our parallel code has been compared with ParaSails. For the symmetric problems tested the two codes provided comparable results. In the nonsymmetric instances our implementation of FSAI–BiCGstab reveals more efficient than ParaSails both in terms of iteration number and total CPU time.

References

1. J. Bear, *Hydraulics of Groundwater*, McGraw-Hill, New York, 1979.
2. M. Benzi, J. K. Cullum, and M. Tůma, *Robust approximate inverse preconditioning for the conjugate gradient method*, SIAM J. Sci. Comput., 22 (2000), pp. 1318–1332.

3. M. Benzi and M. Tůma, *A sparse approximate inverse preconditioner for nonsymmetric linear systems*, SIAM J. Sci. Comput., 19 (1998), pp. 968–994.
4. ——, *A comparative study of sparse approximate inverse preconditioners*, Applied Numerical Mathematics, 30 (1999), pp. 305–340.
5. M. Benzi, J. Marin, and M. Tůma, *A two-level parallel preconditioner based on sparse approximate inverses*, in Iterative Methods in Scientific Computation IV, D. R. Kincaid and A. C. Elster, eds., vol. 5 of IMACS Series in Computational and Applied Mathematics, New Brunswick, New Jersey, USA, 1999, pp. 167–178.
6. L. Bergamaschi, A. Martínez, and G. Pini, *Parallel solution of sparse eigenproblems by simultaneous Rayleigh quotient optimization with FSAI preconditioning*, in Parallel Computing. Software Technology, Algorithms, Architectures & Applications, G. R. Joubert and W. Nagel, eds., Elsevier, North-Holland, 2004, pp. 275–282.
7. L. Bergamaschi and M. Putti, *Efficient parallelization of preconditioned conjugate gradient schemes for matrices arising from discretizations of diffusion equations*, in Proceedings of the Ninth SIAM Conference on Parallel Processing for Scientific Computing, March, 1999. (CD–ROM).
8. E. Chow, *Parallel implementation and practical use of sparse approximate inverse preconditioners with a priori sparsity patterns*, Intl. J. High Perf. Comput. Appl., 15 (2001), pp. 56–74.
9. M. J. Grote and T. Huckle, *Parallel preconditioning with sparse approximate inverses*, SIAM J. Sci. Comput., 18 (1997), pp. 838–853.
10. D. Hysom and A. Pothen, *A scalable parallel algorithm for incomplete factor preconditioning*, SIAM J. Sci. Comput., 22 (2001), pp. 2194–2215.
11. I. E. Kaporin, *New convergence results and preconditioning strategies for the conjugate gradient method*, Numer. Lin. Alg. Appl., 1 (1994), pp. 179–210.
12. L. Yu. Kolotilina, A. A. Nikishin, and A. Yu. Yeremin, *Factorized sparse approximate inverse preconditionings IV. Simple approaches to rising efficiency*, Numer. Lin. Alg. Appl., 6 (1999), pp. 515–531.
13. L. Yu. Kolotilina and A. Yu. Yeremin, *Factorized sparse approximate inverse preconditionings I. Theory*, SIAM J. Matrix Anal. Appl., 14 (1993), pp. 45–58.
14. Z. Li, Y. Saad, and M. Sosonkina, *pARMS: a parallel version of the algebraic recursive multilevel solver*, Numer. Linear Algebra Appl., 10 (2003), pp. 485–509. Preconditioning, 2001 (Tahoe City, CA).
15. A. A. Nikishin and A. Yu. Yeremin, *Prefiltration technique via aggregation for constructing low-density high-quality factorized sparse approximate inverse preconditionings.*, Numer. Linear Alg. Appl., 10 (2003), pp. 235–246.
16. Y. Saad, *A flexible inner-outer preconditioned GMRES algorithm*, SIAM J. Sci. Comput., 14 (1993), pp. 461–469.
17. Y. Saad, *ILUT: A dual threshold incomplete ILU factorization*, Num. Lin. Alg. Appl., 1 (1994), pp. 387–402.
18. H. A. van der Vorst, *Bi-CGSTAB: A fast and smoothly converging variant of BI-CG for the solution of nonsymmetric linear systems*, SIAM J. Sci. Stat. Comput., 13 (1992), pp. 631–644.

Krylov and Polynomial Iterative Solvers Combined with Partial Spectral Factorization for SPD Linear Systems

Luc Giraud[1], Daniel Ruiz[2], and Ahmed Touhami[2,*]

[1] CERFACS, 42 Avenue G. Coriolis, 31057 Toulouse Cedex, France
Luc.Giraud@cerfacs.fr
[2] ENSEEIHT-IRIT, 2 rue Charles Camichel, 31071 Toulouse Cedex, France
{Daniel.Ruiz, Ahmed.Touhami}@enseeiht.fr

Abstract. When solving the Symmetric Positive Definite (SPD) linear system $\mathbf{Ax} = \mathbf{b}$ with the conjugate gradient method, the smallest eigenvalues in the matrix $\mathbf{A}$ often slow down the convergence. Consequently if the smallest eigenvalues in $\mathbf{A}$ could be somehow "removed", the convergence may be improved. This observation is of importance even when a preconditioner is used, and some extra techniques might be investigated to improve furthermore the convergence rate of the conjugate gradient on the given preconditioned system. Several techniques have been proposed in the literature that either consist of updating the preconditioner or enforcing the conjugate gradient to work in the orthogonal complement of an invariant subspace associated with small eigenvalues. In this work, we compare the numerical efficiency, computational complexity, and sensitivity to the accuracy of the spectral information of the techniques presented in [1], [2] and [3]. A more detailed description of these approaches as well as other comparable techniques on a range of standard test problems is available in [4].

Keywords. Chebyshev polynomials, Lanczos method, filtering, deflation, spectral preconditioning, iterative methods.

1 Introduction

The Krylov solver of choice for the solution of large sparse and symmetric positive linear systems of the form

$$\mathbf{Ax} = \mathbf{b} \tag{1}$$

is the conjugate gradient [5], [6], which constructs the kth iterate $\mathbf{x}^{(k)}$ as an element of

$$\mathbf{x}^{(0)} + \mathrm{Span}\{\mathbf{r}^{(0)}, \ldots, \mathbf{A}^{(k-1)}\mathbf{r}^{(0)}\}$$

so that $\left\|\mathbf{x}^{(k)} - \mathbf{x}^*\right\|_{\mathbf{A}}$ is minimized, where $\mathbf{x}^*$ is the exact solution of (1). The convergence of the conjugate gradient is governed by the eigenvalue distribution

* Corresponding author.

M. Daydé et al. (Eds.): VECPAR 2004, LNCS 3402, pp. 637–656, 2005.

in **A** [7]. This is in particular (but not only) highlighted by the bound on the rate of convergence of the conjugate gradient method given by Golub and Loan (cf . [5], Theorem. 10.2.6), which depends on the condition number of **A**. Thus removing the smallest eigenvalues in the matrix **A** can have a beneficial effect on the convergence. Several techniques have been proposed in the literature that either consists of updating the preconditioner or enforcing the conjugate gradient to work in the orthogonal complement of an invariant subspace associated with small eigenvalues.

The aim of this work it to compare several of these techniques in terms of numerical efficiency. Among these techniques we consider first the two-phase approach recently proposed in [1]. Analogously to direct methods, it includes a preliminary phase which we call "partial spectral factorization phase" followed by a solution phase, both only based on numerical tools that are usually exploited in iterative methods.

In a first stage, this algorithm computes (with a predetermined accuracy) a basis for the invariant subspace associated with the smallest eigenvalues in **A**. It uses Chebyshev polynomials as a spectral filtering tool in **A**, to damp in a set of vectors, eigencomponents associated with the largest eigenvalues in **A** and maintain, in the sequence of Krylov vectors, eigencomponents associated with the largest eigenvalues in **A** under a given level of filtering.

Chebyshev iteration can also be used to compute the part of the solution associated with the largest eigenvalues in **A**. The second part of the solution, associated with the smallest eigenvalues in **A**, is obtained with an oblique projection of the initial residual onto the "*near*"-invariant subspace, in order to get those eigencomponents in the solution corresponding to the smallest eigenvalues of **A**, and exploits the precomputed partial spectral factorization of **A**.

In this paper, the computation of an orthonormal basis **W** using the "partial spectral factorization" is exploited in combination with other approaches. In particular, we consider the version of the conjugate gradient algorithm used in [2] for instance, with a starting guess obtained with an oblique projection of the righ-hand side onto the orthogonal complement of **W**. As representative of techniques exploiting the spectral information to update the preconditioner we also consider the approach proposed in [3] that attempts to shift the smallest eigenvalues close to one where most of the eigenvalues of the preconditioned matrix should be located.

The outline of the paper is as follows. In Section 2, we describe the Chebyshev-based partial spectral decomposition. We devote Section 3 to present various solution techniques that exploit the spectral information about the matrix **A**, and indicate how they can be combined with "partial spectral factorization" to construct efficient iterative solvers. In Section 4, we compare these techniques in terms of numerical efficiency and computational complexity on a set of model problems from Matrix Market, or arising from the discretization via finite element technique of some heterogeneous diffusion PDEs in 2D. Finally, we conclude with some open questions and conclusions in Section 5.

2 A Chebyshev-Based Partial Spectral Decomposition

In this section, we introduce a Chebyshev-based filtering technique combined with some Lanczos process that is used to compute a basis for the invariant subspace associated with the smallest eigenvalues in $\mathbf{A}$. To do so, we start with an initial randomly generated vector which we normalize, and we use Chebyshev polynomials in $\mathbf{A}$ to "*damp*", in this starting vector, the eigencomponents associated with all the eigenvalues in some predetermined range. We can fix a positive number $\lambda_{\min}(\mathbf{A}) < \mu < \lambda_{\max}(\mathbf{A})$, and decide to compute all the eigenvectors associated with all the eigenvalues in the range $[\lambda_{\min}(\mathbf{A}), \mu]$. The computation of $\lambda_{\max}(\mathbf{A})$ is usually not too difficult, and in some cases, a sharp upper-bound may be already available through some *a priori* knowledge of the numerical properties of $\mathbf{A}$. If the eigenvalues of $\mathbf{A}$ are well clustered, the number of remaining eigenvalues inside the predetermined range $[\lambda_{\min}(\mathbf{A}), \mu]$, (with reasonable μ like $\lambda_{\max}/100$, or $\lambda_{\max}/10$, for instance) should be small compared to the size of the linear system.

Let us denote by

$$\mathbf{A} = \mathbf{U}\mathbf{\Lambda}\mathbf{U}^T = \mathbf{U}_1\mathbf{\Lambda}_1\mathbf{U}_1^T + \mathbf{U}_2\mathbf{\Lambda}_2\mathbf{U}_2^T$$

the eigendecomposition of the SPD matrix $\mathbf{A}$, $\mathbf{\Lambda}_1$ being the diagonal matrix made with all eigenvalues of $\mathbf{A}$ less than μ and $\mathbf{U}_1$ the unitary matrix whose columns are the corresponding orthonormalized eigenvectors in matrix form, and $\mathbf{\Lambda}_2$ and $\mathbf{U}_2$ the complementary corresponding matrices. Now, consider the translation plus homothetic transformation:

$$\lambda \in \mathbb{R} \longmapsto \omega_\mu(\lambda) = \frac{\lambda_{\max}(\mathbf{A}) + \mu - 2\lambda}{\lambda_{\max}(\mathbf{A}) - \mu},$$

with $\lambda_{\min}(\mathbf{A}) < \mu < \lambda_{\max}(\mathbf{A})$ given above. This transformation maps $\lambda_{\max}(\mathbf{A})$ to -1, μ to 1, and 0 to

$$\omega_\mu(0) = d_\mu = \frac{\lambda_{\max}(\mathbf{A}) + \mu}{\lambda_{\max}(\mathbf{A}) - \mu} > 1.$$

Then, exploiting the optimal properties of Chebyshev polynomials $T_m(x)$ on the interval $[-1, 1]$ (given in Theorem 4.2.1 in [8–page 47]), we can introduce

$$\mathcal{F}_m(\lambda) = \frac{T_m(\omega_\mu(\lambda))}{T_m(d_\mu)}, \tag{2}$$

where $\mathcal{F}_m(\lambda)$ has minimum l_∞ norm on the interval $[\mu, \lambda_{\max}(\mathbf{A})]$ over all polynomial Q_m of degree less than or equal to m satisfying $Q_m(0) = 1$.

The example in Figure 1 shows the values of the Chebyshev polynomial $\mathcal{F}_{16}(x)$ as defined in (2) with $\lambda_{min} = 10^{-16}$, $\lambda_{max} = 1$, and $\mu = 10^{-1}$. On the right side of Figure 1, we plot the value of $\mathcal{F}_{16}(x)$ on the range $[\mu, \lambda_{max}]$, to illustrate how we can manage, with Chebyshev polynomials, a spectral filter on the interval $[\lambda_{max}/10, \lambda_{max}]$ bounded uniformly below $\varepsilon = 10^{-4}$. On the left

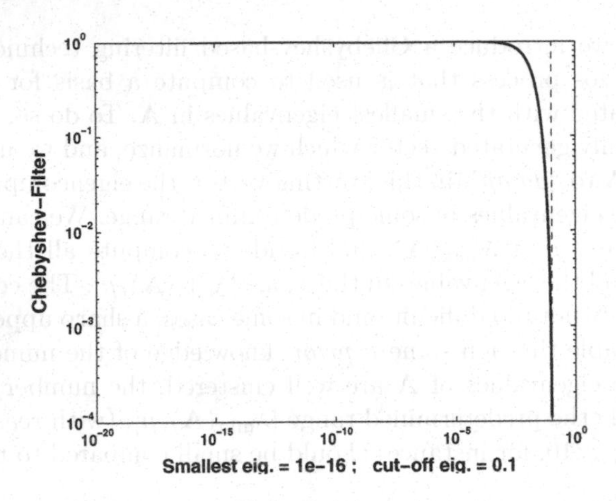

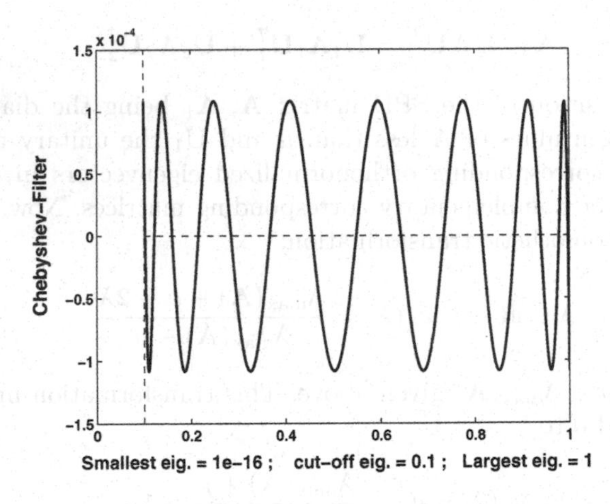

Fig. 1. Use of Chebyshev polynomials in the range $[\mu, \lambda_{max}]$, with $\mu = \lambda_{max}/10$

side, we plot in Log-Log Scale $\mathcal{F}_{16}(x)$ on the interval $[\lambda_{min}, \mu]$. We can see that the fixed value $\mathcal{F}_{16}(0) = 1$ maintains, because of continuity, the values of $\mathcal{F}_{16}(\lambda)$ close to 1 for all $\lambda \leq 10^{-2}$.

Let now $\mathbf{P}$ be a randomly generated and normalized vector, and consider

$$\mathbf{Z} = \mathcal{F}_m(\mathbf{A})\,\mathbf{P} = \mathbf{U}_1\mathcal{F}_m(\mathbf{\Lambda}_1)\mathbf{U}_1^T\,\mathbf{P} + \mathbf{U}_2\mathcal{F}_m(\mathbf{\Lambda}_2)\mathbf{U}_2^T\,\mathbf{P}, \tag{3}$$

with $\mathcal{F}_m$ defined as in (2). For given values of μ, $\lambda_{max}(\mathbf{A})$ and ε, we can fix the degree m of T_m such that $1/|T_m(d_\mu)| < \varepsilon$ on $[\mu, \lambda_{max}(\mathbf{A})]$, out of which we obtain $||\mathcal{F}_m(\mathbf{\Lambda}_2)||_\infty < \varepsilon$, and using (3) we can write

$$||\mathbf{U}_2^T\,\mathbf{Z}||_2 \leq ||\mathcal{F}_m(\mathbf{\Lambda}_2)||_2\,||\mathbf{U}_2^T\,\mathbf{P}||_2 \leq \varepsilon\,||\mathbf{U}_2^T\,\mathbf{P}||_2 \tag{4}$$

Equation (4) explains explicitly how the action of the Chebyshev filtering in $\mathbf{A}$ applied on a set of vectors can reduce below a value ε the eigencomponents, with respect to the invariant subspace $\mathbf{U}_2$ linked to all eigenvalues the range $[\mu, \lambda_{max}(\mathbf{A})]$, relatively to the others. The number of Chebyshev iterations to achieve a given level of filtering ε, is directly related to the rate of convergence of Chebyshev polynomials on the interval $[\mu, \lambda_{max}]$ [8] which depends only on the ratio λ_{max}/μ. As illustrated in Figure 1, 16 Chebyshev iterations are enough to reach a level $\varepsilon = 10^{-4}$ on the interval $[\mu, \lambda_{max}]$ with $\lambda_{max}/\mu = 10$. If we take $\lambda_{max}/\mu = 100$, for instance the number of Chebyshev iterations to reach the same level will then become 50.

Then, after "*filtering*" the starting vector, we obtain an initial vector $\mathbf{W}^{(0)}$ with eigencomponents under the desired level ε for those eigenvalues in the range $[\mu, \lambda_{\max}(\mathbf{A})]$, and relatively much bigger eigencomponents linked with the smallest eigenvalues in $\mathbf{A}$. We then use a Lanczos type of approach to build a "*near*"-invariant basis starting with these filtered vectors. As discussed in [1], the major trouble of the Lanczos iterations is that it deteriorates gradually the gap between the eigencomponents in the Lanczos vectors $\mathbf{W}$, initially set to about ε in $\mathbf{W}^{(0)}$. To circumvent this problem, a few extra Chebyshev filtering iterations are performed at each Lanczos iteration to maintain, under the level ε, the relative gap in those eigencomponents associated with all eigenvalues in the range $[\mu, \lambda_{max}(\mathbf{A})]$ in the newly computed Lanczos vector $\mathbf{W}^{(k+1)}$.

For the details of this particular combination of Lanczos iteration and Chebyshev polynomials, we refer to [1] and [4]. As opposed to classical polynomial precoditioning techniques, the use of Chebyshev filtering polynomials in this context can be viewed as a way to perform implicitly some sort of deflation with respect to the invariant subspace linked with the largest eigenvalues in $\mathbf{A}$, as proposed and discussed for instance in [9], [10], [11], in order to compute those eigenvectors lying in the orthogonal complement of this invariant subspace. Chebyshev filtering polynomials, in that respect, do not present the same nice properties as projectors used commonly in such deflations (which have explicit eigenvalues equal to 0 and 1) but still, they can mimic partly these properties and offer an alternative to achieve the same behavior in the Lanczos iteration as with deflation techniques, but without having to compute effectively the basis vectors associated with this deflation.

This combination of Lanczos technique with Chebyshev filtering enables to compute an approximation of all eigenvalues and eigenvectors in a given range $[\lambda_{min}(\mathbf{A}), \mu[$ without any *a priori* knowledge of their count. Consequently, to be efficient, the method relies on the fact that the spectrum of $\mathbf{A}$ is mostly concentrated in the range $[\mu, \lambda_{max}(\mathbf{A})]$, with reasonable value of μ so that the number of Chebyshev iterations required to reach a given filtering level ε be not too large. In that respect, a preliminary preconditioner applied to $\mathbf{A}$, with classical preconditioning techniques, can be also useful to transform and clusterise better the spectrum in the iteration matrix. The value of μ, as we will see in the following, also controls the reduced condition number $\lambda_{max}(\mathbf{A})/\mu$ that we will implicitly exploit after derivation of a "*near*"-invariant subspace associated with all eigenvalues in $\mathbf{A}$ in the range $]0, \mu]$.

3 Alternative Techniques for the Solution of Linear Systems That Exploit the Partial Spectral Decomposition

In this section, we consider solution techniques that exploit the spectral information about the matrix **A**. The common general idea beneath all such techniques, as we will see in detail in the following paragraphs, is to use some precomputed "*near*"-invariant subspace **W** to build a projection onto the orthogonal complement of **W** in some appropriate norm and to exploit such a projector to maintain, implicitly or explicitly all through the iterations, the computed residuals in this orthogonal complement.

In this way, the expectation is that the iterative solver, like the conjugate gradient method for instance, will behave as if the spectrum of **A** was reduced to only those eigenvalues with eigenvectors not lying in the subspace **W**. This is of interest, in particular, if this subspace **W** approximates those eigenvectors associated to the smallest eigenvalues in **A** since the reduced condition number of **A** restricted to the orthogonal complement of **W** should be much smaller. In this section, we discuss 3 different solution techniques that exploit this general idea, but in a different manner each.

A general idea can be to decompose the solution of (1) in two parts. The first can be obtained directly with an oblique projection of the right hand side **b** onto the orthogonal complement of **W**. The second part of this solution can then be obtained whether by means of Chebyshev iterations, as in § 3.1, or with the use of the conjugate gradient with the first part of this solution as a starting guess, as detailed in § 3.2. These first two techniques make use of the projection with respect to **W** only at the starting point.

The last technique, detailed in § 3.3, aims at building a preconditioner in **A** that shifts, by adding the value 1, those eigenvalues linked to the "*near*"-invariant subspace **W**, and to run a preconditioned conjugate gradient in the usual way to compute the solution of (1).

3.1 Combination of Oblique Projection and Chebyshev Iteration

In this paragraph, we describe how we use the "*near*"-invariant subspace **W** linked to the smallest eigenvalues to compute solution of linear system with **A**.

The idea is to perform an oblique projection of the initial residual ($\mathbf{r}^{(0)} = \mathbf{b} - \mathbf{A}\mathbf{x}^{(0)}$) onto this "*near*"-invariant subspace in order to get the eigencomponents in the solution corresponding to the smallest eigenvalues, viz.

$$\begin{aligned} \mathbf{r}^{(\mathrm{proj})} &= \mathbf{r}^{(0)} - \mathbf{AW}\left(\mathbf{W}^T\mathbf{AW}\right)^{-1}\mathbf{W}^T\mathbf{r}^{(0)}, \\ \text{with}\ \ \mathbf{x}^{(\mathrm{proj})} &= \mathbf{x}^{(0)} + \mathbf{W}\left(\mathbf{W}^T\mathbf{AW}\right)^{-1}\mathbf{W}^T\mathbf{r}^{(0)}. \end{aligned} \tag{5}$$

To compute the remaining part of the solution vector $\mathbf{A}\widetilde{\mathbf{x}} = \mathbf{r}^{(\mathrm{proj})}$, one can then use the classical Chebyshev algorithm with eigenvalue bounds given by μ and $\lambda_{max}(\mathbf{A})$, as explained in § 2.

We get the following algorithm, which we call INIT-CHEBYSHEV.

ALGORITHM: INIT-CHEBYSHEV

Input: A "*near*"-invariant orthonormal basis $\mathbf{W}$ linked to the all eigenvalues in $\mathbf{A}$ in the range $]0, \mu[$

Begin

Chebyshev iteration on the interval $[\mu, \lambda_{max}(\mathbf{A})]$

1. $\alpha_\mu = \dfrac{2}{\lambda_{max}(\mathbf{A}) - \mu}$ and $d_\mu = \dfrac{\lambda_{max}(\mathbf{A}) + \mu}{\lambda_{max}(\mathbf{A}) - \mu}$
2. $\sigma_0 = 1$ and $\sigma_1 = d_\mu$
3. Choose an initial guess $\mathbf{x}^{(0)}$ and $\mathbf{r}^{(0)} = \mathbf{b} - \mathbf{A}\mathbf{x}^{(0)}$
4. $\mathbf{x}^{(1)} = \mathbf{x}^{(0)} + \dfrac{\alpha_\mu}{d_\mu}\mathbf{r}^{(0)}$ and $\mathbf{r}^{(1)} = \mathbf{b} - \mathbf{A}\mathbf{x}^{(1)}$
5. **For** $m = 1, 2, \ldots,$ until $\sigma_m < \delta$ **Do:**
   i. $\sigma_{m+1} = 2\, d_\mu\, \sigma_m - \sigma_{m-1}$
   ii. $\mathbf{x}^{(m+1)} = 2\, \dfrac{\sigma_m}{\sigma_{m+1}} \left(d_\mu \mathbf{x}^{(m)} + \alpha_\mu \mathbf{r}^{(m)} \right) - \dfrac{\sigma_{m-1}}{\sigma_{m+1}}\, \mathbf{x}^{(m-1)}$
   iii. $\mathbf{r}^{(m+1)} = \mathbf{b} - \mathbf{A}\mathbf{x}^{(m+1)}$
6. **EndDo**

Oblique projection onto Span($\mathbf{W}$)

7. $\mathbf{x}^{(\text{proj})} = \mathbf{W}\left(\mathbf{W}^T \mathbf{A} \mathbf{W} \right)^{-1} \mathbf{W}^T\, \mathbf{r}^{(m+1)}$
8. $\mathbf{x}^{(\text{sol})} = \mathbf{x}^{(\text{proj})} + \mathbf{x}^{(m+1)}$

End

Note that, if $\mathbf{W}$ exactly spans the invariant subspace generated by $\mathbf{U}_1$, the Chebyshev iteration and the oblique projection steps in this solution phase can be performed, though sequentially, in any order *a priori*. However, since Span($\mathbf{W}$) is only an approximation of this invariant subspace, we prefer to perform the Chebyshev step first, followed then by the oblique projection, because this helps to increase the accuracy of the oblique projection by "*minimizing*" the influence of the eigencomponents in $\mathbf{U}_2$ relative to those in $\mathbf{U}_1$ in the scalar products $\mathbf{W}^T\mathbf{r}$ in (5). This is of particular interest when the filtering level ε used to compute the basis $\mathbf{W}$, as in § 2, is not close to machine precision.

The first part in INIT-CHEBYSHEV, from step 1 to 6 corresponds to the application of Chebyshev polynomial in $\mathbf{A}$ that attempts to reduce by a factor δ the eigencomponents in $\mathbf{r}^{(0)}$ associated with the eigenvalues in the range $[\mu, \lambda_{\max}(\mathbf{A})]$, providing thus the resulting residual $\mathbf{r}^{(m+1)} = \mathcal{F}_{m+1}(\mathbf{A})\, \mathbf{r}^{(0)}$ and the corresponding update $\mathbf{x}^{(m+1)}$ such that $\mathbf{b} - \mathbf{A}\mathbf{x}^{(m+1)} = \mathbf{r}^{(m+1)}$. The factor δ is *a priori* the same as the level of filtering ε described in § 2, except that it may be fixed on purpose to a value different than ε in this solution phase.

Steps ii and iii are connected together with the relation that,

$$\mathbf{r}^{(m+1)} = \mathcal{F}_{m+1}(\mathbf{A})\,\mathbf{r}^{(0)} = \frac{T_{m+1}(\omega_\mu(\mathbf{A}))\,\mathbf{r}^{(0)}}{T_{m+1}(d_\mu)} = \frac{1}{\sigma_{m+1}} T_{m+1}\Big(d_\mu\,\mathbf{I} - \alpha_\mu\,\mathbf{A}\Big)\,\mathbf{r}^{(0)}$$

where we note $d_\mu = \dfrac{\lambda_{max}+\mu}{\lambda_{max}-\mu}$, $\alpha_\mu = \dfrac{2}{\lambda_{max}-\mu}$ and $\sigma_m = T_m(d_\mu)$ for all $m \geq 0$. In that respect, step iii can also be replaced by the equivalent following three-term recurrent relation,

$$\mathbf{r}^{(m+1)} = 2\,\frac{\sigma_m}{\sigma_{m+1}}\Big(d_\mu \mathbf{r}^{(m)} - \alpha_\mu \mathbf{A}\mathbf{r}^{(m)}\Big) - \frac{\sigma_{m-1}}{\sigma_{m+1}}\,\mathbf{r}^{(m-1)} \tag{6}$$

that corresponds to the Chebyshev iteration giving $\mathbf{r}^{(m+1)}$.

3.2 Conjugate Gradient with an Oblique Projection as a Starting Point

As proposed in [2], once an approximation of the invariant subspace associated with the smallest eigenvalues is obtained, we can use it for the computation of further solutions. As can be seen in Algorithm Init-CG, the idea is to perform an oblique projection of the initial residual onto this invariant subspace in order to get the eigencomponents in the solution corresponding to the smallest eigenvalues, as in (5), and then to perform a classical conjugate gradient to compute the remaining part of the solution vector. Note that we do not project the given matrix in any way, but that we just run the classical conjugate gradient with the original matrix and an initial guess which, we expect, will remove all the troubles that the classical conjugate gradient algorithm would encounter otherwise.

Indeed, in the presence of clusters of eigenvalues at the extreme of the spectrum of **A**, combined with some ill-conditioning, classical conjugate gradient may show a convergence history with sequences of plateau and sharp drops. These plateau (see [7]) are well understood and correspond to the discovery of

Algorithm: Init-CG

Input: A "*near*"-invariant orthonormal basis **W** linked to the all eigenvalues in **A** in the range $]0, \mu[$.

Begin

1. $\mathbf{x}^{(0)} = \mathbf{W}\Big(\mathbf{W}^T\mathbf{A}\mathbf{W}\Big)^{-1}\mathbf{W}^T\mathbf{b}$
2. $\mathbf{x}^{(sol)} = \text{Classical CG}\Big(\mathbf{A}, \mathbf{b}, \mathbf{x}^{(0)}, \text{tol}\Big)$.

End

separate eigenvalues within a cluster. The problem is that these plateau can be rather large, in terms of number of iterations, even when these extreme clusters incorporate only a few eigenvalues. Block conjugate gradient can help to reduce the size of these plateau, but in general can not remove then completely (see [2]).

3.3 A Spectral Low-Rank Update Preconditioner

In this paragraph, we consider the approach proposed in [3] that attempts to remove the effect of the smallest eigenvalues in magnitude in the preconditioned matrix, and can potentially can speed-up the convergence of Krylov solvers.

Roughly speaking, this technique consists in building the preconditioner

$$\mathbf{M} = \mathbf{I} + \mathbf{W}\left(\mathbf{W}^T\mathbf{A}\mathbf{W}\right)^{-1}\mathbf{W}^T \tag{7}$$

that shifts, by adding the value 1, those eigenvalues in **A** linked to the "*near*"-invariant subspace **W**. We can then use this spectral preconditioner **M** to run the preconditioned CG, yielding thus the following algorithm which we call SLRU (Spectral Low Rank Update).

ALGORITHM: SLRU

Input: A "*near*"-invariant orthonormal basis **W** linked to the all eigenvalues in **A** in the range $]0, \mu[$.

Begin

1. $\mathbf{M} = \mathbf{I} + \mathbf{W}\left(\mathbf{W}^T\mathbf{A}\mathbf{W}\right)^{-1}\mathbf{W}^T$.

2. $\mathbf{x}^{(\mathrm{sol})} = \text{Preconditioned CG}\left(\mathbf{A}, \mathbf{b}, \mathbf{x}^{(0)}, \mathbf{M}, \mathrm{tol}\right)$.

End

---

4 Numerical Experiments

In this section, we show some numerical results that illustrate the effectiveness of combining the "partial spectral decomposition" with the different approaches for the solution of sparse symmetric and positive definite (SPD) linear systems, as described in the previous sections.

As we will see in the following, the experiments show that the different algorithms exhibit very similar numerical behavior. This is particularly true when the precomputed "*near*"-invariant basis **W** is obtained with an accuracy close to machine precision. In this case, this numerical behavior corresponds to the one that we would obtain in general with a linear system where the smallest eigenvalue is given by the cut-of value μ in the partial spectral factorization phase,

Table 1. Set of SPD test matrices

Name	Size	Short description
BCSSTK14	1806	Static analyses in structural engineering - Roof of the Omni Coliseum, Atlanta
PDE 2	7969	2D heterogenous diffusion equation discretized by finite element in a L shape region

with a resulting reduced condition number of $\lambda_{max}(\mathbf{A})/\mu$. The purpose of the following sections is to analyse experimentally this in details, and in particular to investigate the impact of much less accurate precomputed $\mathbf{W}$ basis on the actual behavior of these various algorithms.

For all the numerical experiments reported in this section, a first level of left preconditioner $\mathbf{M}_1 = \mathbf{L}\mathbf{L}^T$ is constructed using Incomplete Cholesky factorization IC(t) of $\mathbf{A}$ with t as the threshold. The purpose of this first preliminary preconditioning of the given linear system is to clusterize better the eigenvalues so that the invariant subspace associated with the smallest eigenvalues be of small rank. We denote by $\widehat{\mathbf{b}} = \mathbf{L}^{-1}\mathbf{b}$ and $\widehat{\mathbf{A}} = \mathbf{L}^{-1}\mathbf{A}\mathbf{L}^{-T}$ the matrix symmetrically preconditioned with IC(t) which is SPD and similar to the matrix $\mathbf{M}_1^{-1}\mathbf{A}$.

Table 1 summarizes, for each matrix, the size and the application in which this matrix arises. In particular, matrices BCSSTK are extracted from the Harwell-Boeing collection [12].

In Table 2, we illustrate the effect of Incomplete Cholesky (IC) factorization on the eigenvalue distribution of the preconditioned matrix $\widehat{\mathbf{A}}$. As mentioned before, a clustered spectrum of the preconditioned matrix is usually considered a desirable property for fast convergence of Krylov solvers.

4.1 The Cost of Precomputing the Orthonormal Basis W

The partial spectral factorization (described in § 2) depends mainly on two different parameters which correspond to the choice of the cut-off eigenvalue μ, and of the filtering level ε under which we try to maintain, by means of Chebyshev filtering, those eigencomponents relative to all eigenvalues bigger than μ.

Table 2. Distribution of eigenvalues

Matrix	Unpreconditioned matrix		Preconditioned matrix			Number of eigenvalues below the
	Smallest eig	Largest eig	Threshold IC(t)	Smallest eig	Largest eig	$\mu = \lambda_{max}/10$
BCSSTK14	1	$1.19 \cdot 10^{10}$	10^{-2}	$1.41 \cdot 10^{-2}$	3.13	23
PDE 2	$2.92 \cdot 10^{-9}$	2.17	10^{-2}	$1.79 \cdot 10^{-7}$	1.11	3

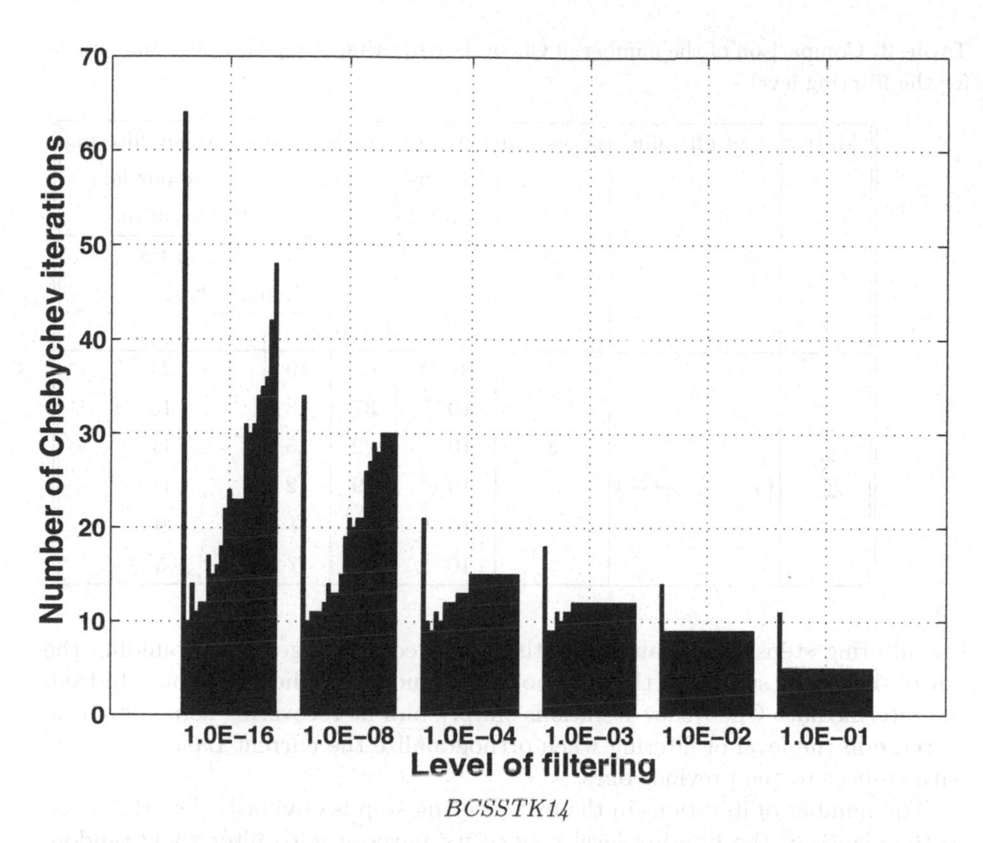

BCSSTK14

Fig. 2. Number of Chebyshev iterations at each Lanczos step during the partial spectral factorization

The cut-off eigenvalue, μ, splits the spectrum of the matrix $\widehat{\mathbf{A}}$ in two subsets and fixes the dimension k of the invariant subspace that will be computed in the partial spectral factorization phase, depending of course on the actual eigenvalue distribution in the matrix $\widehat{\mathbf{A}}$. It also determines the convergence rate of the classical Chebyshev iterations that will be performed, both when filtering the Lanczos vectors and when computing the solution. The rapid change in the Chebyshev rate of convergence with smaller values of μ induces much more Chebyshev filtering steps at each iteration, and this could only be counterbalanced by a very strong reduction in the total number of iterations in the partial spectral factorization phase. In other words, it is worth reducing the value of μ only if there is a very strong clustering of eigenvalues in the spectrum of the iteration matrix $\widehat{\mathbf{A}}$ and if the change in μ helps to reduce by a large amount the dimension of the invariant subspace that will be approximated.

The choice of the level of filtering ε influences directly the numerical quality of the computed "*near*"-invariant basis $\mathbf{W}$ in the Chebyshev-based partial spectral factorization phase. Concerning the number of Chebyshev iterations, we can observe in Figure 2 and Table 3 that these mostly differ in the first and

Table 3. Comparison of the number of Chebyshev filtering steps with different choices for the filtering level

Matrix	Cut-off value μ	#eigs < μ	Value of filtering level ε	Number of Chebyshev filtering iterations in the partial spectral factorization — k-th vector in the Lanczos basis: $k=0$	$k=1$	$k=2$
PDE 2	$\left(\mu = \frac{\lambda_{max}(\mathbf{A})}{10}\right)$	3	10^{-16}	65	19	24
			10^{-8}	37	18	15
			10^{-4}	22	15	11
			10^{-3}	19	12	11
			10^{-2}	15	9	9
			10^{-1}	12	5	5

last filtering steps, and that during the intermediate stages when building the "*near*"-invariant subspace, these do not vary much with the choice of ε. Indeed, the intermediate Chebyshev iterations simply aim at recovering some potential increase in the level of filtering when orthogonalize the current Lanczos vectors with respect to the previous ones.

The number of iterations in the first filtering step is obviously directly linked to the choice of the filtering level ε since its purpose is to filter some random starting set of vectors under that level. At the final stage also, when convergence or "*near*"-invariance with the respect to subspace associated with the smallest eigenvalues of $\widehat{\mathbf{A}}$ is reached in the partial spectral factorization phase, some vectors in the last computed set may become strongly colinear to the orthogonal complement of this invariant subspace, implying thus more Chebyshev filtering iterations. These are the two reasons why smaller values for ε imply more Chebyshev iterations in the first and last filtering stages.

4.2 Comparison of the Different Techniques for the Solution Phase

In this subsection, we compare the numerical convergence behavior of the various algorithms described in §3. We discuss their numerical efficiency, computational complexity and sensitivity to the accuracy of the spectral information. We mention beforehand that, when the basis $\mathbf{W}$ is computed with a filtering level close to machine precision (10^{-16}), all algorithms presented in §3 exhibit the same linear rate of convergence. This linear rate of convergence obtained, which can be observed on the convergence histograms in Figure 3, is directly connected to the cut-off value μ that fixes a posteriori the reduced condition number of the linear system that is solved. This is in agreement with the theory on the numerical behavior of the conjugate gradient algorithm.

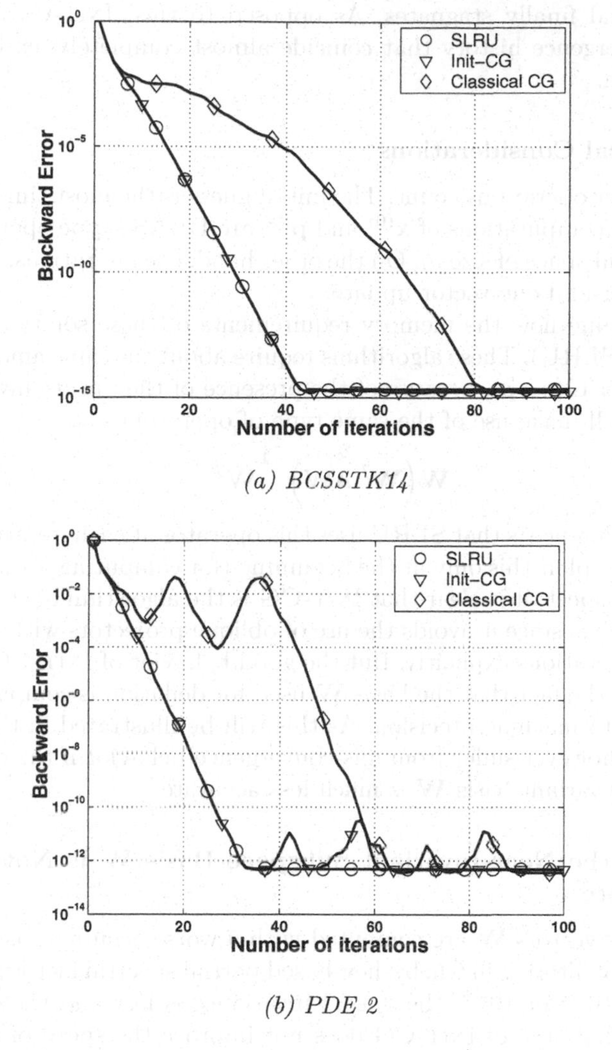

(a) BCSSTK14

(b) PDE 2

Fig. 3. Convergence history for a level of filtering ε equal to 10^{-16}

The stopping criterion used in these tests is the backward error $||\widehat{\mathbf{b}} - \widehat{\mathbf{A}}\mathbf{x}^{(i)}||_2 / ||\widehat{\mathbf{b}}||_2$. In all our runs, we have monitored this backward error down to 10^{-16} when possible. The initial guess is always $\mathbf{x}^{(0)} = 0$ and the right hand side is chosen so that the solution of the initial linear system is the vector of all ones. The block size s is set to 1 and the cut-off eigenvalue μ to $\lambda_{max}(\widehat{\mathbf{A}})/10$.

In Figure 3, we compare the convergence history of the standard conjugate gradient versus INIT-CG and SLRU obtained with a basis $\mathbf{W}$ computed with a level of filtering ε equal to 10^{-16}. The backward error of the standard conjugate gradient method tends to decrease very rapidly in the first few iterations, then decreases more slowly in an intermediate phase during which oscillations may also appear, and finally stagnates. As opposed to this, INIT-CG and SLRU present a convergence history that coincide almost completely, except perhaps after stagnation.

4.3 Practical Considerations

We first consider operations count. The initial guess is the most time-consuming overhead. Both computations of $\mathbf{x}^{(0)}$ and $\mathbf{p}^{(0)}$ are BLAS2 type operations (projection onto a subspace of size p). On the other hand, each iterations, adds merely one dot-product and one vector update.

Let us examine now the memory requirements of these solution techniques (INIT-CG and SLRU). These algorithms require about the same amount of extra storage, of order $\mathcal{O}(n * p)$, because of the presence of the "*near*"-invariant basis $\mathbf{W}$. They also all make use of the same type of operator

$$\mathbf{W}\left(\mathbf{W}^T\mathbf{A}\mathbf{W}\right)^{-1}\mathbf{W}^T.$$

The only difference is that SLRU uses this operator at each iteration, whereas INIT-CG does exploit this only at the beginning (for computing a starting vector $\mathbf{x}^{(0)}$). In that respect, it is clear that INIT-CG is the algorithm of choice because it is less expensive, since it avoids the use of oblique projectors within the conjugate gradient iterations explicitly. But the good behavior of INIT-CG illustrated above is due to the fact that the basis $\mathbf{W}$ used for deflation is computed with an accuracy close to machine precision. As this will be illustrated in the following, INIT-CG may however suffer from mis-convergence behavior if the computation of the "*near*"-invariant basis $\mathbf{W}$ is much less accurate.

4.4 When the Near-Invariant Subspace Basis W Is Not Very Accurate

When the basis vectors $\mathbf{W}$ are computed with a worse accuracy, as for instance when the level of filtering in Chebyshev-based partial spectral factorization phase is around $\varepsilon = 10^{-2}$ or 10^{-1}, the "*near*"-invariance is lost and the oblique projection in the first step of INIT-CG does not improve the speed of convergence of INIT-CG which then behaves like a classical conjugate gradient.

On the left side of Figure 4, we show the impact of varying the filtering level ε (when computing the "*near*"-invariant $\mathbf{W}$ basis) on the convergence of INIT-CG. For example, when the level of filtering ε is equal to 10^{-8}, the phenomenon of plateau occurs again, but at intermediate levels. Indeed, INIT-CG (Classical CG with initial oblique projection) shows two different stages in its numerical behavior. The first stage follows very closely the convergence history observed in Figure 3 when the computation of the $\mathbf{W}$ basis is very accurate. Then, the speed of convergence is disrupted at an intermediate level of the backward error directly linked to the value of the filtering level ε, and the convergence history resembles again that of the classical conjugate gradient on the original system.

The reasons for this is that INIT-CG provides an initial residual to the conjugate gradient that has larger eigencomponents in the orthogonal complement of

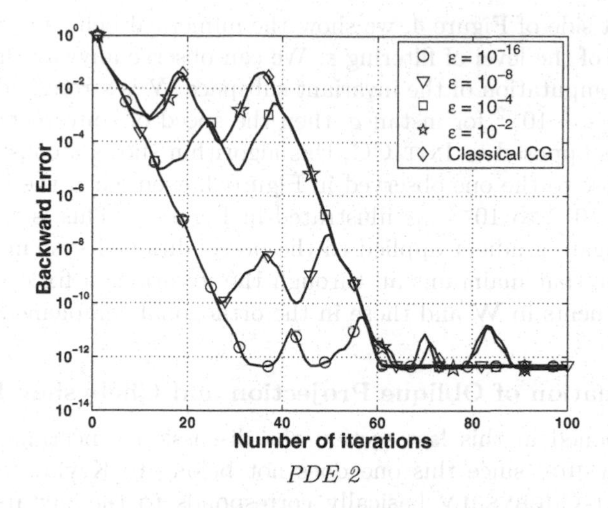

PDE 2

INIT-CG: Conjugate Gradient with an oblique projection as a starting point

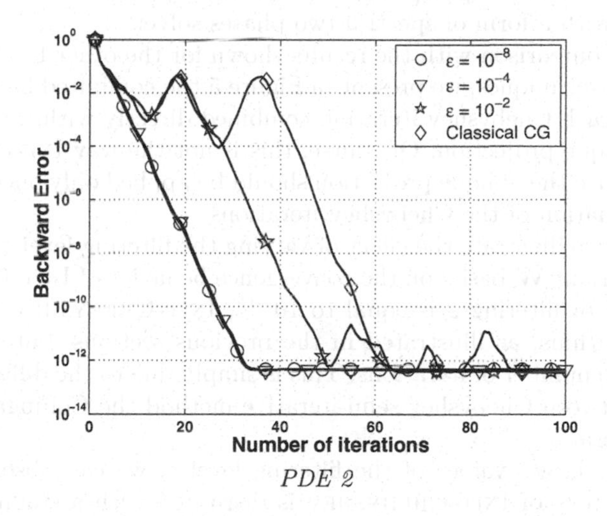

PDE 2

SLRU: Spectral low-rank update preconditioner

Fig. 4. Convergence history of INIT-CG and SLRU for different values of level of filtering ε

the basis **W** and eigencomponents of order ε in the **W** subspace. Then, after the residual norm has been reduced to about the value of ε, the conjugate gradient is left again with a Krylov direction with eigencomponents all of about the same magnitude, and then exhibits the same type of convergence behavior linked to the ill-conditioning of the given matrix. For efficiency, it is clear that one should then the stop the iterations in INIT-CG at a backward error about the filtering level ε before reaching this intermediate plateau.

On the right side of Figure 4, we show the numerical behavior of SLRU for different values of the level of filtering ε. We can observe anyway that if the accuracy in the computation of the invariant subspace $\mathbf{W}$ is strongly deteriorated, as with a value $\varepsilon = 10^{-2}$ for instance, then the speed of convergence may vary slightly. Still, as opposed to INIT-CG, this algorithm shows a constant numerical behavior close to the one observed in Figure 3, even when the filtering level is in the range 10^{-8} to 10^{-2}, as illustrated in Figure 4. This is the benefit of using the conjugate gradient applied to the preconditioned system via spectral low-rank update that maintains all through the iterations a fixed gap between the eigencomponents in $\mathbf{W}$ and those in the orthogonal complement of $\mathbf{W}$.

4.5 Combination of Oblique Projection and Chebyshev Iteration

We have postponed in this last section the discussion concerning the results for INIT-CHEBYSHEV since this one does not belong to Krylov techniques as the others. INIT-CHEBYSHEV basically corresponds to the juxtaposition of a Chebyshev iterative solver and an oblique projection that act independently on two separate subparts in the spectrum of $\mathbf{A}$. In that respect, it is closer to a direct method with a form of spectral two-phases solver.

For sake of comparison with the results shown for the other Krylov based iterative solution techniques, we present in Figure 5 the computed backward error obtained after each Chebyshev iteration combined directly with the application of the final oblique projection. Of course, this is not the way INIT-CHEBYSHEV should work, since the oblique projection should be applied only once at the end, after the termination of the Chebyshev iterations.

In Figure 5, we illustrate the effect of varying the filtering level ε (when computing the invariant $\mathbf{W}$ basis) on the convergence behavior of INIT-CHEBYSHEV. When the level of filtering ε is equal to 10^{-16}, INIT-CHEBYSHEV behaves like the other algorithms, as illustrated in the previous sections, but with a little difference in the number of iterations. This is simply due to the different rates of convergence between Chebyshev semi-iterative method the Conjugate Gradient method in general.

However, for larger values of the filtering level ε, we can observe that the convergence history of INIT-CHEBYSHEV is disrupted with a stagnation of the backward error at a level close to the actual value of ε. The reason for this is simply that the "*near*"-invariant basis $\mathbf{W}$ has eigencomponents of order ε with respect to those eigenvectors corresponding to eigenvalues in the range $[\mu, \lambda_{max}]$. Obviously, this level of information is incorporated explicitly in the oblique projection onto this subspace that is applied as a final stage, and perturbs the eigencomponents in the final residual at about that level. For a more detailed analysis of this phenomenon, we refer to [1].

Nevertheless, it is also possible to iterate on that solution phase, and improve the solution with iterative refinement in the usual way (see [13], for instance). Because we use an oblique projector (5), we may expect that iterative refinement will help to improve the solution only when the filtering level ε is at least smaller than the square root of the condition number of the iteration matrix (see [1]).

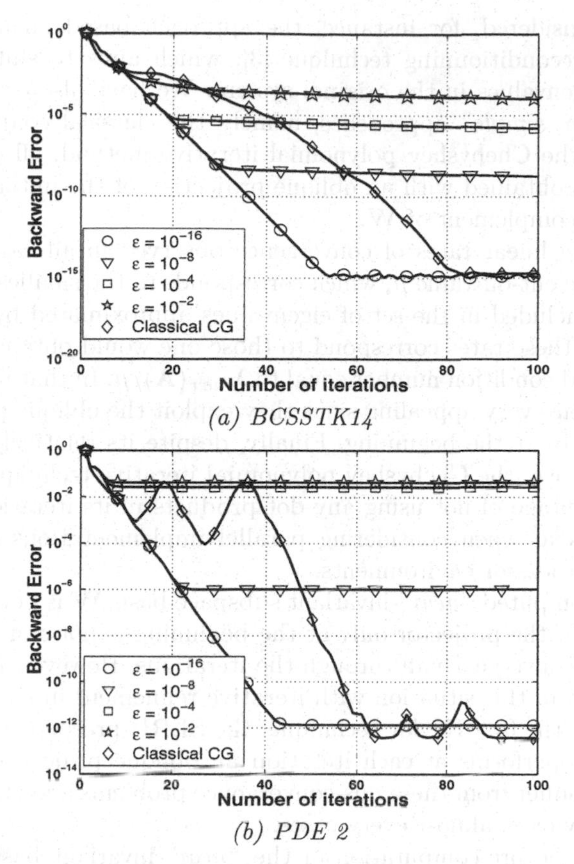

Fig. 5. Convergence history of INIT-CHEBYSHEV for different values of level of filtering ε

It is also worth mentioning that this combination of oblique projection and Chebyshev iteration is based only on kernels commonly used in iterative methods, that enable us to keep the given ill-conditioned matrix in implicit form. It requires only matrix-vector products plus some vector updates, but no dot-products, as opposed to the other Krylov based solution techniques. This remark is also of some importance in the context of parallel computing, and in particular in distributed memory environments.

5 Cost Benefit and Concluding Remarks

In this paper, we have compared various iterative solution techniques that exploit partial spectral information about the matrix **A**. The common general idea in all these techniques is to use some precomputed "*near*"-invariant subspace **W** to build a projection onto the orthogonal complement of **W** in some appropriate norm, and to exploit such a projector to maintain, implicitly or explicitly all through the iterations, the computed residuals in this orthogonal complement.

We have considered, for instance, the approach based on a spectral low-rank update preconditionning technique [3], which aims to shift close to one the smallest eigenvalues in the original system. We have also experienced with two other "*more simple*" approaches, namely the classical conjugate gradient algorithm and the Chebyshev polynomial iterative method, all combined with a starting guess obtained with an oblique projection of the initial residual onto the orthogonal complement of **W**.

The resulting linear rates of convergence observed in all cases are directly connected to the cut-off value μ, which corresponds to the smallest eigenvalue in **A** that is not included in the set of eigenvalues approximated by the basis **W**. More precisely, these rates correspond to those one would obtain with matrices having a reduced condition number equal to $\lambda_{max}(\mathbf{A})/\mu$. In that respect, the last two techniques are very appealing since they exploit the oblique projection with the basis **W** only at the beginning. Finally, despite its relatively higher linear rate of convergence, the Chebyshev polynomial iterative technique presents the additional advantage of not using any dot-products in its iterations, which can be quite beneficial when considering parallel implementations on distributed memory multiprocessor environments.

When the computed "*near*"-invariant subspace basis **W** is not very accurate, the simple use of the projector only at the beginning is not sufficient to ensure a linear rate of convergence all through the iterations. However, it is possible to recover easily from this situation with iterative refinement in the usual way. As opposed to this, the first Krylov technique, the SLRU preconditioned conjugate gradient, which performs at each iteration an oblique projection with the **W** basis, does not suffer from these mis-convergence problems and shows a constant numerical behavior in almost every cases.

Concerning the pre-computation of the "*near*"-invariant basis **W**, we have used the partial spectral factorization algorithm proposed by [1], which involves the use of Chebyshev polynomials as a filtering tool within the Lanczos algorithm. This eigencomputation has a cost that depends on the dimension and on the accuracy of the computed "*near*"-invariant subspace and, to be effective,

Table 4. Number of amortization vectors. The computation of the amortization vectors is relative to SLRU and a tolerance of 10^{-10}

Level (ε)	Number of Amortization Vectors	
	BCSSTK14	PDE 2
10^{-16}	18	4
10^{-8}	14	2
10^{-4}	11	1
10^{-3}	9	1
10^{-2}	7	2
10^{-1}	4	3

the gains obtained in the acceleration of the convergence of the given iterative solvers must compensate in some way the extra cost for the computation of this spectral information. In particular, if one is interested in solving consecutively several linear systems with the same coefficient matrix but different right-hand sides, then the gains obtained can reduce quite substantially the total computational time in the long run.

Finally, we illustrate this last issue experimentally in the case of the SLRU preconditioning technique. In Table 4, we indicate the minimum number of consecutive conjugate gradient runs (preconditioned by means of the SLRU preconditioning technique) before the gains obtained actually counterbalance the price for the preliminary computation of the **W** basis with some given filtering level ε. The cost, evaluated in terms of total number matrix-vector products, of this preliminary partial spectral factorization are detailed in Table 3. We can observe that this extra cost can indeed be overcomed in very few consecutive runs of the iterative solver, depending obviously on the desired level of accuracy when precomputing the invariant subspace, and also provided that this accurary is not too low in which case the SLRU preconditioner might not be so effective, as we can observe for instance when the filtering level is only set to 10^{-1}.

Acknowledgement

We would like to thank Serge Gratton and Mario Arioli for the fruitful discussions during the development of this work.

References

1. Arioli, M., Ruiz, D.: A Chebyshev-based two-stage iterative method as an alternative to the direct solution of linear systems. Technical Report RAL-TR-2002-021, Rutherford Appleton Laboratory, Atlas Center, Didcot, Oxfordshire, OX11 0QX, England (2002)
2. Arioli, M., Ruiz, D.: Block conjugate gradient with subspace iteration for solving linear systems. In Margenov, S.D., Vassilevski, P.S., eds.: Iterative Methods in Linear Algebra, II. Volume 3 in the IMACS Series in Computational and Applied Mathematics. Proceedings of The Second IMACS International Symposium on Iterative Methods in Linear Algebra. (1995)
3. Carpentieri, B., Duff, I.S., Giraud., L.: A class of spectral two-level preconditioners. SIAM Journal on Scientific Computing **25** (2003) 749–765
4. Giraud, L., Ruiz, D., Touhami, A.: A comparative study of iterative solvers exploiting spectral information for SPD linear systems. Technical Report TR/PA/04/40 CERFACS, Toulouse, France (2004). Also Technical Report RT/TLSE/04/03, ENSEEIHT–IRIT, Toulouse, France (2004)
5. Golub, G.H., Van Loan, C.F.: Matrix Computations. Johns Hopkins Studies in the Mathematical Sciences. The Johns Hopkins University Press, Baltimore, MD, USA, third edition (1996)
6. Saad, Y.: Iterative Methods for Sparse Linear Systems. PWS Publishing Company, Boston (1996)

7. van der Sluis, A., van der Vorst, H.A.: The rate of convergence of conjugate gradients. Numer. Math. **48** (1986) 543–560
8. Hageman, L.A., Young, D.M.: Applied Iterative Methods. Academic Press, New York and London (1981)
9. Nabben, R., Vuik, C.: A comparison of deflation and coarse grid correction applied to porous media flow. Report 03-10, Delft University of Technology, Department of Applied Mathematical Analysis, Delft (2003)
10. Nicolaides, R.: Deflation of conjugate gradients with applications to boundary value problems. SIAM Journal on Numerical Analysis **24** (1987) 355–365
11. Saad, Y., Yeung, M., Erhel, J., Guyomarc'h, F.: A deflated version of the conjugate gradient algorithm. SIAM Journal on Scientific Computing **21** (2000) 1909–1926
12. Duff, I.S., Grimes, R.G., Lewis, J.G.: Users' guide for the Harwell-Boeing sparse matrix collection. Technical Report TR/PA/92/86, CERFACS, Toulouse, France (1992). Also RAL Technical Report RAL 92-086.
13. Higham, N.J.: Accuracy and Stability of Numerical Algorithms. SIAM, Philadelphia (1997)

Three Parallel Algorithms for Solving Nonlinear Systems and Optimization Problems*

Jesús Peinado and Antonio M. Vidal

Departamento de Sistemas Informáticos y Computación
Universidad Politécnica de Valencia. Valencia, 46071, Spain
Phone: +(34)-6-3877798. Fax: +(34)-6-3877359
{Jpeinado, avidal}@dsic.upv.es

Abstract. In this work we describe three sequential algorithms and their parallel counterparts for solving nonlinear systems, when the Jacobian matrix is symmetric and positive definite. This case appears frequently in unconstrained optimization problems. Two of the three algorithms are based on Newton's method. The first solves the inner iteration with Cholesky decomposition while the second is based on the inexact Newton methods family, where a preconditioned CG method has been used for solving the linear inner iteration. In this latter case and to control the inner iteration as far as possible and avoid the oversolving problem, we also parallelized several forcing term criteria and used parallel preconditioning techniques. The third algorithm is based on parallelizing the BFGS method. We implemented the parallel algorithms using the SCALAPACK library. Experimental results have been obtained using a cluster of Pentium II PC's connected through a Myrinet network. To test our algorithms we used four different problems. The algorithms show good scalability in most cases.

1 Introduction and Objectives

In many engineering problems the solution is achieved by solving a nonlinear system; sometimes this kind of problems appears as a consequence of a discretization of a PDE; sometimes as a mathematical model problem; and sometimes when working with an unconstrained minimization problem [5]. In some of these cases the Jacobian matrix (Hessian) of the nonlinear system is symmetric; and if the function to minimize is convex, then the Jacobian matrix is also positive definite [5]. In this work we describe three portable sequential and parallel algorithms to solve large scale nonlinear systems where the Jacobian matrix is symmetric and positive definite. Two of them are based on Newton's method [9]. The first algorithm uses the Cholesky [8] direct method to solve the inner linear system and is known as the Newton-Cholesky method. The second algorithm uses a preconditioned CG [7] iterative linear method known as the Newton-CG method. This latter method belongs to the family known as the Inexact Newton methods [9]. We used a preconditioning technique by scaling [8]

* Work supported by Spanish CICYT. Project TIC 2000-1683-C03-03.

M. Daydé et al. (Eds.): VECPAR 2004, LNCS 3402, pp. 657–670, 2005.

the linear system because in some cases this kind of technique helps to improve the convergence. We also implemented several forcing terms criteria [6] to avoid the oversolving problem. The third algorithm is known as BFGS [5]. This method is similar to Broyden's method, but it does not use the *QR* decomposition (as Broyden's) but uses Cholesky decomposition at the first iteration, and for other iterations uses a process called BFGS updating [5], similar to the *QR* updating [8].

We used the analytic Jacobian matrix when known, and finite difference techniques [9] to approximate the Jacobian matrix when not known. Our objective was to use as standard a method as possible. Our algorithms allow us to work with large problems. Furthermore, our algorithms are efficient in terms of computational cost. The algorithms we present here are included in a framework tool where, given a nonlinear system of equations, the best parallel method to solve the system must be chosen. This framework was developed in [13].

In our experiments for this paper we have considered and also parallelized four different test problems.

All our algorithms were implemented using portable standard packages. In the serial algorithms we used the LAPACK [1] library, and the CERFACS codes [7] for the CG algorithm, while the parallel codes use the CERFACS codes and the SCALAPACK [4] and BLACS [16] libraries on top of the MPI [15] communication library. All the codes were implemented using C++ and some routines use FORTRAN 77 algorithms.

Experimental results were obtained using a cluster of Pentium II PC's connected through a Myrinet [11] network. However, other machines could be used due to the portability of the packages and our code. We achieved good results and show that both algorithms are scalable.

2 Solving Nonlinear Systems. Unconstrained Minimization

Solving nonlinear systems is not a trivial task. There are several families of methods to solve this kind of problem. We cannot reference a "perfect" method to solve all the nonlinear systems, because each engineering, physics, or mathematical problem has different characteristics. Below we present three methods for solving nonlinear systems, with a symmetric and positive definite Jacobian matrix.

2.1 Newton's Method. The Cholesky and CG Iterations

A good method to solve the nonlinear system: $F(x) = 0, F \in \Re^n, \; x \in \Re^n$ is Newton's method, because it is powerful and has quadratic local convergence [9]. Newton's method is based on the following algorithm, where J is the Jacobian matrix and F is the function whose root is to be found (k is the iteration number):

$$J(x_k)s_k = -F(x_k), \;\; x_{k+1} = x_k + s_k,$$

$$J(x_k) \in \Re^{n\times n}; \;\; s_k, \; F(x_k) \in \Re^n$$

A linear solver must be used to solve the linear system $J(x_k)s_k = -F(x_k)$. Unconstrained minimization problems [5] also give rise to a nonlinear system. Given: $F: \Re^n \rightarrow \Re$, *twice continuously differentiable*, find

$$x^* \in \Re^n \,/\, F(x^*) = \min_x F(x).$$

Thus to solve the minimization problem it is necessary to compute the gradient, and solve $\nabla F(x^*) = 0$.

We summarize some possible ways of solving this problem:

Newton and Newton-Cholesky Methods. In Newton's method we use *LU* decomposition if *J* is a general matrix, or Cholesky [8] decomposition if *J* is symmetric and positive definite, thus obtaining Newton-Cholesky's method. With Cholesky decomposition *J* is factorized as follows:

$$J = LL^T,\ \ L \text{ is lower triangular}\ \ \ J(x_k) \in \Re^{nxn}.$$

Newton-GMRES and Newton-CG Inexact Newton Methods. In these methods the solution of the Newton step is approximated with an iterative linear method. In an *Inexact Newton Method* [9] each step satisfies:

$$\left\| J(x_k)s_k + F(x_k) \right\|_2 = \eta_k \left\| F(x_k) \right\|_2 \ \ (1).$$

One of the best methods to solve the iterative linear system in the nonsymmetric case is the GMRES [14] method. If the Jacobian matrix is symmetric and positive definite, the linear iteration can be changed by Conjugate Gradient (CG) iteration [9]. There are several improvements to these methods: firstly, the forcing term technique [6] for controlling the precision to solve the linear system at each iteration; secondly, preconditioning techniques for improving convergence [8].

2.2 Forcing Terms. Preconditioning Techniques

As stated above, an inexact Newton method is a kind of Newton method for solving a nonlinear system, where the *k*th iteration, the step from the current approximate solution is required to satisfy (1). But we want to choose a strategy to select η_k as the outer iteration progresses. This term is known as *forcing term*. There are several possible strategies to choose a forcing term, independent of the particular linear solver. Setting to a constant for the entire iteration is often a reasonable strategy, but the choice of that constant depends on the problem. If a constant is too small much effort can be wasted in the initial stages of the iteration. This problem is known as oversolving [6]. Oversolving means that the linear equation for the Newton step is solved to a precision far beyond what is needed to correct the nonlinear equation. We used here three possible alternatives, two obtained from [6] and the third from [9]. For more information, see [13].

Preconditioning and Scaling Techniques

Due to the fact that the preconditioning techniques depend on the problem to be solved, we developed our algorithm so that the user can choose a preconditioner. Our algorithm is ready to use left or right preconditioners. We used scaling techniques [8] by rows (row scaling) and by rows and columns (row-column equilibration). To do this we used routines supplied by the LAPACK and SCALAPACK. Note that in the CG method the preconditioner matrix is the same using right or left preconditioning because the linear system matrix A is symmetric.

2.3 Broyden and BFGS Methods

Another possible way of solving the problem is to use the BFGS (Broyden-Fletcher-Goldfarb-Shano) method [5]. This is a particular version of Broyden's Method [5] that deals with symmetric and positive Jacobian matrices. In this family of methods the Jacobian matrix and the linear decomposition (QR or Cholesky) are only computed at the first iteration. This feature reduces the computing time but the convergence is often made worse. The next Jacobians matrices are approximated using the BFGS updating.

In Broyden's method the Jacobian matrix in the next iteration J_+ is obtained (approximated) by using the current Jacobian matrix J_c. This is $J_+ = Q_+R_+ = Q_cR_c + uv^T$. In the BFGS method the next Jacobian matrix $H_+ = L_+L_+^T$ is computed with $L_+^T = L_c^T + \hat{w}\hat{z}^T L_c^T$. L_c is the Cholesky factor and $H_c = L_cL_c^T$. Computing L_+ can be done in $O(n^2)$. $\hat{w}$ and $\hat{z}$ can be computed from s_+ and $F(x_+)$.

To improve the convergence it is possible to compute (not approximate) the exact Jacobian matrix in some iterations. We applied this technique and improved the convergence in some cases thus achieving a smaller number of iterations.

3 Newton-Cholesky, Newton-CG and BFGS Sequential Methods. Practical Implementation

In this section, we present our Newton-Cholesky, Newton-CG and BFGS serial algorithms. The Cholesky and BFGS codes are based on LAPACK, while the CG code is based on a CERFACS implementation. In the next section we will justify why we chose this code.

All the algorithms give the user the possibility of choosing the following steps for the nonlinear system: the initial guess, the function, the Jacobian matrix, and the stopping criterion. Furthermore, the Newton-CG algorithm gives the user the possibility of choosing the preconditioner and the oversolving criterion for the inner iteration.

Our serial implementations are based on using the BLAS and LAPACK numerical kernels. As stated before the Newton-Cholesky method is similar to the Newton method, but LU decomposition has been substituted by for Cholesky's.

Serial Algorithms Newton-Cholesky/Newton-CG

Choose a starting point x_k

Compute vector $F(x_k)$

While the stopping criterion has not been reached

  Compute the Jacobian matrix $J(x_k)$

  Solve the linear system $J(x_k)s_k = -F(x_k)$ (using Cholesky or CG methods)

  Update the iterate $x_{k+1} = x_k + s_k$

  Compute vector $F(x_k)$

The CG algorithm from CERFACS implements a powerful, efficient and configurable method. Furthermore it has several advantages over other implementations: it leaves the user the possibility of working with numerical kernels (see the next paragraph). And it can be modified to work with SIMD parallel computing. We will explain how we did this in Section 4.

It is known that almost all the Krylov [14] methods are based on Matrix-vector multiplications and dot products. This is an advantage since the CG code we use has a reverse (revcom) communication scheme that allows us to work with the method by using *a driver* routine to compute the method, and when the driver needs a numerical operation, it calls numerical kernels or to a user defined routine, in our case BLAS1 to compute the dot products and BLAS2 to compute the Matrix-vector products. We use the (DSYMV) routine for the Matrix-vector product routine, and (DDOT) to compute the scalar products. The user can also change these routines for others as he wishes. The preconditioning routines can be left to the user or be supplied by us. We used two LAPACK routines to carry out the scaling technique. The (DPOEQU) routine checks if the scaling is necessary, and the (DLAQSY) routine applies the scaling to matrix A if it is necessary. It is very important to note that applying the scaling on the right hand side b, and on the solution of the system x (when computed) must be done by using our routines. With this scheme we can parallelize the entire nonlinear scheme including the CG iteration. We will study this in the next section.

The BFGS code has a common part with the Newton-Cholesky because it is necessary to compute the Cholesky decomposition, but only in the first iteration. For other iterations it is necessary to compute the Cholesky factor in the current iteration L_+ using the Cholesky factor of the previous iteration L_c.

4 Parallel Algorithms

4.1 How to Parallelize the Sequential Algorithms

The parallel version we implemented uses the SCALAPACK library. Within an iteration, the function evaluation, the computation of the Jacobian matrix, the computing of the forcing term, the solution of the linear system including the matrix-vector products and the scalar products and the update of the iterate are parallelized. The evaluation of the function and the computation of the Jacobian matrix depend on the problem to be solved. We analyze this in the case study section. Below we explain how to parallelize the other steps. The SCALAPACK library uses a 2-D block cyclic data distribution.

With this data distribution a good compromise can be reached to parallelize the Cholesky and CG iterations, the different forcing term criteria and the updating of the iterate. All these steps depend on the method used. Furthermore the problem dependant steps must be also paralellized.

We emphasize parallelizing the CG iteration because the Cholesky iteration is already implemented in the SCALAPACK package and is similar to other works using the Newton method previously developed by us [13].

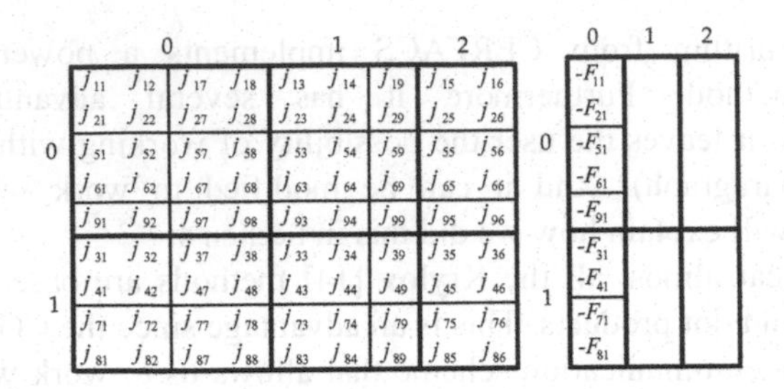

Fig. 1. SCALAPACK block cyclic distribution for the Jacobian matrix and the function

Parallelizing the CG Code

There are two different parts to parallelize: the driver routine, and the reverse communication steps.

Driver routine: The same driver routine [7] is run on all the mesh processors, and some of their data is replicated locally (repeated on all the mesh processors), and some is stored distributed (across the mesh processors) by cyclic blocks. Matrix A, the right hand side of the linear system b, the computed solution of linear system x, and the temporal vectors used for the algorithm, are stored distributed. The other terms are stored replicated locally.

Reverse communication mechanism: this mechanism must be adapted to work in parallel. The precondition steps are problem dependant and are left to the user, but we have to modify the Matrix-vector product and the vector dots steps. To compute the Matrix-vector dot we use the PBLAS routine (`PDSYMV`). A user defined routine could be used. (`PDSYMV`) uses the 2-D cyclic distribution as shown in Figure 1. As already mentioned, the CERFACS CG code is ready [7] to work in an SIMD [10] environment, but SCALAPACK is not fully compliant with this because the right hand side vectors (rhs) are not stored in all the processors. We can solve this by sending the subvector to all the processors of the same row. This can be appreciated in Figure 2.

The other operation is the parallel dot product. The two data vectors are distributed as vector F. See Figure 1. We can compute each subvector product and then do a global sum operation. This operation is optimized in SCALAPACK. Then the result of the dot product goes to all the column processors.

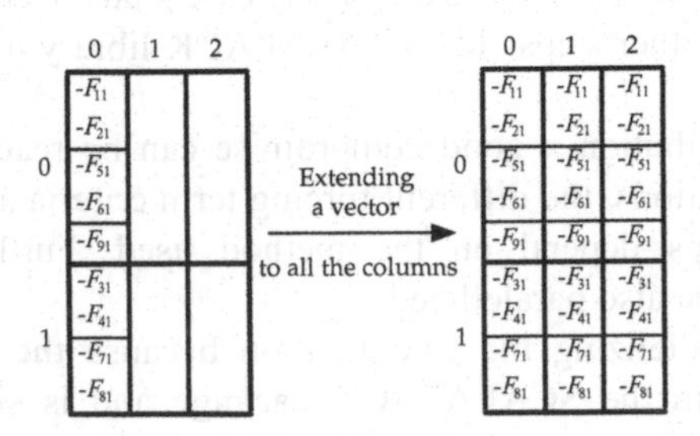

Fig. 2. The column processors must have the rhs vectors to work with the CG driver

To use the scaling techniques shown in Section 2, the parallel SCALAPACK scaling routines analogous to the serial routines in LAPACK must be used: routine (`PDPOEQU`) checks if the scaling is necessary on a distributed matrix. The routine (`PDLAQSY`) applies the scaling on the matrix if necessary. Again as in the serial case, these routines do not scale the right hand side and the computed solution of the linear system. We applied these scalings by developing a parallel routine to carry out this process.

Forcing Term

The operations done to compute the forcing term criterion in the former section are based [13] on computing several 2-norms. Furthermore there is a Matrix-vector product. And some operations are done by using the values computed in the former iteration.

Parallelizing the BFGS Algorithm

The process is complex [13] and it involves many communications. We used LAPACK routines to compute and apply Givens rotations. These are done in each processor (sequential code) involved in computing the Givens rotation. SCALAPACK routines are used to manage the processors involved in BFGS updating. And the following BLACS routines are used to carry out the necessary communications to send/receive the elements of the rotation, exchanging the rows of L: `cdgsum2d, Cdgebs2d` and `Cdgbr2d`.

Nonlinear Iterate Update

When the linear system is solved, solution s is stored in the right hand side vector. Then we only have to broadcast s to all the processors, and update the iterate. Then the update $x_{(k+1)} = x_{(k)} + s_{(k)}$ can be done.

5 Case Studies

In this section we study the four test problems. As mentioned before, the first problem is obtained as a consequence of discretizing a PDE problem, while the others are unconstrained minimization problems.

The Chandrasekhar H-Equation

This problem [9] is used to solve exit distribution problems in radiative transfer. The equation can be discretized. The resulting discrete problems is:

$$F_i(x) = x_i - \left(1 - \frac{c}{2N}\sum_{j=1}^{N}\frac{\mu_i x_j}{\mu_i + \upsilon_j}\right)^{-1} = 0 \qquad i = 1 \ldots n.$$

In this case we will assume that all μ_i terms have the same value. This assumption makes the Jacobian matrix symmetric and positive definite. We parallelized both the computation of the function and the computation of the Jacobian matrix by using the standard SCALAPACK distribution.

Minimizing an H-Ellipse with Respect to a Coordinate

This is a synthetic problem for testing the algorithms developed in this work. We want to find the minimum value of a coordinate of an H-ellipse. An H-ellipse [3] is a special ellipse extended to a $(n+1)$-dimensional space:

$$\frac{x_0^2}{v_0} + \frac{x_1^2}{v_1} + \cdots + \frac{x_n^2}{v_n} = c .$$

For example, to minimize x_0: $x_0 = \pm v_0 \sqrt{c \quad \frac{x_1^2}{v_1} \quad \cdots \quad \frac{x_n^2}{v_n}}$.

We need to find $x^* = (x_1, x_2 ... x_n)^T$ that minimizes the expression. There are two solutions for the above expression. A negative sign expression (where there is a local minimum), and a positive sign expression (where there is a local maximum). Then the function to minimize is:

$$F(x_1, x_2 ... x_n) = v_0 \sqrt{c \quad \frac{x_1^2}{v_1} \quad \cdots \quad \frac{x_n^2}{v_n}}$$

Computing the Minimum of an n-Dimensional Wave

Our idea here is to use a case modeling several mechanical engineering problems by using a general parameterized formula. We want to minimize:

$$F(x) = h + c\sqrt{t}\sin(t) \qquad t = 1 + \sum_{i=1}^{n} \left(\frac{x_i \quad b_i}{a_i} \right)^2$$

Minimizing the Leakage of a Hydraulic Network

This is a real problem [2]. It is a minimization problem whose scheme is very usual in a certain kind of mathematical and engineering problems. The function is a sum of m functions:

$$\min F(x) = \sum_{p=1}^{m} q_p(x) .$$

The goal of our problem is to minimize the pressure on several points of a set of pipes that make a hydraulic network. To do this it is necessary to control and adjust the position of some valves in the network. In our problem the positions of the valves are n entries of vector x. In hydraulic engineering the pressure function [2] for each pipe in the network is as follows:

$$q_p(x) = K_p P_p^{1.1}(x) , P_p = P_p(x,t) \; \forall \; p = 1, 2 \ldots m .$$

The pressure equation for this kind of problem is always known: it depends on valve position and consumption along time (t). To work with several test cases, the pressure equations can be simulated with functions modeling the pressure along time, but these equations only depend on the valve positions (x). The idea is to use functions with curves modeling water consumption. Each pipe has a minimum pressure:

$$P_p(x) \geq Pmin_p .$$

In some cases the Hessian for this problem is symmetric and positive definite. In our example, we used three functions with n valves.

6 Experimental Results

Below is a brief study of the performance our algorithms. The complete study can be found in [13]. We report here the most relevant results. We present the experimental results for each problem we implemented. For each problem we could use a forward finite differences approach, or an analytic Jacobian matrix. We used the forward finite difference formula only in the first case because as we mentioned before it is faster than the analytic formula.

Furthermore we used the techniques for avoiding the oversolving problem as explained above. We also used scaling techniques to improve convergence. Improving the behaviour of the problem using these techniques depends on the problem to be solved. The possible improvement using a forcing term depends on heuristic [6] factors while improvements using scaling depend on the condition number of the linear system matrix. We used both techniques in all the test problems and they have been useful in the H-ellipse problem. The entire test was done using two different problem sizes: $N=1600$ and $N=2000$. In problems with a longer computing time these sizes are adequate because an increase in the number of processors leads to a substantial reduction in computing time. In problems with a shorter computing time we must emphasize the behaviour of our algorithms which obtain good parallel execution times even when the execution times are very short. Finally computing time and speedup is shown for all the case studies.

The hardware used was a cluster of 16 Pentium II/300 MHz PC's with a Myrinet network [11]. We varied the number of processors and type of network topologies.

The Symmetric Chandrasekhar H-Equation

We show the computing time results for the symmetric H-Chandrasekhar equation. We include the best network topology (best mesh type) for each test. All the algorithms reach the convergence for all the tests.

Table 1 shows the good behaviour of our algorithms. It can be seen that the execution times decrease as the number of processors increases. Furthermore there is some influence with variations in network topology. In general it is good practice to use a square mesh. The fastest method was BFGS (BFGS). Newton-CG (NCG) worked better than Newton-Cholesky (NCH) because the linear system is solved very fast with the CG method, because convergence of the linear system is reached with a very few inner iterations. Figure 3 shows the speedup. The results are good and show the algorithms are scalable for this problem.

Table 1. Parallel computing time in seconds. Chandrasekhar problem

Chandrasekhar		Newton-CG		Newton-Cholesky		BFGS	
Proc .Nr.	Best Mesh	1600	2000	1600	2000	1600	2000
1		1575,12	3067,88	1654,14	3226,60	131,51	243,31
2	1x2	810,35	1581,72	844,36	1647,12	67,40	138,64
4	2x2	448,62	869,35	440,39	851,58	39,60	70,98
6	3x2	295,44	538,03	311,71	566,00	28,90	50,68
8	8x1	232,37	445,59	255,58	485,49	22,65	39,05
9	3x3	205,23	381,58	217,11	401,89	22,21	38,37
10	5x2	182,36	351,12	193,82	370,74	19,14	33,01
12	3x4	148,16	291,79	161,76	304,03	20,03	32,70
16	4x4	114,86	216,73	119,93	229,70	14,66	24,61

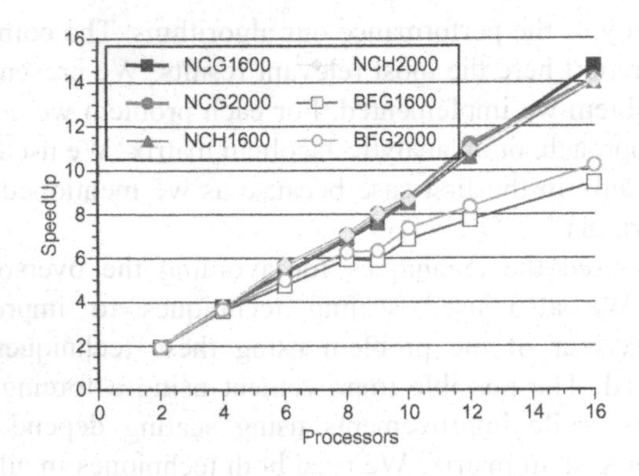

Fig. 3. Speedup. Chandrasekhar problem

Minimizing an H-Ellipse with Respect to a Coordinate

We show the computing time results for this problem. We used the Newton-CG algorithms with (and without) a forcing term (Fterm) criterion and with scaling (Equ) and without scaling (NoEqu). BFGS does not converge. The Newton-Cholesky converges in all the cases. The Newton-CG algorithm only converges when using one of the improvements above. Using both of them together does not improve the results.

Table 2. Parallel computing time in seconds. H-Ellipse problem

H-elipse	NewtonCG (NoEqu,Fterm)		Newton-CG (Equ)		Newton-Cholesky	
Proc. Nr.	1600	2000	1600	2000	1600	2000
1	319,5	616,2	14,16	22,09	69,94	130,98
2	162,13	326,09	7,77	12,07	31,51	58,38
4	75,47	150,36	3,96	6,15	18,02	32,24
6	57,47	113,39	2,73	4,17	13,8	24,08
8	42,77	84,05	2,06	3,18	11,25	19,54
9	37,45	67,44	1,87	2,78	9,64	16,78
10	48,84	90,21	1,76	2,66	11,36	19,56
12	33,79	61,37	1,42	2,14	8,46	14,48
16	24,1	45,1	1,05	1,62	6,40	10,82

We do not include the best mesh type parameter because it is different for each algorithm for a given number of processors. The complete test can be consulted in [13] and it can appreciated that the algorithms work better with square (or almost square) meshes.

Table 2 shows the behaviour of our algorithms. It can be appreciated that the execution times decrease as the number of processors increases. The best method is Newton-CG with scaling (NCGFE), then Newton-Cholesky (NCH) and finally Newton-CG (NCGF) with a forcing term criterion. Good scaling is crucial (if necessary) in Newton-CG because as mentioned when explaining the method, decreasing the condition number of the linear system matrix improves convergence. In Figure 4 we show the speedup for this problem:

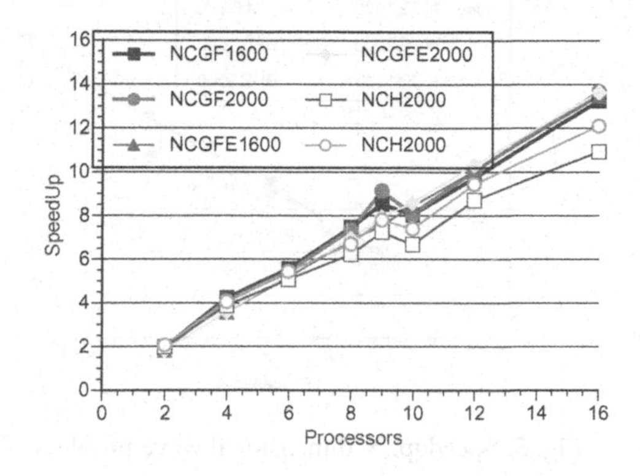

Fig. 4. Speedup. H-ellipse problem

Computing the Minimum of an n-Dimensional Wave

We show the results for this problem in Table 3. Both algorithms converge for all the problems. No improvements were found when using a forcing term or the scaling technique. The computing time results are as follows:

Table 3. Parallel computing time in seconds. Minimum of a n n-dimensional wave

Newton-CG	Newton-CG		Newton-Cholesky		BFGS	
Proc. Nr.	1600	2000	1600	2000	1600	2000
1	8,33	12,68	52,51	123,64	40,17	72,84
2	4,41	6,88	23,22	55,49	21,75	38,86
4	2,21	3,53	13,42	30,30	13,19	22,60
6	1,54	2,36	10,33	22,77	10,64	17,91
8	1,15	1,8	8,44	18,52	8,61	14,00
9	1,05	1,61	7,24	15,93	9,43	15,69
10	0,94	1,46	7,82	16,92	8,52	13,85
12	0,81	1,26	6,07	13,20	7,80	12,73
16	0,6	0,93	4,86	10,36	7,16	11,67

As with the above problems, we can appreciate that computing time decreases when the number of processors is increased. Again the best topology varies when changing the algorithm and the problem size, although the most efficient results are obtained when using square or almost square (vertical) 2D-meshes. See [13].

The Newton-CG works better because it performs a very few internal iterations. Note the behaviour of the algorithm working with low computing times that are difficult to parallelize. BGFS does not perform better due to the communications generated in the BFGS updating. Our algorithms and our problem implementation are scalable although the BFGS method is less scalable. The speedup results can be observed in Figure 5.

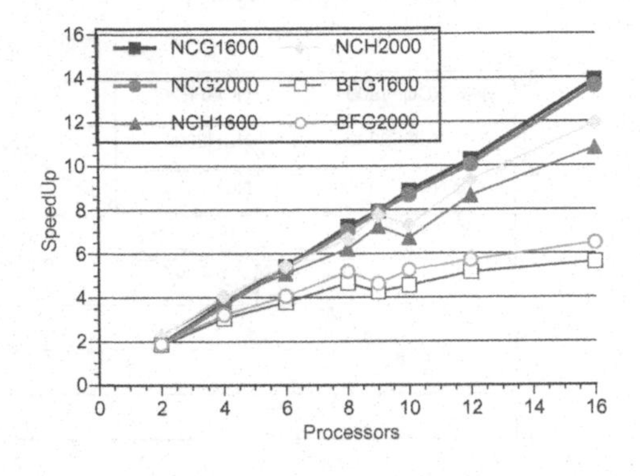

Fig. 5. Speedup. N-dimensional wave problem

Minimizing the Leakage of a Hydraulic Network

We show the results for the Hydraulic engineering problem in Table 4. The algorithm converges for the entire test. There is no improvement when using a forcing term criterion or the scaling technique. We include the results only computing the exact Jacobian (BFGS_0) at the first iteration, or computing it every 6 iterations (BFGS_6). The numerical results are:

Table 4. Parallel computing time in seconds. Hydraulic problem

Hydraulic	Newton-CG		Newton-Cholesky		BFGS(0)		BFGS(6)	
Proc. Nr.	1600	2000	1600	2000	1600	2000	1600	2000
1	138,75	189,18	229,23	342,90	93,39	153,30	94,99	161,48
2	75,82	104,11	114,76	169,73	68,61	107,76	45,01	76,12
4	38,88	53,1	61,89	89,96	36,91	57,13	23,45	38,12
6	24,06	36,11	44,78	64,70	33,83	52,34	19,13	29,31
8	20,13	27,27	35,40	50,55	24,17	37,55	14,78	22,28
9	18,2	23,61	31,20	43,96	26,35	39,94	13,80	20,92
10	16,37	22,12	30,27	43,12	24,43	36,31	13,33	20,59
12	13,71	18,13	25,04	35,44	22,57	34,77	11,39	17,77
16	9,36	13,74	18,64	26,52	20,29	35,39	9,35	13,68

Below we include the speedup for this problem. There are some oscillations with the speedup in the Newton-CG method when solving *N=2000*. We studied this behaviour and found it was due to variations in the iterations of the CG method when changing the number of processors. These variations could be due to the error produced when computing the parallel scalar products. This is a known issue when working with parallel inexact Newton Methods [12]. In BFGS it is better to compute the exact Jacobian matrix every 6 iterations.

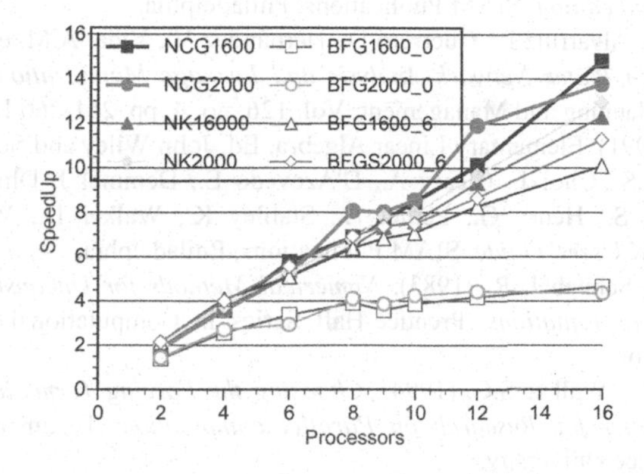

Fig. 6. Speedup. Hydraulic problem

7 Conclusions

We have developed and compared three efficient methods to solve nonlinear systems when the Jacobian matrix is symmetric and positive definite. Our methods work with any nonlinear problem. The user only has to supply the starting guess, the function and the Jacobian matrix (both parallelized if possible). If the user only knows the function, the methods can compute the Jacobian matrix by using a forward finite difference technique. If we work with optimization problems, the function must be changed by the gradient, and the Jacobian matrix by the Hessian.

We worked with four problems, two synthetic and two real problems. We obtained good performance when working with the parallel version of our algorithms, and all test problems were scalable for all the methods. The best method depends on the problem to be solved. BFGS is fast but does not always converge. Newton-CG works faster than Cholesky's only when the CG linear method performs a few iterations.

In our implementation it is also important to note the portability of the code because all the software used, the LAPACK/SCALAPACK libraries and CG (from CERFACS) standard codes, and PVM/MPI parallel environments is standard and portable

Acknowledgments

We want to thank Professor Luc Giraud from CERFACS (Parallel Algorithms Project) for his invaluable help in explaining to us the difficulties and characteristics of his GMRES and CG codes. Without his help, it would have been impossible to develop this work.

References

[1] Anderson E., Bai Z., Bischof C., Demmel J., Dongarra J., Du Croz J., Greenbaum A., Hammarling S., Mackenney A., Ostrouchov S., Sorensen D. (1995). *LAPACK Users' Guide. Second edition.* SIAM Publications, Philadelphia.

[2] Alonso J.M., Alvarruiz F., Guerrero D., Hernández V., Vidal A.M. et al. (2000). *Parallel Computing in Water Network Analysis and Leakage Minimizations*. Journal of Water Resources Planning and Management. Vol. 126, no. 4, pp. 251-260 ISSN 0733-9496.

[3] Anton H. (1991). Elementary Linear Algebra. Ed. John Wiley and Sons. New York

[4] Blackford L.S., Choi J., Cleary A., D'Azevedo E., Demmel J. Dhillon I., Dongarra J., Hammarling S., Henry G., Petitet A., Stanley K., Walker D., Whaley R.C. (1997). *SCALAPACK Users' Guide*. SIAM Publications, Philadelphia.

[5] Dennis J.E., Schnabel, R. (1983). *Numerical Methods for Unconstrained Optimization and Nonlinear Equations.* Prentice-Hall Series in Computational Mathematics. Cleve Moler Advisor.

[6] Einsestat S.C., Walker S.C. (1994*). Choosing the Forcing Terms in an Inexact Newton Method. Center for Research on Parallel computation.* Technical Report CRPRPC-TR94463. Rice University.

[7] Frayssé V., Giraud L. (2000). *A Set of Conjugate Gradient routines for Real and Complex Arithmetics*. CERFACS Technical Report TR/PA/00/47.

[8] Golub G. H., Van Loan C. F. (1996). *Matrix Computations (third edition)*. John Hopkins University Press.

[9] Kelley C.T. (1995). *Iterative Methods for Linear and Nonlinear Equations*. Frontiers in Applied Mathematics, SIAM Publications, Philadelphia..

[10] Kumar R., Grama A., Gupta A., Karypis G. (1994). *Introduction to Parallel Computing: Design and Analysis of Algorithms*. The Benjamin Cumimngs Publishing Company.

[11] Myrinet.*Myrinet overview*.http://www.myri.com/

[12] Peinado J, Vidal A.M. (2002) *A Parallel Newton-GMRES Algorithm for Solving Large Scale Nonlinear Systems*. VecPar '2002. Selected Papers and Invited Talks. Lecture Notes in Computer Science. Ed. Springer. LNCS 2565 pp.328-342.

[13] Peinado J. (2003). Ph.D. *Thesis. Parallel Resolution of Nonlinear Systems.* September 2003. DSIC-UPV.

[14] Saad Y., Schultz M. (1986). *GMRES: A generalized minimal residual algorithm for solving nonsymmetric linear systems*. SIAM J. Sci. Stat, Comput. 7, 856-869.

[15] Skjellum A. (1994). *Using MPI: Portable Programming with Message-Passing Interface*. MIT Press Group.

[16] Whaley R.C. (1994). *(BLACS): Analysis and Implementation Across Multiple Parallel Architectures.* Computer Science Dept. Technical Report CS 94-234, University of Tennesee, Knoxville.

Numerical Integration of the Differential Riccati Equation: A High Performance Computing Approach

Enrique Arias[1] and Vicente Hernández[2]

[1] Departamento de Informática, Universidad Castilla-La Mancha,
Avda. España s/n,02071-Albacete, Spain
earias@info-ab.uclm.es
Tel: +34-967-599200 Ext: 2597 Fax: +34-967-599224

[2] Departamento de Sistemas Informátios y Computación,
Universidad Politécnica de Valencia,
Camino de Vera s/n, 46022-Valencia, Spain
vhernand@dsic.upv.es
Tel: +34-96-3877350 Ext: 73560 Fax: +34-96-3877359

Abstract. This paper presents a **High Performance Computing Approach (HPC)** to the code appeared in [1] called DRESOL, for the **Numerical Integration of the Differential Riccati Equation**. This equation arises in the application of quadratic optimization for motion control to the feedback control of robotic manipulators. In this paper the main changes carried out in the DRESOL package and the new block oriented subroutines for computing the Sylvester and Lyapunov equations in order to obtain a sequential HPC implementation are described. From this new sequential implementation parallel algorithms for distributed memory platforms have been also carried out.

Keywords: Numerical Methods, Parallel Computing.

1 Introduction

During the last years, the availability of more powerful computers at a low cost has allowed great advances in the theory and application of non linear control [2, 3, 4, 5].

This paper is concerned with the application of parallel computing for solving the differential Riccati equation (DRE). This equation arises on several applications of different fields of science and engineering, such as the quadratic optimization for motion control to the feedback control of robotic manipulators.

It is possible to find in the literature different methods for solving the DRE [6, 7]: direct integration methods [8], methods based on Bernoulli substitutions [9, 10], methods based on superposition rule of Riccati solutions [11, 12], numerical integration of the Riccati equation based on a Newton iteration method [13, 1], symplectic methods [15, 16], methods based on Magnus series [17], the Mobius scheme [18], etc.

M. Daydé et al. (Eds.): VECPAR 2004, LNCS 3402, pp. 671–684, 2005.

The work presented in this paper has been based on the implementation developed by Luca Dieci [1] for solving the DRE by means of numerical integration.

This paper is organized as follows. First, an exhaustive study of the DRESOL package has been carried out in order to identify the main computational kernels to be enhanced. Section 2 describes the DRESOL package and several changes are proposed from a High Performance Computing (HPC) point of view. These changes can be clasified into minor changes and important changes. Minor changes are related to intensive use of standard linear algebra libreries such as Basic Linear Algebra Subroutines (BLAS) [20] and Linear Algebra PACKage (LAPACK) [21]. Important changes consist in new block oriented algorithms. In particular, new routines to solve the Sylvester and Lyapunov equations have been carried out under a block-oriented philosophy in order to improve performance.

Once a better sequential implementation was available, a distributed implementation has been carried out using Parallel Basic Linear Algebra Subroutines (PBLAS) [22] and Scalable Linear Algebra PACKage (ScaLAPACK) [23] for computations and Message Passing Interface (MPI) [24] for comunication and synchronization purposes. Details on the parallel implementation are introduced in Section 4. Section 5 presents the cases of study and some experimental results. Finally, some conclusions and future work are outlined in Section 6.

2 Numerical Integration of the Differential Riccati Equation

In this Section, the methodology introduced in [1] by Luca Dieci for solving the Differential Riccati Equation (DRE) by numerical integration is presented.

DRE means the expression of a time-dependent change of variables which decouples a linear system of Ordinary Differential Equations (ODEs). Then, let $A(t)$ be the matrix of the underlying linear system. This matrix is a continuous matrix function and all the quantities are time dependent. Let's consider the two-point boundary value problems (TPBVPs) with separated boundary conditions (BCs), where the partitioning conforms to the BCs, and the matrix B_{12} is invertible

$$\dot{y} = \begin{bmatrix} \dot{y}_1 \\ \dot{y}_2 \end{bmatrix} = \begin{bmatrix} A_{11} & A_{12} \\ A_{21} & A_{22} \end{bmatrix} \begin{bmatrix} y_1 \\ y_2 \end{bmatrix}, \; 0 < t < T,$$

$$[B_{11}, B_{12}] \begin{bmatrix} y_1 \\ y_2 \end{bmatrix}_{t=0} = \beta_1, \; [B_{21}, B_{22}] \begin{bmatrix} y_1 \\ y_2 \end{bmatrix}_{t=T} = \beta_2,$$

where $A_{11} \in R^{n \times n}$, $A_{22} \in R^{m \times m}$, $y_1 \in R^n$, $y_2 \in R^m$, and the constant matrices $B_{12} \in R^{m \times m}$, $B_{21} \in R^{n \times n}$, $B_1 \in R^m, B_2 \in R^n$. It is possible to consider the time-dependent change of variable

$$S(t) = \begin{pmatrix} I_n & 0 \\ X(t) & I_n \end{pmatrix}, \quad X(t) \in R^{n \times m}.$$

Assuming a new variable $w = S^{-1}(t)y$, the system is partly decoupled resulting the following structure

$$\dot{w} = \hat{A}(t)w, \tag{1}$$

$$\hat{A}(t) = S^{-1}AS - S^{-1}\dot{S} = \begin{pmatrix} A_{11} + A_{12}X & A_{12} \\ 0 & A_{22} - XA_{12} \end{pmatrix} = \begin{pmatrix} \hat{A}_{11} & A_{12} \\ 0 & \hat{A}_{22} \end{pmatrix}, \tag{2}$$

plus BCs for w. This problem has solution if and only if $X(t)$ satisfies the DRE

$$\dot{X} = A_{21} + A_{22}X - XA_{11} - XA_{12}X = F(t, X),\ \ X(0) = X_0, \tag{3}$$

and initial values X_0. By requiring

$$X_0 = -B_{12}^{-1}B_{11}, \tag{4}$$

then the BCs in (1) are also decoupled

$$(0, B_{12})w(0) = \beta_1,\ \ (B_{21} + B_{22}X(T), B_{22})w(T) = \beta_2. \tag{5}$$

So, the original BVP (Boundary Value Problem) can be solved according to the following strategy:

1. Solve equations (3,4) and the w_2 – *equation* in (2,5) from 0 to T.
2. Solve w_1 – *equation* in (2,5) from T to 0 and recover $y = S(t)w$.

All the possible strategies need to integrate (3), and the numerical integration of this equation is the main purpose of this section, presented in [1].

A particular case of the Riccati equation (3), is the *symmetric* DRE

$$\dot{X} = A_{21} - A_{11}^T X - XA_{11} - XA_{12}X,\ \ X(0) = X_0. \tag{6}$$

This kind of equation (6) appears when the matrix $A(t)$ is the particular Hamiltonian

$$A(t) = \begin{pmatrix} A_{11} & A_{12} \\ A_{21} & A_{11}^T \end{pmatrix}$$

being $n = m$, $A_{12} = A_{21}^T$, $A_{21} = A_{21}^T$ semidefinite positive, and the solution is $X = X^T$. This *symmetric* DRE can be found in several application such as optimal control.

A common way to deal with the numerical integration of DRE is to rewrite it as an nm-vector differential equation, and then use any available initial value problem (IVP) software.

In [1] implicit schemes are used to solve the discrete version of the DRE according to the following property:

Property. Discretization by an implicit scheme of a polynomial differential equation reduces it to a polynomial algebraic equation of the same degree.

In particular, the DRE reduces to an Algebraic Riccati Equation (ARE) after discretization. Then, this ARE is solved at each integration step by means of Newton's type iteration.

The implementation appeared in [1] is referred as DRESOL package. This implementation is based on the IVP software package LSODE developed by Hindsmarsh [14].

Several methods have been implemented for solving the AREs. However, in the context of stiff DREs, the best choice for solving the associated ARE obtained by using implicit schemes are Newton's or quasi-Newton's methods.

In both cases, at each iteration step a Sylvester equation has to be solved

$$G_{11}Y - YG_{22} = H, \tag{7}$$

where Y and $H \in R^{n \times m}$, $G_{11} \in R^{n \times n}$, and $G_{22} \in R^{m \times m}$. H, G_{11} y G_{22} change at each iteration if Newton's method is used, while G_{11} and G_{22} stay constant if quasi-Newton's method is used.

A standard solution for solving the equation (7) is the Bartels-Stewart algorithm [19]. By using this approach, the matrices G_{11} and G_{22} are both reduced, via orthogonal U and V, to Schur form, obtaining the equivalent system

$$(U^T G_{11} U)(U^T Y V) - (U^T Y V)(V^T G_{22} V) = U^T H V, \tag{8}$$

Once the quasi-triangular system (8) is solved for $U^T Y V$, then Y is easily recovered.

3 DRESOL Package: A High Performance Computing Version

In this Section, the changes carried out in the DRESOL package are described. The aim of these changes is to adapt, from a implementation point of view, the original DRESOL package to the current architectures, taking into account the memory hierarchy. Also, the changes introduced keep the following two important features:

- **Portability:** Standard Linear Algebra libraries have been used to achieve portable algorithms.
- **Efficiency:** Good performance has been achieved by improving certain routines. Enhancements in the routines consist on developing block oriented algorithms. In this way, the memory hierarchies of the modern computers can be efficiently exploited.

The DRESOL Package consists on the following subroutines and functions:

- **dresol and drequs**: They both constitute merely an interface between the user's driver and the differential Riccati equation solver, subroutine *drequ*.
- **intdy**: It computes an interpolated value of the y vector (which is the unrolled Riccati matrix x) at $t = tout$.
- **stode**: This is the core integrator, which does one step of the integration and the associated error control and eigenvalues checking.
- **cfode**: It sets all method coefficients and tests constants.

- **f, doblks, and foftex**: They compute the right hand side of the Riccati equation by maximizing efficiency.
- **prepj, jac**: They manage computation and preprocess of the Jacobian matrix and the Newton iteration matrix $p = i - hl_0j$. There are several important technical details here. These routines call the routines *axpxbd, atxxad, orthes, ortran, hqr3, exchng, qrstep* and *split* to perform an ordered eigenvalue decomposition.
- **solsy**: It manages solution of linear system in the Chord iteration and in the quasi-Newton iteration. It calls the routines *axpxbs, shrslv, atxxas, symslv and sysslv* to solve the decomposed Sylvester equations.
- **ewset**: It sets the error weight vector *ewt* before each step.
- **vnorm**: It computes the weighted max norm of a vector.
- **srcom**: It is a user-callable routine to save and restore the contents of the internal common blocks.
- **r1mach**: It computes the unit roundoff in a machine-independent manner.
- **xerrwv, xsetun, and xsetf**: They handle the printing of all error messages and warnings. *xerrwv* is machine-dependent.

These routines do not use any linear algebra library in order to improve performance, such as BLAS [20] or LAPACK [21]. Only, some EISPACK [25] routines were used. But nowadays, there are new implementations, more robust and more efficient.

Two types of changes can be considered:

- The first one is related to the intensive use of BLAS and LAPACK.
- The second one consists on new block oriented algorithms.

As an example of the first type of changes, consider the following piece of code that belongs to **subroutine f()**

```
      do 20 j=1,nnq
         do 20 i=1,nnp
            prec=0.e0
            do 15 k=1,nnq
               prec=prec-xsave(i,k)*como(k,j)
 15         continue
            effe(i,j)=prec
 20   continue
```

This block can be replaced by a BLAS subroutine

```
       alpha=-1.0
       beta=0.0
       call sgemm("N","N",nnp,nnq,nnq,alpha,xsave,nnp,como,n,
     $              beta,effe,nnp)
```

Also, in the following piece of code, which belongs to **subroutine prepj()**, an important change can be carried out

```
call orthes(na,n,nlow,nup,a,ort)
call ortran(na,n,nlow,nup,a,ort,u)
call hqr3(a,u,n,nlow,nup,eps,er,ei,itype,na,nu)
```

This piece of code can be replaced by a LAPACK subroutine

```
     call sgees('V','N',dummy,n,a,na,sdim,er,ei,u,nu,
    $           ort,lort,dummy,iflag)
```

More significant changes have been introduced in **subroutine atxxas()** and **subroutine axpxbs()**. In these routines, a Lyapunov and a Sylvester equation of quasi-triangular systems is respectively solved, by using the so-called Bartels-Stewart algorithm [19].

Let's consider the solution of the quasi-triangular Sylvester equation

$$AX + XB = C, \tag{9}$$

where $A \in R^{m\times m}$, $B \in R^{n\times n}$, X, $C \in R^{m\times n}$, being A and B in real Schur form The real Schur form is a triangular form by blocks of dimension 1×1 o 2×2, according to the eigenvalues. That is, if the eigenvalue is real, then the algorithm works with a 1×1 block. On the other hand, when the eigenvalue is complex, the algorithm works with a 2×2 block. The matrices A, B, X and C are partitioned in blocks of dimension $r \times s$, r, $s \in 1, 2$. Then, the following partition is obtained

$$A = \begin{pmatrix} A_{11} & \cdots & A_{1p} \\ & \ddots & \vdots \\ 0 & & A_{pp} \end{pmatrix}, \; B = \begin{pmatrix} B_{11} & \cdots & B_{1q} \\ & \ddots & \vdots \\ 0 & & B_{qq} \end{pmatrix},$$

$$X = \begin{pmatrix} X_{11} & \cdots & X_{1q} \\ \vdots & \ddots & \vdots \\ X_{p1} & \cdots & X_{pq} \end{pmatrix}, \; C = \begin{pmatrix} C_{11} & \cdots & C_{1q} \\ \vdots & \ddots & \vdots \\ C_{p1} & \cdots & C_{pq} \end{pmatrix}.$$

So, matrix A is partitioned into $p \times p$ blocks, matrix B is partitioned into $q \times q$ blocks and matrices X and C are partitioned into $p \times q$ blocks, where all the diagonal blocks have dimension 1×1 or 2×2.

A non-diagonal block A_{ij} has dimension $r \times s$, being the block A_{ii} of dimension $r \times r$ and the block A_{jj} of dimension $s \times s$.

Matching the block (i, j), $1 \leq i \leq p$, $1 \leq j \leq q$, at both sides of the equation (9) the following expression is obtained

$$\sum_{k=i}^{p} \sum_{l=1}^{j} \left(A_{ik} X_{kl} + X_{kl} B_{lj} \right) = C_{ij}, \tag{10}$$

from which the following $p \times q$ Sylvester equations are obtained

$$A_{ii} X_{ij} + X_{ij} B_{jj} = \hat{C}_{ij}, \tag{11}$$

```
subroutine stsyl-version1 ( m, n, A, B, X, C )
    for i = p : -1 : 1
        for j = 1 : q
            C_{1:i,j} = C_{1:i,j} - ( A_{1:i,i} X_{i,1:j-1} + X_{i,1:j-1} B_{1:j-1,j} )
            trsyl( |A_ii|, |B_jj|, A_ii, B_jj, X_ij, C_ij )
            C_{1:i-1,j} = C_{1:i-1,j} - ( A_{1:i-1,i}(X_ij) + X_ij B_jj )
        end_for
    end_for
end stsyl-version1
```

Fig. 1. Algorithm to solve the quasi-triangular Sylvester equation. Version 1

```
subroutine stsyl-version2 ( m, n, A, B, X, C )
    for i = m/ib : -1 : 1
        for j = 1 : n/jb
            C_{1:i,j} = C_{1:i,j} - ( A_{1:i,i} X_{i,1:j-1} + X_{i,1:j-1} B_{1:j-1,j} )
            trsyl( |A_ii|, |B_jj|, A_ii, B_jj, X_ij, C_ij )
            C_{1:i-1,j} = C_{1:i-1,j} - ( A_{1:i-1,i} X_ij + X_ij B_jj )
        end_for
    end_for
end stsyl-version2
```

Fig. 2. Block-oriented algorithm to solve the quasi-triangular Sylvester equation. Version 2

$\forall i = 1, \ldots, p$, $\forall j = 1, \ldots, q$, with

$$\hat{C}_{ij} = C_{ij} - \sum_{\substack{k=i,l=1 \\ (k,l)\neq(i,j)}}^{p,j} \left(A_{ik} X_{kl} + X_{kl} B_{lj} \right). \tag{12}$$

The Sylvester equations (11) are at most of order 2, and then they can be solved using the *trsyl* LAPACK subroutine. Before solving these Sylvester equations, one needs to update the block C_{ij} according to the expression (12). The algorithm to solve the quasi-triangular Sylvester equation is shown in Figure 1.

Although the *stsyl-version1* algorithm is block-oriented, the size of these blocks is at most 2×2. This block size is too reduced in order to take benefits of the memory hierarchy of current computers, in particular, the cache memory. From the *stsyl-version1* algorithm is possible to derive the block oriented algorithm of Figure 2.

The block-oriented algorithm *stsyl-version2* is similar to the algorithm *stsyl-version1*, except that the block sizes in the first case are ib and jb, and in the second case either 1×1 or 2×2, depending upon whether the eigenvalues are real or complex.

A similar analysis can be done to obtain a block-oriented algorithm to solve the quasi-triangular Lyapunov equation.

4 Parallel Implementations

The parallel implementation has been done in a multicomputer or distributed memory platform (cluster of PC's).

Basically, clusters are different computers interconnected by means of a network. Each computer has its own memory. Communications between processors are done by message passing through the network. Thus, in these platforms, the bottleneck is the network.

The features of the platform have been taken into account in order to make appropriate and efficient implementations, keeping the good features in all software implementations such as portability, efficiency and robustness. To accomplish these features, standard libraries have been used for computation, communication and synchronization purposes.

In the distributed memory implementation, PBLAS [22] and ScaLAPACK [23] have been used for computation. The message passing interface is provided by the environment MPI [24]. Then, the distributed memory implementation consists on replacing calls to BLAS in the sequential implementation, with calls to PBLAS, and calls to LAPACK, with calls to ScaLAPACK. However, there are not parallel routines to compute the Sylvester (*pstsyl*) or Lyapunov (*pstlyc*) equations by blocks, as noted in the previous section. So, new parallel block oriented routines have been developed for these cases.

Figure 3 shows the parallel block oriented algorithm to solve a quasi-triangular Lyapunov equation. In this code, the updating steps are done in parallel. Mean-

Subroutine pstlyc (n, A, X, C)
    To communicate diagonal blocks of matrix A
    for $i = 1 : p$
        $X_{i,1:i-1} = (X_{1:i-1,i})^T$
        for $j = i : p$
            /* Update I in parallel: block $C_{i:j,j}$ by psgemm */
            $C_{i:j,j} = C_{i:j,j} - X_{i,1:j-1}A_{1:j-1,j}$
            /* To solve Lyapunov equation by processor (i,j) */
            if idfila==i **and** idcolumna==j **then**
              trsyl($|A_{ii}|$, $|A_{jj}|$, A_{ii}^T, A_{jj}, X_{ij}, C_{ij})
            end if
            /* Update II en parallel: block $C_{i+1:j,j}$ by psgemm*/
            $C_{i+1:j,j} = C_{i+1:j,j} - A_{i,i+1:j}^T X_{ij}$
        end for
    end for
end subroutine

Fig. 3. Parallel block oriented algorithm for solving the quasi-triangular Lyapunov equation while, the solution of the Lyapunov equation is done by only one processor due to the fact that only one processor has all data blocks without additional communications. This circumstance has been motivated by two factors. The first one is the block oriented implementation. The other one is the initial communications of the diagonal blocks of matrix A involved in the Lyapunov equation.

The parallel implementation carried out can be considered incomplete because a distributed memory implementation of a Schur factorization routine (with reordering facilities) has not been developed, due to the fact that this task was out of the scope of this work.

5 Experimental Results

The experimental results have been obtained in a Cluster of PC's. This platform belongs to the High Performance Networking and Computing Group (GRyCAP) [26] from the Universidad Politécnica de Valencia. The cluster consists on 12 nodes (one is used as a front end) IBM xSeries 330 SMP under the Linux RedHat 7.1 operating system. Each node has 2 Pentium III processors at 866 Mhz with 512 MB of main memory and 256 KB of cache memory. The nodes are interconnected by means of a Gigabit Ethernet. In order to consider a true distributed memory platform, only one of the two processors at each node has been used, due to the fact that the communications between processors at each node use threads and not message passing. So, 11 processors can be used in fact for test purposes.

The case of study has been extracted from [1], and consists on the following matrices:

$$A11 = A22 = 0;$$

$$A12 = A21 = \alpha I_n.$$

where n is the variable dimension of the problem and $\alpha > 0$, $0 \leq t \leq T = 10$. Initial values for vector X_0 are ones. The relative and absolute tolerance has the value $RTOL = ATOL = 10^{-3}$.

In our tests the value of n could be 128, 256, 512 and 1024. The value of the variable α is 10.

Another parameter to take into account is *icontr* which allows to solve a symmetric (*icontr* $= 0$) or non-symmetric (*icontr* $= 1$) Differential Riccati Equation. This parameter permits to take benefits of the symmetry of the Differential Riccati Equation. In terms of computation, it means to solve a Lyapunov equation (symmetric case) or a Sylvester equation (non-symmetric case).

Due to the fact that the DRESOL package has a lot of routines, it gets difficult to test all of them. So, the following measures have been considered:

(a) Time to perform one iteration of the DRESOL package to solve a symetric Differential Riccati Equation.
(b) Time to perform one iteration of the DRESOL package to solve a non-symetric Differential Riccati Equation.
(c) Time to solve a quasi-triangular Lyapunov equation.

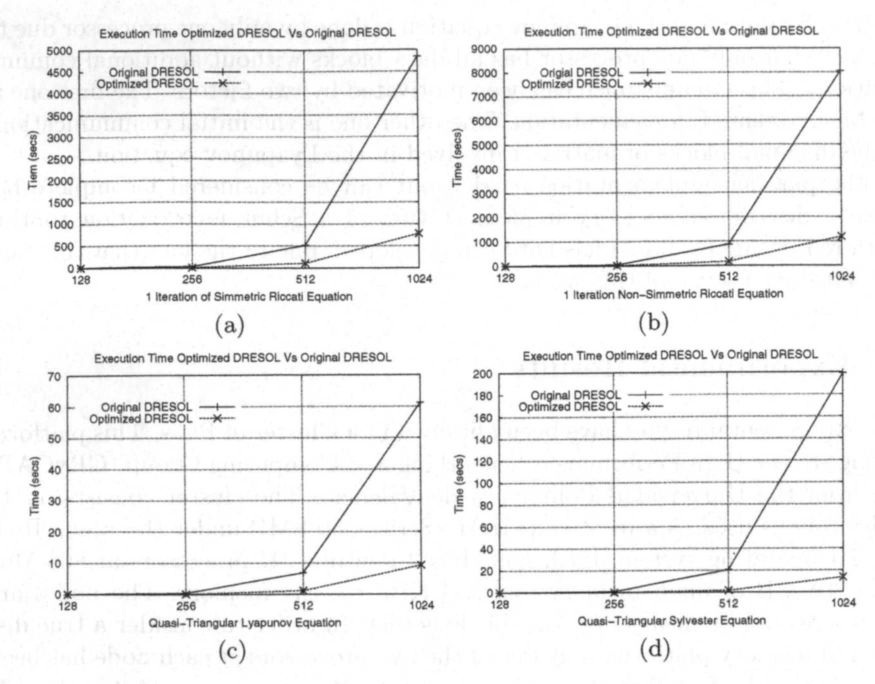

Fig. 4. Comparison between execution times of original DRESOL package and the optimized one. Sequential version. (a) Test a. (b) Test b. (c) Test c. (d) Test d

(d) Time to solve a quasi-triangular Sylvester equation.

The graphics in Figure 4 show the execution time obtained by the original DRESOL package and the optimized package in a single processor (sequential version).

The experimental results show a spectacular reduction in the execution time between the original DRESOL package and the optimized package developed in this work. As the dimension of the problem increases, the improvement of the execution time gets better and better. This is due to the fact that the new implementation is block-oriented. Thus, it is able to take benefits of the memory hierarchy of current computers. In particular, the data flow between main memory and cache memory decreases drastically, thus obtaining better performance. It is also remarkable the fact that better performance is obtained by exploiting the symmetry of the problem, being almost twice faster.

The following parameters have been considered to run the test set into the parallel architecture:

- Block size: This parameter is very important in order to take benefits of the cache memory. The following block sizes have been considered: 8, 16, 32, 64 and 128.
- Number of processors: The values considered are 2, 4, and 8 processors.
- Topology of the grid in ScaLAPACK:

- 2 processors: 2×1 and 1×2.
- 4 processors: 4×1, 1×4 and 2×2.
- 8 processors: 8×1, 1×8, 4×2 and 2×4.
- 16 processors: 16×1, 1×16, 4×4, 8×2 and 2×8.

The performace obtained in the parallel implementations are evaluated in terms of:

- Execution time: It is the time spent for solving the problem.
- Speed-up: It is defined as the ratio of the time taken to solve a problem on a processor to the time required to solve the same problem on a parallel computer with p identical processors.
- Efficiency: It is a measure of the fraction of time for which a processor is usefully employed; it is defined as the ratio of the speed-up to the number of processors.

The graphics in Figure 5 show the execution time obtained by the optimized DRESOL package and the distributed memory version. The execution time has been dramatically reduced by using 2 processors. This reduction becomes less significant by using 4 and 8 processors when the time to perform one iteration is considered. This circumstance is due to the fact that the Schur factorization is sequentially computed, thus performance decreases. Considering the time to

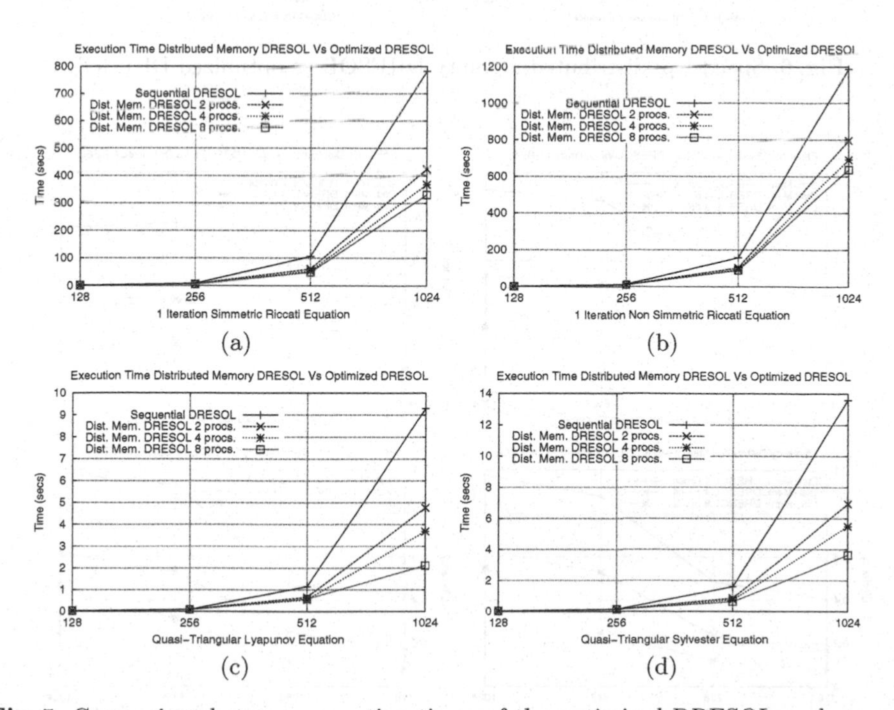

Fig. 5. Comparison between execution times of the optimized DRESOL package and the distributed memory version. (a) Test a. (b) Test b. (c) Test c. (d) Test d

solve a Sylvester or Lyapunov equation, it is possible to appreciate a significant reduction in the execution time by using also 4 and 8 processors, because there

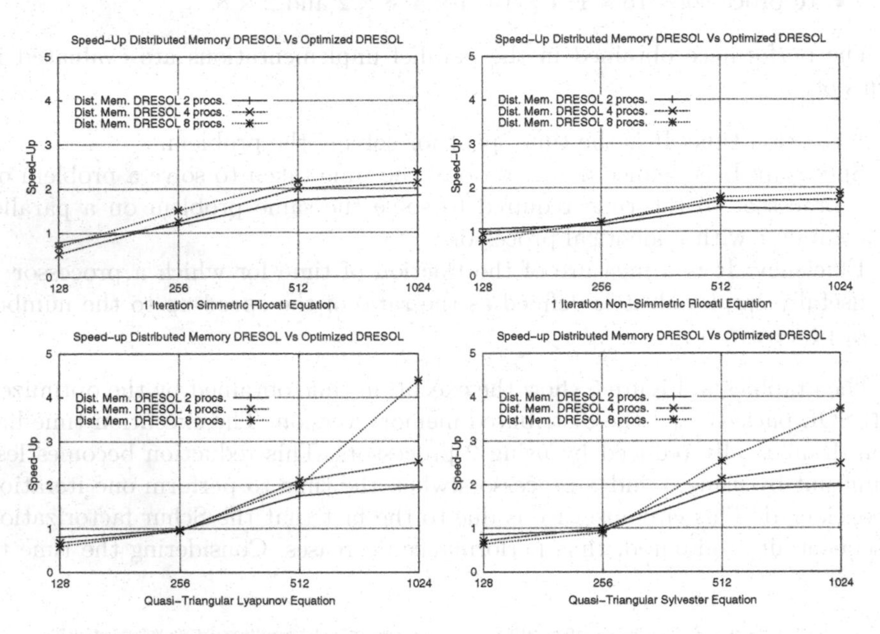

Fig. 6. Speed-Up: distributed memory DRESOL vs optimized DRESOL

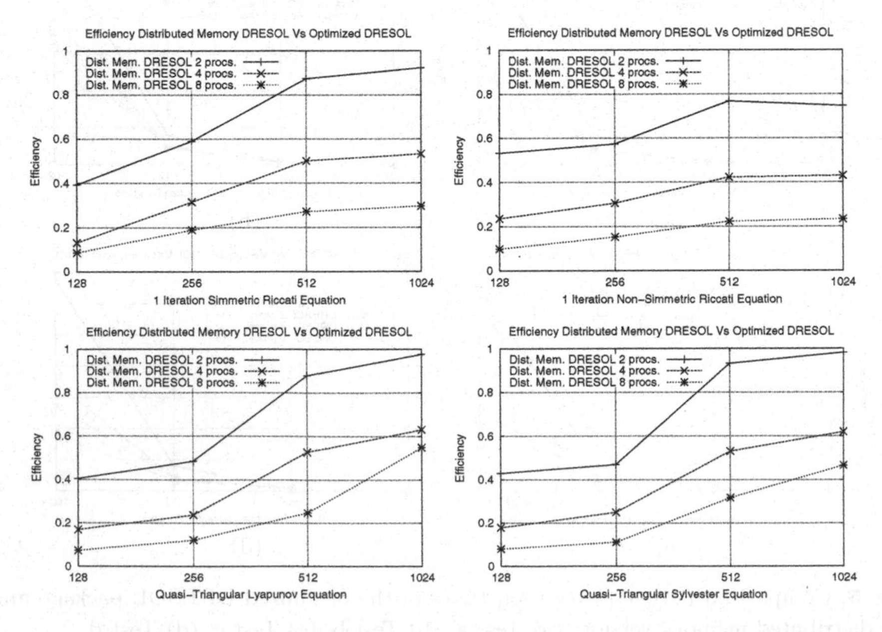

Fig. 7. Efficiency: distributed memory DRESOL vs optimized DRESOL

is not influence of the Schur factorization. Again, better times are obtained by exploting the structure of the problem.

Figure 6 and 7 show the comparison between the optimized DRESOL package and the distributed memory version, in terms of speed-up and efficiency, respectively. In the cases of considering the time to perform one iteration of the code, good speed-up and efficiency are obtained for 2 processors, but not for 4 and 8 processors, due to the circsumstance previously explained. If it is considered the case of solving a Sylvester or Lyapunov equation, performance, in terms of speed-up and efficiency, increases with the size of the problem till almost the theorical values. Also, performance for 4 and 8 processors increases with respect to the size of the problem.

In any case, the times obtained by the sequential optimized implementation are very small, so any reduction of execution time by using the parallel implementation involves a huge amount of work. This amount of work is justified when high dimension problems have to be dealt with in real time.

6 Conclusions

In this paper a High Performance Computing Approach for solving the Differential Riccati Equation by Numerical Integration has been introduced, based on the original package called DRESOL by Luca Dieci [1].

This HPC approach has been developed in two phases. The first one consists on an exhaustive study of the original DRESOL package in order to identify the main computational kernels. From this study, a new sequential optimized version of DRESOL has been carried out. From this sequential optimized version, a new parallel implementation has been developed on a distributed memory architecture.

The sequential and the parallel implementations have been carried out taking into account three desirable features in high quality software: Portability, Efficiency and Robustness. Portability has been obtained by using standard libraries for computations (BLAS, LAPACK, PBLAS and ScaLAPACK) and for communications and synchronizations (MPI). The block oriented implementations improve the efficiency of the implementation by taking benefits of the memory hierarchy. Finally, the numerical methods employed in the implementations guarantee the robustness of the software.

With respect to the experimental results, the sequential optimized version dramatically reduces the execution time obtained by the original version. The parallel implementations improve the performance of the new sequential optimized version, achieving speed-up nearly to the theorical maximum.

References

1. Luca Dieci, *Numerical Integration of the diferential Riccati equation and some related issues*, SIAM J. on Mumerical Anaysis, v. 29, n. 3, pp. 781-815, 1992.
2. Isidori,A.. *Nonlinear Control Systems: An Introduction.* Springer-Verlag. 1989.
3. Spong, M.W. and Vidyasagar, M.. *Robot dynamics and control.* John Whiley and Sons. 1989.

4. SLOTINE, J. J. AND W. LI. *Applied Nonlinear Control.* PrenticeHall. 1984.
5. ASTROM, K. J. AND WITTENMARK, B.. *Computer-Controlled Systems: Theory and Design.* Prentice-Hall, Upper Saddle River, New Jersey, 1997.
6. C. S. KENNEY AND R. B. LEIPNIK. *Numerical Integration of the Differential Matrix Riccati Equation.* IEEE Transactions on Automatic Control, pp. 962-979, 1985.
7. CHIU H. CHOI. *A survey of numerical methods for solving matrix Riccati differential equations.* Southeastcon'90 Proceedings. pp. 696–700, 1990.
8. LAINIOTIS, D. G.. *Partitioned Riccati solutions and integration-free doubling algorithms.* IEEE Trans. on Automatic Control, 21, pp. 677–689, 1976.
9. LEIPNIK, R. B.. *A canonical form and solution for the matrix Riccati differential equation.* Bull Australian Math. Soc., 26, pp. 355–261, 1985.
10. DAVISON, E. J. AND MAKI, M. C.. *The numerical solution of the matrix Riccati differential equation.* IEEE Trans. on Automatic Control, 18, pp. 71–73, 1973.
11. RAND, D. W. AND WINTERNITZ, P.. *Nonlinear Superposition Principles: a new numerical method for solvinf matrix Riccati equations.* Comp. Phys. Commun., 33, pp. 305-328, 1984.
12. SORINE, M AND WINTERNITZ, P.. *Superposition laws for the solution of differential Riccati equations.* IEEE Trans. on Automatic Control, 30, pp. 266–272, 1985.
13. CHIU H. CHOI AND ALAN J. LAUB. *Efficient Matrix-Valued Algorithms for Solving Stiff Riccati Differential Equations.* IEEE Trans. on Automatic Control, 35-7, pp. 770-776, 1990.
14. A.C. HINDMARSH. *Brief Description of ODEPACK – A Systematized Collection of ODE Solvers*, http://www.netlib.org/odepack/doc
15. FENG KANG AND QIN MENG-ZHAO. *The sumplectic methods for the computation of Hamiltonian equations.* Proceedings of the 1st Chinese Conference for Numerical Methods for PDE's, pp. 1–37, 1987.
16. J. M. SANZ-SERNA, *Symplectic integrators for Hamiltonian Problems: an overview*, Acta Numerica, v. 1, pp. 243-286, 1992.
17. ISERLES, A. AND NORSETT, S. P.. *On the solution of linear differential equations in Lie groups.* Technical report, Cambrigde University, 1997.
18. JEREMY SCHIFF AND S. SHNIDER, *A natural approach to the numerical integration of riccati differential equations*, SIAM, v. 36, n. 5, pp. 1392-1413, 1999.
19. R.H. BARTELS AND G.W. STEWART, *Solution of the Matrix Equation $AX+XB=C$: Algorithm 432*, CACM, v. 15, pp. 820-826, 1972.
20. DONGARRA, J. AND DU CROZ, J. AND HAMMARLING, S. AND HANSON, R. J.. *An Extended Set of FORTRAN Basic Linear Algebra Subroutines.* ACM Trans. Math. Software, 1988. http://www.netlib.org/blas.tgz
21. ANDERSON, E. AND OTHERS. *LAPACK Users' Guide.* SIAM, Philadelphia, PA, 1994. http://www.netlib.org/lapack.
22. J. CHOI AND J. DONGARRA AND S. OSTROUCHOV AND A. PETITET AND D. WALKER, *A Proposal for a Set of Parallel Basic Linear Algebra Subprograms*, Department of Computer Science, University of Tennessee, Technical Report UT-CS-95-292, May, 1995.
23. BLACKFORD ET AL., *ScaLAPACK Users' Guide*, SIAM, Philadelphia, PA, 1997.
24. W. GROPP AND E. LUSK AND A. SKJELLUM, *Using MPI: Portable Parallel Programming with the Message-Passing Interface*, MIT Press, 1994.
25. B. T. SMITH, ET AL, *Matrix Eigensystems Routines - EISPACK Guide*, Springer Lecture Notes in Computer Science, 1976.
26. HIGH PERFORMANCE NETWORKING AND COMPUTING GROUP. *(GRYCAP).* http://www.grycap.upv.es.

An Efficient and Stable Parallel Solution for Non-symmetric Toeplitz Linear Systems*

Pedro Alonso[1], José M. Badía[2], and Antonio M. Vidal[1]

[1] Universidad Politécnica de Valencia,
cno. Vera s/n, 46022 Valencia, Spain
{palonso, avidal}@dsic.upv.es
[2] Universitat Jaume I, Campus de Riu Sec,
12071 Castellón de la Plana, Spain
badia@icc.uji.es

Abstract. In this paper, we parallelize a new algorithm for solving non–symmetric Toeplitz linear systems. This algorithm embeds the Toeplitz matrix in a larger structured matrix, then transforms it into an embedded Cauchy–like matrix by means of trigonometric modifications. Finally, the algorithm applies a modified QR transformation to triangularize the augmented matrix. The algorithm combines efficiency and stability. It has been implemented using standard tools and libraries, thereby producing a portable code. An extensive experimental analysis has been performed on a cluster of personal computers. Experimental results show that we can obtain efficiencies that are similar to other fast parallel algorithms, while obtaining more accurate results with only one iterative refinement step in the solution.

Keywords: Non–symmetric Toeplitz, Linear Systems, Displacement Structure, Cauchy–like matrices, Parallel Algorithms, Clusters.

1 Introduction

In this paper, we present a new parallel algorithm for the solution of Toeplitz linear systems such as

$$Tx = b\,, \tag{1}$$

is presented, where the Toeplitz matrix $T \in \mathrm{R}^{n\times n}$ has the form $T = (t_{ij}) = (t_{i-j})_{i,j=0}^{n-1}$, and where $b, x \in \mathrm{R}^n$ are the independent and the solution vector, respectively.

Many *fast* algorithms that solve this problem can be found in the literature; that is, algorithms whose computational cost is $O(n^2)$. Almost all these algorithms produce poor results unless strongly regular matrices are used; that is, matrices whose leading submatrices are all well–conditioned. Several methods

* Supported by Spanish projects CICYT TIC 2000-1683-C03-03 2003-08238-C02-02.

M. Daydé et al. (Eds.): VECPAR 2004, LNCS 3402, pp. 685–698, 2005.

have been proposed to improve the solution of (1), including the use of *look-ahead* or refinement techniques [1, 2].

It is difficult to obtain efficient parallel versions of *fast* algorithms, because they have a reduced computational cost and they also have many dependencies among fine–grain operations. These dependencies produce many communications, which are a critical factor in obtaining efficient parallel algorithms, especially on distributed memory computers. This problem could partially explain the small number of parallel algorithms that exist for dealing with Toeplitz linear systems. For instance, parallel algorithms that use systolic arrays to solve Toeplitz systems can be found in [3, 4]. Other parallel algorithms deal only with positive definite matrices (which are strongly regular) [5], or with symmetric matrices [6]. There also exist parallel algorithms for shared memory multiprocessors [7, 8, 9] and several parallel algorithms have recently been proposed for distributed architectures [10].

If we apply some of the techniques to improve the stability of the sequential algorithms that solve (1), it is more difficult to obtain efficient parallel versions. In this work, we combine the advantages of two important results from [2, 10] in order to derive a parallel algorithm that exploits the special structure of the Toeplitz matrices and is both efficient and stable.

Another of our main goals is to offer an efficient parallel algorithm for general purpose architectures, i.e. clusters of personal computers. The algorithm presented in this paper is portable because it is based on the libraries LAPACK [11] and ScaLAPACK [12] that are sequential and parallel, respectively.

In [10], we proposed two parallel algorithms for the solution of Toeplitz linear systems. The first algorithm, called PAQR, performs a QR decomposition of T. This decomposition is modified by a correcting factor that increases the orthogonality of the factor Q, so that the algorithm is backward stable and the solution is more accurate. The main drawback of this algorithm is that it produces poor speed–up's for more than two processors. In the second algorithm, called PALU, the Toeplitz matrix is transformed into a Cauchy–like matrix by means of discrete trigonometric transforms. This algorithm obtains better speed–up's than the PAQR algorithm, but produces results that are not very accurate. In this paper, we present a new parallel algorithm that combines the efficiency of the PALU algorithm with the stability and precision of the PAQR algorithm.

The structure of this paper is the following. In Section 2, we present the sequential algorithm. In Section 3 we show how to perform the triangular decomposition of a Cauchy–like matrix. In Section 4, we describe the parallel algorithm. The experimental results are shown in Section 5, and our conclusions are presented in the last section.

2 The Sequential Algorithm

The algorithm proposed and parallelized in this paper performs a *modified* QR decomposition in order to solve the Toeplitz system (1). This *modified* QR decomposition is based on the idea set out in [2] and applied in the PAQR parallel algorithm mentioned in the previous section. The *modified* QR decomposition can be described as follows.

An augmented matrix M is constructed with the following decomposition

$$M = \begin{pmatrix} T^T T & T^T \\ T & 0 \end{pmatrix} = \begin{pmatrix} R^T & 0 \\ Q & \Delta \end{pmatrix} \begin{pmatrix} R & Q^T \\ 0 & -\Delta^T \end{pmatrix}, \qquad (2)$$

where R^T and Δ are lower triangular matrices and Q is a square matrix. Using exact arithmetic, the factors Q and R in (2) form the QR decomposition of T ($T = QR$); however, the computed factor Q may actually lose its orthogonality. By introducing the correcting factor Δ, the product $(\Delta^{-1}Q)$ is almost orthogonal. We will refer to the decomposition $T = \Delta(\Delta^{-1}Q)R$ as *modified* QR decomposition. This factorization is then used to obtain x in the linear system (1) by using $x = R^{-1}(Q^T\Delta^{-T})\Delta^{-1}b$.

The matrix M is a *structured* matrix, which means that a *fast* algorithm can be used to compute the triangular factorization shown in (2). One of these algorithms is the Generalized Schur Algorithm. In [10], we proposed a parallel version of the Generalized Schur Algorithm, but its scalability for more than two processors is not very good. This fact is mainly due to the form of the so–called *displacement matrices* used. To improve its efficiency, we apply the same transformation technique used in the PALU algorithm to M in order to work with diagonal displacement matrices. This kind of displacement matrix allows us to avoid a large number of the communications during the application of the parallel algorithm.

Structured matrices are characterized by the *displacement rank* property [13]. We start from the following displacement representation of matrix M (2) with respect to a displacement matrix F,

$$\nabla_{M,F} = FM - MF = GHG^T, \quad F = Z_0 \oplus Z_1, \qquad (3)$$

where $Z_\epsilon \in \mathrm{R}^{n\times n}$ is a zero matrix whose first superdiagonal and subdiagonal are equal to one and whose first and last diagonal entries are equal to ϵ. The rank of $\nabla_{M,F}$ (3) is 8. Matrices $G \in \mathrm{R}^{2n\times 8}$ and $H \in \mathrm{R}^{8\times 8}$ are called a *generator pair*.

First, we define

$$\hat{\mathcal{S}} = \sin\frac{ij\pi}{n+1}, \; i,j = 1,\ldots,n, \quad \text{and} \quad \mathcal{S} = \sqrt{\frac{2}{n+1}}\,\hat{\mathcal{S}},$$

as the unnormalized and the normalized Discrete Sine Transforms (DST), respectively. Matrix $\mathcal{S}$ is symmetric and orthogonal. Matrix Z_0 can be diagonalized by means of the DST $\mathcal{S}$; that is, $\mathcal{S}Z_0\mathcal{S} = \Lambda_0$, where Λ is diagonal.

In the same way, we define

$$\hat{\mathcal{C}} = \xi_j \cos\frac{(2i+1)j\pi}{2n}, \; i,j = 0,\ldots,n-1, \quad \text{and} \quad \mathcal{C} = \sqrt{\frac{2}{n}}\,\hat{\mathcal{C}},$$

where $\xi_j = 1/\sqrt{2}$ for $j = 0$, and $\xi_j = 1$ otherwise, as the unnormalized and the normalized Discrete Cosine Transforms (DCT), respectively. Matrix $\mathcal{C}$ is non–symmetric and orthogonal. Matrix Z_1 can be diagonalized by means of the DCT $\mathcal{C}$; that is, $\mathcal{C}^T Z_1 \mathcal{C} = \Lambda_1$, where Λ_1 is diagonal.

There exist four distinct versions of DST and DCT which are numbered as I, II, III and IV [14, 15]. Transforms DST and DCT, as defined in this paper, correspond to DST-I and DCT-II, respectively. We also exploit *superfast* algorithms that perform the transformation of a vector by DST or DCT in $O(n \log n)$ operations.

Now, we construct the following transformation

$$S = \begin{pmatrix} \mathcal{S} & 0 \\ 0 & \mathcal{C} \end{pmatrix} . \tag{4}$$

Pre– and post–multiplying equation (3) by S leads us to a displacement representation of M with diagonal displacement matrices.

$$\begin{aligned} S^T(FM - MF)S &= S^T(GHG^T)S \,, \\ (S^TFS)(S^TMS) - (S^TMS)(S^TFS) &= (S^TG)H(S^TG)^T \,, \\ \Lambda\hat{C} - \hat{C}\Lambda &= \hat{G}H\hat{G}^T \,, \end{aligned} \tag{5}$$

where $\Lambda = \left(\Lambda_0 \oplus \Lambda_1 \right)$, and $\hat{C}$ has the form

$$\hat{C} = \begin{pmatrix} C^TC & C^T \\ C & 0 \end{pmatrix} = \begin{pmatrix} \mathcal{S} & 0 \\ 0 & \mathcal{C}^T \end{pmatrix} \begin{pmatrix} T^TT & T^T \\ T & 0 \end{pmatrix} \begin{pmatrix} \mathcal{S} & 0 \\ 0 & \mathcal{C} \end{pmatrix} ,$$

with $C = \mathcal{C}^T T \mathcal{S}$. It is easy to see that, by means of DST and DCT, we obtain a Cauchy–like linear system from the Toeplitz linear system (1)

$$(\mathcal{C}^T T \mathcal{S})(\mathcal{S}x) = (\mathcal{C}^T b) \;\rightarrow\; C\hat{x} = \hat{b} \,. \tag{6}$$

Matrices C and $\hat{C}$ are called *Cauchy–like* matrices and are also structured. The advantage of using *Cauchy–like* matrices is that the diagonal form of the displacement matrix Λ allows us to avoid a great number of communications.

We apply these ideas in the following algorithm to solve (1) in three steps.

1. The computation of generator G (3) and $\hat{G} = S^TG$, where S is defined in (4).
2. The computation of the triangular decomposition of matrix $\hat{C}$,

$$\hat{C} = \begin{pmatrix} C^TC & C^T \\ C & 0 \end{pmatrix} = \begin{pmatrix} \hat{R}^T & 0 \\ \hat{Q} & \hat{\Delta} \end{pmatrix} \begin{pmatrix} \hat{R} & \hat{Q}^T \\ 0 & -\hat{\Delta}^T \end{pmatrix} , \tag{7}$$

   The embedded factors $\hat{Q}$, $\hat{R}$ and $\hat{\Delta}$ represent the modified QR decomposition of matrix C, $C = \hat{\Delta}(\hat{\Delta}^{-1}\hat{Q})\hat{R}$.
3. The computation of the solution of the system (6)

$$\hat{x} = \hat{R}^{-1}(\hat{Q}^T\hat{\Delta}^{-T})\hat{\Delta}^{-1}\hat{b} \,, \tag{8}$$

   and the solution of the Toeplitz system (1), $x = \mathcal{S}\hat{x}$.

The form of generator pair (G, H) (3) can be derived analytically as follows: Given the following partitions of T and $T^T T$,

$$T = \begin{pmatrix} t_0 & v^T \\ u & T_0 \end{pmatrix} = \begin{pmatrix} T_0 & \bar{v} \\ \bar{u}^T & t_0 \end{pmatrix}, \quad T^T T = \begin{pmatrix} s_0 & w_0^T \\ w_0 & S_0 \end{pmatrix} = \begin{pmatrix} S_1 & w_1 \\ w_1^T & s_1 \end{pmatrix},$$

where a vector $\bar{x}$ represents the vector $x = \begin{pmatrix} x_0 & x_1 & \cdots & x_{n-2} & x_{n-1} \end{pmatrix}^T$ with the elements placed in the reverse order $\bar{x} = \begin{pmatrix} x_{n-1} & x_{n-2} & \cdots & x_1 & x_0 \end{pmatrix}^T$, and given the following partitions for the resulting vectors u, v, w_0 and w_1, in the above partitions,

$$u = \begin{pmatrix} t_1 \\ u' \end{pmatrix} = \begin{pmatrix} u'' \\ t_{n-1} \end{pmatrix}, \quad v = \begin{pmatrix} t_{-1} \\ v' \end{pmatrix} = \begin{pmatrix} v'' \\ t_{-n+1} \end{pmatrix},$$
$$w_0 = \begin{pmatrix} \alpha \\ w_0' \end{pmatrix}, \quad w_1 = \begin{pmatrix} w_1' \\ \beta \end{pmatrix},$$

we obtain the following generator pair

$$G = \begin{pmatrix} 0 & 0 & 0 & t_{-1} & 1 & 0 & t_{n-1} & 0 \\ w_0' & w_1' & \bar{u}' & v' & \hat{0} & \hat{0} & \bar{u}'' & v'' \\ 0 & 0 & t_1 & 0 & 0 & 1 & 0 & t_{-n+1} \\ t_0 + t_1 & 0 & 0 & 1 & 0 & 0 & 0 & 1 \\ u' & \bar{v}' & \hat{0} & \hat{0} & \hat{0} & \hat{0} & \hat{0} & \hat{0} \\ 0 & t_0 + t_{-1} & 1 & 0 & 0 & 0 & 1 & 0 \end{pmatrix}, \quad H = \begin{pmatrix} & I_4 \\ -I_4 & \end{pmatrix},$$

where I_4 is the identity of order 4. A demonstration of the form of the generator can be found in [16]. The computation of $\hat{G}$ involves $O(n \log(n))$ operations.

Once the generator has been computed, a *fast* triangular factorization ($(O(n^2)$ operations) of matrix $\hat{C}$ (7) can be performed by means of the process explained in the following section.

3 Triangular Decomposition of Symmetric Cauchy–Like Matrices

For the discussion of this section, let us start with the following displacement representation of a symmetric Cauchy–like matrix $C \in \mathrm{R}^{n \times n}$,

$$\Lambda C - C \Lambda = G H G^T, \tag{9}$$

where Λ is diagonal and where (G, H) is the generator pair. For the sake of convenience, we have used the same letters as in the previous section in this representation of a general Cauchy–like matrix.

Generally, the displacement representation (9) arises from other displacement representations, like the displacement representation of a symmetric Toeplitz matrix or another symmetric structured matrix. Matrix C is not explicitly formed in order to reduce the computational cost; matrix C is implicitly known by means of matrices G, H and Λ.

From equation (9), it is clear that any column of C, $C_{:,j}$, $j = 0, \ldots, n-1$, of C can be obtained by solving the Sylvester equation

$$\Lambda C_{:,j} - C_{:,j}\lambda_{j,j} = GHG_{j,:}^T ,$$

and that the $(i,j)^{th}$ element of C can be computed as

$$C_{i,j} = \frac{G_{i,:}HG_{j,:}^T}{\lambda_{i,i} - \lambda_{j,j}} , \tag{10}$$

for all i, j with $i \neq j$, that is, for all the off–diagonal elements of C. We can use this assumption because all the elements of the diagonal matrix Λ (5) used in our case are different. With regard to the diagonal elements of C, assume that they have been computed a priori.

Given the first column of C, we use the following partition of C and Λ,

$$C = \begin{pmatrix} d & c^T \\ c & C_1 \end{pmatrix} , \quad \Lambda = \begin{pmatrix} \lambda & 0 \\ 0 & \hat{\Lambda} \end{pmatrix} ,$$

with $C_1, \hat{\Lambda} \in \mathrm{R}^{(n-1)\times(n-1)}$, $c \in \mathrm{R}^{n-1}$ and $d, \lambda \in \mathrm{R}$, to define the following matrix X,

$$X = \begin{pmatrix} 1 & 0 \\ l & I_{n-1} \end{pmatrix} , \quad X^{-1} = \begin{pmatrix} 1 & 0 \\ -l & I_{n-1} \end{pmatrix} ,$$

where $l = c/d$. By applying $X^{-1}(.)X^{-T}$ to equation (9) we have,

$$\begin{gathered} X^{-1}(\Lambda C - C\Lambda)X^{-T} = \\ (X^{-1}\Lambda X)(X^{-1}CX^{-T}) - (X^{-1}CX^{-T})(X^T\Lambda X^{-T}) = \\ \begin{pmatrix} \lambda & 0 \\ \hat{\Lambda}l - \lambda l & \hat{\Lambda} \end{pmatrix} \begin{pmatrix} d & 0 \\ 0 & C_{sc} \end{pmatrix} - \begin{pmatrix} d & 0 \\ 0 & C_{sc} \end{pmatrix} \begin{pmatrix} \lambda & l^T\hat{\Lambda} - \lambda l^T \\ 0 & \hat{\Lambda} \end{pmatrix} = \\ (X^{-1}G)H(X^{-1}G)^T , \end{gathered}$$

where $C_{sc} = C_1 - (cc^T)/d$ is the Schur complement of C with respect to d. At this point, we have the first column of L, $\begin{pmatrix} 1 & l^T \end{pmatrix}^T$, and the first diagonal entry, d, of the LDL^T decomposition of $C = LDL^T$, with L being a lower unit triangular factor and D diagonal.

Equating the (2, 2) position in the above equation, we have the displacement representation of the Schur complement of C with respect to its first element,

$$\hat{\Lambda}C_{sc} - C_{sc}\hat{\Lambda} = G_1HG_1^T ,$$

where G_1 is the portion of $X^{-1}G$ from the second row down. The process can now be repeated on the displacement equation of the Schur complement C_{sc} to get the second column of L and the second diagonal element of D. After n steps, the LDL^T factorization of C is obtained.

Now, we show how to apply the LDL^T decomposition to matrix $\hat{C} = S^T MS$ (7). The LDL^T factorization of $\hat{C}$ can be represented as

$$\hat{C} = \begin{pmatrix} C^T C & C^T \\ C & 0 \end{pmatrix} = \begin{pmatrix} \hat{\bar{R}}^T & 0 \\ \hat{\bar{Q}} & \hat{\bar{\Delta}} \end{pmatrix} \begin{pmatrix} \hat{\bar{D}}_1 & 0 \\ 0 & -\hat{\bar{D}}_2 \end{pmatrix} \begin{pmatrix} \hat{\bar{R}} & \hat{\bar{Q}}^T \\ 0 & \hat{\bar{\Delta}}^T \end{pmatrix} , \tag{11}$$

where $\hat{\bar{R}}^T$ and $\hat{\bar{\Delta}}$ are lower unit triangular matrices, and $\hat{\bar{D}}_1$ and $\hat{\bar{D}}_2$ are diagonal.

From (7) and (11), we have $\hat{R} = \hat{\bar{D}}_1^{\frac{1}{2}} \hat{\bar{R}}$, $\hat{Q} = \hat{\bar{Q}} \hat{\bar{D}}_1^{\frac{1}{2}}$ and $\hat{\Delta} = \hat{\bar{\Delta}} \hat{\bar{D}}_2^{\frac{1}{2}}$, and the solution of system (1) is $x = \mathcal{S} \hat{\bar{R}}^{-1} \hat{\bar{Q}}^T \hat{\bar{\Delta}}^{-T} \hat{\bar{D}}_2^{-1} \hat{\bar{\Delta}}^{-1} \hat{b}$.

Diagonal entries of Cauchy–like matrix $C^T C$ cannot be computed by means of (10), so we need an algorithm to obtain them without explicitly computing all the Cauchy–like matrix entries by means of discrete transforms. An algorithm for the computation of these entries exists and has a computational cost of $O(n \log n)$ operations. An algorithm for the computation of the entries of $\mathcal{O}^T T^T T \mathcal{O}$, where $\mathcal{O}$ is any of the existing orthogonal trigonometric transforms, can be found in [15]. A more specific algorithm can be found in [16] for the case in which $\mathcal{O} = \mathcal{S}$.

4 The Parallel Algorithm

Most of the cost of the parallel algorithm is incurred during the second step, that is, during the triangular decomposition of the Cauchy–like matrix $\hat{C}$. The triangularization process deals with the $2n \times 8$ entries of the generator of $\hat{C}$. Usually, $2n \gg 8$ and the operations performed in the triangularization process can be carried out independently on each row of the generator. In order to get the maximum efficiency in the global parallel algorithm, we have chosen the best data distribution for the triangular factorization step.

The generator $\hat{G}$ of the displacement representation of $\hat{C}$,

$$\Lambda \hat{C} - \hat{C} \Lambda = \hat{G} H \hat{G}^T ,$$

is partitioned into $2n/\nu$ blocks of size $\nu \times 8$, and they are cyclically distributed onto an array of p processors, denoted by P_k, for $k = 0, \ldots, p-1$, in such a way that block $\hat{G}_i$, $i = 0, \ldots, 2n/\nu - 1$, belongs to processor $P_{i \bmod p}$ (Fig. 1). For simplicity, in the presentation we assume that $2n \bmod \nu = 0$, although we do not have this restriction in the implemented algorithm. This distribution is performed and managed by ScaLAPACK tools on a one–dimensional mesh of p processors.

The lower unit triangular factor,

$$L = \begin{pmatrix} \hat{\bar{R}}^T & 0 \\ \hat{\bar{Q}} & \hat{\bar{\Delta}} \end{pmatrix} , \tag{12}$$

obtained by the triangular factorization algorithm, is partitioned into square blocks of size $\nu \times \nu$ forming a two–dimensional mesh of $(2n/\nu) \times (2n/\nu)$ blocks.

P_0	$\hat{G}_0$	L_{00}
P_1	$\hat{G}_1$	L_{10} L_{11}
P_2	$\hat{G}_2$	L_{20} L_{21} L_{22}
P_0	$\hat{G}_3$	L_{30} L_{31} L_{32} L_{33}
P_1	$\hat{G}_4$	L_{40} L_{41} L_{42} L_{43} L_{44}
$\vdots$	$\vdots$	$\vdots$ $\vdots$ $\vdots$ $\vdots$ $\vdots$ $\ddots$

Fig. 1. Example of data distribution for $p = 3$

Let $L_{i,j}$ be the $(i,j)^{th}$ block of L; then, blocks $L_{i,j}$, $j = 0, \ldots, 2n/\nu - 1$, belong to processor $P_{i \bmod p}$ (Fig. 1). The diagonal elements of matrix D (11) are stored in the diagonal entries of L (all diagonal entries of L are equal to one).

The block size ν influences the efficiency of the parallel algorithm. Large values of ν produce a low number of large messages, but the workload is unbalanced among the processors, while small values of ν produce a higher number of smaller messages, but also a better workload balance. Furthermore, the effect of the block size depends on the hardware platform used, and so the block size must be chosen by experimental tuning.

Once we know the data distribution among the processors, we briefly describe the parallel version of the algorithm. The generator is computed in the first step. The computation of the generator involves a Toeplitz matrix–vector product and the translation to the Cauchy–like form by means of the trigonometric transforms. The trigonometric transforms are applied with subroutines from the library BIHAR [17, 18]. Currently, there are no efficient parallel implementations of these routines for distributed memory architectures so the trigonometric transforms will be applied sequentially in this first step of the parallel algorithm. This first step has a low weight in the total execution time if the problem size plus one is not a prime number, as will be shown in Section 5.

The block triangular factorization can be described as an iterative process of $2n/\nu$ steps. When the iteration k starts (Fig. 2), blocks $\hat{G}_i$ such that $i < k$ have already been zeroed, and blocks $L_{i,j}$, such that $j < k$ for all i, have also been computed. Then, during the iteration k, the processor that contains block

$$
\begin{array}{c|llll}
\hat{\mathbf{0}} & L_{00} & & & \\
\vdots & \vdots & \ddots & & \\
\hat{\mathbf{0}} & L_{(k-1)0} & \cdots & L_{(k-1)(k-1)} & \\
\hat{G}_k & L_{k0} & \cdots & L_{k(k-1)} & L_{kk} \\
\hat{G}_{k+1} & L_{(k+1)0} & \cdots & L_{(k+1)(k-1)} & L_{(k+1)k} \\
\hat{G}_{k+2} & L_{(k+2)0} & \cdots & L_{(k+2)(k-1)} & L_{(k+2)k} \\
\vdots & \vdots & \vdots & \vdots & \vdots
\end{array}
$$

Fig. 2. Iteration example

$\hat{G}_k$ computes $L_{k,k}$, zeroes $\hat{G}_k$ and broadcasts the suitable information to the rest of the processors. Then, the processors that contain blocks of $\hat{G}$ and L with indexes that are greater than k, compute blocks $L_{i,k}$ and update blocks $\hat{G}_i$, for $i = k+1, \ldots, 2n/\nu - 1$. Blocks $L_{i,j}$, with $j > k$ for all i, have not yet been referenced at this iteration.

Finally, the third step of the algorithm is carried out by means of calls to PBLAS routines in order to solve three triangular systems and one matrix–vector product in parallel. The parallel algorithm is summarized in Algorithm 1.

Algorithm 1. (Algorithm QRMC). *Given a non-symmetric Toeplitz matrix* $T \in R^{n\times n}$ *and an independent vector* $b \in R^n$, *this algorithm returns the solution vector* $x \in R^n$ *of the linear system* $Tx = b$ *using a* column *array of p processors.*

For all P_i, $i = 0, \ldots, p-1$, *do*

1. *Each processor computes the full generator matrix* $\hat{G}$ *(5) and stores its own row blocks in order to form the distributed generator (Fig. 1).*
2. *All processors compute the triangular factor* L *(12) in parallel using the parallel triangular decomposition algorithm explained above. Factor* L *will be distributed as shown in Fig. 1.*
3. *All processors compute* $\hat{x}$ *by means of (8) using parallel routines of PBLAS, and* P_0 *computes the solution* $x = S x$ *of the linear system.*

In order to improve the precision of the results, an iterative refinement can be applied by repeating the third step of the algorithm.

5 Experimental Results

All experimental analyse were carried out in a cluster with 12 nodes connected by a Gigabit Ethernet. Each node is a bi–processor board IBM xSeries 330 SMP composed of two Intel Pentium III at 866 MHz with 512 Mb of RAM and 256 Kb of cache memory.

The first analysis concerns the block size ν. The main conclusion of that analysis is that the worst results are obtained with small values of ν. The results improve as ν grows, arriving to the optimum when $\nu \approx 50$. Thus, the time spent by the parallel algorithm grows very slowly as we approach the maximum value of $\nu = 2n/p$ (only one block per processor). This performance behaviour is independent of both the number of processors and the matrix size. A full experimental analysis of the influence of the block size can be found in [16].

The second experimental analysis deals with the effect of the three main steps of the parallel algorithm on its total cost. Table 1 shows the time in seconds spent on each of these steps. First of all, it can be observed that the time used for the calculus of the generator tends to grow with the problem size, but it shows large changes for certain values of n. More specifically, the execution time of the trigonometric transform routines depends on the size of the prime factors of $(n + 1)$. That is, for problem sizes $n = 2800, 3000, 4000$, $(n + 1)$ is a prime number and the divide–and–conquer techniques used in trigonometric transform

Table 1. Execution time of each step of the parallel algorithm on one processor

n	primes of $(n+1)$	generator calculus	factorization step	system solution	total time
2000	$3 \times 23 \times 29$	0.32 (9%)	2.76 (81%)	0.32 (9%)	3.24
2200	$3^4 \times 71$	0.38 (9%)	3.47 (82%)	0.39 (9%)	4.43
2400	7^4	0.42 (8%)	4.52 (84%)	0.47 (9%)	5.34
2600	$3^2 \times 17^2$	0.48 (7%)	5.49 (84%)	0.55 (8%)	6.54
2800	2801	1.58 (17%)	6.81 (75%)	0.74 (8%)	9.04
3000	3001	1.79 (17%)	7.77 (75%)	0.84 (8%)	10.65
3200	$3 \times 11 \times 97$	0.67 (6%)	9.48 (86%)	0.82 (7%)	10.62
3400	19×179	0.76 (6%)	10.62 (86%)	0.94 (8%)	12.43
3600	13×277	0.85 (6%)	12.49 (87%)	1.05 (7%)	14.26
3800	$3 \times 7 \times 181$	0.89 (6%)	14.05 (87%)	1.17 (7%)	16.15
4000	4001	2.99 (15%)	15.96 (78%)	1.50 (7%)	20.24

routines cannot be exploited. However, with the exception of these cases, the weight (in parenthesis) of this step with respect to the total time is low. The most costly step is the second one, the "factorization step". The third step has a low impact on the total time. If we need to apply an iterative refinement step, the time for the "system solution" doubles the time shown in Table 1.

Our third experimental analysis is about the precision of the results obtained with the parallel algorithm. To perform this analysis we use the forward or relative error, defined as $\|x - \tilde{x}\|/\|x\|$, where x and $\tilde{x}$ are the exact and the computed solution respectively. In order to obtain the relative errors, we have computed the independent vector $b = Tx$ with a known solution x and with all entries equal to 1.

The matrix used for the experimental analysis of the precision of the solution and the stability of the algorithm is

$$T = T_0(\epsilon) + \epsilon T_1 ,$$

where $T_0(\epsilon) = \left(t_{|i-j|}\right)_{i,j=0}^{n-1}$ is a symmetric Toeplitz matrix with $t_i = (1/2)^i$, for all $i \neq 0$, and $t_0 = \epsilon$. When $\epsilon \ll 1$, all leading principal submatrices of size $3k+1$, $k = 0, \ldots, (n-1)/3$ are near–singular. We have used a value of $\epsilon = 10^{-14}$, thus, matrix T_0 is a non–strongly regular matrix. Classical *fast* algorithms like Levinson [19] and *superfast* algorithms like Bitmead–Anderson [20] fail or may produce poor results with non–strongly regular matrices. These problems occur even with well–conditioned matrices, at least for non–symmetric and symmetric indefinite matrices [21]. The only solution is to apply some additional look–ahead or refinement techniques in order to stabilize the algorithm and to improve the accuracy of the solution. Matrix T_1 is a randomly formed Toeplitz matrix with entries belonging to $[0, 1]$. We use matrix ϵT_1 to obtain a non–symmetric Toeplitz matrix T with features that are similar to T_0.

We will compare the relative error of the QRMC algorithm with two other parallel algorithms. The first algorithm, called QRMS (QR Modified decomposi-

Table 2. Forward error

n	QRMS	LUC	QRMC
2000	5.72×10^{-13}	3.12×10^{-11}	2.78×10^{-10}
2300	1.74×10^{-12}	2.30×10^{-11}	4.62×10^{-10}
2600	6.76×10^{-13}	3.93×10^{-11}	1.71×10^{-09}
2900	4.25×10^{-13}	9.77×10^{-11}	1.05×10^{-09}
3200	2.02×10^{-12}	6.29×10^{-12}	6.20×10^{-10}
3500	3.10×10^{-13}	9.72×10^{-11}	1.88×10^{-09}
3800	1.11×10^{-12}	2.31×10^{-11}	2.87×10^{-10}

Table 3. Forward error with one iteration of iterative refinement

n	QRMS	LUC	QRMC
2000	6.07×10^{-16}	4.03×10^{-13}	5.28×10^{-16}
2300	5.19×10^{-16}	8.55×10^{-13}	4.89×10^{-16}
2600	6.15×10^{-16}	4.04×10^{-13}	3.36×10^{-16}
2900	6.67×10^{-16}	5.15×10^{-12}	3.41×10^{-16}
3200	6.67×10^{-16}	1.43×10^{-12}	5.92×10^{-16}
3500	7.03×10^{-16}	3.42×10^{-12}	2.69×10^{-16}
3800	5.16×10^{-16}	6.13×10^{-13}	6.72×10^{-16}

tion with the Generalized Schur Algorithm), corresponds to the PAQR algorithm mentioned in the introduction and in Section 2 of this paper. The second algorithm, called LUC (LU factorization of a Cauchy–like translation of the Toeplitz matrix), is described in [10]. Basically, this algorithm performs a transformation of the Toeplitz system into a Cauchy–like system and computes a LU factorization of the resulting matrix.

Table 2 shows that we did not obtain the expected precision in the solution, that is, a relative error close to the relative error obtained with the QRMS algorithm. This is due to the errors induced by the discrete trigonometric transforms. The transforms $\mathcal{S}$ and $\mathcal{C}$ have an undesirable impact on the computation of the generator and on the solution obtained by the parallel algorithm.

However, the results in Table 3 show that the solution can be greatly improved by means of only one step of iterative refinement. This refinement uses matrices $\hat{\bar{R}}^T$, $\hat{\bar{Q}}$, and $\hat{\bar{\Delta}}$ that form the L (12). A similar refinement technique can be applied to the solution obtained with the LUC algorithm, but the precision does not improve as much as with the QRMC algorithm. In [16], we performed a similar analysis using the backward error, and we arrived to similar conclusions.

Finally, we analyze the execution time of the QRMC algorithm with several processors and different problem sizes (Fig. 3). It can be seen that the time decreases as the number of processors increases.

Fig. 4 allows us to compare the execution time of the three algorithms for a fixed problem size. The difference in time is mainly due to the size of the generator used in each case. While LUC works on a generator of size $n \times 4$,

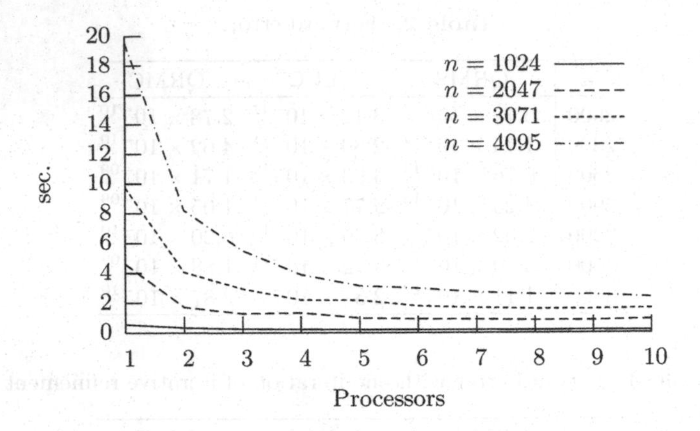

Fig. 3. Time in seconds of the QRMC algorithm

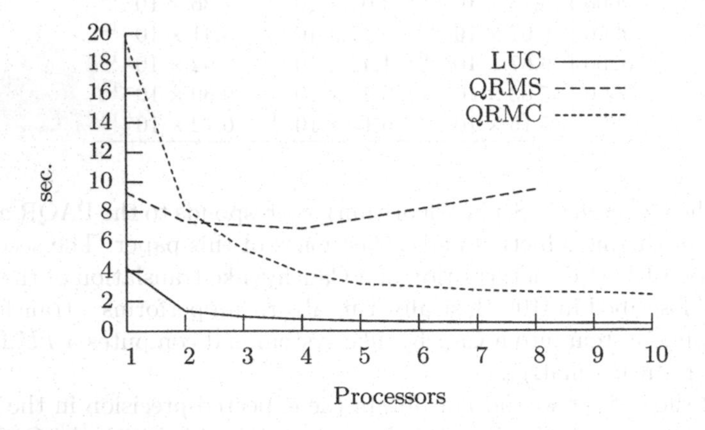

Fig. 4. Time in seconds of the three algorithms for $n = 4095$

QRMS works on a generator of size $2n \times 6$ and QRMC works on a generator of size $2n \times 8$. The embedding technique used by these last two algorithms to improve the stability and the precision of their results increases the number of rows from n to $2n$. The number of columns is related to the rank of the displacement representation. Also, the last step of the QRMS and QRMC algorithms involves the solution of three triangular systems and a matrix–vector product, while LUC only solves two triangular systems. In addition, the time shown in Fig. 4 for QRMC includes one step of the iterative refinement solution, which is not performed in the other two parallel algorithms.

The LUC and QRMC algorithms have very similar behaviours. The execution time decreases very fast when few processors are used, and decreases more slowly when the number of processors used increases. In the QRMS algorithm, the time is only reduced when two processors, or even four processors are used for large problems. For more processors, the time increases due to the effect of the communication cost. Furthermore, QRMS can only be executed on $p = 2^i$ processors, for any integer i. Our goal when developing the QRMC algorithm was to improve QRMS by means of the Cauchy translation and the results show that we have reduced the communication cost, obtaining an efficiency that is similar to the LUC algorithm, while keeping the precision of the results obtained with the QRMS algorithm. As an example, in the problem in Fig. 4, the relative error is 2.33×10^{-11} for LUC, 1.48×10^{-12} for QRMS, and 3.13×10^{-16} for QRMC using only one iteration of iterative refinement in this last case.

6 Conclusions

In this paper, we describe the parallelization of a new algorithm for solving non–symmetric Toeplitz systems. The parallel algorithm combines the efficiency of the LUC algorithm and the stability of the QRMS algorithms presented in [10].

This algorithm embeds the original Toeplitz matrix into an augmented matrix in order to improve its stability and the precision of the results. The application of only one step of iterative refinement to the solution produces very precise results. However, as can be observed in [10], this technique alone produces a non-scalable parallel algorithm due to the high communication cost.

In this paper, we solve this problem by applying trigonometric transforms that convert the original Toeplitz matrix into a Cauchy–like matrix. This transformation allows us to deal with diagonal displacement matrices, which greatly reduces the communication cost. The efficiency obtained with this method is similar to other less accurate parallel algorithms, while maintaining the accuracy of the solution.

An experimental analysis of the algorithm was performed in a cluster of personal computers. Standard tools and libraries, both sequential and parallel, were used. This has produced a code that is portable to different parallel architectures.

The experimental results show the precision and efficiency of the new parallel algorithm with few processors (< 10). The scalability of the parallel algorithm is quite good, considering the low cost of the sequential algorithm that we are parallelizing, $O(n^2)$. The analysis also shows the negative effect of some steps of the algorithm in certain cases. Future work should be done to address this problem.

Acknowledgments

The authors are indebted to Prof. Raymond Hon–fu Chan for his reference to several useful papers and Prof. Daniel Potts for his useful hints and explanations about the computation of diagonal entries of a symmetric Cauchy–like matrix.

References

1. R. W. Freund. A look-ahead Schur-type algorithm for solving general Toeplitz systems. *Zeitschrift für Angewandte Mathe. und Mechanik*, 74:T538–T541, 1994.

2. S. Chandrasekaran and Ali H. Sayed. A fast stable solver for nonsymmetric Toeplitz and quasi-Toeplitz systems of linear equations. *SIAM Journal on Matrix Analysis and Applications*, 19(1):107–139, January 1998.
3. S. Y. Kung and Y. H. Hu. A highly concurrent algorithm and pipelined architecture for solving Toeplitz systems. *IEEE Trans. Acoustics, Speech and Signal Processing*, ASSP-31(1):66, 1983.
4. D. R. Sweet. The use of linear-time systolic algorithms for the solution of toeplitz problems. Technical Report JCU-CS-91/1, Department of Computer Science, James Cook University, January 1 1991. Tue, 23 Apr 1996 15:17:55 GMT.
5. I. Ipsen. Systolic algorithms for the parallel solution of dense symmetric positive-definite Toeplitz systems. Technical Report YALEU/DCS/RR-539, Department of Computer Science , Yale University, May 1987.
6. D. J. Evans and G. Oka. Parallel solution of symmetric positive definite Toeplitz systems. *Parallel Algorithms and Applications*, 12(9):297–303, September 1998.
7. I. Gohberg, I. Koltracht, A. Averbuch, and B. Shoham. Timing analysis of a parallel algorithm for Toeplitz matrices on a MIMD parallel machine. *Parallel Computing*, 17(4–5):563–577, July 1991.
8. K. Gallivan, S. Thirumalai, and P. Van Dooren. On solving block toeplitz systems using a block schur algorithm. In Jagdish Chandra, editor, *Proceedings of the 23rd International Conference on Parallel Processing. Volume 3: Algorithms and Applications*, pages 274–281, Boca Raton, FL, USA, August 1994. CRC Press.
9. S. Thirumalai. *High performance algorithms to solve Toeplitz and block Toeplitz systems*. Ph.d. thesis, Graduate College of the University of Illinois at Urbana-Champaign, 1996.
10. Pedro Alonso, José M. Badía, and Antonio M. Vidal. Parallel algorithms for the solution of toeplitz systems of linear equations. In *Proceedeings of the Fifth International Conference On Parallel Processing and Applied Mathmatics*, Czestochowa, Poland, September 2003 (to appear in *Lecture Notes in Computer Science*).
11. E. Anderson et al. *LAPACK Users' Guide*. SIAM, Philadelphia, 1995.
12. L.S. Blackford et al. *ScaLAPACK Users' Guide*. SIAM, Philadelphia, 1997.
13. Thomas Kailath and Ali H. Sayed. Displacement structure: Theory and applications. *SIAM Review*, 37(3):297–386, September 1995.
14. Georg Heinig and Adam Bojanczyk. Transformation techniques for Toeplitz and Toeplitz-plus-Hankel matrices. I. transformations. *Linear Algebra and its Applications*, 254(1–3):193–226, March 1997.
15. Daniel Potts and Gabriele Steidl. Optimal trigonometric preconditioners for non-symmetric Toeplitz systems. *Linear Algebra and its Applications*, 281(1–3):265–292, September 1998.
16. Pedro Alonso, José M. Badía, and Antonio M. Vidal. An efficient and stable parallel solution for non–symmetric Toeplitz linear systems. Technical Report II-DSIC-08/2004, Universidad Politécnica de Valencia, 2004.
17. P. N. Swarztrauber. *Vectorizing the FFT's*. Academic Press, New York, 1982.
18. P. N. Swarztrauber. FFT algorithms for vector computers. *Parallel Computing*, 1(1):45–63, August 1984.
19. N. Levinson. The Wiener RMS (Root Mean Square) error criterion in filter design and prediction. *Journal of Mathematics and Physics*, 25:261–278, 1946.
20. Robert R. Bitmead and Brian D. O. Anderson. Asymptotically fast solution of Toeplitz and related systems of linear equations. *Linear Algebra and its Applications*, 34:103–116, 1980.
21. James R. Bunch. Stability of methods for solving Toeplitz systems of equations. *SIAM Journal on Scientific and Statistical Computing*, 6(2):349–364, April 1985.

Partial Spectral Information from Linear Systems to Speed-Up Numerical Simulations in Computational Fluid Dynamics

C. Balsa[1], J.M.L.M. Palma[1], and D. Ruiz[2]

[1] Faculdade de Engenharia da Universidade do Porto,
Rua Dr. Roberto Frias s/n, 4200-465 Porto, Portugal
{cbalsa, jpalma}@fe.up.pt

[2] Ecole Nationale Supérieure d'Electrotechnique, d'Electronique,
d'Informatique, d'Hydraulique et des Télécommunications de Toulouse,
2, rue Charles Camichel, BP 7122 - F 31071 Toulouse Cedex 7, France
ruiz@enseeiht.fr

Abstract. It was observed that all the different linear systems arising in an iterative fluid flow simulation algorithm have approximately constant invariant subspaces associated with their smallest eigenvalues. For this reason, we propose to perform one single computation of the eigenspace associated with the smallest eigenvalues, at the beginning of the iterative process, to improve the convergence of the Krylov method used in subsequent iterations of the fluid flow algorithm by means of this precomputed partial spectral information. The *Subspace Inverse Iteration Method* with *Stabilized Block Conjugate Gradient* is our choice for computing the spectral information, which is then used to remove the effect of the smallest eigenvalues in two different ways: either building a spectral preconditioner that shifts these eigenvalues from almost zero close to the unit value, or performing a deflation of the initial residual in order to remove parts of the solution corresponding to the smallest eigenvalues. Under certain conditions, both techniques yield a reduction of the number of iterations in each subsequent runs of the Conjugate Gradient algorithm.

1 Introduction

The starting point for this work was the iterative solution of multiple systems of linear equations, all Symmetric and Positive Definite (SPD), that occur in a Computational Fluid Dynamic (CFD) code based on the SIMPLE fluid flow algorithm [1]. This is what has been called a segregated approach, where the mass and momentum conservation equations are solved in sequence and in an iterative manner, as separate equation systems.

In every iteration i of the SIMPLE algorithm, a new system of linear equations

$$A_i x = b_i \tag{1}$$

M. Daydé et al. (Eds.): VECPAR 2004, LNCS 3402, pp. 699–715, 2005.

must be solved, where both the coefficient matrix A_i and its right-hand side b_i are updated. We have observed that, despite of the updating of A_i and b_i between iterations, the characteristics of the invariant subspaces associated with the smallest eigenvalues in these varying linear systems did not change much. This is very important, because it is well-known that eigenvalues near the origin slow down the convergence of Krylov based methods like CG [2] (for instance, in the case of SPD systems). The idea was then to take advantage of this near-stability of the invariant subspace associated with the smallest eigenvalues, computing it only once and using it to improve the convergence of the CG method in subsequent iterations of the SIMPLE fluid flow algorithm.

The paper is organized as follows. In section 2 we analyze the characteristics and the origin of the systems to be solved. Section 3 includes a short explanation of the techniques being used to evaluate and cancel the effect of the smallest eigenvalues. In section 4 is concerned with the results and their discussion. Finally, in section 5, we present the costs and benefits of the proposed technique and conclude with some observations about the experiments and a list of other issues, still requiring further analysis are analyzed.

2 The Test Problem

As a basis for our study, we used the coefficient matrices of the pressure p which results from the classical problem of the flow inside a square box with a sliding lid, for a Reynolds number ($Re = U_{lid}L/\nu$) of 100, based on lid velocity (U_{lid}) and box size (L =1 m).

2.1 Fluid Flow Equations

The fluid flow governing equations are the continuity,

$$\frac{\partial u}{\partial x} + \frac{\partial v}{\partial y} = 0 , \tag{2}$$

and the Navier-Stokes equations,

$$\frac{\partial(uu)}{\partial x} + \frac{\partial(uv)}{\partial y} = -\frac{1}{\rho}\frac{\partial p}{\partial x} + \nu\left(\frac{\partial^2 u}{\partial x^2} + \frac{\partial^2 u}{\partial y^2}\right) \tag{3a}$$

$$\frac{\partial(uv)}{\partial x} + \frac{\partial(vv)}{\partial y} = -\frac{1}{\rho}\frac{\partial p}{\partial y} + \nu\left(\frac{\partial^2 v}{\partial x^2} + \frac{\partial^2 v}{\partial y^2}\right) \tag{3b}$$

which establish the principles of mass (2) and momentum conservation (3) in a two-dimensional stationary flow of an incompressible constant-property Newtonian fluid. u and v are the velocity components along the x and y Cartesian coordinate directions, p is the pressure and ρ and ν are the density and the fluid's kinematic viscosity.

The value of the unknowns u, v and p was determined using the SIMPLE fluid flow algorithm [1], where the momentum and mass conservation principles are enforced by alternately solving, until convergence, equations (3) and a Poisson-type pressure-correction equation derived by algebraic manipulation of the continuity (2) and momentum (or velocity) equations (3). The advective and diffusive terms are discretized in a non-staggered grid [3], using hybrid and second order central differencing, in a finite volume approach [4], which yields a five-diagonal coefficient matrix. The hybrid scheme switches between upwind and central differencing schemes, depending on whether the Reynolds number based on the numerical cell size is higher or lower than 2.

The total number of global iterations performed by the SIMPLE algorithm, to reach convergence, is between 100 and 450, depending on the kind of formulation and of the used under-relaxation factors [5]. In our test problem it converge in 250 iterations.

2.2 Properties of the Coefficient Matrices

The domain was discretized with a regular mesh of 120×120 nodes, with Dirichlet's boundary conditions. The result was a five-diagonal matrix of dimension $N = 13924$, with $NZ = 69148$ non-zero elements. In table 1, we present some properties of the coefficient matrices A_1, of the momentum and pressure correction system, solved in the first global iteration.

The test systems used in this work are derived from the pressure-correction equation. As we can see in table 1, these systems are ill-conditioned, compared with the velocity systems, because their smallest eigenvalues are much closer to the origin. The linear systems involving the velocities appear to be rather well conditioned, even with other grid sizes that we have tried in some preliminary tests. The pressure systems are positive definite because we considered Dirichlet's boundary conditions for the pressure-correction equation. In the case of Neumann conditions on all boundaries, these systems would have been singular and more difficult to solve.

Some of the eigenvalues of the pressure system, computed with ARPACK [6], are presented in table 2. These values belong to the systems solved at the first global iteration ($A_1 x = b_1$), and to the same system preconditioned with M_1,

Table 1. Properties of the coefficient matrix of velocity and pressure with Dirichlet boundary conditions

System	Momentum u	Pressure p
N	13924	13924
NZ	69148	69148
$\|\|A\|\|_\infty$	$1.057e-01$	$1.006e-02$
$\|\|A\|\|_2$	$9.971e-02$	$1.005e-02$
$\kappa_2(A)$	$4.233e+00$	$3.131e+05$
$\kappa_\infty(A)$	$5.812e+00$	$3.130e+01$

Table 2. Eigenvalues of the pressure system solved in the first global iteration, before and after preconditioning

Eigenvalue	A_1	$M_1^{-1}A_1$
λ_1	$3.211006e-08$	$4.428912e-05$
λ_2	$8.877776e-07$	$1.232523e-03$
λ_3	$9.572325e-07$	$1.328789e-03$
λ_4	$1.769538e-06$	$2.473183e-03$
λ_5	$3.550269e-06$	$4.915760e-03$
λ_{10}	$7.985443e-06$	$1.100848e-02$
λ_{15}	$1.164407e-05$	$1.611350e-02$
λ_{20}	$1.591812e-05$	$2.193773e-02$
λ_{max}	$1.005278e-02$	$1.226147e-00$
κ_2	$3.130730e+05$	$2.768510e+04$

Table 3. Cosine of the angles between the invariant subspaces associated with the 20 smallest eigenvalues

$min\{SVD(V_l^T V_p)\}$	$M_l = M_p = I$	$M_l = IC(A_l)$, $M_p = IC(A_p)$	$M_l = M_p = IC(A_1)$
$min\{SVD(V_1^T V_{10})\}$	0.9999999211680	0.9999999106341	0.9999999123822
$min\{SVD(V_1^T V_{100})\}$	0.9999999704539	0.9999999236925	0.9999999693204
$min\{SVD(V_1^T V_{200})\}$	0.9999999025570	0.9999999240168	0.9999999946030

the standard Incomplete Cholesky (IC). As we can see, the IC preconditioner clusters the eigenvalues near one, but it is not very effective in decreasing the spectral condition number, which changes by one order of magnitude only.

In table 3, we present the minimum cosine of the angles between the two subspaces generated by the eigenvectors associated with the 20 smallest eigenvalues of $M_l A_l$ and $M_p A_p$ are presented. The indices l and p refer to the global iterations of the SIMPLE algorithm and M_l and M_p are the corresponding IC preconditioners. The computation is made using the Singular Value Decomposition (SVD), as proposed in [2], where $SVD(V_l^T V_p)$ gives the cosines of the angles between the two subspaces $Range(V_l)$ and $Range(V_p)$, and its minimum value denotes the maximum angle. The eigenvectors, located in the columns of the matrices V_l and V_p, have been computed with ARPACK. Despite the changes in the systems in each global iteration, the spectral characteristics of the coefficient matrices A_i do not change much from the first to the last global iteration. Indeed, as we can see in table 3, the angles between the invariant subspaces associated with the smallest eigenvalues stay very close to zero.

3 Methods to Obtain and Use the Spectral Information

As mentioned before, we are concerned with the consecutive solution of several linear systems with different right-hand sides and different coefficient matrices but having spectral characteristics approximately constant. To benefit from this, we propose a two phase approach that firstly computes a part of the spectral information associated with the smallest eigenvalues in the first system, and which afterwards uses this information to compute the solution of the remaining systems.

In this section, we describe the method for computing the eigenvalues and the eigenvectors of the matrix A_1, which is based on the Inverse Subspace Iteration. Next, we discuss two techniques, both exploiting the precomputed invariant subspace, that can be used to improve the CG convergence: first, the deflation of the starting guess and second, the use of a spectral preconditioner.

3.1 Computation of the Invariant Subspaces

To compute the spectral information we have implemented the Inverse Subspace iteration in conjunction with the stabilized Block Conjugate Gradient method (BlockCG) to compute the required sets of multiple solutions, as proposed in [7] where this combination is called BlockCGSI algorithm. This algorithm computes s eigenvectors associated to the s smallest eigenvalues of the preconditioned pressure matrix $M^{-1}A$, at the first SIMPLE iteration.

Algorithm 1 The BlockCGSI algorithm

$Z^{(0)}$ =RANDOM(n,s)
M-Orthonormalize $Z^{(0)}$:
  $V^{(0)}R_0 = Z^{(0)}$ such that $V^{(0)^T}MV^{(0)} = I_{s\times s}$
for $k = 1$ to m **do**
  Use BlockCG with block size s to solve:
    $M^{-1}AZ^{(k)} = V^{(k-1)}$
  M-Orthonormalize $Z^{(k)}$:
    $Q^{(k)}R_k = Z^{(k)}$ such that $Q^{(k)^T}MQ^{(k)} = I_{s\times s}$
  Set $\beta_k = Q^{(k)^T}AQ^{(k)}$
  Diagonalize β_k:
    $\beta_k = U_k\Delta_k U_k^T$
    where $U_k^T = U_k^{-1}$
    and Δ_k =Diag$(\delta_1, ..., \delta_s)$ (Ritz values)
  Set $V^{(k)} = Q^{(k)}U_k$ (Ritz vectors)
end for

In algorithm 1, the BlockCG solves the s linear systems $M^{-1}Az_j^{(k)} = v_j^{(k-1)}$, with $j = 1, ..., s$, simultaneously, where the matrix A is preconditioned with M (the standard IC preconditioner) in a symmetrical way in order to make the coefficient matrix similar to a Symmetric and Positive Definite (SPD) one. The Block Conjugate Gradient algorithm under concern is a numerically stable variant [8] that avoids the numerical problems that can occur when some of the s systems are about to converge.

The block solution matrix $Z^{(k)}$, corresponding to the s right-hand sides (columns of $V^{(k-1)}$), is computed with an accuracy determined by a threshold value ϵ, which is an input of the algorithm. The other input values are m, the number of inverse iterations of the BlockCGSI algorithm, and s, the block size, which defines the size of the computed invariant subspace. These parameters can not be set in an optimal manner a-priori because this depends on the kind of problem and on the desired final accuracy. This depends also how we take advantage of the computed spectral information in the following consecutive solutions. In section 4, we investigate experimentally the impact of varying the values of m and s in the context of the current problem.

3.2 Deflation

Once the invariant subspace linked to the smallest eigenvalues of the linear system solved in the first global iteration is obtained, we can use it for the computation of the solutions in each of the following global iterations. The idea is to use this spectral information to remove the effect of the poor conditioning in these linear systems.

One of the possible methods is to perform a deflation on the initial residual. This enables to obtain the eigencomponents of the solution corresponding to the smallest eigenvalues. With this initial starting guess, we expect that the CG will converge to the remaining part of the solution very quickly, since the difficulties caused by the smallest eigenvalues have been swallowed. In this way, the Conjugate Gradient should reach linear convergence immediately [9].

In each global iteration i of the SIMPLE algorithm, we run the Conjugate Gradient to solve $M_1^{-1}A_i x = M_1^{-1}b_i$ with the starting guess

$$x^{(0)} = V^{(m)} \Delta_m^{-1} V^{(m)^T} b_i, \tag{4}$$

where $V^{(m)}$ and Δ_m are the matrices of the s Ritz vectors and values, obtained after m inverse iterations of the BlockCGSI algorithm on the preconditioned matrix $M_1^{-1}A_1$.

3.3 Spectral Low Rank Update (SLRU) Preconditioner

Another way of exploiting spectral information of the coefficient matrix is to perform a deflation at each CG iteration, instead of just at the beginning. This approach, proposed in [10], is called Spectral Low Rank Update (SLRU) preconditioning.

The computation of the solution of the systems $M_1^{-1}A_i x = M_1^{-1}b_i$, is obtained by running the CG algorithm on an equivalent system $MA_i x = Mb_i$, where the preconditioner M is given by

$$M = M_1^{-1} + V^{(m)} \Delta_m^{-1} V^{(m)^T}. \tag{5}$$

This preconditioner will shift the smallest eigenvalues in the coefficient matrix $M_1^{-1}A_i$ close to one (see [10]). In some cases, it can be interesting to shift the smallest eigenvalues close to some predetermined value λ (with $\lambda = \lambda_{max}$, for instance), in which case the spectral preconditioner must be set to

$$M = M_1^{-1} + \lambda V^{(m)} \Delta_m^{-1} V^{(m)^T}. \tag{6}$$

This is not useful in our test problem, since the spectrum is previously clustered near one with the first level of preconditioning M_1 (see table 2).

In comparison with the deflation on the starting guess, presented in equation (4), the SLRU preconditioner implies additional work at each CG iteration.

4 Experiments

In this section, we illustrate the potential of the methods presented above, in the context of the SIMPLE algorithm. We analyse the computation phase of the spectral information and its consequences on the acceleration of the CG convergence. All the tests have been performed on a Personal Computer with an AMD® Athlon® Processor, with clock frequency of 1.6 MHz and 250 MB of RAM memory. All runs have been performed with Matlab® R12.

We compute, using algorithm 1, the s smallest eigenvalues, with s varying from 5 and 20, in the system $A_1 x = b_1$ and we then run the CG algorithm on all the sample systems. In each run at the classical Conjugate Gradient algorithm, preconditioned with Incomplete Cholesky, we compare the performance improvements obtained with the two different techniques introduced above, namely the SLRU preconditioner and the deflation of the smallest eigenvalues in the starting guess.

4.1 Monitoring the BlockCGSI Algorithm

The main parameters of algorithm 1, that must be carefully chosen, are the threshold value (ϵ) for the stopping criterion of the BlockCG algorithm, the dimension of the computed invariant subspace (block size s), and the number of inverse iterations (m).

The accuracy in the computation of the multiple solutions required in the subspace inverse iterations, obtained by means of the stabilized BlockCG, is set by the threshold value ϵ used in the stopping criterion. The choice of this stopping criterion is indeed a crucial point because it can lead to a great amount of unnecessary extra work, proportional to the block size s. We recall that the linear systems that we need to solve at each inverse iteration is of the type

$$M^{-1} A z = v,$$

which is equivalent to the symmetrized system $R^{-T} A R^{-1} R z = R v$, assuming that the preconditioner is factorized as $M = R^T R$. Based on the ideas developed by Rigal and Gaches [11], we have chosen the normwise backward error

$$\omega_1 = \frac{||Rv - R^{-T} A z||_2}{||R^{-T} A R^{-1}||_2 ||Rz||_2 + ||Rv||_2}, \tag{7}$$

where z denotes the current iterate approximating the solution of the system above, to monitor the convergence and to stop the iterations when this backward error is smaller than some threshold ϵ. Assuming also that the preconditioned matrix $R^{-T}AR^{-1}$ has a maximum eigenvalue close to one – which is the case in general with classical preconditioning techniques, and can at any rate be achieved with some additional scaling factor – and considering that the right-hand side vector v is M-orthogonal, we also mention that this backward error measure can be simplified to

$$\omega_1 = \frac{||v - M^{-1}Az||_M}{||z||_M + 1}, \tag{8}$$

since the preconditioner M is supposed to be SPD.

The reasons for this choice is that it is the measure of the residual norms in the symmetrized system $R^{-T}AR^{-1}Rz = Rv$ that bounds effectively the error on the approximated eigenvalues. Indeed, at each inverse subspace iteration, the Ritz values $\delta_1^{(k)}, ..., \delta_s^{(k)}$ (diagonal elements of Δ_k) ranged in increasing order, and the corresponding Ritz vectors $v_1^{(k)}, ..., v_s^{(k)}$ (columns of $V^{(k)}$), approximate the eigenvalues $\lambda_1, ..., \lambda_s$ and the eigenvectors $u_1, ..., u_s$, of $M^{-1}A$. The error bound on each eigenpair, given by [12], is:

$$|\lambda_j - \delta_j^{(k)}| \leq \frac{||Av_j^{(k)} - \delta_j^{(k)} M v_j^{(k)}||_{M^{-1}}}{||Mv_j^{(k)}||_{M^{-1}}}, \tag{9}$$

which also corresponds to

$$|\lambda_j - \delta_j^{(k)}| \leq ||M^{-1}Av_j^{(k)} - \delta_j^{(k)} v_j^{(k)}||_M \tag{10}$$

and incorporates the same matrix-norm as in the backward error measure (8). In that respect, we may expect a better agreement between the choice of the threshold value ϵ used to stop the BlockCG iterations and the final level of accuracy that can be reached in the approximated eigenvalues.

To start with the illustration of the different issues discussed above, we first show in figure 1 the convergence history of the backward error ω_1 within the BlockCG run at the first inverse iteration in algorithm 1. We indicate in the X-axis the number of matrix-vectors multiplications, which correspond to the iteration count multiplied by the block size s (fixed to $s = 5$ and $s = 10$ in the two plots). The convergence history that is shown is relative to the solution vector z_1 associated with the smallest eigenvalues λ_1, because its computation is the most sensitive to round-off error propagation. Just for sake of comparison, we also plot the convergence history of the classical standard relative residual ratio:

$$\omega_2 = \frac{||M^{-1}Az - v||_2}{||v||_2}. \tag{11}$$

We can observe that the choice of the criterion based on (11) can increase drastically the number of BlockCG iterations for the same fixed level ϵ. For example, with a threshold value $\epsilon = 10^{-4}$ and a block size $s = 10$, the use of a residual

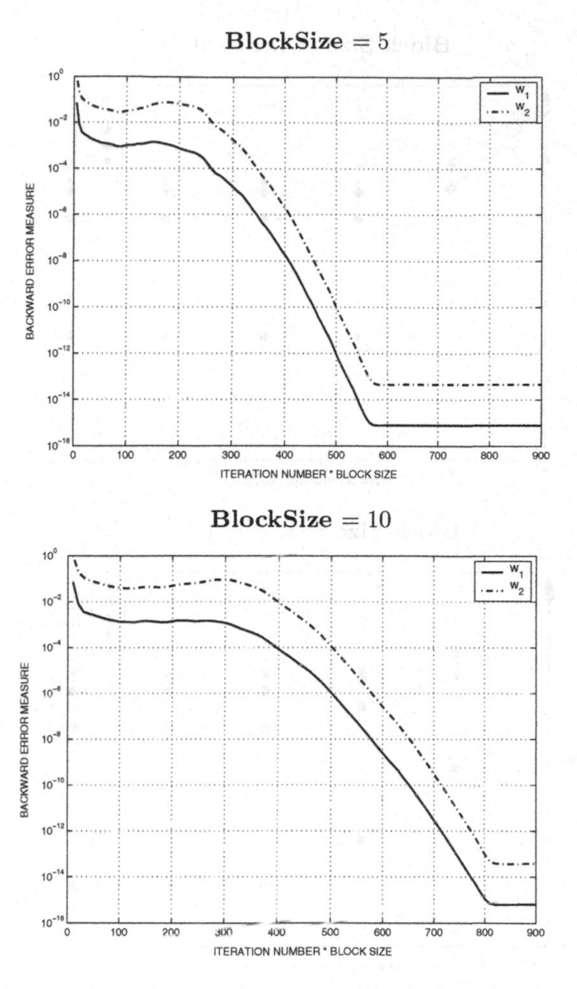

Fig. 1. Two different residual measure for the convergence of the BlockCG with block size 5 and 10

norm ω_2 instead of ω_1 increases the total number of matrix-vector multiplications from 400 to 500 (see figure 1). It is also important to note that the amount of matrix-vector multiplications does not grow proportionally with the block size s.

One of the most crucial issues in these experiments is the good adequacy between the threshold value ϵ and the level of stagnation of the error bound (10) in the inverse iterations. In figure 2, we can actually see that this error bound decreases very fast as the number of inverse iterations grows, and stagnates around a level very close to the threshold value ϵ. This justifies experimentally the discussion we had before concerning our choice for the backward error measure (8), and gives the potential for an easy control of the numerical behavior of algorithm 1. Finally, the stagnation of this bound after a certain number of

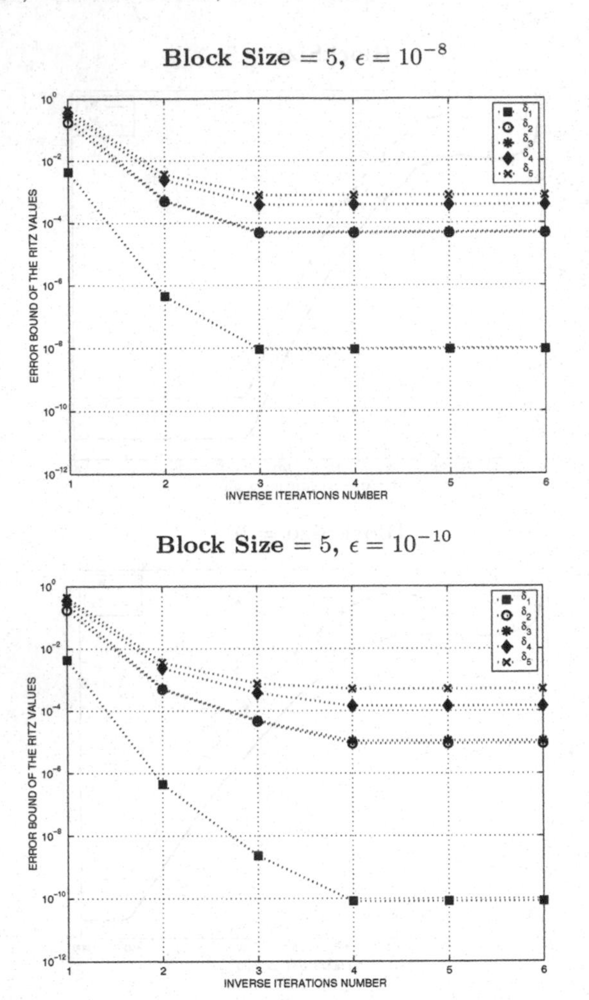

Fig. 2. Evolution of the error bound (10) associated to the five smallest eigenvalues, computed with the BlockCGSI algorithm

inverse iterations is a direct consequence of the level of accuracy in the computed solutions. For example, if we want to approximate the five smallest eigenvalues with a threshold value $\epsilon = 10^{-10}$, the BlockCGSI algorithm stagnates after four inverse iterations (see figure 2), and after this stage, the BlockCG needs only one iteration to reach the level ϵ (see table 4). This occurs simply because there is no more refinement possible with the same ϵ.

Dividing equation (10) by $\delta_j^{(k)}$ instead of λ_j, we can get an estimate of the relative residual associated to λ_j, which indicates the number of correct digits in each approximated eigenvalue $\delta_j^{(k)}$. In table 5, we give the value of this estimate corresponding to $\delta_1^{(k)}$ (obtained with a block size $s = 5$), as a function of the number of inverse iterations and for different values of the threshold ϵ. For

Table 4. Total number of matrix-vector products performed by the BlockCGSI algorithm as a function of the number of inverse iterations

	Number of BlockCG Iterations * Block Size															
	$\epsilon = 10^{-4}$				$\epsilon = 10^{-6}$				$\epsilon = 10^{-8}$				$\epsilon = 10^{-10}$			
Inverse	Block Size				Block Size				Block Size				Block Size			
Iterations	5	10	15	20	5	10	15	20	5	10	15	20	5	10	15	20
1	260	400	480	560	345	500	600	600	410	580	600	600	455	600	600	600
2	50	70	90	120	215	310	360	420	305	430	495	600	365	510	600	600
3	10	10	15	20	10	20	15	20	165	220	210	260	255	350	390	460
4	5	10	15	20	5	10	15	20	5	10	15	20	120	130	15	100
5	5	10	15	20	5	10	15	20	5	10	15	20	5	10	15	20
6	5	10	15	20	5	10	15	20	5	10	15	20	5	10	15	20

Table 5. Estimate of the relative residual associated with the smallest eigenvalue λ_1 as a function of the number of inverse iterations and different threshold values

Inverse	$\|\|M^{-1}Av_1^{(k)} - \delta_1^{(k)}v_1^{(k)}\|\|_M/\delta_1^{(k)}$			
Iterations	$\epsilon = 10^{-4}$	$\epsilon = 10^{-6}$	$\epsilon = 10^{-8}$	$\epsilon = 10^{-10}$
1	$6.71101e+01$	$6.71317e+01$	$6.71317e+01$	$6.71317e+01$
2	$2.25445e+00$	$2.35220e-02$	$1.47802e-02$	$1.02337e-02$
3	$1.60105e+00$	$1.86759e-02$	$2.81274e-04$	$5.23205e-05$
4	$1.60105e+00$	$1.86759e-02$	$2.15853e-04$	$1.85662e-06$
5	$1.60105e+00$	$1.86759e-02$	$2.15853e-04$	$1.85662e-06$
6	$1.60105e+00$	$1.86759e-02$	$2.15853e-04$	$1.85662e-06$

instance, with $\epsilon = 10^{-10}$, the BlockCGSI algorithm gives an approximation of the smallest eigenvalue with six correct digits after four inverse iterations.

Concerning the choice of the block size s, we must consider different conflicting aspects in which there is some compromise to be reached. On the one hand, the more eigenvectors associated to the smallest eigenvalues are approximated, the faster should be the convergence of the CG algorithm in the solution phase, because we expect that the CG algorithm will work as if the condition number was reduced to about $\lambda_{\max}/\lambda_{s+1}$. But on the other hand, as s grows, the work to compute the spectral information also grows. However, and this has been mentioned before, this computational work does not grow proportionally with the block size, and it is thus important to find out the appropriate number of eigenvalues to approximate.

The last important aspect concerning the choice of the block size s, is that the convergence of the inverse iterations in the BlockCGSI algorithm can also be improved with larger values of s. Indeed, the subspace computed in the BlockCGSI algorithm converges toward an invariant subspace of the iteration matrix with a rate that depends on the relative gap between the approximated eigenvalues and the smallest of the remaining ones. As it was shown in [7], this rate of convergence is governed by the relation:

$$|\lambda_j - \delta_j^{(k)}| \sim \left(\frac{\lambda_j}{\lambda_{s+1}}\right)^k, \quad 1 \le j \le s, \tag{12}$$

and it may be useful in some cases to increase the block size s just to benefit from a better gap in the relation above. This technique is also denoted *"Guard Vectors"*. Nevertheless, the relation (12) relies on the fact that the solutions in each inverse iteration are obtained as exactly as possible. This is not the case in our experiments where we actually play with rather large values for the threshold ϵ, expecting to reduce quite reasonably the total amount of work for very improvements in the solution phase, as this will be illustrated in the following sections.

4.2 Improving the Convergence of the CG Algorithm

Based on the precomputed spectral information, we can improve the convergence of the CG algorithm at each of the consecutive runs in the SIMPLE algorithm.

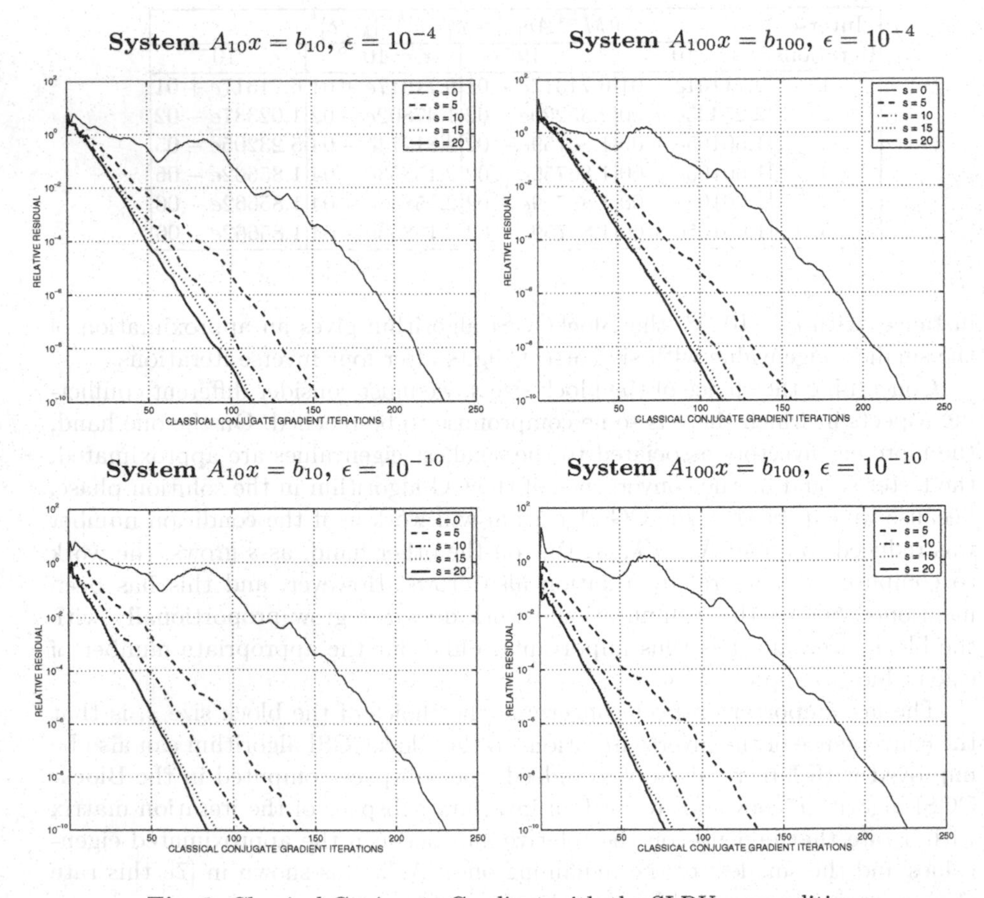

Fig. 3. Classical Conjugate Gradient with the SLRU preconditioner

To illustrate this, we have selected two pressure-correction systems coming from the global iterations 10 and 100 in the SIMPLE algorithm, and have solved these systems with the CG algorithm accelerated by means of the SLRU preconditioner or the deflation of the initial solution vector. The backward error is based on w_2, equation (11), mainly for reasons of simplicity in the discussion and also because very similar improvements can be observed with other type of residual measures that can be found in literature (see [13] for instance).

The results show that, as expected, the spectral information calculated in the first global iteration is valid during the whole iterative process of the SIMPLE algorithm. In previous experiments, we observed that the reduction in the number of CG iterations is similar to the one obtained in the case of the two systems $A_{10}x = b_{10}$ and $A_{100}x = b_{100}$ presented here.

The Incomplete Cholesky preconditioner clusters the eigenvalues near one but still leaves a few ones remote (see table 2). These are shifted to the right of the unity by the SLRU preconditioner introduced in equation (5), yielding a lower condition number and, therfore, an improved CG performance. However,

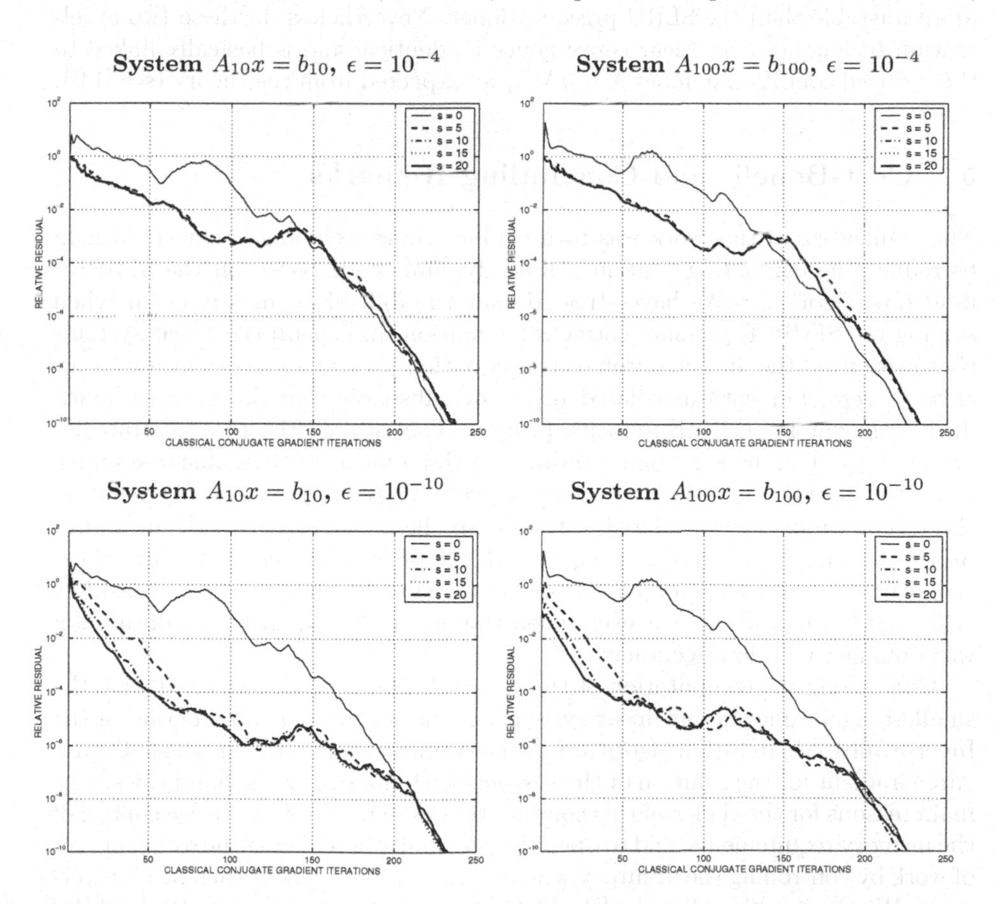

Fig. 4. Classical Conjugate Gradient with the deflation of the starting guess

we can observe in figure 3 that, after removing the 10 smallest eigenvalues, no much better improvements are obtained on the convergence rate. This occurs because the reduced condition number $\lambda_{\max}/\lambda_{s+1}$ remains at the level of 10^2 when $s \geq 10$ (see table 2).

An important property of the SLRU preconditioner is that it does not need a high accuracy in the spectral information to be efficient. As we can observe in figure 3, the curves obtained with $\epsilon = 10^{-4}$ and $\epsilon = 10^{-10}$ are very similar, despite the fact that δ_1 has no correct digit in the first case and 6 correct digits in the second case (see table 5).

The alternative approach that we have experimented to remove the effect of the smallest eigenvalues is the deflation technique. As it was mentioned before, the idea is to include in the starting guess the information corresponding to the smallest eigenvalues. As opposed to the SLRU preconditioner the deflation of the starting guess shows a linear convergence only during the first few iterations, and requires also that the spectral information be computed with better accuracy (see figure 4). This suggests that the deflation technique is numerically more unstable than the SLRU preconditioner. Nevertheless, in these two acceleration techniques, the linear convergence is identical and is basically linked to the reduced condition number $\lambda_{\max}/\lambda_{s+1}$ as expected from the theory (see [14]).

5 Cost-Benefit and Concluding Remarks

The framework of this work was to introduce some techniques that could help to reduce the computing time in a fluid dynamic code based on the SIMPLE fluid flow algorithm. We have stressed that the difficulties mostly occur when solving the SIMPLE pressure-correction equation. In general the linear systems that arise from the discretization of this equation present some eigenvalues very close to zero, but with associated invariant subspaces that do not vary much during the whole SIMPLE iterative process. This opened the way to strategies based on spectral information for improving the solution of these linear systems. In that respect, we have focused on two closely related approaches: (1) deflating the eigencomponents associated with the smallest eigenvalues with an appropriate starting guess, or (2) using the SLRU preconditioner that shifts these eigenvalues away from zero. The latter appeared to be numerically more stable, achieving linear convergence, even when the pre-computed spectral information was obtained with low accuracy.

Concerning the computation of the spectral information associated with the smallest eigenvalues in our linear systems, we have chosen a combination of the Inverse Subspace Iteration algorithm with a stabilized version of the Block Conjugate Gradient for the solution of the systems with multiple right-hand sides. The main reasons for the choice of this combination were to enable a precise control of the memory requirements, and to open the possibility of reducing the total amount of work by controlling the accuracy when solving the systems at each inverse iteration. We also mention that the BlockCG algorithm can incorporate BLAS 3 [15] operations and therefore be efficiently implemented on modern computers.

Table 6. Costs and benefits due to CG with the SLRU prconditioner

s	ϵ	Eq. Iter. (costs)	CG Its	CG Time	Benefits Its	Benefits Time	SIMPLE Amortization
5	10^{-4}	320	79	7.65	74	4.03	5
	10^{-10}	1195	53	5.18	100	6.50	12
10	10^{-4}	470	62	8.65	91	3.03	6
	10^{-10}	1590	38	5.32	115	6.36	14
15	10^{-4}	580	56	9.32	97	2.36	6
	10^{-10}	1590	30	4.96	123	6.72	13
20	10^{-4}	680	54	9.96	99	1.72	7
	10^{-10}	1760	29	5.32	124	6.36	15

This eigencomputation has a cost that depends on the dimension and on the accuracy of the computed invariant subspace. Once this information is obtained, we can use it to accelerate the CG in the successive global iterations in the SIMPLE algorithm. The success of all this implies that the initial cost can be recovered by the reduction of the number of CG iterations in the following runs.

In tables 6 and 7 the cost-benefit of the two proposed techniques are illustrated. The system used is $A_{100}x = b_{100}$, and we stop the CG iterations when the relative residual norm $||r^{(k)}||_2/||r^{(0)}||_2$ is below 10^{-4}. This is a value which is not too low and is normally used in this type of simulation. In this context, the classical Conjugate Gradient algorithm converges in 153 iterations and 11.68 seconds. In these tables, we indicate first the total number of matrix-vector multiplications performed to get the spectral information, which we denote by "*Equivalent Iterations*", i.e. the cost of computing the spectral information. In each case, the number of CG iteration and the time to converge are indicated, as well as the benefits with respect to the convergence without any acceleration (153 iterations and 11.68 sec.). Finally, the "*SIMPLE Amortization*" denotes the number of global iterations after which the extra cost in terms of matrix-vector multiplications for computing the spectral information is actually fully recovered. For instance, in table 6, for $s = 10$ and $\epsilon = 10^{-10}$, 1590 equivalent iterations are needed for the spectral pre-computation, out of which CG convergence is achieved in 38 iterations, i.e. a reduction of 75% compared to the run without spectral preconditioning. The 1590 extra equivalent iterations are payed back after 14 global iterations, and in the remaining global iterations of the SIMPLE algorithm, the total time spent in the solver will be halved.

As we can see in table 7, the efficient use of the deflation technique requires a greater initial cost, since it needs higher accuracy in the computation of the spectral information. However, the deflation technique just implies an initial extra operation, and is less costly than the SLRU preconditioner. For example, in the same situation ($s = 10$ and $\epsilon = 10^{-10}$), the deflation leads to the same reduction in iterations, but in terms of computing time, the reduction is approximately of 80% and bigger compared to the SLRU preconditioner. Therefore, depending on the total number of global iterations that will be performed, the deflation technique can still be a good option.

Table 7. Costs and benefits to CG with deflation of the starting guess

s	ϵ	Eq. Iter. (costs)	CG		Benefits		SIMPLE Amortization
			Its	Time	Its	Time	
5	10^{-8}	880	54	3.19	99	8.49	9
	10^{-10}	1195	53	3.16	100	8.52	12
10	10^{-8}	1230	46	2.83	107	8.85	12
	10^{-10}	1590	39	2.35	114	9.33	14
15	10^{-8}	1350	46	2.83	107	8.85	13
	10^{-10}	1590	38	2.35	113	9.33	15
20	10^{-8}	1460	46	2.83	107	8.85	14
	10^{-10}	1760	38	2.35	113	9.33	16

Finally, the success of this approach also depends on the appropriate combined monitoring of the inverse iterations and of the BlockCG convergence within these iterations. We have proposed and validated experimentally a stopping criterion for the BlockCG algorithm that presents good numerical properties to achieve such a goal. Of course, we still need to investigate this issue from the theoretical point of view, aiming to build an algorithm that can adapt automatically to the properties of the given matrices and compute invariant subspaces with some a-priori fixed accuracy.

The results presented here show the effectiveness in reducing the computing time in what can be considered as a not too difficult fluid flow calculation. Our plan is also to investigate the potential of such an approach in the context of more complex numerical simulations, where the stability of spectral properties of the linear systems will be not so strong.

References

1. Patankar, S.V., Spalding, B.B.: A calculation procedure for heat, mass and momentum transfer in three-dimensional parabolic flows. International Journal of Heat and Mass Transfer **15** (1972) 1787–1806
2. Golub, G.H., Loan, C.F.V.: Matrix Computation. The Johns Hopkins University Press, Baltimore and London (1983)
3. Rhie, C.M., Chow, W.L.: Numerical study of the turbulent flow past an airfoil with trailing edge separation. AIAA Journal **21** (1983) 1525–1532
4. Ferziger, J.H., Perić, M.: Computational Methods for Fluid Dynamics. 3rd edn. Springer (2002)
5. McGuirk, J.J., Palma, J.M.L.M.: The efficiency of alternative pressure-correction formulations for incompressible turbulent flow problems. Computers & Fluids **22** (1993) 77–87
6. Lehoucq, R., Sorensen, D., Yang, C.: Arpack users' guide: Solution of large scale eigenvalue problems with implicitly restarted arnoldi methods (1997)
7. Arioli, M., Ruiz, D.: Block conjugate gradient with subspace iteration for solving linear systems. In: Iterative Methods in Linear Algebra, Second IMACS Symposium on Iterative Metohds in Linear Algebra, Blagoevgrad, Bulgaria, S. Margenov and P. Vassilevski (eds.) (June 17-20, 1995) 64–79

8. Arioli, M., Duff, I., Ruiz, D., Sadkane, M.: Block lanczos techniques for accelerating the block cimmino method. Journal on Scientific Computing **16** (November, 1995) 1478–1511
9. Van der Sluis, A., van der Vorst, H.A.: The rate of convergence of Conjugate Gradients. Numerical Mathematics **48** (1986) 543–560
10. Carpentieri, B., Duff, I.S., Giraud., L.: A class of spectral two-level preconditioners. Technical Report TR/PA/02/55, CERFACS (2002) Also Technical Report RAL-TR-2002-020. To appear in SIAM J. Scientific Computing.
11. Rigal, J.L., Gaches, J.: On the compatibility of a given solution with the data of a linear system. Journal of the ACM **14** (July 1967) 543–548
12. Parlett, B.N.: The Symmetric Eigenvalue Problem. SIAM, Philadclphia (1998)
13. Arioli, M.: A stopping criterion for conjugate gradient algorithm in a finite element method. Numerical Mathematics, DOI 10.1007/s00211-003-0500-y (2003) 1–24
14. L. Giraud, D. Ruiz, A. Touhami: A comparative study of iterative solvers exploiting spectral information for SPD systems. Technical Report RT/TLSE/04/03, ENSEEIHT - IRIT (2004)
15. Dongarra, J., Decroz, J., Hammarlung, S., Duff, I.: A set of level 3 basic linear algebra subprograms. ACM Trans. Math. Software **16** (1990) 1–28

Parallel Newton Iterative Methods Based on Incomplete LU Factorizations for Solving Nonlinear Systems

Josep Arnal[1], Héctor Migallón[2], Violeta Migallón[1], and José Penadés[1]

[1] Departamento de Ciencia de la Computación e Inteligencia Artificial, Universidad de Alicante, E-03071 Alicante, Spain
{arnal, violeta, jpenades}@dccia.ua.es

[2] Departamento de Física y Arquitectura de Computadores, Universidad Miguel Hernández, 03202 Elche, Alicante, Spain
hmigallon@umh.es

Abstract. Parallel iterative algorithms based on the Newton method and on two of its variations, the Shamanskii method and the Chord method, for solving nonlinear systems are proposed. These algorithms also use techniques from the non–stationary multisplitting methods. Concretely, in order to construct the multisplitting, ILU factorizations are considered. Convergence properties of these parallel methods are studied for H–matrices. Computational results, on a distributed multiprocessor IBM RS/6000 SP, that show the effectiveness of these methods are included to illustrate the theoretical results.

Topics: Numerical methods (nonlinear algebra), Parallel and distributed computing.

1 Introduction

In this paper we consider the problem of solving a nonlinear system of the form

$$F(x) = 0, \quad F : \mathbb{R}^n \to \mathbb{R}^n. \tag{1}$$

Considering that a solution $x^\star$ exists, we can use for solving the nonlinear system (1) the classical Newton method (cf. [12]). Given an initial vector $x^{(0)}$, this method produces the following sequence of vectors

$$x^{(\ell+1)} = x^{(\ell)} - x^{(\ell+\frac{1}{2})}, \quad \ell = 0, 1, \ldots, \tag{2}$$

where $x^{(\ell+\frac{1}{2})}$ is the solution of the linear system

$$F'(x^{(\ell)})z = F(x^{(\ell)}), \tag{3}$$

and $F'(x)$ denotes the Jacobian matrix.

Iterative methods can be used for the solution of (3). In this case we are in the presence of a Newton iterative method. Descriptions of these methods can

M. Daydé et al. (Eds.): VECPAR 2004, LNCS 3402, pp. 716–729, 2005.

be found, e.g., in [12] and [15]. In order to generate efficient algorithms to solve nonlinear system (1) on a parallel computer, in [1] we constructed a parallel Newton iterative algorithm to solve the general nonlinear system (1) that uses non–stationary multisplitting models to approximate the linear system (3).

To construct these methods, let us consider for each x, a multisplitting of $F'(x)$, $\{M_k(x), N_k(x), E_k\}_{k=1}^{p}$, that is, a collection of splittings

$$F'(x) = M_k(x) - N_k(x), \quad 1 \leq k \leq p,$$

and diagonal nonnegative weighting matrices E_k which add to the identity. Let us further consider a sequence of integers $q(\ell, s, k)$, $\ell = 0, 1, 2, \ldots$, $s = 1, 2, \ldots, m_\ell$, $1 \leq k \leq p$, called non–stationary parameters. Following [1] or [2], the linear system (3) can be approximated by $x^{(\ell+\frac{1}{2})}$ as follows

$$\begin{aligned}
x^{(\ell+\frac{1}{2})} &= z^{(m_\ell)}, \\
z^{(s)} &= H_{\ell,s}(x^{(\ell)})z^{(s-1)} + B_{\ell,s}(x^{(\ell)})F(x^{(\ell)}), \quad s = 1, 2, \ldots, m_\ell, \\
H_{\ell,s}(x) &= \omega \sum_{k=1}^{p} E_k \left(M_k^{-1}(x)N_k(x)\right)^{q(\ell,s,k)} + (1-\omega)I, \qquad (4) \\
B_{\ell,s}(x) &= \omega \sum_{k=1}^{p} E_k \sum_{j=0}^{q(\ell,s,k)-1} \left(M_k^{-1}(x)N_k(x)\right)^j M_k^{-1}(x) \\
&= \omega \sum_{k=1}^{p} E_k \left(I - (M_k^{-1}(x)N_k(x))^{q(\ell,s,k)}\right) (F'(x))^{-1},
\end{aligned}$$

where $\omega > 0$, is the relaxation parameter and $z^{(0)} = 0$. Thus

$$x^{(\ell+\frac{1}{2})} = \left(\sum_{i=1}^{m_\ell - 1} \prod_{j=i+1}^{m_\ell} H_{\ell,j}(x^{(\ell)})B_{\ell,i}(x^{(\ell)}) + B_{\ell,m_\ell}(x^{(\ell)}) \right) F(x^{(\ell)}),$$

where $\prod_{j=i+1}^{m_\ell} H_{\ell,j}(x)$ denotes the product of the matrices $H_{\ell,j}(x)$ in the order $H_{\ell,m_\ell}(x)H_{\ell,m_\ell-1}(x) \ldots H_{\ell,i+1}(x)$. Therefore, from (2) the relaxed non–stationary parallel Newton iterative method can be written as follows

$$x^{(\ell+1)} = G_{\ell,m_\ell}(x^{(\ell)}), \qquad (5)$$

where $G_{\ell,m_\ell}(x) = x - A_{\ell,m_\ell}(x)F(x)$, and

$$A_{\ell,m_\ell}(x) = \left(\sum_{i=1}^{m_\ell - 1} \prod_{j=i+1}^{m_\ell} H_{\ell,j}(x)B_{\ell,i}(x) + B_{\ell,m_\ell}(x) \right). \qquad (6)$$

We point out that we have introduced a relaxation parameter in the same sense as was done in [10] for solving linear systems. Particularly, if $\omega = 1$, this method reduces to that obtained in [1]. On the other hand, this method extends the parallel Newton method introduced by White [17].

The experiments displayed in [1] show the good behaviour of these methods. The convergence was shown in two cases, when the Jacobian matrix is either monotone or H–matrix. Concretely, in the last case, H–compatible splittings of the Jacobian matrix were considered. In Section 3 we analyze the convergence of these methods when incomplete LU (ILU) factorizations are used in order to obtain the multisplitting of the Jacobian matrix, actually H–matrix. Moreover, we describe two acceleration techniques for the non–stationary parallel Newton iterative method (5) based on both the Chord method and the Shamanskii method. In Section 4 we present some numerical experiments, which illustrate the performance of these algorithms. We would like to point out that an ILU factorization of an H–matrix is not necessarily an H–compatible splitting. Therefore, in this paper we provide new convergence results of our methods for H–matrices. Previously, in Section 2 we present some notation, definitions and preliminary results which we refer to later on.

2 Preliminary Results

A matrix A is said to be a nonsingular M–matrix if A has all nonpositive off–diagonal entries and it is monotone, i.e., $A^{-1} \geq O$. For any matrix $A = (a_{ij}) \in \mathbb{R}^{n\times n}$, we define its comparison matrix $\langle A\rangle = (\alpha_{ij})$ by $\alpha_{ii} = |a_{ii}|$, $\alpha_{ij} = -|a_{ij}|$, $i \neq j$. The matrix A is said to be an H–matrix if $\langle A\rangle$ is a nonsingular M–matrix. The splitting $A = M - N$ is called a regular splitting if $M^{-1} \geq O$ and $N \geq O$; the splitting is an H–compatible splitting if $\langle A\rangle = \langle M\rangle - |N|$; see e.g., [5] or [16].

Let $x > 0$, we consider the vector norm

$$\|y\|_x = \inf\{\beta > 0 \;:\; -\beta x \leq y \leq \beta x\}. \tag{7}$$

This vector norm is monotonic and satisfies $\|\,|B|x\,\|_x = \|B\|_x$, where $\|B\|_x$ denotes the matrix norm of B corresponding to the norm defined in (7).

Iterative methods can be classified according to their rate of convergence (see e.g., [12]). In this paper we use the following rates of convergence. Let $\{x^{(\ell)}\}_{\ell=0}^{\infty}$ be a sequence of iterates in $\mathbb{R}^n$ which converges to $x^\star \in \mathbb{R}^n$, then the iterates converge Q-linearly (or linearly) to $x^\star$ if there exists $c \geq 0$ such that $\|x^{(\ell+1)} - x^\star\| \leq c\|x^{(\ell)} - x^\star\|$. They converge Q–quadratically (or quadratically) to $x^\star$ if there exists $c \geq 0$ such that $\|x^{(\ell+1)} - x^\star\| \leq c\|x^{(\ell)} - x^\star\|^2$. They converge Q–superlinearly (or superlinearly) to $x^\star$ if $\lim_{\ell\to\infty} \frac{\|x^{(\ell+1)}-x^\star\|}{\|x^{(\ell)}-x^\star\|} = 0$. Finally, the iterates converge R–(quadratically, superlinearly, linearly) if and only if $\|x^{(\ell)} - x^\star\| \leq b_\ell$, where the sequence $b_0, b_1, b_2, \ldots$ converges Q–(quadratically, superlinearly, linearly) to 0.

The following theorems show the existence of ILU factorizations of matrices.

Theorem 1. *Let A be an $n \times n$ M–matrix, then for every zero pattern subset S of $S_n = \{(i,j) : i \neq j,\ 1 \leq i,j \leq n\}$, there exist a unit lower triangular matrix $L = (l_{ij})$, an upper triangular matrix $U = (u_{ij})$, and a matrix $N = (n_{ij})$ with $l_{ij} = u_{ij} = 0$ if $(i,j) \in S$ and $n_{ij} = 0$ if $(i,j) \notin S$, such that $A = LU - N$ is a regular splitting of A. Moreover, the factors L and U are unique.*

Proof. See e.g., [13].

Theorem 2. *Let A be an $n \times n$ H–matrix, then for every zero pattern subset S of $S_n = \{(i,j) : i \neq j,\ 1 \leq i,j \leq n\}$, there exist a unit lower triangular matrix $L = (l_{ij})$, an upper triangular matrix $U = (u_{ij})$, and a matrix $N = (n_{ij})$ with $l_{ij} = u_{ij} = 0$ if $(i,j) \in S$ and $n_{ij} = 0$ if $(i,j) \notin S$, such that $A = LU - N$. The factors L and U not only are unique but are H–matrices.*

Proof. See e.g., Theorem 2.5 and Lemma 2.6 of [8].

The following theorem reviews a basic property of the ILU factorizations for H–matrices.

Theorem 3. *Let A be an $n \times n$ H–matrix. Let $A = LU - N$ and $\langle A \rangle = \hat{L}\hat{U} - \hat{N}$ be the ILU factorizations of A and $\langle A \rangle$ corresponding to a zero pattern subset S of $S_n = \{(i,j) : i \neq j,\ 1 \leq i,j \leq n\}$, respectively. Then $|(LU)^{-1}N| \leq (\hat{L}\hat{U})^{-1}\hat{N}$.*

Proof. The proof can be found in [11].

3 Convergence

In order to obtain a multisplitting of the Jacobian matrix $F'(x)$, we can perform ILU factorizations of this matrix.

This entails for each k, $1 \leq k \leq p$, a decomposition of the form $F'(x) = M_k(x) - N_k(x)$, where $M_k(x) = L_k(x)U_k(x)$ and the matrices $L_k(x)$ and $U_k(x)$ are unit lower triangular and upper triangular matrices, respectively, and $N_k(x)$ is the residual or error of the factorization. This incomplete factorization is rather easy to compute. A general algorithm for building ILU factorizations can be derived by performing Gaussian elimination and dropping some elements in predetermined nondiagonal positions (see e.g., [13]).

As was done in [1], to study the convergence of the iterative scheme (5) when $M_k(x) = L_k(x)U_k(x)$, we need to make the following assumptions.

(i) There exists $r_0 > 0$ such that F is differentiable on $S_0 \equiv \{x \in \mathbb{R}^n : \|x - x^\star\| < r_0\}$,
(ii) the Jacobian matrix at $x^\star$, $F'(x^\star)$, is nonsingular,
(iii) there exists $\vartheta > 0$ such that for $x \in S_0$, $\|F'(x) - F'(x^\star)\| \leq \vartheta \|x - x^\star\|$.
(iv) There exist $t_k > 0$, $1 \leq k \leq p$, such that for $x \in S_0$, $\|M_k(x) - M_k(x^\star)\| \leq t_k \|x - x^\star\|$.
(v) $M_k(x^\star)$, $1 \leq k \leq p$, are nonsingular.
(vi) There exits $0 \leq \alpha < 1$, such that, for each positive integer s and $\ell = 0, 1, \ldots$, $\|H_{\ell,s}(x^\star)\| \leq \alpha$, where $H_{\ell,s}(x^\star)$ is defined in (4).

Theorem 4. *Let assumptions (i)–(vi) hold and $F(x^\star) = 0$. Let $\{m_\ell\}_{\ell=0}^{\infty}$ be a sequence of positive integers, and define*

$$m = \max\left[\{m_0\} \cup \left\{ m_\ell - \sum_{i=0}^{\ell-1} m_i \ : \ \ell = 1, 2, \ldots \right\}\right]. \tag{8}$$

Suppose that $m < +\infty$, Consider that the splittings $F'(x^\star) = M_k(x^\star) - N_k(x^\star)$, $1 \le k \le p$, satisfy $\|M_k^{-1}(x^\star)N_k(x^\star)\| < 1$. Then, there exist $r > 0$ and $c < 1$ such that, for $x^{(0)} \in S \equiv \{x \in \mathbb{R}^n : \|x - x^\star\| < r\}$, the sequence of iterates defined by (5) converges to $x^\star$ and satisfies $\|x^{(\ell+1)} - x^\star\| \le c^{m_\ell}\|x^{(\ell)} - x^\star\|$.

Proof. The proof follows in a similar way as the proof of Theorem 2 in [1] and it can be found in [2].

The following result shows the convergence when $F'(x^\star)$ is an H–matrix and the multisplitting of $F'(x^\star)$ is obtained from ILU factorizations. Note that when $F'(x^\star)$ is an H–matrix, if we express $F'(x^\star) = D - B$, with $D = \text{diag}(F'(x^\star))$, then $|D|$ is nonsingular and $\rho(|D|^{-1}|B|) < 1$ (see e.g., [6]).

Theorem 5. *Let assumptions (i)–(iv) hold and $F(x^\star) = \dot{0}$. Let $\{m_\ell\}_{\ell=0}^{\infty}$, be a sequence of positive integers, and define m as in (8). Suppose that $m < +\infty$. Let $F'(x^\star)$ be an H–matrix. Let $F'(x^\star) = M_k(x^\star) - N_k(x^\star)$, $1 \le k \le p$, with $M_k(x^\star) = L_k(x^\star)U_k(x^\star)$, be the ILU factorization of $F'(x^\star)$ corresponding to a zero pattern subset S_k of $S_n = \{(i,j) : \ i \ne j, \ 1 \le i,j \le n\}$. Moreover, assume that the relaxation parameter ω, satisfies $\omega \in \left(0, \frac{2}{1+\rho}\right)$, with $\rho = \rho(|D|^{-1}|B|)$, then, there exist $r > 0$ and $c < 1$ such that, for $x^{(0)} \in S \equiv \{x \in \mathbb{R}^n : \ \|x - x^\star\| < r\}$, the sequence of iterates defined by (5) converges to $x^\star$ and satisfies*

$$\|x^{(\ell+1)} - x^\star\| \le c^{m_\ell}\|x^{(\ell)} - x^\star\|. \tag{9}$$

Proof. From Theorem 4 it follows that is suffices to prove that there exists $\alpha < 1$ such that $\|M_k^{-1}(x^\star)N_k(x^\star)\| \le \alpha$, $1 \le k \le p$, and $\|H_{\ell,s}(x^\star)\| \le \alpha$, $\ell = 0, 1, 2, \ldots$, $s = 1, 2, \ldots, m_\ell$, for some matrix norm. Let $\langle F'(x^\star)\rangle = \hat{L}_k(x^\star)\hat{U}_k(x^\star) - \hat{N}_k(x^\star)$, $1 \le k \le p$, be the ILU factorization of $\langle F'(x^\star)\rangle$ corresponding to the zero pattern subset S_k of S_n. Then, taking into account Theorem 3, it obtains

$$\begin{aligned} |M_k(x^\star)^{-1}N_k(x^\star)| &= |(L_k(x^\star)U_k(x^\star))^{-1}N_k(x^\star)| \\ &\le (\hat{L}_k(x^\star)\hat{U}_k(x^\star))^{-1}\hat{N}_k(x^\star). \end{aligned} \tag{10}$$

Since $\langle F'(x^\star)\rangle = |D| - |B| = |D|(I - |D^{-1}||B|)$, using (10) some manipulations yield

$$\begin{aligned} |M_k(x^\star)^{-1}N_k(x^\star)| &\le I - (\hat{L}_k(x^\star)\hat{U}_k(x^\star))^{-1}\langle F'(x^\star)\rangle \\ &= I - (\hat{L}_k(x^\star)\hat{U}_k(x^\star))^{-1}|D|(I - J), \end{aligned} \tag{11}$$

where $J = |D|^{-1}|B|$.

Consider the vector $u = (1, 1, \ldots, 1)^T \in \mathbb{R}^n$. Since J is nonnegative, the matrix $J + \epsilon uu^T$ is irreducible for all $\epsilon > 0$, and then (see e.g., [16]) there exists a positive Perron vector x_ϵ such that

$$(J + \epsilon uu^T)x_\epsilon = \rho_\epsilon x_\epsilon, \tag{12}$$

where $\rho_\epsilon = \rho(J + \epsilon uu^T)$. The continuity of the spectral radius ensures that there exists ϵ_0 such that $\rho_\epsilon < 1$, for all $0 < \epsilon \leq \epsilon_0$.

Since $\langle F'(x^\star)\rangle$ is an M–matrix, by Theorem 1, $(\hat{L}_k(x^\star)\hat{U}_k(x^\star))^{-1} \geq O$ and therefore $(\hat{L}_k(x^\star)\hat{U}_k(x^\star))^{-1}|D|\epsilon uu^T$ is nonnegative. Moreover, it is easy to see that $|D^{-1}| \leq (\hat{L}_k(x^\star)\hat{U}_k(x^\star))^{-1}$, $1 \leq k \leq p$. Then, from (11) and (12) it follows

$$\begin{aligned} |M_k(x^\star)^{-1}N_k(x^\star)|x_\epsilon &\leq \left[I - (\hat{L}_k(x^\star)\hat{U}_k(x^\star))^{-1}|D|\left(I - (J + \epsilon uu^T)\right)\right] x_\epsilon \\ &= x_\epsilon - (\hat{L}_k(x^\star)\hat{U}_k(x^\star))^{-1}|D|\left(1 - \rho_\epsilon\right) x_\epsilon \leq \rho_\epsilon x_\epsilon. \end{aligned} \tag{13}$$

Then, by the definition of $H_{\ell,s}(x)$ in (4) we obtain the following inequalities

$$\begin{aligned} |H_{\ell,s}(x^\star)|x_\epsilon &\leq \omega \sum_{k=1}^{p} E_k |(M_k(x^\star)^{-1}N_k(x^\star)|^{q(\ell,s,k)} x_\epsilon + |1-\omega| x_\epsilon \\ &\leq \omega \sum_{j=1}^{r} E_j \rho_\epsilon^{q(\ell,s,k)} x_\epsilon + |1-\omega| x_\epsilon \\ &\leq (\omega\rho_\epsilon + |1-\omega|)\, x_\epsilon, \quad \ell = 0, 1, \ldots, \quad s = 1, 2, \ldots, m_\ell. \end{aligned}$$

Moreover, as $0 < \omega < 2/(1+\rho)$, then $\omega\rho + |1-\omega| < 1$; the continuity of the spectral radius guarantees that there exists ϵ_1 such that $\alpha_\epsilon = \omega\rho_\epsilon + |1-\omega| < 1$ for all $0 < \epsilon \leq \epsilon_1$. Therefore, for all $0 < \epsilon \leq \min\{\epsilon_0, \epsilon_1\}$, $|H_{\ell,s}(x^\star)|x_\epsilon < \alpha_\epsilon x_\epsilon$, where $x_\epsilon > 0$. Thus, using the matrix norm induced by the vector norm (7), one obtains

$$\|H_{\ell,s}(x^\star)\|_{x_\epsilon} \leq \alpha_\epsilon < 1, \; \ell = 0, 1, \ldots, \quad s = 1, 2, \ldots, m_\ell. \tag{14}$$

Therefore, by (13) and (14), setting $\alpha = \max\{\rho_\epsilon, \alpha_\epsilon\}$ the proof is complete.

We want to point out that, when ILU factorizations are used in order to construct the multisplitting, the Jacobian matrix must be factored at each nonlinear iteration of the non–stationary parallel Newton iterative method (5). One approach to reduce the cost of each iteration is to consider the following sequence of iterates

$$x^{(\ell+1)} = \hat{G}_{\ell,m_\ell}(x^{(\ell)}, x^{(0)}), \tag{15}$$

where $\hat{G}_{\ell,m_\ell}(x, y) = x - A_{\ell,m_\ell}(y)F(x)$ and $A_{\ell,m_\ell}(y)$ is defined in (6). This method, based on the Chord method (see e.g., [12]), will be called the non–stationary parallel Chord iterative method. The only difference in implementation from the non–stationary parallel Newton iterative method is that the computation and, therefore, the obtaining of the multisplitting of the Jacobian matrix is done before the nonlinear iteration begins. This technique could be interesting to reduce the computational time. The difference in the iteration itself is that another approximation to $F'(x^{(\ell)})$ is used. Another technique consists of alternating a Newton step with a sequence of Chord steps. In this case we can describe the transition from $x^{(\ell)}$ to $x^{(\ell+1)}$ by

$$y^{(1)} = G_{\ell,m_\ell}(x^{(\ell)}),\ \ y^{(j+1)} = \hat{G}_{\ell,m_\ell}(y^{(j)}, x^{(\ell)}),\ j = 1, 2, \ldots, \tau - 1,\ \ x^{(\ell+1)} = y^{(\tau)}.$$

This method, based on the Shamanskii method (see e.g., [12]), will be called the non–stationary parallel Shamanskii iterative method. Note that $\tau = 1$ is the non–stationary parallel Newton iterative method and $\tau = \infty$ is the corresponding Chord method. It is not difficult to see that the convergence results of Theorem 4, and therefore, those of Theorem 5, remain valid for both variations of the Newton method described above.

Remark 1. The expression (9) in Theorem 5 implies that, if we choose $m_\ell = \jmath$, where $\jmath$ is a positive integer constant, then this non–stationary Newton method converges Q–linearly; if we choose $m_\ell = \ell$ or $m_\ell = 2^\ell$ then the method converges Q–superlinearly; and if $m_\ell = 2^\ell$ then the convergence of the method is also R–quadratic.

Remark 2. In [1] and [2] convergence properties of non–stationary parallel Newton iterative methods were studied when the Jacobian matrix is monotone and the splittings are weak regular. From Theorem 1 it follows that the splittings obtained from ILU factorizations of an M–matrix are regular splittings and then weak regular splittings. Therefore, Theorem 3 of [1] and Theorem 3 of [2] give convergence results for the sequence of iterates defined by (5) when the Jacobian matrix is an M–matrix and ILU factorizations of this matrix are used. On the other hand, these theorems are also valid for the Chord and Shamanskii variations.

Remark 3. One can use p different relaxation parameters $\omega_1, \omega_2, \ldots, \omega_p$, associated with each of the splittings. In this case, our convergence results remain valid and the proofs are almost identical.

4 Numerical Experiments

We have implemented the methods described in this paper on a distributed multiprocessor IBM RS/6000 SP with 8 nodes. These nodes are 120 MHz Power2 Super Chip and are connected through a high performance switch with latency time of 40 microseconds and a bandwidth of 110 Mbytes per second. The parallel environment has been managed using the MPI library of parallel routines [7]. Moreover, we have used the BLAS routines [4] for vector computations and the SPARSKIT [14] routines for handling sparse matrices.

In order to illustrate the behaviour of the above algorithms, we have considered a classic nonlinear elliptic partial differential equation, known as the Bratu problem. In this problem, heat generation from a combustion process is balanced by heat transfer due to conduction. The model problem is given as

$$\nabla^2 u - \lambda e^u = 0, \tag{16}$$

where u is the temperature and λ is a constant known as the Frank–Kamenetskii parameter (see e.g., [3]). There are two possible steady–state solutions to this problem for a given value of λ. One solution is close to $u = 0$ and it is easy to obtain. A close starting point is needed to converge to the other solution. For our model case, we consider a 3D square domain of unit length and $\lambda = 6$.

In order to solve equation (16) using the finite difference method, we consider a grid in the 3D domain of d^3 nodes equally spaced by $h = \frac{1}{d+1}$. This discretization yields a nonlinear system of the form $Ax + \Phi(x) = b$, where $\Phi : \mathbb{R}^n \to \mathbb{R}^n$ is a nonlinear diagonal mapping (i.e., the ith component Φ_i of Φ is a function only of x_i). The Jacobian matrix of $F(x) = Ax + \Phi(x) - b$ is a sparse matrix of order d^3 and the typical number of nonzero elements per row of this matrix is seven, with fewer in rows corresponding to boundary points of the physical domain.

In order to explain the splittings used in our experiments, we can express the Jacobian of $F(x) = Ax + \Phi(x) - b$ in a general form as follows

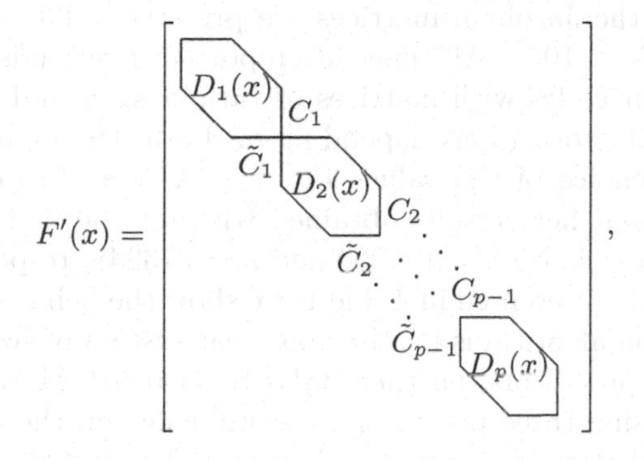

where $D_k(x)$, $1 \le k \le p$, are matrices of size $n_k \times n_k$ such that $\sum_{k=1}^{p} n_k = n$, and n is the size of the problem to be solved. That is, the block diagonal matrix $\mathrm{diag}(D_1(x), D_2(x), \ldots, D_p(x))$ consists of diagonal blocks of $F'(x)$.

Let us further consider an ILU factorization of each matrix $D_k(x)$, $1 \le k \le p$, that is $D_k(x) = L_k(x)U_k(x) - R_k(x)$. With this notation, in our experiments, we consider the splittings $F'(x) = M_k(x) - N_k(x)$ determined by $M_k(x) = \mathrm{diag}(I, \ldots, I, L_k(x)U_k(x), I, \ldots, I)$, $1 \le k \le p$. Each diagonal weighting matrix E_k has ones in the entries corresponding to the diagonal block $D_k(x)$ and zero otherwise.

Let us denote by ILU(S) the incomplete LU factorization associated with the zero pattern subset S of $S_n = \{(i,j) : i \ne j,\ 1 \le i,j \le n\}$. In particular, when $S = \{(i,j) : a_{ij} = 0\}$, the incomplete factorization with null fill–in is known as ILU(0). To improve the quality of the factorization, many strategies for altering the pattern have been proposed. In the "level of fill–in" factorizations [9], ILU(κ), a level of fill–in is recursively attributed to each fill–in position from the levels of its parents. Then, the positions of level lower than κ are removed from S. In the experiments reported here, we have used these ILU(κ) factorizations for the matrices $D_k(x)$, $1 \le k \le p$ defined above. We have modified the SPARSKIT routine which obtains the ILU(κ) factorization in order to improve the factorizations of successive matrices with the same sparsity pattern.

Table 1. Parallel Newton algorithms, $n = 373248$, $p = 3$

		$m_\ell = 1$		$m_\ell = 3$		$m_\ell = 7$		$m_\ell = \ell$	
Level	q	Iter.	Time	Iter.	Time	Iter.	Time	Iter.	Time
0	1	2827	1513.9	943	778.8	405	569.6	76	451.3
	3	964	784.2	322	520.1	139	446.5	45	**429.1**
	5	587	653.9	197	486.1	85	**439.6**	35	438.2
	6	493	625.8	165	**480.6**	72	445.9	32	444.3
	11	277	**583.1**	93	493.2	41	479.6	24	525.1
	15	207	587.9	70	519.2	31	513.3	21	538.5
1	1	1390	1685.2	464	739.8	200	471.4	54	406.9
	2	732	1021.4	245	521.6	106	378.4	39	**320.5**
	5	315	631.0	106	402.8	46	**342.4**	26	344.6
	8	208	540.4	70	**391.1**	31	365.3	21	365.5
	15	122	**509.3**	42	426.2	19	419.7	17	473.6
	18	105	514.7	36	441.7	16	432.1	15	456.9

In all the experiments we have considered $n_k = n/p$, $1 \leq k \leq p$, such that the block structure of the Jacobian matrices are preserved. The stopping criterion used was $\|F(x^{(\ell)})\|_2 < 10^{-7}$. All times are reported in seconds.

We have run our codes with matrices of various sizes and different levels of fill–in for the ILU factorizations depending on both, the number of processors used (p) and the choice of the values n_k, $1 \leq k \leq p$. In order to focus our discussion, we present here results obtained with $d = 50$ and $d = 72$ that lead to nonlinear systems of size $n = 125000$ and $n = 373248$, respectively.

In the first results, presented in Table 1, we show the behaviour of the parallel non–relaxed Newton algorithms for the nonlinear system of size $n = 373248$, for a fixed number of processors, on the IBM RS/6000 SP. More specifically, this table illustrates, using three processors, the influence on the execution time of the number of local steps performed ($q(\ell, k) = q$, $1 \leq k \leq p$), for several values of linear steps m_ℓ and different levels of fill-in, $\kappa = 0$ and $\kappa = 1$.

It can be observed that the number of global iterations decreases as the number of local steps q increases. Therefore, if this reduction of global iterations balances the realization of more local updates per processor, less execution time will be observed. As it can be observed, the computational time starts to decrease as the number of local updates q increases up to some optimal value of q after which the time starts to increase. This behaviour is independent of the choice of m_ℓ. However, the optimal value of q decreases when m_ℓ increases. This fact can also be seen in Figure 1. In this figure we show the execution time to solve the nonlinear system of size $n = 125000$ using two processors. In this figure several incomplete LU factorizations are considered for $m_\ell = 7$ and $m_\ell = \ell$.

We would like to remark that (see Table 1 and Figure 1), the best results were obtained setting incomplete LU factorizations with level 1 of fill–in, independently of the value of m_ℓ. The levels of fill–in ILU(κ) refer to the amount of fill–in allowed during the incomplete factorization. Using $\kappa = 0$ gives upper and lower triangular factors L and U such that $L + U$ has the same sparsity pattern as the original matrix. Using $\kappa = n$ gives the full LU factorization of A without pivoting. For given values of m_ℓ and q, increasing the level of fill–in κ, provides a better quality method in terms of its rate of convergence (that is, in terms of reducing the number of global iterations required), but at the cost of increas-

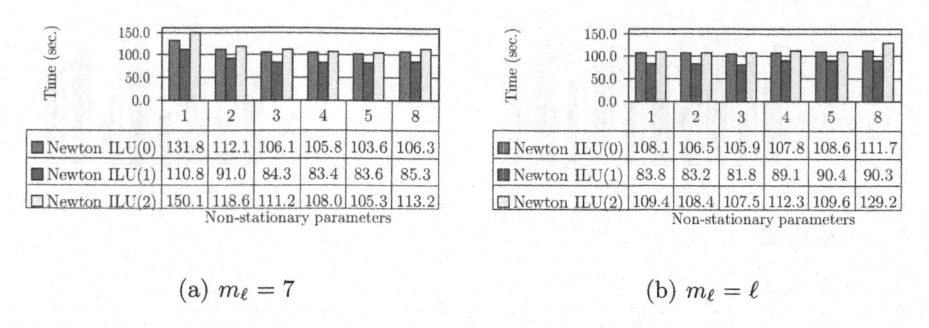

Time (sec.)
150.0
100.0
50.0
0.0
1 2 3 4 5 8
Newton ILU(0) 131.8 112.1 106.1 105.8 103.6 106.3
Newton ILU(1) 110.8 91.0 84.3 83.4 83.6 85.3
Newton ILU(2) 150.1 118.6 111.2 108.0 105.3 113.2
Non-stationary parameters
Time (sec.)
150.0
100.0
50.0
0.0
1 2 3 4 5 8
Newton ILU(0) 108.1 106.5 105.9 107.8 108.6 111.7
Newton ILU(1) 83.8 83.2 81.8 89.1 90.4 90.3
Newton ILU(2) 109.4 108.4 107.5 112.3 109.6 129.2
Non-stationary parameters

(a) $m_\ell = 7$ (b) $m_\ell = \ell$

Fig. 1. Parallel Newton algorithms, $n = 125000$, $p = 2$

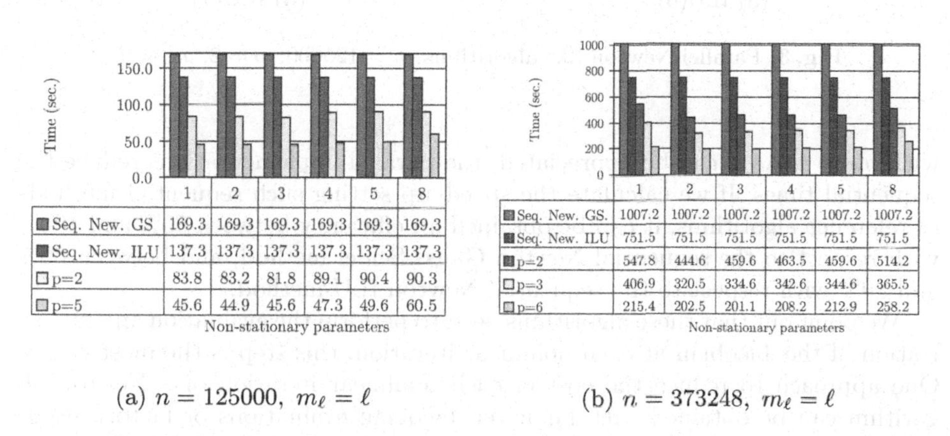

Time (sec.)
150.0
100.0
50.0
0.0
1 2 3 4 5 8
Seq. New. GS 169.3 169.3 169.3 169.3 169.3 169.3
Seq. New. ILU 137.3 137.3 137.3 137.3 137.3 137.3
p=2 83.8 83.2 81.8 89.1 90.4 90.3
p=5 45.6 44.9 45.6 47.3 49.6 60.5
Non-stationary parameters
Time (sec.)
1000
800
600
400
200
0
1 2 3 4 5 8
Seq. New. GS. 1007.2 1007.2 1007.2 1007.2 1007.2 1007.2
Seq. New. ILU 751.5 751.5 751.5 751.5 751.5 751.5
p=2 547.8 444.6 459.6 463.5 459.6 514.2
p=3 406.9 320.5 334.6 342.6 344.6 365.5
p=6 215.4 202.5 201.1 208.2 212.9 258.2
Non-stationary parameters

(a) $n = 125000$, $m_\ell = \ell$ (b) $n = 373248$, $m_\ell = \ell$

Fig. 2. Parallel Newton algorithms for different number of processors versus the sequential Newton Gauss-Seidel and the sequential Newton ILU(1) algorithms with $m_\ell = \ell$

ing storage. Therefore, it can be seen that, for $\kappa > 1$ this reduction of global iterations does not balance the increase of the computational cost obtained by increasing the level of fill–in.

On the other hand, in the above table and figures it can also be observed that the optimal choice for the values m_ℓ, $\ell = 0, 1, \ldots$, is $m_\ell = \ell$. From a theoretical point of view, if we choose $m_\ell = \ell$ then the method converges Q–superlinearly, and in this case, the faster convergence in terms of its rate of convergence has effect on the execution time.

Figure 2 illustrates the influence of the number of processors used for several values of q in relation to the size of the matrix. Figure 2(a) corresponds to the matrix of size $n = 125000$ and Figure 2(b) corresponds to the matrix of size $n = 373248$. In both cases $\kappa = 1$ (ILU(1)) and $m_\ell = \ell$ are considered. Let us highlight some observations of the results in Figures 2(a) and 2(b). We have compared the results of these figures with both the well–known sequential Newton Gauss–Seidel method and the sequential Newton ILU method [12], for several values of m_ℓ. In both cases, the best sequential methods where obtained

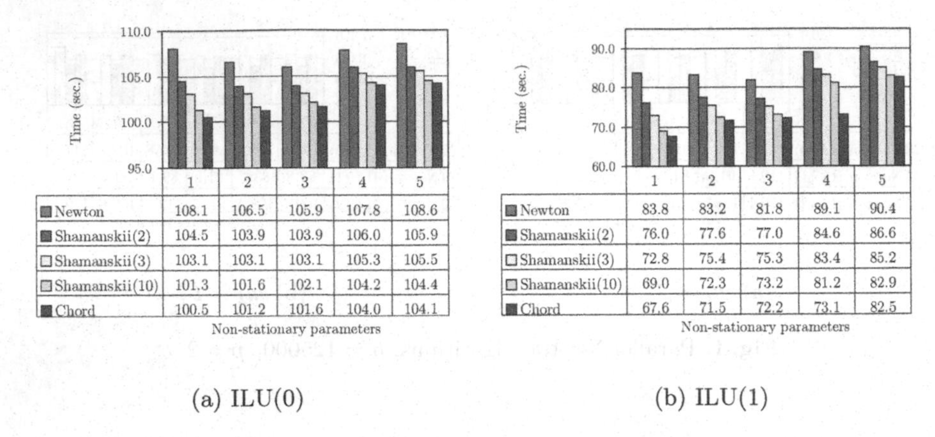

(a) ILU(0) (b) ILU(1)

Fig. 3. Parallel Newton like algorithms, $n = 125000$, $p = 2$, $m_\ell = \ell$

with $m_\ell = \ell$. As it can be appreciated, the parallel implementations reduce the sequential times. If we calculate the speed–up setting such sequential methods as reference algorithms, it can be obtained an efficiency between 80% and 100% with respect to the sequential Newton Gauss-Seidel method, and between 60% and 80% with respect to the sequential Newton ILU method.

We point out that these algorithms need to perform the evaluation and factorization of the Jacobian at each nonlinear iteration; this step is the most costly. One approach to reduce the cost of each nonlinear iteration of a Newton algorithm can be obtained with far fewer Jacobian evaluations or factorizations (Shamanskii method), or with only one Jacobian evaluation, before the nonlinear iteration begins (Chord method).

We continue with an analysis of the effects of the use of these two Newton like methods in the parallel algorithms discussed above. Figure 3 shows the behaviour of these variations of the parallel Newton method studied here for the nonlinear system of size $n = 125000$ using two processors and $m_\ell = \ell$. In this figure, Shamanskii(τ) indicates that the Jacobian matrix is updated each τ nonlinear iterations. Figure 3(a) displays the execution time of these algorithms when ILU(0) factorizations are used, while Figure 3(b) corresponds to ILU(1) factorizations. The results obtained indicate that the parallel Chord methods are the best option for this problem. and the best results were obtained, again, with level 1 of fill–in (i.e., using ILU(1)). Table 2 also illustrates this fact and it is representative of other runs we have performed. In this table we have obtained the execution time of the parallel Chord iterative algorithms for the problem of size $n = 373248$. Concretely, this table displays the results obtained depending on the number of processors used ($p = 2, 3, 6$), for several values of linear steps ($m_\ell = 1, 3, 7, \ell$) and different levels of fill–in ($\kappa = 0, 1, 2$). We want to point out that the best times were obtained, as in the previous experiments, for $m_\ell = \ell$. Besides, this table shows that, for $m_\ell = \ell$, a good choice of the non–stationary parameters is one a little greater or equal than one (between 1 and 3). For this

Table 2. Parallel Chord algorithms, $n = 373248$

		$m_\ell = 1$			$m_\ell = 3$			$m_\ell = 7$			$m_\ell = \ell$		
Level	q	$p=2$	$p=3$	$p=6$	$p=2$	$p=3$	$p=6$	$p=2$	$p=3$	$p=6$	$p=2$	$p=3$	$p=6$
0	1	918.4	691.9	485.5	680.2	505.2	340.8	613.8	452.3	300.3	**579.0**	430.0	283.8
	2	731.0	538.1	360.4	608.6	439.1	279.4	576.0	412.6	257.4	579.2	**404.9**	253.8
	3	671.9	491.6	321.4	588.5	422.5	262.2	**568.1**	**404.5**	246.6	579.4	415.6	**246.4**
	4	644.7	476.1	305.7	581.7	**421.4**	**257.6**	568.2	406.5	**246.2**	605.1	419.8	246.6
	5	631.0	469.6	**301.3**	**578.0**	424.7	261.1	570.1	413.2	248.8	604.9	427.7	252.8
	6	622.8	**468.8**	301.8	578.6	428.3	265.9	568.5	417.4	255.9	609.3	434.4	259.7
	9	**615.3**	476.4	316.3	589.7	445.9	286.2	583.2	445.1	278.6	613.3	483.0	299.6
1	1	536.0	404.8	288.9	422.0	313.0	214.0	390.5	288.3	192.6	425.3	287.8	184.5
	2	459.3	342.2	236.9	399.5	292.5	192.5	**385.1**	**280.7**	**180.5**	**402.4**	**285.2**	**184.3**
	3	437.3	327.5	222.2	**395.7**	**291.6**	**188.7**	390.6	284.6	181.1	417.3	287.2	185.6
	4	430.6	**324.3**	**220.1**	399.1	296.0	192.8	396.8	290.8	185.9	426.6	316.3	194.5
	5	**428.0**	326.3	222.7	404.6	303.7	198.1	401.3	299.7	194.3	426.9	321.1	200.4
2	1	577.7	383.3	269.2	515.7	**318.7**	212.9	**478.6**	**301.2**	**197.2**	**434.1**	**306.7**	**192.8**
	2	557.2	357.5	245.5	**500.3**	320.8	**211.3**	485.9	312.1	202.1	573.8	321.3	211.1
	3	613.8	**355.7**	**242.9**	551.1	330.0	217.7	546.1	323.5	213.1	666.3	347.5	219.7
	4	**525.5**	360.2	247.6	568.9	341.3	227.2	722.0	341.4	223.6	683.1	371.2	239.6

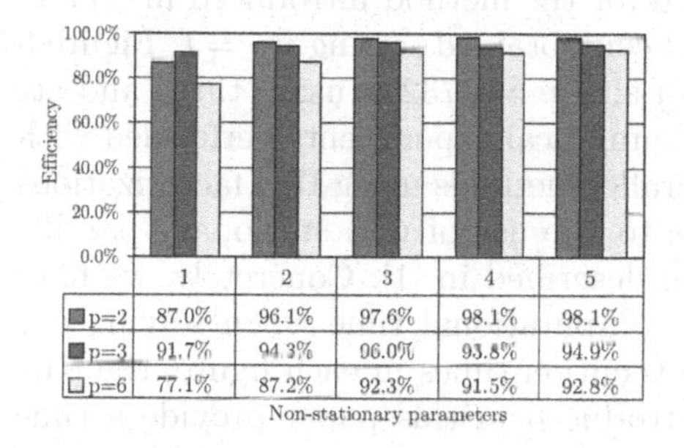

(a) Sequential algorithm: the same algorithm executed on a single processor

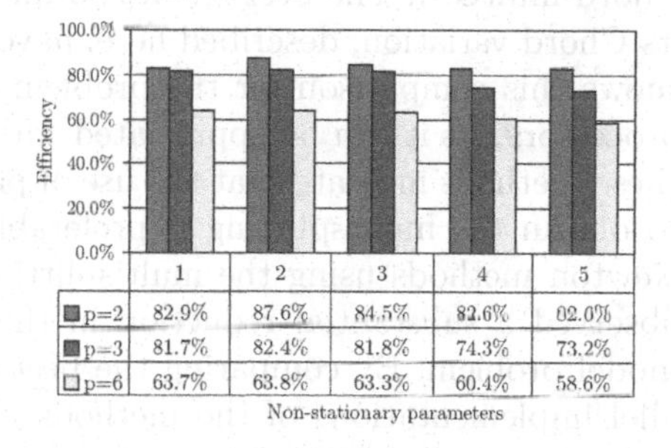

(b) Sequential algorithm: Chord ILU(1) algorithm

Fig. 4. Efficiency of the parallel Chord algorithms setting as reference different sequential algorithms, $n = 373248$, $m_\ell = \ell$

value of m_ℓ, Figure 4 shows the efficiency of the best results of Table 2, that is when ILU(1) is used. Concretely, Figure 4(a) displays the efficiency with respect to the same algorithm executed on a single processor, and Figure 4(b) displays the efficiency of these parallel algorithms over the fastest serial algorithm we have obtained, that is, the Chord variation of the Newton ILU(1) algorithm. Clearly, these efficiencies show a good degree of parallelism of the algorithms discussed here.

On the other hand, we have compared the methods introduced here with those corresponding to the algorithms presented in [1] (denoted here by New-multGS). The splittings used in this case are the same as the ones used in [1]. That is, a Gauss-Seidel type multisplitting is used. Moreover, we have implemented the corresponding Chord variation of those methods (denoted here

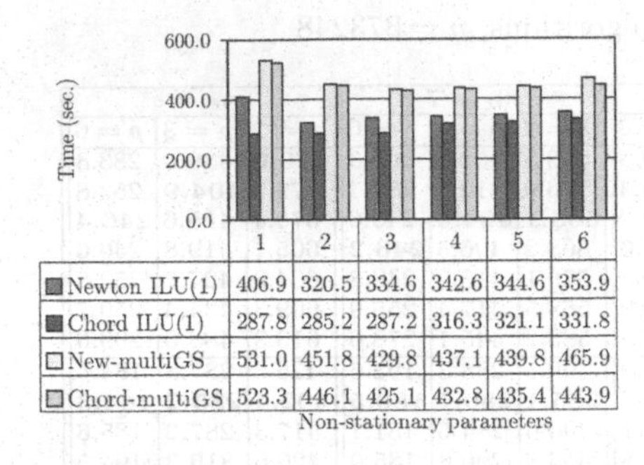

	1	2	3	4	5	6
Newton ILU(1)	406.9	320.5	334.6	342.6	344.6	353.9
Chord ILU(1)	287.8	285.2	287.2	316.3	321.1	331.8
New-multiGS	531.0	451.8	429.8	437.1	439.8	465.9
Chord-multiGS	523.3	446.1	425.1	432.8	435.4	443.9

(a) Number of processors, $p = 3$

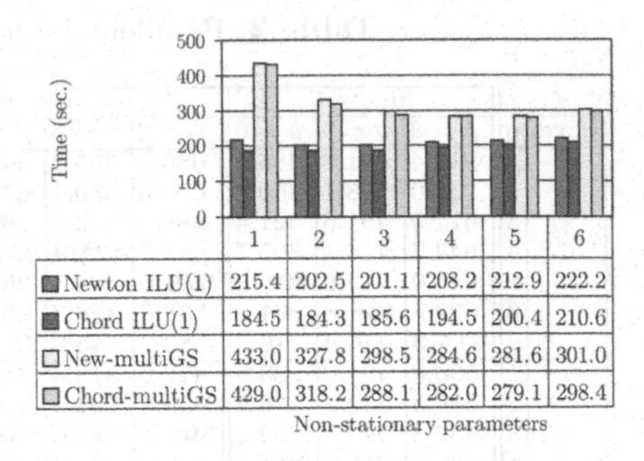

	1	2	3	4	5	6
Newton ILU(1)	215.4	202.5	201.1	208.2	212.9	222.2
Chord ILU(1)	184.5	184.3	185.6	194.5	200.4	210.6
New-multiGS	433.0	327.8	298.5	284.6	281.6	301.0
Chord-multiGS	429.0	318.2	288.1	282.0	279.1	298.4

(b) Number of processors, $p = 6$

Fig. 5. Parallel Newton like algorithms, $n = 373248$

Chord-multGS). The best results obtained for the method introduced in [1] and its Chord variation, described here, have been obtained setting $m_\ell = \ell$. Figure 5 shows this comparison for the problem of size $n = 373248$ using three and six processors. As it can be appreciated, the numerical experiments performed with these methods indicate that the use of parallel methods using ILU factorizations to obtain the multisplitting is preferable to the use of non–stationary parallel Newton methods using the multisplitting described in [1]. Concretely, we have observed a substantial reduction in the computational time when solving our model problem. By comparing the best execution times in each figure, the parallel implementations of the methods introduced in this paper provide a time reduction of about 26%-57%.

Finally, we want to mention that we can also accelerate the convergence by using a relaxation parameter $\omega \neq 1$. The experiments performed have shown time reductions up to 25%-30% when comparing to the best non-relaxed method. For example, the best time in Figure 5(a) is 285.2 sec. (Chord ILU(1), $m_\ell = \ell$, $q = 2$); however, the relaxed version always reduces the execution time for $\omega \in (1, 1.5]$, obtaining the best result (197.6 sec.) for $\omega = 1.5$.

References

[1] Arnal, J., Migallón, V., Penadés, J.: Non–stationary parallel Newton iterative methods for nonlinear problems. Lectures Notes in Computer Science **1981** (2001) 380–394.

[2] Arnal, J., Migallón, V., Penadés, J.: Relaxed parallel Newton algorithms for nonlinear systems. In Proceedings of the 2002 Conference on Computational and Mathematical Methods on Science and Engineering (CMMSE–2002), volume I, (2002) 51–60.

[3] Averick, B. M., Carter, R. G., More, J. J., Xue, G.: The MINPACK–2 Test Problem Collection. Technical Report MCS-P153-0692, Mathematics and Computer Science Division, Argonne National Laboratory (1992).

[4] BLAS: Basic Linear Algebra Subroutine, Available on line from http://www.netlib.org/index.html.

[5] Berman, A., Plemmons, R. J.: Nonnegative Matrices in the Mathematical Sciences. Academic Press, New York, third edition (1979). Reprinted by SIAM, Philadelphia (1994).

[6] Frommer, A., Mayer, G.: Convergence of relaxed parallel multisplitting methods. Linear Algebra and its Applications **119** (1989) 141–152.

[7] Gropp, W., Lusk, E., Skjellum, A.: Using MPI: Portable Parallel Programming. MIT Press, Cambridge, MA (1994).

[8] Kim, S. W., Yun, J. H.: Block ILU factorization preconditioners for a block–tridiagonal H–matrix. Linear Algebra and its Applications **317** (2000) 103–125.

[9] Langtangen, H. P.: Conjugate gradient methods and ILU preconditioning of non-symmetric matrix systems with arbitrary sparsity patterns. International Journal for Numerical Methods in Fluids **9** (1989) 213–233.

[10] Mas, J., Migallón, V., Penadés, J., Szyld, D. B.: Non–stationary parallel relaxed multisplitting methods. Linear Algebra and its Applications **241/243** (1996) 733–748.

[11] Messaoudi, A.: On the stability of the incomplete LU factorizations and characterizations of H–matrices. Numerische Mathematik **69** (1995) 321–331.

[12] Ortega, J. M., Rheinboldt, W. C.: Iterative Solution of Nonlinear Equations in Several Variables. Academic Press, San Diego (1970).

[13] Saad, Y.: Iterative Methods for Sparse Linear Systems. PWS Publishing Company (1996). Also available in http://www.cs.umn.edu/~saad/books.html.

[14] Saad, Y.: SPARSKIT: A basic tool kit for sparse matrix computation. Technical Report 90–20. Research Institute for Advanced Computer Science, NASA Ames Research Center, Moffet Field, CA (1990). Second version of SPARSKIT available on line from http://www-users.cs.umn.edu/~saad/software/SPARSKIT/sparskit.html

[15] Sherman, A.: On Newton–iterative methods for the solution of systems of nonlinear equations. SIAM Journal on Numerical Analysis **15** (1978) 755–771.

[16] Varga, R. S.: Matrix Iterative Analysis. Prentice Hall, Englewood Cliffs, New Jersey (1962).

[17] White, R. E.: Parallel algorithms for nonlinear problems. SIAM Journal on Algebraic Discrete Methods **7** (1986) 137–149.

Author Index

Lecture Notes in Computer Science

For information about Vols. 1–3369

please contact your bookseller or Springer

Vol. 3525: A.E. Abdallah, C.B. Jones, J.W. Sanders (Eds.), Communicating Sequential Processes. XIV, 321 pages. 2005.

Vol. 3510: T. Braun, G. Carle, Y. Koucheryavy, V. Tsaoussidis (Eds.), Wired/Wireless Internet Communications. XIV, 366 pages. 2005.

Vol. 3503: S.E. Nikoletseas (Ed.), Experimental and Efficient Algorithms. XIV, 605 pages. 2005.

Vol. 3501: B. Kégl, G. Lapalme (Eds.), Advances in Artificial Intelligence. XV, 458 pages. 2005. (Subseries LNAI).

Vol. 3500: S. Miyano, J. Mesirov, S. Kasif, S. Istrail, P. Pevzner, M. Waterman (Eds.), Research in Computational Molecular Biology. XVII, 632 pages. 2005. (Subseries LNBI).

Vol. 3492: P. Blache, E. Stabler, J. Busquets, R. Moot (Eds.), Logical Aspects of Computational Linguistics. X, 363 pages. 2005. (Subseries LNAI).

Vol. 3489: G.T. Heineman, J.A. Stafford, H.W. Schmidt, K. Wallnau, C. Szyperski, I. Crnkovic (Eds.), Component-Based Software Engineering. XI, 358 pages. 2005.

Vol. 3488: M.-S. Hacid, N.V. Murray, Z.W. Raś, S. Tsumoto (Eds.), Foundations of Intelligent Systems. XIII, 700 pages. 2005. (Subseries LNAI).

Vol. 3467: J. Giesl (Ed.), Term Rewriting and Applications. XIII, 517 pages. 2005.

Vol. 3465: M. Bernardo, A. Bogliolo (Eds.), Formal Methods for Mobile Computing. VII, 271 pages. 2005.

Vol. 3463: M. Dal Cin, M. Kaâniche, A. Pataricza (Eds.), Dependable Computing - EDCC 2005. XVI, 472 pages. 2005.

Vol. 3461: P. Urzyczyn (Ed.), Typed Lambda Calculi and Applications. XI, 433 pages. 2005.

Vol. 3459: R. Kimmel, N.A. Sochen, J. Weickert (Eds.), Scale Space and PDE Methods in Computer Vision. XI, 634 pages. 2005.

Vol. 3456: H. Rust, Operational Semantics for Timed Systems. XII, 223 pages. 2005.

Vol. 3455: H. Treharne, S. King, M. Henson, S. Schneider (Eds.), ZB 2005: Formal Specification and Development in Z and B. XV, 493 pages. 2005.

Vol. 3454: J.-M. Jacquet, G.P. Picco (Eds.), Coordination Models and Languages. X, 299 pages. 2005.

Vol. 3453: L. Zhou, B.C. Ooi, X. Meng (Eds.), Database Systems for Advanced Applications. XXVII, 929 pages. 2005.

Vol. 3452: F. Baader, A. Voronkov (Eds.), Logic for Programming, Artificial Intelligence, and Reasoning. XI, 562 pages. 2005. (Subseries LNAI).

Vol. 3450: D. Hutter, M. Ullmann (Eds.), Security in Pervasive Computing. XI, 239 pages. 2005.

Vol. 3449: F. Rothlauf, J. Branke, S. Cagnoni, D.W. Corne, R. Drechsler, Y. Jin, P. Machado, E. Marchiori, J. Romero, G.D. Smith, G. Squillero (Eds.), Applications of Evolutionary Computing. XX, 631 pages. 2005.

Vol. 3448: G.R. Raidl, J. Gottlieb (Eds.), Evolutionary Computation in Combinatorial Optimization. XI, 271 pages. 2005.

Vol. 3447: M. Keijzer, A. Tettamanzi, P. Collet, J.v. Hemert, M. Tomassini (Eds.), Genetic Programming. XIII, 382 pages. 2005.

Vol. 3444: M. Sagiv (Ed.), Programming Languages and Systems. XIII, 439 pages. 2005.

Vol. 3443: R. Bodik (Ed.), Compiler Construction. XI, 305 pages. 2005.

Vol. 3442: M. Cerioli (Ed.), Fundamental Approaches to Software Engineering. XIII, 373 pages. 2005.

Vol. 3441: V. Sassone (Ed.), Foundations of Software Science and Computational Structures. XVIII, 521 pages. 2005.

Vol. 3440: N. Halbwachs, L.D. Zuck (Eds.), Tools and Algorithms for the Construction and Analysis of Systems. XVII, 588 pages. 2005.

Vol. 3439: R.H. Deng, F. Bao, H. Pang, J. Zhou (Eds.), Information Security Practice and Experience. XII, 424 pages. 2005.

Vol. 3437: T. Gschwind, C. Mascolo (Eds.), Software Engineering and Middleware. X, 245 pages. 2005.

Vol. 3436: B. Bouyssounouse, J. Sifakis (Eds.), Embedded Systems Design. XV, 492 pages. 2005.

Vol. 3434: L. Brun, M. Vento (Eds.), Graph-Based Representations in Pattern Recognition. XII, 384 pages. 2005.

Vol. 3433: S. Bhalla (Ed.), Databases in Networked Information Systems. VII, 319 pages. 2005.

Vol. 3432: M. Beigl, P. Lukowicz (Eds.), Systems Aspects in Organic and Pervasive Computing - ARCS 2005. X, 265 pages. 2005.

Vol. 3431: C. Dovrolis (Ed.), Passive and Active Network Measurement. XII, 374 pages. 2005.

Vol. 3429: E. Andres, G. Damiand, P. Lienhardt (Eds.), Discrete Geometry for Computer Imagery. X, 428 pages. 2005.

Vol. 3427: G. Kotsis, O. Spaniol (Eds.), Wireless Systems and Mobility in Next Generation Internet. VIII, 249 pages. 2005.

Vol. 3423: J.L. Fiadeiro, P.D. Mosses, F. Orejas (Eds.), Recent Trends in Algebraic Development Techniques. VIII, 271 pages. 2005.

Vol. 3422: R.T. Mittermeir (Ed.), From Computer Literacy to Informatics Fundamentals. X, 203 pages. 2005.

Vol. 3421: P. Lorenz, P. Dini (Eds.), Networking - ICN 2005, Part II. XXXV, 1153 pages. 2005.

Vol. 3420: P. Lorenz, P. Dini (Eds.), Networking - ICN 2005, Part I. XXXV, 933 pages. 2005.

Vol. 3419: B. Faltings, A. Petcu, F. Fages, F. Rossi (Eds.), Constraint Satisfaction and Constraint Logic Programming. X, 217 pages. 2005. (Subseries LNAI).

Vol. 3418: U. Brandes, T. Erlebach (Eds.), Network Analysis. XII, 471 pages. 2005.

Vol. 3416: M. Böhlen, J. Gamper, W. Polasek, M.A. Wimmer (Eds.), E-Government: Towards Electronic Democracy. XIII, 311 pages. 2005. (Subseries LNAI).

Vol. 3415: P. Davidsson, B. Logan, K. Takadama (Eds.), Multi-Agent and Multi-Agent-Based Simulation. X, 265 pages. 2005. (Subseries LNAI).

Vol. 3414: M. Morari, L. Thiele (Eds.), Hybrid Systems: Computation and Control. XII, 684 pages. 2005.

Vol. 3412: X. Franch, D. Port (Eds.), COTS-Based Software Systems. XVI, 312 pages. 2005.

Vol. 3411: S.H. Myaeng, M. Zhou, K.-F. Wong, H.-J. Zhang (Eds.), Information Retrieval Technology. XIII, 337 pages. 2005.

Vol. 3410: C.A. Coello Coello, A. Hernández Aguirre, E. Zitzler (Eds.), Evolutionary Multi-Criterion Optimization. XVI, 912 pages. 2005.

Vol. 3409: N. Guelfi, G. Reggio, A. Romanovsky (Eds.), Scientific Engineering of Distributed Java Applications. X, 127 pages. 2005.

Vol. 3408: D.E. Losada, J.M. Fernández-Luna (Eds.), Advances in Information Retrieval. XVII, 572 pages. 2005.

Vol. 3407: Z. Liu, K. Araki (Eds.), Theoretical Aspects of Computing - ICTAC 2004. XIV, 562 pages. 2005.

Vol. 3406: A. Gelbukh (Ed.), Computational Linguistics and Intelligent Text Processing. XVII, 829 pages. 2005.

Vol. 3404: V. Diekert, B. Durand (Eds.), STACS 2005. XVI, 706 pages. 2005.

Vol. 3403: B. Ganter, R. Godin (Eds.), Formal Concept Analysis. XI, 419 pages. 2005. (Subseries LNAI).

Vol. 3402: M. Daydé, J.J. Dongarra, V. Hernández, J.M.L.M. Palma (Eds.), High Performance Computing for Computational Science - VECPAR 2004. XVIII, 732 pages. 2005.

Vol. 3401: Z. Li, L.G. Vulkov, J. Waśniewski (Eds.), Numerical Analysis and Its Applications. XIII, 630 pages. 2005.

Vol. 3399: Y. Zhang, K. Tanaka, J.X. Yu, S. Wang, M. Li (Eds.), Web Technologies Research and Development - APWeb 2005. XXII, 1082 pages. 2005.

Vol. 3398: D.-K. Baik (Ed.), Systems Modeling and Simulation: Theory and Applications. XIV, 733 pages. 2005. (Subseries LNAI).

Vol. 3397: T.G. Kim (Ed.), Artificial Intelligence and Simulation. XV, 711 pages. 2005. (Subseries LNAI).

Vol. 3396: R.M. van Eijk, M.-P. Huget, F. Dignum (Eds.), Agent Communication. X, 261 pages. 2005. (Subseries LNAI).

Vol. 3395: J. Grabowski, B. Nielsen (Eds.), Formal Approaches to Software Testing. X, 225 pages. 2005.

Vol. 3394: D. Kudenko, D. Kazakov, E. Alonso (Eds.), Adaptive Agents and Multi-Agent Systems II. VIII, 313 pages. 2005. (Subseries LNAI).

Vol. 3393: H.-J. Kreowski, U. Montanari, F. Orejas, G. Rozenberg, G. Taentzer (Eds.), Formal Methods in Software and Systems Modeling. XXVII, 413 pages. 2005.

Vol. 3392: D. Seipel, M. Hanus, U. Geske, O. Bartenstein (Eds.), Applications of Declarative Programming and Knowledge Management. X, 309 pages. 2005. (Subseries LNAI).

Vol. 3391: C. Kim (Ed.), Information Networking. XVII, 936 pages. 2005.

Vol. 3390: R. Choren, A. Garcia, C. Lucena, A. Romanovsky (Eds.), Software Engineering for Multi-Agent Systems III. XII, 291 pages. 2005.

Vol. 3389: P. Van Roy (Ed.), Multiparadigm Programming in Mozart/Oz. XV, 329 pages. 2005.

Vol. 3388: J. Lagergren (Ed.), Comparative Genomics. VII, 133 pages. 2005. (Subseries LNBI).

Vol. 3387: J. Cardoso, A. Sheth (Eds.), Semantic Web Services and Web Process Composition. VIII, 147 pages. 2005.

Vol. 3386: S. Vaudenay (Ed.), Public Key Cryptography - PKC 2005. IX, 436 pages. 2005.

Vol. 3385: R. Cousot (Ed.), Verification, Model Checking, and Abstract Interpretation. XII, 483 pages. 2005.

Vol. 3383: J. Pach (Ed.), Graph Drawing. XII, 536 pages. 2005.

Vol. 3382: J. Odell, P. Giorgini, J.P. Müller (Eds.), Agent-Oriented Software Engineering V. X, 239 pages. 2005.

Vol. 3381: P. Vojtáš, M. Bieliková, B. Charron-Bost, O. Sýkora (Eds.), SOFSEM 2005: Theory and Practice of Computer Science. XV, 448 pages. 2005.

Vol. 3380: C. Priami (Ed.), Transactions on Computational Systems Biology I. IX, 111 pages. 2005. (Subseries LNBI).

Vol. 3379: M. Hemmje, C. Niederee, T. Risse (Eds.), From Integrated Publication and Information Systems to Information and Knowledge Environments. XXIV, 321 pages. 2005.

Vol. 3378: J. Kilian (Ed.), Theory of Cryptography. XII, 621 pages. 2005.

Vol. 3377: B. Goethals, A. Siebes (Eds.), Knowledge Discovery in Inductive Databases. VII, 190 pages. 2005.

Vol. 3376: A. Menezes (Ed.), Topics in Cryptology – CT-RSA 2005. X, 385 pages. 2005.

Vol. 3375: M.A. Marsan, G. Bianchi, M. Listanti, M. Meo (Eds.), Quality of Service in Multiservice IP Networks. XIII, 656 pages. 2005.

Vol. 3374: D. Weyns, H.V.D. Parunak, F. Michel (Eds.), Environments for Multi-Agent Systems. X, 279 pages. 2005. (Subseries LNAI).

Vol. 3372: C. Bussler, V. Tannen, I. Fundulaki (Eds.), Semantic Web and Databases. X, 227 pages. 2005.

Vol. 3371: M.W. Barley, N. Kasabov (Eds.), Intelligent Agents and Multi-Agent Systems. X, 329 pages. 2005. (Subseries LNAI).

Vol. 3370: A. Konagaya, K. Satou (Eds.), Grid Computing in Life Science. X, 188 pages. 2005. (Subseries LNBI).